PROBLEMS AND SOLUTIONS IN ELECTROMAGNETIC THEORY

PROBLEMS AND SOLUTIONS IN ELECTROMAGNETIC THEORY

C. M. LERNER

A WILEY-INTERSCIENCE PUBLICATION
JOHN WILEY & SONS
New York • Chichester • Brisbane • Toronto • Singapore

Library of Congress Cataloging in Publication Data:

Lerner, C. M.
Problems and solutions in electromagnetic theory.

"A Wiley-Interscience publication."
1. Electromagnetic theory—Problems, exercises, etc. I. Title.

QC670.75.L47 1983 530.1'41'076 83-6548
ISBN 0-471-88678-5

Printed in the United States of America

10 9 8 7 6 5 4 3 2 1

Preface

This is a book consisting of 425 problems in Electromagnetic Theory with completed, thoroughly worked out solutions for every problem. The level of the problems vary from that of a freshman level (approximately one-third of the book is on this level), to that of an advanced graduate level (approximately one-third of the book is on this level). The other third of the book consists of the intermediate level problems, which deal with work commensurate with typical junior-senior level course in the field.

The intent of this book is to explain the physical principles and ideas of electromagnetic theory -- along with the subtle, and not so subtle, mathematical manipulations and fundamental useful techniques -- through the use of problem solving. There are conceptions that can only be learned by working through many problems in electricity and magnetism, such as those contained in this book. Frequently, when attempting to learn electricity and magnetism, students of physics do not understand the physical concept until they have been shown samples of its use through problem solving and have subsequently worked out many problems on their own. The reason for this is that even when a student thinks that they understand the physics of a situation, they often find that having a "general feeling" for how to solve a problem is not enough, and to actually get a result that makes sense they run into mnay obstacles along the way that they may not have anticipated. They find that they have to know how to think of the "next step" needed to complete a problem. Thus, the purpose of this book is to try to clearly illustrate the use of the physical principles of electromagnetic theory in many varied and useful examples.

Therefore, this book should appeal to a large audience of both students, who will use the problems as a guide for a more intensive understanding of electromagnetic theory, and also for the teacher, who will use the book as a means of finding innovative and interesting methods and examples for explaining this field of physics in a more comprehensive and understandable nature. One may believe perhaps that, in fact, the problems are worked out in too much detail and does not, therefore, allow the readers to proceed to think on their own. This is not the case here. These problems include only what I consider to be a sufficient number of steps or explanations to convey intelligibly the physical problem without the reader having to guess at what is going on. This in no way means that the problems are worked out in such detail so there would not have to be any thinking on the part of the reader. The book is written so that it will appeal to all students, at any level of sophistication in the application of the physical principles of electromagnetic theory to the understanding of the problems involved. In addition, the book undoubtedly will be extremely useful to graduate students in their study for preliminary and comprehensive examinations offered as a requirement in most graduate programs granting advanced degrees in physics, chemistry, and engineering.

I owe special thanks and outstanding appreciation and admiration to the fascinating people whose ideas form the basis of this book: Maxwell, Lorentz, Einstein, Feynman, Hawking, and Dirac.

As I have been told by countless number of E & M specialists, "Here is all the background you need to solve and possess a complete understanding of all these problems:

$$\nabla \cdot \mathbf{D} = 4\pi\rho, \quad \nabla \times \mathrm{H} = \frac{4\pi}{c}\ \mathrm{j} + \frac{1}{c}\frac{\partial \mathbf{D}}{\partial t}$$

$$\nabla \cdot \mathbf{B} = 0, \quad \nabla \times \mathrm{E} + \frac{1}{c}\frac{\partial \mathbf{B}}{\partial t} = 0."$$

I have already tried it on this basis, but I do not want to influence you by delving into my findings.

Cathy M. Lerner

Cambridge, Massachusetts
June 1985

Contents

PROBLEMS AND SOLUTIONS IN ELECTROMAGNETIC THEORY

Chapter 1

PROBLEM 1.1 Prove that

(a)
$$\oint_s \phi \mathbf{n}\, da = \oint_v \nabla\phi\, dv$$

(b)
$$\oint_s \mathbf{F}(\mathbf{G}\cdot\mathbf{n})\, da = \int_v \mathbf{F}\,\mathrm{div}\,\mathbf{G}\, dv + \int_v (\mathbf{G}\cdot\nabla)\mathbf{F}\, dv$$

[*Hint:* Use the divergence theorem.]

Solution:

(a)
$$\oint_s \phi\mathbf{n}\, da = \int_v \nabla\phi\, dv$$

Let $\mathbf{k}$ be some constant vector and ϕ an arbitrary scalar function. By the divergence theorm (Gauss)

$$\int_{\text{closed } s} \mathbf{k}\phi\cdot\mathbf{n}\, da = \int_v \nabla\cdot(\mathbf{k}\phi)\, d^3r$$

Also, $\nabla\cdot(\mathbf{k}\phi) = \phi\nabla\cdot\mathbf{k} + \mathbf{k}\cdot\nabla\phi$. However, $\nabla\cdot\mathbf{k} = 0$ since $\mathbf{k} = \text{constant}$.

$$\therefore \int_s \mathbf{k}\phi\cdot\mathbf{n}\, da = \int_v \mathbf{k}\cdot\nabla\phi\, d^3r, \quad \mathbf{k}\cdot\int_s \phi\mathbf{n}\, da = \mathbf{k}\cdot\int_v \nabla\phi\, d^3r$$

Since $\mathbf{k}$ is *arbitrary*

$$\int_s \phi\mathbf{n}\, da = \int_v \nabla\phi\, d^3r \quad \text{Q.E.D.}$$

Note: This suggests the following, coordinate-independent definition of $\nabla\phi$:

$$\nabla\phi = \lim_{v\to 0} \frac{1}{v}\int_s \phi\mathbf{n}\, da$$

(b) Let $\mathbf{k}$ be an arbitrary constant vector. By the divergence theorem

$$\int_{\text{closed } s} [(\mathbf{k}\cdot\mathbf{F})\mathbf{G}]\cdot\mathbf{n}\, da = \int_v \nabla\cdot[(\mathbf{k}\cdot\mathbf{F})\mathbf{G}]\, d^3r$$

$$= \int_v [\mathbf{G} \cdot \nabla(\mathbf{k} \cdot \mathbf{F}) + (\mathbf{k} \cdot \mathbf{F})\nabla \cdot \mathbf{G}]\, d^3r$$

Note

$$\mathbf{G} \cdot \nabla(\mathbf{k} \cdot \mathbf{F}) = \sum_{i=1}^{3} \mathbf{G} \cdot \nabla(k_i F_i) = \sum_i k_i (\mathbf{G} \cdot \nabla) F_i = \mathbf{k} \cdot (\mathbf{G} \cdot \nabla)\mathbf{F}$$

$$\therefore \quad \mathbf{k} \cdot \int \mathbf{F}\mathbf{G} \cdot \mathbf{n}\, da = \mathbf{k} \cdot \int [\mathbf{F}\nabla \cdot \mathbf{G} + (\mathbf{G} \cdot \nabla)\mathbf{F}]\, d^3r$$

But **k** is arbitrary,

$$\therefore \quad \int \mathbf{F}\mathbf{G} \cdot \mathbf{n}\, da = \int [\mathbf{F}\nabla \cdot \mathbf{G} + (\mathbf{G} \cdot \nabla)\mathbf{F}]\, d^3r$$

PROBLEM 1.2 Let $\mathbf{F}(\mathbf{r})$ be a vector field whose value is specified on some closed boundary and let $\mathbf{C}(\mathbf{r})$ and $d(\mathbf{r})$ be the given curl and divergence of **F** in the region enclosed by the boundary.

(a) Show that $\mathbf{F}(\mathbf{r})$ is unique inside the boundary [i.e., the boundary values, $\mathbf{C}(\mathbf{r})$ and $d(\mathbf{r})$ determine $\mathbf{F}(\mathbf{r})$ inside the boundary].
(b) If $\mathbf{F}(\mathbf{r})$ goes to zero faster than f/r^2 (f constant) as $r \to \infty$ and $\mathbf{C}(\mathbf{r})$ and $d(\mathbf{r})$ go to zero at infinity, from your knowledge of electrostatics and magnetostatics, give an explicit expression for $\mathbf{F}(\mathbf{r})$ in terms of integrals involving **C** and d.
(c) By using vector analytic methods and δ-function techniques show that your "guess" for **F** leads to $\nabla \times \mathbf{F} = \mathbf{C}$, $\nabla \cdot \mathbf{F} = d$.

Solution:

(a) We make use of the following equations:

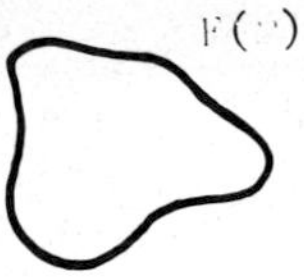

$$\nabla \times \mathbf{F}(\mathbf{r}) = \mathbf{C}(\mathbf{r}) \quad \text{and} \quad \nabla \cdot \mathbf{F}(\mathbf{r}) = d(\mathbf{r})$$

We want to show that there cannot be two **F**'s both satisfying these equations. First, we will assume there are and then we show that this cannot be true. Assume $\exists$ $\mathbf{F}_1$ and $\mathbf{F}_2$ both satisfying all conditions. Then

$$\nabla \times \mathbf{F}_1 = \nabla \times \mathbf{F}_2 = \mathbf{C}(\mathbf{r}) \quad \Rightarrow \quad \nabla \times (\mathbf{F}_1 - \mathbf{F}_2) = 0$$

$$\nabla \cdot \mathbf{F}_1 = \nabla \cdot \mathbf{F}_2 = d(\mathbf{r}) \quad \Rightarrow \quad \nabla \cdot (\mathbf{F}_1 - \mathbf{F}_2) = 0$$

Let $\mathbf{G} = \mathbf{F}_1 - \mathbf{F}_2$

$$\nabla \times \mathbf{G} = 0 \quad \Rightarrow \quad \mathbf{G} = -\nabla\phi \quad (\text{if curl } \mathbf{G} = 0)$$

$\nabla \cdot G = 0 \quad \Rightarrow \quad G = \nabla^2 \phi \quad$ (like $\nabla \cdot E = 0$)

$\nabla \cdot E = \frac{\rho}{\varepsilon_0} = -\nabla^2 \phi$

If $\quad \nabla \cdot E = 0$ then $\nabla^2 \phi = 0$.

Now drawing this (since it looks like $\nabla \cdot B = 0$), we see that $G \cdot \hat{n} = 0$.

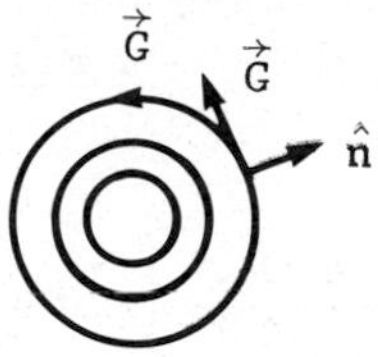

Define $u = \phi \nabla \phi$. Then

$$\nabla \cdot \mathbf{u} = \nabla \cdot (\phi \nabla \phi) = \underbrace{\phi \nabla^2 \phi}_{=0} + |\nabla \phi|^2 = (\nabla \phi)^2$$

Now to make use of $G \;\; \mathbf{n} = 0 = -\nabla \phi \cdot \mathbf{n}$, use the divergence theorem

$$\int_{vol} \nabla \cdot \mathbf{F}\, dv = \oint_S \mathbf{F} \cdot \hat{\mathbf{n}}\, da$$

$$\int_{vol} \nabla \cdot \mathbf{u}\, dv = \int_{vol} \nabla \cdot \phi \nabla \phi\, dv = \int_S \phi \underbrace{\nabla \phi \cdot \hat{\mathbf{n}}}_{=0}\, da = 0$$

$$\int_{vol} (\nabla \phi)^2\, dv = 0 \Rightarrow (\nabla \phi)^2 = 0 \text{ inside volume.}$$

$\therefore \quad \nabla \phi = -G = F_2 - F_1 = 0$

So F_2 and F_1 must both be equal. Thus $\mathbf{F}(\mathbf{r})$ is unique (inside the boundary).

(*Note:* The proof shows only the normal component of **F** completely determines the solution, so only it should be specified; anything more than this and $\nabla \times \mathbf{F}$ and $\nabla \cdot \mathbf{F}$ inside actually overdetermines **F**.)

(b) $\mathbf{F}(\mathbf{r}) \to 0$ faster than f/r^2 as $r \to \infty$, and $\mathbf{C}(\mathbf{r})$ and $d(\mathbf{r})$ go to 0 at infinity (compare to boundary condition on **B** and **E**). We want an expression for $\mathbf{F}(\mathbf{r})$ in terms of integrals involving **C** and d.

$$\nabla \times \mathbf{F}(\mathbf{r}) = \mathbf{C}(\mathbf{r}) \qquad \text{(i)}$$

$$\nabla \cdot \mathbf{F}(\mathbf{r}) = d(\mathbf{r}) \qquad \text{(ii)}$$

Equation (ii) is like

$$\nabla \cdot E = \frac{\rho(\mathbf{r})}{\varepsilon_0} \to E = \frac{1}{4\pi\varepsilon_0} \int \frac{\rho(\mathbf{r}')(\mathbf{r} - \mathbf{r}')\, dv'}{|\mathbf{r} - \mathbf{r}'|^3}$$

and Equation (i) is like

$$\nabla \times \mathbf{B} = \mu \cdot \mathbf{j} \rightarrow B = \frac{\mu_0}{4\pi} \int \frac{\mathbf{j}(\mathbf{r}') \times (\mathbf{r} - \mathbf{r}')\, dv'}{|\mathbf{r} - \mathbf{r}'|^3}$$

when the boundary is at infinity. So

$$\mathbf{F} = \frac{1}{4\pi} \left[\int \frac{d(\mathbf{r}')(\mathbf{r} - \mathbf{r}')\, dv'}{|\mathbf{r} - \mathbf{r}'|^3} + \int \frac{\mathbf{C}(\mathbf{r}') \times (\mathbf{r} - \mathbf{r}')\, dv'}{|\mathbf{r} - \mathbf{r}'|^3} \right]$$

(c)

$$\nabla \cdot \mathbf{F} = \frac{1}{4\pi} \int \nabla \cdot \left(\frac{\mathbf{C}(\mathbf{r}') \times (\mathbf{r} - \mathbf{r}')}{|\mathbf{r} - \mathbf{r}'|^3} \right) dv' + \frac{1}{4\pi} \int d(\mathbf{r}') \nabla \cdot \frac{(\mathbf{r} - \mathbf{r}')}{|\mathbf{r} - \mathbf{r}'|^3}\, dv'$$

The second term,

$$-\nabla |\mathbf{r} - \mathbf{r}'|^{-1} = \frac{\mathbf{r} - \mathbf{r}'}{|\mathbf{r} - \mathbf{r}'|^3},$$

so

$$\nabla \cdot \frac{(\mathbf{r} - \mathbf{r}')}{|\mathbf{r} - \mathbf{r}'|^3} = -\nabla \cdot \frac{\nabla 1}{|\mathbf{r} - \mathbf{r}'|} = -\nabla^2 \frac{1}{|\mathbf{r} - \mathbf{r}'|}$$

From

$$\nabla^2 \frac{1}{|\mathbf{r} - \mathbf{r}'|} = -4\pi\, \delta(\mathbf{r} - \mathbf{r}')$$

we get

$$\int \frac{1}{4\pi} d(\mathbf{r}') \nabla \cdot \frac{(\mathbf{r} - \mathbf{r}')}{|\mathbf{r} - \mathbf{r}'|^3}\, dv' = \frac{1}{4\pi} \int d(\mathbf{r}') \frac{\nabla^2 (-1)}{|\mathbf{r} - \mathbf{r}'|}\, dv'$$

$$= \frac{1}{4\pi} \int d(\mathbf{r}')\, [+4\pi\, \delta(\mathbf{r} - \mathbf{r}')]\, dv'$$

$$= \int d(\mathbf{r}')\, \delta\, (\mathbf{r} - \mathbf{r}')\, dv'$$

$$= d(\mathbf{r})$$

The first term,

$$\nabla \cdot \left(\frac{\mathbf{C}(\mathbf{r}') \times (\mathbf{r} - \mathbf{r}')}{|\mathbf{r} - \mathbf{r}'|^3} \right) = \nabla \cdot \left(\mathbf{C}(\mathbf{r}') \times \nabla \frac{1}{|\mathbf{r} - \mathbf{r}'|} \right)$$

$$= \nabla \frac{1}{|\mathbf{r} - \mathbf{r}'|} \cdot \underbrace{\nabla \times \mathbf{C}(\mathbf{r}')}_{=0} - \mathbf{C}(\mathbf{r}') \cdot \underbrace{\nabla \times \nabla \frac{1}{|\mathbf{r} - \mathbf{r}'|}}_{=0}$$

Here, the first term on the right-hand side is zero because C(r') does not depend on r, and the second term is zero because curl $\nabla\phi = 0$. So

$$\nabla \cdot \mathbf{F} = d(\mathbf{r}) .$$

$$\nabla \times \mathbf{F} = \frac{1}{4\pi}\left[\int \nabla \times \left(\mathbf{C}(\mathbf{r}') \times \frac{(\mathbf{r}-\mathbf{r}')}{|\mathbf{r}-\mathbf{r}'|^3}\right) dv' + \int \nabla \times \left(\frac{d(\mathbf{r}')(\mathbf{r}-\mathbf{r}')\,dv'}{|\mathbf{r}-\mathbf{r}'|^3}\right)\right]$$

$$= \frac{1}{4\pi}\left[\int \nabla \times \left(\mathbf{C}(\mathbf{r}') \times \nabla \frac{1}{|\mathbf{r}-\mathbf{r}'|}\right) dv' + \int d(\mathbf{r}')\nabla \times \left(\nabla \frac{1}{|\mathbf{r}-\mathbf{r}'|}\, dv'\right)\right]$$

The last term is zero since $\nabla \times \nabla\phi = 0$.

$$\nabla \times \mathbf{C}(\mathbf{r}') \times \nabla \frac{1}{|\mathbf{r}-\mathbf{r}'|} = \mathbf{C}(\mathbf{r}')\nabla \cdot \nabla \frac{1}{|\mathbf{r}-\mathbf{r}'|} - \nabla \frac{\nabla 1}{|\mathbf{r}-\mathbf{r}'|}\ \nabla \cdot \mathbf{C}(\mathbf{r}')$$

$$+ \left(\nabla \frac{1}{|\mathbf{r}-\mathbf{r}'|} \cdot \nabla\right)\mathbf{C}(\mathbf{r}') - [\mathbf{C}(\mathbf{r}') \cdot \nabla]\ \frac{\nabla 1}{|\mathbf{r}-\mathbf{r}'|}$$

Here the second and third terms on the right-hand side are zero because C is independent of r. Therefore,

$$\nabla \times \mathbf{C}(\mathbf{r}') \times \nabla \frac{1}{|\mathbf{r}-\mathbf{r}'|} = \mathbf{C}(\mathbf{r})\ \nabla^2 \frac{1}{|\mathbf{r}-\mathbf{r}'|} - [\mathbf{C}(\mathbf{r}') \cdot \nabla]\ \frac{1}{|\mathbf{r}-\mathbf{r}'|}$$

Now using the identity

$$\oint_s \mathbf{F}(\mathbf{G} \cdot \mathbf{n})\, da = \int_v \mathbf{F}\, \mathrm{div}\, \mathbf{G}\, dv + \int_v (\mathbf{G} \cdot \nabla)\mathbf{F}\, dv$$

$$\int_{\substack{\text{vol}\\ \text{at } \infty}} [\mathbf{C}(\mathbf{r}') \cdot \nabla]\frac{1}{|\mathbf{r}-\mathbf{r}'|} = \oint_{\substack{s\\ \text{at } \infty}} \frac{1}{|\mathbf{r}-\mathbf{r}'|}\ (\mathbf{C}(\mathbf{r}') \cdot \hat{\mathbf{n}})\, da$$

$$- \int_{\text{vol}} \frac{1}{|\mathbf{r}-\mathbf{r}'|}\ (\nabla \cdot \mathbf{C}(\mathbf{r}'))\, dv$$

Note: For the last integral of the surface at infinity, use the fact that as $\mathbf{r}'\to\infty$, $\mathbf{C}(\mathbf{r}')\to 0$ faster than $1/\mathbf{r}'$. The last volume integral integration can be carried out using the fact that $\nabla \times \mathbf{F}$ and div $(\nabla \times \mathbf{F})$ are zero; also C(r') is independent of $\nabla(\mathbf{r})$. So these two terms equal 0 and we are left with

$$\nabla \times \left(\mathbf{C}(\mathbf{r}') \times \nabla \frac{1}{|\mathbf{r}-\mathbf{r}'|}\right) = -4\pi\,\delta(\mathbf{r}-\mathbf{r}')\mathbf{C}(\mathbf{r}') + \text{something that integrates to zero}$$

$$\nabla \times \mathbf{F} = \int -4\pi\,\delta\,(\mathbf{r}-\mathbf{r}')\mathbf{C}(\mathbf{r}')\, dv = \mathbf{C}(\mathbf{r})$$

PROBLEM 1.3 Given that

$$A(r) = \frac{\mu_0}{4\pi} \int \frac{J(r')}{|r - r'|} dv'$$

where the integration is over all space and J(r) goes to zero faster than $1/r'$ as $|r'| \to \infty$, show that

$$\nabla \cdot A = 0, \quad \nabla \times A = \frac{\mu_0}{4\pi} \int \frac{J(r') \times (r - r')}{|r - r'|^3} dv', \quad \nabla^2 A = -\mu_0 J$$

Solution:

$$\nabla \cdot A = \frac{\mu_0}{4\pi} \int \nabla \cdot \frac{J(r')}{|r - r'|} dv'$$

$$= \frac{\mu_0}{4\pi} \int \left[\frac{1}{|r - r'|} \underbrace{\nabla \cdot J(r')}_{=0} + J(r') \cdot \nabla \frac{1}{|r - r|} \right] dv'$$

Note: $\nabla \cdot J(r') = 0$ since ∇ operates on r only. Therefore,

$$\nabla \cdot A = \frac{-\mu_0}{4\pi} \int J(r') \cdot \nabla' \frac{1}{|r - r'|} dv'$$

$$= -\frac{\mu_0}{4\pi} \int [\nabla' \cdot J(r') \frac{1}{|r - r'|} - \frac{1}{|r - r'|} \underbrace{\nabla' \cdot J(r')}_{=0}] dv'$$

Also $\nabla' \cdot J(r') = 0$ by conservation of charge, so

$$\nabla \cdot A = -\frac{\mu_0}{4\pi} \int_{S_\infty} \frac{J(r') \cdot n'}{|r - r'|} ds' = 0$$

since integrand $< 1/r'$.

$$\nabla \times A = \frac{\mu_0}{4\pi} \int \nabla \times \frac{J(r')}{|r - r'|} dv'$$

$$= \frac{\mu_0}{4\pi} \int [+J(r') \times \frac{(r - r')}{|r - r'|^3} + \frac{1}{|r - r'|} \underbrace{\nabla \times J(r')}_{=0}]$$

$\nabla \times J(r') = 0$ because we are operating with unprimed ∇. So,

$$\nabla^2 A = \frac{\mu_0}{4\pi} \int J(r') \nabla^2 \frac{1}{|r - r'|} dv'$$

$$= \frac{\mu_0}{4\pi} \int -4\pi \delta (r - r') J(r') dv' = -\mu_0 J(r)$$

PROBLEM 1.4 Consider the curvilinear coordinate system given by the following transformation:

$$x = \xi_1 \xi_2, \quad y = \sqrt{(\xi_1^2 - d^2)(1 - \xi_2^2)}, \quad z = \xi_3$$

where $\xi_1 \geq |d|$, $|\xi_2| \leq 1$, and $-\infty \leq \xi_3 \leq \infty$. (This represents elliptic cylindrical coordinates).

(a) Sketch (qualitatively) the projection of a few constant ξ_1 surfaces and constant ξ_2 surfaces onto a plane of constant ξ_3.
(b) Show that this coordinate system is orthogonal.
(c) Calculate the scale coefficients h_1, h_2, and h_3.
(d) Calculate the unit vectors $\hat{\mathbf{a}}_1$, $\hat{\mathbf{a}}_2$, and $\hat{\mathbf{a}}_3$.

[*Hint:* There is no need to solve for ξ_1, ξ_2, and ξ_3 in parts (b), (c) and (d) of this problem.]

Solution:

(a) We substitute $\xi_2 = x/\xi_1$ into the expression for y:

$$y = \sqrt{(\xi_1^2 - d^2)(1 - x^2/\xi_1^2)} \rightarrow \frac{x^2}{\xi_1^2} + \frac{y^2}{(\xi_1^2 - d^2)} = 1$$

Since $\xi_1 \geq |d|$, the curves of $\xi_1 = \text{constant}$ are ellipses. Next we substitute $\xi_1 = x/\xi_2$ in the expression for y. This yields

$$\frac{x^2}{\xi_2^2} - \frac{y^2}{1 - \xi_2^2} = d^2$$

Since $|\xi_2| < 1$ this equation yields hyperbolas.

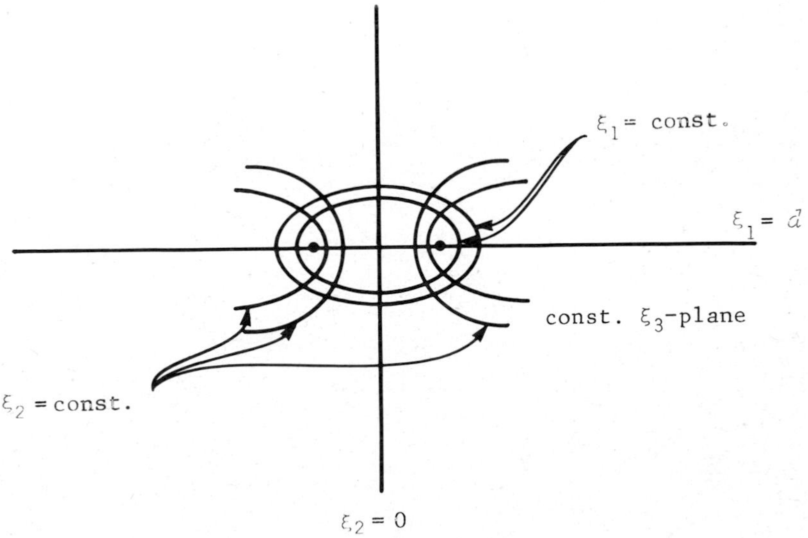

(b) $$\frac{\partial \mathbf{r}}{\partial \xi_2} \cdot \frac{\partial \mathbf{r}}{\partial \xi_1} = \frac{\partial x}{\partial \xi_2}\frac{\partial x}{\partial \xi_1} + \frac{\partial y}{\partial \xi_2}\frac{\partial y}{\partial \xi_1} + \frac{\partial z}{\partial \xi_2}\frac{\partial z}{\partial \xi_1}$$

$$= \left[\xi_2\xi_1 - \frac{\xi_2\xi_1(1-\xi_2^2)(\xi_1^2-d^2)}{(\xi_1^2-d^2)(1-\xi_2^2)} + 0 \times 0 \right] = 0 \quad \text{[see part (c)]}$$

$$\frac{\partial \mathbf{r}}{\partial \xi_2} \cdot \frac{\partial \mathbf{r}}{\partial \xi_3} = \left[\frac{\partial x}{\partial \xi_2} \times 0 + \frac{\partial y}{\partial \xi_2} \times 0 + 0 \times \frac{\partial z}{\partial \xi_3} \right] = 0$$

Similarly $\dfrac{\partial \mathbf{r}}{\partial \xi_1}\,\dfrac{\partial \mathbf{r}}{\partial \xi_3} = 0$

(c) $$\frac{\partial x}{\partial \xi_1} = \xi_2, \quad \frac{\partial x}{\partial \xi_2} = \xi_1, \quad \frac{\partial y}{\partial \xi_1} = \frac{\xi_1(1-\xi_2^2)}{\sqrt{(\xi_1^2-d^2)(1-\xi_2^2)}}$$

$$\frac{\partial y}{\partial \xi_2} = -\frac{(\xi_1^2-d^2)\xi_2}{\sqrt{(\xi_1^2-d^2)(1-\xi_2^2)}}, \quad \frac{\partial y}{\partial \xi_3} = 0, \quad \frac{\partial x}{\partial \xi_3} = 0$$

$$\frac{\partial z}{\partial \xi_1} = \frac{\partial z}{\partial \xi_2} = 0, \quad \frac{\partial z}{\partial \xi_3} = 1$$

$$h_1 = \left[\left(\frac{\partial x}{\partial \xi_1}\right)^2 + \left(\frac{\partial y}{\partial \xi_1}\right)^2 + \left(\frac{\partial z}{\partial \xi_1}\right)^2 \right]^{\frac{1}{2}} = \left[\xi_2^2 + \frac{\xi_1^2(1-\xi_2^2)}{(\xi_1^2-d^2)(1-\xi_2^2)} + 0 \right]^{\frac{1}{2}}$$

$$= \left[\frac{\xi_1^2 - \xi_2^2 d^2}{\xi_1^2 - d^2} \right]^{\frac{1}{2}}$$

$$h_2 = \left[\left(\frac{\partial x}{\partial \xi_2}\right)^2 + \left(\frac{\partial y}{\partial \xi_2}\right)^2 + \left(\frac{\partial z}{\partial \xi_2}\right)^2 \right]^{\frac{1}{2}} = \left[\xi_1^2 + \frac{(\xi_1^2-d^2)\xi_2^2}{(\xi_1^2-d^2)(1-\xi_2^2)} \right]^{\frac{1}{2}}$$

$$= \left[\frac{\xi_1^2 - \xi_2^2 d^2}{1 - \xi_2^2} \right]^{\frac{1}{2}}$$

$$h_3 = \left[0 + 0 + \left(\frac{\partial z}{\partial \xi_3}\right)^2 \right]^{\frac{1}{2}} = 1$$

(d) $$\hat{\mathbf{a}}_1 = \frac{1}{h_1}\frac{\partial \vec{\mathbf{r}}}{\partial \xi_1} = \left[\frac{\xi_1^2 - d^2}{\xi_1^2 - \xi_2^2 d^2} \right]^{\frac{1}{2}} \left[\hat{\mathbf{i}}\xi_2 + \hat{\mathbf{j}}\, \frac{(1-\xi_2^2)\xi_1}{\sqrt{(\xi_1^2-d^2)(1-\xi_2^2)}} \right]$$

$$\hat{\mathbf{a}}_2 = \frac{1}{h_2}\frac{\partial \vec{\mathbf{r}}}{\partial \xi_2} = \left[\frac{1-\xi_2^2}{\xi_1^2 - \xi_2^2 d^2}\right]^{\frac{1}{2}} \left[\hat{\mathbf{i}}\xi_1 + \hat{\mathbf{j}}\,\frac{(d^2-\xi_2^2)\,\xi_2}{\sqrt{(\xi_1^2 - d^2)(1-\xi_2^2)}}\right]$$

$$\hat{\mathbf{a}}_3 = \hat{\mathbf{k}}$$

PROBLEM 1.5 Obtain explicit expressions for $|r - r'|$ in Cartesian, Spherical Polar, and Cylindrical coordinates.

Cartesian

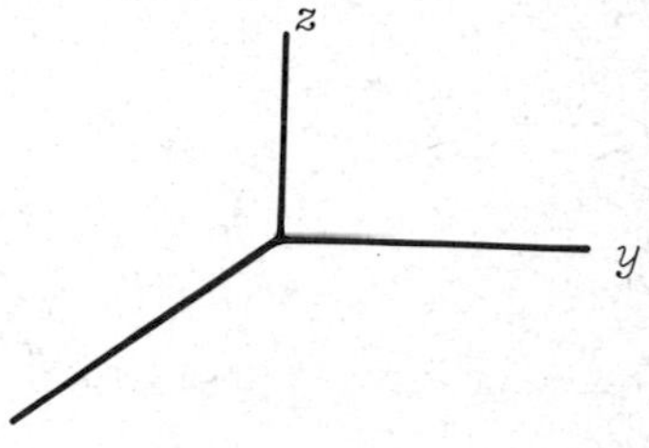

$$\mathbf{r} = \hat{\mathbf{i}}x + \hat{\mathbf{j}}y + \hat{\mathbf{k}}z$$
$$\mathbf{r}' = \hat{\mathbf{i}}x' + \hat{\mathbf{j}}y' + \hat{\mathbf{k}}z'$$
$$\mathbf{r} - \mathbf{r}' = \hat{\mathbf{i}}(x-x') + \hat{\mathbf{j}}(y-y') + \hat{\mathbf{k}}(z-z')$$
$$|\mathbf{r}-\mathbf{r}'| = [(x-x')^2 + (y-y')^2 + (z-z')^2]^{\frac{1}{2}}$$

Spherical polar

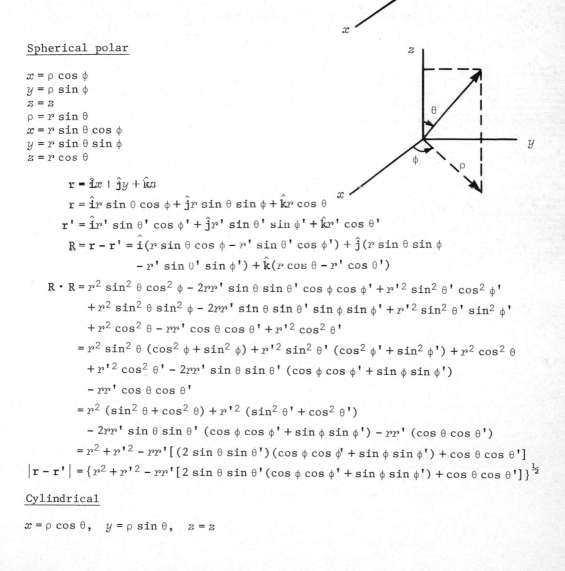

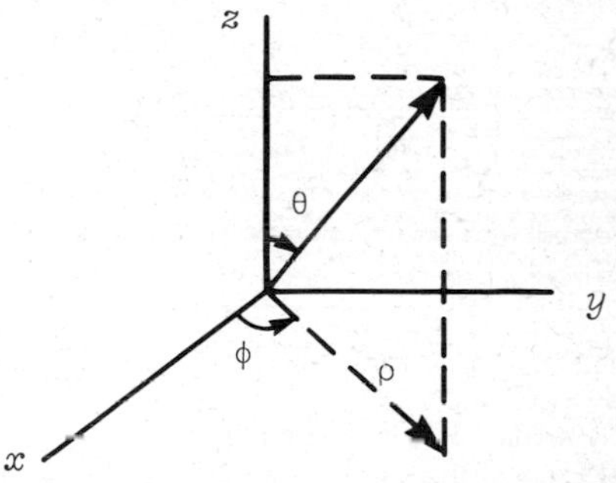

$$x = \rho\cos\phi$$
$$y = \rho\sin\phi$$
$$z = z$$
$$\rho = r\sin\theta$$
$$x = r\sin\theta\cos\phi$$
$$y = r\sin\theta\sin\phi$$
$$z = r\cos\theta$$

$$\mathbf{r} = \hat{\mathbf{i}}x + \hat{\mathbf{j}}y + \hat{\mathbf{k}}z$$
$$\mathbf{r} = \hat{\mathbf{i}}r\sin\theta\cos\phi + \hat{\mathbf{j}}r\sin\theta\sin\phi + \hat{\mathbf{k}}r\cos\theta$$
$$\mathbf{r}' = \hat{\mathbf{i}}r'\sin\theta'\cos\phi' + \hat{\mathbf{j}}r'\sin\theta'\sin\phi' + \hat{\mathbf{k}}r'\cos\theta'$$
$$\underline{\mathbf{R}} = \mathbf{r} - \mathbf{r}' = \hat{\mathbf{i}}(r\sin\theta\cos\phi - r'\sin\theta'\cos\phi') + \hat{\mathbf{j}}(r\sin\theta\sin\phi - r'\sin\theta'\sin\phi') + \hat{\mathbf{k}}(r\cos\theta - r'\cos\theta')$$

$$\begin{aligned}
\mathbf{R}\cdot\mathbf{R} &= r^2\sin^2\theta\cos^2\phi - 2rr'\sin\theta\sin\theta'\cos\phi\cos\phi' + r'^2\sin^2\theta'\cos^2\phi' \\
&\quad + r^2\sin^2\theta\sin^2\phi - 2rr'\sin\theta\sin\theta'\sin\phi\sin\phi' + r'^2\sin^2\theta'\sin^2\phi' \\
&\quad + r^2\cos^2\theta - rr'\cos\theta\cos\theta' + r'^2\cos^2\theta' \\
&= r^2\sin^2\theta(\cos^2\phi + \sin^2\phi) + r'^2\sin^2\theta'(\cos^2\phi' + \sin^2\phi') + r^2\cos^2\theta \\
&\quad + r'^2\cos^2\theta' - 2rr'\sin\theta\sin\theta'(\cos\phi\cos\phi' + \sin\phi\sin\phi') \\
&\quad - rr'\cos\theta\cos\theta' \\
&= r^2(\sin^2\theta + \cos^2\theta) + r'^2(\sin^2\theta' + \cos^2\theta') \\
&\quad - 2rr'\sin\theta\sin\theta'(\cos\phi\cos\phi' + \sin\phi\sin\phi') - rr'(\cos\theta\cos\theta') \\
&= r^2 + r'^2 - rr'[(2\sin\theta\sin\theta')(\cos\phi\cos\phi' + \sin\phi\sin\phi') + \cos\theta\cos\theta']
\end{aligned}$$

$$|\mathbf{r}-\mathbf{r}'| = \{r^2 + r'^2 - rr'[2\sin\theta\sin\theta'(\cos\phi\cos\phi' + \sin\phi\sin\phi') + \cos\theta\cos\theta']\}^{\frac{1}{2}}$$

Cylindrical

$x = \rho\cos\theta,\quad y = \rho\sin\theta,\quad z = z$

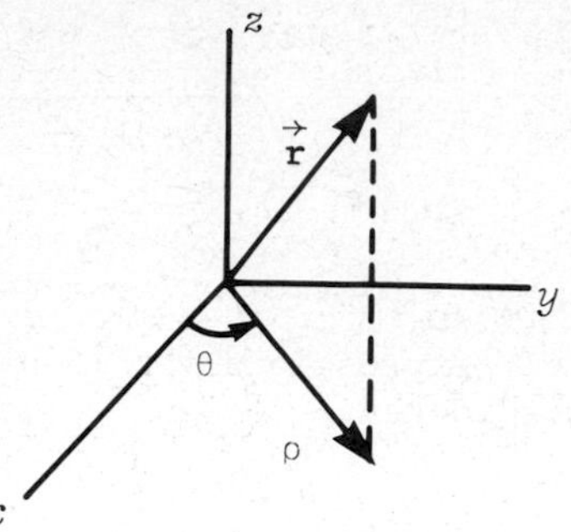

$$\mathbf{r} = \hat{\mathbf{i}}x + \hat{\mathbf{j}}y + \hat{\mathbf{k}}z$$
$$\mathbf{r} = \hat{\mathbf{i}}\rho\cos\theta + \hat{\mathbf{j}}\rho\sin\theta + \hat{\mathbf{k}}z$$
$$\mathbf{r}' = \hat{\mathbf{i}}\rho'\cos\theta' + \hat{\mathbf{j}}\rho'\sin\theta' + \hat{\mathbf{k}}z'$$
$$\mathbf{R} = \mathbf{r} - \mathbf{r}' = \hat{\mathbf{i}}(\rho\cos\theta - \rho'\cos\theta') + \hat{\mathbf{j}}(\rho\sin\theta - \rho'\sin\theta') + \hat{\mathbf{k}}(z - z')$$

$$\mathbf{R}\cdot\mathbf{R} = \rho^2\cos^2\theta - 2\rho\rho'\cos\theta\cos\theta' + \rho'^2\cos^2\theta' + \rho^2\sin^2\theta - 2\rho\rho'\sin\theta\sin\theta' + \rho'^2\sin^2\theta' + z^2 - 2zz' + z'^2$$

$$\mathbf{R}\cdot\mathbf{R} = \rho^2 + \rho'^2 - 2\rho\rho'(\cos\theta\cos\theta' + \sin\theta\sin\theta') + z^2 - 2zz' + z'^2$$

$$|\mathbf{r} - \mathbf{r}'| = [\rho^2 + \rho'^2 - 2\rho\rho'(\cos\theta\cos\theta' + \sin\theta\sin\theta') + z^2 - 2zz' + z'^2]^{\frac{1}{2}}$$

If $\mathbf{r}' = 0$

$$|\mathbf{r} - \mathbf{r}'| = \sqrt{\rho^2 + z^2}$$

PROBLEM 1.6 Calculate $\nabla^2(1/|\mathbf{r} - \mathbf{r}'|)$ explicitly in Cartesian coordinates.

Solution:

$$\nabla^2 \frac{1}{|\mathbf{r} - \mathbf{r}'|} = \nabla^2[(x - x')^2 + (y - y')^2 + (z - z')^2]^{-\frac{1}{2}}$$

$$\frac{\partial}{\partial x}[(x - x')^2 + (y - y')^2 + (z - z')^2]^{-\frac{1}{2}} = -\tfrac{1}{2}[(x - x')^2 + (y - y')^2 + (z - z')^2]^{-3/2}\,2(x - x')(1)$$

$$\frac{\partial}{\partial x}\frac{1}{|\mathbf{r} - \mathbf{r}'|} = (x' - x)[(x - x')^2 + (y - y')^2 + (z - z')^2]^{-3/2}$$

$$\frac{\partial^2}{\partial x^2}\frac{1}{|\mathbf{r} - \mathbf{r}'|} = (x' - x)(-3/2)[(x - x')^2 + (y - y')^2 + (z - z')^2]^{-5/2}(2)(x - x') - [(x - x')^2 + (y - y')^2 + (z - z')^2]^{-3/2}$$

$$\frac{\partial^2}{\partial x^2}\frac{1}{|\mathbf{r} - \mathbf{r}'|} = -3(x' - x)(x - x')[(x - x')^2 + (y - y')^2 + (z - z')^2]^{-5/2} - [(x - x')^2 + (y - y')^2 + (z - z')^2]^{-3/2}$$

$$\frac{\partial^2}{\partial y^2}\frac{1}{|\mathbf{r}-\mathbf{r}'|} = -3(y'-y)(y-y')[(x-x')^2+(y-y')^2+(z-z')^2]^{-5/2}$$
$$-[(x-x')^2+(y-y')^2+(z-z')^2]^{-3/2}$$

$$\frac{\partial^2}{\partial z^2}\frac{1}{|\mathbf{r}-\mathbf{r}'|} = -3(z'-z)(z-z')[(x-x')^2+(y-y')^2+(z-z')^2]^{-5/2}$$
$$-[(x-x')^2+(y-y')^2+(z-z')^2]^{-3/2}$$

$$\left(\frac{\partial^2}{\partial x^2}+\frac{\partial^2}{\partial y^2}+\frac{\partial^2}{\partial z^2}\right)\frac{1}{|\mathbf{r}-\mathbf{r}'|} = (3x'^2+3x^2-6xx'+3y'^2+3y^2-6yy'+3z'^2+3z^2-6zz'$$
$$-3x^2+6xx'-3x'^2-3y^2+6yy'-3y'^2-3z^2+6zz'-3z'^2)/$$
$$[(x-x')^2+(y-y')^2+(z-z')^2]^{-5/2}$$

$$\nabla^2\frac{1}{|\mathbf{r}-\mathbf{r}'|} = 0$$

PROBLEM 1.7 Show $\nabla^2(1/r) = 0$, $r \neq 0$; use spherical coordinates.

Solution:

$$\nabla^2\left(\frac{1}{r}\right) = \frac{1}{r^2}\left[\frac{\partial}{\partial r}r^2\left(-\frac{1}{r^2}\right)\right] = \frac{1}{r^2}\frac{\partial}{\partial r}(-1) = 0 \quad \text{if } r \neq 0$$

PROBLEM 1.8 Prove $\iint_s \hat{\mathbf{n}} \times \nabla\phi \, da = \int_c \phi \, d\mathbf{s}$.

Solution: First prove Stokes' theorem,

$$\iint_s (\nabla \times \mathbf{f}) \cdot \hat{\mathbf{n}} \, da = \oint_c \mathbf{f} \cdot d\mathbf{s}$$

The surface s is divided into a large number of cells. The surface of the ith cell is Δs_i and the curve bounding it is c_i:

$$\oint_c \mathbf{f} \cdot d\mathbf{s} = \sum_i \oint_{c_i} \mathbf{f} \cdot d\mathbf{s}$$

Let the number of cells become infinite:

$$\oint_c \mathbf{f} \cdot d\mathbf{s} = \lim_{\Delta s_i \to 0} \sum_i \frac{1}{\Delta s_i} \oint_{c_i} \mathbf{f} \cdot d\mathbf{s} \, \Delta s_i$$

$$= \iint_s (\nabla \times \mathbf{F}) \cdot \hat{\mathbf{n}} \, da \quad \text{Q.E.D.}$$

Now use Stokes' theorem to prove

$$\iint_s \hat{\mathbf{n}} \times \nabla\phi \, da = \oint_c \phi \, d\mathbf{s}$$

Let $\hat{\varphi} = \phi\mathbf{a}$ where a is an arbitrary constant vector

$$\oint_C \hat{\varphi} \cdot d\mathbf{s} = \iint_S (\nabla \times \phi\mathbf{a}) \cdot \hat{\mathbf{n}}\, da$$

$$= \iint_S (\nabla\phi \times \mathbf{a}) \cdot \hat{\mathbf{n}}\, da$$

because $\phi\nabla \times \mathbf{a} = 0$, or

$$\mathbf{a} \cdot \oint_C \phi\, d\mathbf{s} = \mathbf{a} \cdot \iint_S \hat{\mathbf{n}} \times \nabla\phi\, da$$

$$\mathbf{a} \cdot \left(\oint_C \phi\, d\mathbf{s} - \iint_S \hat{\mathbf{n}} \times \nabla\phi\, da\right) = 0$$

Since a has arbitrary direction

$$\oint_C \phi\, d\mathbf{s} = \iint_S \hat{\mathbf{n}} \times \nabla\phi\, da \quad \text{Q.E.D.}$$

PROBLEM 1.9 Show that if we set $\mathbf{A} = \nabla V + \nabla \times (\mathbf{a}U)$, where $\nabla^2 V = \nabla^2 U = 0$ and $\partial a_\beta / \partial x_\alpha = c\,\delta_{\alpha\beta}$:

(a) A satisfies the equation $\nabla \cdot \mathbf{A} = 0$.
(b) $\nabla \times (\nabla \times \mathbf{A}) = 0$.

Solution: Given $\mathbf{A} = \nabla V + \nabla \times (\mathbf{a}U)$ where $\nabla^2 V = \nabla^2 U = 0$ and $\partial a_\beta / \partial r_\alpha = c\delta_{\alpha\beta}$ we show the following:

(a) $\nabla \cdot \mathbf{A} = 0$

$\nabla \cdot \mathbf{A} = \nabla \cdot (\nabla V + \nabla \times (\mathbf{a}U))$

$$= \underbrace{\nabla^2 V}_{=0} + \nabla \cdot \begin{vmatrix} \hat{\mathbf{i}} & \hat{\mathbf{j}} & \hat{\mathbf{k}} \\ \frac{\partial}{\partial x} & \frac{\partial}{\partial y} & \frac{\partial}{\partial z} \\ a_x U & a_y U & a_z U \end{vmatrix}$$

$$= \nabla \cdot \left[\hat{\mathbf{i}}\left(\frac{\partial}{\partial y}(a_z U) - \frac{\partial}{\partial z}(a_y U)\right) + \hat{\mathbf{j}}\left(\frac{\partial}{\partial z}(a_x U) - \frac{\partial}{\partial x}(a_z U)\right)\right.$$

$$\left. + \hat{\mathbf{k}}\left(\frac{\partial}{\partial x}(a_y U) - \frac{\partial}{\partial y}(a_y U)\right)\right]$$

$$= \nabla \cdot \left[\hat{\mathbf{i}}\left(a_z \frac{\partial U}{\partial y} - a_y \frac{\partial U}{\partial z}\right) + \hat{\mathbf{j}}\left(a_x \frac{\partial U}{\partial z} - a_z \frac{\partial U}{\partial x}\right) + \hat{\mathbf{k}}\left(a_y \frac{\partial U}{\partial x} - a_x \frac{\partial U}{\partial y}\right)\right]$$

$$= a_z \frac{\partial^2 U}{\partial y \partial x} - a_y \frac{\partial^2 U}{\partial z \partial x} + a_x \frac{\partial^2 U}{\partial z \partial y} - a_z \frac{\partial^2 U}{\partial x \partial y} + a_y \frac{\partial^2 U}{\partial x \partial z} - a_x \frac{\partial^2 U}{\partial y \partial z}$$

$= 0$ Q.E.D.

(b) $\nabla \times (\nabla \times \mathbf{A}) = 0$. First look at $\nabla \times \mathbf{A}$:

$$\underbrace{\nabla \times \nabla V}_{=0} + \nabla \times (\nabla \times (\mathbf{a}U))$$

$$= \nabla \times \left[\hat{\mathbf{i}}\left(a_z\frac{\partial U}{\partial y} - a_y\frac{\partial U}{\partial z}\right) + \hat{\mathbf{j}}\left(a_x\frac{\partial U}{\partial z} - a_z\frac{\partial U}{\partial x}\right) + \hat{\mathbf{k}}\left(a_y\frac{\partial U}{\partial x} - a_x\frac{\partial U}{\partial y}\right)\right]$$

$$= \begin{vmatrix} \hat{\mathbf{i}} & \hat{\mathbf{j}} & \hat{\mathbf{k}} \\ \frac{\partial}{\partial x} & \frac{\partial}{\partial y} & \frac{\partial}{\partial z} \\ \left(a_z\frac{\partial U}{\partial y} - a_y\frac{\partial U}{\partial z}\right) & \left(a_x\frac{\partial U}{\partial z} - a_z\frac{\partial U}{\partial x}\right) & \left(a_y\frac{\partial U}{\partial x} - a_x\frac{\partial U}{\partial y}\right) \end{vmatrix}$$

$$= \hat{\mathbf{i}}\left[a_y\frac{\partial^2 U}{\partial x\partial y} + c - a_x\frac{\partial^2 U}{\partial y^2} - a_x\frac{\partial^2 U}{\partial z^2} + a_z\frac{\partial^2 U}{\partial x\partial z} + c\right]$$

$$+ \hat{\mathbf{j}}\left[a_z\frac{\partial^2 U}{\partial y\partial z} + c - a_y\frac{\partial^2 U}{\partial z^2} - a_y\frac{\partial^2 U}{\partial x^2} + a_x\frac{\partial^2 U}{\partial y\partial z} + c\right]$$

$$+ \hat{\mathbf{k}}\left[a_x\frac{\partial^2 U}{\partial z\partial x} + c - a_z\frac{\partial^2 U}{\partial x^2} - a_z\frac{\partial^2 U}{\partial y^2} + a_y\frac{\partial^2 U}{\partial z\partial y} + c\right]$$

Now take the curl of this expression:

$$\begin{vmatrix} \hat{\mathbf{i}} & \hat{\mathbf{j}} & \hat{\mathbf{k}} \\ \frac{\partial}{\partial x} & \frac{\partial}{\partial y} & \frac{\partial}{\partial z} \\ 2c + a_y\frac{\partial^2 U}{\partial x\partial y} + a_z\frac{\partial^2 U}{\partial x\partial z} & 2c + a_z\frac{\partial^2 U}{\partial y\partial z} + a_x\frac{\partial^2 U}{\partial y\partial x} & 2c + a_x\frac{\partial^2 U}{\partial z\partial x} + a_y\frac{\partial^2 U}{\partial z\partial y} \\ -a_x\left(\frac{\partial^2 U}{\partial y^2} + \frac{\partial^2 U}{\partial z^2}\right) & -a_y\left(\frac{\partial^2 U}{\partial z^2} + \frac{\partial^2 U}{\partial x^2}\right) & -a_z\left(\frac{\partial^2 U}{\partial x^2} + \frac{\partial^2 U}{\partial y^2}\right) \end{vmatrix}$$

$$= \hat{\mathbf{i}}\left[a_x\frac{\partial^3 U}{\partial x\,\partial y\,\partial z} + a_y\frac{\partial^3 U}{\partial z\,\partial y^2} + c - a_z\frac{\partial}{\partial y}\left(\frac{\partial^2 U}{\partial x^2} + \frac{\partial^2 U}{\partial y^2}\right)\right.$$

$$\left. - a_z\frac{\partial^3 U}{\partial y\,\partial z^2} - c - a_x\frac{\partial^3 U}{\partial x\,\partial y\,\partial z} + a_y\frac{\partial}{\partial z}\left(\frac{\partial^2 U}{\partial z^2} + \frac{\partial^2 U}{\partial x^2}\right)\right]$$

$$+ \hat{\mathbf{j}}\left[a_y\frac{\partial^3 U}{\partial x\,\partial y\,\partial z} + a_z\frac{\partial^3 U}{\partial x\,\partial z^2} + c - a_x\frac{\partial}{\partial z}\left(\frac{\partial^2 U}{\partial y^2} + \frac{\partial^2 U}{\partial z^2}\right)\right.$$

$$- a_x \frac{\partial^3 U}{\partial z\, \partial x^2} - c - a_y \frac{\partial^3 U}{\partial x\, \partial y\, \partial z} + a_z \frac{\partial}{\partial x}\left(\frac{\partial^2 U}{\partial x^2} + \frac{\partial^2 U}{\partial y^2}\right)\Bigg]$$

$$+ \hat{\mathbf{k}}\left[a_z \frac{\partial^3 U}{\partial x\, \partial y\, \partial z} + a_x \frac{\partial^3 U}{\partial y\, \partial x^2} + c - a_y \frac{\partial}{\partial x}\left(\frac{\partial^2 U}{\partial z^2} + \frac{\partial^2 U}{\partial x^2}\right)\right.$$

$$\left. - a_y \frac{\partial^3 U}{\partial x\, \partial y^2} - c - a_z \frac{\partial^3 U}{\partial x\, \partial y\, \partial z} + a_x \frac{\partial}{\partial y}\left(\frac{\partial^2 U}{\partial y^2} + \frac{\partial^2 U}{\partial z^2}\right)\right]$$

$$= \hat{\mathbf{i}}\left[a_y \frac{\partial}{\partial z}\left(\frac{\partial^2 U}{\partial y^2} + \frac{\partial^2 U}{\partial z^2} + \frac{\partial^2 U}{\partial x^2}\right) - a_z \frac{\partial}{\partial y}\left(\frac{\partial^2 U}{\partial x^2} + \frac{\partial^2 U}{\partial y^2} + \frac{\partial^2 U}{\partial z^2}\right)\right]$$

$$+ \hat{\mathbf{j}}\left[a_z \frac{\partial}{\partial x}\left(\frac{\partial^2 U}{\partial z^2} + \frac{\partial^2 U}{\partial x^2} + \frac{\partial^2 U}{\partial y^2}\right) - a_x \frac{\partial}{\partial z}\left(\frac{\partial^2 U}{\partial y^2} + \frac{\partial^2 U}{\partial z^2} + \frac{\partial^2 U}{\partial x^2}\right)\right]$$

$$+ \hat{\mathbf{k}}\left[a_x \frac{\partial}{\partial y}\left(\frac{\partial^2 U}{\partial x^2} + \frac{\partial^2 U}{\partial y^2} + \frac{\partial^2 U}{\partial z^2}\right) - a_y \frac{\partial}{\partial x}\left(\frac{\partial^2 U}{\partial z^2} + \frac{\partial^2 U}{\partial x^2} + \frac{\partial^2 U}{\partial y^2}\right)\right]$$

$= 0$ Q.E.D.

since in the last statement each of the terms in parentheses equals zero.

PROBLEM 1.10 Define or describe rectangular (R), cylindrical (C), and spherical (S) for the following:

(a) Coordinate Systems and Differential Geometry

*Coordinate systems
*Coordinate transformations
*Line elements
*Components of metric tensor
*Differential distances along coordinate lines
*Differential areas on coordinate surfaces
*Differential volume elements
*Illustration of generalized curvilinear coordinates

(b) Vector Analysis

*Unit vectors
*General vector representation
*Vector addition and subtraction
*Dot product
*Cross product
*Dot and cross products of unit vectors
*Dot products between unit vectors in different systems
*Method for changing coordinate representation of a vector
*Line element in vector notation
*Grad $U = \nabla U$
*Div $\mathbf{F} = \nabla \cdot \mathbf{F}$
*Curl $\mathbf{F} = \nabla \times \mathbf{F}$
*Div Grad U = Laplacian $U = \nabla^2 U$
*Divergence theorem, Stokes' theorem, and Green's theorem.

Solution:

Coordinate Systems

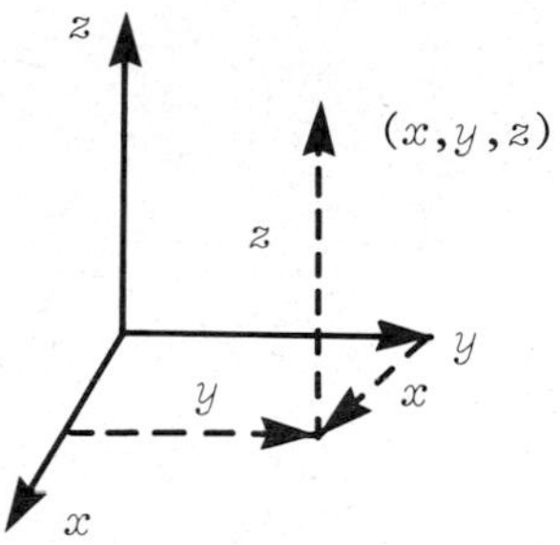

Rectangular
(R)

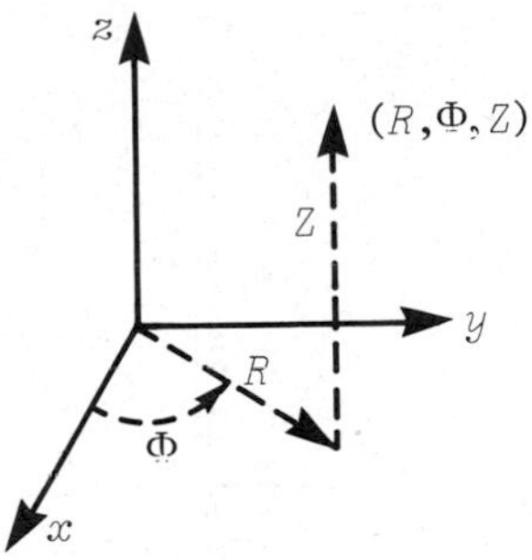

Cylindrical
(C)

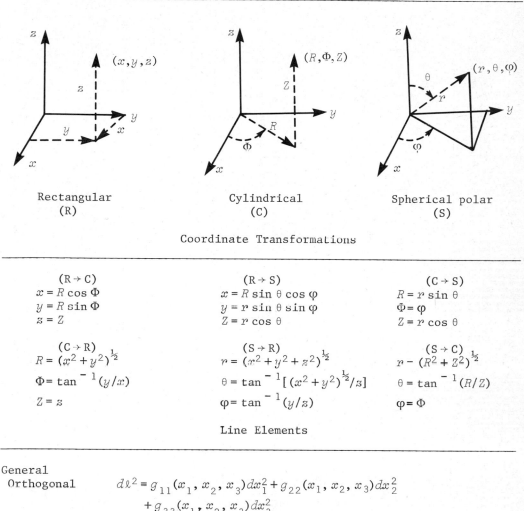

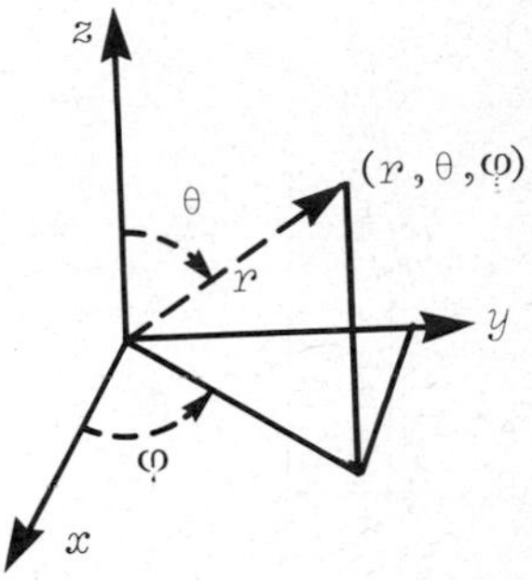

Spherical polar
(S)

Coordinate Transformations

(R → C)	(R → S)	(C → S)
$x = R\cos\Phi$	$x = R\sin\theta\cos\varphi$	$R = r\sin\theta$
$y = R\sin\Phi$	$y = r\sin\theta\sin\varphi$	$\Phi = \varphi$
$z = Z$	$Z = r\cos\theta$	$Z = r\cos\theta$
(C → R)	(S → R)	(S → C)
$R = (x^2 + y^2)^{1/2}$	$r = (x^2 + y^2 + z^2)^{1/2}$	$r = (R^2 + Z^2)^{1/2}$
$\Phi = \tan^{-1}(y/x)$	$\theta = \tan^{-1}[(x^2 + y^2)^{1/2}/z]$	$\theta = \tan^{-1}(R/Z)$
$Z = z$	$\varphi = \tan^{-1}(y/z)$	$\varphi = \Phi$

Line Elements

General Orthogonal $\quad d\ell^2 = g_{11}(x_1, x_2, x_3)dx_1^2 + g_{22}(x_1, x_2, x_3)dx_2^2 + g_{33}(x_1, x_2, x_3)dx_3^2$

Rectangular (R) $\quad d\ell^2 = dx^2 + dy^2 + dz^2$

Cylindrical (C) $\quad d\ell^2 = dR^2 + R^2\, d\Phi^2 + dZ$

Spherical (S) $\quad d\ell^2 = dr^2 + r^2\, d\theta^2 + r^2 \sin^2\theta\, d\varphi^2$

Components of Metric Tensor

General Orthogonal	x_1	x_2	x_3	$g_{11}(x_1, x_2, x_3)$	$g_{22}(x_1, x_2, x_3)$	$g_{33}(x_1, x_2, x_3)$
Rectangular (R)	x	y	z	1	1	1
Cylindrical (C)	R	Φ	Z	1	R^2	1
Spherical (S)	r	θ	φ	1	r^2	$r^2\sin^2\theta$

Differential Distances Along Coordinate Lines

General Orthogonal Coordinate Varied	General Orthogonal Element of Length	Rectangular (R) Coordinate Varied	Rectangular (R) Element of Length	Cylindrical (C) Coordinate Varied	Cylindrical (C) Element of Length	Spherical (S) Coordinate Varied	Spherical (S) Element of Length
x_1	$d\ell_1 = g_{11}^{1/2}\,dx_1$	x	$d\ell_x = dx$	R	$d\ell_R = dR$	r	$d\ell_r = dr$
x_2	$d\ell_2 = g_{22}^{1/2}\,dx_2$	y	$d\ell_y = dy$	Φ	$d\ell_\Phi = R\,d\Phi$	θ	$d\ell_\theta = r\,d\theta$
x_3	$d\ell_3 = g_{33}^{1/2}\,dx_3$	z	$d\ell_z = dz$	Z	$d\ell_z = dZ$	φ	$d\ell_\varphi = r\sin\theta\,d\varphi$

Differential Areas on Coordinate Surfaces

General Orthogonal Constant Coordinate	General Orthogonal Element of Area	Rectangular (R) Constant Coordinate	Rectangular (R) Element of Area	Cylindrical (C) Constant Coordinate	Cylindrical (C) Element of Area	Spherical (S) Constant Coordinate	Spherical (S) Element of Area
x_1	$dA_1 = d\ell_2 d\ell_3$ --------------- $(g_{22}g_{33})^{1/2}dx_2dx_3$	x	$dA_x = d\ell_y d\ell_z$ ------------ $dy\,dz$	R	$dA_R = d\ell_\Phi d\ell_Z$ ------------- $R\,d\Phi\,dZ$	r	$dA_r = d\ell_\theta d\ell_\varphi$ ------------ $r^2 \sin\theta\,d\theta\,d\varphi$
x_2	$dA_2 = d\ell_1 d\ell_3$ --------------- $(g_{11}g_{33})^{1/2}dx_1dx_3$	y	$dA_y = d\ell_x d\ell_z$ ------------ $dx\,dz$	Φ	$dA_\Phi = d\ell_R d\ell_Z$ -------------- $dR\,dZ$	θ	$dA_\theta = d\ell_r d\ell_\varphi$ ------------ $r \sin\theta\,dr\ \ \varphi$
x_3	$dA_3 = d\ell_1 d\ell_2$ --------------- $(g_{11}g_{22})^{1/2}dx_1dx_2$	z	$dA_z = d\ell_x d\ell_y$ ------------ $dx\,dy$	Z	$dA_Z = d\ell_R d\ell_\Phi$ -------------- $R\,dR\,d\varphi$	φ	$dA_\varphi = d\ell_r d\ell_\theta$ ------------ $r\,dr\,d\theta$

Differential Volume Elements

General Orthogonal	$dv = d\ell_1 d\ell_2 d\ell_3 = (g_{11}g_{22}g_{33})^{\frac{1}{2}} dx_1 dx_2 dx_3$
Rectangular	$dv = d\ell_x d\ell_y d\ell_z = dx\, dy\, dz$
Cylindrical	$dv = d\ell_R d\ell_\Phi d\ell_Z = R\, dR d\Phi dZ$
Spherical	$dv = d\ell_r d\ell_\theta d\ell_\varphi = r^2 \sin\theta\, dr\, d\theta\, d\varphi$

Illustration of Generalized Curvilinear Coordinates

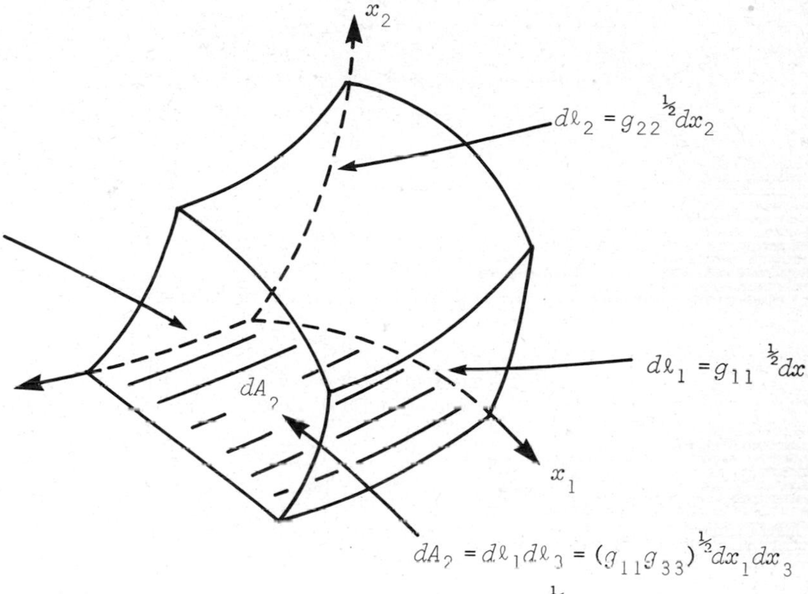

$$dv = d\ell_1 d\ell_2 d\ell_3 = (g_{11}g_{22}g_{33})^{\frac{1}{2}} dx_1 dx_2 dx_3$$

Unit Vectors

General Orthogonal	$\mathbf{i}_1$	$\mathbf{i}_2$	$\mathbf{i}_3$
Rectangular	$\mathbf{i}$	$\mathbf{j}$	$\mathbf{k}$
Cylindrical	$\mathbf{i}_R$	$\mathbf{i}_\Phi$	$\mathbf{i}_Z$
Spherical	$\mathbf{i}_r$	$\mathbf{i}_\theta$	$\mathbf{i}_\varphi$

At each point in space there are three mutually orthogonal unit vectors. Each unit vector is tangent to a coordinate line passing through the point and indicates the direction of positive increase of the coordinate line at the point.

General Vector Representation

General Orthogonal	$\mathbf{A}(x_1, x_2, x_3) = \hat{\mathbf{i}}_1 A_1(x_1, x_2, x_3) + \hat{\mathbf{i}}_2 A_2(x_1, x_2, x_3) + \hat{\mathbf{i}}_3 A_3(x_1, x_2, x_3)$
Rectangular	$\mathbf{A}(x, y, z) = \hat{\mathbf{i}} A_x(x, y, z) + \hat{\mathbf{j}} A_y(x, y, z) + \hat{\mathbf{k}} A_z(x, y, z)$
Cylindrical	$\mathbf{A}(R, \Phi, Z) = \hat{\mathbf{i}}_R A_R(R, \Phi, Z) + \hat{\mathbf{i}}_\Phi A_\Phi(R, \Phi, Z) + \hat{\mathbf{i}}_Z A_Z(R, \Phi, Z)$
Spherical	$\mathbf{A}(r, \theta, \varphi) = \hat{\mathbf{i}}_r A_r(r, \theta, \varphi) + \hat{\mathbf{i}}_\theta A_\theta(r, \theta, \varphi) + \hat{\mathbf{i}}_\varphi A_\varphi(r, \theta, \varphi)$

Vector Addition and Subtraction

$$\begin{aligned} \mathbf{C} = \mathbf{A} + \mathbf{B} &= (\hat{\mathbf{i}}_1 A_1 + \hat{\mathbf{i}}_2 A_2 + \hat{\mathbf{i}}_3 A_3) + (\hat{\mathbf{i}}_1 B_1 + \hat{\mathbf{i}}_2 B_2 + \hat{\mathbf{i}}_3 B_3) \\ &= \hat{\mathbf{i}}_1 (A_1 + B_1) + \hat{\mathbf{i}}_2 (A_2 + B_2) + \hat{\mathbf{i}}_3 (A_3 + B_3) \\ &= \hat{\mathbf{i}}_1 C_1 + \hat{\mathbf{i}}_2 C_2 + \hat{\mathbf{i}}_3 C_3 \end{aligned}$$

where

$$C_1 = A_1 + B_1, \quad C_2 = A_2 + B_2, \quad C_3 = A_3 + B_3$$

Subtraction is performed in an analogous manner.

Dot Product $\mathbf{A} \cdot \mathbf{B} = AB \cos [A,B]$

General Orthogonal	$\mathbf{A} \cdot \mathbf{B} = A_1 B_1 + A_2 B_2 + A_3 B_3$
Rectangular	$\mathbf{A} \cdot \mathbf{B} = A_x B_x + A_y B_y + A_z B_z$
Cylindrical	$\mathbf{A} \cdot \mathbf{B} = A_R B_R + A_\Phi B_\Phi + A_Z B_Z$
Spherical	$\mathbf{A} \cdot \mathbf{B} = A_r B_r + A_\theta B_\theta + A_\varphi B_\varphi$

The square of the magnitude of a vector is given by

$$A^2 = \mathbf{A} \cdot \mathbf{A} = A_1^2 + A_2^2 + A_3^2$$

$$\mathbf{A} \cdot \mathbf{B} = \mathbf{B} \cdot \mathbf{A}$$

Cross Product $\mathbf{C} = \mathbf{A} \times \mathbf{B}$ $C = AB \sin [A,B]$

General Orthogonal	$\mathbf{A} \times \mathbf{B} = \hat{\mathbf{i}}_1 (A_2 B_3 - A_3 B_2) + \hat{\mathbf{i}}_2 (A_3 B_1 - A_1 B_3) + \hat{\mathbf{i}}_3 (A_1 B_2 - A_2 B_1)$
Rectangular	$\mathbf{A} \times \mathbf{B} = \hat{\mathbf{i}}(A_y B_z - A_z B_y) + \hat{\mathbf{j}}(A_z B_x - A_x B_z) + \hat{\mathbf{k}}(A_x B_y - A_y B_x)$
Cylindrical	$\mathbf{A} \times \mathbf{B} = \hat{\mathbf{i}}_R (A_\Phi B_Z - A_Z B_\Phi) + \hat{\mathbf{i}}_\Phi (A_Z B_R - A_R B_Z) + \hat{\mathbf{i}}_Z (A_R B_\Phi - A_\Phi B_R)$
Spherical	$\mathbf{A} \times \mathbf{B} = \hat{\mathbf{i}}_r (A_\theta B_\varphi - A_\varphi B_\theta) + \hat{\mathbf{i}}_\theta (A_\varphi B_r - A_r B_\varphi) + \hat{\mathbf{i}}_\varphi (A_r B_\theta - A_\theta B_r)$

The cross product can be expressed as a determinant

General Orthogonal	Rectangular	Cylindrical	Spherical
$\mathbf{A}\times\mathbf{B} = \begin{vmatrix} \hat{\mathbf{i}}_1 & \hat{\mathbf{i}}_2 & \hat{\mathbf{i}}_3 \\ A_1 & A_2 & A_3 \\ B_1 & B_2 & B_3 \end{vmatrix}$	$\mathbf{A}\times\mathbf{B} = \begin{vmatrix} \hat{\mathbf{i}} & \hat{\mathbf{j}} & \hat{\mathbf{k}} \\ A_x & A_y & A_z \\ B_x & B_y & B_z \end{vmatrix}$	$\mathbf{A}\times\mathbf{B} = \begin{vmatrix} \hat{\mathbf{i}}_R & \hat{\mathbf{i}}_\Phi & \hat{\mathbf{i}}_Z \\ A_R & A_\Phi & A_Z \\ B_R & B_\Phi & B_Z \end{vmatrix}$	$\mathbf{A}\times\mathbf{B} = \begin{vmatrix} \hat{\mathbf{i}}_r & \hat{\mathbf{i}}_\theta & \hat{\mathbf{i}}_\varphi \\ A_r & A_\theta & A_\varphi \\ B_r & B_\theta & B_\varphi \end{vmatrix}$

$$\mathbf{A}\times\mathbf{B} = -\,\mathbf{A}\times\mathbf{B}$$

The vector $\mathbf{C} = \mathbf{A}\times\mathbf{B}$ will be perpendicular to the plane containing **A** and **B**. It will point in the direction that a right-handed screw would advance if it were screwed through the plane by rotating it in the direction established by rotating **A** (first vector) into **B** (second vector) through the smaller angle.

Dot and Cross Products of Unit Vectors

In each coordinate system $\hat{\mathbf{i}}_p \cdot \hat{\mathbf{i}}_q = 1$ if $p = q$, 0 if $p \neq q$

In each coordinate system the cross product of the unit vectors can be found by the scheme:

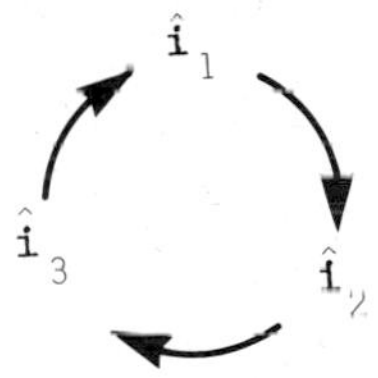

Thus $\hat{\mathbf{i}}_1 \times \hat{\mathbf{i}}_2 = \hat{\mathbf{i}}_3$; $\hat{\mathbf{i}}_3 \times \hat{\mathbf{i}}_1 = \hat{\mathbf{i}}_2$; $\hat{\mathbf{i}}_2 \times \hat{\mathbf{i}}_3 = \hat{\mathbf{i}}_1$. Reversing the order introduces a minus sign.

Dot Products Between Unit Vectors in Different Systems

Cylindrical <—> Rectangular

$\hat{\mathbf{i}}_R \cdot \hat{\mathbf{i}} = \cos\Phi$ $\qquad$ $\hat{\mathbf{i}}_\Phi \cdot \hat{\mathbf{i}} = -\sin\Phi$ $\qquad$ $\hat{\mathbf{i}}_Z \cdot \hat{\mathbf{i}} = 0$

$\hat{\mathbf{i}}_R \cdot \hat{\mathbf{j}} = \sin\Phi$ $\qquad$ $\hat{\mathbf{i}}_\Phi \cdot \hat{\mathbf{j}} = \cos\Phi$ $\qquad$ $\hat{\mathbf{i}}_Z \cdot \hat{\mathbf{j}} = 0$

$\hat{\mathbf{i}}_R \cdot \hat{\mathbf{k}} = 0$ $\qquad$ $\hat{\mathbf{i}}_\Phi \cdot \hat{\mathbf{k}} = 0$ $\qquad$ $\hat{\mathbf{i}}_Z \cdot \hat{\mathbf{k}} = 1$

Spherical <-> Cylindrical

$\hat{i}_r \cdot \hat{i}_R = \sin\theta$	$\hat{i}_\theta \cdot \hat{i}_R = \cos\theta$	$\hat{i}_\varphi \cdot \hat{i}_R = 0$
$\hat{i}_r \cdot \hat{i}_\Phi = 0$	$\hat{i}_\theta \cdot \hat{i}_\Phi = 0$	$\hat{i}_\varphi \cdot \hat{i}_\Phi = 1$
$\hat{i}_r \cdot \hat{i}_Z = \cos\theta$	$\hat{i}_\theta \cdot \hat{i}_Z = -\sin\theta$	$\hat{i}_\varphi \cdot \hat{i}_Z = 0$

Spherical <-> Rectangular

$\hat{i}_r \cdot \hat{i} = \sin\theta\cos\varphi$	$\hat{i}_\theta \cdot \hat{i} = \cos\theta\cos\varphi$	$\hat{i}_\varphi \cdot \hat{i} = -\sin\varphi$
$\hat{i}_r \cdot \hat{j} = \sin\theta\sin\varphi$	$\hat{i}_\theta \cdot \hat{j} = \cos\theta\sin\varphi$	$\hat{i}_\varphi \ \hat{j} = \cos\varphi$
$\hat{i}_r \cdot \hat{k} = \cos\theta$	$\hat{i}_\theta \cdot \hat{k} = -\sin\theta$	$\hat{i}_\varphi \cdot \hat{k} = 0$

Method for Changing Coordinate Representation of a Vector

To change A from rectangular to cylindrical coordinates first equate:

$$\hat{i}A_x + \hat{j}A_y + \hat{k}A_z = \hat{i}_R A_R + \hat{i}_\Phi A_\Phi + \hat{i}_Z A_Z,$$

where A_x, A_y, and A_z are assumed known.

Dot with $\hat{i}_R$ and obtain

$$\hat{i}_R \cdot \hat{i}A_x + \hat{i}_R \cdot \hat{j}A_y + \hat{i}_R \cdot \hat{k}A_z = A_R.$$

Dot with $\hat{i}_\Phi$ and obtain

$$\hat{i}_\Phi \cdot \hat{i}A_x + \hat{i}_\Phi \cdot \hat{j}A_y + \hat{i}_\Phi \cdot \hat{k}A_z = A_\Phi.$$

Dot with $\hat{i}_Z$ and obtain

$$\hat{i}_Z \cdot \hat{i}A_x + \hat{i}_Z \cdot \hat{j}A_y + \hat{i}_Z \cdot \hat{k}A_z = A_Z.$$

Evaluate dot products to complete transformation.

Line Element in Vector Notation

General Orthogonal	$d\vec{\ell} = \hat{i}_1[g_{11}(x_1, x_2, x_3)]^{\frac{1}{2}}dx_1 + \hat{i}_2[g_{22}(x_1, x_2, x_3)]^{\frac{1}{2}}dx_2 + \hat{i}_3[g_{33}(x_1, x_2, x_3)]^{\frac{1}{2}}dx_3$
Rectangular	$d\vec{\ell} = \hat{i}dx + \hat{j}dy + \hat{k}dz$
Cylindrical	$d\vec{\ell} = \hat{i}_R dR + \hat{i}_\Phi R d\Phi + \hat{i}_Z dZ$
Spherical	$d\vec{\ell} = \hat{i}_r dr + \hat{i}_\theta r d\theta + \hat{i}_\varphi r \sin\theta d\varphi$

Grad $U = \nabla U$

General Orthogonal	$\hat{\mathbf{i}}_1(g_{11})^{-\frac{1}{2}}\dfrac{\partial U}{\partial x_1}+\hat{\mathbf{i}}_2(g_{22})^{-\frac{1}{2}}\dfrac{\partial U}{\partial x_2}+\hat{\mathbf{i}}_3(g_{33})^{-\frac{1}{2}}\dfrac{\partial U}{\partial x_3}$
Rectangular	$\hat{\mathbf{i}}\dfrac{\partial U}{\partial x}+\hat{\mathbf{j}}\dfrac{\partial U}{\partial y}+\hat{\mathbf{k}}\dfrac{\partial U}{\partial z}$
Cylindrical	$\hat{\mathbf{i}}_R\dfrac{\partial U}{\partial R}+\hat{\mathbf{i}}_\Phi\dfrac{1}{R}\dfrac{\partial U}{\partial \Phi}+\hat{\mathbf{i}}_Z\dfrac{\partial U}{\partial Z}$
Spherical	$\hat{\mathbf{i}}_r\dfrac{\partial U}{\partial r}+\hat{\mathbf{i}}_\theta\dfrac{1}{r}\dfrac{\partial U}{\partial \theta}+\hat{\mathbf{i}}_\varphi\dfrac{1}{r\sin\theta}\dfrac{\partial U}{\partial \varphi}$

Div $\mathbf{F} = \nabla \cdot \mathbf{F}$

General Orthogonal	$(g_{11}g_{22}g_{33})^{-\frac{1}{2}}\left(\dfrac{\partial}{\partial x_1}[F_1(g_{22}g_{33})^{\frac{1}{2}}]+\dfrac{\partial}{\partial x_2}[F_2(g_{33}g_{11})^{\frac{1}{2}}]+\dfrac{\partial}{\partial x_3}[F_3(g_{11}g_{22})^{\frac{1}{2}}]\right)$
Rectangular	$\dfrac{\partial}{\partial x}F_x+\dfrac{\partial}{\partial y}F_y+\dfrac{\partial}{\partial z}F_z$
Cylindrical	$\dfrac{1}{R}\dfrac{\partial}{\partial R}(RF_R)+\dfrac{1}{R}\dfrac{\partial}{\partial \Phi}F_\Phi+\dfrac{\partial}{\partial Z}F_Z$
Spherical	$\dfrac{1}{r^2}\dfrac{\partial}{\partial r}(r^2F_r)+\dfrac{1}{r\sin\theta}\dfrac{\partial}{\partial\theta}(\sin\theta\, F_\theta)+\dfrac{1}{r\sin\theta}\dfrac{\partial}{\partial\varphi}F_\varphi$

Curl $\mathbf{F} = \nabla \times \mathbf{F}$

General Orthogonal

$$\hat{\mathbf{i}}_1(g_{22}g_{33})^{-\frac{1}{2}}\left(\frac{\partial}{\partial x_2}[F_3(g_{33})^{\frac{1}{2}}]-\frac{\partial}{\partial x_3}[F_2(g_{22})^{\frac{1}{2}}]\right)$$
$$+\hat{\mathbf{i}}_2(g_{33}g_{11})^{-\frac{1}{2}}\left(\frac{\partial}{\partial x_3}[F_1(g_{11})^{\frac{1}{2}}]-\frac{\partial}{\partial x_1}[F_3(g_{33})^{\frac{1}{2}}]\right)$$
$$+\hat{\mathbf{i}}_3(g_{11}g_{22})^{-\frac{1}{2}}\left(\frac{\partial}{\partial x_1}[F_2(g_{22})^{\frac{1}{2}}]-\frac{\partial}{\partial x_2}[F_1(g_{11})^{\frac{1}{2}}]\right)$$
$$=(g_{11}g_{22}g_{33})^{-\frac{1}{2}}\begin{vmatrix}\hat{\mathbf{i}}_1(g_{11})^{\frac{1}{2}} & \hat{\mathbf{i}}_2(g_{22})^{\frac{1}{2}} & \hat{\mathbf{i}}_3(g_{33})^{\frac{1}{2}}\\ \dfrac{\partial}{\partial x_1} & \dfrac{\partial}{\partial x_2} & \dfrac{\partial}{\partial x_3}\\ F_1(g_{11})^{\frac{1}{2}} & F_2(g_{22})^{\frac{1}{2}} & F_3(g_{33})^{\frac{1}{2}}\end{vmatrix}$$

Determinant should be expanded on first row.

Rectangular $\hat{\mathbf{i}}\left(\dfrac{\partial}{\partial y}F_z-\dfrac{\partial}{\partial z}F_y\right)+\hat{\mathbf{j}}\left(\dfrac{\partial}{\partial z}F_x-\dfrac{\partial}{\partial x}F_z\right)+\hat{\mathbf{k}}\left(\dfrac{\partial}{\partial x}F_y-\dfrac{\partial}{\partial y}F_x\right)$

Cylindrical $\hat{i}_R\left[\frac{1}{R}\frac{\partial}{\partial \Phi}F_Z - \frac{\partial}{\partial Z}F_\Phi\right] + \hat{i}_\Phi\left[\frac{\partial}{\partial Z}F_R - \frac{\partial}{\partial R}F_Z\right] + \hat{i}_Z\,\frac{1}{R}\left[\frac{\partial}{\partial R}(RF_\Phi) - \frac{\partial}{\partial \Phi}F_R\right]$

Spherical $\hat{i}_r\,\frac{1}{r\sin\theta}\left[\frac{\partial}{\partial\theta}(\sin\theta F_\varphi) - \frac{\partial}{\partial\varphi}F_\theta\right] + \hat{i}_\theta\left[\frac{1}{r\sin\theta}\frac{\partial}{\partial\varphi}F_r - \frac{1}{r}\frac{\partial}{\partial r}(rF_\varphi)\right]$

$$+\hat{i}_\varphi\,\frac{1}{r}\left[\frac{\partial}{\partial r}(rF_\theta) - \frac{\partial}{\partial\theta}F_r\right]$$

$$\nabla^2 U = \text{Div Grad } U = \text{Laplacian } U$$

General Orthogonal $(g_{11}g_{22}g_{33})^{-1}\left\{\frac{\partial}{\partial x_1}\left[(g_{22}g_{33}/g_{11})^{\frac{1}{2}}\frac{\partial U}{\partial x_1}\right]\right.$

$$\left.+\frac{\partial}{\partial x_2}\left[(g_{33}g_{11}/g_{22})^{\frac{1}{2}}\frac{\partial U}{\partial x_2}\right] + \frac{\partial}{\partial x_3}\left[(g_{11}g_{22}/g_{33})^{\frac{1}{2}}\frac{\partial U}{\partial x_3}\right]\right\}$$

Rectangular $\frac{\partial^2 U}{\partial x\,\partial x} + \frac{\partial^2 U}{\partial y\,\partial y} + \frac{\partial^2 U}{\partial z\,\partial z}$

Cylindrical $\frac{1}{R}\frac{\partial}{\partial R}(R\frac{\partial U}{\partial R}) + \frac{1}{R^2}\frac{\partial^2 U}{\partial\Phi\partial\Phi} + \frac{\partial^2 U}{\partial Z\partial Z}$

Spherical $\frac{1}{r^2}\frac{\partial}{\partial r}(r^2\frac{\partial U}{\partial r}) + \frac{1}{r^2\sin\theta}\frac{\partial}{\partial\theta}(\sin\theta\frac{\partial U}{\partial\theta}) + \frac{1}{r^2\sin^2\theta}\frac{\partial^2 U}{\partial\varphi\partial\varphi}$

$$\oint_s \mathbf{F}\cdot\hat{\mathbf{n}}\,da = \int_v \text{div } \mathbf{F}\,dv \quad \text{(Divergence theorem)}$$

$$\oint_c \mathbf{F}\cdot d\mathbf{l} = \int_s \text{curl } \mathbf{F}\cdot\hat{\mathbf{n}}\,da \quad \text{(Stokes' theorem)}$$

$$\int_v (\tau\nabla^2\mu - \mu\nabla^2\tau)\,dv = \oint_s (\tau \text{ grad } \mu - \mu \text{ grad } \tau)\cdot\hat{\mathbf{n}}\,da \quad \text{(Green's theorem)}$$

PROBLEM 1.11 Determine the diverngence and curl of the following vector:

$$\mathbf{A} = \hat{\mathbf{r}}r^2\cos\theta + \hat{\theta}r + \hat{\varphi}\sin\theta$$

Solution:

$$\mathbf{A} = \hat{\mathbf{r}}r^2\cos\theta + \hat{\theta}r + \hat{\varphi}\sin\theta$$

Divergencc

$$\nabla\cdot\mathbf{A} = \frac{1}{r^2}\frac{\partial(r^2A_r)}{\partial r} + \frac{1}{r\sin\theta}\frac{\partial(A_\theta\sin\theta)}{\partial\theta} + \frac{1}{r\sin\theta}\frac{\partial A_\varphi}{\partial\varphi}$$

$$\nabla\cdot\mathbf{A} = \frac{1}{r^2}\frac{\partial}{\partial r}(r^4\cos\theta) + \frac{1}{r\sin\theta}\frac{\partial}{\partial\theta}(r\sin\theta) + \frac{1}{r\sin\theta}\frac{\partial(\sin\theta)}{\partial\varphi}$$

$$\nabla\cdot\mathbf{A} = 4r\cos\theta + \cot\theta$$

Curl

$$\nabla \times \mathbf{A} = \begin{vmatrix} \dfrac{\hat{r}}{r^2 \sin\theta} & \dfrac{\hat{\theta}}{r\sin\theta} & \dfrac{\hat{\varphi}}{r} \\ \dfrac{\partial}{\partial r} & \dfrac{\partial}{\partial \theta} & \dfrac{\partial}{\partial \varphi} \\ r^2\cos\theta & r^2 & r\sin^2\theta \end{vmatrix}$$

$$\nabla \times \mathbf{A} = \frac{\hat{r}}{r^2 \sin\theta}(2r\sin\theta\cos\theta) - \frac{\hat{\theta}}{r\sin\theta}(\sin^2\theta - 0) + \frac{\hat{\varphi}}{r}(2r + r^2\sin\theta)$$

$$\nabla \times \mathbf{A} = \hat{r}\,\frac{2\cos\theta}{r} - \hat{\theta}\,\frac{\sin\theta}{r} + \hat{\varphi}\,(2 + r\sin\theta)$$

PROBLEM 1.12 If A is any vector field with continuous derivatives, div (curl A) = 0 or, using the "del" notation, $\nabla \cdot (\nabla \times \mathbf{A}) = 0$. Here are two different ways to prove it:

(a) (Uninspired straightforward calculation in a particular coordinate system): Using the formula for ∇ in Cartesian coordinates, work out the string of second partial derivatives that $\nabla \cdot (\nabla \times \mathbf{A})$ implies.

(b) (With the divergence theorem and Stokes' theorem, no coordinates are needed): Consider the surface S in the figure, a balloon almost cut in two, which is bounded by the closed curve C. Think about the line interval, over a curve like C, of any vector field. Then invoke Stokes' and Gauss' theorems with suitable arguments.

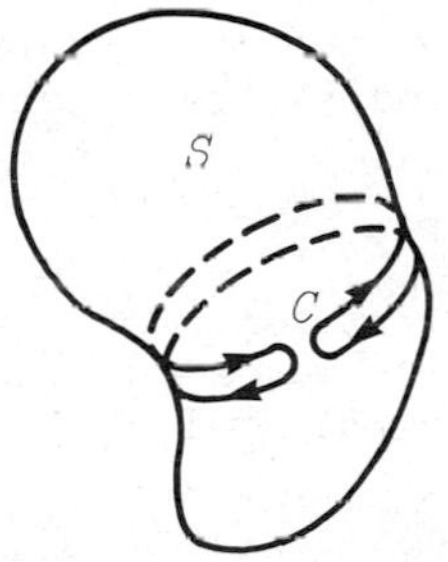

Solution:

(a) Using a straightforward calculation:

$$\nabla \cdot (\nabla \times \mathbf{A}) = \frac{\partial}{\partial x}\left(\frac{\partial A_z}{\partial y} - \frac{\partial A_y}{\partial z}\right) + \frac{\partial}{\partial y}\left(\frac{\partial A_x}{\partial z} - \frac{\partial A_z}{\partial x}\right) + \frac{\partial}{\partial z}\left(\frac{\partial A_y}{\partial x} - \frac{\partial A_x}{\partial y}\right)$$

$$= \frac{\partial^2 A_z}{\partial x\,\partial y} - \frac{\partial^2 A_y}{\partial x\,\partial z} + \frac{\partial^2 A_x}{\partial y\,\partial z} - \frac{\partial^2 A_z}{\partial y\,\partial x} + \frac{\partial^2 A_y}{\partial z\,\partial x} - \frac{\partial^2 A_x}{\partial z\,\partial y} = 0$$

(b) Using Stokes' and Gauss' theorems:

$$\oint_C \mathbf{a} \cdot d\mathbf{S} = \int_S (\nabla \times \mathbf{A}) \cdot d\mathbf{a}$$

If C is allowed to be brought to a single line so that S encloses v, then

$$\int_S (\nabla \times \mathbf{A}) \quad d\mathbf{a} = \int_v \nabla \cdot (\nabla \times \mathbf{A}) dv$$

But, in that case, $\oint \mathbf{A} \cdot d\mathbf{S}$ is the integral out and back along the same line and must be zero. So,

$$\oint_C \mathbf{A} \cdot d\mathbf{S} = 0 = \int_S (\nabla \times \mathbf{A}) \cdot d\mathbf{a} = \int_v \nabla \cdot (\nabla \times \mathbf{A}) dv$$

and, since v is arbitrary, this must be true for all integrands, that is,

$$\nabla \cdot (\nabla \times \mathbf{A}) \equiv 0$$

PROBLEM 1.13

(a) By explicitly calculating the components of $\nabla \times \mathbf{E}$, show that the vector function $E_x = 6xy$, $E_y = 3x^2 - 3y^2$, $E_z = 0$ is a possible electrostatic field. (Of course, you may also prove it in another way by finding a scalar function of which it is the gradient.) Evaluate the divergence of this field.

(b) Draw "field lines" for the vector function $\mathbf{A} = -y\hat{\mathbf{x}} + x\hat{\mathbf{y}}$ in the xy plane. Calculate the curl of $\mathbf{A}$ and sketch a vector indicating its direction. Calculate the integral

$$\oint \mathbf{A} \cdot d\boldsymbol{\ell}$$

over the closed curve $x^2 + y^2 = 1$, $z = 0$. Show that Stokes' theorem holds, by calculating the surface integral of $\nabla \times \mathbf{A}$ over the surface bounded by this curve.

Solution:

(a) $E_x = 6xy, \quad E_y = 3x^2 - 3y^2, \quad E_z = 0$:

$$\nabla \times \mathbf{E} = \left(\frac{\partial(0)}{\partial y} - \frac{\partial(3x^2 - 3y^2)}{\partial z}\right)\hat{\mathbf{x}} + \left(\frac{\partial(6xy)}{\partial z} - \frac{\partial(0)}{\partial x}\right)\hat{\mathbf{y}} + \left(\frac{\partial(3x^2 - 3y^2)}{\partial x} - \frac{\partial(6xy)}{\partial y}\right)\hat{\mathbf{z}}$$

$$= 0\hat{\mathbf{x}} + 0\hat{\mathbf{y}} + (6x - 6x)\hat{\mathbf{z}} = 0$$

$$\nabla \cdot \mathbf{E} = \frac{\partial(6xy)}{\partial x} + \frac{\partial(3x^2 - 3y^2)}{\partial y} + \frac{\partial(0)}{\partial z} = 6y - 6y = 0$$

(b) For $\mathbf{A} = -y\hat{\mathbf{x}} + x\hat{\mathbf{y}}$,

$$\nabla \times \mathbf{A} = \left(\frac{\partial(x)}{\partial x} - \frac{\partial(-y)}{\partial y}\right)\hat{\mathbf{z}} = 2\hat{\mathbf{z}}$$

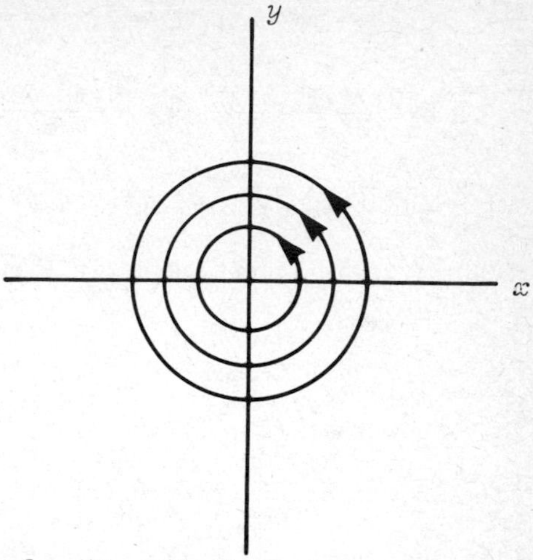

Since $\mathbf{A} = -y\hat{\mathbf{x}} + x\hat{\mathbf{y}}$ is the equation for a line element along a circle of radius

$r = \sqrt{x^2 + y^2}$, $\mathbf{A}$ and $d\boldsymbol{\ell}$ are parallel and

$$\oint \mathbf{A} \cdot d\boldsymbol{\ell} = \int_0^{2\pi} r^2 d\theta = 2\pi r^2 = 2\pi$$

with $r^2 = 1$. With $d\mathbf{a} = da\hat{\mathbf{z}}$

$$\int (\nabla \times \mathbf{A}) \cdot d\mathbf{a} = 2\int da \text{ (over a circle of unit radius)}$$

$$= 2\pi r^2 = 2\pi$$

PROBLEM 1.14 Calculate the curl and the divergence of each of the following vector fields. If the curl turns out to be zero, try to discover a scalar function ϕ of which the vector field is the gradient:

(a) $F_x = x + y$; $F_y = -x + y$; $F_z = -2z$.

(b) $G_x = 2y$; $G_y = 2x + 3z$; $G_z = 3y$.

(c) $H_x = x^2 - z^2$; $H_y = 2$; $H_z = 2xz$.

Solution:

(a) For $F_x = x + y$; $F_y = -x + y$; $F_z = -2z$:

$$\nabla \times \mathbf{F} = \left(\frac{\partial(-2z)}{\partial x} - \frac{\partial(-x+y)}{\partial z}\right)\hat{\mathbf{x}} + \left(\frac{\partial(x+y)}{\partial z} - \frac{\partial(-2z)}{\partial x}\right)\hat{\mathbf{y}} + \left(\frac{\partial(-x+y)}{\partial x} - \frac{\partial(x+y)}{\partial y}\right)\hat{\mathbf{z}}$$

$$= 0\hat{\mathbf{x}} + 0\hat{\mathbf{y}} + (-1-1)\hat{\mathbf{z}} = -2\hat{\mathbf{z}}$$

(b) For $G_x = 2y$; $G_y = 2x + 3z$; $G_z = 3y$:

$$\nabla \times \mathbf{G} = \left(\frac{\partial(3y)}{\partial y} - \frac{\partial(2x+3z)}{\partial z}\right)\hat{\mathbf{x}} + \left(\frac{\partial(2y)}{\partial z} - \frac{\partial(3y)}{\partial x}\right)\hat{\mathbf{y}} + \left(\frac{\partial(2x+3z)}{\partial x} - \frac{\partial(2y)}{\partial y}\right)\hat{\mathbf{z}}$$

$$= (3-3)\hat{\mathbf{x}} + 0\hat{\mathbf{y}} + (2-2)\hat{\mathbf{z}} = 0$$

Try $\int \mathbf{G} \cdot d\boldsymbol{\ell}$ for $d\boldsymbol{\ell} = \hat{\mathbf{x}}dx$, $\hat{\mathbf{y}}dy$, $\hat{\mathbf{z}}dz$. $\int G_x d_x = 2xy + f(y,z)$, $\int G_y d_y = 2xy + 3yz + g(x,z)$, $\int G_z d_z = 3yz + h(x,y)$. From this it is clear that $\phi = 2xy + 3yz$ satisfies all three equations.

(c) For $H_x = x^2 - z^2$; $H_y = 2$; $H_z = 2xz$:

$$\nabla \times \mathrm{H} = \left(\frac{\partial(2xy)}{\partial z} - \frac{\partial(2)}{\partial y}\right)\hat{\mathrm{x}} + \left(\frac{\partial(x^2 - z^2)}{\partial z} - \frac{\partial(2xz)}{\partial x}\right)\hat{\mathrm{y}} + \left(\frac{\partial(2)}{\partial x} - \frac{\partial(x^2 - z^2)}{\partial y}\right)\hat{\mathrm{z}}$$

$$= 0\hat{\mathrm{x}} + (-2z - 2z)\hat{\mathrm{y}} + 0\hat{\mathrm{z}} = -4z\hat{\mathrm{y}}$$

PROBLEM 1.15 Accept as definition that $P_n(x) \equiv (1/2^n n!)D^n(x^2 - 1)^n$, where $D \equiv d/dx$.

(a) Show that $P'_{n+1}(x) - P'_{n-1}(x) = (2n+1)P_n(x)$

(b) Show that $(2n+1)x\, P_n(x) = (n+1)P_{n+1}(x) + nP_{n-1}(x)$

(c) Show that $nP_n(x) = xP_n(x) - P'_{n-1}(x)$

(d) Show that $(1 - x^2)P'_n(x) = nP_{n-1} - nxP_n$

(c) and (d) may be derived from (a) and (b). Therefore, we have obtained the following results:

$$P'_{n+1}(x) - P'_{n-1}(x) = (2n+1)P_n(x) \qquad \text{(i)}$$

$$(2n+1)x\, P_n(x) = (n+1)P_{n+1} + nP_{n-1} \qquad \text{(ii)}$$

$$nP_n(x) = xP'_n - P'_{n-1} \qquad \text{(iii)}$$

$$(1 - x^2)P'_n(x) = nP_{n-1}(x) - nxP_n(x) \qquad \text{(iv)}$$

(e) Now show that the Legendre's polynomials satisfy Legendre's equation.

Solution:

(a) $$P'_{n+1}(x) - P'_{n-1}(x) = \frac{1}{2^{n+1}(n+1)!} D^{n+2}(x^2 - 1)^{n+1} - \frac{1}{2^{n-1}(n-1)!} D^n(x^2 - 1)^{n-1}$$

$$= \frac{1}{2^{n+1}(n+1)!} D^{n+1}\{(n+1)(x^2 - 1)^n 2x\} - \frac{1}{2^{n-1}(n-1)!} D^n(x^2 - 1)^{n-1}$$

$$= \frac{1}{2^n n!} D^{n+1}\{x(x^2 - 1)^n\} - \frac{1}{2^{n-1}(n-1)!} D^n(x^2 - 1)^{n-1}$$

$$= \frac{1}{2^n n!} D^n\{(x^2 - 1)^n + n(x^2 - 1)^{n-1}2x^2\} - \frac{1}{2^{n-1}(n-1)!} D^n(x^2 - 1)^{n-1}$$

$$= \frac{1}{2^n n!} D^n(x^2 - 1)^n + \frac{1}{2^{n-1}(n-1)!} D^n\{x^2(x^2 - 1)^{n-1}\}$$

$$- \frac{1}{2^{n-1}(n-1)!} D^n(x^2 - 1)^{n-1}$$

$$= \frac{1}{2^n n!} D^n (x^2 - 1)^n + \frac{1}{2^{n-1}(n-1)!} D^n\{x^2(x^2 - 1)^{n-1} - (x^2 - 1)^{n-1}\}$$

$$= \frac{1}{2^n n!} D^n (x^2 - 1)^n + \frac{2n}{2^n n!} D^n (x^2 - 1)^n$$

$$= (1 + 2n)P_n(x)$$

$$\therefore\ P'_{n+1}(x) - P'_{n-1}(x) = (2n+1)P_n(x) \text{ as required.}$$

(b) We first note that if f is an arbitrary function, then

$$D^n(fx) = x(D^n f) + n(D^{n-1} f) \quad \text{(v)}$$

This can be proved either by induction or by the Liebnitz formula

$$D^n[f(x)s(x)] = \sum_{m=0}^{n} \frac{n!}{m!(n-m)!} (D^{n-m} f)(d^m s)$$

Also,

$$P_{n+1}(x) = \frac{1}{2^{n+1}(n+1)!} D^{n+1}(x^2 - 1)^{n+1} = \frac{1}{2^n n!} D^n[x(x^2 - 1)^n] \quad \text{(vi)}$$

Using (v) with $f = (x^2 - 1)^n$, we can therefore write

$$(2n+1)x\ P_n(x) = \frac{(2n+1)}{2^n n!} xD^n(x^2 - 1)^n = \frac{2n+1}{2^n n!} \{D^n[x(x^2 - 1)^n] - nD^{n-1}(x^2 - 1)^n\}$$

$$= \frac{(n+1)}{2^n n!} D^n[x(x^2 - 1)^n] + \frac{n}{2^n n!} D^n[x(x \quad - 1)^n] - \frac{2n^2}{2^n n!} D^{n-1}(x^2 - 1)^n$$

$$- \frac{n}{2^n n!} D^{n-1}(x^2 - 1)^n$$

$$= (n+1)P_{n+1} + \frac{1}{2^n(n-1)!} D^n[(x^2 - 1)^n + nx(x^2 - 1)^{n-1} 2x]$$

$$- \frac{n}{2^{n-1}(n-1)!} D^{n-1}(x^2 - 1)^n - \frac{1}{2^n(n-1)!} D^{n-1}(x^2 - 1)^n$$

$$= (n+1)P_{n+1} + \frac{1}{2^n(n-1)!} D^n(x^2 - 1)^n + \frac{n}{2^{n-1}(n-1)!} D^{n-1}[x^2(x^2 - 1)^{n-1}]$$

$$- \frac{n}{2^{n-1}(n-1)!} D^{n-1}(x^2 - 1)^n - \frac{1}{2^n(n-1)!} D^{n-1}(x^2 - 1)^n$$

$$= (n+1)P_{n+1} + \frac{n}{2^{n-1}(n-1)!} D^{n-1}[x^2(x^2-1)^{n-1} - (x^2-1)^n]$$

$$= (n+1)P_{n+1} + \frac{n}{2^{n-1}(n-1)!} D^{n-1}[(x^2-1)^{n-1}(x^2-x^2+1)]$$

$$= (n+1)P_{n+1} + \frac{n}{2^{n-1}(n-1)!} D^{n-1}(x^2-1)^{n-1} = (n+1)P_{n+1} + nP_{n-1},$$

that is,

$$(2n+1)x\, P_n(x) = (n+1)P_{n+1} + nP_{n-1} \quad \text{as required.}$$

(c) Differentiate Eq. (ii),

$$(2n+1)P_n + (2n+1)x\, P'_n = (n+1)P'_{n+1} + nP'_{n-1}$$

From (i),

$$(n+1)P'_{n+1} = (2n+1)(n+1)P_n(x) + (n+1)P'_{n-1}$$

Substituting,

$$(2n+1)P_n + (2n+1)x\, P'_n = (2n+1)(n+1)P_n(x) + (n+1)P'_{n-1} + nP'_{n-1}$$

that is,

$$(2n+1)P_n + (2n+1)x\, P'_n = (2n+1)(n+1)P_n(x) + (2n+1)P'_{n-1}$$

$$\therefore \quad P_n + P'_n = nP_n(x) + P_n(x) + P'_{n-1}(x)$$

$$\therefore \quad nP_n(x) = xP'_n - P'_{n-1}(x) \quad \text{as required.}$$

(d) To prove this result, we start with Eqs. (i) and (iii)

$$nP_n(x) = xP'_n - P'_{n-1}(x) \qquad \text{(iii)}$$

$$-(2n+1)P_n(x) = P'_{n+1} + P'_{n-1}(x) \qquad \text{(i)}$$

adding $-(n+1)P_n = xP'_n - P'_{n+1}$, that is:

$$P'_{n+1} = xP'_n + (n+1)P_n$$

or equivalently

$$P'_n = xP'_{n-1} + nP_{n-1} \qquad \text{(vii)}$$

Now, multiply Eq. (iii) by x:

$$nxP_n(x) = x^2P'_n - xP'_{n-1}$$

Therefore,

$$(1 - x^2)P'_n = P'_n - xP'_{n-1} - nxP_n$$

$$= nP_{n-1} - nxP_n \quad \text{by Eq. (vii)}$$

which is the required result. Therefore,

$$(1 - x^2)P'_n(x) = nP_{n-1}(x) - nxP_n(x)$$

(e) To show that the Legendre's polynomials satisfy Legendre's equation, we differentiate Eq. (iv),

$$\frac{d}{dx}\{(1 - x^2)P'_n(x)\} = nP'_{n-1}(x) - nP_n x - nxP'_n(x)$$

Using Eq. (iii), this becomes

$$\frac{d}{dx}\{(1 - x^2)P'_n(x)\} = nxP'_n - n^2P_n(x) - nP_n(x) - nxP'_n(x)$$

$$\therefore \frac{d}{dx}\{(1 - x^2)P'_n(x)\} + n(n+1)P_n(x) = 0 \quad \text{as required!}$$

PROBLEM 1.16

(a) Prove the orthogonality properties of the Legendre polynomials.
(b) Expand an arbitrary function $f(x)$ defined in the interval $[-1, 1]$.

Solution: Consider

$$P_m\left[\frac{d}{dx}(1 - x^2)P'_n(x) + n(n+1)P_n(x)\right] = 0$$

$$P_n\left[\frac{d}{dx}(1 - x^2)P'_m(x) + m(m+1)P_m(x)\right] = 0$$

Subtracting and integrating from -1 to $+1$, we obtain

$$\int_{-1}^{1} P_m \frac{d}{dx}\left[(1 - x^2)P'_n\right]dx - \int_{-1}^{1} P_n \frac{d}{dx}\{(1 - x^2)P'_m\}dx$$

$$= m(m+1) - n(n+1)\int_{-1}^{1} P_m(x)P_n(x)dx$$

The left-hand side of the equation is clearly zero, if one integrates by parts $[(x^2 - 1) = 0$ at $x = \pm 1]$. Therefore, we have

$$m(m+1) - n(n+1)\int_{-1}^{1} P_m(x)P_n(x)dx = 0$$

So, if $m \neq n$, the integral must vanish. Therefore,

$$\int_{-1}^{1} P_m(x) P_n(x)\,dx = 0 \quad \text{if } m \neq n$$

We now need to evaluate $\int_{-1}^{1} P_n^2(x)\,dx$. Consider:

$$(2^n n!)^2 \int_{-1}^{+1} P_n^2(x)\,dx = \int_{-1}^{+1} D^n(x^2-1) D^n(x^2-1)^n\,dx$$

$$= D^{n-1}(x^2-1)^n D^n(x^2-1)^n \Big|_{-1}^{+1} - \int_{-1}^{+1} D^{n-1}(x^2-1) D^{n+1}(x^2-1)^n\,dx$$

$$= -\int_{-1}^{+1} D^{n-1}(x^2-1)^n D^{n+1}(x^2-1)^n\,dx,$$

since $D^{n-1}(x^2-1)^n$ contains (x^2-1) as a factor repeating the procedure n times.

$$(2^n n!)^2 \int_{-1}^{+1} P_n^2\,dx = (-1)^n \int_{-1}^{+1} (x^2-1) D \quad (x^2-1)^n\,dx = (-1)^n (2n)! \int_{-1}^{+1} (x^2-1)^n\,dx$$

where we have used $D^{2n}(x^2-1)^n = (2n)!$

$$\int_{-1}^{+1} P_n^2(x)\,dx = \frac{(-1)^n\,(2n)!}{(2^n n!)^2} \int_{-1}^{+1} (x^2-1)^n\,dx = \frac{(2n)!}{2^{2n}(n!)^2} \int_{-1}^{+1} (1-x^2)^n\,dx$$

But

$$\int_{-1}^{+1} (1-x^2)^n\,dx = 2\int_0^1 (1-x^2)^n\,dx = \int_0^1 (1-t)^n t^{-\frac{1}{2}}\,dt = \frac{\Gamma(n+1)\Gamma(\frac{1}{2})}{\Gamma(n+{}^3/_2)}$$

where $\Gamma(x)$ is the gamma function,

$$\Gamma(x) \equiv \int_0^\infty t^{x-1} e^{-t}\,dt,\ \Gamma(x) = (x-1)\Gamma(x-1);\ \Gamma(n) = (n-1)! \text{ if } n \text{ is an integer}$$

$$\therefore \int_{-1}^{+1} P_n^2(x)\,dx = \frac{(2n)!}{2^{2n}(n!)^2} \; \underbrace{\frac{n!\Gamma(\frac{1}{2})}{(n+\frac{1}{2})(n-\frac{1}{2})\ldots(\frac{3}{2})\frac{1}{2}\Gamma(\frac{1}{2})}}_{N+1 \text{ factors}} = \frac{(2n)!\,n!}{2^{2n}(n!)^2} \; \frac{2^{n+1}}{(2n+1)(2n-1)\cdot 3}$$

but

$$\begin{aligned}(2n+1)! &= \{(2n+1)(2n-1)\ldots 5\cdot 3\}\{(2n)(2n-2)\ldots 2\}\\ &= \{(2n+1)(2n-1)\ldots 5\cdot 3\}\,2^n\{n(n-1)(n-2)\ldots 1\}\\ &= \{(2n+1)(2n-1)\ldots 5\cdot 3\}\,2^n n!\end{aligned}$$

Therefore,

$$\int_{-1}^{1} P_n^2(x)\,dx = \frac{(2n)!2}{2^n n!}\,\frac{2^n n!}{(2n+1)!} = \frac{2(2n)!}{(2n+1)!} = \frac{2}{2n+1}$$

Finally, we express the orthogonality of the Legendre polynomials in the form

$$\int_{-1}^{+1} P_m(x)P_n(x)\,dx = \frac{2}{2n+1}\,\delta_{nm} \tag{i}$$

where

$$\delta_{nm} = \begin{cases} 0 & \text{if } n \neq m \\ 1 & \text{if } n = m \end{cases}$$

Let

$$f(x) = \sum_{n=0}^{\infty} A_n P_n(x)$$

Multiplying this equation by $P_m(x)$ and integrating, we obtain

$$\int_{-1}^{+1} f(x)P_m(x)\,dx = \sum_{n=0}^{\infty} A_n \int_{-1}^{+1} P_n(x)P_m(x) = \sum_{n=0}^{\infty} A_n \frac{1}{2n+1}\,\delta_{mn}$$

$$= A_m \frac{2}{2m+1}$$

that is,

$$f(x) = \sum_{n=0}^{\infty} A_n P_n(x), \quad -1 \le x \le 1 \tag{ii}$$

$$A_n = \frac{2n+1}{2}\int_{-1}^{+1} f(x)P_n(x)$$

PROBLEM 1.17 Evaluate $\int_{-1}^{+1} x^2 P_n P_m\,dx$

Solution:

$$(2n+1)x\,P_n(x) = (n+1)P_{n+1} + nP_{n-1}$$

or

$$xP_n = \left(\frac{n+1}{2n+1}\right)P_{n+1} + \left(\frac{n}{2n+1}\right)P_{n-1} \tag{i}$$

Therefore,

$$x^2 P_n = \left(\frac{n+1}{2n+1}\right) P_{n+1} + \left(\frac{n}{2n+1}\right) P_{n-1} \tag{ii}$$

Using Eq. (i) again

$$xP_{n+1} = \left(\frac{n+2}{2n+3}\right) P_{n+2} + \left(\frac{n+1}{2n+3}\right) P_n$$

$$xP_{n-1} = \left(\frac{n}{2n-1}\right) P_n + \left(\frac{n-1}{2n-1}\right) P_{n-2}$$

So that Eq. (ii) becomes

$$x^2 P_n = \left(\frac{n+1}{2n+1}\right)\left[\left(\frac{n+2}{2n+3}\right) P_{n+2} + \left(\frac{n+1}{2n+3}\right) P_n\right] + \left(\frac{n}{2n+1}\right)\left[\frac{n}{2n-1} P_n + \left(\frac{n-1}{2n-1}\right) P_{n-2}\right]$$

$$= \frac{(n+1)(n+2)}{(2n+1)(2n+3)} P_{n+2} + \left[\frac{(n+1)^2}{(2n+1)(2n+3)} + \frac{n^2}{(2n+1)(2n-1)}\right] P_n$$

$$+ \frac{n(n-1)}{(2n+1)(2n-1)} P_{n-2}$$

Therefore,

$$\int_{-1}^{+1} x^2 P_n P_m = \frac{(n+1)(n+2)}{(2n+1)(2n+3)} \int_{-1}^{+1} P_{n+2} P_m + \left[\frac{(n+1)^2}{(2n+1)(2n+3)}\right.$$

$$\left. + \frac{n^2}{(2n+1)(2n-1)}\right] \int_{-1}^{+1} P_n P_m dx + \frac{n(n-1)}{(2n+1)(2n-1)} \int_{-1}^{+1} P_{n-2} P_m$$

So, the integral $\int_{-1}^{1} x^2 P_n P_m$ is equal to zero except for

(a) $m = n + 2$, then

$$\int x^2 P_n P_{n+2} = \frac{(n+1)(n+2)}{(2n+1)(2n+3)} \frac{2}{2(n+2)+1} = \frac{2(n+1)(n+2)}{(2n+1)(2n+3)(2n+5)}$$

(b) $m = n$, then

$$\int x^2 P_n P_n = \frac{2}{(2n+1)^2} \left(\frac{(n+1)^2}{2n+3} + \frac{n^2}{(2n-1)}\right)$$

[When $m = n - 2$, the integral is also nonzero. But this is case (a), with m and n interchanged.]

PROBLEM 1.18 Consider the function

$$F(x,a) = \frac{1}{\sqrt{2\pi}} \frac{\exp[-(x^2/2a^2)]}{a}$$

Plot $F(x,a)$ for $a = 1$, 0.4, and 0.1. What is the area under the curve as a function of a? Show that if we define

$$\Delta(x - x_0) \equiv \lim_{a \to 0} F(x - x_0, a)$$

then,

$$G(\xi) = \int_{-\infty}^{+\infty} \Delta(\xi - x) G(x)\, dx$$

where G is an arbitrary function. Identify the function $(\xi - x)$.

Solution:

$$F(x,a) = \frac{1}{\sqrt{2\pi}a} e^{-x^2/2a^2}$$

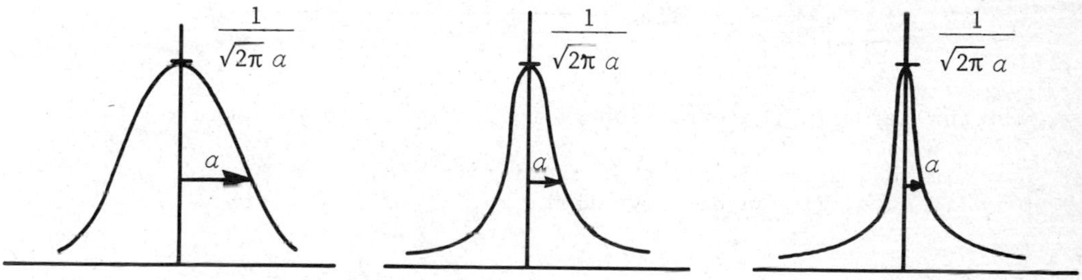

→ Decreasing a

$$\int_{-\infty}^{+\infty} \frac{1}{\sqrt{2\pi}a} e^{-x^2/2a^2} dx = \frac{1}{\sqrt{2\pi}a} \int_{-\infty}^{+\infty} e^{-x^2/2a^2} dx;$$

Use

$$\int_0^{\infty} e^{-\alpha x^2} dx = \tfrac{1}{2}\sqrt{\pi/\alpha}$$

Therefore, with $\alpha = \frac{1}{2}a^2$,

$$\int_{-\infty}^{+\infty} e^{-x^2/2a^2} dx = 2 \int_0^{\infty} e^{-x^2/2a^2} dx = \sqrt{\pi 2a^2} = \sqrt{2\pi}a$$

Therefore,

$$\int_{-\infty}^{+\infty} \frac{1}{\sqrt{2\pi}a} e^{-x^2/2a^2} dx = 1$$

which is known as the normalized Gaussian function. Notice that when $a \to 0$, the function becomes nonvanishing only in a small neighborhood of $x = 0$. Its value at 0 increases infinitely, but the area under the curve is always 1, as proved above. Let

$$\Delta(x - x_0) \equiv \lim_{a \to 0} F(x - x_0, a) = \lim_{x \to 0} \frac{1}{\sqrt{2\pi}\, a} e^{-(x - x_0)^2/2a^2}$$

Assume $G(x)$ can be expanded in Taylor's series about x_0, so that

$$\int_{-\infty}^{+\infty} \Delta(x - x_0) G(x) dx = \int_{-\infty}^{+\infty} \Delta(x - x_0) \sum_{n=0}^{\infty} \frac{(x - x_0)^n}{n!} \frac{dG^n(x_0)}{dx^n}$$

$$= G(x_0) \lim_{a \to 0} \int_{-\infty}^{+\infty} \frac{e^{-(x - x_0)^2/2a^2}}{\sqrt{2\pi}\, a} dx$$

$$+ \sum_{n=1}^{\infty} \frac{dG^n(x_0)}{dx^n} \lim_{a \to 0} \int_{-\infty}^{+\infty} \frac{(x - x_0)^n}{n!} \frac{e^{-(x - x_0)^2/2a^2}}{\sqrt{2\pi}\, a} dx$$

Now, all the terms in the expansion go to zero for the following:

(a) If n is odd, this is trivial (function is odd)
(b) If n is even, it can be shown that

$\int_{-\infty}^{+\infty} Y^n e^{-\alpha Y^2} dx$ when $n \geq 2$ is proportional to α^{-k}, $k \geq 3/2$, that is, proportional to a^m, with $m \geq 3$. Therefore, as $a \to 0$, integral vanishes.

Then,

$$\int_{-\infty}^{+\infty} \Delta(x - x_0) G(x) dx = G(x_0) \lim_{a \to 0} \int_{-\infty}^{+\infty} \frac{e^{-(x - x_0)^2/2a^2}}{\sqrt{2\pi}\, a} = G(x_0) \cdot 1 = G(x_0)$$

$$\Delta(x - x_0) \equiv \delta(x - x_0)$$

Note: One can understand why it is true that

$$\int_{-\infty}^{+\infty} G(x) \lim_{a \to 0} e^{-(x - x_0)^2/2a^2} = G(x_0) \int_{-\infty}^{+\infty} \frac{e^{-(x - x_0)^2/2a^2}}{\sqrt{2\pi}\, a} dx \quad \text{when } a \to 0$$

If one realizes that as $a \to 0$, $\{(e^{-(x - x_0)^2/2a^2})/\sqrt{2\pi}\, a\}$ is only *nonvanishing* in a small neighborhood of x_0, where, unless $G(x)$ has discontinuities at x_0, we may assume $G(x)$ to be constant. In the limit $a \to 0$, this becomes an exact statement.

Chapter 2

PROBLEM 2.1 Two free point charges $+q$ and $+4q$ are a distance ℓ apart. A third charge is so placed that the entire system is in equilibrium. Find the location, magnitude, and sign of the third charge.

Solution: We know that the third charge has to be in the middle and negative.

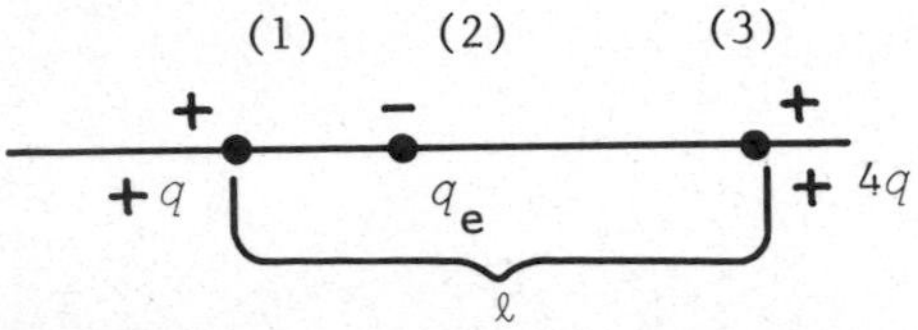

Force between (1) and (2) = force between (3) and (2) = force between (1) and (3):

$$F_{3\to 2} = \frac{1}{4\pi\varepsilon_0}\frac{q_e 4q}{x^2} = \frac{1}{4\pi\varepsilon_0}\frac{4qq}{\ell^2} = \frac{1}{4\pi\varepsilon_0}\frac{qq_e}{(\ell - x)^2}$$

$$\frac{q_e 4q}{4\pi\varepsilon_0 x^2} = \frac{qq_e}{4\pi\varepsilon_0(\ell - x)^2},\quad \frac{4}{x^2} = \frac{1}{(\ell - x)^2},\quad \frac{x^2}{4} = (\ell - x)^2$$

$$x^2 = 4(\ell - x)^2$$

$$x^2 = 4(\ell^2 - 2\ell x + x^2)$$

$$= 4\ell^2 - 8\ell x + 4x^2 - x^2$$

$$3x^2 - 8\ell x + 4\ell^2 = 0$$

$$a = 3,\ b = -8\ell,\ c = 4\ell^2 \quad \text{in} \quad x = \frac{-b \pm \sqrt{b^2 - 4ac}}{2a}$$

$$x = \frac{8\ell \pm \sqrt{64\ell^2 - 4(3)(4\ell^2)}}{6} = \frac{8\ell \pm 4\ell}{6}$$

$x = 2\ell$ or ${}^2/_3\,\ell$, but we know that the charge is in the middle, so x cannot be greater than ℓ, so $x = {}^2/_3\,\ell$. Therefore,

$$\frac{1}{4\pi\varepsilon_0}\frac{4qq}{\ell^2} = \frac{1}{4\pi\varepsilon_0}\frac{q_e 4q}{({}^4/_9\,\ell^2)}$$

$$q = \frac{q_e}{{}^4/_9},\quad q_e = \frac{-4}{9}q$$

PROBLEM 2.2 What is the force on the charge in the lower left-hand corner of the square in the figure? Assume that $q = 1.0 \times 10^{-7}$ C and $a = 5.0$ cm.

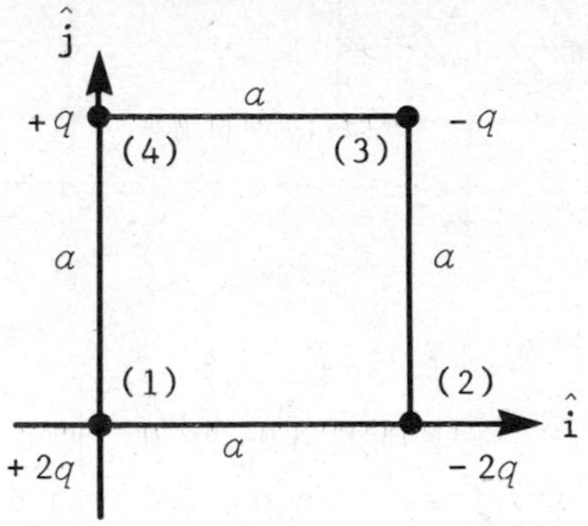

Solution:

$$\mathbf{F}_1 = \mathbf{F}_{2\to 1} + \mathbf{F}_{3\to 1} + \mathbf{F}_{4\to 1}$$

$$\mathbf{F}_{2\to 1} = \frac{1}{4\pi\varepsilon_0}\frac{(2q)(2q)}{a^2}\,\hat{\mathbf{i}}$$

$$\mathbf{F}_{4\to 1} = \frac{1}{4\pi\varepsilon_0}\frac{(q)(2q)}{a^2}\,(-\hat{\mathbf{j}})$$

$$\mathbf{F}_{3\to 1} = \frac{1}{4\pi\varepsilon_0}\frac{(q)(2q)}{[(2a^2)^{\frac{1}{2}}]^2}\cos\theta\hat{\mathbf{i}} + \frac{1}{4\pi\varepsilon_0}\frac{(q)(2q)}{[(2a^2)^{\frac{1}{2}}]^2}\sin\theta\hat{\mathbf{j}}$$

$$\mathbf{F}_1 = \frac{1}{4\pi\varepsilon_0}\frac{4q^2}{a^2}\hat{\mathbf{i}} + \frac{1}{4\pi\varepsilon_0}\frac{2q^2}{2a^2}\cos\theta\hat{\mathbf{i}} + \frac{1}{4\pi\varepsilon_0}\frac{q^2}{a^2}\sin\theta\hat{\mathbf{j}} - \frac{1}{4\pi\varepsilon_0}\frac{2q^2}{a^2}\hat{\mathbf{j}}$$

$$\mathbf{F}_1 = \hat{\mathbf{i}}\left[\left(9\times 10^9\ \frac{\mathrm{N\cdot m^2}}{\mathrm{C^2}}\right)(4)\left(\frac{1\times 10^{-7}\,\mathrm{C}}{0.05\,\mathrm{m}}\right)^2 + \left(9\times 10^9\ \frac{\mathrm{N\cdot m^2}}{\mathrm{C^2}}\right)\left(\frac{1\times 10^{-7}\,\mathrm{C}}{0.05\,\mathrm{m}}\right)^2\cos 45^\circ\right]$$

$$+\hat{\mathbf{j}}\left[\left(9\times 10^9\ \frac{\mathrm{N\cdot m^2}}{\mathrm{C^2}}\right)\left(\frac{1\times 10^{-7}\,\mathrm{C}}{0.05\,\mathrm{m}}\right)^2\sin 45^\circ - \left(9\times 10^9\ \frac{\mathrm{N\cdot m^2}}{\mathrm{C^2}}\right)(2)\left(\frac{1\times 10^{-7}\,\mathrm{C}}{0.05\,\mathrm{m}}\right)^2\right]$$

$$\mathbf{F}_1 = \hat{\mathbf{i}}\left[\left(9\times 10^9\ \frac{\mathrm{N\cdot m^2}}{\mathrm{C^2}}\right)(4)\left(4\times 10^{-12}\ \frac{\mathrm{C^2}}{\mathrm{m^2}}\right)\right.$$

$$\left. + \left(9\times 10^9\ \frac{\mathrm{N\cdot m^2}}{\mathrm{C^2}}\right)\left(4\times 10^{-12}\ \frac{\mathrm{C^2}}{\mathrm{m^2}}\right)(0.707)\right]$$

$$+\hat{\mathbf{j}}\left[\left(9\times 10^9\ \frac{\mathrm{N\cdot m^2}}{\mathrm{C^2}}\right)\left(4\times 10^{-12}\ \frac{\mathrm{C^2}}{\mathrm{m^2}}\right)(0.707)\right.$$

$$\left. - \left(9\times 10^9\ \frac{\mathrm{N\cdot m^2}}{\mathrm{C^2}}\right)(2)\left(4\times 10^{-12}\ \frac{\mathrm{C^2}}{\mathrm{m^2}}\right)\right]$$

$$\mathbf{F} = (0.169\ \mathrm{N})\hat{\mathbf{i}} - (4.7\times 10^{-12}\mathrm{N})\hat{\mathbf{j}}$$

$F_x = +0.17$ N $F_y = -0.05$ N.

PROBLEM 2.3 A "dipole" is formed from a rod of length $2a$ and two charges, $+q$ and $-q$. Two such dipoles are oriented as shown in the figure.

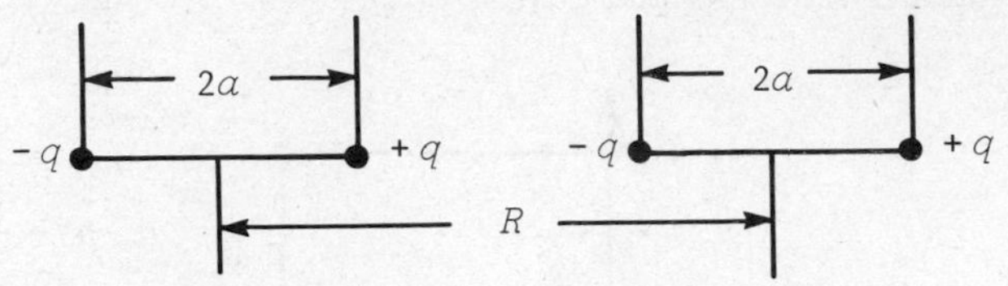

Their centers are separated by the distance R.

(a) Calculate the force exerted on the left dipole.
(b) For $R \gg a$, show that the magnitude of the force exerted on the left dipole is approximately given by $F = 3p^2/2\pi\varepsilon_0 R^4$, where $p = 2qa$ is the "dipole moment."

Solution:

(a) For the force exerted on the left dipole:

$$\mathbf{F}_1 = \mathbf{F}_{3\to 1} + \mathbf{F}_{4\to 1}$$

$$= \frac{1}{4\pi\varepsilon}\frac{(q)(q)}{r^2}(-)\hat{\mathbf{i}} + \frac{1}{4\pi\varepsilon}\frac{(q)(q)}{r^2}\hat{\mathbf{i}}$$

$$\mathbf{F}_1 = \frac{-\hat{\mathbf{i}}}{4\pi\varepsilon}\frac{(q)(q)}{R^2} + \mathbf{i}\frac{1}{4\pi\varepsilon}\frac{(q)(q)}{(R+2a)^2}$$

$$\mathbf{F}_2 = \mathbf{F}_{3\to 2} + \mathbf{F}_{4\to 2}$$

$$= \frac{1}{4\pi\varepsilon}\frac{(q)(q)}{r^2}\hat{\mathbf{i}} + \frac{1}{4\pi\varepsilon}\frac{(q)(q)}{r^2}(-)\hat{\mathbf{i}}$$

$$\mathbf{F}_2 = \frac{\hat{\mathbf{i}}}{4\pi\varepsilon}\frac{(q)(q)}{(R-2a)^2} - \frac{\hat{\mathbf{i}}}{4\pi\varepsilon}\frac{(q)(q)}{(R)^2}$$

$$\mathbf{F}_L = \mathbf{F}_1 + \mathbf{F}_2$$

$$\mathbf{F}_L = \frac{-\hat{\mathbf{i}}}{4\pi\varepsilon}\frac{q^2}{R^2} + \frac{\hat{\mathbf{i}}}{4\pi\varepsilon}\frac{q^2}{(R+2a)^2} + \frac{\hat{\mathbf{i}}}{4\pi\varepsilon}\frac{q^2}{(R-2a)^2} - \frac{\hat{\mathbf{i}}}{4\pi\varepsilon}\frac{q^2}{R^2}$$

$$= \frac{\hat{\mathbf{i}}q^2}{4\pi\varepsilon}\left(\frac{1}{(R+2a)^2} + \frac{1}{(R-2a)^2}\right) - \frac{\hat{\mathbf{i}}q^2}{4\pi\varepsilon}\left(\frac{1}{R^2} + \frac{1}{R^2}\right)$$

$$= \frac{\hat{\mathbf{i}}q^2}{4\pi\varepsilon}\left(\frac{2R^2 + 8a^2}{(R+2a)^2(R-2a)^2}\right) - \frac{\hat{\mathbf{i}}q^2}{4\pi\varepsilon}\left(\frac{2}{R^2}\right)$$

$$= \frac{\hat{\mathbf{i}}q^2}{4\pi\varepsilon}\left(\frac{2R^2 + 8a^2}{(R+2a)^2(R-2a)^2} - \frac{2}{R^2}\right)$$

$$= \frac{\hat{\mathbf{i}}q^2}{2\pi\varepsilon}\left(\frac{R^2+4a^2}{(R+2a)^2(R-2a)^2} - \frac{1}{R^2}\right)$$

$$\mathbf{F}_L = \frac{\hat{\mathbf{i}}q^2}{2\pi\varepsilon}\left(\frac{R^2+4a^2}{(R^2-4a^2)^2} - \frac{1}{R^2}\right)$$

(b) For $R >> a$ the magnitude of force on the left dipole is $\sim F = 3p^2/2\pi\varepsilon_0 R^4$, where $p = (q)(2a)$ is the dipole moment.

$$F = \frac{q^2}{2\pi\varepsilon}\left(\frac{R^2+4a^2}{(R^2-4a^2)^2} - \frac{1}{R^2}\right)$$

$$= \frac{q^2}{2\pi\varepsilon}\left(\frac{(R^2)(R^2+4a^2)}{(R^2)(R^2-4a^2)^2} - \frac{1(R^2-4a^2)^2}{R^2(R^2-4a^2)^2}\right)$$

$$= \frac{q^2}{2\pi\varepsilon}\left(\frac{R^2(R^2+4a^2) - (R^2-4a^2)^2}{R^2(R^2-4a^2)^2}\right)$$

$$= \frac{q^2}{2\pi\varepsilon}\left(\frac{R^4+4R^2a^2-(R^4-8a^2R^2+16a^4)}{R^2(R^2-4a^2)^2}\right) - \frac{q^2}{2\pi\varepsilon}\left(\frac{R^4+4R^2a^2-R^4+8a^2R^2-16a^4}{R^2(R^2-4a^2)^2}\right)$$

$$= \frac{q^2}{2\pi\varepsilon}\left(\frac{12R^2a^2-16a^4}{R^2(R^2-4a^2)^2}\right) = \left(\frac{1}{2\pi\varepsilon}\right)\frac{(3)(4)q^2a^2R^2-16q^2a^4}{R^2(R^2-4a^2)^2}$$

$$= \left(\frac{1}{2\pi\varepsilon}\right)\frac{3p^2R^2-16q^2a^2}{R^2(R^2-4a^2)^2}$$

$$F = \frac{1}{2\pi\varepsilon}\,\frac{3p^2-16q^2a^2/R^2}{(R^2-4a^2)^2}$$

For $R >> a$

$$F = \frac{1}{2\pi\varepsilon}\,\frac{3p^2}{R^4}$$

PROBLEM 2.4 You are given two metal spheres mounted on portable insulating supports. Find a way to give them equal and opposite charges. You may use a glass rod rubbed with silk but may not touch it to the spheres. Do the spheres have to be of equal size for your method to work?

Solution: Bring two neutral spheres together and touch them; then bring up the glass rod rubbed with silk (with a net positive charge), and one side will become like the negative side of a dipole, and the other like the positive side. Then take the rod away and pull the spheres apart. The spheres do not have to be of equal size for this to work; all the following situations will leave the spheres with equal charge.

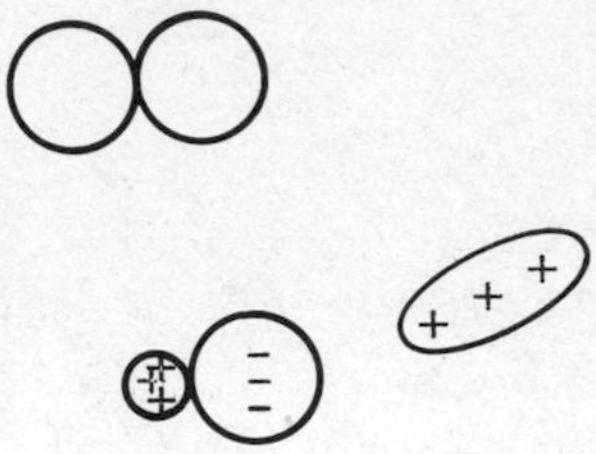

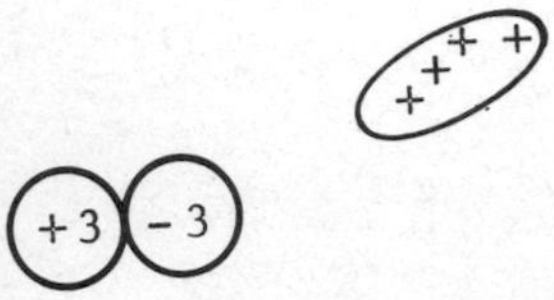

Here an equal amount of electrons can be attracted to the right as always; therefore, if the spheres are of equal size, the charge will be the greatest it can be in that situation.

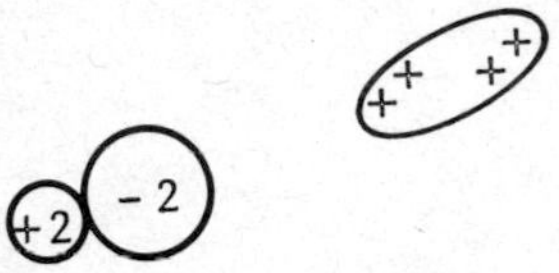

Here not as many electrons are in the little sphere, so less can be pulled right.

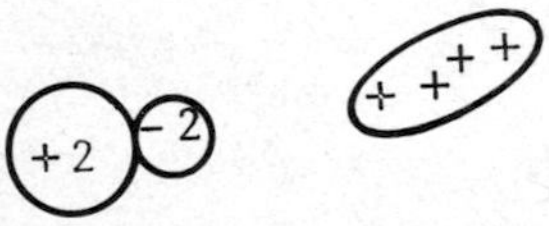

Here not as many electrons can go into the little ball before the repulsion force stopping the e^-'s from going right will be equal to the attractive force of the rod, so less go over then in equal spheres.

PROBLEM 2.5 A charged rod attracts bits of dry cork dust, which, after touching the rod, often jump violently away from it. Explain.

Solution:

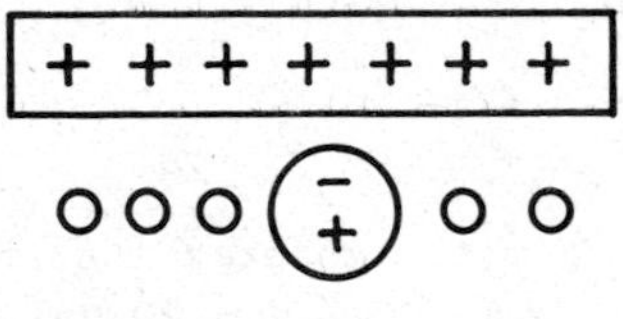

The particles (1) are attracted to the rod, but as soon as they touch it, the e^- will go onto the rod, leaving (1) with a net positive charge next to a positive rod, so it will be repelled.

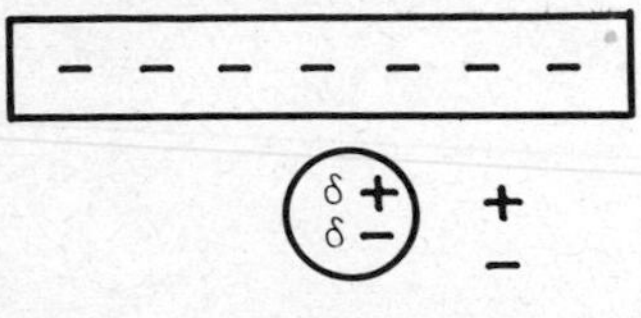

$\delta+$ = extra little bit of positive charge
$\delta-$ = extra little bit of negative charge

Here the same thing will happen: the cork dust will be attracted to the rod, but then the e^- from the rod will go onto the dust, leaving it with a net negative charge near a negatively charged rod, so it will be repelled away from the rod.

PROBLEM 2.6 If a charged glass rod is held near one end of an insulated uncharged metal rod as in the Figure, electrons are drawn to one end, as shown. Why does the flow of electrons cease? There is an almost inexhaustible supply of them in the metal rod.

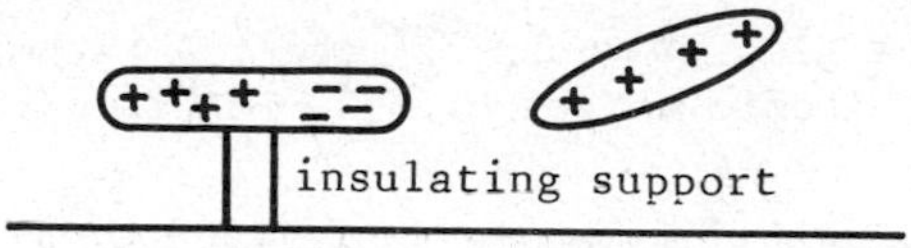

Solution:

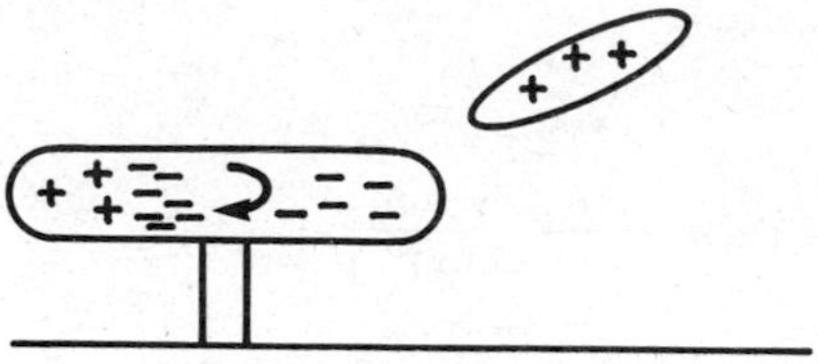

After a certain number of electrons are attracted to the right-hand side, additional electrons that want to move right will be attracted by the positive rod, but, at the same time, they will be repelled by the electrons already there (on the right side); therefore, eventually the force of repulsion on an electron from the electrons already on the right-hand side will equal the force of attraction on the electron from the positively charged rod, and the electrons will remain where they are.

PROBLEM 2.7

(a) A positively charged glass rod attracts a suspended object. Can we conclude that the object is negatively charged?
(b) A positively charged glass rod repels a suspended object. Can we conclude that the object is positively charged?

Solution:

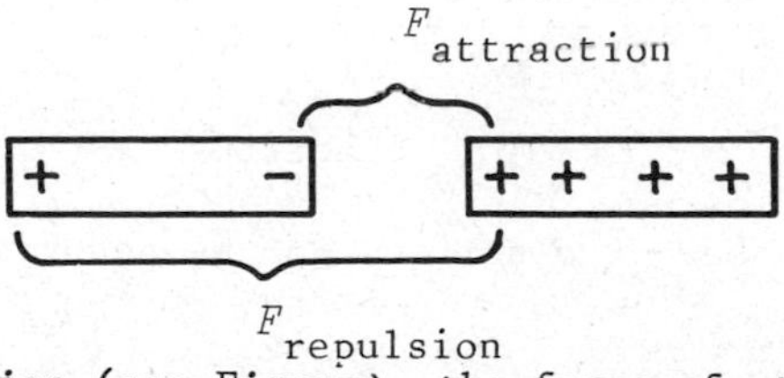

(a) In this situation (see Figure), the force of attraction is greater than the force of repulsion because the distance between the rod and the negative charge is smaller than the distance between the rod and the positive charge. If r is smaller from $F = (1/4\pi\varepsilon)\ (q_1 q_2 / r^2)$, the force will be greater, so the glass rod will attract the suspended object even though it is not negatively charged. So we cannot conclude that the object is negatively charged.
(b) Yes.

PROBLEM 2.8 Estimate roughly the number of coulombs of positive charge in a glass of water. Assume the volume to be 500 cm^3.

Solution: To find how many atoms there are in 500 cm^3 of water, first note that 1 g water = 1 cm^3 water, so that the mass of the volume of water is 500 g, because the density of water is 1 g/cm^3. Relate mass to number of atoms by

$$\frac{N}{N_0} = \frac{m}{M},$$

where N = number of molecules, m = mass of water, M = atomic weight of water, N_0 = Avogadro's number. Atomic weight of H_2O = 18.016 g/mole, therefore,

$$N = \frac{mN_0}{M} = \frac{(6.02 \times 10^{23} \text{ molecules/mole})(500 \text{ cm}^3)(1 \text{ g/cm}^3)}{(18.016 \text{ g/mole})}$$

$$N = 1.67 \times 10^{25} \text{ molecules}$$

So there are

$$1.67 \times 10^{25} \text{ molecules} \left(\frac{10 \text{ protons}}{\text{molecule}}\right)\left(1.6 \times 10^{-19} \frac{\text{C}}{\text{proton}}\right)$$

$$= 2.67 \times 10^{7} \text{ C}.$$

PROBLEM 2.9

(a) How many electrons would have to be removed from a penny to leave it with a charge of $+10^{-7}$ C?
(b) To what fraction of the electrons in the penny does this correspond?
(c) What would the force between two such pennies place 1.0 m apart?

Solution:

(a) If you take away -1×10^{-7} C of e^-, it will leave the penny with 1×10^{-7} C charge, since it was neutral to start:

$$(-1 \times 10^{-7} \text{ C})\left(\frac{1e^-}{-1.6 \times 10^{-19} \text{ C}}\right)$$

$$= 6.25 \times 10^{11} \text{ electrons}$$

(b) If N is the total number of electrons in the penny and

$$\frac{N}{N_0} = \frac{m}{M},$$

where N = number of atoms, N_0 = Avogadro's number, m = mass of Cu, M = atomic weight of Cu, then

$$N = \frac{(6 \times 10^{23} \text{ atoms/mole})(3.1 \text{ g})}{64 \text{ g/mole}}$$

$$= 2.9 \times 10^{22} \text{ atoms in a Cu penny}$$

Each atom has 29 electrons, so

$$\frac{6.25 \times 10^{11} \text{ electrons}}{(29 \text{ electrons/atom})(2.9 \times 10^{22} \text{ atoms})} = 7.43 \times 10^{-13}$$

(c) Neglecting gravitational force,

$$F = \frac{1}{4\pi\varepsilon} \frac{q_1 q_2}{r^2}$$

$$F = \frac{(9 \times 10^9\ N \cdot m^2/C^2)(1 \times 10^{-7}\ C)(1 \times 10^{-7}\ C)}{(1\ m^2)} = 9.0 \times 10^{-5}\ N$$

PROBLEM 2.10 What would be the force of attraction between two 1.0 C charges separated by a distance of 1.0 m? By 1.0 mile?

Solution:

(a)

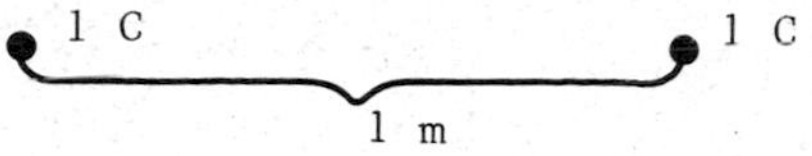

$$F = \frac{1}{4\pi\varepsilon_0} \frac{q_1 q_2}{r^2} = \frac{(9 \times 10^9\ N \cdot m^2/C^2)(1\ C)(1\ C)}{1\ m^2} = 9 \times 10^9\ N \text{ at } 1 \text{ m}$$

(b)

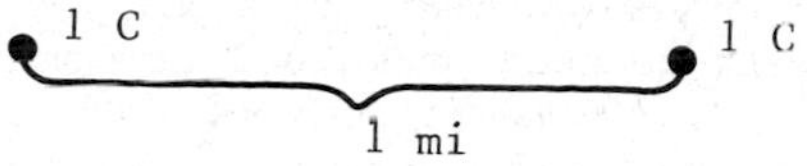

1 mi = 5280 ft

$$1 \text{ mi} \quad \frac{5280 \text{ ft}}{\text{mi}} \quad \frac{0.3048 \text{ m}}{\text{ft}} = 1609.34 \text{ m}$$

$$F = \frac{(9 \times 10^9\ N \cdot m^2/C^2)\ (1\ C)(1\ C)}{(1609.34\ m^2)} = 3.47 \times 10^3\ N \text{ at } 1 \text{ mi}$$

PROBLEM 2.11 What is the force of attraction between a single sodium ion and an adjcent chlorine ion in a salt crystal if their separation is 2.82×10^{-10} m?

Solution: Charge on Na ion = + 1 = 1.6×10^{-19} C. Charge on Cl ion = -1.6×10^{-19} C.

$$F = \frac{1}{4\pi\varepsilon} \frac{q_1 q_2}{r^2} = \frac{(9 \times 10^9\ N \cdot m^2/C^2)(1.6 \times 10^{-19}\ C)(1.6 \times 10^{-19}\ C)}{[(2.82 \times 10^{-10})^2\ m^2]}$$

$$= 2.89 \times 10^{-9}\ N \text{ (attractive)}$$

PROBLEM 2.12 An electric field E with an average magnitude of about 150 N/C points downward in the earth's atmosphere. We wish to "float" a sulfur sphere weighing 1.0 lb in this field by charging it.

(a) What charge (sign and magnitude) must be used?
(b) Why is the experiment not practical? Give a qualitative reason to prove your point.

Solution:

(a)

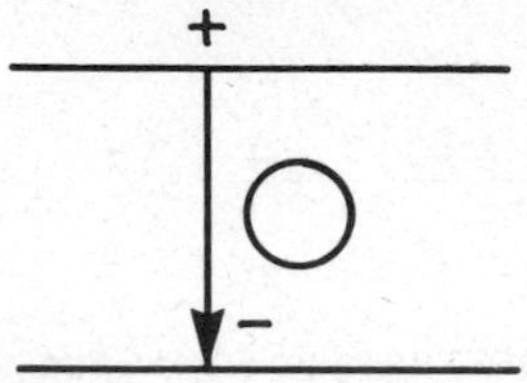

Use negative charge.

$$E = \frac{F}{q}$$

$$F = qE$$

$$mg = qE$$

$$q = \frac{(1\ \text{lb.})(0.45\ \text{kg/lb})(9.8\ \text{m/s}^2)}{(150\ \text{N/C})}$$

$$q = -0.030\ \text{C}$$

(b) Why is the experiment not practical? Eventually, the forces would pull the sphere apaart. (The electric field and the sphere's weight, each pulling in a different direction, would cause the sphere to break apart; the sphere would blow up because of mutual Coulomb repulsion.)

If the sphere was negatively charged, it would be repelled; but at the same time the positive charge in the sphere would want to go in the direction of the electric field, so the force between them would be going opposite to the way the negative charge was going, and the sphere would break apart.

PROBLEM 2.13

(a) Sketch qualitatively the lines of force associated with two separated point charges q and $-2q$.
(b) Three charges are arranged in an equilateral triangle as in the figure. What is the direction of the electric field acting on the charge $+q$?
(c) Sketch qualitatively the lines of force associated with a thin, circular, uniformly charged disk of radius R. (*Hint*: Consider as limiting cases points very close to the surface and points far from it.) Show the lines only in a plane containing the axis of the disk.

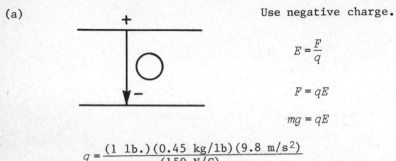

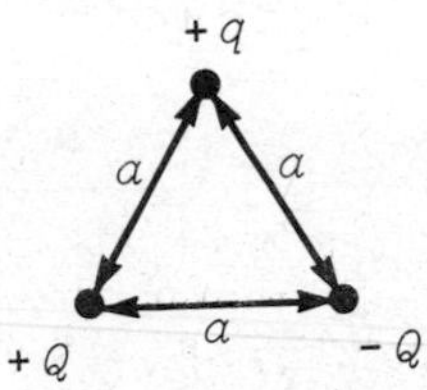

Solution:

(a)

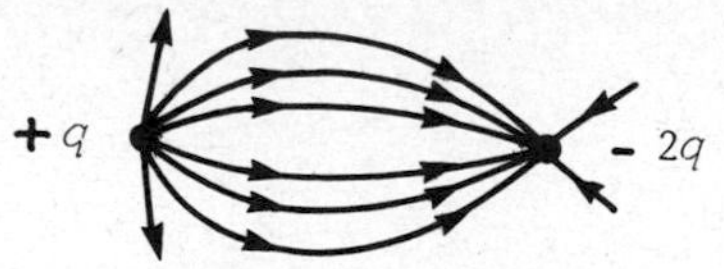

(b)

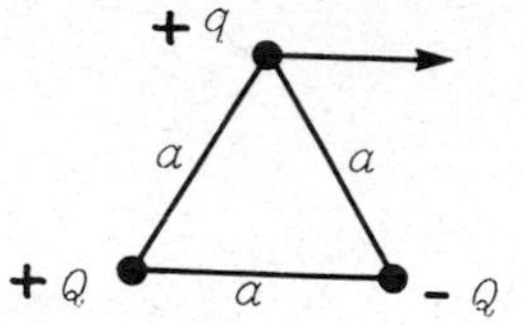

(c)

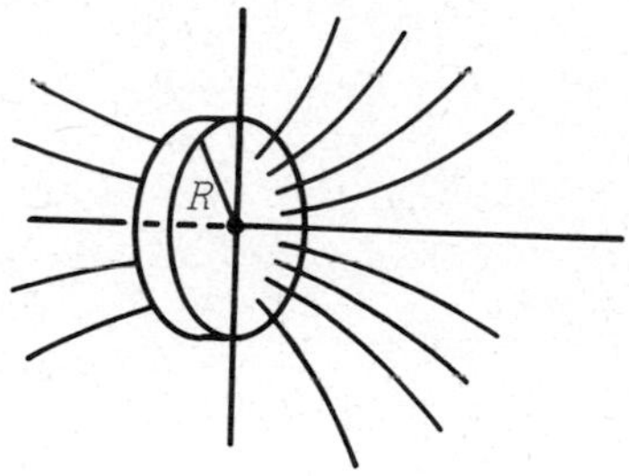

PROBLEM 2.14 Equal and opposite charges of magnitude 2.0×10^{-7} C are 15 cm apart.

(a) What is the magnitude and direction of E at a point midway between the charges?
(b) What force (magnitude and direction) would act on an electron placed there?

Solution:

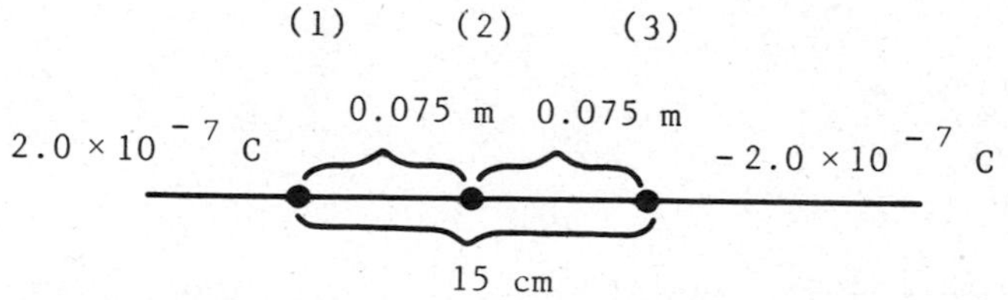

(a) $E = E_1 + E_2$

$$E_2 = \frac{\hat{i}}{4\pi\varepsilon} \frac{(2 \times 10^{-7}\ C)}{5.625 \times 10^{-3}\ m^2} + \frac{\hat{i}}{4\pi\varepsilon} \frac{(2 \times 10^{-7}\ C)}{5.625 \times 10^{-3}\ m^2}$$

$$= (\hat{i}) \left(9 \times 10^9\ \frac{N \cdot m^2}{C^2}\right) \frac{(2 \times 10^{-7}\ C)}{5.625 \times 10^{-3}\ m^2} + (\hat{i}) \left(9 \times 10^9\ \frac{N \cdot m^2}{C^2}\right) \frac{(2 \times 10^{-7}\ C)}{5.625 \times 10^{-3}\ m^2}$$

$$= \hat{\imath}\left(3.2 \times 10^5 \ \frac{N}{C}\right) + \hat{\imath}\left(3.2 \times 10^5 \ \frac{N}{C}\right)$$

$E_2 = 6.4 \times 10^5 \ \frac{N}{C}$ toward negative charge.

(b) $F = \frac{1}{4\pi\varepsilon} \frac{q_1 q_2}{r^2}, \quad q_e = -1.6 \times 10^{-19} \ C$

$$F_2 = F_{1 \to 2} + F_{3 \to 2}$$

$$F_2 = \frac{-\hat{\imath}}{4\pi\varepsilon} \frac{(2 \times 10^{-7} \ C)(1.6 \times 10^{-19} \ C)}{5.625 \times 10^{-3} \ m^2} + \frac{-\hat{\imath}}{4\pi\varepsilon} \frac{(2 \times 10^{-7} \ C)(1.6 \times 10^{-19} \ C)}{5.625 \times 10^{-3} \ m^2}$$

$$F_2 = -\hat{\imath}\left(9 \times 10^9 \ \frac{N \cdot m^2}{C^2}\right) \frac{(2 \times 10^{-7} \ C)(1.6 \times 10^{-19} \ C)}{5.625 \times 10^{-3} \ m^2}$$

$$-\hat{\imath}\left(9 \times 10^9 \ \frac{N \cdot m^2}{C^2}\right) \frac{(2 \times 10^{-7} \ C)(1.6 \times 10^{-19} \ C)}{5.625 \times 10^{-3} \ m^2}$$

$$= \hat{\imath}(1.024 \times 10^{-13} \ N)$$

$F_2 = 1.0 \times 10^{-13}$ N toward positive charge.

PROBLEM 2.15 Two point charges of magnitude $+2.0 \times 10^{-7}$ C and $+8.5 \times 10^{-8}$ C are 12 cm apart.

(a) What electric field does each produce at the site of the other?
(b) What force acts on each?

Solution:

(a)

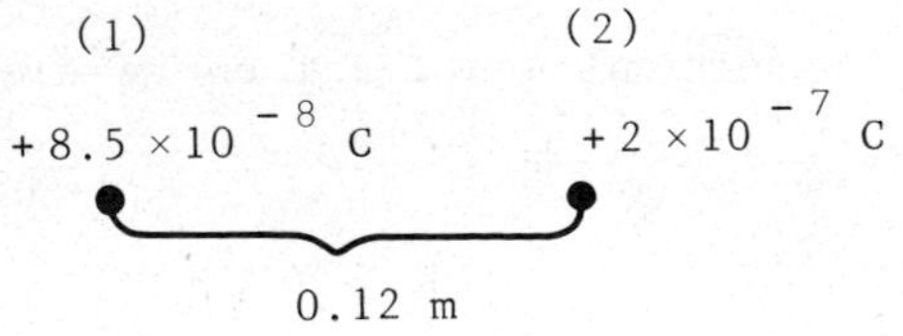

$$E_{at(1)} = \frac{-\hat{\imath}}{4\pi\varepsilon} \frac{(2 \times 10^{-7} \ C)}{(0.12 \ m)^2} = 1.25 \times 10^5 \ \frac{N}{C} \text{ in } -\hat{\imath} \text{ direction}$$

$$E_{at(2)} = \frac{+\hat{\imath}}{4\pi\varepsilon} \frac{(8.5 \times 10^{-8} \ C)}{(0.12 \ m)^2} = 5.31 \times 10^4 \ \frac{N}{C} \text{ in } +\hat{\imath} \text{ direction}$$

$$F_{2 \to 1} = \frac{-\hat{\imath}}{4\pi\varepsilon} \frac{(8.5 \times 10^{-8} \ C)(2 \times 10^{-7} \ C)}{(0.12 \ m)^2} = 1.063 \times 10^{-2} \ N \text{ in } -\hat{\imath} \text{ direction}$$

$$F_{1 \to 2} = \frac{-\hat{\imath}}{4\pi\varepsilon} \frac{(8.5 \times 10^{-8} \ C)(2 \times 10^{-7} \ C)}{(0.12 \ m)^2} = 1.063 \quad 10^{-2} \ N \text{ in } +\hat{\imath} \text{ direction}$$

PROBLEM 2.16 Two charges $q_1(2.1 \times 10^{-8}$ coul), and $q_2(-4q_1)$ are placed at a distance a(50 cm) apart. Find the point along the straight line joining the two charges at which the electric field is zero.

Solution:

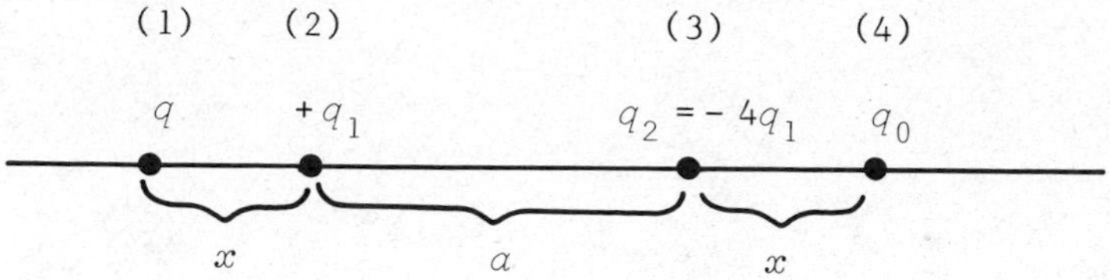

q_1 = positive test charge which cannot be on the right side of q_2 by observation, because the distance from point 2 to point 4 is too small, and the charge at 2 is so large that it is not possible to have $E = 0$ there. Therefore, the point at which $E = 0$ must be to the left of q_1.

$$E_3 = E_{1\to 3} + E_{2\to 3}$$

$$E_3 = \frac{\hat{i}}{4\pi\varepsilon}\frac{q_1}{(a+x)^2} \frac{-\hat{i}}{4\pi\varepsilon}\frac{(-4q_1)}{x^2} = 0$$

$$= \frac{\hat{i}}{4\pi\varepsilon}\left(\frac{q_1}{(a+x)^2} - \frac{4q_1}{x^2}\right) = 0$$

$$\frac{q_1}{(a+x)^2} = \frac{4q_1}{x^2}$$

$$(a+x)^2 = \frac{x^2}{4}$$

$$a^2 + 2ax + x^2 - \frac{1}{4}x^2 = 0$$

$$\frac{3}{4}x^2 + 2ax + a^2 = 0$$

$$x^2 + \frac{8}{3}ax + \frac{4}{3}a^2 = 0$$

$$a = \frac{3}{4}, \quad b = 2a, \quad c = a^2$$

$$x = \frac{-b \pm \sqrt{b^2 - 4ac}}{2a}$$

$$x = \frac{-2a \pm \sqrt{4a^2 - 4(3/4)(a^2)}}{2(3/4)}$$

$$x = \frac{-2a \pm a}{3/2} \qquad x = \frac{-a}{3/2} \qquad x = \frac{-3a}{3/2}$$

$$x = -\frac{2}{3}a \quad \text{or} \quad x = -2a$$

If 3 is on the left side

$$E_3 = \frac{-\hat{i}}{4\pi\varepsilon}\frac{q_1}{x^2} + \frac{\hat{i}}{4\pi\varepsilon}\frac{4q_1}{(a+x)^2} = \frac{\hat{i}}{4\pi\varepsilon}\left(\frac{4q_1}{(a+x)^2} - \frac{q_1}{x^2}\right) = 0$$

$$\frac{4q_1}{(a+x)^2} = \frac{q_1}{x^2}$$

$$\frac{1}{4}(a+x)^2 = x^2$$

$$\frac{1}{4}(a^2 + 2ax + x^2) = x^2$$

$$\frac{1}{4}a^2 + \frac{1}{2}ax + \frac{1}{4}x^2 - x^2 = 0$$

$$\frac{1}{4}a^2 + \frac{1}{2}ax - \frac{3}{4}x^2 = 0$$

$$-\frac{3}{4}x^2 + \frac{1}{2}ax + \frac{1}{4}a^2 = 0$$

$$-3x^2 + 2ax + a^2 = 0, \quad p = -3, \quad q = 2a, \quad r = a^2$$

$$x = \frac{-q \pm \sqrt{q^2 - 4pr}}{2p}$$

$$x = \frac{-2a \pm \sqrt{4a^2 - 4(-3)(a^2)}}{2(-3)}$$

$$x = \frac{-2a \pm 4a}{-6}$$

$$x = a \quad \text{or} \quad x = \frac{2a}{-6} = -\frac{1}{3}a$$

Where $x = a$ is only positive distance. So point is 50 cm to the left of q, and 100 cm to the left of q_2.

PROBLEM 2.17 Figure shows an electric quadrupole consisting of two electric dipoles. Show that if $r >> a$, the value of E at point P is $E = 3Q/4\pi\varepsilon_0 r^4$, where $Q = 2qa^2$. (Q is called the quadrupole moment of the charge distribution.

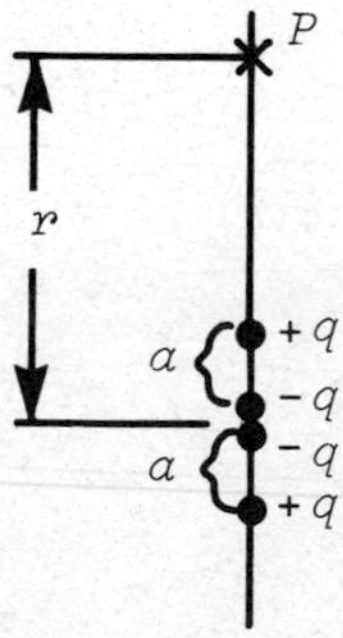

Solution:

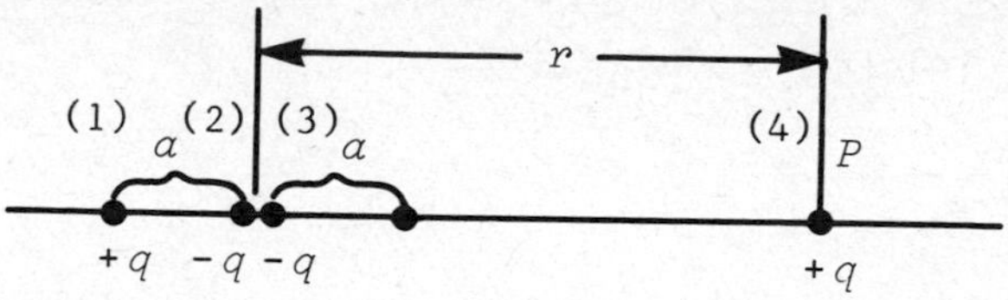

$$E_p = E_{1\to p} + E_{2\to p} + E_{3\to p} + E_{4\to p}$$

$$E_p = \frac{\hat{i}}{4\pi\varepsilon}\,\frac{q}{(r+a)^2} + \frac{-\hat{i}}{4\pi\varepsilon}\,\frac{q}{r^2} + \frac{-\hat{i}}{4\pi\varepsilon}\,\frac{q}{r^2} + \frac{\hat{i}}{4\pi\varepsilon}\,\frac{q}{(r-a)^2}$$

$$E_p = \frac{\hat{i}q}{4\pi\varepsilon}\left(\frac{1}{(r+a)^2} + \frac{1}{(r-a)^2}\right) - \frac{\hat{i}q}{4\pi\varepsilon}\left(\frac{1}{r^2} + \frac{1}{r^2}\right)$$

$$E_p = \frac{\hat{i}q}{4\pi\varepsilon}\left(\frac{(r-a)^2 + (r+a)^2}{(r+a)^2(r-a)^2}\right) - \frac{\hat{i}q}{4\pi\varepsilon}\left(\frac{2}{r^2}\right)$$

$$E_p = \frac{\hat{i}q}{4\pi\varepsilon}\left(\frac{2r^2 + 2a^2}{(r+a)^2(r-a)^2}\right) - \frac{\hat{i}q}{4\pi\varepsilon}\left(\frac{2}{r^2}\right)$$

$$E_p = \frac{q}{4\pi\varepsilon}\left[\frac{2r^2 + 2a^2}{(r+a)^2(r-a)^2} - \frac{2}{r^2}\right] = \frac{q}{4\pi\varepsilon}\left[\frac{(r^2)(2r^2 + 2a^2)}{r^2(r+a)^2(r-a)^2} - \frac{2(r+a)^2(r-a)^2}{r^2(r+a)^2(r-a)^2}\right]$$

$$E_p = \frac{q}{4\pi\varepsilon}\left[\frac{(r^2)(2r^2 + 2a^2) - (2)(r+a)^2(r-a)^2}{r^2(r+a)^2(r-a)^2}\right]$$

$$E_p = \frac{q}{4\pi\varepsilon}\left[\frac{2r^4 + 2r^2a^2 - 2(r^4 - 2a^2r^2 + a^4)}{r^2(r+a)^2(r-a)^2}\right]$$

$$E_p = \frac{q}{4\pi\varepsilon}\left[\frac{2r^4 + 2r^2a^2 - 2r^4 + 4a^2r^2 - 2a^4}{r^2(r+a)^2(r-a)^2}\right]$$

$$E_p = \frac{q}{4\pi\varepsilon}\left[\frac{6r^2a^2 - 2a^4}{r^2(r+a)^2(r-a)^2}\right]$$

for $r >> a$,

$$E_p = \left(\frac{1}{4\pi\varepsilon}\right)\frac{6r^2qa^2 - 2qa^2a^2}{r^2(r+a)^2(r-a)^2}$$

$$E_p = \frac{1}{4\pi\varepsilon} \frac{3Qr^2/r^2 - Qa^2/r^2}{[r^2(r+a)^2(r-a)^2]r^{-2}}$$

$$E_p = \frac{1}{4\pi\varepsilon} \frac{3Q}{r^4}$$

PROBLEM 2.18 A thin glass rod is bent into a semicircle of radius R. A charge $+Q$ is uniformly distributed along the upper half and a charge $-Q$ is uniformly distributed along the lower half, as shown in the figure. Find the electric field E at P, the center of the semicircle.

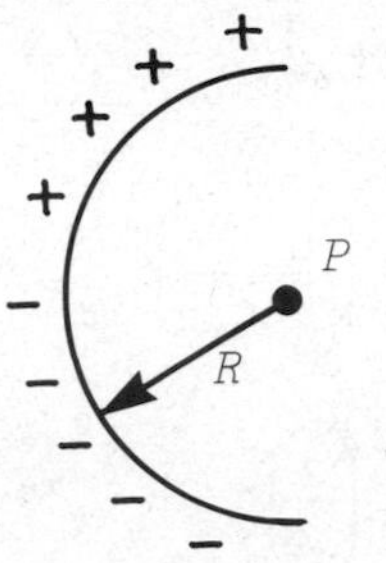

Solution:

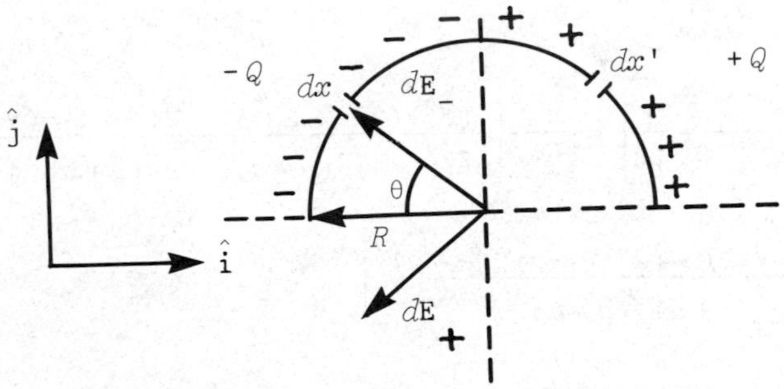

$dx = rd\theta$
$dq = \lambda rd\theta$ = charge on dx

$$\lambda = \frac{Q}{\pi r}$$

$$\mathbf{E}_T = \mathbf{E}_- + \mathbf{E}_+$$

$$d\mathbf{E}_- = \frac{-\hat{\mathbf{i}}}{4\pi\varepsilon} \frac{dq}{R^2} \cos\theta + \frac{\hat{\mathbf{j}}}{4\pi\varepsilon} \frac{dq}{R^2} \sin\theta$$

$$d\mathbf{E}_+ = \frac{-\hat{\mathbf{i}}}{4\pi\varepsilon} \frac{dq'}{R^2} \cos\theta + \frac{-\hat{\mathbf{j}}}{4\pi\varepsilon} \frac{dq'}{R^2} \sin\theta'$$

$$d\mathbf{E}_- = \frac{-\hat{\mathbf{i}}}{4\pi\varepsilon} \frac{\lambda R d\theta}{R^2} \cos\theta + \frac{\hat{\mathbf{j}}}{4\pi\varepsilon} \frac{\lambda R d\theta}{R^2} \sin\theta$$

$$E_- = \frac{-\hat{i}\lambda}{4\pi\varepsilon R}\int_0^{\pi/2} \cos\theta\, d\theta + \frac{\hat{j}\lambda}{4\pi\varepsilon R}\int_0^{\pi/2} \sin\theta\, d\theta$$

$$E_- = \frac{-\hat{i}\lambda}{4\pi\varepsilon R} + \frac{\hat{j}\lambda}{4\pi\varepsilon R}$$

$$dE_+ = \frac{-\hat{i}}{4\pi\varepsilon}\,\frac{\lambda' R\,d\theta'}{R^2}\cos\theta' - \frac{\hat{j}}{4\pi\varepsilon}\,\frac{\lambda' R\,d\theta'}{R^2}\sin\theta'$$

$$E_+ = \frac{-\hat{i}\lambda'}{4\pi\varepsilon R}\int_0^{\pi/2} \cos\theta'\,d\theta' - \frac{\hat{j}\lambda'}{4\pi\varepsilon R}\int_0^{\pi/2} \sin\theta'\,d\theta'$$

$$E_+ = \frac{-\hat{i}\lambda'}{4\pi\varepsilon R} - \frac{\hat{j}\lambda'}{4\pi\varepsilon R}$$

$$E_T = E_- + E_+ = \frac{-\hat{i}\lambda}{4\pi\varepsilon R} + \frac{\hat{j}\lambda}{4\pi\varepsilon R} - \frac{\hat{i}\lambda'}{4\pi\varepsilon R} - \frac{\hat{j}\lambda'}{4\pi\varepsilon R}$$

$$= \frac{-\hat{i}}{4\pi\varepsilon R}(\lambda + \lambda') + \frac{\hat{j}}{4\pi\varepsilon R}(\lambda - \lambda'), \quad \lambda = \lambda' = 2Q/\pi r$$

$$= \frac{-\hat{i}}{4\pi\varepsilon R}\left(\frac{4Q}{\pi r}\right) + \frac{\hat{j}}{4\pi\varepsilon R}\left(\frac{2Q}{\pi r} - \frac{2Q}{\pi r}\right)$$

$$E = \frac{-\hat{i}Q}{\pi^2 R^2 \varepsilon}, \quad E = \frac{Q}{\pi^2 \varepsilon R^2} \text{ in } -\hat{i} \text{ direction.}$$

PROBLEM 2.19 A point charge of Q_2 C is located on the x axis a distance a from another point charge of Q_1 C, as shown in the figure. Calculate the force on Q_2 and then calculate the work to move Q_2 from a to a distance b from Q_1.

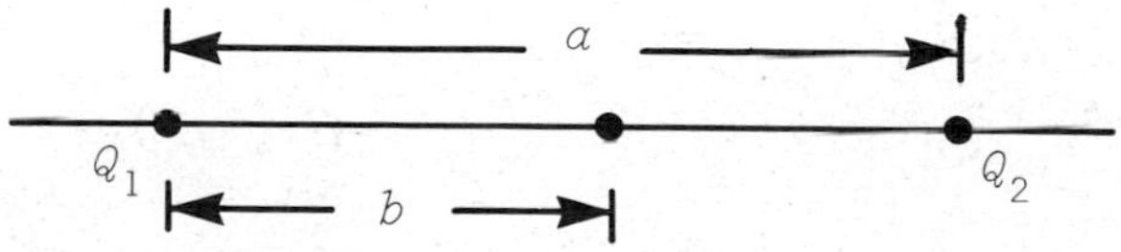

Solution:

(a) $$F_{21} = F_{Q_1 \to Q_2} = \frac{1}{4\pi\varepsilon}\,\frac{Q_1 Q_2}{a^2}$$

(b) $$W = \int F dx = \int_a^b \frac{1}{4\pi\varepsilon}\,\frac{Q_1 Q_2}{x^2}(-)\,dx$$

$$= \frac{-Q_1 Q_2}{4\pi\varepsilon}\int_a^b \frac{1}{x^2}\,dx$$

$$= \frac{-Q_1 Q_2}{4\pi\varepsilon}(-)\frac{1}{x}\Big|_a^b$$

$$= \left(\frac{-Q_1Q_2}{4\pi\varepsilon}\right)\left(-\frac{1}{b}\right) - \left[\left(\frac{-Q_1Q_2}{4\pi\varepsilon}\right)\left(-\frac{1}{a}\right)\right]$$

$$= \frac{+Q_1Q_2}{4\pi\varepsilon} \cdot \frac{1}{b} - \frac{Q_1Q_2}{4\pi\varepsilon} \cdot \frac{1}{a}$$

$$W = \frac{Q_1Q_2}{4\pi\varepsilon}\left(\frac{1}{b} - \frac{1}{a}\right)$$

PROBLEM 2.20 A thin circular ring of 3 cm radius has a total charge of 10^{-3} C uniformly distributed on it.

(a) What is the force on a charge of 10^{-2} C at its center?
(b) What would be the force on this charge if it were placed at a distance of 4 cm from the ring, along its axis?

Solution:

(a) Here $r = 0.3$ m and the total charge $= 1 \times 10^{-3}$ C.

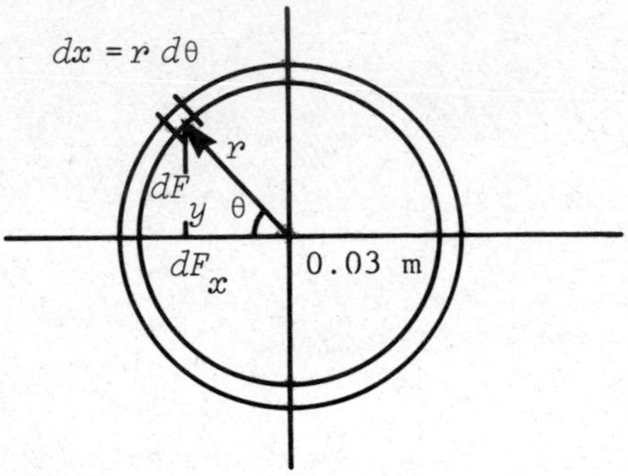

$$\lambda = \frac{1 \times 10^{-3}\ \text{C}}{2\pi(.03\ \text{m})} = 5.3 \times 10^{-3}\ \frac{\text{C}}{\text{m}}$$

$$dq = \lambda dx$$

$$dx = r d\theta$$

$$dF = \frac{1}{4\pi\varepsilon_0} \frac{Q_1 dq}{r^2}$$

$$dq = \lambda r d\theta$$

$$dF = dF_x + dF_y$$

$$d\mathbf{F} = \hat{\imath}\, \frac{1 \cdot Q}{4\pi\varepsilon_0} \frac{dq\, Q}{r^2} \cos\theta - \hat{\jmath}\, \frac{1 \cdot Q}{4\pi\varepsilon} \frac{dq\, Q}{r^2} \sin\theta$$

$$d\mathbf{F} = \hat{\imath}\, \frac{1 \cdot Q}{4\pi\varepsilon} \frac{\lambda r d\theta}{r^2} \cos\theta - \hat{\jmath}\, \frac{1 \cdot Q}{4\pi\varepsilon} \frac{\lambda r d\theta}{r^2} \sin\theta$$

$$\mathbf{F} = \int d\mathbf{F} = \frac{\hat{\mathbf{i}}\lambda r Q}{r^2 4\pi\varepsilon} \int_0^{2\pi} \cos\theta \, d\theta - \frac{\hat{\mathbf{j}} Q\lambda r}{4\pi\varepsilon r^2} \int_0^{2\pi} \sin\theta \, d\theta$$

$$= \frac{\hat{\mathbf{i}}\lambda\, Q}{4\pi\varepsilon r} \sin\theta \Big|_0^{2\pi} - \frac{\hat{\mathbf{j}} Q\, \lambda}{4\pi\varepsilon r} (-\cos\theta) \Big|_0^{2\pi}$$

$$= \frac{\hat{\mathbf{i}}\lambda Q}{4\pi\varepsilon r} \sin 2\pi - \frac{\hat{\mathbf{i}}\lambda Q}{4\pi\varepsilon} \sin 0 + \frac{\hat{\mathbf{j}} Q\lambda}{4\pi\varepsilon r} \cos 2\pi - \frac{\hat{\mathbf{j}} Q\lambda}{4\pi\varepsilon r} \cos 0$$

$$\mathbf{F} = \frac{\hat{\mathbf{j}} Q\lambda}{4\pi\varepsilon r} - \frac{\hat{\mathbf{j}} Q\lambda}{4\pi\varepsilon r} = 0$$

(b) Here total charge $Q - 1 \times 10^{-3}$ C and $Q_0 = 1 \times 10^{-2}$ C.

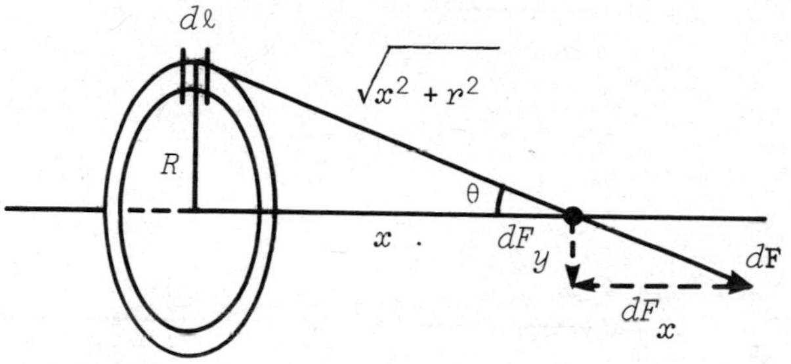

$$F = \frac{1}{4\pi\varepsilon} \frac{Q_1 Q_2}{r^2}$$

$$\lambda = \frac{Q_R}{2\pi r}$$

$$dq = \lambda \, d\,\ell$$

$$dF = dF_x + dF_y$$

$$dF_x = dF \cos\theta$$

$$dF_y = dF \sin\theta$$

$$dF = \hat{\mathbf{i}} \frac{1}{4\pi\varepsilon_0} \frac{Q dq}{(x^2 + r^2)} \cos\theta$$

$$d\mathbf{F} = \hat{\mathbf{i}} \frac{1}{4\pi\varepsilon} \frac{Q\, \lambda\, d\ell}{(x^2 + r^2)} \cos\theta$$

$$\mathbf{F} = \frac{\hat{\mathbf{i}} Q\lambda \cos\theta}{4\pi\varepsilon (x^2 + r^2)} \int_0^{2\pi r} d\ell$$

$$\mathbf{F} = \frac{\hat{\mathbf{i}} Q\lambda \cos\theta \, 2\pi r}{4\pi\varepsilon (x^2 + r^2)}$$

$$F = \hat{i}\left(9\times10^{9}\ \frac{N\cdot m^{2}}{C^{2}}\right)(1\times10^{-2}\ C)$$

$$\left(5.3\times10^{-3}\ \frac{C}{m}\right)\left(\frac{0.04}{\sqrt{0.04^{2}+0.03^{2}}}\right)(2\pi)\left(\frac{0.03\ m}{0.04^{2}\ m^{2}+0.03^{2}\ m^{2}}\right)$$

$$F = \hat{i}\ 2.88\times10^{7}\ N$$

PROBLEM 2.21 The linear charge density on the rod shown in the figure, 2 m long, is given by $\mu = \mu_0 + 2x$ C/m where x is the distance measured from one end of the rod. What is the total charge on the rod?

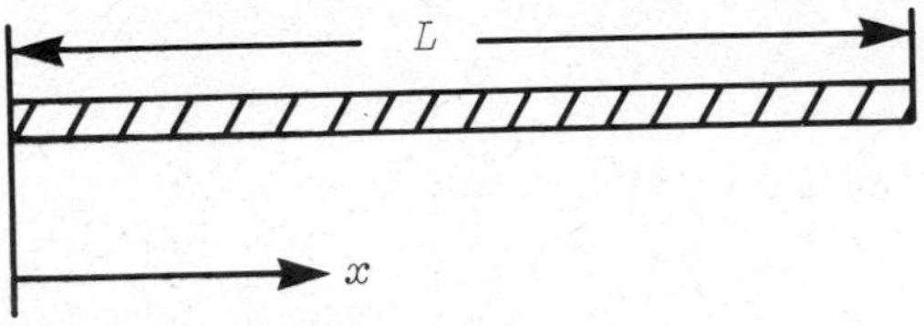

Solution:

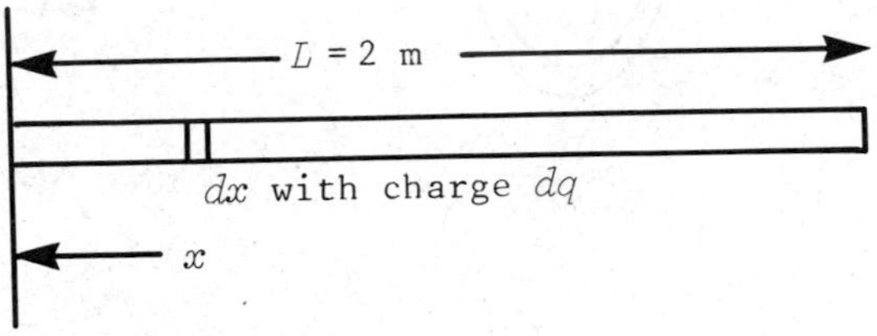

$$dq = (\mu_0 + 2x)\,dx$$

$$\mu = \frac{dq}{dx}$$

$$dx = \mu\ dx$$

$$\int dx = \int \mu\ dx$$

$$Q = \int_0^2 (\mu_0 + 2x)\,dx$$

$$= \mu_0 x + \frac{2x^2}{2}\Bigg|_0^2$$

$\therefore$ Total charge $Q = (2\mu_0 + 4)$C

Problem 2.22 Three point charges $+Q_1$, $-Q_2$, and $+Q_3$ area equally spaced along a line as shown in the figure. If the magnitudes of Q_1 and Q_2 are equal, what must be the magnitude of Q_3 in order that the net force on Q_1 be zero?

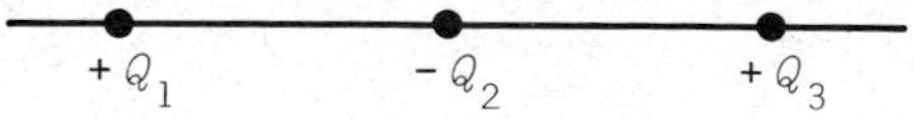

Solution:

$|Q_1| = |Q_2|$

$$F_{a_1} = 0 = F_{2\to 1} + F_{3\to 1} = F_{12} + F_{13}$$

(where, in general, $a \to b$ = on b, by a)

$$F_{12} = \frac{1}{4\pi\varepsilon}\,\frac{Q_1^2}{r^2}$$

$$F_{13} = \frac{1}{4\pi\varepsilon}\,\frac{Q_1 Q_3}{(2r)^2}$$

$$\frac{1}{4\pi\varepsilon}\,\frac{Q_1^2}{r^2} + \frac{1}{4\pi\varepsilon}\,\frac{Q_1 Q_3}{4r^2} = 0$$

$$\frac{KQ_1^2}{r^2} = -\frac{KQ_1 Q_3}{4r^2}$$

$$Q_1 = \frac{-Q_3}{4},\quad Q_3 = -4Q_1,\quad Q_3 = -4(-Q_2),\quad Q_3 = 4Q_2,\quad Q_1 = -Q_2.$$

PROBLEM 2.23 A rectangular plate 4 cm by 6 cm has a uniform surface charge density along the long dimension (x direction) but a variable surface charge density along the short dimension (y direction). The charge density is given by $\sigma = 10^{-6} y$ C/m^2. Determine the total charge on the plate.

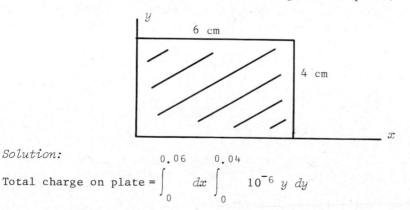

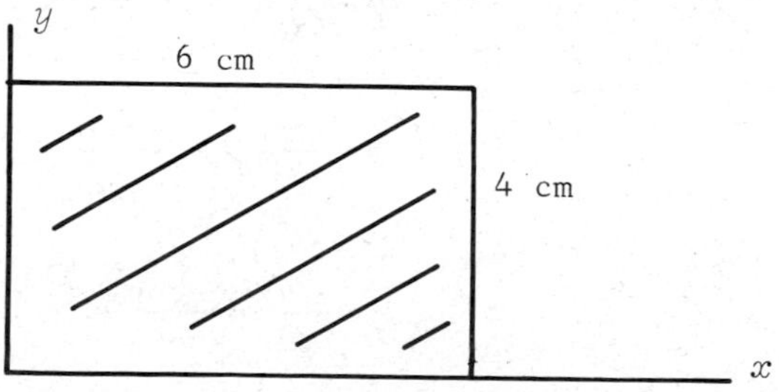

Solution:

$$\text{Total charge on plate} = \int_0^{0.06} dx \int_0^{0.04} 10^{-6}\, y\, dy$$

$$= \frac{10^{-6}y^2}{2}\Bigg|_0^{0.04}$$

$$= \frac{10^{-6}\ (0.04)^2}{2}$$

$$= \int_0^{0.06} 8 \times 10^{-10}\ dx$$

$$Q = (8 \times 10^{-10})(0.06) = 4.8 \times 10^{-11}\ \text{C}$$

PROBLEM 2.24

(a) Find the magnitude E_r of the radial component of the electric field due to a dipole.
(b) For what values of θ is E_r zero?

Solution:

(a)

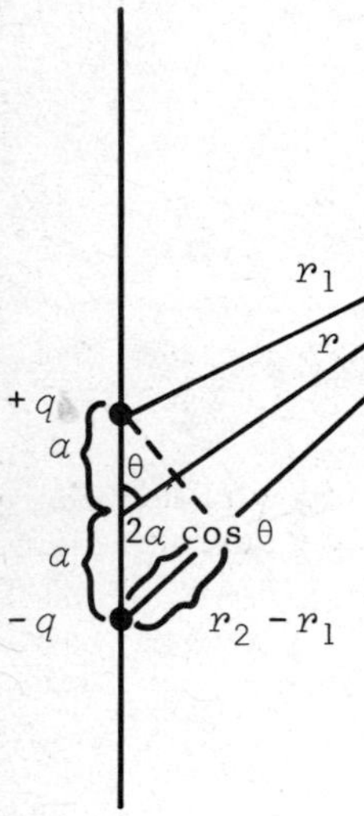

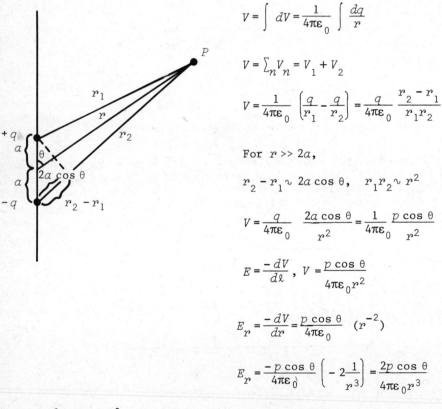

$$V = \int dV = \frac{1}{4\pi\varepsilon_0}\int \frac{dq}{r}$$

$$V = \sum_n V_n = V_1 + V_2$$

$$V = \frac{1}{4\pi\varepsilon_0}\left(\frac{q}{r_1} - \frac{q}{r_2}\right) = \frac{q}{4\pi\varepsilon_0}\ \frac{r_2 - r_1}{r_1 r_2}$$

For $r >> 2a$,

$$r_2 - r_1 \sim 2a\cos\theta, \quad r_1 r_2 \sim r^2$$

$$V = \frac{q}{4\pi\varepsilon_0}\ \frac{2a\cos\theta}{r^2} = \frac{1}{4\pi\varepsilon_0}\ \frac{p\cos\theta}{r^2}$$

$$E = \frac{-dV}{d\ell}, \quad V = \frac{p\cos\theta}{4\pi\varepsilon_0 r^2}$$

$$E_r = \frac{-dV}{dr} = \frac{p\cos\theta}{4\pi\varepsilon_0}\ (r^{-2})$$

$$E_r = \frac{-p\cos\theta}{4\pi\varepsilon_0}\left(-2\frac{1}{r^3}\right) = \frac{2p\cos\theta}{4\pi\varepsilon_0 r^3}$$

(b) At 90° and 270°, but E is not 0 at these angles.

PROBLEM 2.25 Two conducting spheres, one of radius 6.0 cm and the other of radius 12 cm, each have a charge of 3.0×10^{-6} C and are very far apart. If the spheres are connected by a conducting wire, find

(a) the direction of motion and the magnitude of the charge transferred,
(b) the final charge on and potential of each sphere.

Solution:

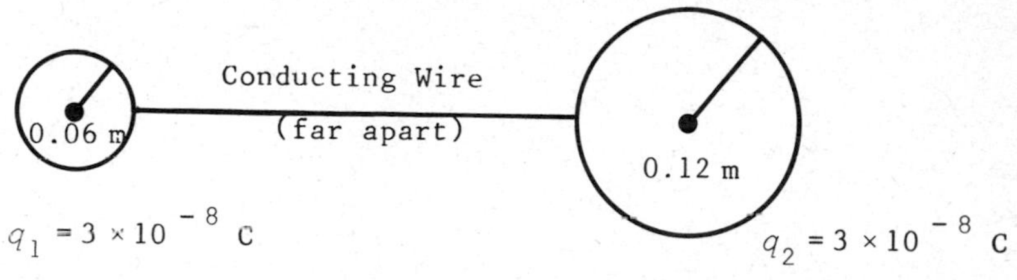

(a)

$$V_1 = \frac{1}{4\pi\varepsilon}\frac{q}{r} = \frac{(9 \times 10^9\ \mathrm{N \cdot m^2/C^2})(3 \times 10^{-8}\ \mathrm{C})}{(0.06\ \mathrm{m})} = 4.5 \times 10^3\ V$$

$$V_2 = \frac{(9 \times 10^9\ \mathrm{N \cdot m^2/C^2})(3 \times 10^{-8}\ \mathrm{C})}{0.12\ \mathrm{m}} = 2.25 \times 10^3\ V$$

Direction of motion toward 0.12 m radius ball. They want to be at the same potential, $V_1 = V_2$. Therefore,

$$\frac{1}{4\pi\varepsilon}\frac{q_1}{r_1} = \frac{1}{4\pi\varepsilon}\frac{q_2}{r_2}$$

$$\frac{q_1}{r_1} = \frac{q_2}{r_2}$$

$$q_1 = \frac{q_2 r_1}{r^2}$$

$$q_1 + q_2 = 6 \times 10^{-8}\ \mathrm{C}$$

$$q_1 = 6 \times 10^{-8}\ \mathrm{C} - 4 \times 10^{-8}\ \mathrm{C}$$

$$q_1 = 2 \times 10^{-8}\ \mathrm{C}$$

Magnitude of charge transferred $= 1 \times 10^{-8}$ C

$$q_1 + \frac{q_2 r_1}{r_2} = 6 \times 10^{-8}\ \mathrm{C}$$

$$q_2\left(1 + \frac{r_1}{r_2}\right) = 6 \times 10^{-8}\ \mathrm{C}$$

$$q_2 = \frac{6 \times 10^{-8}\ \mathrm{C}}{1 + 0.06/0.12} = 4 \times 10^{-8}\ \mathrm{C}$$

(b) The final charge and potential of each sphere are calculated as follows:

Smaller sphere

$$q = 2 \times 10^{-8}\ \text{C}$$

$$V = \frac{1}{4\pi\varepsilon}\frac{q}{r} = \frac{(9 \times 10^{9}\ \text{N} \cdot \text{m}^2/\text{C}^2)(2 \times 10^{-8}\ \text{C})}{0.06\ \text{m}} = 3 \times 10^{3}\ \text{V}$$

Larger sphere

$$q = 4 \times 10^{-8}\ \text{C}$$

$$V = \frac{(9 \times 10^{9}\ \text{N} \cdot \text{m}^2/\text{C}^2)(4 \times 10^{-8}\ \text{C})}{0.12\ \text{m}} = 3 \times 10^{3}\ \text{V}$$

PROBLEM 2.26 The electric potential varies along the x axis as shown in the figure. For each of the intervals shown (ignore the behavior at the end points of the intervals), find the x component of the electric field.

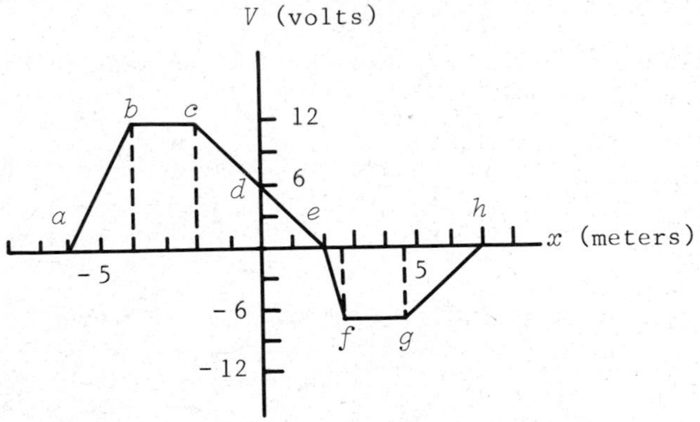

Solution:

(a - b) $V = (12/2)x$, $V = 6x$, $E_x = \frac{-\partial V}{\partial x} = -6\ \frac{\text{volts}}{\text{meter}}$

(b - c) $V = 0x$, $E_x = 0.0\ \frac{\text{volts}}{\text{meter}}$

(c - d) $V = -(6/2)x + b$, $V = -3x + b$, $E_x = \frac{-\partial V}{\partial x} = \frac{-\partial(-3x)}{\partial x} = 3\ \frac{\text{volts}}{\text{meter}}$

(c - e) same, $+3\ \frac{\text{volts}}{\text{meter}} = E_x$

(c - e) $+3\ \frac{\text{volts}}{\text{meter}} = E_x$

(e - f) $V = (7.5/0.5)x$, $E_x = +15\ \frac{\text{volts}}{\text{meter}}$

(f - g) $0.0\ \frac{\text{volts}}{\text{meter}}$

(*g* − *h*) $V = (+7.5/2.5)x, \quad E_x = -3\,\frac{\text{volts}}{\text{meter}}$

PROBLEM 2.27 Two charges $q(+2.0 \times 10^{-6}$ C) are fixed in space a distance $d(2.0$ cm) apart, as shown in the figure.

(a) What is the electric potential at point *C*?
(b) A third charge $q(+2.0 \times 10^{-6}$ C) is brought very slowly from infinity to *C*. How much work must be done?
(c) What is the potential energy *U* of the configuration when the third charge is in place?

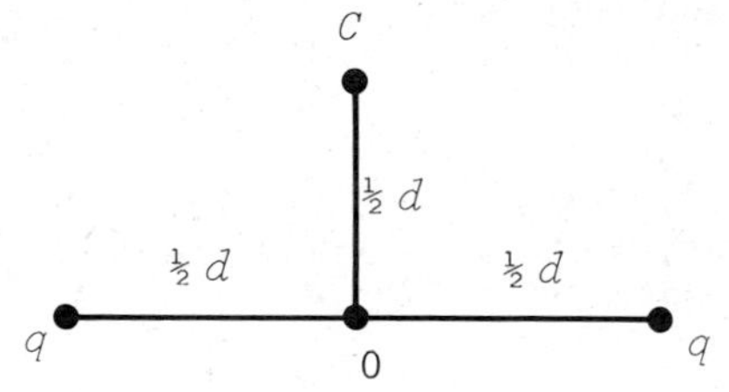

Solution:

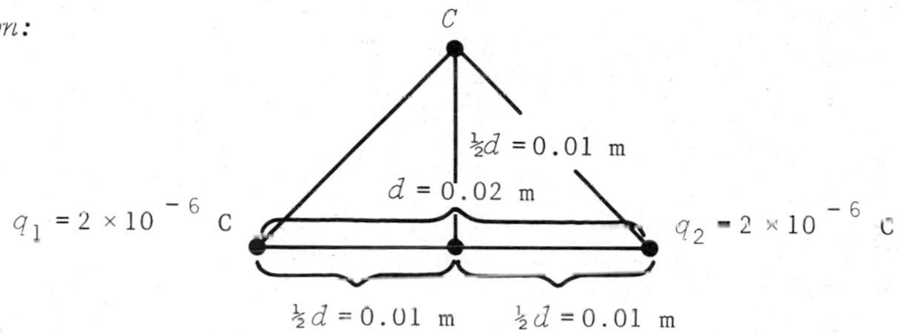

(a) $$V_C = \frac{1}{4\pi\varepsilon}\frac{q_1}{r} + \frac{1}{4\pi\varepsilon}\frac{q_2}{r} = \frac{2q}{4\pi\varepsilon r}$$

$$r = \sqrt{(0.01\ \text{m})^2 + (0.01\ \text{m})^2}$$

$$r = 0.014\ \text{m}$$

$$V_C = \frac{2(2 \times 10^{-6}\ \text{C})(9 \times 10^9\ \text{N} \cdot \text{m}^2/\text{C}^2)}{0.014\ \text{m}} = 2.57 \times 10^6\ \text{J/C}$$

$$= 2.57 \times 10^6\ \text{V}$$

(b) The work required to assemble the system by bringing in *q* from an infinite distance is the electric potential energy:

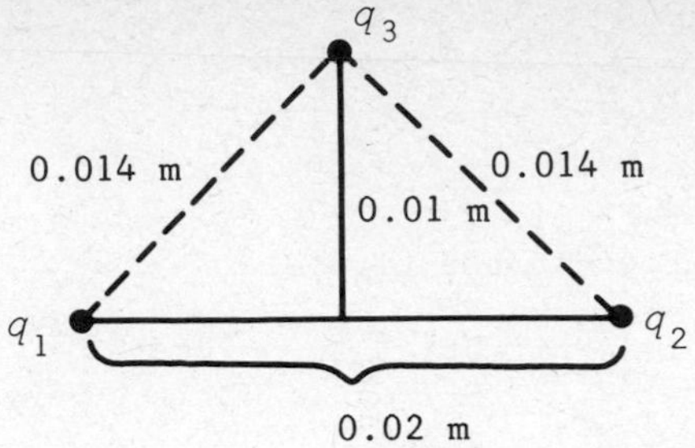

The E potential at the original site of $q_3 = 2.57 \times 10^6$ V. If q_3 is moved in from infinity to the original distance r, from the definition of potential, $V = W/q$. The work done must be $W = U = Vq = (2.57 \times 10^6 \text{ V})(2 \times 10^{-6} \text{ C}) = 5.14$ J.

(c) The electric potential energy (U) of a system of point charges is the work required to assemble this system of charges by bringing them in from an infinite distance:

$$U = U_{12} + U_{13} + U_{23}$$

$$V = W/q$$

$$U = W = Vq$$

$$U = \frac{q^2}{4\pi\varepsilon r_{12}} + \frac{q^2}{4\pi\varepsilon r_{13}} + \frac{q^2}{4\pi\varepsilon r_{23}}$$

$$U = W = \frac{q^2}{4\pi\varepsilon r}$$

$$U = \frac{q^2}{4\pi\varepsilon}\left(\frac{1}{r_{12}} + \frac{1}{r_{13}} + \frac{1}{r_{23}}\right)$$

$$U = (2 \times 10^{-6} \text{ C})^2 \left(9 \times 10^9 \frac{\text{N} \cdot \text{m}^2}{\text{C}^2}\right)\left(\frac{1}{0.02 \text{ m}} + \frac{1}{0.014 \text{ m}} + \frac{1}{0.014 \text{ m}}\right)$$

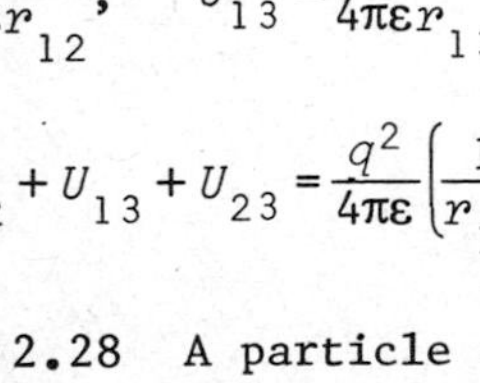

$$U = 6.94 \text{ J}$$

$$q_1 = q_2 = q_3$$

$$U_{12} = \frac{q_1 q_2}{4\pi\varepsilon r_{12}}, \quad U_{13} = \frac{q_1 q_3}{4\pi\varepsilon r_{13}}, \quad U_{23} = \frac{q_2 q_3}{4\pi\varepsilon r_{23}}$$

$$U = U_{12} + U_{13} + U_{23} = \frac{q^2}{4\pi\varepsilon}\left(\frac{1}{r_{12}} + \frac{1}{r_{13}} + \frac{1}{r_{23}}\right)$$

PROBLEM 2.28 A particle of mass m, positive charge q, and initial kinetic energy K.E.$_i$ is projected from infinity directly toward a heavy nucleus of charge Q. How close to the center of the nucleus is the particle when it comes instantaneously to rest?

Solution:

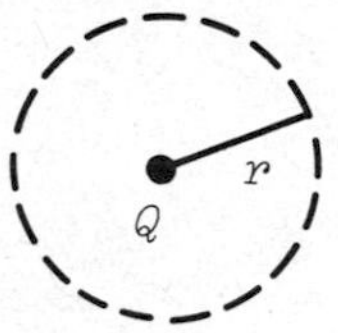

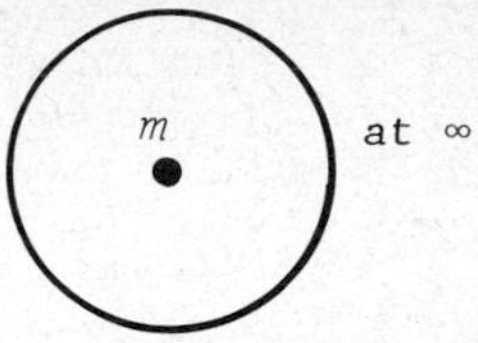

initial K.E. $= \frac{1}{2} mv^2$

The particle comes to rest when the potential energy (P.E.) = kinetic energy (K.E.).

$$\text{P.E. for nucleus} = \frac{1}{4}\frac{Qq}{r}\frac{mv^2\, 4\pi\varepsilon}{2Qq}$$

$$V = \frac{W}{q}, \quad W = U = Vq, \quad W = \frac{Qq}{4\pi\varepsilon r},$$

$$\frac{1}{r} = \frac{(\text{K.E.}_i)\, 4\pi\varepsilon}{Qq}$$

$$r = \frac{Qq}{(\text{K.E.}_i)\, 4\pi\varepsilon}$$

PROBLEM 2.29 Two small metal spheres of mass m_1(5.0 g) and mass m_2(10 g) carry equal positive charges $q(5.0 \times 10^{-6}$ C). The spheres are connected by a massless string of length d(1.0 m), which is much greater than the sphere radius.

(a) What is the electrostatic potential energy of the system?
(b) The string is cut. At that instant what is the acceleration of each of the spheres?
(c) A long time after the string is cut, what is the velocity of each sphere? Ignore gravitational effects.

Solution:

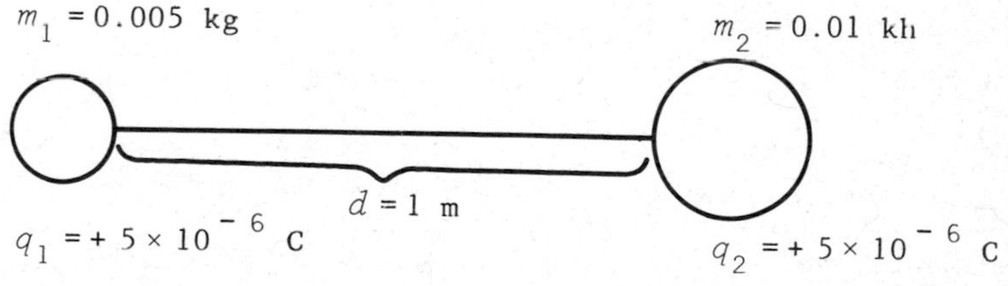

(a) What is the electrostatic potential energy of the system? This is the work required to assemble this system of charges by bringing them in from an infinite distance. q_2 goes to infinity and is at rest. The electric potential at the original site of q_2 caused by q_1 is

$$V = \frac{1}{4\pi\varepsilon}\frac{q_1}{r}$$

Now q_2 is moved in from infinity to the original distance r, the work required is (from the definition of potential) $W = Vq_2$. This work is the electric potential energy (U) of the system $q_1 + q_2$:

$$U = W = \frac{1}{4\pi\varepsilon_0}\frac{q_1q_2}{r_{12}}$$

$$U = W = \frac{1}{4\pi\varepsilon_0}\frac{q_1q_2}{r_{12}} = \frac{(9 \times 10^9\ \text{N}\cdot\text{m}^2/\text{C}^2)(5 \times 10^{-6}\ \text{C})^2}{1\ \text{m}}$$

$$= 0.225\ \text{J}$$

(b) $F_{2\to 1} = F_{1\to 2}$

$F_{2\to 1} = m_1a_1$, $F_{1\to 2} = m_2a$

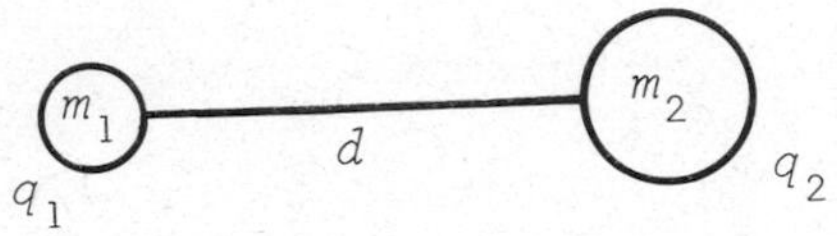

$$F_{1\to 2} = F_{2\to 1} = \frac{1}{4\pi\varepsilon}\frac{q_1q_2}{r^2} = \frac{(9 \times 10^9\ \text{N}\cdot\text{m}^2/\text{C}^2)(5 \times 10^{-6}\ \text{C})^2}{1\ \text{m}^2}$$

$$= 0.225\ \text{N}$$

$F = m_1a_1$, $F = m_2a_1$

$$a_1 = \frac{F}{m_1} = \frac{0.225\ \text{kg}\cdot\text{m/s}^2}{0.005\ \text{kg}} = 45\ \text{m/s}^2$$

$$a_2 = \frac{F}{m_2} \frac{0.225\ \text{kg}\cdot\text{m/s}^2}{0.01\ \text{kg}} = 22.5\ \text{m/s}^2$$

(c) <u>Before</u>

K.E. = 0, maximum P.E.
P.E. = E_{tot} = 0.225 J

Momentum = 0

After a long time, there will be a constant velocity.

$-m_1v_1 = m_2v_2$

$\frac{1}{2}m_1v_1^2 + \frac{1}{2}m_2v_2^2 = 0.225$ J

$m_2v_2 = -m_1v_1$

<u>After</u>

Maximum, all K.E., P.E. = 0
K.E. = E_{tot} = 0.225 J

K.E. = $\frac{1}{2}mv_1^2 - \frac{1}{2}mv_2^2$

Total momentum = 0

$0 = m_1v_1 + m_2v_2$

$-m_1v_1 = m_2v_2$

$$v_2 = \frac{-m_1 v_1}{m_2}$$

$$\tfrac{1}{2} m_1 v_1^2 + \tfrac{1}{2} m_2 \left(\frac{m_1^2 v_1^2}{m^2}\right) = 0.225 \text{ J}$$

$$\tfrac{1}{2} m_1 v_1^2 \left(1 + \frac{m_1}{m_2}\right) = 0.225 \text{ J}$$

$$v_1 = \sqrt{\frac{2(0.225 \text{ J})}{m_1(1 + m_1/m_2)}} = \sqrt{\frac{2(0.225 \text{ J})}{(0.005 \text{ kg})(1 + 0.005 \text{ kg}/0.01 \text{ kg})}}$$

$$v_1 = 7.75 \text{ m/s}$$

$$-m_1 v_1 = m_2 v_2$$

$$v_1 = \frac{-m_1 v_1}{m_2} = \frac{-(0.005 \text{ kg})(7.75 \text{ m/s})}{(0.01 \text{ kg})}$$

$$v_2 = -3.88 \text{ m/s}$$

PROBLEM 2.30

(a) A particle of charge Q is kept in a fixed position at a point P and a second particle of mass m, having the same charge Q, is initially held at rest a distance r_1 from P. The second particle is then released and is repelled from the first one. Find the speed at the instant it is a distance r_2 from P. Let $Q = 3.1 \times 10^{-6}$ C, $m = 2.0 \times 10^{-5}$ kg, $r_1 = 9.0 \times 10^{-4}$ m, and $r_2 = 25 \times 10^{-4}$ m.

(b) A particle having a charge of -2.0×10^{-9} C is acted on by a downward electric force of 3.0×10^{-6} N in a uniform electric field.

(1) What is the magnitude of the electric field?
(2) What are the magnitude and direction of the electric force exerted on a proton placed in this field?
(3) What is the gravitational force on the proton?
(4) What is the ratio of the electric to the gravitational forces in this case?

Solution:

(a)

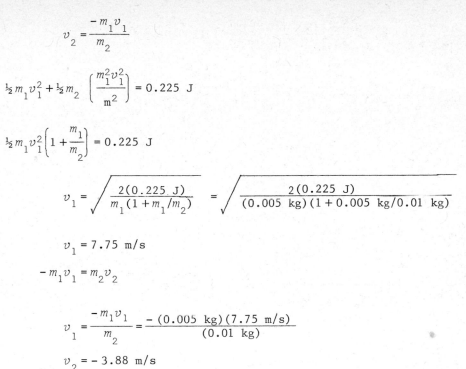

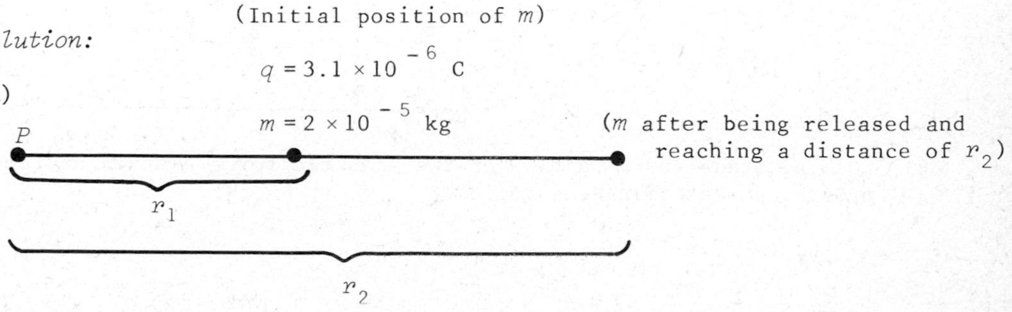

$$V_{\text{at } r_2} = \frac{1}{4\pi\varepsilon}\frac{q}{r} = \left(9 \times 10^9 \ \frac{\text{N} \cdot \text{m}^2}{\text{C}^2}\right)\left(\frac{3.1 \times 10^{-6} \ \text{C}}{25 \times 10^{-4} \ \text{m}}\right) = 1.17 \times 10^7 \ \frac{\text{J}}{\text{C}}$$

$$V_{\text{at } r_1} = \frac{1}{4\pi\varepsilon}\frac{(3.1 \times 10^{-6} \ \text{C})}{9 \times 10^{-4} \ \text{m}} = \left(9 \times 10^9 \ \frac{\text{N} \cdot \text{m}^2}{\text{C}^2}\right)\left(\frac{3.1 \times 10^{-6} \ \text{C}}{9 \times 10^{-4} \ \text{m}}\right)$$

$$V_{\text{at } r_1} = 3.1 \times 10^7 \ \frac{\text{J}}{\text{C}}$$

$$\Delta\text{P.E.} = 1.93 \times 10^7 \ \text{V}$$

$$K_f = qV$$

$$K_f = (1.93 \times 10^7 \ \text{V})(3.1 \times 10^{-6} \ \text{C}) = 59.83 \ \text{V} \cdot \text{C}$$

$$K_f = \tfrac{1}{2} m v_f^2$$

$$v_f = \sqrt{\frac{2K_f}{m}} = \sqrt{\frac{2(59.83) \ \text{kg} \cdot \text{m}^2/\text{s}^2}{2 \times 10^{-5} \ \text{kg}}}$$

$$v_f = 2.45 \times 10^3 \ \text{m/s}$$

(b)

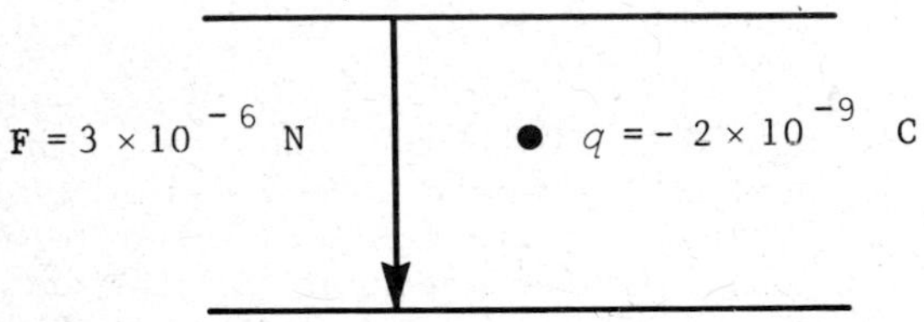

(1) The magnitude of the electric field is

$$E = \frac{\text{force}}{\text{size of charge (in C)}} = \frac{3 \times 10^{-6} \ \text{N}}{2 \times 10^{-9} \ \text{C}} = 1.5 \times 10^3 \ \frac{\text{N}}{\text{C}}$$

in the upward direction.

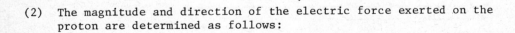

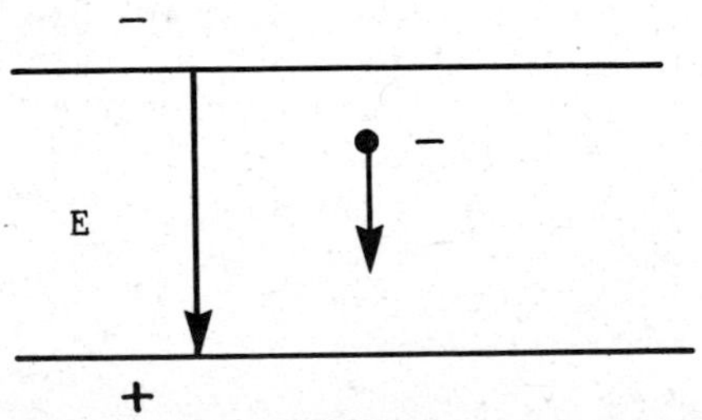

(2) The magnitude and direction of the electric force exerted on the proton are determined as follows:

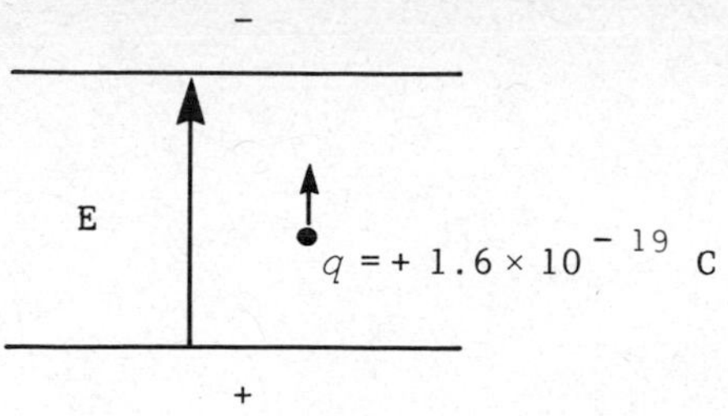

$$E = \frac{F}{q}$$

$$F = Eq$$

$$F = (1.5 \times 10^3 \ \frac{N}{C})(1.6 \times 10^{-19} \ C)$$

$$F = 2.4 \times 10^{-16} \ N, \text{ directed upward.}$$

(3) The gravitational force on the proton is

$$F = (1.67261 \times 10^{-27} \ kg)(9.8 \ m/s^2) = 1.64 \times 10^{-26} \ N$$

(4) The ratio of the electric to gravitational force is

$$\frac{E\ell}{G} = \frac{2.4 \times 10^{-16} \ N}{1.64 \times 19^{-24} \ N} = 1.47 \times 10^8 : 1 \ (!)$$

PROBLEM 2.31

(a) Between two parallel, flat, conducting surfaces of spacing d (1.0 cm) and potential difference V (1.0×10^4 V), an electron is projected from one plate directly toward the second. What is the initial velocity of the electron if it comes to rest just at the surface of the second plate?

(b) A uniform electric field exists in a region between two oppositely charged plates. An electron is released from rest at the surface of the negatively charged plate and strikes the surface of the opposite plate, 2.0 cm away in a time 1.5×10^{-8} s.

(1) What is the speed of the electron as it strikes the second plate?
(2) What is the magnitude of the electric field **E**?

Solution:

(a)

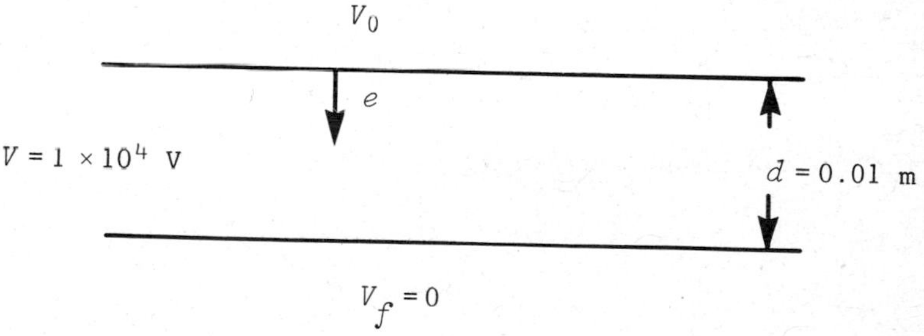

$$(\text{K.E.}) = qV, \quad V = Ed, \quad E = \frac{V}{d}$$

$$V'^2 = V_0'^2 + 2ax$$

$$V'^2 = V_0'^2 + 2\frac{q}{m} Ed$$

$$\tfrac{1}{2} mV'^2 = \tfrac{1}{2} mV_0'^2 + qEd$$

$$(\text{K.E.})_f = (\text{K.E.})_0 + qEd$$

$$0 = (\text{K.E.})_0 + qEd$$

$$(\text{K.E.})_0 = qV$$

$$(\text{K.E.})_0 = (1.6 \times 10^{-19}\ \text{C})(1 \times 10^4\ \text{V})$$

$$(\text{K.E.})_0 = 1.6 \times 10^{-15}\ \text{J}$$

$$(\text{K.E.})_0 = \tfrac{1}{2} m V_0'^2$$

$$V_0' = \sqrt{\frac{2(\text{K.E.})_0}{m}} = \sqrt{\frac{2(1.6 \times 10^{-15})\ \text{kg} \cdot \text{m}^2/\text{s}^2}{9.11 \times 10^{-31}\ \text{kg}}}$$

$$V_0' = 5.9 \times 10^7\ \text{m/s}$$

(b)

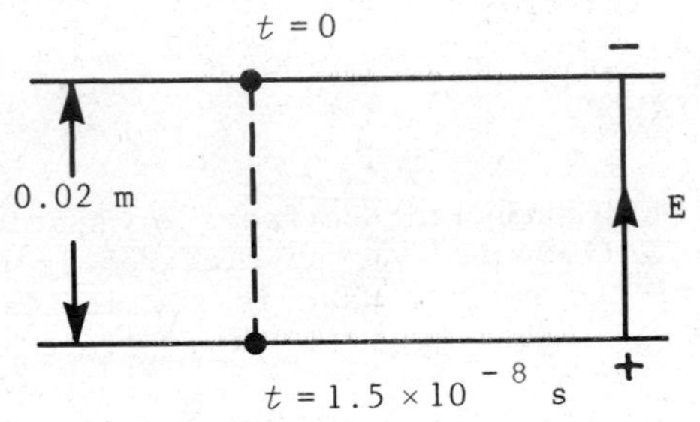

(1) The speed of the electron as it strikes the second plate is determined as follows:

$$E = \frac{F}{q}, \quad F = qE$$

$$F = ma, \text{ so } ma = aE$$

$$a = qE/m(\text{const.})$$

for constant acceleration,

$$V'^2 = V_0'^2 + 2ax$$

$$\tfrac{1}{2} m V'^2 = \tfrac{1}{2} m V_0'^2 + qEd$$

$$\text{K.E.}_f = \text{K.E.}_i + qEd$$

$$(\text{mass } e^- = 9.11 \times 10^{-31}\ \text{kg})$$

$$\text{K.E.}_f = qEd = (1.6 \times 10^{-19}\ \text{C})(E)(0.02\ \text{m})$$

$$\tfrac{1}{2}(9.11 \times 10^{-31}\ \text{kg})\ V_f'^2 = (1.6 \times 10^{-19})(E)(0.2\ \text{m})$$

$$V_f'^2 = 7.03 \times 10^9\ \frac{\text{C} \cdot \text{m}}{\text{kg}}\ E$$

$$E = 1.42 \times 10^{-10}\ V_f^2$$

$$(0.02\ \text{m}) = \tfrac{1}{2}\,\frac{qE}{m}\,t^2$$

$$0.02\ \text{m} = \frac{(\tfrac{1}{2})(1.6 \times 10^{-19}\ \text{C})(1.42 \times 10^{-10})(V_f^2)(-1.5 \times 10^{-8}\ \text{s})^2}{(9.11\ 10^{-31}\ \text{kg})}$$

$$V_f^2 = 7.14 \times 10^{12}, \quad V_f' = 2.67 \times 10^6\ \text{m/s}$$

(2) The magnitude of the electric field is

$$E = 1.41 \times 10^{-10}\ V_f'^2 = 1.42 \times 10^{-10}(7.14 \times 10^{12}) = 1.01 \times 10^3\ \text{N/C}$$

PROBLEM 2.32 An alpha particle is accelerated through a potential difference of 1.0×10^6 V in an electrostatic generator.

(a) What kinetic energy does it acquire?
(b) What kinetic energy would a proton acquire under these same circumstances?
(c) Which particle would acquire the greater speed, starting from rest?

Solution:

(a) The electric potential energy of the system is reduced by qV (because this is the work that an external agent would have to do to restore the system to its original condition). The decrease in potential energy appears as K.E. of the particle,

$$\text{K.E.} = qV$$

$$\text{K.E.} = (1 \times 10^6\ \text{V})(3.2 \times 10^{-19}\ \text{C})$$

$$\text{K.E.} = 3.2 \times 10^{-13}\ \text{J}$$

(b) $\text{K.E.} = (1 \times 10^6\ \text{V})(1.6 \times 10^{-19}\ \text{C}) = 1.6 \times 10^{-13}\ \text{J}$.

(c) The proton (because it is lighter $\rightarrow \text{K.E.} = \tfrac{1}{2}mv^2$).

PROBLEM 2.33 A positive charge q is distributed uniformly throughout a nonconducting spherical volume of radius R. Show that the potential a distance a from the center, where $a < R$, is given by

$$V = \frac{q(3R^2 - a^2)}{8\pi\varepsilon_0 R^3}$$

Solution:

$$\mathbf{E}_{\text{outside}} = \frac{1}{4\pi\varepsilon}\,\frac{q}{a^2}\,\hat{\mathbf{r}}, \quad \mathbf{E}_{\text{inside}} = \hat{\mathbf{r}}\,\frac{\rho a}{3\varepsilon} = \hat{\mathbf{r}}\,\frac{qa}{4\pi R^3\varepsilon}, \quad \rho = \frac{q}{{}^4\!/_3\pi R^3}$$

$$V(a) - V(\infty) = -\int_{\infty}^{R} \frac{1}{4\pi\varepsilon}\frac{q}{a^2}\,\hat{\mathbf{r}} - \int_{R}^{a} \frac{1}{4\pi R^3\varepsilon}\,qa\,\hat{\mathbf{r}}$$

$$V(a) = \frac{-q}{4\pi\varepsilon}\int_{\infty}^{R} \frac{1}{a^2}\,\hat{\mathbf{r}} - \frac{1}{4\pi\varepsilon R^3}\int_{R}^{a} a\hat{\mathbf{r}}$$

$$= \frac{-q}{4\pi\varepsilon}\left(-\frac{1}{a}\right)\Bigg|_{\infty}^{R} - \frac{q}{4\pi\varepsilon R^3}\left(\tfrac{1}{2}a^2\right)\Bigg|_{R}^{a}$$

$$= \frac{+q}{4\pi\varepsilon}\frac{1}{R} - \frac{q}{8\pi\varepsilon R^3}(a^2 - R^2)$$

$$= \frac{q(2R^2 - a^2 + R^2)}{8\pi\varepsilon R^2} = \frac{+q(3R^2 - a^2)}{8\pi\varepsilon R^3}$$

PROBLEM 2.34 A spherical drop of water carrying a charge of 3×10^{-11} C has a potential of 500 V at its surface.

(a) What is the radius of the drop?
(b) If two such drops, of the same charge and radius, combine to form a single spherical drop, what is the potential at the surface of the new drop so formed?

Solution:

(a) $$V = \frac{1}{4\pi\varepsilon}\frac{q}{r}$$

$$4\pi\varepsilon r V = q$$

$$r = \frac{q}{4\pi\varepsilon V} = \frac{(9 \times 10^9)(3 \times 10^{-11})}{500} = 5.4 \times 10^{-4} \text{ m}$$

(b) $$\text{V} = \frac{1}{4\pi\varepsilon}\frac{2q}{r_2} = \frac{(9 \times 10^9)(2)(3 \times 10^{-11})}{(6.8 \times 10^{-4})}$$

$$\text{V} = 7.94 \times 10^2 \text{ V}$$

PROBLEM 2.35 A point charge has $q = +1.0 \times 10^{-6}$ C. Consider point A, which is 2.0 m distant, and point B, which is 1.0 m distant, in a direction diametrically opposite, as in figure a.

(a) What is the potential difference $V_A - V_B$?
(b) Repeat if points A and B are located as in figure b.

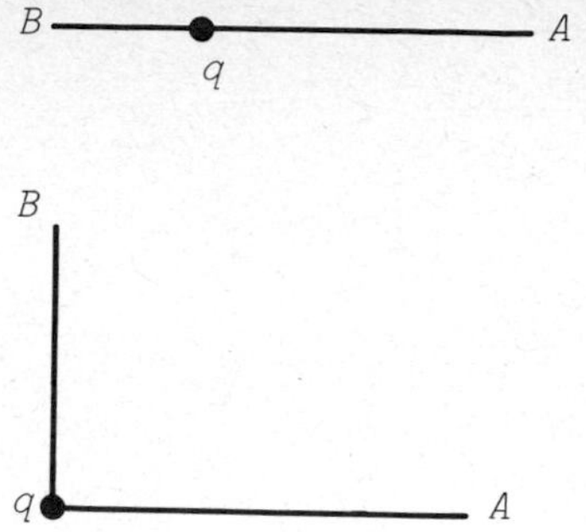

Solution:

(a)

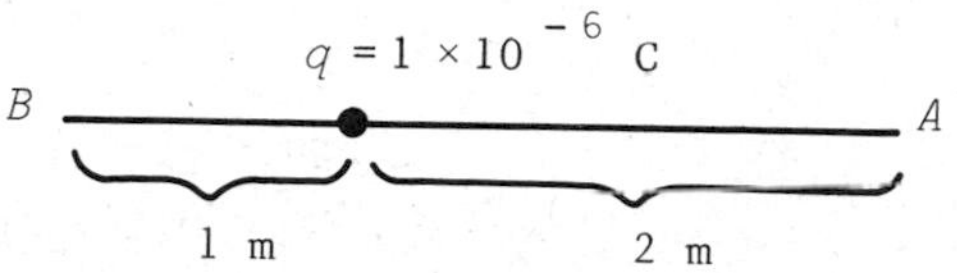

$$|V_B - V_A| = \left| \frac{-q}{4\pi\varepsilon_0} \int_{r_A}^{r_B} \frac{1}{r^2}\,dr = \frac{q}{4\pi\varepsilon_0}\left(\frac{1}{r_B} - \frac{1}{r_A}\right) \right|$$

$$= |(1 \times 10^{-6})(9 \times 10^{9})(1 - \tfrac{1}{2})|$$

$$= |4.5 \times 10^{3}\ \text{V}| \quad \text{and } V_A > V_B \text{ so,} \quad V_A - V_B = -4500\ \text{V}$$

(b) Same, does not depend on path.

PROBLEM 2.36 As shown in the figure two large, flat, horizontal metal plates are a distance a (2.0 cm) apart and are maintained at a potential difference V (500 V), the lower plate being positive. A beam of electrons is introduced midway between the plates moving parallel to them at speed v (2.0×10^{7} m/s). At what horizontal distance d will the beam hit the positive plate?

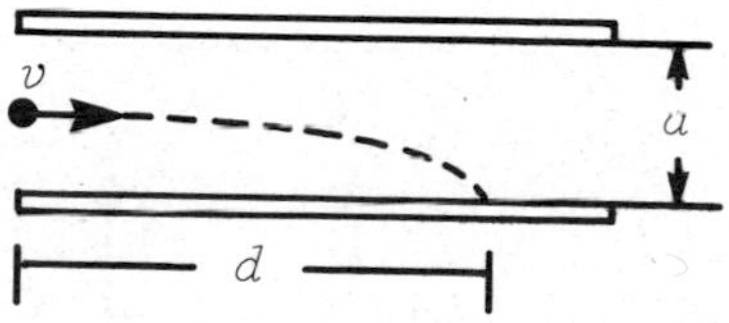

Solution:

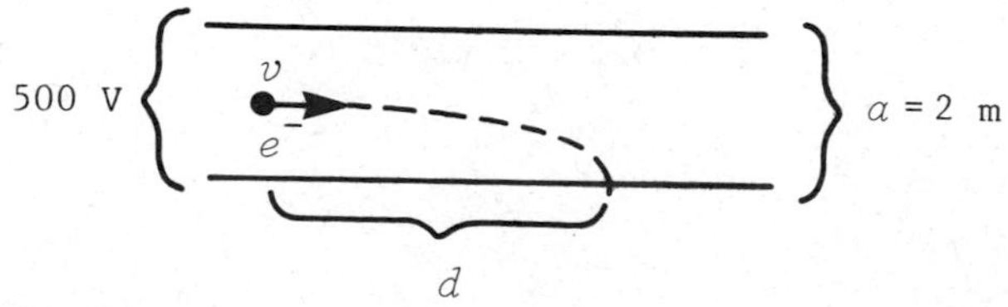

$v^2 = v_0^2 + 2ax$, $F = qE = ma$

$0 = (2 \times 10^{7}\ \text{m/s})^2 - 2ax$, $a = qE/m$, $V = Ed$, $E = V/d$

$$a = \frac{qV}{dm} = \frac{(-1.6 \times 10^{-19}\ \text{C})(500\ \text{N} \cdot \text{m/C})}{(0.02\ \text{m})(9.1 \times 10^{-31}\ \text{kg})}$$

$$a = 4.4 \times 10^{15}\ \text{m/s}^2$$

$$2ax = (2 \times 10^{7}\ \text{m/s})^2$$

$$x = \frac{(2 \times 10^{7}\ \text{m/s})^2}{2(4.4 \times 10^{15}\ \text{m/s}^2)}$$

$$x = 4.6 \times 10^{-2}\ \text{m}$$

$$x = 4.6\ \text{cm}$$

PROBLEM 2.37 A charge of 10^{-8} C can be produced by simple rubbing. To what potential would such a charge raise an insulated conducting sphere of 10 cm radius?

Solution:

$$q = 1 \times 10^{-8}\ \text{C}$$

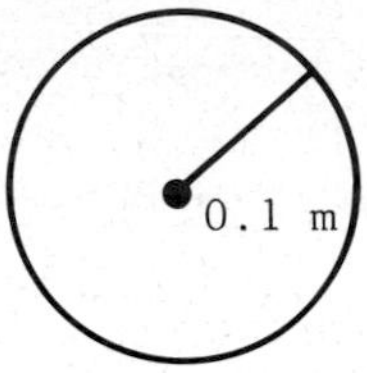

$$V = \frac{1}{4\pi\varepsilon_0}\frac{q}{r} = \frac{(9 \times 10^{9}(\text{N} \cdot \text{m}^2/\text{C}^2))(1 \times 10^{-8}\ \text{C})}{0.1\ \text{m}} = 9 \times 10^{2}\ \text{V}$$

PROBLEM 2.38 An infinite charged sheet has a surface charge density σ of $1.0 \times 10^{-7}\ \text{C/m}^2$. How far apart are the equipotential surfaces whose potentials differ by 5.0 V?

Solution:

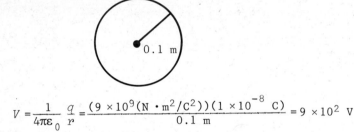

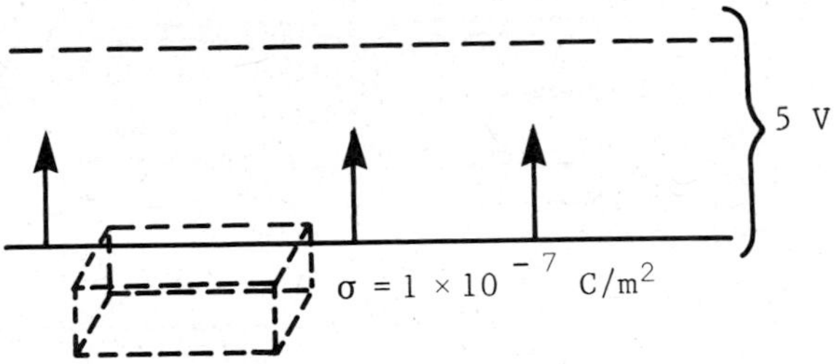

$$V_B - V_A = W_{AB}/q_0, \quad W_{AB} = Fd = q_0 Ed$$

$$V_B - V_A = q_0 Ed/q_0, \quad V = Ed = 5\ \text{V}, \quad d = 5\ \text{V}/E$$

$$EA = q/\varepsilon_0$$

$$E(2A) = \sigma A/\varepsilon_0$$

$$E = \frac{\sigma}{2\varepsilon} = \frac{1 \times 10^{-7}\ \mathrm{C/m^2}}{2(8.55 \times 10^{-12}\ \mathrm{C^2/N \cdot m^2})} = 5.8 \times 10^{3}\ \mathrm{N/C}$$

$$d = \frac{5\ \mathrm{N \cdot m/C}}{5.8 \times 10^{3}\ \mathrm{N/C}} \quad 8.6 \times 10^{-4}\ \mathrm{m}$$

$$d = 0.86\ \mathrm{mm}$$

PROBLEM 2.39 An oil drop of mass m (3.27×10^{-10} g) remains stationary between two horizontal parallel plates of separation d (2.00 cm) and potential V (2.00×10^{4} V), the electric force balancing the gravitational force. What is the charge on the *oil* drop?

Solution:

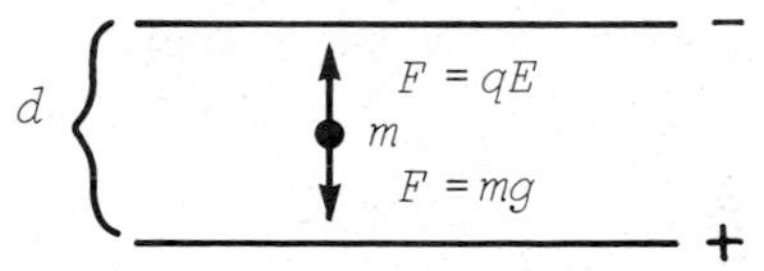

$$V = Ed$$

$$W = fd = qEd$$

$$f = qE = q(V/d)$$

$$V = \frac{W}{q_0} = \frac{fd}{q_0}$$

$$q_0 = \frac{Fd}{V} = \frac{(3.27 \times 10^{-13}\ \mathrm{kg})(9.80\ \mathrm{m/s^2})(0.02\ \mathrm{m})}{2 \times 10^{4}\ \mathrm{V}}$$

$$q_0 = 3.2 \times 10^{-18}\ \mathrm{C}$$

PROBLEM 2.40 The phenomenon of hydration is important in the chemistry of aqueous solutions. This refers to the fact that an ion in solution gathers around itself a cluster of water molecules, which cling to it rather tightly. [See, for example, G. C. Pimentel, ed., *Chemistry, an Experimental Science*, p. 314 (Freeman, San Francisco, 1963); or L. Pauling, *General Chemistry*, p. 205 (Freeman, San Francisco, 1953).] The force of attraction between a dipole and a point charge is responsible for this. Estimate the energy required to separate an ion carrying a single charge e from a water molecule, assuming that initially the ion is located 1.5 Å from the effective location of the H_2O dipole. (This distance is actually a rather ill-defined quantity since the water molecule, viewed from close up, is a charge distribution, not an infinitesimal dipole.) Which part of the water molecule will be found nearest to a negative ion?

Solution:

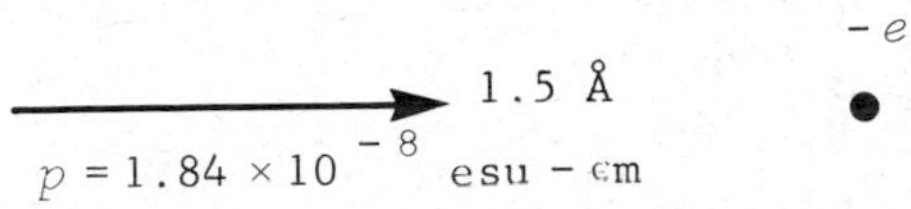

The positive on H end of the H_2O molecule will be pointed toward the negative ion. If we can assume the field to be that of a "true" dipole,

$$W = \int \mathbf{F} \cdot d\mathbf{s} = e \int E_z d_z = 2ep \int_{1.5}^{\infty} \frac{dz}{d^3}$$

$$W = ep \left(-\frac{1}{z^2} \right)_{1.5}^{\infty} = \frac{4.8 \times 10^{-10} \times 1.84 \times 10^{-18}}{(1.5 \times 10^{-8})^2}$$

$$W = 3.93 \times 10^{-12} \text{ erg} = 3.93 \times 10^{-19} \text{ J} = 2.45 \text{ eV}$$

PROBLEM 2.41 A voltage V is applied between two coaxial conducting cylinders of radius a and b $(a < b)$. Neglecting end effects, show that if the outer cylinder is grounded, the potential in the space between the cylinders is

$$\phi(r) = V \frac{\ln b - \ln r}{\ln b - \ln a} .$$

Solution:

$$\phi_b - \phi_r = -\int_r^b E\, dr$$

$$\phi_r - \phi_b = \int_r^b E\, dr = \frac{\lambda}{2\pi\varepsilon} \int_r^b \frac{dr}{r} = \frac{\lambda}{2\pi\varepsilon} [\ln b - \ln r]$$

$$\phi_a - \phi_b = V = \frac{\lambda}{2\pi\varepsilon} [\ln b - \ln a]$$

$$\lambda = \frac{2\pi\varepsilon (V)}{\ln b - \ln a}$$

$$\phi(r) - 0 = \frac{2\pi\varepsilon V}{2\pi\varepsilon} \frac{\ln b - \ln r}{\ln b - \ln a} = \frac{\ln b/r}{\ln b/a} V$$

PROBLEM 2.42 Find an expression for the magnitude of the electric field at a distance x along the perpendicular bisector of a uniformly charged rod of length a and total charge q, see figure.

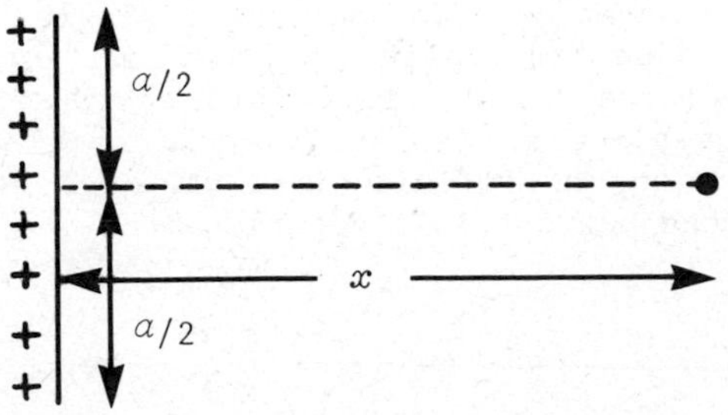

Solution:

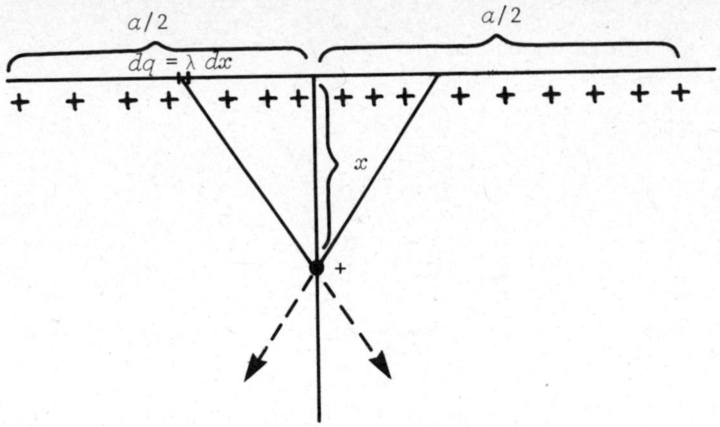

So E is down

$$dE = \frac{1}{4\pi\varepsilon}\,\frac{dq}{r^2}$$

$$dE = \frac{1}{4\pi\varepsilon}\,\frac{\lambda d\ell}{\left(\sqrt{\ell^2 + x^2}\right)^2}$$

$$\int dE = \int_{a/2}^{a/2} \frac{1}{4\pi\varepsilon}\,\frac{\lambda d\ell}{(x^2 + \ell^2)}$$

$$E = \frac{\lambda}{4\pi\varepsilon}\int_{-a/2}^{a/2} \frac{d\ell}{x^2 + \ell^2}$$

Note:

$$\int \frac{dx}{c^2 + x^2} = \frac{1}{c}\,\tan^{-1}\frac{x}{c}$$

so

$$E = \frac{\lambda}{4\pi\varepsilon}\left(\frac{1}{x}\,\tan^{-1}\frac{\ell}{x}\right)_{-a/2}^{a/2}$$

$$E = \frac{\lambda}{2\pi\varepsilon}\left[\frac{1}{x}\,\tan^{-1}\frac{a}{2x} - \frac{1}{x}\,\tan^{-1}\left(-\frac{a}{2x}\right)\right]$$

PROBLEM 2.43 What is the magnitude of the dipole moment of each of the charge distribution in parts (*a*), (*b*), and (*c*) of the figure? What is the direction of the dipole moment vector **p**?

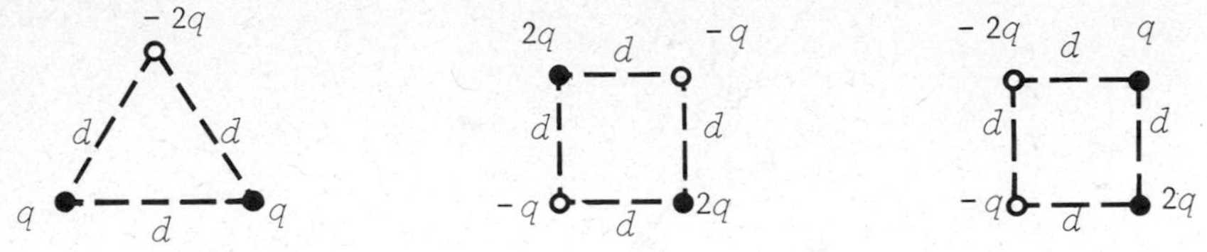

Solution:

(a)

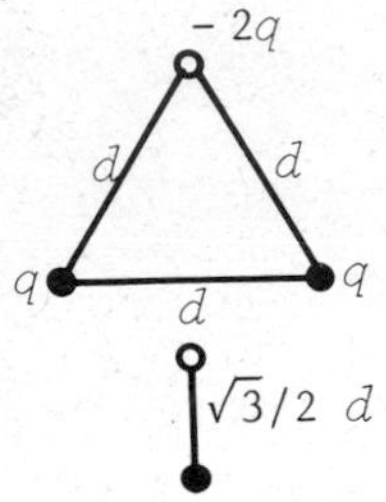

No x component to **p**, since it looks like:

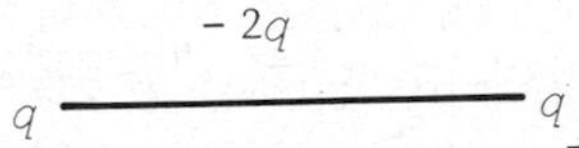

$p_y = (2q)(\sqrt{3}/2\ d) = \sqrt{3}\ qd$ downward.

(b) 0

(c)

$p_x = 3qd$ to right

$p_y = qd$ downward

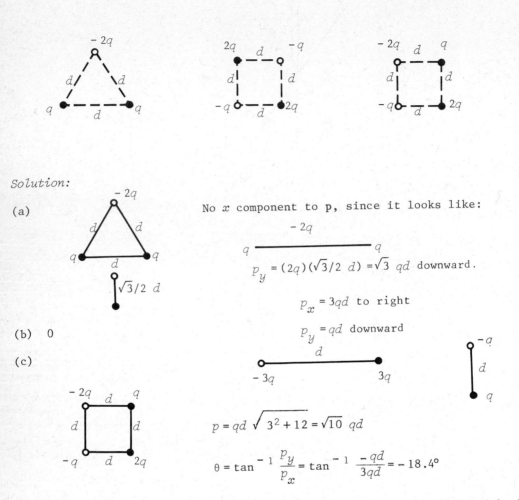

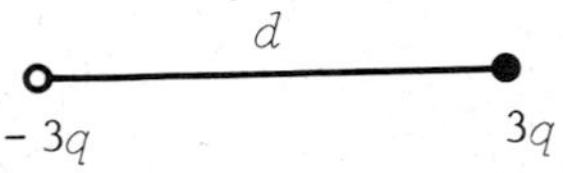

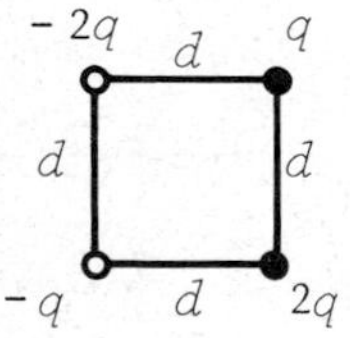

$$p = qd\sqrt{3^2 + 1^2} = \sqrt{10}\ qd$$

$$\theta = \tan^{-1}\frac{p_y}{p_x} = \tan^{-1}\frac{-qd}{3qd} = -18.4°$$

PROBLEM 2.44 A hydrogen chloride molecule is located at the origin with the H-Cl line along the z axis and Cl uppermost. What is the direction of the electric field, and its strength in statvolts/cm, at a point 10 Å up from the origin, on the z axis? At a point 10 Å out from the origin on the y axis?

Solution:

Here $\mathbf{p} = -1.03 \times 10^{-18}\ \hat{z}$ esu-cm and for a point at $z = 10$ Å, $\theta = 0°$,

$$E_y = 0$$

$$E_z = \frac{1.03 \times 10^{-18}}{(10^{-7}\ \text{cm})^3}(3 \times 1 - 0) = 2.06 \times 10^3\ \frac{\text{esu}}{\text{cm}}\ \text{down}$$

For a point at $y = 10$ Å, $\theta = 90°$

$$E_y = 0$$

$$E_z = \frac{1.03 \times 10^{-18}}{(10^{-7})^3}(3 \times 0 - 1) = 1.03 \times 10^3 \ \frac{\text{esu}}{\text{cm}^2} \ \text{up}$$

PROBLEM 2.45

(a) Write down the differential equations, boundary conditions, and relationships between various vector fields involved in the solution of problems in the electrostatics of isotropic, homogeneous, linear dielectric media.

(b) A dielectric sphere of permittivity ε and radius R is placed in a uniform electric field. Calculate the electrostatic potential and the electric field, both inside and outside the sphere.

Solution:

(a) $\nabla \cdot \mathbf{D} = \rho_f$

$\nabla \times \mathbf{E} = 0$

$\mathbf{D} = \varepsilon_0 \mathbf{E} + \mathbf{P} = \varepsilon \mathbf{E}$

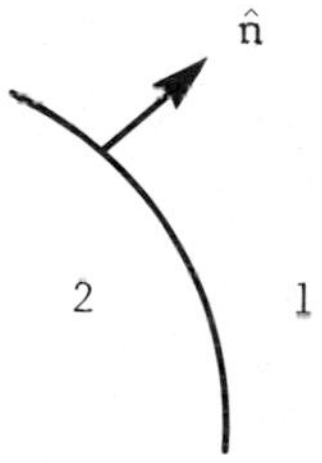

$B_{1n} = B_{2n}$

$D_{2n} - D_{1n} = \sigma$

$E_{1t} = E_{2t}$

From regularity at origin and infinity; and potential continuous across boundary.

(b)

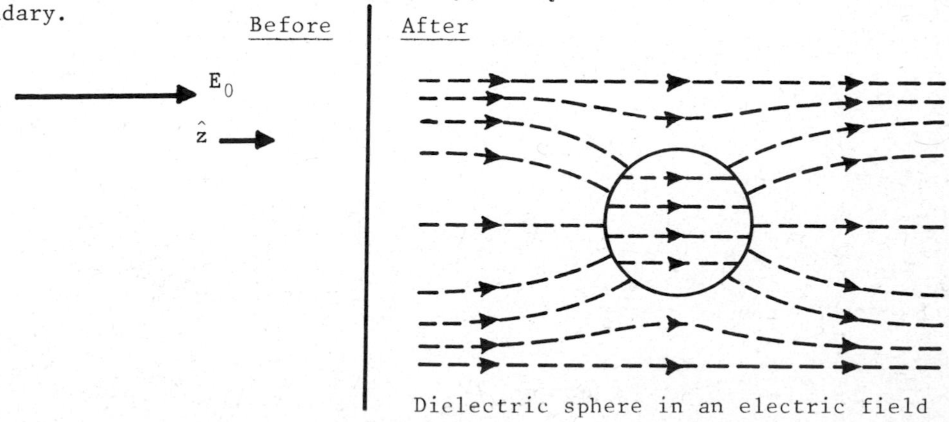

Dielectric sphere in an electric field

For spherical coordinates we have,

$$u(r,\theta) = \sum_{n=0}^{\infty} A_n r^n P_n \cos\theta + \sum_{n=0}^{\infty} B_n r^{-(n+1)} P_n(\cos\theta),$$

n	P_n
0	1
1	$\cos\theta$
2	$\frac{1}{2}(3\cos^2\theta - 1)$

Match:

(1) Regularity at origin and at infinity
(2) Boundary conditions
(3) Condition given as $r \to \infty$, $\mathbf{E} \to E_0\hat{\mathbf{z}}$, $\mathbf{E} = -\dfrac{\partial U}{\partial Z}$, $U(r,\theta) = -E_0 z$

$$= -E_0 r \cos\theta$$

$$U_{\text{in}} = A_0 + \frac{B_0}{r} + A_1 r \cos\theta + \frac{B_1}{r^2}\cos\theta$$

$$U_{\text{out}} = A_0' + \frac{B_0'}{r} + A_1' r \cos\theta + \frac{B_1'}{r^2}\cos\theta$$

at $r = 0$, all B's $= 0$, regularity at origin.

$$U_{\text{in}} = A + A_1 r \cos\theta$$

at $r = \infty$, $U_{\text{out}} = -E_0 r\cos\theta = A_0' = -E_0 r\cos\theta$, so $A_0' = 0$.

$$U_{\text{out}} = \frac{B_0'}{r} - E_0 r\cos\theta + \frac{B_1'}{r^2}\cos\theta$$

potential continuous across boundary gives

$$A_0 + A_1 R\cos\theta = \frac{B_0'}{R} + \left(\frac{B_1'}{R^2} - E_0 R\right)\cos\theta$$

Equating coefficients of the orthogonal polynomials (Legendre polynomials) on both sides

$A_0 = B_0'/R$; we can set $B_0' = 0$ if sphere carries no net charge so

$B_1' = 0$, $A_0 = 0$

(1) $A_1 R = \dfrac{B_1'}{R^2} - E_0 R$

$$U_{\text{in}} = A_1 r\cos\theta$$

$$U_{\text{out}} = -E_0 r\cos\theta + \frac{B_1'}{r^2}\cos\theta$$

$$D_{1n} = D_{2n} + \varepsilon \left.\frac{\partial U_{in}}{\partial r}\right|_{r=R} = +\varepsilon_0 \left.\frac{\partial U_{out}}{\partial r}\right|_{r=R}$$

$$\varepsilon A_1 \cos\theta = \varepsilon_0(-)E_0 \cos\theta - 2\varepsilon_0 \frac{B_1'}{R^3} \cos\theta$$

(2) $$\varepsilon A_1 = -E_0\varepsilon_0 - 2\varepsilon_0 \frac{B_1'}{R^3}$$

$$A_1 = \frac{B_1'}{R^3} - E_0, \quad A_1 = -E_0 \frac{\varepsilon_0}{\varepsilon} - \frac{2\varepsilon_0}{\varepsilon} \frac{B_1'}{R^3}$$

$$\frac{B_1'}{R^3} - E_0 = -E_0 \frac{\varepsilon_0}{\varepsilon} - \frac{2\varepsilon_0}{\varepsilon} \frac{B_1'}{R^3}$$

$$\frac{B_1'}{R^3}\left(1 + \frac{2\varepsilon_0}{\varepsilon}\right) = E_0\left(1 - \frac{\varepsilon_0}{\varepsilon}\right)$$

$$B_1' = E_0 R^3 \frac{1 - \varepsilon_0/\varepsilon}{1 + 2\varepsilon_0/\varepsilon} = E_0 R^3 \left(\frac{\varepsilon - \varepsilon_0}{\varepsilon + 2\varepsilon_0}\right)$$

$$A_1 = E_0 \left(\frac{\varepsilon - \varepsilon_0}{\varepsilon + 2\varepsilon_0} - 1\right)$$

$$A_1 = E_0 \left(\frac{\varepsilon - \varepsilon_0}{\varepsilon + 2\varepsilon_0} - \frac{\varepsilon + 2\varepsilon_0}{\varepsilon + 2\varepsilon_0}\right) = E_0 \left(\frac{\varepsilon - \varepsilon_0 - \varepsilon - 2\varepsilon_0}{\varepsilon + 2\varepsilon_0}\right)$$

$$A_1 = \frac{E_0(-3\varepsilon_0)}{\varepsilon + 2\varepsilon_0}$$

$$U_{in} = A_1 r \cos\theta, \quad \text{where} \quad A_1 = \frac{-3E_0\varepsilon_0}{\varepsilon + 2\varepsilon_0}$$

$$U_{out} = -E_0 r \cos\theta + \frac{B_1' \cos\theta}{r^2}, \quad \text{where} \quad B_1' = E_0 R^3 \left(\frac{\varepsilon - \varepsilon_0}{\varepsilon + 2\varepsilon_0}\right)$$

$\mathbf{E} = -\nabla U$, using spherical coordinates take the derivative,

$$-\mathbf{E}_{out} = \hat{\mathbf{r}} \frac{\partial U_{out}}{\partial r} + \hat{\theta} \frac{1}{r} \frac{\partial U_{out}}{\partial \theta} = \hat{\mathbf{r}} \left[E_0 \cos\theta + B_1' \cos\theta \left(\frac{-2}{r^3}\right)\right]$$

$$+ \hat{\theta} \frac{1}{r} \left(-E_0 r + \frac{B_1'}{r^2}\right)(-\sin\theta)$$

$$-\mathbf{E}_{in} = \hat{\mathbf{r}}\,\frac{\partial U_{in}}{\partial r} + \hat{\theta}\,\frac{1}{r}\,\frac{\partial U_{in}}{\partial \theta}$$

$$-\mathbf{E}_{in} = \hat{\mathbf{r}}\,A_1 \cos\theta + \hat{\theta}\,\frac{1}{r}\,A_1\,(-\sin\theta)$$

PROBLEM 2.46 A sphere of radius 20 cm has a volume charge density which varies with radius according to $\rho = 10^{-6}r^2$ C/m^3.

(a) Determine the total charge within the sphere.
(b) Determine the charge contained in the spherical shell between $r = 10$ cm and $r = 20$ cm.

Solution:

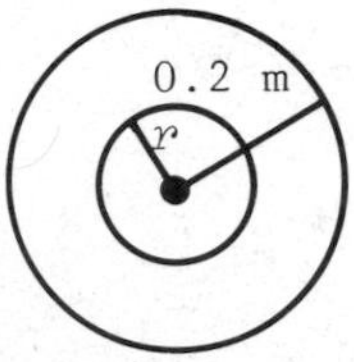

For $\rho = 1 \times 10^{-6} r^2$ C/m^3

(a) $$\int_0^{0.2} \rho 4\pi r^2 dr = \int_0^{0.2} (4 \times 10^{-6})(r^2)(\pi) r^2 dr$$

$$= \int_0^{0.2} (1.256 \times 10^{-5}) r^4 dr$$

$$= \frac{1.256 \times 10^{-5}}{5}\,[0.2^5 - 0^5] = 8.04 \times 10^{-10}\ \text{C}$$

(b) $$\int_{0.1}^{0.2} (1.256 \times 10^{-5}) r^4 dr$$

$$= \frac{1.256 \times 10^{-5}}{5}\,[0.2^5 - 0.1^5] = 7.787 \times 10^{-10}\ \text{C}$$

PROBLEM 2.47 Calculate Φ_E through a hemisphere of radius R. The field of **E** is uniform and is parallel to the axis of the hemisphere.

Solution:

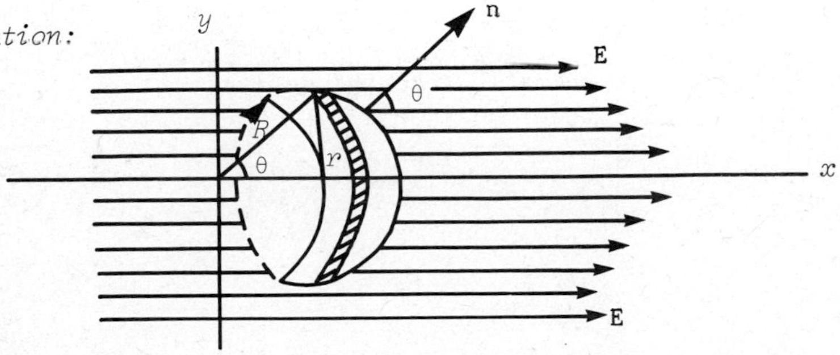

$\mathbf{E} \cdot \hat{\mathbf{n}} = En \cos\theta$

$\sin = \frac{r}{R} \Rightarrow r = R \sin\theta$

$$dA = 2\pi R \sin\theta \, (R d\theta)$$

$$\phi = \oint \mathbf{E} \cdot \hat{\mathbf{n}} \, dA$$

$$\phi = \oint E n \cos\theta \, dA$$

$$\phi = \oint E n \cos\theta \; 2\pi R^2 \sin\theta \, d\theta$$

$$= E \, 2\pi R^2 \oint_0^{\pi/2} \cos\theta \sin\theta \, d\theta$$

$$\int \sin^n ax \cos ax \, dx = \frac{1}{a(n+1)} \sin^{n+1} ax, \quad n \neq -1$$

$$\int \sin\theta \cos\theta \, d\theta = \tfrac{1}{2} \sin^2\theta$$

$$\phi = E \pi R^2 \left[\sin^2(\pi/2) - \sin^2 0\right] = \pi E R^2$$

PROBLEM 2.48 A having a charge of -2.0×10^{-9} C is acted on by a downward electric force of 3.0×10^{-6} N in a uniform electric field.

(a) What is the magnitude of the electric field?
(b) What are the magnitude and direction of the electric force exerted on a proton placed in this field?
(c) What is the gravitational force on the proton?
(d) What is the ratio of the electric to the gravitational forces in this case?

Solution:

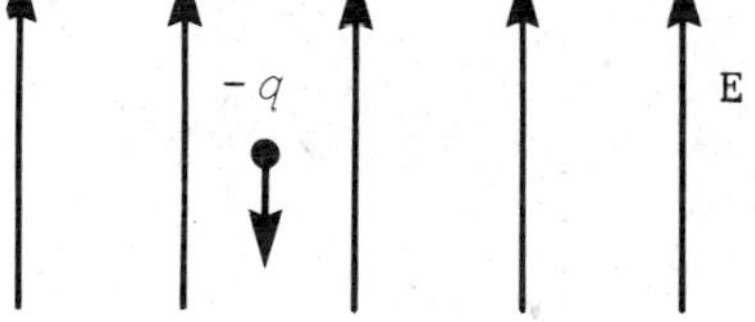

$$q = -2 \times 10^{-9} \text{ C}$$

$$\mathbf{F}_E = 3 \times 10^{-6} \text{ N, down}$$

(a) $$\mathbf{E} = \frac{3 \times 10^{-6} \text{ N}}{2 \times 10^{-9} \text{ C}} = 1.5 \times 10^3 \text{ N/C}$$

(b) $$F_{up} = (1.6 \times 10^{-19} \text{C})(1.5 \times 10^3 \text{ N/C}) = 2.4 \times 10^{-16} \text{ N}$$

(c) $$(1.67 \times 10^{-27} \text{ kg})(9.8 \text{ m/s}^2) = 1.64 \times 10^{-26} \text{ N, down}$$

(d) $$\frac{2.4 \times 10^{-16} \text{ N}}{1.64 \times 10^{-26} \text{ N}} = 1.46 \times 10^{10}$$

PROBLEM 2.49 A charge density defined by $\rho = K/r$ exists for a spherical region of space defined by $0 < r < R$.

(a) Find the electric field intensity at all points in space.
(b) Find the electric potential at all points in space.
(c) Show that the potential found in part b satisfies the equation $\nabla^2\phi = -\rho/\varepsilon$ for both $r < R$ and $r > R$.

Solution:

(a) $\rho = \dfrac{K}{r} \quad 0 < r < R$

$r < R$

$$\oint \mathbf{D} \cdot d\mathbf{s} = \int \rho \, dv$$

$$D4\pi r^2 = \int_0^r \frac{K}{r} 4\pi r^2 dr$$

$$= \int_0^r K4\pi r \, dr$$

$$= \frac{K4\pi r^2}{2}$$

$$D4\pi r^2 = 2\pi K r^2$$

$$\varepsilon_0 2E = K$$

$$E = \frac{K}{2\varepsilon_0}$$

$r > R$

$$E = \frac{1}{4\pi\varepsilon} \frac{Q_T}{r^2}$$

$$Q_T = \int_0^R \rho \, dv = K4\pi \frac{R^2}{2} = 2\pi K R^2$$

$$E = \frac{1}{4\pi\varepsilon} \frac{2\pi K R^2}{r^2}$$

$$E = \frac{KR^2}{2\varepsilon r^2}$$

(b)

For points outside ($r > R$)

$$\phi_r - \phi_\infty = -\int_\infty^r \mathbf{E} \cdot d\mathbf{r} = \int_r^\infty E_{\text{outside}} \, dr = \int_r^\infty \frac{KR^2}{2\varepsilon r^2} dr = \frac{KR^2}{2\varepsilon} \int_r^\infty r^{-2} dr$$

$$= \frac{-KR^2}{2\varepsilon}\left(\frac{1}{r}\right)_r^\infty = \frac{-KR^2}{2\varepsilon}\left(0-\frac{1}{r}\right) = \frac{KR^2}{2\varepsilon r}$$

For points inside $(r < R)$

$$\phi_r - \phi_\infty = -\int_\infty^r \mathbf{E}\cdot d\mathbf{r} = \int_r^\infty E_{\text{inside}}\,dr + \int_R^\infty E_{\text{outside}}\,dr$$

$$= \int_r^R \frac{K}{2\varepsilon_0}\,dr + \int_R^\infty \frac{KR^2}{2\varepsilon r^2}\,dr = \frac{K}{2\varepsilon_0}(R-r) + \frac{-KR^2}{2\varepsilon}\left(\frac{1}{\infty}-\frac{1}{R}\right)$$

$$= \frac{K}{2\varepsilon_0}(R-r) + \frac{KR^2}{2\varepsilon_0 R} = \frac{KR}{2\varepsilon_0} - \frac{Kr}{2\varepsilon_0} + \frac{KR}{2\varepsilon_0} = \frac{2KR}{2\varepsilon_0} - \frac{Kr}{2\varepsilon_0}$$

$$= \frac{KR}{\varepsilon_0} - \frac{Kr}{2\varepsilon_0} = \frac{2KR - Kr}{2\varepsilon_0} = \frac{K(2R-r)}{2\varepsilon_0}$$

$$= \frac{K}{\varepsilon}\left(R - \frac{r}{2}\right)$$

(c)

For $r > R$

$$\nabla^2\phi = \frac{-\rho}{\varepsilon}, \quad \rho = 0$$

$\nabla^2\phi = 0$, $\phi(r)$, ϕ is a function of r

$$\nabla^2\phi = \frac{1}{r^2}\frac{\partial}{\partial r}\left(r^2\frac{\partial\phi}{\partial r}\right) = 0$$

$$\frac{1}{r^2} \neq 0 \Rightarrow \frac{\partial}{\partial r}\left(r^2\frac{\partial\phi}{\partial r}\right) = 0 \Rightarrow r^2\frac{\partial\phi}{\partial r} = \text{constant} = b$$

$r^2\dfrac{\partial\phi}{\partial r} = b$, but $\phi(r)$, therefore

$$r^2\frac{d\phi}{dr} = b$$

$$\int d\phi = \int br^{-2}dr$$

$$\phi(r) = \frac{-b}{r} + c$$

The boundary conditions are

at $r = \infty$, $\phi = 0$

at $r = r'$, $r' > R$, $\phi = \dfrac{KR^2}{2\varepsilon r'}$

at $r = r'$, $\phi = \frac{1}{4\pi\varepsilon}\frac{Q_T}{r'}$

$$= \frac{1}{4\pi\varepsilon}\frac{2\pi KR^2}{r'}$$

$$= \frac{KR^2}{2\varepsilon r'}$$

$$\phi(\infty) = \frac{-b}{\infty} + c = 0 \Rightarrow c = 0$$

$$\phi(r) = \frac{-b}{r}$$

$$\phi(r') = \frac{-b}{r'} = \frac{KR^2}{2\varepsilon r'} \Rightarrow b = \frac{-KR^2}{2\varepsilon}$$

$$\phi(r) = \frac{-(-KR^2/2\varepsilon)}{r} + 0$$

$$\phi(r) = \frac{KR^2}{2\varepsilon r}$$

Also substituting,

$$\nabla^2\phi = \frac{-\rho}{\varepsilon}, \quad \text{outside } (r > R) \quad \rho = 0, \quad \phi(r > R) = \frac{KR^2}{2\varepsilon r}$$

$$\nabla^2\phi = 0$$

$$\nabla^2\phi = \frac{1}{r^2}\frac{\partial}{\partial r}\left(r^2\frac{\partial\phi}{\partial r}\right) = \frac{1}{r^2}\frac{\partial}{\partial r}\left[(r^2)\left(\frac{KR^2}{2\varepsilon}\right)\left(-\frac{1}{r^2}\right)\right]$$

$$= -\frac{1}{r^2}\frac{\partial}{\partial r}\frac{KR^2}{2\varepsilon} = -\frac{1}{r^2}(0) = 0$$

<u>For $r < R$</u>

$$\phi(r < R) = \frac{K}{\varepsilon}\left(R - \frac{r}{2}\right) = \frac{K}{\varepsilon}R - \frac{K}{\varepsilon}\frac{r}{2}$$

$$\nabla^2\phi = \frac{1}{r^2}\frac{\partial}{\partial r}\left(r^2\frac{\partial\phi}{\partial r}\right) = \frac{1}{r^2}\frac{\partial}{\partial r}\left[(r^2)\left(\frac{-K}{2\varepsilon}\right)\right] = \frac{-K}{2\varepsilon r^2}\frac{\partial}{\partial r}r^2$$

$$= \frac{-K2r}{2\varepsilon r^2} = \frac{-K}{\varepsilon r} = -(K/r)/\varepsilon = \frac{-\rho}{\varepsilon}$$

$[\rho = K/r]$

PROBLEM 2.50 Consider the rectangular charge distribution in the figure.

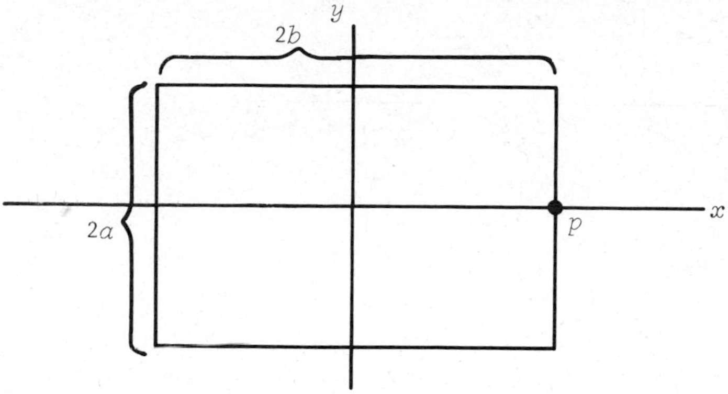

Assume the surface charge density is given by $\sigma = kxy$, where k is constant, and find the electric potential at point P located on the x axis.

Solution:

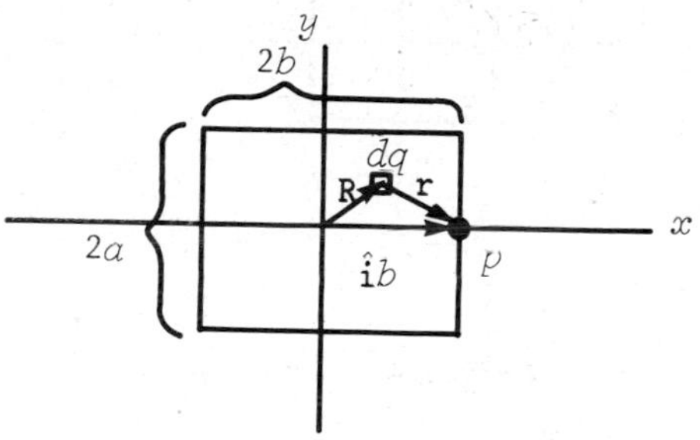

$$dq = \sigma\, dx\, dy = kxy\, dx\, dy$$

$$\sigma = kxy, \quad k \text{ constant}$$

$$d\phi = \frac{\sigma\, dx\, dy}{4\pi\varepsilon r}$$

$$\mathbf{R} = \hat{\mathbf{i}}x + \hat{\mathbf{j}}y$$

$$\mathbf{R} + \mathbf{r} = \hat{\mathbf{i}}b$$

$$\mathbf{r} = \hat{\mathbf{i}}b - (\hat{\mathbf{i}}x + \hat{\mathbf{j}}y) = \hat{\mathbf{i}}(b - x) + \hat{\mathbf{j}}y$$

$$r = \sqrt{(b - x)^2 + y^2}$$

$$d\phi = \frac{1}{4\pi\varepsilon} \frac{Kxy\, dx\, dy}{\sqrt{(b - x)^2 + y^2}}$$

$$\phi = \int_{x=-b}^{b} \int_{y=-a}^{a} \frac{Kxy\, dx\, dy}{4 \sqrt{(b - x)^2 + y^2}}$$

$$\phi = \frac{Kx\, dx}{4\pi\varepsilon} \int_{x=-b}^{b} \int_{y=-a}^{a} \frac{y}{\sqrt{(b - x)^2 + y^2}}$$

where the integral over y equals $\int_y = \frac{Kx\,dx}{4\pi\varepsilon}\left[(b-x)^2+y^2\right]^{\frac{1}{2}}\Big|_{-a}^{a}$

$$\int_y = \frac{Kx\,dx}{4\pi\varepsilon}\sqrt{(b-x)^2+a^2} - \frac{Kx\,dx}{4\pi\varepsilon}\sqrt{(b-x)^2+a^2} = 0$$

$$\phi = \int_{x=-b}^{b} \frac{Kx\,dx}{4\pi\varepsilon}(0) = 0$$

PROBLEM 2.51 Determine the electric potential produced by the rectangular charge distribution $\sigma = kxy$ (on the rectangle in the figure), at arbitrary points in space that is correct to the quadrapole term. Using the potential, find the electric field intensity at these same points.

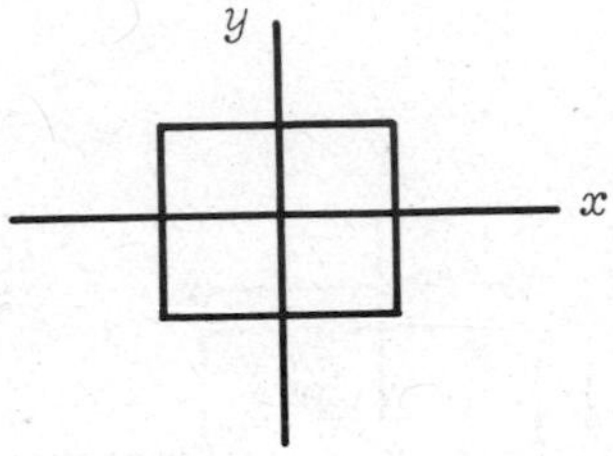

Solution:

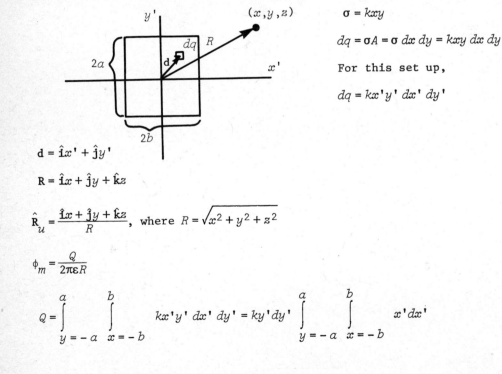

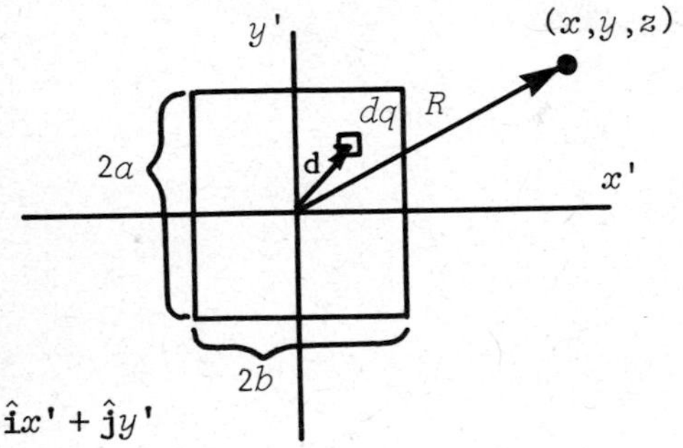

$\sigma = kxy$

$dq = \sigma A = \sigma\,dx\,dy = kxy\,dx\,dy$

For this set up,

$dq = kx'y'\,dx'\,dy'$

$\mathbf{d} = \hat{\mathbf{i}}x' + \hat{\mathbf{j}}y'$

$\mathbf{R} = \hat{\mathbf{i}}x + \hat{\mathbf{j}}y + \hat{\mathbf{k}}z$

$\hat{\mathbf{R}}_u = \frac{\hat{\mathbf{i}}x + \hat{\mathbf{j}}y + \hat{\mathbf{k}}z}{R}$, where $R = \sqrt{x^2+y^2+z^2}$

$$\phi_m = \frac{Q}{2\pi\varepsilon R}$$

$$Q = \int_{y=-a}^{a}\int_{x=-b}^{b} kx'y'\,dx'\,dy' = ky'dy'\int_{y=-a}^{a}\int_{x=-b}^{b} x'dx'$$

where the integral over x equals

$$\frac{x'^2}{2}\Bigg|_{-b}^{b} = \frac{b^2}{2} - \frac{b^2}{2} = 0$$

$$\phi_m = \frac{0}{2\pi\varepsilon R}$$

$$\phi_m = 0$$

$$Q = \int_{y'=-a}^{a} ky'dy'(0) = 0$$

$$\phi_d = \frac{1}{4\pi\varepsilon R^2}\int (\mathbf{d}\cdot\hat{\mathbf{R}}_u)dq$$

$$\mathbf{d}\cdot\hat{\mathbf{R}}_u = (\hat{\mathbf{i}}x' + \hat{\mathbf{j}}y')\cdot\left(\hat{\mathbf{i}}\,\frac{x}{R} + \hat{\mathbf{j}}\,\frac{y}{R} + \hat{\mathbf{k}}\,\frac{z}{R}\right) = \frac{x'x}{R}\quad\frac{y'y}{R}$$

$$\phi_d = \frac{1}{4\pi\varepsilon R^3}\int (x'x + y'y)(kx'y'dx'dy')$$

$$= \frac{1}{4\pi\varepsilon R^3}\int kx'^2xy'dx'dy' + ky'^2yx'dx'dy'$$

$$= \frac{k}{4\pi\varepsilon R^3}\int_{y=-a}^{a}\int_{x=-b}^{b} y'xx'^2dx'dy' + y'^2yx'dx'dy'$$

$$= \frac{k}{4\pi\varepsilon R^3}\int_{y=-a}^{a} y'x\left(\frac{x'^3}{3}\Bigg|_{-b}^{b}\right)dy' + yy'^2\left(\frac{x'^2}{2}\right)\Bigg|_{-b}^{b} dy'$$

where the last term equals zero.

$$\phi_d = \frac{k}{4\pi\varepsilon R^3}\int_{y=-a}^{+a} \frac{2b^3}{3}\,xy'dy' = \frac{k2b^3x}{4\pi\varepsilon R^3 3}\left(\frac{y'^2}{2}\Bigg|_{-a}^{a}\right) = 0$$

where the term in parentheses equals zero.

So $\qquad \phi_d = 0$

$$\phi_q = \frac{1}{4\pi\varepsilon R^3} \int (\mathbf{d} \cdot \hat{\mathbf{R}}_u)^2 dq$$

$$(\mathbf{d} \cdot \hat{\mathbf{R}}_u)^2 = \left(\frac{xx'}{R} + \frac{yy'}{R}\right)^2 = \left(\frac{xx'}{R} + \frac{yy'}{R}\right)\left(\frac{xx'}{R} + \frac{yy'}{R}\right)$$

$$= \frac{x^2x'^2}{R^2} + \frac{2xyx'y'}{R^2} + \frac{y^2y'^2}{R^2} = (x^2x'^2 + 2xyx'y' + y^2y'^2)\left(\frac{1}{R^2}\right)$$

$$\phi_q = \frac{1}{4\pi\varepsilon R^5} \int (x^2x'^2 + 2xyx'y' + y^2y'^2)(kx'y')dx'dy'$$

$$= \frac{k}{4\pi\varepsilon R^5} \int (x^2x'^2x'y' + 2xyx'y'x'y' + y^2y'^2x'y')dx'dy'$$

$$= \frac{k}{4\pi\varepsilon R^5} \int_{y=-a}^{a} \int_{x=-b}^{b} x^2x'^3y' + 2xyx'^2y'^2 + y^2x'y'^3dx'dy'$$

where the integral over x equals

$$\int_x = x^2y' \underbrace{\left(\frac{x'^4}{4}\right)\Bigg|_{-b}^{b}}_{=0} + 2xyy'^2 \left(\frac{x'^3}{3}\Bigg|_{-b}^{b}\right) + y^2y'^3 \underbrace{\left(\frac{x'^2}{2}\Bigg|_{-b}^{b}\right)}_{=0}$$

$$\int_x = 2xyy'^2\left(\frac{2b^3}{3}\right)$$

$$\int_x = \frac{4}{3}\, xyy'^2b^3$$

$$= \frac{kb^34xy}{4\pi\varepsilon R^5 3} \int_{y=-a}^{a} y'^2dy' = \frac{kxyb^3}{3\pi\varepsilon R^5}\left(\frac{y'^3}{3}\Bigg|_{-a}^{a}\right) = \frac{kxyb^32a^3}{9\pi\varepsilon R^5}$$

$$\phi_q = \frac{2kxyb^3a^3}{9\pi\varepsilon R^5} \tag{1}$$

$$\mathbf{E} = -\nabla\phi$$

$$\phi_{\text{approx.}} = \phi_m + \phi_d + \phi_q = 0 + 0 + \phi_q = \frac{2kxyb^3a^3}{9\pi\varepsilon R^5}$$

$$E = -\left(\hat{\imath}\,\frac{\partial}{\partial x} + \hat{\jmath}\,\frac{\partial}{\partial y} + \hat{k}\,\frac{\partial}{\partial z}\right)\left(\frac{2ka^3b^3}{9\pi\varepsilon}\;\frac{xy}{(x^2+y^2+z^2)^{5/2}}\right)$$

$$= \frac{-\,2kb^3a^3}{9\pi\varepsilon}\left(\hat{\imath}\,\frac{\partial}{\partial x} + \hat{\jmath}\,\frac{\partial}{\partial y} + \hat{k}\,\frac{\partial}{\partial z}\right)[xy(x^2+y^2+z^2)^{-\,5/2}]$$

where $\dfrac{-\,2kb^3a^3}{9\pi\varepsilon} = c = \text{a constant}$

$$\mathrm{E} = c\,\{\hat{\imath}[(xy)(-\tfrac{5}{2})(x^2+y^2+z^2)^{-7/2}(2x) + (x^2+y^2+z^2)^{-5/2}(y)]$$

$$+\hat{\jmath}[(xy)(-\tfrac{5}{2})(x^2+y^2+z^2)^{-7/2}(2y) + (x^2+y^2+z^2)^{-5/2}(x)]$$

$$+\hat{k}[(xy)(-\tfrac{5}{2})(x^2+y^2+z^2)^{-7/2}(2z) + 0]\}$$

$$= c\left[\hat{\imath}\left(\frac{-\,5x^2y}{(x^2+y^2+z^2)^{7/2}} + \frac{y}{(x^2+y^2+z^2)^{5/2}}\right)\right.$$

$$\left.+\hat{\jmath}\left(\frac{-\,5y^2x}{(x^2+y^2+z^2)^{7/2}} + \frac{x}{(x^2+y^2+z^2)^{5/2}}\right) + \hat{k}\left(\frac{-\,5xyz}{(x^2+y^2+z^2)^{7/2}}\right)\right]$$

$$\mathrm{E} = \frac{+\,2kb^3a^3}{9\pi\varepsilon}\left[\hat{\imath}\left(\frac{+\,5x^2y(x^2+y^2+z^2)^{5/2} + y(x^2+y^2+z^2)^{7/2}}{(x^2+y^2+z^2)^{6}}\right)\right. \qquad (2)$$

$$\left.+\hat{\jmath}\left(\frac{+\,5y^2x(x^2+y^2+z^2)^{5/2} + x(x^2+y^2+z^2)^{7/2}}{(x^2+y^2+z^2)^{6}}\right) + \hat{k}\left(\frac{5xyz}{(x^2+y^2+z^2)^{7/2}}\right)\right]$$

PROBLEM 2.52 It is found experimentally that the electric field in a large region of the earth's atmosphere is directed vertically downward. At an altitude of 300 m the field is 60 N/C and at an altitude of 200 m it is 100 N/C. Find the net amount of charge contained in a cube 100 m on an edge extending from a height of 200 m to a height of 300 m.

Solution:

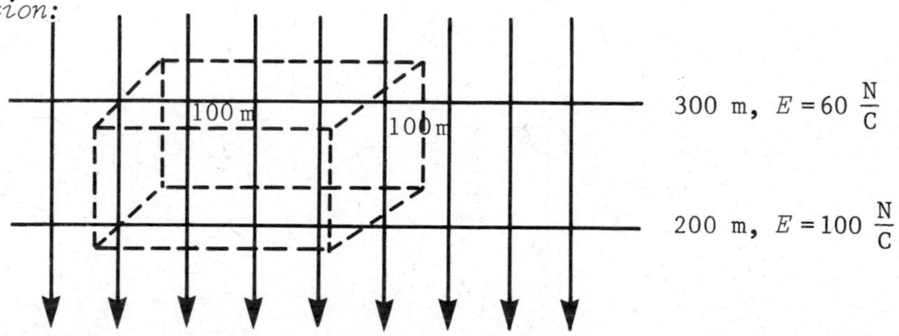

$$\varepsilon_0 \phi_E = \varepsilon_0 \oint \mathbf{E} \cdot \hat{\mathbf{n}}\, dA = q_0$$

$$\text{Total flux} = \phi = \phi_{in} - \phi_{out} = \frac{q_0}{\varepsilon_0}$$

Therefore,

$$\varepsilon_0 (EA_{in} - EA_{out}) = q_0$$

$$\varepsilon_0 \left(-60\ \frac{\text{N}}{\text{C}}\right)(100\ \text{m})^2 - \varepsilon_0 \left(-100\ \frac{\text{N}}{\text{C}}\right)(100\ \text{m}^2) = q_0$$

$$\left(8.85 \times 10^{-12}\ \frac{C^2}{\text{N} \cdot \text{m}^2}\right)\left(-60\ \frac{\text{N}}{\text{C}}\right)(1 \times 10^4\ \text{m}^2)$$

$$-\left(8.85 \times 10^{-12}\ \frac{C^2}{\text{N} \cdot \text{m}^2}\right)\left(-1 \times 10^2\ \frac{\text{N}}{\text{C}}\right)(1 \times 10^4\ \text{m}^2) = q$$

$$-5.3 \times 10^{-6}\ \text{C} + 8.85 \times 10^{-6}\ \text{C} = q$$

$$q = +3.55 \times 10^{-6}\ \text{C}$$

PROBLEM 2.53 A thin glass rod is bent into a semicircle of radius R. The linear charge density varies according to $\lambda = \lambda_0 \sin\theta$, where θ varies from 0 to π. Find the electric field intensity at the center of the semicircle.

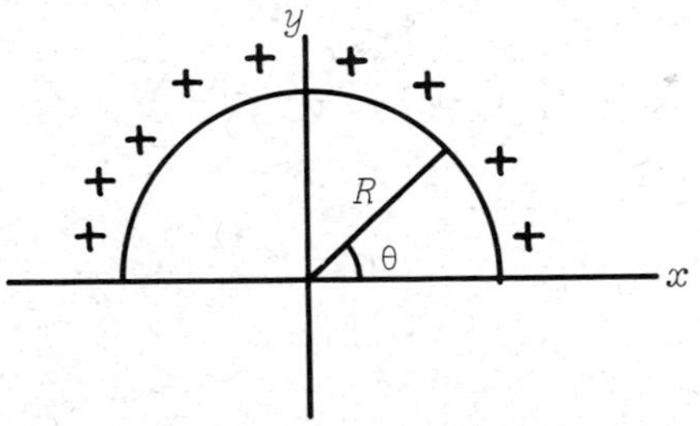

Solution:

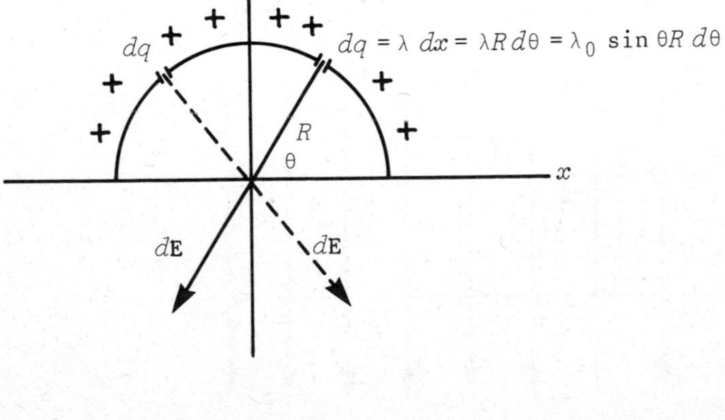

$$d\mathbf{E} = \frac{1}{4\pi\varepsilon} \frac{dq}{r^2}$$

$$d\mathbf{E} = \hat{\mathbf{i}}\,\frac{1}{4\pi\varepsilon}\,\frac{\lambda_0 \sin\theta R\,d\theta}{R^2}\cos\theta - \hat{\mathbf{j}}\,\frac{1}{4\pi\varepsilon}\,\frac{\lambda_0 \sin\theta R\,d\theta}{R^2}\sin\theta,$$

the first term $\to 0$ by symmetry.

$$d\mathbf{E} = -\hat{\mathbf{j}}\,\frac{\lambda_0 \sin^2\theta R\,d\theta}{4\pi\varepsilon R^2}$$

$$\mathbf{E} = \frac{-\hat{\mathbf{j}}\lambda_0}{4\pi\varepsilon R}\int_0^\pi \sin^2\theta\,d\theta = \frac{-\hat{\mathbf{j}}\lambda_0}{4\pi\varepsilon R}\,[\tfrac{1}{2}\theta - \tfrac{1}{4}\sin^2\theta]_0^\pi$$

$$\mathbf{E} = \frac{-\hat{\mathbf{j}}\lambda_0}{4\pi\varepsilon R}\left(\frac{\pi}{2}\right)$$

$$\mathbf{E} = -\hat{\mathbf{j}}\,\frac{\lambda_0}{8\varepsilon R}$$

PROBLEM 2.54 A total charge Q is distributed uniformly throughout the volume of a sphere of radius R. The electric field (radial) is

$$E = \begin{cases} \dfrac{Q}{r^2}, & r > R \\[2ex] \dfrac{Q}{R^3}r, & r < R \end{cases}$$

By calculating the line integral from infinity in to r, determine the potential for (a) $r > R$; (b) $r - R$; (c) $r < R$; (d) $r = 0$.

Solution:

(a) $$\phi_r - \phi_\infty = -\int_\infty^r \frac{Q}{r^2}\,dr = +\frac{Q}{r} = \phi$$

(b) $$\frac{Q}{R} = \phi$$

(c) $$\phi_r - \phi_\infty = -\int_\infty^R \mathbf{E}\cdot d\mathbf{r} - \int_R^r \frac{Q}{R^3}\,r\,dr = \frac{Q}{R} - \frac{Q}{R^3}\left(\frac{r^2}{2} - \frac{R^2}{2}\right)$$

$$\phi = \frac{Q}{R} - \frac{Q}{2R^3}\,(r^2 - R^2)$$

$$\phi = \frac{Q}{R}\left(1 - \frac{1}{2R^2}\,(r^2 - R^2)\right)$$

(d) $$\phi = \frac{2Q}{2R}\quad\frac{Q(+R^2)}{2R^3} = \frac{3Q}{2R}$$

PROBLEM 2.55 Show that for frequencies at which vacuum tubes are ordinarily operated, the vacuum displacement current is negligible compared with the current carried by electrons. Take, for example, a tube with cathode and plate area each 1 cm^2, space 5 mm apart, passing 10 mA (rms) at a frequency of 10^6 cps (Hz). Assume parallel-plate geometry and that the cathode is at fixed potential and the rms ac voltage across the tube is 100 V. What is the magnitude of the rms displacement current through the tube?

Solution:

$$I_d = \int \mu_0\varepsilon_0 \frac{\partial E}{\partial t}\, dA = \mu_0\varepsilon_0 \frac{\partial E}{\partial t}\, \pi r^2$$

$$E = \frac{V}{d} = \frac{100}{5\times 10^{-3}} = 2\times 10^4 \text{ V/m}$$

If $E = E_0 \sin \omega t$

$$\frac{\partial E}{\partial t} = \omega E_0 \cos \omega t$$

$$I_d = \mu_0\varepsilon_0\omega_0 E_0 \pi r^2$$

where $\pi r^2 = A$

$$I_d = (4\pi\times 10^{-7})(8.85\times 10^{-12})(2\pi)10^6(2\times 10^4)(10^{-4})$$

$$I_d = 1.4\times 10^{-10} \text{ A}$$

PROBLEM 2.56 Calculate the relaxation time (characteristic decay time) of the charge in a homogeneous conductor. (*Hint:* Combine the conductivity equation, the equation of continuity, and the source equation for the electric displacement to give a differential equation for ρ, and integrate.)

Solution:

Continuity equation

$$\frac{\partial \rho}{\partial t} + \nabla\cdot \mathbf{j} = 0$$

Conduction equation

$$\mathbf{j} = \sigma \mathbf{E}$$

therefore

$$\frac{\partial \rho}{\partial t} + \sigma\nabla\cdot \mathbf{E} = 0$$

Source equation for D

$$\nabla\cdot \mathbf{D} = \rho$$

Supplementary Condition

$$\mathbf{D} = \varepsilon \mathbf{E}$$

so that

$$\nabla \cdot \mathbf{D} = \varepsilon \nabla \cdot \mathbf{E} = \rho$$

therefore

$$\frac{\partial \rho}{\partial t} + \frac{\sigma}{\varepsilon} \rho = 0$$

$$\int_{\rho_0}^{\rho} \frac{d\rho'}{\rho'} = \int_0^t -\frac{\sigma}{\varepsilon}\, dt$$

$$\rho = \rho_0 \, e^{(-\sigma/\varepsilon)t}$$

$$\therefore \quad \text{relaxation time} = \frac{\varepsilon}{\sigma}$$

PROBLEM 2.57 Calculate the potential when an insulated uncharged conducting sphere is placed in a constant external electric field. Calculate the surface charge distribution of the sphere.
(*Hint:* Use method of images).

Solution:

If there are two charges $\pm Q$ located at positions $z = \mp R$, then in a region near the origin whose dimensions are very small compared to R there is an approximately constant electric field

$$\mathbf{E}_0 \simeq \frac{Q}{2\pi\varepsilon_0 R^2}\,\hat{\mathbf{k}}$$

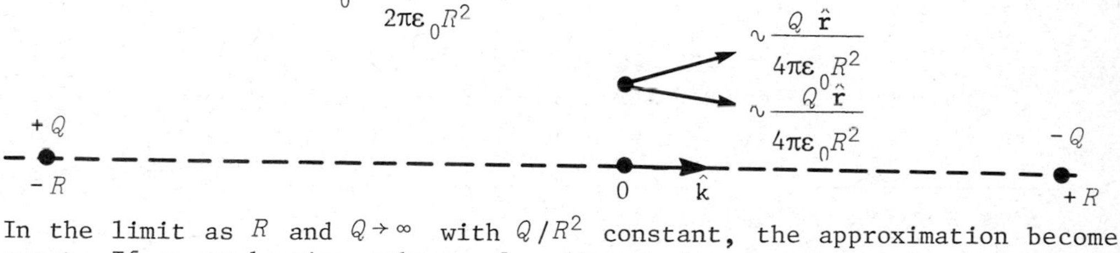

In the limit as R and $Q \to \infty$ with Q/R^2 constant, the approximation becomes exact. If a conducting sphere of radius a is placed at the origin, the potential will be that due to the charges $\pm Q$ at $\mp R$ and their images. First consider the images required for the charge $-Q$ at R,

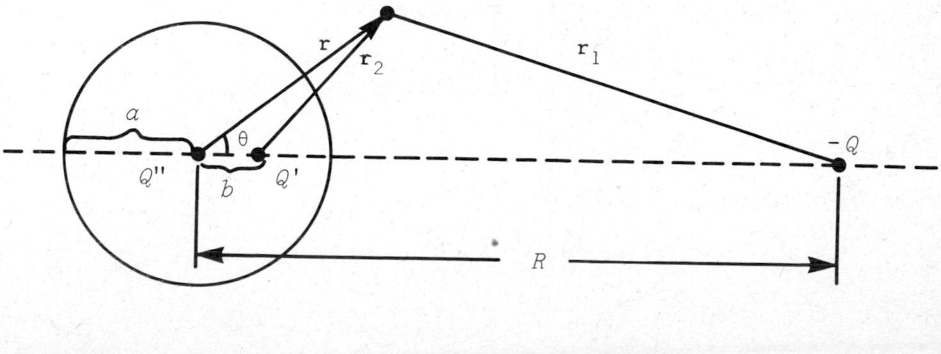

The charge Q' is such that the combination Q' and $-Q$ make the sphere a surface of zero potential. The charge Q'' changes the potential of the sphere to fit the boundary condition:

$\phi(r,\theta,\phi)$ due to Q' and $-Q$

$$\phi(r,\theta,\phi) = \frac{-Q}{4\pi\varepsilon_0 r_1} + \frac{Q'}{4\pi\varepsilon_0 r_2}$$

$$= \frac{1}{4\pi\varepsilon_0}\left(\frac{-Q}{(r^2+R^2-2rR\cos\theta)^{\frac{1}{2}}} + \frac{Q'}{(r^2+b^2-2rb\cos\theta)^{\frac{1}{2}}}\right)$$

$\phi(a,\theta,\phi) = 0$ for all θ and ϕ

This will be so if $b = a^2/R$ and $Q' = (a/R)Q$. Since the net charge on the sphere is zero,

$$Q'' = -Q' = -\frac{a}{R}Q$$

Applying the same line of reasoning to the charge $+Q$ at $-R$, the correct image charges are:

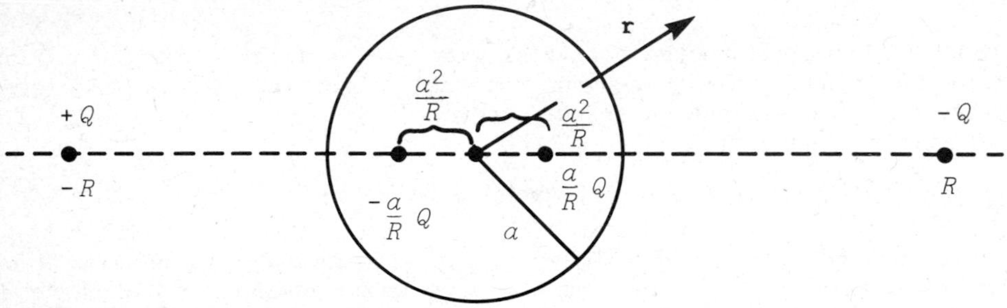

The Q'''s associated with the $+Q$ and the $-Q$ are equal and opposite, therefore the image charge at the origin is zero:

$$\phi(\mathbf{r}) = \frac{1}{4\pi\varepsilon_0}\left(\frac{Q}{(r^2+R^2+2rR\cos\theta)^{\frac{1}{2}}} - \frac{Q}{(r^2+R^2-2rR\cos\theta)^{\frac{1}{2}}}\right.$$

$$\left. - \frac{aQ}{R\,[r^2+a^4/R^2+(2a^2r/R)\cos\theta]^{\frac{1}{2}}} + \frac{aQ}{R[r^2+a^4/R^2-(2a^2r/R)\cos\theta]^{\frac{1}{2}}}\right)$$

In the first two terms $R \gg r$

$$(r^2+R^2+2rR\cos\theta)^{-\frac{1}{2}} \simeq \frac{1}{R}\left(1+\frac{2r}{R}\cos\theta\right)^{-\frac{1}{2}}$$

$$\simeq \frac{1}{R}\left(1-\frac{1}{2}\frac{2r}{R}\cos\theta\right) = \frac{1}{R}\left(1-\frac{r}{R}\cos\theta\right)$$

Similarly,

$$(r^2+R^2-2rR\cos\theta)^{-\frac{1}{2}} \simeq \frac{1}{R}\left(1+\frac{r}{R}\cos\theta\right)$$

In the second two terms $r >> R$,

$$\left(r^2 + \frac{a^4}{R^2} + \frac{2a^2 r}{R}\cos\theta\right)^{-\frac{1}{2}} \simeq \frac{1}{r}\left(1 + \frac{2a^2}{Rr}\cos\theta\right)^{-\frac{1}{2}}$$

$$\simeq \frac{1}{r}\left(1 - \frac{1}{2}\frac{2a^2}{Rr}\cos\theta\right) = \frac{1}{r}\left(1 - \frac{a^2}{Rr}\cos\theta\right)$$

Similarly,

$$\left(r^2 + \frac{a^4}{R^2} - \frac{2a^2 r}{R}\cos\theta\right)^{-\frac{1}{2}} \simeq \frac{1}{r}\left(1 + \frac{a^2}{Rr}\cos\theta\right)$$

$$\therefore \quad \phi(\mathbf{r}) \simeq \frac{1}{4\pi\varepsilon_0}\left(-\frac{2Q}{R^2}\, r\cos\theta + \frac{a^3 Q}{r^2 R^2}\cos\theta\right)$$

in the limit as $r \to \infty$. In that limit $2Q/R^2 \to E_0$, the applied uniform field,

$$\phi(\mathbf{r}) = -\frac{E_0}{4\pi\varepsilon_0}\left(r - \frac{a^3}{r^2}\right)\cos\theta \tag{1}$$

Surface charge $\sigma = \varepsilon_0 E(a)$

$$\sigma = -\varepsilon_0 \left.\frac{\partial\phi}{\partial r}\right|_a = +\frac{E_0}{4\pi}\left.\left(1 + 2\frac{a^3}{r^3}\right)\right|_a \cos\theta$$

$$\sigma = \frac{3E_0\cos\theta}{4\pi} \tag{2}$$

PROBLEM 2.58 Calculate the potential inside an infinite rectangular wave-guide with grounded conducting walls at

$x = 0, a$ and $y = 0, b$

in the presence of a line charge λ (C/m) inside the guide and parallel to it at

$x = c \qquad y = d$

Solution:

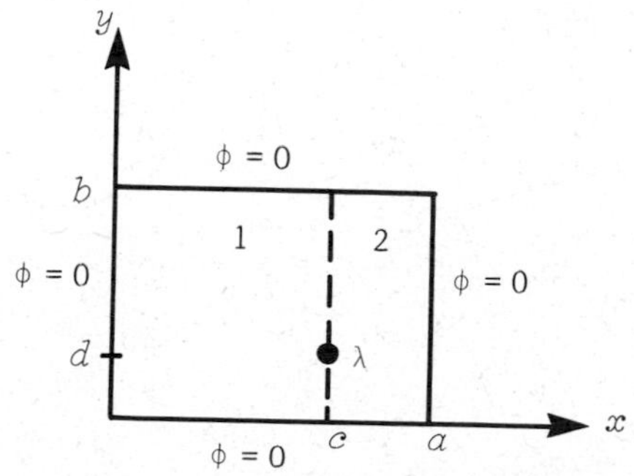

Let $\phi(x,y) = X(x)Y(y)$, between the conductors except at (c,d)

$$\nabla^2\phi = 0$$

$$Y\frac{d^2X}{dx^2} + X\frac{d^2Y}{dy^2} = 0$$

$$\frac{1}{X}\frac{d^2X}{dx^2} + \frac{1}{Y}\frac{d^2Y}{dy^2} = 0$$

Since it is possible to vary y without affecting X, each term must be constant:

$$\frac{d^2Y}{dy^2} + k^2Y = 0 \qquad \frac{d^2X}{dx^2} - k^2X = 0$$

$$Y = A' \sin ky = B' \cos ky$$

$$X = C'e^{kx} + D'e^{-kx}$$

$$Y = 0 \text{ at } y = 0 \text{ and } y = b$$

$$\therefore \quad Y = A' \sin\frac{n\pi y}{b}$$

$$X = C'e^{n\pi x/b} + D'e^{-n\pi x/b}$$

$$X = 0 \text{ at } x = 0 \text{ and } x = a$$

These conditions cannot be simultaneously satisfied by either term in X so that the solution must be written separately for the region $0 \leq x \leq c$ (region 1) and $c \leq x \leq a$ (region 2):

$$X_1 = C_1e^{n\pi x/b} + D_1e^{-n\pi x/b}, \quad 0 \leq x \leq c$$

$$X_1 = 0 \text{ at } x = 0$$

$$0 = C_1 + D_1$$

$$\therefore \quad D_1 = -C_1$$

$$X_1 = C_1(e^{n\pi x/b} - e^{-n\pi x/b})$$

$$X_2 = C_2e^{n\pi x/b} + D_2e^{-n\pi x/b}, \quad c \leq r \leq a$$

$$X_2 = 0 \text{ at } x = a$$

$$0 = C_2e^{n\pi a/b} + D_2e^{-n\pi a/b}$$

$$0 = C_2e^{2n\pi a/b} + D_2$$

$$X_2 = C_2(e^{n\pi x/b} - e^{2n\pi a/b}\, e^{-n\pi x/b})$$

$$X_2 = C_2(e^{n\pi x/b} - e^{n\pi(2a-x)/b})$$

$$\phi_1 = \sum_{n=1}^{\infty} A_n \sin\frac{n\pi y}{b}\,(e^{n\pi x/b} - e^{-n\pi x/b}), \quad 0 \le x \le c$$

$$\phi_2 = \sum_{n=1}^{\infty} B_n \sin\frac{n\pi y}{b}\,(e^{n\pi x/b} - e^{n\pi(2a-x)/b}), \quad c \le x \le a$$

$\phi_1 = \phi_2$ at $x = c$

$$\sum_{n=1}^{\infty} A_n \sin\frac{n\pi y}{b}\,(e^{n\pi c/b} - e^{-n\pi c/b}) = \sum_{n=1}^{\infty} B_n \sin\frac{n\pi y}{b}\,(e^{n\pi c/b} - e^{n\pi(2a-c)/b})$$

$$A_n\left(\frac{e^{2n\pi c/b} - 1}{e^{n\pi c/b}}\right) = B_n\left(\frac{e^{2n\pi c/b} - e^{2n\pi a/b}}{e^{n\pi c/b}}\right)$$

$$A_n = B_n \frac{e^{2n\pi c/b} - e^{2n\pi a/b}}{e^{2n\pi c/b} - 1} \triangleq B_n g$$

at $x = c$, $\eta(y) = \lambda\delta(y - d)$

$$\left[\left(\frac{\partial\phi_1}{\partial x}\right)_{x=c} - \left(\frac{\partial\phi_2}{\partial x}\right)_{x=c}\right] = \frac{\eta(y)}{\varepsilon_0} = \frac{\lambda\delta(y-d)}{\varepsilon_0}$$

$$\sum_{n=1}^{\infty} B_n g \sin\frac{n\pi y}{b}\left(\frac{n\pi}{b}\right)(e^{n\pi c/b} + e^{-n\pi c/b})$$

$$-\sum_{n=1}^{\infty} B_n \sin\frac{n\pi y}{b}\left(\frac{n\pi}{b}\right)(e^{n\pi c/b} + e^{n\pi(2a-c)/b}) = \frac{\lambda\delta(y-d)}{\varepsilon_0}$$

Multiply both sides through by $\sin m\pi y/b$,

$$\sum_{n=1}^{\infty} B_n \sin\frac{n\pi y}{b}\sin\frac{m\pi y}{b}\,\frac{n\pi}{b}\,[e^{n\pi c/b}(g-1) + e^{-n\pi c/b}(g - e^{2n\pi a/b})]$$

$$= \frac{\lambda\delta(y-d)}{\varepsilon_0}\sin\frac{m\pi y}{b}$$

Integrating from 0 to b,

$$B_m \frac{b}{2}\frac{m\pi}{b}\,[e^{m\pi c/b}(g-1) + e^{-m\pi c/b}(g - e^{2m\pi a/b})] = \frac{\lambda}{\varepsilon_0}\sin\frac{m\pi d}{b}$$

back to n,

$$B_n = \frac{2\lambda}{n\pi\varepsilon_0} \frac{\sin(n\pi d/b)}{[e^{n\pi c/b}(g-1) + e^{-n\pi c/b}(g - e^{2n\pi a/b})]}$$

$$\phi_1 = \sum_{n=1}^{\infty} \frac{2\lambda \sin(n\pi d/b)}{n\pi\varepsilon} \frac{(g \sin(n\pi y/b))(e^{n\pi x/b} - e^{-n\pi x/b})}{[e^{n\pi c/b}(g-1) + e^{-n\pi c/b}(g - e^{2n\pi a/b})]}$$

$$\phi_1 = \sum_{n=1}^{\infty} \frac{2\lambda \sin(n\pi x/b)}{n\pi\varepsilon_0} \frac{(\sin \pi y/b)(e^{n\pi x/b} - e^{n\pi(2a-x)/b})}{[e^{n\pi c/b}(g-1) + e^{-n\pi c/b}(g - e^{2n\pi a/b})]}$$

where

$$g = \frac{e^{2n\pi c/b} - e^{2n\pi a/b}}{e^{2n\pi c/b} - 1}$$

PROBLEM 2.59 Given a thin circular disk of dielectric material having uniform polarization P parallel to its axis, find the electrostatic potential at a point of observation on the axis of the disk, whose distance x from the disk is large compared to the thickness of the disk but not large compared to the diameter of the disk:

for $x > 0$,

$$\phi = \frac{Pt}{2\varepsilon_0}\left(1 - \frac{x}{(R^2 + x^2)^{\frac{1}{2}}}\right)$$

for $x < 0$,

$$\phi = -\frac{Pt}{2\varepsilon_0}\left(1 + \frac{x}{(R^2 + x^2)^{\frac{1}{2}}}\right)$$

Solution:

The situation we have is this:

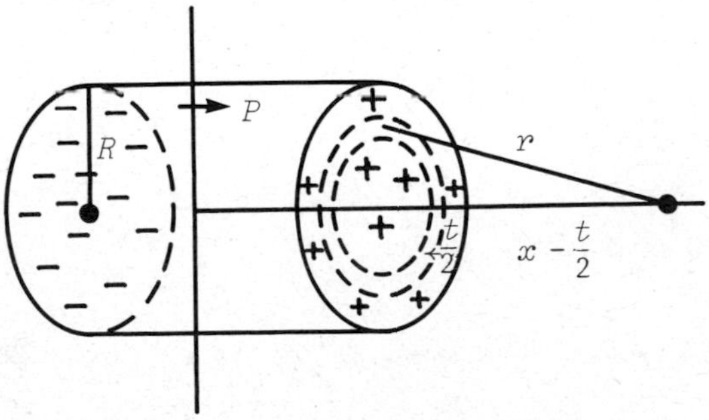

The charge density on either face is $\sigma = P$. For the + face

$$d\phi_+ = \frac{1}{4\pi\varepsilon_0}\frac{dq}{r} = \frac{1}{4\pi\varepsilon_0}\frac{P2\pi a da}{[a^2 + (x - t/2)^2]^{\frac{1}{2}}}$$

$$\phi_+ = \frac{P}{2\varepsilon_0}\int_0^R \frac{a da}{[a^2 + (x - t/2)^2]^{\frac{1}{2}}} = \frac{P}{2\varepsilon_0}[a^2 + (x - t/2)^{\frac{1}{2}}]\Big|_0^R$$

$$\phi_+ = \frac{P}{2\varepsilon_0}\{[R^2 + (x - t/2)^2]^{\frac{1}{2}} \pm (x - t/2)\}$$

where the plus or minus sign results from taking the square root. We have to choose the proper sign to give the correct result at known positions.

As $x \to \infty$, $\phi_+ \to 0$; choosing the - sign does this for all positive values of x but does not work for negative values of x, since at $x = -\infty$, $\phi \neq 0$. So if we are to the right of the sheet,

$$\phi_+ = \frac{P}{2\varepsilon_0}\{[R^2 + (x - t/2)^2]^{\frac{1}{2}} - (x - t/2)\}$$

and if we are to the left of the sheet,

$$\phi_+ = \frac{P}{2\varepsilon_0}\{[R^2 + (x - t/2)^2]^{\frac{1}{2}} + (x - t/2)\}$$

Similarly for the points to the right of the negative sheet,

$$\phi_- = -\frac{P}{2\varepsilon_0}\{[R^2 + (x + t/2)^2]^{\frac{1}{2}} - (x + t/2)\}$$

and for points to the left of the negative sheet,

$$\phi_- = -\frac{P}{2\varepsilon_0}\{[R^2 + (x + t/2)^2]^{\frac{1}{2}} + (x + t/2)\}$$

So for $x > t/2$, the total potential is

$$\phi = \frac{P}{2\varepsilon_0}\{[R^2 + (x - t/2)^2]^{\frac{1}{2}} - (x - t/2) - [R^2 + (x + t/2)^2]^{\frac{1}{2}} + (x + t/2)\}$$

$$\phi = \frac{P}{2\varepsilon_0}\{[R^2 + (x - t/2)^2]^{\frac{1}{2}} - [R^2 + (x + t/2)^2]^{\frac{1}{2}} + t\}$$

and for $-t/2 < x < t/2$,

$$\phi = \frac{P}{2\varepsilon_0}\{[R^2 + (x - t/2)^2]^{\frac{1}{2}} + (x - t/2) - [R^2 + (x + t/2)^2]^{\frac{1}{2}} + (x + t/2)\}$$

$$\phi = \frac{P}{2\varepsilon_0}\{[R^2 + (x - t/2)^2]^{\frac{1}{2}} - [R^2 + (x + t/2)^2]^{\frac{1}{2}} + 2x\}$$

and for $x < -t/2$,

$$\phi = \frac{P}{2\varepsilon_0}\{[R^2 + (x - t/2)^2]^{\frac{1}{2}} + (x - t/2) - [R^2 + (x + t/2)^2]^{\frac{1}{2}} - (x + t/2)\}$$

$$\phi = \frac{P}{2\varepsilon_0}\{[R^2 + (x - t/2)^2]^{\frac{1}{2}} - [R^2 + (x + t/2)^2]^{\frac{1}{2}} - t\}$$

For $x >> t$,

$$(x - t/2)^2 \simeq x^2 - xt, \quad (x + t/2)^2 \simeq x^2 + xt$$

and if $R >> t$, also,

$$(R^2 + x^2 - xt)^{\frac{1}{2}} = (R^2 + x^2)^{\frac{1}{2}}\left(1 - \frac{xt}{(R^2 + x^2)}\right)^{\frac{1}{2}} \simeq (R^2 + x^2)^{\frac{1}{2}} - \frac{xt}{2(R^2 + x^2)^{\frac{1}{2}}}$$

and

$$(R^2 + x^2 + xt)^{\frac{1}{2}} = (R^2 + x^2)^{\frac{1}{2}} + \frac{xt}{2(R^2 + x^2)^{\frac{1}{2}}}$$

So with this for $x > t/2$,

$$\phi = \frac{P}{2\varepsilon_0}\left((R^2 + x^2)^{\frac{1}{2}} - \frac{xt}{2(R^2 + x^2)^{\frac{1}{2}}} - (R^2 + x^2)^{\frac{1}{2}} - \frac{xt}{2(R^2 + x^2)^{\frac{1}{2}}} + t\right)$$

$$\phi = \frac{Pt}{2\varepsilon_0}\left(1 - \frac{x}{(R^2 + x^2)^{\frac{1}{2}}}\right)$$

PROBLEM 2.60 An object having a mass of 10 g and a charge $+8 \times 10^{-5}$ C is placed in an electric field given in N/C by $E = \hat{\imath}3 \times 10^3 - \hat{\jmath}6 \times 10^2$.

(a) What are the magnitude and direction of the force on the object?
(b) If the object starts from rest at the origin, what will be its coordinates after 3 s?

Solution:

(a)

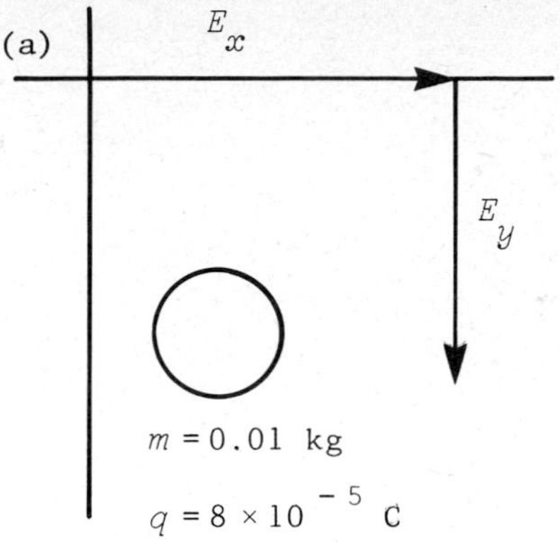

$$F = qE$$

$$F = (8 \times 10^{-5}\ \text{C})(\hat{\mathbf{i}}3 \times 10^3 - \hat{\mathbf{j}}6 \times 10^2)$$

$$F = (8 \times 10^{-5}\ \text{C})\hat{\mathbf{i}}(3 \times 10^3) - (8 \times 10^{-5}\ \text{C})\hat{\mathbf{j}}(6 \times 10^2)$$

$$F = 2.4 \times 10^{-1}\hat{\mathbf{i}} - 4.8 \times 10^{-2}\hat{\mathbf{j}}$$

$$F_{mag} = \sqrt{(0.24)^2 + (0.048)^2} = 0.245\ \text{N}$$

$$\text{Direction} \Rightarrow \frac{-0.48}{0.24} = \tan\theta$$

at $-11.3°$

(b)

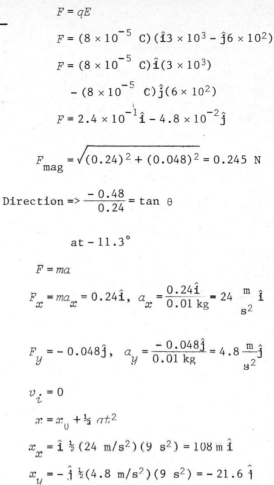

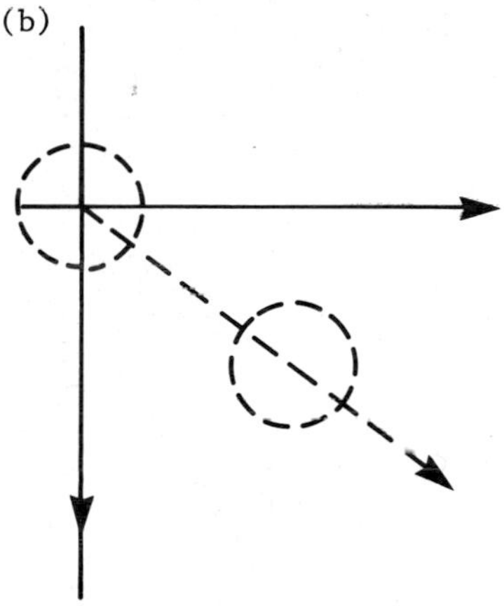

$$F = ma$$

$$F_x = ma_x = 0.24\hat{\mathbf{i}},\quad a_x = \frac{0.24\hat{\mathbf{i}}}{0.01\ \text{kg}} = 24\ \frac{\text{m}}{\text{s}^2}\hat{\mathbf{i}}$$

$$F_y = -0.048\hat{\mathbf{j}},\quad a_y = \frac{-0.048\hat{\mathbf{j}}}{0.01\ \text{kg}} = 4.8\frac{\text{m}}{\text{s}^2}\hat{\mathbf{j}}$$

$$v_i = 0$$

$$x = x_0 + \tfrac{1}{2}at^2$$

$$x_x = \hat{\mathbf{i}}\ \tfrac{1}{2}(24\ \text{m/s}^2)(9\ \text{s}^2) = 108\ \text{m}\ \hat{\mathbf{i}}$$

$$x_y = -\hat{\mathbf{j}}\ \tfrac{1}{2}(4.8\ \text{m/s}^2)(9\ \text{s}^2) = -21.6\ \hat{\mathbf{j}}$$

PROBLEM 2.61

(a) What is the acceleration of an electron in a uniform electric field of 1.0×10^6 N/C?

(b) How long would it take for the electron starting from rest to attain one-tenth the speed of light?

Solution:

(a) $F = qE = ma$

$$a = \frac{qE}{m} = \frac{(1.6 \times 10^{-19}\ \text{C})(1.0 \times 10^6\ \text{N/C})}{(9.1 \times 10^{-31}\ \text{kg})}$$

$$= 1.76 \times 10^{17}\ \text{N/kg}$$

$$a = 1.8 \times 10^{17}\ \text{m/s}^2$$

(b) $v = at$

$$t = \frac{v}{a} = \frac{0.1(3.0 \times 10^8 \text{ m/s})}{1.8 \times 10^{17} \text{ m/s}^2}$$

$$t = 1.67 \times 10^{-10} \text{ s}$$

PROBLEM 2.62 In an early run (1911), Millikan observed that the following measured charges, among others, appeared at different times on a single drop:

6.563×10^{-19} C	13.13×10^{-19} C	19.71×10^{-19} C
8.204×10^{-19} C	16.48×10^{-19} C	22.89×10^{-19} C
11.50×10^{-19} C	18.08×10^{-19} C	26.13×10^{-19} C

What value for the elementary charge e can be deduced from these data?

Solution: Find differences and see that all are multiples of smallest difference.

Difference	Charge
	6.563×10^{-19} C
1.641	
	8.204×10^{-19} C
3.300	
	11.50×10^{-19} C
1.630	
	13.13×10^{-19} C
3.350	
	16.48×10^{-19} C
1.600	
	18.08×10^{-19} C
1.630	
	19.71×10^{-19} C
3.180	
	22.89×10^{-19} C
3.240	
	26.13×10^{-19} C

Therefore, 1.625×10^{-19} C = average of all smallest numerical distances.

PROBLEM 2.63 Oil Drop Experiment. R. A. Millikan set up an apparatus (see figure for this problem) in which a tiny, charged oil drop, placed in an electric field E, could be "balanced" by adjusting E until the electric force on the drop was equal and opposite to its weight. If the radius of the drop is 1.64×10^{-4} cm and E at balance is 1.92×10^5 N/C, (a) what charge is on the drop? (b) Why did Millikan not try to balance electrons in his apparatus instead of oil drops? The density of the oil is 0.851 g/cm^3. (Millikan first measured the electronic charge in this way. He measured the drop radius by observing the limiting speed that the drops attained when they fell in air with the electric field turned off. He charged the oil drops by irradiating them with X-rays.)

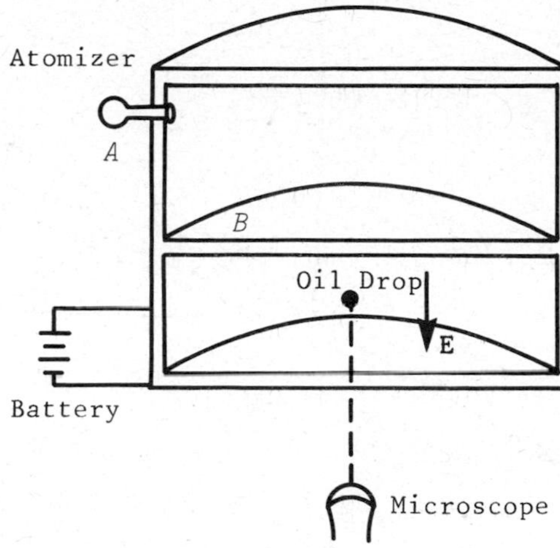

Solution:

(a) E at balance is 1.92×10^5 N/C.

Radius of drop = $(1.64 \times 10^{-4}\text{ cm})\left(\frac{1\text{ m}}{100\text{ cm}}\right) = 1.64 \times 10^{-6}$ m

Volume of drop = $\frac{4}{3}\pi(1.64 \times 10^{-6}\text{ m})^3 = 1.85 \times 10^{-17}\text{ m}^3$, (since so small, can consider it a sphere)

Density of oil = 0.851 g/cm^3

Mass of drop = $(1.85 \times 10^{-17}\text{ m}^3)(8.51 \times 10^2\text{ kg/m}^3)$

$$m = 1.57 \times 10^{-14}\text{ kg}$$

$$F = qE = ma \Rightarrow q = \frac{ma}{E} = \frac{mg}{E} = \frac{\frac{4}{3}\pi r^3 \rho g}{E}$$

$$q = \frac{(1.57 \times 10^{-14}\text{ kg})(9.8\text{ m/s}^2)}{1.92 \times 10^5\text{ N/C}}$$

$$q = 8.02 \times 10^{-19}\text{ C}$$

$$\frac{8.02 \times 10^{-19}\text{ C}}{1.6 \times 10^{-19}\text{ C}/e^-} = 5e^-$$

(b) One cannot see electrons; also E would be too small for practical considerations.

PROBLEM 2.64 Find the spherical components of the electric field intensity produced by an electric dipole located symmetrically about the origin and lying along the x-axis at points that are far compared to the separation d of the two charges.

Solution:

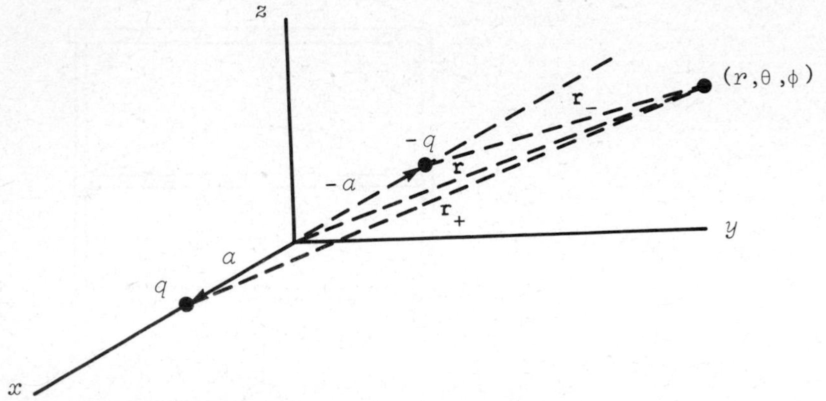

$$\phi = \frac{1}{4\pi\varepsilon_0}\frac{q}{r_+} - \frac{1}{4\pi\varepsilon_0}\frac{q}{r_-}$$

$$\phi = \frac{q}{4\pi\varepsilon_0}\left(\frac{1}{r_+} - \frac{1}{r_-}\right) = \frac{q}{4\pi\varepsilon_0}\left(\frac{r_- - r_+}{r_+ r_-}\right)$$

$$\mathbf{r} = \hat{\mathbf{i}}a + \mathbf{r}_+ = -\hat{\mathbf{i}}a + \mathbf{r}_-$$

$$\mathbf{r}_+ = \mathbf{r} - \hat{\mathbf{i}}a, \quad \mathbf{r}_- = \mathbf{r} + \hat{\mathbf{i}}a$$

$$r_+ = (r^2 + a^2 - 2a\hat{\mathbf{i}} \cdot \mathbf{r})^{\frac{1}{2}}$$

$$r_- = (r^2 + a^2 + 2a\hat{\mathbf{i}} \cdot \mathbf{r})^{\frac{1}{2}}$$

Using the polar expression for $\hat{\mathbf{r}}$,

$$r_+ = (r^2 + a^2 - 2ar \sin\theta \cos\phi)^{\frac{1}{2}}, \quad r_- = (r^2 + a^2 + 2ar \sin\theta \cos\phi)^{\frac{1}{2}}$$

$$\phi = \frac{q}{4\pi\varepsilon_0}\frac{r_- - r_+}{r_+ r_-}\frac{r_- + r_+}{r_- + r_+} = \frac{q}{4\pi\varepsilon_0}\frac{r_-^2 - r_+^2}{r_+ r_-(r_- + r_+)}$$

$$\phi \simeq \frac{q}{4\pi\varepsilon_0}\frac{4ar \sin\theta \cos\phi}{2r^3}$$

$$\mathbf{E} = -\nabla\phi = -\left(\hat{\mathbf{r}}\frac{\partial}{\partial r} + \hat{\theta}\frac{1}{r}\frac{\partial}{\partial\theta} + \hat{\varphi}\frac{1}{r\sin\theta}\frac{\partial}{\partial\phi}\right)\left(\frac{p}{4\pi\varepsilon_0}\frac{\sin\theta\cos\phi}{r^2}\right)$$

$$\mathbf{E} = -\hat{\mathbf{r}}\frac{p}{4\pi\varepsilon_0}(-2)\frac{\sin\theta\cos\phi}{r^3} - \hat{\theta}\frac{1}{r}\frac{p}{4\pi\varepsilon_0}\frac{\cos\theta\cos\phi}{r^2}$$

$$-\hat{\varphi}\frac{1}{r\sin\theta}\frac{p}{4\pi\varepsilon_0}\frac{\sin\theta\,(-\sin\phi)}{r^2}$$

$$\mathbf{E} = \hat{\mathbf{r}}\,\frac{p}{2\pi\varepsilon_0}\frac{\sin\theta\cos\phi}{r^3} - \hat{\theta}\,\frac{p\cos\theta\cos\phi}{4\pi\varepsilon_0 r^3} + \hat{\varphi}\,\frac{p\sin\phi}{4\pi\varepsilon_0 r^3}$$

PROBLEM 2.65 One model of hydrogen is of a point charge $+q$ surrounded by a spherically symmetric distribution of negative charge of density $\rho(r) = -be^{-2r/a_1}$, where a_1 is a constant to 5.28×10^{-11} m.

(a) Determine the value of b.
(b) Determinc the potential at points in space a distance r from $+q$ by direct integration over the charge distribution.

Solution:

(a) $Q = \int \rho\, dv = -e$

$$-e = \int_0^\infty -be^{-2r/a_1}\,4\pi r^2\,dr = -4\pi b\int_0^\infty r^2 e^{-2r/a_1}\,dr$$

$b = 3.46\times10^{11}\ \mathrm{C/m^3}$

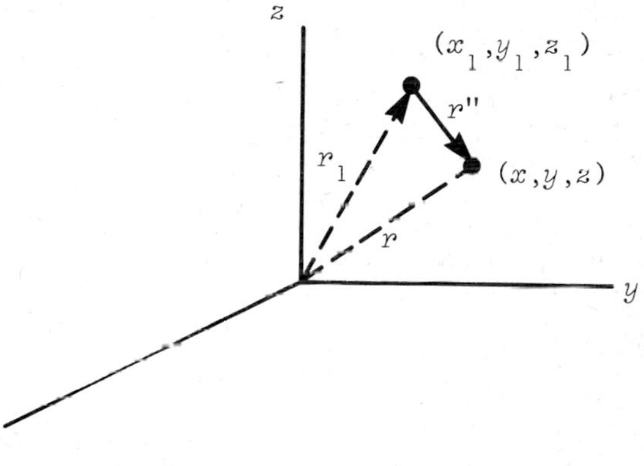

(b) $dV = \dfrac{1}{4\pi\varepsilon_0}\dfrac{dq}{r''}$

$$dq = \rho(r)r^2\sin\theta\,d\theta\,d\phi\,dr$$

$$r'' = |\mathbf{r}-\mathbf{r}_1| = (r^2 + r_1^2 - 2\mathbf{r}\cdot\mathbf{r}_1)^{\frac{1}{2}} = (r^2 + r_1^2 - 2xx' - 2yy' - 2zz')^{\frac{1}{2}}$$

But

$$x_1 = r_1\sin\theta\cos\phi,\quad y' = r_1\sin\theta\sin\phi,\quad z' = r_1\cos\theta$$

$$dV = \frac{(-1/4\pi\varepsilon_0)be^{-2r_1/a_1}\,4\pi r_1^2\,dr_1}{(r^2 + r_1^2 - 2xr_1\sin\theta\cos\phi - 2yr_1\sin\theta\sin\phi - 2zr_1\cos\theta)^{\frac{1}{2}}}$$

$$V = \frac{q}{4\pi\varepsilon_0 r} + V_-$$

PROBLEM 2.66 A long hollow conducting cylinder of radius a has a linear charge density $+Q$ C/m. A second cylinder of radius b $(b > a)$ is coaxial with the first and has a linear charge density $-2Q$ C/m.

(a) What is the electric field intensity at points $r > b$?
(b) What is the potential difference between the cylinder?

Solution:

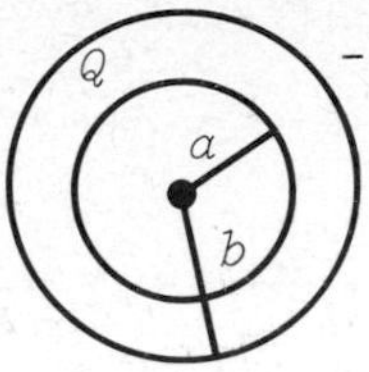

$$E2\pi rL = \frac{1}{\varepsilon_0}(-QL)$$

(a) $E = \dfrac{Q}{2\pi\varepsilon_0 r}$, directed in

(b) $a < r < b$,

$$\mathbf{E} = \frac{Q}{2\pi\varepsilon_0 r}\,\hat{\mathbf{r}}$$

$$V(b) - V(a) = -\int_a^b \mathbf{E}\cdot d\mathbf{r} = -\frac{Q}{2\pi\varepsilon_0}\ln(b/a)$$

PROBLEM 2.67 A dielectric sphere of radius R has a radial polarization that varies with distance from the center of the sphere according to $\mathbf{P} = a\mathbf{r}$ where a is constant. Find the electrostatic potential at the center of the sphere.

Solution:

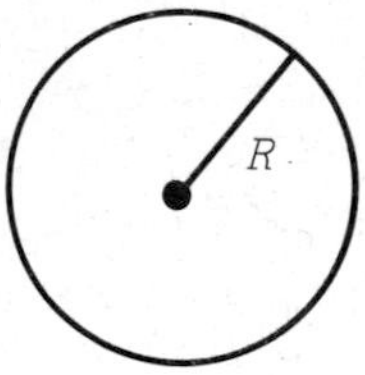

$$\mathbf{P} = ar\hat{\mathbf{r}}$$

Outside sphere

$E = 0$ outside because total bound charge of dielectric is zero (and here no free charge).

$$\mathbf{P} = ar\hat{\mathbf{r}}$$

$$\sigma_b = \mathbf{P}\cdot\hat{\mathbf{r}} = ar$$

$$\rho_b = -\nabla\cdot\mathbf{P} = \frac{-1}{r^2}\frac{\partial(r^2 ar)}{\partial r} = -\frac{1}{r^2}\frac{\partial ar^3}{\partial r} = -\frac{1}{r^2}3ar^2 = -3a$$

Inside sphere

$$\varepsilon E 4\pi r^2 = -3a((4/3)\pi r^3)$$

$$\mathbf{E} = \frac{-ar}{\varepsilon_0}\hat{\mathbf{r}}$$

$$\phi_{\text{center}} - \phi_\infty = -\int_\infty^0 \mathbf{E}\cdot d\boldsymbol{\ell} = +\int_\infty^R 0 + \int_R^0 \frac{+ar}{\varepsilon_0}\,dr = \frac{+a}{\varepsilon_0}\left(\frac{r^2}{2}\right)_R^0$$

$$\phi_{\text{center}} = \frac{-aR^2}{2\varepsilon_0}$$

PROBLEM 2.68 Notice that the vector potential **A** is related to the magnetic field **B** the way **B** is related to the current density **J**. That is, curl **A** = **B**, while curl **B** = $4\pi\mathbf{J}/c$. What statement about **A** corresponds to the statement that the line integral of **B** around any closed path equals $4\pi/c$ times the enclosed current? Consider the magnetic field of a long cylindrical rod carrying a current uniformly distributed over its cross section and parallel to its axis. Here $B = 2Ir/R^2c$ inside and $2I/rc$ outside. Now, using the analogy just mentioned, find a vector potential to go with the field of an infinitely long solenoid. Prove that your **A** leads to the correct **B** inside and outside the solenoid and that div **A** = 0.

Solution:

$$\int \text{curl}\,\mathbf{B}\cdot d\mathbf{a} = \frac{4\pi}{c}\int \mathbf{J}\cdot d\mathbf{a} = \oint \mathbf{B}\cdot d\boldsymbol{\ell}$$

therefore,

$$\int \text{curl}\,\mathbf{A}\cdot d\mathbf{a} = \int \mathbf{B}\cdot d\mathbf{a} = \oint \mathbf{A}\cdot d\boldsymbol{\ell}$$

By analogy $4\pi\mathbf{J}/c \to \mathbf{B}$; for the wire,

$$B(\text{inside}) = \frac{2Ir}{R^2c} = \frac{2\pi J}{c}r$$

$$B(\text{outside}) = \frac{2I}{rc} = \frac{2\pi J}{c}\frac{R^2}{r}$$

Hence for the solenoid,

$$A(\text{inside}) = \frac{B}{2}r$$

$$A(\text{outside}) = \frac{BR^2}{2}\frac{1}{r}$$

and A is in the form of concentric circles. For $\mathbf{B} = B\hat{\mathbf{z}}$,

$$\mathbf{A}(\text{inside}) = \frac{B}{2}\,(-y\hat{\mathbf{x}} + x\hat{\mathbf{y}})$$

$$\mathbf{A}(\text{outside}) = \frac{BR^2}{2}\left(\frac{-y\hat{\mathbf{x}} + x\hat{\mathbf{y}}}{x^2+y^2}\right)$$

$$\nabla \cdot \mathbf{A} = 0, \text{ inside}$$

$$\nabla \cdot \mathbf{A} = BR^2\left[\frac{xy}{(x^2+y^2)^2} + \frac{-xy}{(x^2+y^2)^2}\right] = 0, \text{ outside}$$

$$\nabla \times \mathbf{A} = B\left(-\frac{\partial(-y)}{\partial y} + \frac{\partial(x)}{\partial x}\right)\hat{\mathbf{z}} = B\hat{\mathbf{z}}, \text{ inside}$$

$$\nabla \times \mathbf{A} = \frac{BR^2}{2}\left[-\frac{\partial}{\partial y}\left(\frac{-y}{x^2+y^2}\right) + \frac{\partial}{\partial x}\left(\frac{x}{x^2+y^2}\right)\right]$$

$$= \frac{BR^2}{2}\left(\frac{1}{x^2+y^2} - \frac{2y^2}{(x^2+y^2)^2} + \frac{1}{x^2+y^2} - \frac{2x^2}{(x^2+y^2)^2}\right)$$

$$= BR^2\left(\frac{1}{x^2+y^2} - \frac{x^2+y^2}{(x^2+y^2)^2}\right) = 0, \text{ outside}$$

PROBLEM 2.69 Starting from the general expression for the electrostatic potential at some point r in space due to a surface charge distribution σ, develop the expression for the potential due to a flat disk of radius R having uniform charge density σ on its surface. Assume the disk is in the $z=0$ plane and that its center coincides with the origin. Leave your answer in the form of an integral in cylindrical coordinates but be sure to display all the variables and limits of integration explicitly. Remember that the potential at an arbitrary point is being asked for.

Solution:

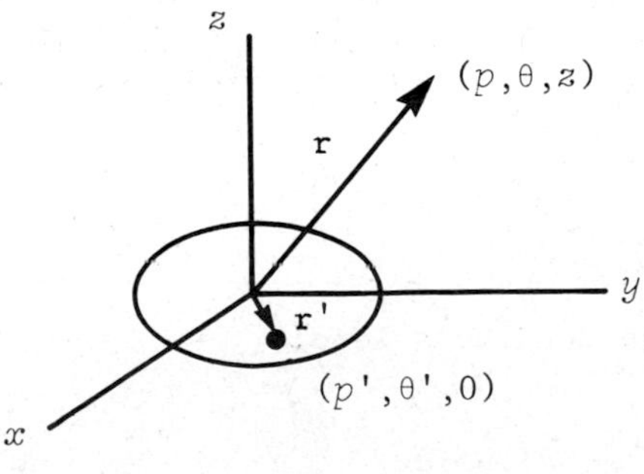

$$4\pi\varepsilon_0 U(\mathbf{r}) = \int_S \frac{\sigma(r')\,ds'}{|\mathbf{r}-\mathbf{r}'|}$$

$$= \int \frac{\sigma(\mathbf{r}')\,ds'}{\sqrt{r^2 - 2\mathbf{r}\cdot\mathbf{r}' + r'^2}}$$

$$x = \rho\cos\theta,\quad y = \rho\sin\theta,\quad z = z$$

therefore,

$$\mathbf{r}\cdot\mathbf{r}' = xx' + yy' + z\cdot 0$$

$$= \rho\rho' \cos\theta \cos\theta' + \rho\rho' \sin\theta \sin\theta'$$

$$= \rho\rho' \cos(\theta - \theta')$$

$$4\pi\varepsilon_0 U(\rho,\theta,z) = \sigma \int_0^R \int_0^{2\pi} \frac{\rho'\,d\rho'\,d\theta}{\sqrt{\rho^2 + z^2 - 2\rho\rho' \cos(\theta - \theta') + \rho'^2}}$$

PROBLEM 2.70 Consider two infinitely long concentric cylindrical surfaces of radii a and b respectively $(a < b)$, whose axes coincide with the z axis. The potential on the inner cylinder is given by $U(a, \theta) = V_a \sin\theta$ and on the outer cylinder $U(b, \theta) = V_b \cos 3\theta$, where θ is the polar angle (measured from x axis, say). Obtain an expression for $U(r, \theta)$ in the region between the two cylinders.

Solution:

The general solution of Laplace's equation in cylindrical coordinates is (for $(\partial U/\partial z) = 0$)

$$4\pi\varepsilon_0 U(r,\theta) = E \ln r + \sum_{n=0}^{\infty} (A_n r^n + B_n r^{-n}) \sin n\theta + \sum_{n=0}^{\infty} (C_n r^n + D_n r^{-n}) \cos n\theta$$

But clearly only the $\sin\theta$ and $\cos 3\theta$ terms are needed to satisfy the boundary conditions in this case. Try

$$4\pi\varepsilon_0 U(r,\theta) = (A_1 r + B_1/r)\sin\theta + (A_3 r^3 + B_1/r^3)\cos 3\theta$$

at $r = b$,

$$(A_1 b + B_1/b)\sin\theta + (A_3 b^3 + B/b^3)\cos 3\theta = V_b \cos 3\theta$$

at $r = a$,

$$(A_1 a + B_1/a)\sin\theta + (A_3 a^3 + B/a^3)\cos 3\theta = V_a \sin\theta$$

by orthogonality:

$$A_1 b + B_1/b = 0, \quad A_3 b^3 + B/b^3 = V_b$$

$$A_1 a + B_1/a = V_a, \quad A_3 a^3 + B/a^3 = 0$$

$$B_1 = -A_1 b^2, \quad A_1 = \frac{V_a}{a - b^2/a}$$

$$B_3 = -A_3 a^6, \quad A_3 = \frac{V_b}{b^3 - a^6/b^3}$$

$$4\pi\varepsilon_0 U(r,\theta) = \left(\frac{V_a ar}{a^2 - b^2} - \frac{V_a b^2}{a^2 - b^2}\frac{a}{r}\right)\sin\theta + \left(\frac{V_b b^3 r^3}{b^6 - a^6} - \frac{V_b b^3 a^6}{b^6 - a^6}\frac{1}{r^3}\right)\cos 3\theta$$

PROBLEM 2.71 A very thin rod of length L and uniform linear charge density λ is placed along the x axis with its center at the origin. Calculate the x component of the electric field at an arbitrary point (x,y,z) in space.

Solution:

$$\mathbf{E} = \frac{1}{4\pi\varepsilon_0}\int \frac{\rho(\mathbf{r}')(\mathbf{r}-\mathbf{r}')}{|\mathbf{r}-\mathbf{r}'|^3}\,dv' \quad \frac{1}{4\pi\varepsilon_0}\int \frac{\lambda(s)[\mathbf{r}-\mathbf{r}'(s)]}{|\mathbf{r}-\mathbf{r}'|^3}\,ds$$

$$= \frac{\lambda}{4\pi\varepsilon_0}\int_{-L/2}^{L/2} \frac{(\mathbf{r}-\mathbf{r}')}{|\mathbf{r}-\mathbf{r}'|}\,dx$$

$$E_x = \frac{\lambda}{4\pi\varepsilon_0}\int_{-L/2}^{L/2} \frac{(x-x')\,dx}{[(x-x')^2+(y-0)^2+(z-0)^2]^{3/2}}$$

$$= \frac{\lambda}{4\pi\varepsilon_0}\int_{-L/2}^{L/2} d\left(\frac{1}{[(x-x')^2+y^2+z^2]^{\frac{1}{2}}}\right)$$

$$= \frac{\lambda}{4\pi\varepsilon_0}\left(\frac{1}{\sqrt{[(x-L/2)^2+y^2+z^2]^{\frac{1}{2}}}} - \frac{1}{\sqrt{(x+L/2)^2+y^2+z^2}}\right)$$

PROBLEM 2.72 Consider two concentric spherical shells of radius r_a and r_c, respectively. The volume between the shells is filled with a dielectric of dielectric constant K_1 out to r_b and the remainder is filled with another dielectric of dielectric constant K_2. Take $r_a < r_b < r_c$. A potential difference of U_0 is imposed between the conductors.

(a) Calculate the electric field in the two regions between the conductors.
(b) Calculate all surface and volume polarization charge densities (σ_p and ρ_p).

All answers are to be expressed as functions of r_a, r_b, r_c, r (distance from the origin), U_0 , and the dielectric constants K_1 and K_2 or the corresponding permittivities.

Solution: Clearly the potentials in the two regions involve only the $1/r$ and constant terms since nothing depends on the angles. Thus, try

$$U = \frac{A_1}{r}, \quad r_a \le r \le r_b;$$

$$U = \frac{A_2}{r}, \quad r_b \le r \le r_c$$

(we don't need the constant). The transverse component of E is automatically continuous at $r = r_b$ because $E_t \equiv 0$ everywhere. We still need $D_{1n} = D_{2n}$ (since no free charge is present at the interface). Thus,

$$\varepsilon_1 \frac{A_1}{r_b^2} = \varepsilon_2 \frac{A_2}{r_b^2} \to A_2 = \frac{\varepsilon_1}{\varepsilon_2} A_1 = \frac{K_1}{K_2} A_1 .$$

We have one more boundary condition:

$$U(r_c) - U(r_a) = U_0, \quad A_1 \left(\frac{K_1}{K_2}\frac{1}{r_c} - \frac{1}{r_a}\right) = U_0, \quad A_1 \left(\frac{K_1 r_a - K_2 r_c}{K_2 r_c r_a}\right) = U_0$$

$$A_1 = \frac{K_2 U_0 r_a r_c}{K_1 r_a - K_2 r_c}, \quad A_2 = \frac{K_1 U_0 r_a r_c}{K_1 r_a - K_2 r_c}$$

therefore,

$$\mathbf{E} = \begin{cases} \dfrac{K_2 U_0 r_a r_c}{K_1 r_a - K_2 r_c} \dfrac{\mathbf{r}}{r^3}, & r_a \le r \le r_b \\ \\ \dfrac{K_1 U_0 r_a r_c}{K_1 r_a - K_2 r_c} \dfrac{\mathbf{r}}{r^3}, & r_b \le r \le r_c \end{cases}$$

(b) $\mathbf{P} = \chi \mathbf{E} = \varepsilon_0 (K - 1)\mathbf{E}$. There is no volume charge density here because

$$\nabla \cdot \mathbf{P} = \frac{1}{r^2} \frac{\partial}{\partial r} (r^2 P_r)$$

and $P_r \sim \text{const.}/r^2$. At $r = r_a$,

$$+\hat{\mathbf{n}} \cdot \mathbf{P} = \sigma_p = -P_{1r}$$

$$= \frac{\varepsilon_0 U_0 r_a r_c}{K_1 r_a - K_2 r_c} \frac{K_2 (K_1 - 1)}{r_a^2}$$

therefore, $\nabla \cdot \mathbf{P} \equiv 0$. At $r = r_b$,

$$\hat{\mathbf{n}} \cdot \mathbf{P} = \sigma_p = +P_{2r} = +\frac{\varepsilon_0 U_0 r_a r_c}{K_1 r_a - K_2 r_c} \frac{K_1 (K_2 - 1)}{r_c^2}$$

There is a σ_p at $r = r_b$,

$$\sigma_p = P_{1r} - P_{2r}$$

$$\sigma_p = \frac{\varepsilon_0 U_0 r_a r_c}{K_1 r_a - K_2 r_c} \frac{1}{r_b^2} [(K_1 - 1)K_2 - (K_2 - 1)K_1]$$

$$\sigma_p = \frac{\varepsilon_0 U_0 r_a r_c}{K_1 r_a - K_2 r_c} \frac{1}{r_b^2} (K_1 - K_2)$$

PROBLEM 2.73 Show that if a charge is placed within a spherical cavity made in a conducting material, the charge will be attracted to the inner surface of the cavity with a force

$$F = \frac{q^2 ar}{4\pi\varepsilon_0 (a^2 - r^2)^2}$$

where q is the charge, a is the radius of the cavity, and r is the distance to the charge from the center of the cavity.

Solution:

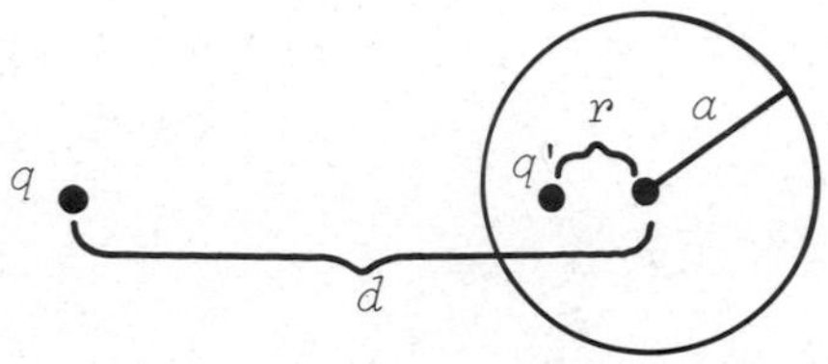

To create the equipotential surface of radius a,

$$q_1 = -\frac{a}{d} q, \quad r = \frac{a^2}{d}$$

Then

$$F = \frac{1}{4\pi\varepsilon} \frac{qq_1}{(d-r)^2} = \frac{1}{4\pi\varepsilon} \frac{(d/a)q_1^2}{(d-r)^2}$$

But $d = a^2/r$, so

$$F = \frac{1}{4\pi\varepsilon} \frac{(1/a)(a^2/r)q_1^2}{(a^2/r - r)^2} = \frac{1}{4\pi\varepsilon} \frac{q_1^2 ar}{(a^2 - r^2)^2}$$

PROBLEM 2.74 Show that the force experienced by a charge q placed at a distance r from the center of an uncharged, insulated, conducting sphere of radius a is

$$F = -\frac{q^2 a^3 (2r^2 - a^2)}{4\pi\varepsilon_0 r^3 (r^2 - a^2)^2}$$

Solution:

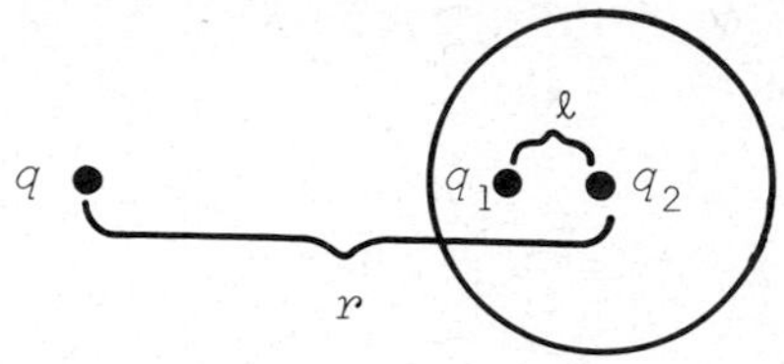

$$q_1 = -\frac{a}{r}q, \quad q_2 = -q_1 = \frac{a}{r}q$$

$$\ell = \frac{a^2}{r}$$

$$F = \frac{1}{4\pi\varepsilon}\;\frac{qq_1}{(r-\ell)^2} + \frac{qq_2}{r^2}$$

$$F = \frac{1}{4\pi\varepsilon}\;\frac{-q^2(a/r)}{(r-a^2/r)^2} + \frac{q(a/r)q}{r^2}$$

$$F = \frac{q^2(a/r)}{4\pi\varepsilon}\left(-\frac{r^2}{(r^2-a^2)^2} + \frac{1}{r^2}\right)$$

$$F = \frac{q^2}{4\pi\varepsilon}\;\frac{a}{r}\left(\frac{-r^4+r^4-2r^2a^2+a^4}{r^2(r^2-a^2)^2}\right)$$

$$F = -\frac{q^2a^3}{4\pi\varepsilon r^3}\left(\frac{2r^2-a^2}{(r^2-a^2)^2}\right)$$

PROBLEM 2.75 A particle with electric charge q is released (from rest) at the distance x_0 from the surface of a large, grounded, conducting plate. The particle is attracted by the plate, and moves toward it. What is the kinetic energy of the particle as a function of its distance x from the plate? (Neglect any energy loss by radiation.) Do you see anything unphysical about your answer?

Solution:

In order to create an equipotential plane we can replace the plane by a point charge $-q$ located the same distance behind the plane as the charge $+q$ is in front of the plane. As a function of position then, the force on the q is

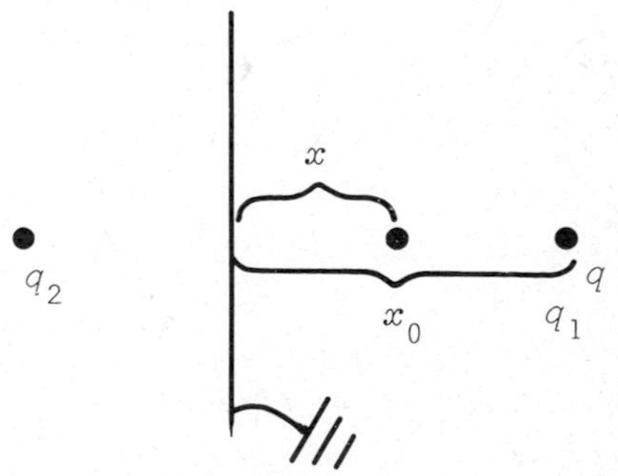

$$\mathbf{F} = -\hat{\mathbf{i}}\,\frac{q_1q_2}{4\pi\varepsilon_0(2x)^2}$$

As it moves from x_0 to x, the work done is

$$W = \int_{x_0}^{x} \mathbf{F} \cdot d\mathbf{r} = K.E._f - K.E.$$

$$K.E. = \int_{x_0}^{x} \frac{q^2}{4\pi\varepsilon_0 (2x)^2} (-dx) = -\frac{q^2}{4\pi\varepsilon_0} \int_{x_0}^{x} \frac{dx}{x^2}$$

$$K.E. = +\frac{q^2}{16\pi\varepsilon_0} \left.\frac{1}{x}\right|_{x_0}^{x} = \frac{q^2}{16\pi\varepsilon_0} \left(\frac{1}{x} - \frac{1}{x_0}\right)$$

At $x = 0$, the kinetic energy would be infinite. Here at $x = 0$, $v = c$, which is impossible.

PROBLEM 2.76 A point charge q is placed at a distance d from an infinite conducting plane having a hemispherical bump of radius a directly in front of q. Show that the point charge q is attracted toward the plane with the force

$$F = \frac{q^2}{16\pi\varepsilon_0 d^2} \left(1 + \frac{16a^3d^5}{(d^4 - a^4)^2}\right)$$

Solution:

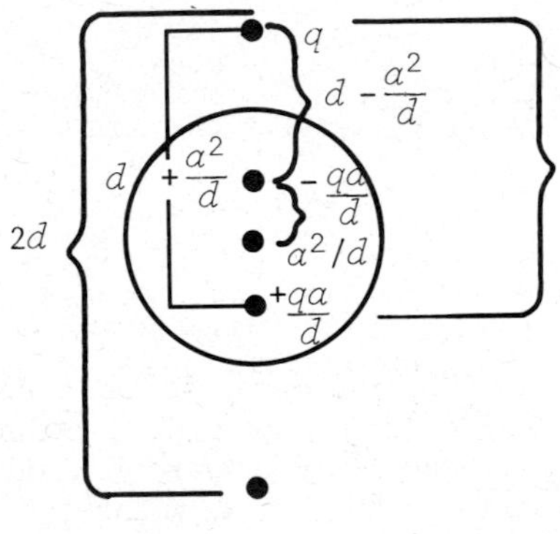

$$F = \frac{1}{4\pi\varepsilon} \left(\frac{q(-qa/d)}{(d - a^2/d)^2} + \frac{q(qa/d)}{(d + a^2/d)^2} + \frac{q(-q)}{(2d)^2}\right)$$

$$= \frac{1}{4\pi\varepsilon} \left(\frac{-q^2a}{d(d - a^2/d)^2} + \frac{q^2a}{d(d + a^2/d)^2} - \frac{q^2}{4d^2}\right)$$

$$F = \frac{q^2}{4\pi\varepsilon} \left(\frac{-a(4d^2)(d + a^2/d)^2}{(4d^2)d(d - a^2/d)^2(d + a^2/d)^2} + \frac{a(4d^2)(d - a^2/d)^2}{d(d + a^2/d)^2(4d^2)(d - a^2/d)^2}\right.$$

$$\left. - \frac{d(d + a^2/d)^2(d - a^2/d)}{d(4d^2)(d + a^2/d)^2(d - a^2/d)^2}\right)$$

$$= \frac{q^2}{4\pi\varepsilon}\left(\frac{-4ad^2(d+a^2/d)^2 + 4ad^2(d-a^2/d)^2 - d(d+a^2/d)^2(d-a^2/d)^2}{4d^3(d-a^2/d)^2(d+a^2/d)^2}\right)$$

$$= \frac{q^2}{16\pi\varepsilon d^2}\left(\frac{-4ad^2(d^2+2a^2+a^4/d^2) + 4ad^2(d^2-2a^2+a^4/d^2) - d(d^4-2a^4+a^8/d^4)}{d(d-a^2/d)^2(d+a^2/d)^2}\right)$$

$$= \frac{q^2}{16\pi\varepsilon d^2}\left(\frac{-4ad^4 - 8a^3d^2 - 4a^5 + 4ad^4 - 8a^3d^2 + 4a^5 - d^5 + 2a^4d - a^8/d^3}{d^5 - 2a^4d + a^8/d^3}\right)$$

$$= \frac{q^2}{16\pi\varepsilon d^2}\left(\frac{-16a^3d^2 - d^5 + 2a^4d - a^8/d^3}{d^5 - 2a^4d + a^8/d^3}\right)$$

$$= \frac{q^2}{16\pi\varepsilon d^2}\left(\frac{-16a^3d^5 - d^8 + 2a^4d^4 - a^8}{d^8 - 2a^4d^4 + a^8}\right)$$

$$= \frac{q^2}{16\pi\varepsilon d^2}\left(\frac{-16a^3d^5}{(d^4-a^4)^2} - \frac{d^8 - 2a^4d^4 + a^8}{d^8 - 2a^4d^4 + a^8}\right)$$

$$F = \frac{-q^2}{16\pi\varepsilon d^2}\left(1 + \frac{16a^3d^5}{(d^4-a^4)^2}\right)$$

PROBLEM 2.77 According to Thomson's model, a hydrogen atom may be imagined as a sphere made of uniformly distributed positive charge q at the center of which a negative point charge $-q$ (the electron) is embedded. Show that if the electron is displaced from its equilibrium position, it will execute simple harmonic vibrations through the center of the "atom" with the frequency f given by

$$f^2 = \frac{q^2}{16\pi^3\varepsilon_0 a^3 m}$$

where a is the radius of the atom (positive sphere) and m is the mass of the electron.

Solution:

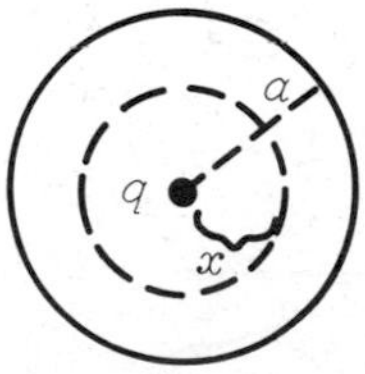

At the position of the electron we need E in order to get the force:

$$\oint \mathbf{E}\cdot d\mathbf{s} = \frac{1}{\varepsilon} Q$$

$$E 4\pi x^2 = \frac{1}{\varepsilon}\,\rho\,{}^4/_3\,\pi x^3$$

$$E = \frac{\rho x}{3\varepsilon}$$

So $F_R = -(\rho x/3\varepsilon)\,q$ is the restoring force on the proton. Since $F_R = m\ddot{x}$, the equation of the electron is

$$m\ddot{x} + \frac{\rho q}{3\varepsilon}\,x = 0$$

From the form of this equation $\omega^2 = \rho q/3\varepsilon/m$

$$f^2 = \frac{1}{4\pi^2}\,\frac{\rho q}{3\varepsilon m}$$

But $\rho = 3q/4\pi a^3$, therefore,

$$f^2 = \frac{1}{4\pi^2}\,\frac{q}{3\varepsilon m}\,\frac{3q}{4\pi a^3}$$

$$f^2 = \frac{q^2}{16\pi^3 \varepsilon a^3 m}$$

$$f \sim 5 \times 10^{-16}\ \text{Hz}$$

PROBLEM 2.78 In figure (a) below, assume that both charges are positive.

(a) Show that E at point P in figure (a), assuming $r \gg a$, is given by

$$E = \frac{1}{4\pi\varepsilon_0}\,\frac{2q}{r^2}$$

(b) What is the direction of **E**?

(c) Is it reasonable that E should vary as r^{-2} here and as r^{-3} for the dipole of figure (a)?

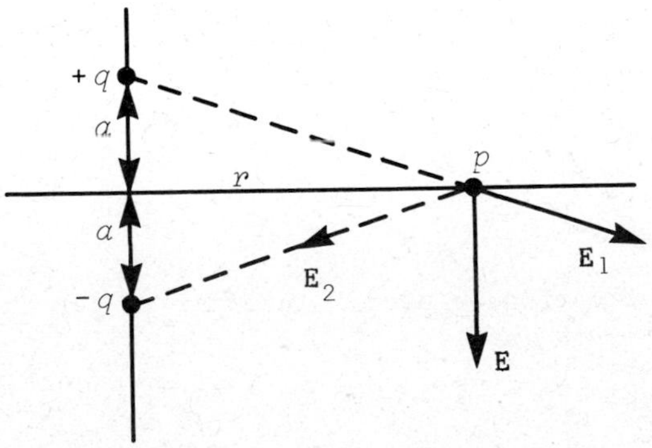

Figure (a)

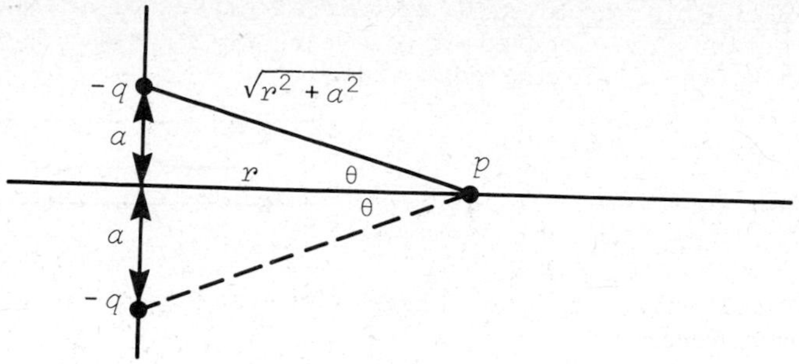

Figure (b)

Solution:

(a)
$$\mathbf{E}_p = \mathbf{E}_{1 \to p} + \mathbf{E}_{2 \to p}$$

$$= \frac{\hat{\mathbf{i}}}{4\pi\varepsilon} \frac{q}{r^2 + a^2} \cos\theta + \frac{\hat{\mathbf{i}}}{4\pi\varepsilon} \frac{q}{r^2 + a^2} \cos\theta$$

therefore

$$E_p = \frac{1}{4\pi\varepsilon} \frac{2q}{r^2}$$

or

$$E_p = \frac{1}{4\pi\varepsilon} \frac{q}{r^2 + a^2} \frac{r}{(r^2 + a^2)^{\frac{1}{2}}} + \frac{1}{4\pi\varepsilon} \frac{q}{r^2 + a^2} \frac{r}{(r^2 + a^2)^{\frac{1}{2}}}$$

$$E_p = \frac{1}{4\pi\varepsilon} \frac{2qr}{(r^2 + a^2)^{3/2}} \qquad \text{if } r >> a$$

$$E_p = \frac{1}{4\pi\varepsilon} \frac{2qr}{r^3} = \frac{1}{4\pi\varepsilon} \frac{2q}{r^2}$$

(b) Direction of $\mathbf{E}$ in $+\hat{\mathbf{i}}$ direction.

(c) The dipole in figure (c) is two equal and opposite charges placed close to each other so that their separate fields at distant points almost, but not quite, cancel; therefore, we can see why $E(r)$ for a dipole varies as $1/r^3$. For a point charge E drops off more slowly, proportional to $1/r^2$.

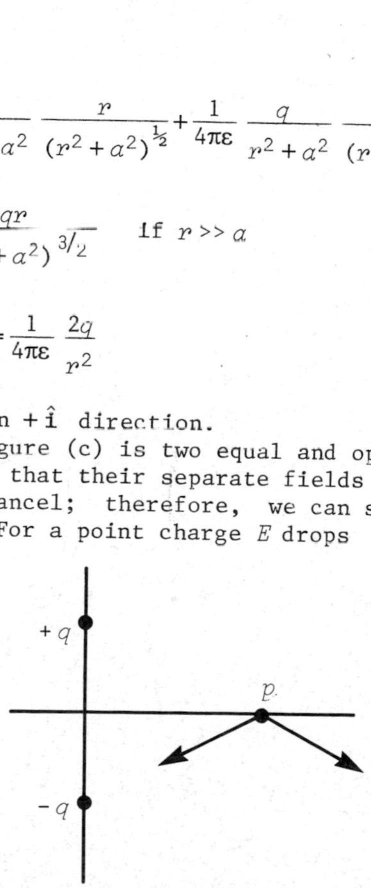

Figure (c)

PROBLEM 2.79 Parallel Plate Problem. Two parallel plates of infinite extent in the x and z directions are located at $y = 0$ and $y = a$, as in the diagram. The plates are kept at the following potentials:

$$\phi(x, y = a) = \phi(|x| > d, y = 0) = 0$$

$$\phi(|x| \leq d, y = 0) = V$$

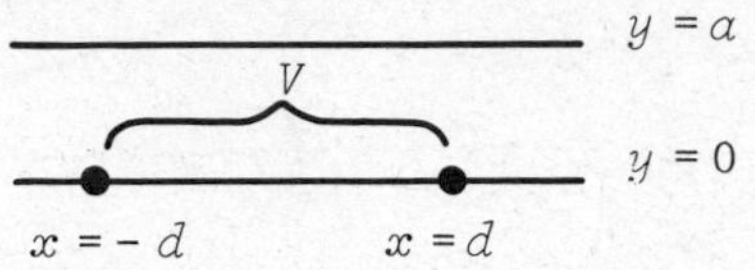

Determine $\phi(xy)$ at all points in the charge-free region between the plates by the following procedure:

(a) Choose the separation constant q^2 for Laplace's equation so that the x dependence is in terms of $\sin qx$ and $\cos qx$. Show that $\phi(x,y)$ is then given by a double integral over q and x' of known functions of x and y.

(b) Assuming that $\phi(x,y)$ is independent of the order of the preceding q and x' integrations, evaluate the q integral via contour integration. Be careful in determining the poles.

(c) Finally, evaluate the x' integral to give expressions for $\phi(x,y)$ for $|x| > d$ and for $|x| \leq d$. Use the expansion of $1 - y/a$ in a Fourier sine series to simplify the expression for $\phi(|x| \leq d, y)$.

Note: This problem should not be solved by choosing q^2 so that the y dependence is in terms of $\sin qy$ and $\cos qy$: the contour integral of part (b) is a key ingredient of this problem.

Solution:

(a)

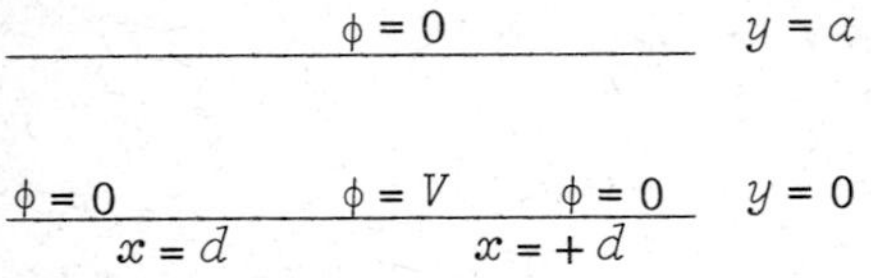

$$\phi(x, y = a) = \phi(|x| > d, y = 0) = 0$$

$$\phi(|x| \leq d, y = 0) = V$$

Solve Laplace's equation

$$\frac{\partial^2 \phi}{\partial x^2} + \frac{\partial^2 \phi}{\partial y^2} = 0$$

for all points in the charge-free region between the plates, except for $-d \leq x \leq d$, $y = 0$; let

$$\phi(x, y) = X(x)Y(y)$$

$$Y \frac{\partial^2 X}{\partial x^2} + X \frac{\partial^2 Y}{\partial y^2} = 0$$

$$\frac{1}{X} \frac{\partial^2 X}{\partial x^2} + \frac{1}{Y} \frac{\partial^2 Y}{\partial y^2} = 0$$

where the first term on the left equals $-q^2$ and the second term equals q^2.

$$\frac{1}{X}\frac{\partial^2 X}{\partial x^2} = -q^2, \quad \frac{\partial^2 X}{\partial x^2} + q^2 X = 0$$

$$\frac{1}{Y}\frac{\partial^2 Y}{\partial y^2} = q^2, \quad \frac{\partial^2 Y}{\partial y^2} - q^2 Y = 0$$

Therefore,

$$X(x) = \sin qx, \ \cos qx \tag{1}$$

$$Y(y) = e^{\pm qy}, \ \sinh qy, \ \cosh qy$$

So $\phi(x, y) = X(x)Y(y)$ with $X(x)$ and $Y(y)$ given in Eq. (1). Now apply the boundary conditions, when $|x| \le d$, $Y(0) = 1$, and $Y(a) = 0$:

$$Y(y) = Ae^{+qy} + Be^{-qy}$$

$$Y(0) = A + B = 1$$

$$A + B = 1, \quad B = 1 - A$$

$$Y(a) = Ae^{qa} + Be^{-qa} = Ae^{qa} + e^{-qa} - Ae^{-qa} = 0$$

Therefore,

$$A(e^{qa} - e^{-qa}) = -e^{-qa}$$

$$A = \frac{-e^{-qa}}{e^{qa} - e^{-qa}}$$

$$B = 1 - A = 1 + \frac{e^{-qa}}{e^{qa} - e^{-qa}} = \frac{e^{qa}}{e^{qa} - e^{-qa}}$$

$$Y(y) = \frac{-e^{-qa}e^{+qy}}{e^{qa} - e^{-qa}} + \frac{e^{qa}e^{-qy}}{e^{qa} - e^{-qa}} = \frac{e^{q(a-y)} - e^{-q(a-y)}}{e^{qa} - e^{-qa}}$$

$$Y(y) = \frac{\sinh q(a-y)}{\sinh qa}$$

$$\phi(x, y) = \frac{1}{\sqrt{2\pi}}\int_{-\infty}^{\infty} A_q Y_q(y) \cos qx \, dq + \frac{1}{\sqrt{2\pi}}\int_{-\infty}^{\infty} B_q Y_q(y) \sin q \, dq$$

$$\phi(x, y) = \frac{1}{\sqrt{2\pi}}\int_{-\infty}^{\infty} Y_q(y)(A_q \cos qx + B_q \sin qx)dq$$

$$A_q = \frac{1}{\sqrt{2\pi}} \int_{-d}^{d} dx' \phi(x', 0) \cos qx'$$

$$B_q = \frac{1}{\sqrt{2\pi}} \int_{-d}^{d} dx' \phi(x', 0) \sin qx'$$

and

$$\phi(x', 0) = V$$

So

$$A_q = \frac{1}{\sqrt{2\pi}} \int_{-d}^{d} dx' V \cos qx'$$

$$B_q = \frac{1}{\sqrt{2\pi}} \int_{-d}^{d} dx' V \sin qx'$$

Therefore,

$$\phi(x, y) = \frac{1}{\sqrt{2\pi}} \int_{-\infty}^{\infty} dq \, \frac{\sinh q(a - y)}{\sinh qa} \left(\frac{1}{\sqrt{2\pi}} \int_{-d}^{d} dx' V \cos qx' \cos qx \right.$$

$$\left. + \frac{1}{\sqrt{2\pi}} \int_{-d}^{d} dx' V \sin qx' \sin qx \right)$$

$$\phi(x, y) = \frac{V}{2\pi} \int_{-\infty}^{\infty} dq \, \frac{\sinh q(a - y)}{\sinh qa} \int_{-d}^{d} dx' (\cos qx' \cos qx + \sin qx' \sin qx)$$

(b)

$$\phi(x, y) = \int_{-\infty}^{\infty} A'_q \, dq \, e^{iqx} e^{qy} + B'_q \, e^{iqx - qy}$$

$$\phi(x, a) = \int_{-\infty}^{\infty} dq \, e^{iqx} \, (A'_q \, e^{qa} + B'_q \, e^{-qa}) = 0$$

$$A'_q = -e^{-2qa} \, B'_q$$

$$B'_q = -e^{2qa} \, A'_q$$

$$\phi(x, 0) = \int_{-\infty}^{\infty} dq \, e^{iqx} \, (A'_q + B'_q)$$

$$A'_q + B'_q = \frac{1}{2\pi}\int_{-\infty}^{\infty} dx'\, e^{-iqx'} V$$

$$A'_q = \frac{V}{2\pi(1 - e^{2qa})}\int_{-\infty}^{\infty} dx'\, e^{-iqx'}$$

$$B'_q = \frac{V}{2\pi(1 - e^{-2qa})}\int_{-\infty}^{\infty} dx'\, e^{-iqx'}$$

$$\phi(x, y) = \frac{-V}{2\pi}\int_{-\infty}^{\infty} dq\, e^{iqx}\int_{-d}^{d} dx'\, e^{-iqx'}\, \frac{\sinh q(y - a)}{\sinh qa}$$

$$\phi(x, y) = \int_{-d}^{d} dx'\, \frac{V}{2\pi}\int_{-\infty}^{\infty} dq\, e^{iq(x - x')}\, \frac{\sinh q(a - y)}{\sinh qa}$$

$\sinh qa = 0$ when $e^{qa} = e^{-qa}$, so $e^{2qa} = 1$. So $2qa = i2\pi n$ or $q = n\pi i/a \rightarrow$ poles (n an integer).

For q integral

$$\text{Residue at } q = \frac{in\pi}{a} = \lim_{q \rightarrow n\pi i/a} e^{iq(x - x')}\, \frac{(q - n\pi i/a)\sinh q\,(a - y)}{\sinh qa}$$

$$= e^{-n\pi(x - x')/a}\, \sinh\left(\frac{\pi i n}{a}(a - y)\right) \lim_{q \rightarrow n\pi i/a} \left(\frac{q - in\pi/a}{\sinh qa}\right)$$

$$= e^{-n\pi(x - x')/a}\, i\,\sin\left(\frac{n\pi}{a}(a - y)\right) \lim_{q - n\pi i/a \rightarrow 0} \frac{q - in\pi/a}{\sinh qa}$$

$$= e^{-n\pi(x - x')/a}\, i\,\sin\left(\frac{n\pi}{a}(a - y)\right) \lim_{q - n\pi i/a \rightarrow 0} \frac{q - n\pi i/a}{(-1)^n \sinh(q - n\pi i/a)a}$$

$$= e^{-n\pi(x - x')/a}\, i\,\sin\left(\frac{n\pi}{a}(a - y)\right) \lim_{q - n\pi i/a \rightarrow 0} \frac{(-1)^n (q - n\pi i/a)}{(q - n\pi i/a)a + [(q - n\pi i/a)a]^3/3! + \cdots}$$

$$\text{Residue at } q = in\pi/a = \frac{i(-1)^n}{a}\, e^{-n\pi(x - x')/a}\, \sin\left(\frac{n\pi}{a}(a - y)\right)$$

For C_1

$$\oint = +2\pi i \sum_{n=1}^{\infty} \text{Res}_{q = +n\pi i/a}$$

$$\int_{C_1} dq\, e^{iq(x-x')} \frac{\sinh q(a-y)}{\sinh qa}, \quad 0 < \theta < \pi, \quad q = q_0 e^{i\theta}$$

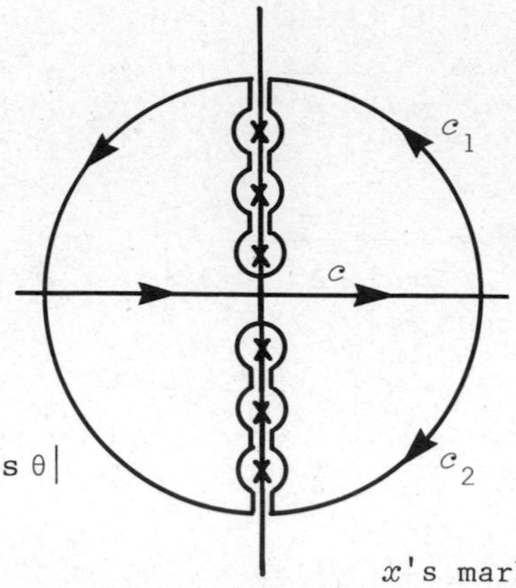

x's mark poles

$$\left|\int_{C_1}\right| \le \int_{\theta > 0} |dq|\,|e^{iq(x-x')}|\,\left|\frac{\sinh q(a-y}{\sinh qa}\right|$$

$$|dq| = q_0 d\theta$$

$$|e^{iq(x-x')}| = e^{-q_0 \sin\theta (x-x')}$$

$$\left|\frac{\sinh q(a-y)}{\sinh qa}\right| \sim \frac{e^{|q(a-y)\cos\theta|}}{e^{|q_0 a \cos\theta|}} = e^{-q_0 y|\cos\theta|}$$

Therefore,

$$\int_{C_1} = 0 \quad \text{for } x - x' > 0$$

For C_2

$$\oint = -2\pi i \sum_{n=1}^{\infty} \text{Res}_{q=-n\pi i/a}$$

$$\int_{C_2} dq\, e^{iq(x-x')} \frac{\sinh q(a-y)}{\sinh qa}, \quad 0 > \theta > \pi, \quad \text{real } q_0 \to \infty$$

$$\left|\int_{C_2}\right| \le \int_{\theta < 0} |dq|\,|e^{iq(x-x')}|\,\left|\frac{\sinh q(a-y)}{\sinh qa}\right|$$

Therefore,

$$\int_{C_2} = 0 \text{ for } x - x' < 0$$

$$I = \int_{-\infty}^{\infty} dq\, e^{iq(x-x')} \frac{\sinh q(a-y)}{\sinh qa}$$

So

$$I = \frac{2\pi}{a} \sum_{n=1}^{\infty} (-1)^{n+1} \sin\left[\frac{n\pi}{a}(a-y)\right] e^{-n\pi|x-x'|/a} = q \text{ integral}$$

Therefore,

$$\phi(x, y) = \int_{-d}^{d} \frac{V}{2\pi}\, dx' \int_{-\infty}^{\infty} dq\, e^{iq(x-x')} \frac{\sinh q(a-y)}{\sinh qa}$$

$$\phi(x, y) = \frac{V}{2\pi} \int_{-d}^{d} dx' \frac{2\pi}{a} \sum_{n=1}^{\infty} (-1)^{n+1} \sin\left[\frac{n\pi}{a}(a-y)\right] e^{-n\pi|x-x'|/a}$$

$$\phi(x, y) = \frac{V}{a} \int_{-d}^{d} dx' \sum_{n=1}^{\infty} (-1)^{n+1} \sin\left[\frac{n\pi}{a}(a-y)\right] e^{-n\pi|x-x'|/a}$$

(c) $X_n = \int_{-\infty}^{\infty} dx'\, Ve^{-n\pi|x-x'|/a}$ for $|x'| \leq d$

$X_n = 0$ for $|x'| > d$

For $x < -d$

$$X_n = V \int_{-d}^{d} dx'\, e^{n\pi(x-x')/a} = \frac{2Va}{n\pi} e^{n\pi x/a} \sinh n\pi d/a$$

For $x > d$

$$X_n = V \int_{-d}^{d} dx' e^{-n\pi(x-x')/a} = \frac{2Va}{n\pi} e^{-n\pi x/a} \sinh n\pi d/a$$

For $-d < x < d$

$$X_n = V \int_{-d}^{x} dx' e^{-n\pi(x-x')/a} + V \int_{x}^{d} e^{n\pi(x-x')/a}\, dx'$$

$$= \frac{Va}{n\pi} [(1 - e^{-n\pi(x+d)/a}) - (e^{n\pi(x-d)/a} - 1)]$$

$$X_n = \frac{2Va}{n\pi} (1 - e^{-n\pi d/a} \cosh n\pi x/a)$$

so for $|x| > d$,

$$\phi(x, y) = \sum_{n=1}^{\infty} (-1)^n \sin\left[\frac{n\pi}{a}(y-a)\right] \frac{2V}{n\pi} e^{-n\pi|x|/a} \sinh n\pi d/a$$

and for $|x| \leq d$,

$$\phi(x, y) = \sum_{n=1}^{\infty} (-1)^n \sin\left[\frac{n\pi}{a}(y - a)\right] \frac{2V}{n\pi} (1 - e^{-n\pi d/a} \cosh n\pi x/a)$$

Now use expansion of $1 - y/a$ in a Fourier sine series to simplify the expression for $\phi(|x| \leq d, y)$:

$$\phi(x, y) = \frac{2V}{\pi} \sum_{n=1}^{\infty} \frac{1}{n} \sin(n\pi y/a)(1 - e^{-n\pi d/a} \cosh n\pi x/a)$$

and

$$1 - y/a = \sum_{n=1}^{\infty} \frac{2}{n\pi} \sin n\pi y/a$$

So

$$\phi(x, y) = V(1 - y/a) - \frac{2V}{\pi} \sum_{n=1}^{\infty} \frac{1}{n} e^{-n\pi d/a} \cosh\left(\frac{n\pi x}{a}\right) \sin\left(\frac{n\pi y}{a}\right) \quad \text{for } |x| \leq d$$

Chapter 3

PROBLEM 3.1 Charge is distributed uniformly throughout an infinitely long cylinder of radius R:

(a) Show that E at a distance r from the cylinder axis $(r < R)$ is given by $E = \rho r/2\varepsilon_0$, where ρ is the density of charge.
(b) What result do you expect for $r > R$?

Solution:

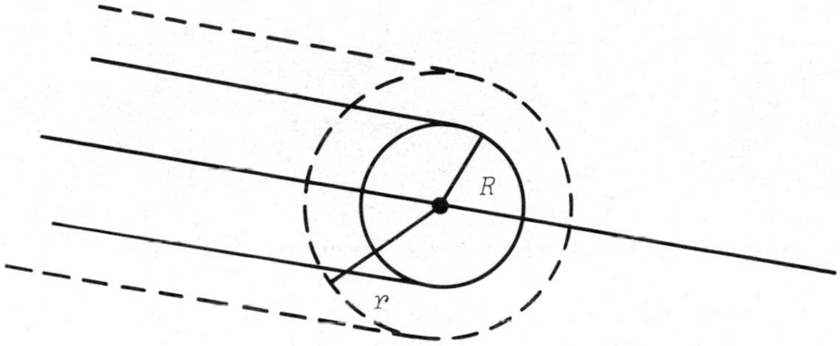

(a) $\underline{r < R}$

$$EA = \frac{Q_T}{\varepsilon_0}$$

$$\varepsilon_0 E(2\pi r h) = \pi r^2 h\rho$$

$$E = \frac{\rho r}{2\varepsilon_0}$$

(b) $\underline{r > R}$

$$\varepsilon_0 E(2\pi r h) = \rho\pi R^2 h$$

$$E = \frac{\rho R^2}{2\varepsilon_0 r}$$

PROBLEM 3.2 A long conducting cylinder of length ℓ, carrying a total charge $+q$, is surrounded by a conducting cylindrical shell of total charge $-2q$, as shown in cross section below. Use Gauss's law to find the following:

(a) The electric field at points outside the conducting shell.
(b) The distribution of charge on the conducting shell.
(c) The electric field in the region between the cylinders.

Solution:

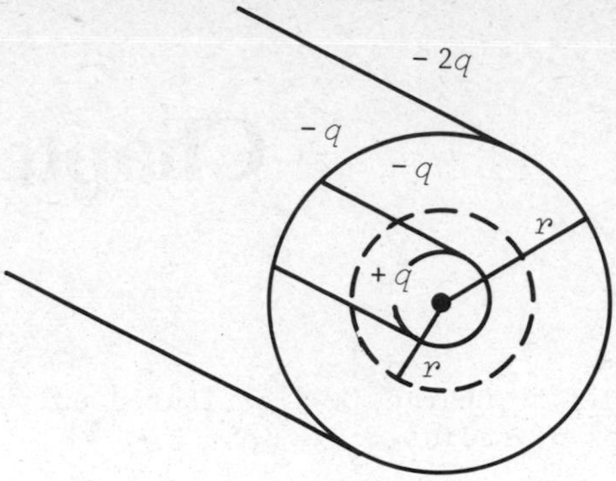

(a)

$$EA = \frac{Q_T}{\varepsilon_0}$$

$$E(2\pi r\ell) = \frac{q - 2q}{\varepsilon_0}$$

$$E = \frac{-q}{\varepsilon_0 (2\pi r\ell)}$$

where the minus sign means that E is directed radially inward. Total charge $= +q - q - q = -q$.

(b) $-q$ on both inner and outer surfaces.

(c)

$$E(2\pi r\ell) = \frac{+q}{\varepsilon_0}$$

$$E = \frac{+q}{2\pi r\ell\varepsilon_0}$$

where the plus sign indicates that E is directed radially outward.

PROBLEM 3.3 An infinite line of charge produces a field of 4.5×10^4 N/C at a distance of 2.0 m. Calculate the linear charge density.

Solution:

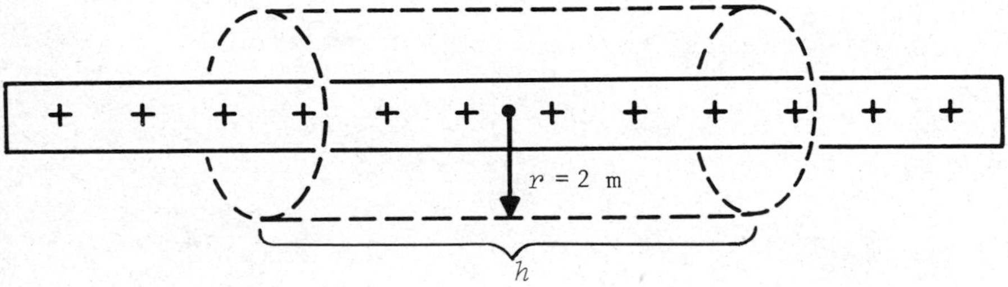

$$\lambda = \frac{\text{charge}}{\text{length}}$$

$$\varepsilon_0 \oint \mathbf{E} \cdot \hat{\mathbf{n}}\, da = Q_T$$

$2\pi rh$ is the area of surface

$$E \oint dA = \frac{Q_T}{\varepsilon_0}$$

$$E(2\pi rh) = \frac{Q_T\ \lambda h}{\varepsilon_0}$$

$$2\pi\varepsilon_0 Er = \lambda = 2\pi(8.85 \times 10^{-12}\ C^2/N \cdot m^2)(4.5 \times 10^4\ N/C)(2m)$$

$$\lambda = 5.0 \times 10^{-6}\ C/m$$

PROBLEM 3.4 Two large nonconducting sheets of positive charge face each other as in the figure. What is E at points (a) to the left of the sheets, (b) between them, and (c) to the right of the sheets? Assume the same surface charge density σ for each sheet. The separation of the sheets is much less than their dimensions. Do not consider points near the edges of the sheets. (*Hint:* E at any point is the vector sum of the separate electric fields set up by each sheet.)

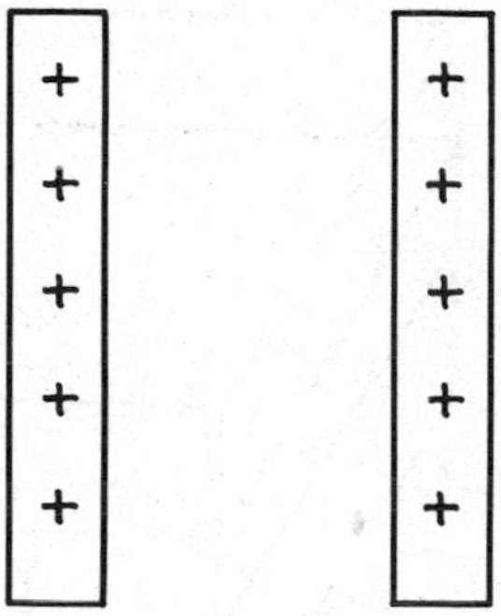

Solution:

(a) $$\oint \mathbf{E} \cdot \hat{\mathbf{n}}\, dA = \frac{Q_T}{\varepsilon_0}$$

$$E\pi r^2 = \frac{\sigma\pi r^2}{\varepsilon_0}$$

$$E = \frac{\sigma}{\varepsilon_0}$$

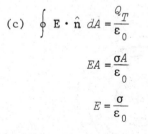

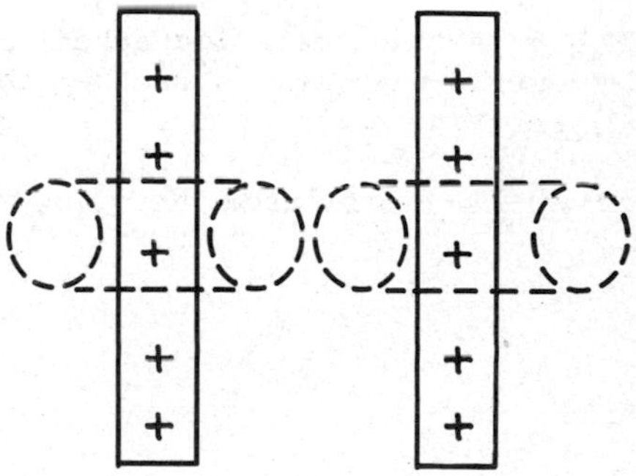

(b) between them $E = 0$.

(c) $$\oint \mathbf{E} \cdot \hat{\mathbf{n}}\, dA = \frac{Q_T}{\varepsilon_0}$$

$$EA = \frac{\sigma A}{\varepsilon_0}$$

$$E = \frac{\sigma}{\varepsilon_0}$$

PROBLEM 3.5 A nonconducting plane slab of thickness d has a uniform volume charge density ρ. Find the magnitude of the electric field at all points in space both (a) inside and (b) outside the slab.

Solution:

(a) $$\int \mathbf{E} \cdot \hat{\mathbf{n}}\, dA = \frac{Q}{\varepsilon_0}$$

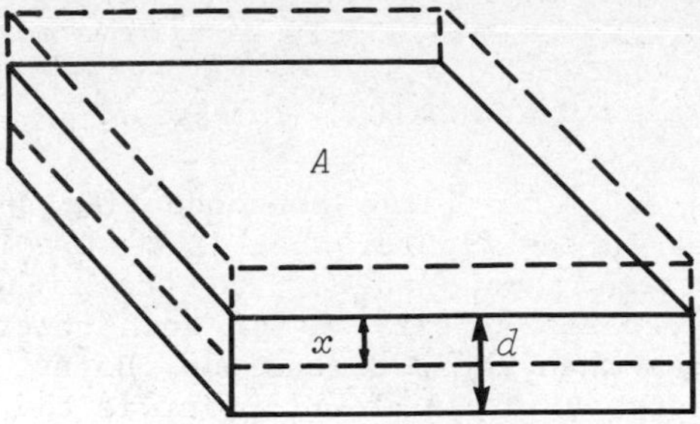

$$EA = \frac{Q}{\varepsilon_0}$$

$$EA = \frac{\rho A x}{\varepsilon_0}$$

$$E = \frac{\rho x}{\varepsilon_0}$$

(b) $\mathbf{E} \cdot \hat{\mathbf{n}}$ on side $= 0$

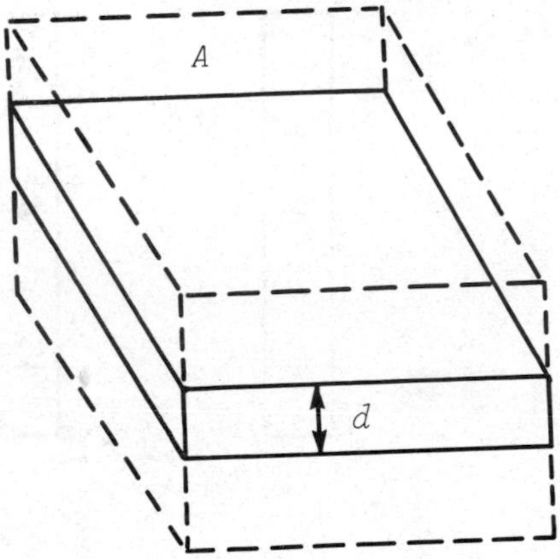

$$EA = \frac{Q}{\varepsilon_0}$$

$$2EA = \frac{\rho A d}{\varepsilon_0}$$

$$E = \frac{\rho d}{2\varepsilon_0}$$

PROBLEM 3.6 A thin metallic spherical shell of radius a carries a charge q_a. Concentric with it is another thin metallic spherical shell of radius b ($b > a$) carrying a charge q_b. Use Gauss's law to find the electric field at radial points r where (a) $r < a$; (b) $a < r < b$; (c) $r > b$. (d) How is the charge on each shell distributed between the inner and outer surfaces of that shell?

Solution:

(a) $\underline{r < a}$

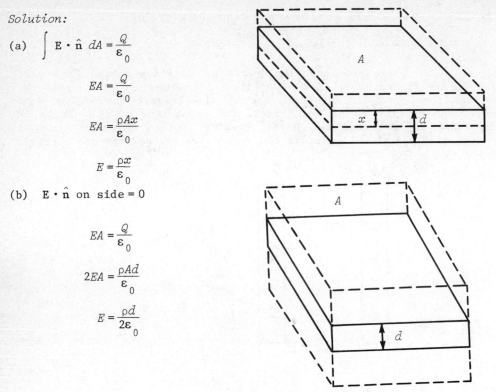

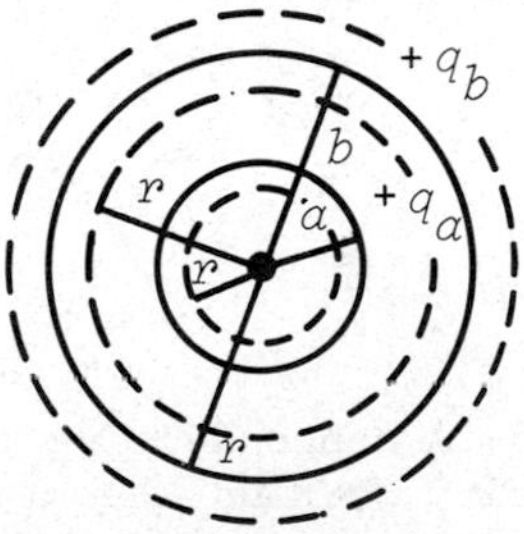

$$\oint \mathbf{E} \cdot \hat{\mathbf{n}}\, dA = \frac{Q_T}{\varepsilon_0}$$

$$E\, dA = \frac{Q_T}{\varepsilon_0}, \text{ but } Q_T = 0, \text{ therefore}$$

$$E = 0$$

(b) $\underline{a < r < b}$

$$\phi_E = \oint \mathbf{E} \cdot \hat{\mathbf{n}}\, dA = \frac{Q_T}{\varepsilon_0}$$

$$EA = \frac{Q_T}{\varepsilon_0}$$

$$E = \frac{q_a}{4\pi r^2 \varepsilon_0}$$

(c) $\underline{r > b}$

$$\oint \mathbf{E} \cdot \hat{\mathbf{n}}\, dA = \frac{Q_T}{\varepsilon_0}$$

$$EA = \frac{Q_T}{\varepsilon_0}$$

$$E = \frac{q_a + q_b}{4\pi r^2 \varepsilon_0}$$

(d) Inner shell: inside $q = 0$
outside: q_a, because it is conductor
outer shell: inside $= -q_a$
outside $= q_a + q_b$

therefore, $q_a + q_b - q_a = q_b$

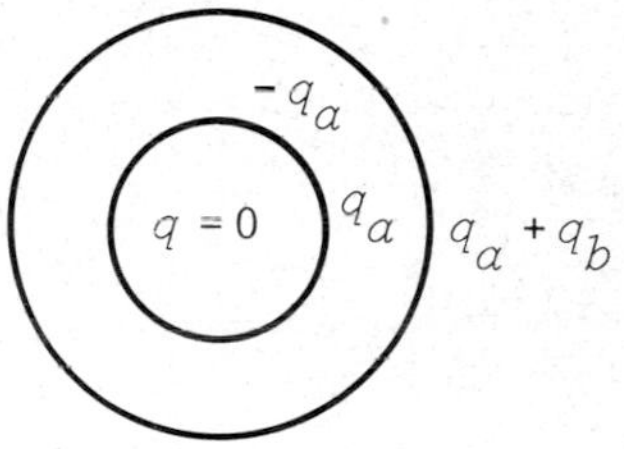

PROBLEM 3.7 The figure shows a spherical nonconducting shell of charge of uniform charge density ρ. Plot E for distances r from the center of the shell ranging from 0 to 30 cm. Assume that $\rho = 1.0 \times 10^{-6}$ C/m^3, $a = 10$ cm, and $b = 20$ cm.

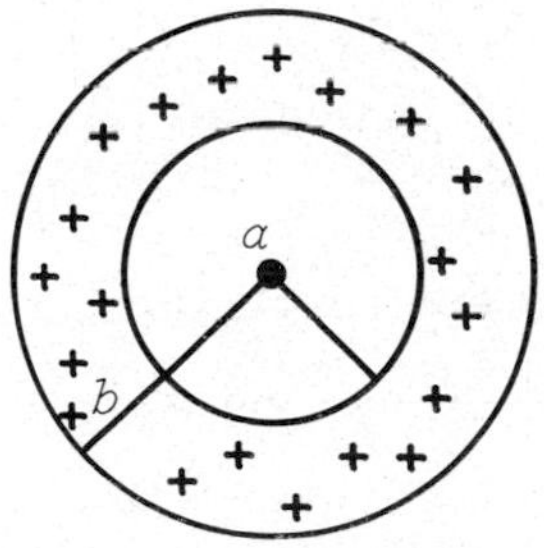

Solution:

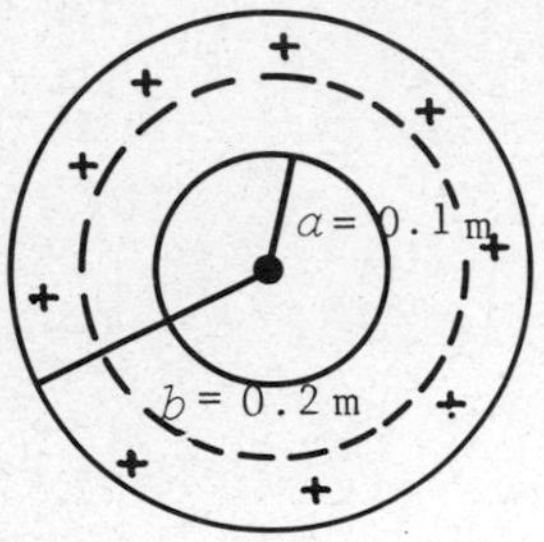

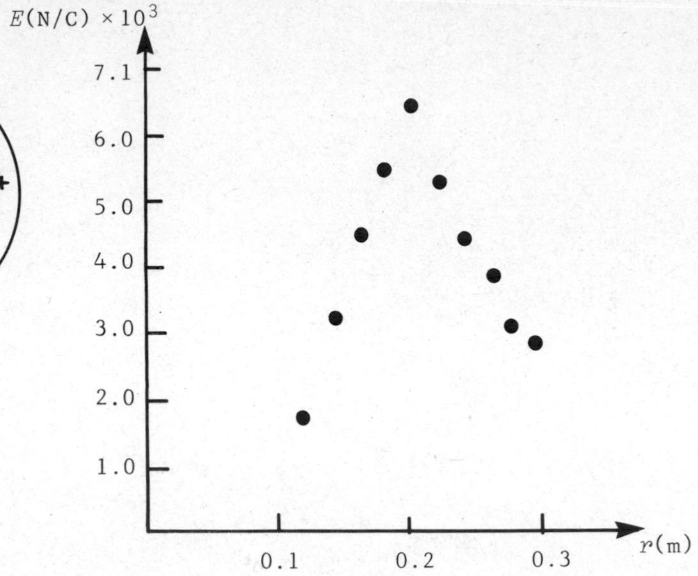

$$\phi_E = \oint \mathbf{E} \cdot \hat{\mathbf{n}}\, dA = \frac{Q_T}{\varepsilon_0}$$

$$\oint E\, dA = \frac{Q_T}{\varepsilon_0}$$

$$\varepsilon_0 E(4\pi r^2) = (^4/_3\, \pi r^3 - {^4/_3}\, \pi a^3)$$

$$E = \frac{(^1/_3\, \rho)4\pi(r^3 - a^3)}{3\varepsilon_0(4\pi r^2)}$$

$$E = \frac{\rho(r^3 - a^3)}{3\varepsilon_0 r^2}$$

$$E(r) = \frac{1 \times 10^{-6}\ (r^3 - 0.001)}{2.66 \times 10^{-11}\ r^2}$$

$$E(r) = \frac{1 \times 10^{-6}\ r^3 - 1 \times 10^{-9}}{2.66 \times 10^{-11}\ r^2}$$ (inside shell from a to b)

$$\phi = \int \mathbf{E} \cdot \hat{\mathbf{n}}\, dA = \frac{Q_T}{\varepsilon_0}$$

$$\varepsilon_0 E(4\pi r^2) = Q_T$$

$$E = \frac{Q_T}{4\pi r^2 \varepsilon_0} = \frac{\rho(4/3\ \pi b^3 - 4/3\ \pi a^3)}{4\pi r^2 \varepsilon_0}$$

$$E = \frac{1/3\ (1 \times 10^{-6})4\pi(0.2^3 - 0.1^3)}{(4\pi)(8.85 \times 10^{-12})r^2}$$

$$E(r) = \frac{263.65}{r^2} \quad \text{(outside)}$$

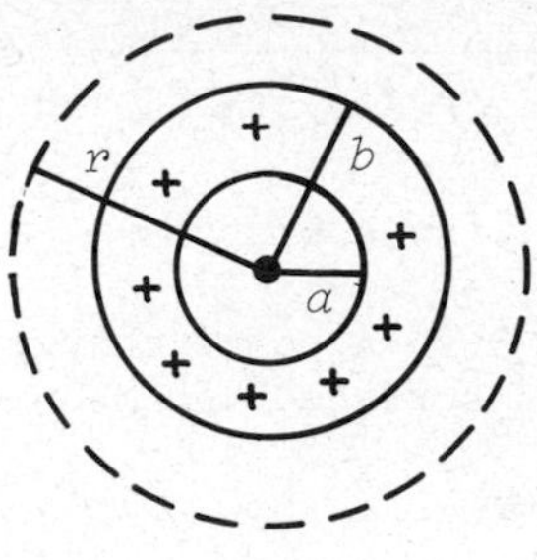

PROBLEM 3.8 A thin-walled metal sphere has a radius of 25 cm and carries a charge of 2.0×10^{-7} C. Find E for a point (a) inside the sphere, (b) just outside the sphere, and (c) 3.0 m from the center of the sphere.

Solution:

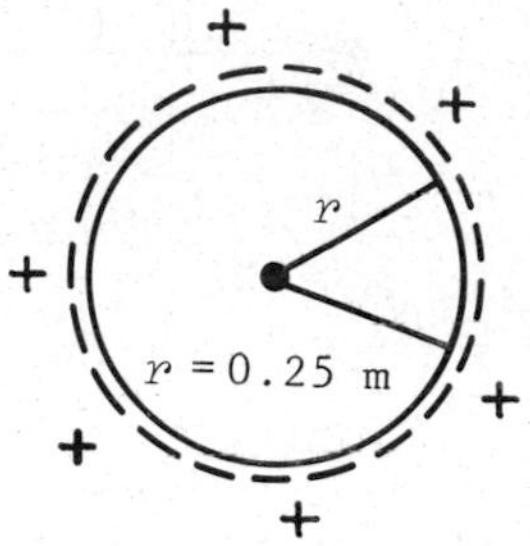

(a) $E = 0$, because an excess charge placed on an insulated conductor, resides on its outer surface.

(b) $$\Phi = \oint \mathbf{E} \cdot \hat{\mathbf{n}}\ dA$$

$$\phi_E = \oint E\ dA = \frac{Q_T}{\varepsilon_0}$$

$$E(4\pi r^2) = \frac{Q_T}{\varepsilon_0}$$

$$E = \frac{Q_T}{4\pi r^2 \varepsilon_0} = \frac{2 \times 10^{-7}\ \text{C}}{4\pi(0.25\ \text{m}^2)(8.85 \times 10^{-12}\ \text{C}^2/\text{N} \cdot \text{m}^2)}$$

$$= 2.88 \times 10^4\ \text{N/C}$$

(c) $$\phi = \oint E\ dA = \frac{Q_T}{\varepsilon_0}$$

$$E(4\pi r^2) = \frac{Q}{\varepsilon_0}$$

$$E = \frac{Q}{4\pi r^2 \varepsilon}$$

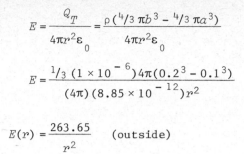

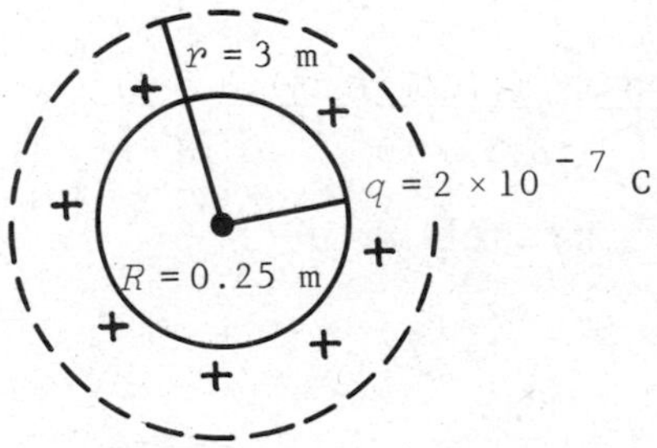

$$= \frac{2 \times 10^{-7}\ \text{C}}{4\pi(3\ \text{m})^2(8.85 \times 10^{-12}\ \text{C}^2/\text{N}\cdot\text{m}^2)}$$

$$= 199.8\ \text{N/C}$$

PROBLEM 3.9 A point charge of 1.0×10^{-6} C is at the center of a cubical Gaussian surface 0.50 m on edge. What is Φ_E for the surface?

Solution:

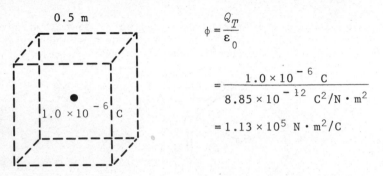

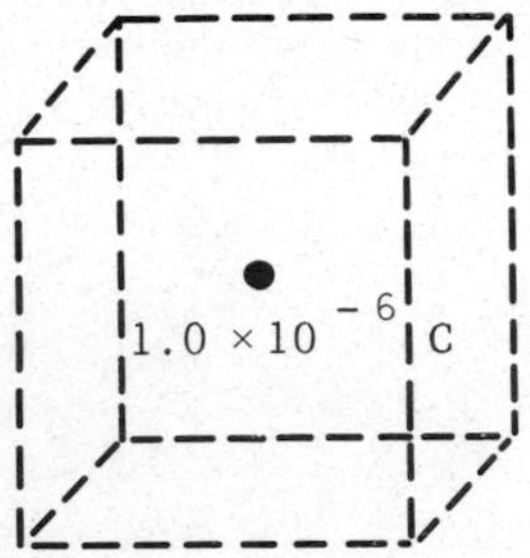

$$\phi = \frac{Q_T}{\varepsilon_0}$$

$$= \frac{1.0 \times 10^{-6}\ \text{C}}{8.85 \times 10^{-12}\ \text{C}^2/\text{N}\cdot\text{m}^2}$$

$$= 1.13 \times 10^5\ \text{N}\cdot\text{m}^2/\text{C}$$

PROBLEM 3.10 A uniformly charged conducting sphere of 1.0 m diameter has a surface charge density of 8 C/m². What is the total electric flux leaving the surface of the sphere?

Solution:

$$\phi_E = \oint \mathbf{E}\cdot\hat{\mathbf{n}}\ dA = Q_T/\varepsilon \text{ and } \sigma = 8\ \text{C/m}^2.$$

$$\phi_E = \oint \mathbf{E}\cdot\hat{\mathbf{n}}\ dA$$

$$= \oint E\ dA = \frac{Q_T}{\varepsilon_0}$$

$$= E \oint dA = E(4\pi r^2) = \frac{Q_T}{\varepsilon_0}$$

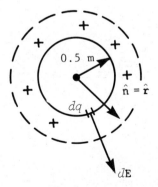

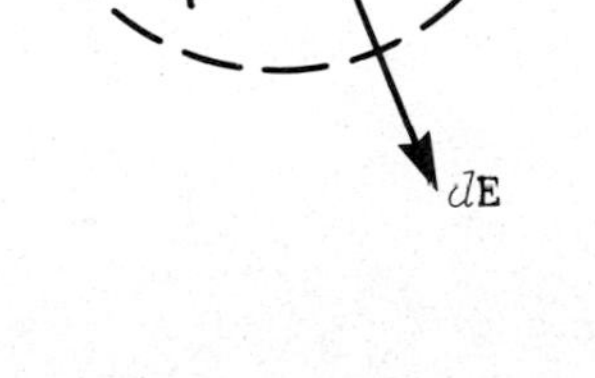

$$\phi_E = \frac{\sigma A}{\varepsilon_0} = \frac{(8\ \text{C/m}^2)(4\pi)(0.5\ \text{m})^2}{8.85 \times 10^{-12}\ \text{C}^2/\text{N}\cdot\text{m}^2}$$

$$\phi_E = 2.84 \times 10^{12}\ \text{N}\cdot\text{m}^2/\text{C}$$

PROBLEM 3.11 Charge on an originally uncharged insulated conductor is separated by holding a positively charged rod nearby (see figure). What can you learn from Gauss's law about the flux for the five Gaussian surfaces shown?

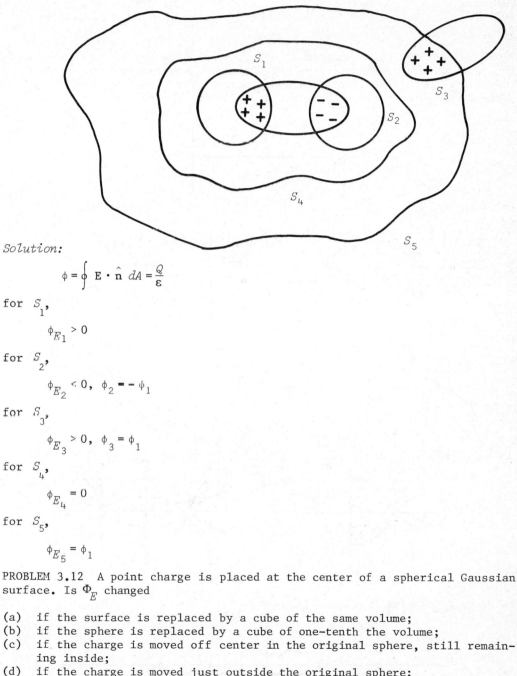

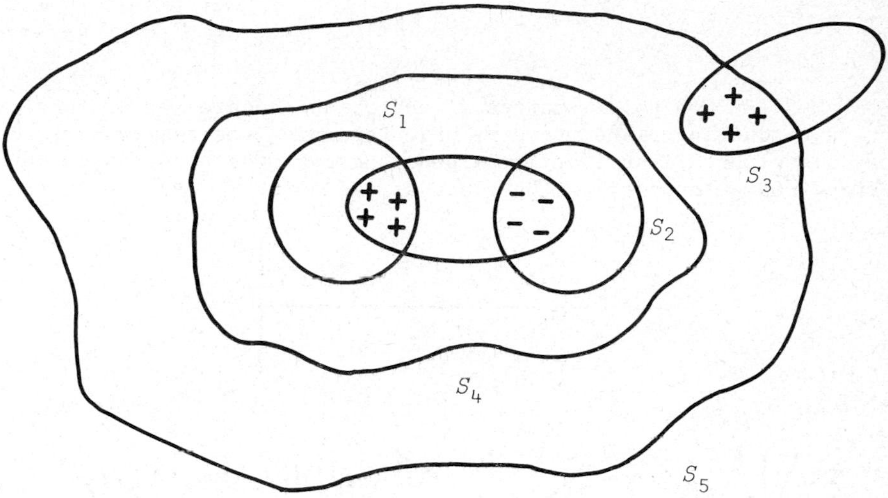

Solution:

$$\phi = \oint \mathbf{E} \cdot \hat{\mathbf{n}}\, dA = \frac{Q}{\varepsilon}$$

for S_1,

$$\phi_{E_1} > 0$$

for S_2,

$$\phi_{E_2} < 0, \ \phi_2 = -\phi_1$$

for S_3,

$$\phi_{E_3} > 0, \ \phi_3 = \phi_1$$

for S_4,

$$\phi_{E_4} = 0$$

for S_5,

$$\phi_{E_5} = \phi_1$$

PROBLEM 3.12 A point charge is placed at the center of a spherical Gaussian surface. Is Φ_E changed

(a) if the surface is replaced by a cube of the same volume;
(b) if the sphere is replaced by a cube of one-tenth the volume;
(c) if the charge is moved off center in the original sphere, still remaining inside;
(d) if the charge is moved just outside the original sphere;
(e) if a second charge is placed near, and outside, the original sphere;
(f) if a second charge is placed inside the Gaussian surface?

Solution:

(a) No
(b) No
(c) No
(d) Yes
(e) No
(f) Yes

PROBLEM 3.13 Two point charges q_A and q_B are located at distances r_A and r_B from a conducting plane where $r_B > r_A$. Determine the resultant force on q_B if they lie along a line that is perpendicular to the plate. Also determine the surface charge density in the plate at the point χ.

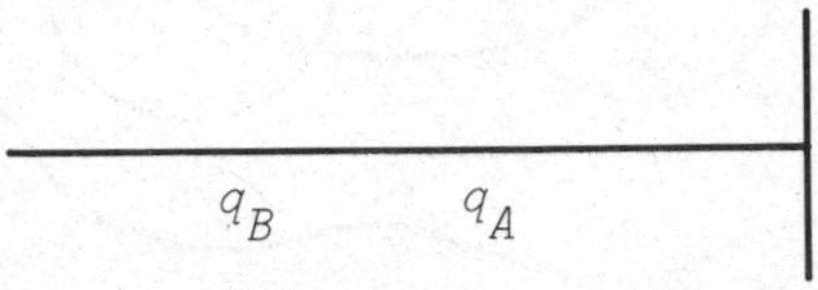

Solution:

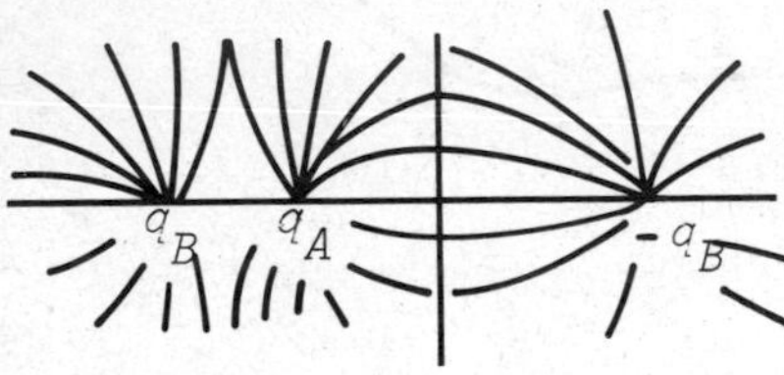

(a)

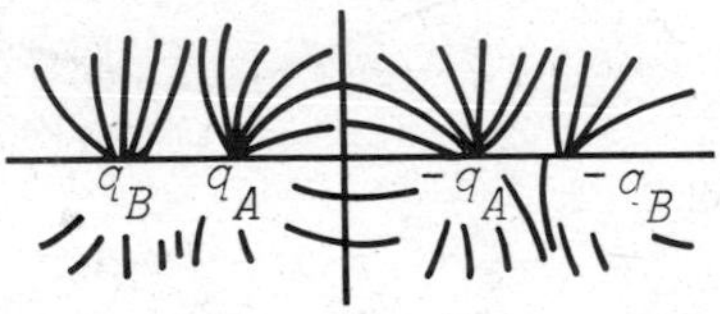

(b)

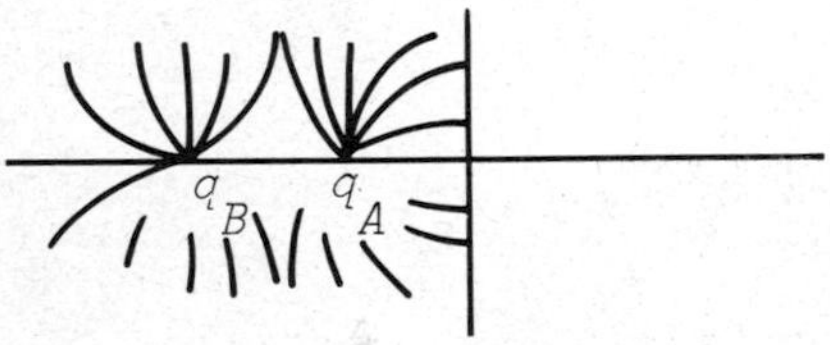

(c)

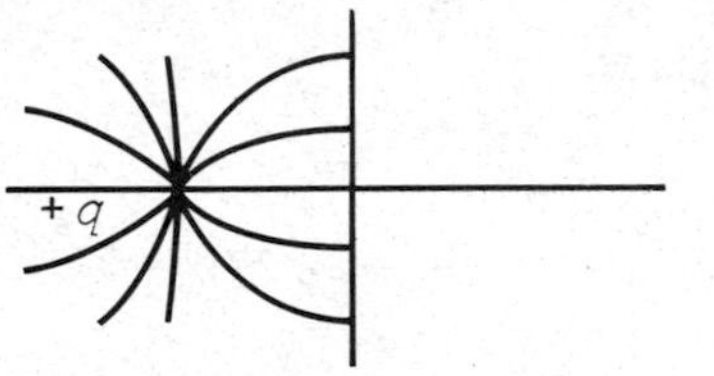

(d)

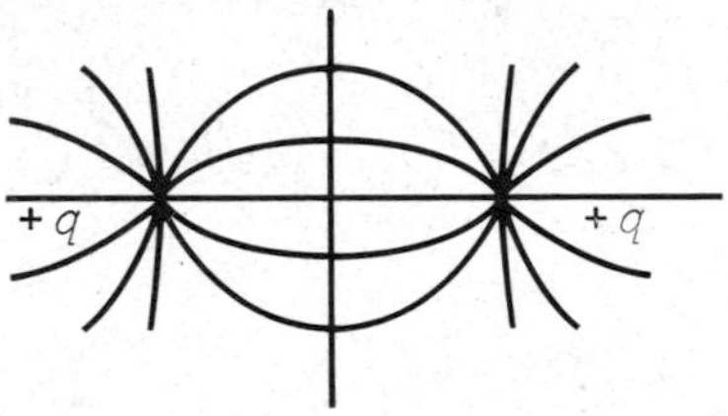

(e)

Refer to figure for Problem 3.13 above,

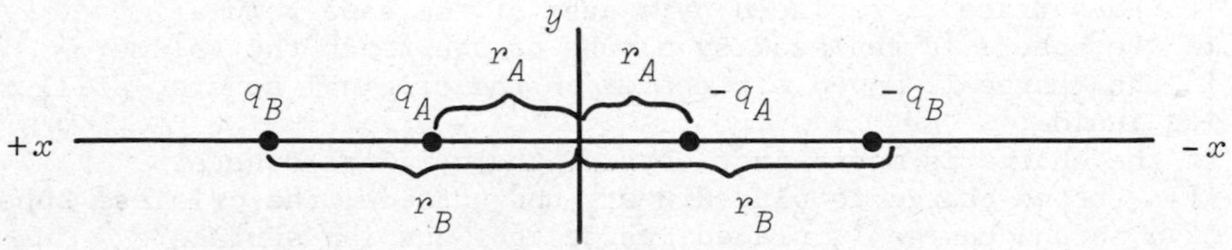

$$\mathbf{F}_{\text{on } q_B} = \frac{q_B q_A}{4\pi\varepsilon(r_B - r_A)^2}(\hat{\mathbf{i}}) + \frac{q_B q_A}{4\pi\varepsilon(r_B + r_A)^2}(-\hat{\mathbf{i}}) + \frac{q_B q_B}{4\pi\varepsilon(2r_B)^2}(-\hat{\mathbf{i}})$$

$$\mathbf{F} = \hat{\mathbf{i}}\,\frac{q_B}{4\pi\varepsilon}\left[\frac{q_A}{(r_B - r_A)^2} - \frac{q_A}{(r_B + r_A)^2} - \frac{q_B}{4r_B^2}\right]$$

Now to find surface charge density at point χ.

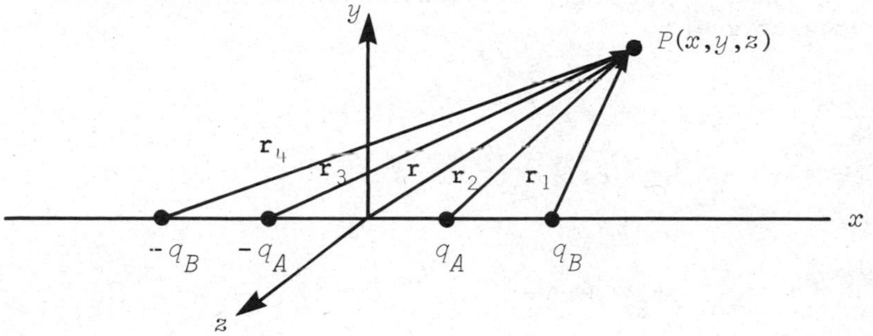

$$\mathbf{r} = \hat{\mathbf{i}}x + \hat{\mathbf{j}}y + \hat{\mathbf{k}}z$$

$$r = \sqrt{x^2 + y^2 + z^2}$$

$$\hat{\mathbf{i}}r_B + \mathbf{r}_1 = \mathbf{r}$$

$$\mathbf{r}_1 = \mathbf{r} - \hat{\mathbf{i}}r_B = \hat{\mathbf{i}}(x - r_B) + \hat{\mathbf{j}}y + \hat{\mathbf{k}}z$$

$$r_1 = \sqrt{(x - r_B)^2 + y^2 + z^2}$$

$$\hat{\mathbf{i}}r_A + \mathbf{r}_2 = \mathbf{r}$$

$$\mathbf{r}_2 = \hat{\mathbf{i}}(x - r_A) + \hat{\mathbf{j}}y + \hat{\mathbf{k}}z$$

$$r_2 = \sqrt{(x - r_A)^2 + y^2 + z^2}$$

$$-r_A\hat{\mathbf{i}} + \mathbf{r}_3 = \mathbf{r}$$

$$\mathbf{r}_3 = \hat{\mathbf{i}}(x + r_A) + \hat{\mathbf{j}}y + \hat{\mathbf{k}}z$$

$$r_3 = \sqrt{(x + r_A)^2 + y^2 + z^2}$$

$$-r_B\hat{\mathbf{i}} + \mathbf{r}_4 = \mathbf{r}$$

$$\mathbf{r}_4 = \hat{\mathbf{i}}(x + r_B) + \hat{\mathbf{j}}y + \hat{\mathbf{k}}z$$

$$r_4 = \sqrt{(x + r_B)^2 + y^2 + z^2}$$

$$\phi(x,y,z) = \frac{1}{4\pi\varepsilon}\left(\frac{q_A}{r_2} - \frac{q_A}{r_3} + \frac{q_B}{r_1} - \frac{q_B}{r_4}\right)$$

$$\phi(x,y,z) = \frac{q_A}{4\pi\varepsilon}\{[(x-r_A)^2+y^2+z^2]^{-\frac{1}{2}} - [(x+r_A)^2+y^2+z^2]^{-\frac{1}{2}}\}$$
$$+\frac{q_B}{4\pi\varepsilon}\{[(x-r_B)^2+y^2+z^2]^{-\frac{1}{2}} - [(x+r_B)^2+y^2+z^2]^{-\frac{1}{2}}\}$$

$\sigma = -\varepsilon_0 (\partial\phi/\partial x)_{x=0}$, using $+x$ because that is the outward normal to the conductor surface.

$$\sigma = \frac{-\varepsilon_0 q_A}{4\pi\varepsilon}\{-\tfrac{1}{2}[(x-r_A)^2+y^2+z^2]^{-3/2}\,2(x-r_A)$$
$$-(-\tfrac{1}{2})[(x+r_A)^2+y^2+z^2]^{-3/2}\,2(x+r_A)\}$$

$$+\frac{-\varepsilon_0 q_B}{4\pi\varepsilon}\{-\tfrac{1}{2}[(x-r_B)^2+y^2+z^2]^{-3/2}\,2(x-r_B)$$
$$-(-\tfrac{1}{2})[(x+r_B)^2+y^2+z^2]^{-3/2}\,2(x+r_B)\}$$

$$\sigma(x,y,z) = \frac{q_A}{4\pi}\left(\frac{x-r_A}{[(x-r_A)^2+y^2+z^2]^{3/2}} - \frac{x+r_A}{[(x+r_A)^2+y^2+z^2]^{3/2}}\right)$$

$$+\frac{q_B}{4\pi}\left(\frac{x-r_B}{[(x-r_B)^2+y^2+z^2]^{3/2}} - \frac{x+r_B}{[(x+r_B)^2+y^2+z^2]^{3/2}}\right)$$

At $x = 0$,

$$\sigma(y,z) = \frac{q_A}{4\pi}\left(\frac{-2r_A}{(r_A^2+y^2+z^2)^{3/2}}\right) + \frac{q_B}{4\pi}\left(\frac{-2r_B}{(r_B^2+y^2+z^2)^{3/2}}\right)$$

$$\sigma(y,z) = \frac{-q_A}{2\pi}\left(\frac{r_A}{(r_A^2+y^2+z^2)^{3/2}}\right) - \frac{q_B}{2\pi}\left(\frac{r_B}{(r_B^2+y^2+z^2)^{3/2}}\right)$$

at point χ, $y = 0$ and $z = 0$,

$$\sigma(0,0) = \frac{-q_A r_A}{2\pi\, r_A^3} - \frac{q_B r_B}{2\pi\, r_B^3}$$

$$\sigma(\text{at point } \chi) = \frac{-1}{2\pi}\left(\frac{q_A}{r_A^2} + \frac{q_B}{r_B^2}\right)$$

A quicker way to find $\sigma(\chi)$ would be to find E at the origin only, which is all you need to solve the problem.

PROBLEM 3.14 Two infinite, grounded conducting planes are at $x = \pm a/2$. A point charge $+Q$ is placed between the planes, where $-a/2 < x' < a/2$. Calculate the Green's function for the distribution and the induced surface charge density on each plane.

Solution:

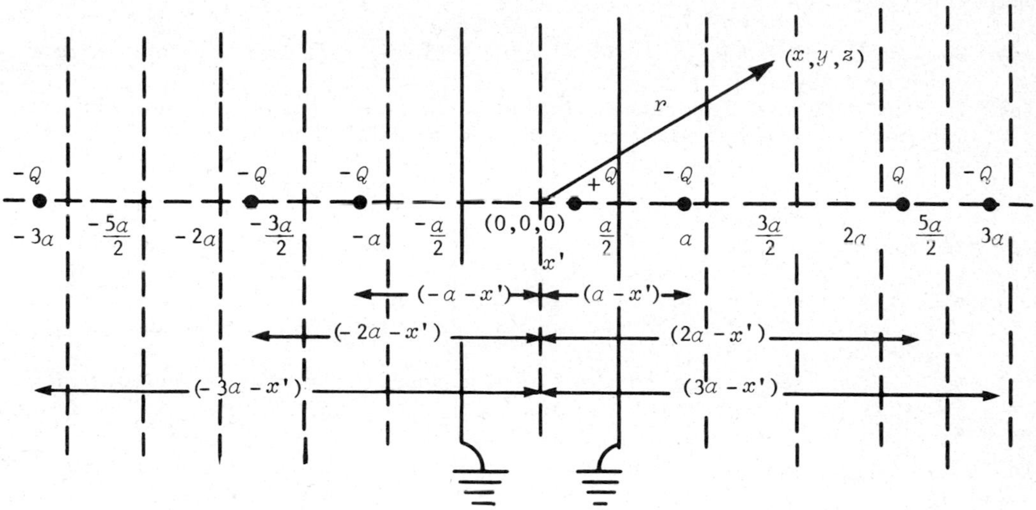

In order for the potential to be zero at both plates simultaneously, we need an infinite number of image charges. The potential due to a unit charge and its images is just the Green's function appropriate for Dirichlet boundary conditions:

$$G(\mathbf{r},\mathbf{r}') = \sum_{n=-\infty}^{\infty} \frac{(-1)^n}{4\pi\varepsilon_0 |\mathbf{r} - [na + (-1)^n x']\hat{\mathbf{i}}|}$$

$$G(x,y,z,x') = \frac{1}{4\pi\varepsilon_0} \sum_{n=-\infty}^{\infty} \frac{(-1)^n}{(\{x - [na + (-1)^n x']\}^2 + y^2 + z^2)^{\frac{1}{2}}}$$

$$\phi(\mathbf{r}) = \frac{Q}{4\pi\varepsilon_0} \sum_{n=-\infty}^{\infty} \frac{(-1)^n}{(\{x - [na + (-1)^n x']\}^2 + y^2 + z^2)^{\frac{1}{2}}}$$

$$\frac{\partial\phi}{\partial x} = \frac{-Q}{4\pi\varepsilon_0} \sum_{n=-\infty}^{\infty} \frac{(-1)^n \{x - [na + (-1)^n x']\}}{(\{x - [na + (-1)^n x']\}^2 + y^2 + z^2)^{3/2}}$$

$$\sigma = \varepsilon_0 E \Big|_{x = \pm a/2} = -\varepsilon_0 \frac{\partial\phi}{\partial x}\Big|_{x = \pm a/2}$$

$$\sigma = +\frac{Q}{4\pi} \sum_{n=-\infty}^{\infty} \frac{(-1)^n \{\pm a/2 - [na + (-1)^n x']\}}{(\{\pm a/2 - [na + (-1)^n x']\}^2 + y^2 + z^2)^{3/2}}$$

where $+a/2$ gives the charge distribution on the plate at $x=+a/2$ and $-a/2$ gives σ on the plate at $x=-a/2$.

PROBLEM 3.15 An observer with a device for measuring the electric field E is located some distance away from a fixed point charge q. A short length of uncharged metal pipe is lowered by an insulating rope until it surrounds the point charge.

(a) How does this affect the electric field measured by the distant observer? Explain.
(b) If you occupy a laboratory inside a large copper box, can you tell whether charges are being moved around outside? Explain.

Solution:

(a) As long as the length of the pipe is small compared to the distance from it to the observer, it will have a negligible effect on the electric field. The outer surface charge on it will be equal to that of the surrounded point charge, and it, in turn, will appear as almost a "point" change at long distances.
(b) No. The surface charge on the closed conducting surface will move until there is no electric field inside.

PROBLEM 3.16 A spherical conductor A contains two spherical cavities. The total charge on the conductor itself is zero. However, there is a point charge q_b at the center of one cavity and q_c at the center of the other. A considerable distance r away is another charge q_d. What force acts on each of the four objects: A, q_b, q_c, q_d? Which answers, if any, are only approximate, and depend on r being relatively large?

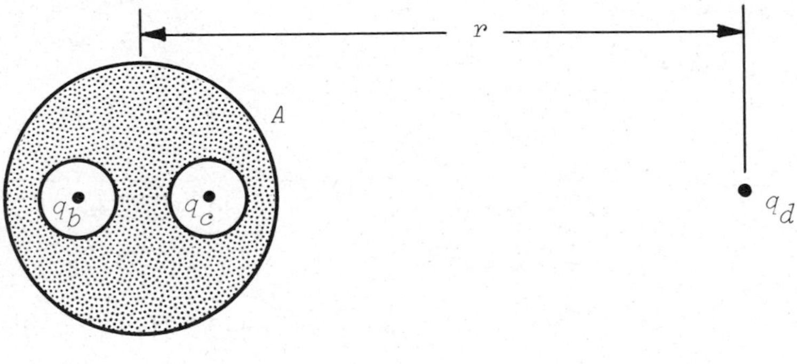

Solution:

No force acts on q_b or q_c. The conductor A shields them from each other and from q_d, and they are each at the center of a spherical cavity. The surface charge on A will equal q_b+q_c. In the absence of q_d, the surface charge density will be uniform. If $r \gg R_A$, this will still be approximately true and

$$F_{Aq_d} \sim \frac{q_d(q_b+q_c)}{r^2}$$

PROBLEM 3.17 By solving the problem of the point charge and the plane conductor we have, in effect, solved every problem that can be constructed from it by superposition. For instance, suppose we have a straight wire, uniformly charged with 10^3 esu per centimeter of length, running parallel to the earth at a height of 5 m. What is the field strength at the surface of the earth, immediately below the wire? What is the electrical force acting on the wire per unit of length?

Solution: The image of the long positively charged wire a distance h above the plane is a long negatively charged wire a distance h below the plane. So, at the plane,

$$E = E_n = -2\,(2\lambda/r)\cos\theta$$

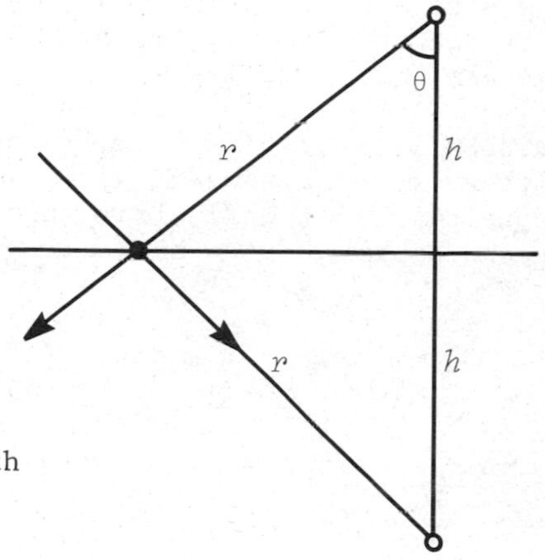

At $r = h$, $\cos\theta = 1$ and

$$E = \frac{-4\lambda}{h} = \frac{-4(10^3\ \text{esu/cm})}{500\ \text{cm}}$$

$$= \sim 8\,\frac{\text{statvolts}}{\text{cm}}$$

$$\frac{F}{\ell} = \frac{q}{\ell}\,E = \lambda\left(\frac{2\lambda}{2h}\right) = \frac{\lambda^2}{h}\ \text{toward the earth}$$

$$= \frac{(10^3\ \text{esu/cm})^2}{500\ \text{cm}} = 2\times 10^3\ \text{dyn/cm}$$

PROBLEM 3.18 A charge Q is located h cm above a conducting plane, just as in the figure. Asked to predict the amount of work that would have to be done to move this charge out to infinite distance from the plane, one student says that it is the same as the work required to separate to infinite distance two charges Q and $-Q$ which are initially $2h$ cm apart, hence $W = Q^2/2h$. Another student calculates the force that acts on the charge as it is being moved and integrates $F\,dx$, but gets a different answer. What did the second student get, and who is right?

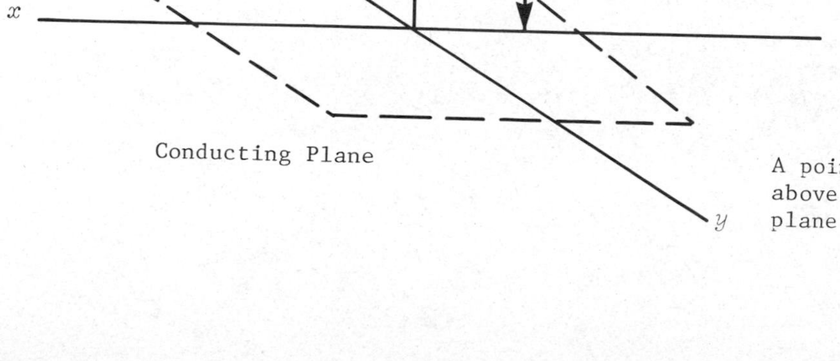

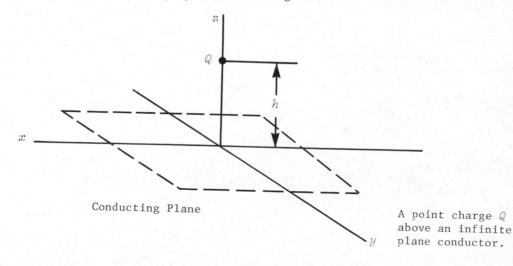

A point charge Q above an infinite plane conductor.

Solution: From the image-charge method,

$$F = \frac{q^2}{r^2} = \frac{q^2}{4h^2}$$

$$W = \int_h^{\infty} F\, dh = \left(\frac{-q^2}{4h}\right)_h^{\infty} = \frac{q^2}{4h}$$

This method is correct. The motion of the actual charge not only increases the distance, it reduces the induced surface charge, so the force and the work decrease faster. Or, a motion Δh of the charge results in a change $2\Delta h$ between charge and image.

PROBLEM 3.19 A 100-pF capacitor is charged to 100 V. After the charging battery is disconnected, the capacitor is connected in parallel to another capacitor. If the final voltage is 30 V, what is the capacitance of the second capacitor? How much energy was lost, and what happened to it?

Solution:

$$Q = C_1\phi_1 = 10^{-10}\ \text{F} \times 100\ \text{V} = 10^{-8}\ \text{C}$$

$$= (C_1 + C_2)\phi_2 = 30(10^{-10} + C_2)$$

$$C_2 = 3.33 \times 10^{-10} - 10^{-10} = 2.33 \times 10^{-10}\ \text{F} = 233\ \text{pF}$$

$$U_1 = \tfrac{1}{2}C_1\phi_1^2 = 5 \times 10^{-7}\ \text{J}$$

$$U_2 = \tfrac{1}{2}(C_1 + C_2)\phi_2^2 = 1.5 \times 10^{-7}\ \text{J}$$

The loss is I^2R loss as the charge is moved.

PROBLEM 3.20 What is the capacitance C of a capacitor that consists of two concentric spherical shells, the inner of radius r_1, the outer of radius r_2? Check your result by relating to the case of a flat parallel-plate capacitor the limiting case in which $r_2 - r_1 << r_1$.

Solution:

$$\phi = Q\left(\frac{1}{r_1} - \frac{1}{r_2}\right) = \frac{Q}{C}$$

$$C = r_1 r_2/(r_2 - r_1)$$

Let

$$s = r_2 - r_1$$

$$C = r_1(r_1 + s)/s \simeq r_1^2/s \quad \text{for } s <<< r_1$$

But

$$A = 4\pi r_1^2$$

or

$$C \simeq A/4\pi s$$

which checks.

PROBLEM 3.21 Let ϕ_{12} be the potential difference between the plates of the spherical capacitor consisting of two concentric spherical shells, the inner of radius r_1, the outer of radius r_2. Find an expression for the electric field, as a function of radius. Calculate the total field energy, $\int (E^2/8\pi)dv$, and show it comes out equal to $\frac{1}{2}C\phi_{12}{}^2$.

Solution:

$$E = \frac{Q}{r^2}, \quad r_1 < r < r_2$$

$$U = \frac{1}{8\pi} \int E^2 \, dv = -\frac{Q^2}{8\pi} \int \frac{r^2 \sin\theta \, dr \, d\theta \, d\phi}{r^4}$$

$$= \frac{Q^2}{2} \int_{r_1}^{r_2} \frac{dr}{r^2} = -\frac{Q^2}{2r} \Bigg|_{r_1}^{r_2} = \tfrac{1}{2}Q^2 \left(\frac{1}{r_1} - \frac{1}{r_2} \right)$$

$$= \tfrac{1}{2} \frac{Q^2}{C} = \tfrac{1}{2}C\phi_{12}{}^2$$

PROBLEM 3.22 Calculate the force that acts on one plate of a parallel-plate capacitor. The potential difference between the plates is 10 statvolts and the plates are squares 20 cm on a side with a separation of 3 cm. If the plates are insulated so that the charge cannot change, how much external work could be done by letting the plates come together? Does this equal the energy that was initially stored in the electric field?

Solution:

$$F = qE = \sigma AE = E^2A/4\pi \neq \text{function of } s$$

$$E \simeq \frac{\Delta d}{s} = \frac{10}{3}, \quad A = 400 \text{ cm}^2$$

$$F = \left(\frac{10}{3} \right)^2 \frac{400}{4\pi} = 354 \text{ dyn}$$

Since F is constant

$$W = Fs = 354 \times 3 = 1061 \text{ erg}$$

$$W = \frac{E^2}{4\pi} (As) = 2 \times \text{energy in field}$$

PROBLEM 3.23

(a) Find the capacitance of a capacitor that consists of two coaxial cylinders, of radii a and b, and length L. Assume $L >> b-a$, so that end corrections may be neglected. Check your result by showing that if the gap between the cylinders, $b-a$, is very small compared to the radius, your formula reduces to one that could have been obtained by using the formula for the parallel-plate capacitor.

(b) A cylinder of 2.00-in. outer diameter hangs, with its axis vertical, from one arm of a beam balance. The lower portion of the hanging cylinder is surrounded by a stationary cylinder, coaxial, with inner diameter 3.00 in. Calculate the magnitude of the force tending to pull the hanging cylinder further down when the potential difference between the two cylinders is 5 kV.

Solution:

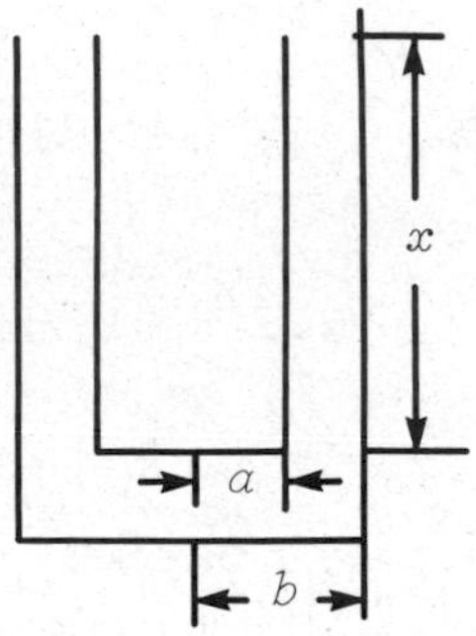

$$E = \frac{2\lambda}{r} = \frac{2Q}{Lr}, \quad a < r < b$$

$$\phi_{ba} = \frac{2Q}{L}\int_a^b \frac{dr}{r} = \frac{2Q}{L}\ln\left(\frac{b}{a}\right) = \frac{Q}{C}$$

$$C = \frac{L}{2\ln(b/a)}, \quad \text{if } b-a = s <<< a$$

$$C = \frac{L}{2\ln(1+s/a)} \simeq \frac{L}{2\,s/a} = \frac{La}{2s}$$

But $A = 2\pi aL$, so $C \simeq A/4\pi s$.

(b)

$$U = \frac{1}{8\pi}\int E^2\,dv, \quad E = \frac{2\lambda}{r} = \frac{2Q}{xr}$$

$$= \frac{1}{8\pi}\frac{4Q^2}{x^2}\int \frac{r dr\,d\theta\,dx}{r^2}\ \frac{4\times 2\pi}{8\pi}\frac{Q^2}{x}\ln\left(\frac{b}{a}\right)$$

But $Q = C\phi = \dfrac{x}{2\ln(b/a)}$

$$U = \frac{\phi^2 x}{4\ln(b/a)}$$

$$F = \frac{dU}{dx} = \frac{\phi^2}{4\ln(b/a)}\ \frac{(5000/300\text{ statvolts})^2}{4\ln(3/2)}$$

$$F = 171\text{ dyn}$$

PROBLEM 3.24 In the field of the point charge over the plane, as in the figure, if you follow a field line that starts out from the point charge in a horizontal direction, that is, parallel to the plane, where does it meet the surface of the conductor? (You'll need Gauss's law and a simple integration.) Some field lines for the charge above the plane are shown in the figure. The field strength at the surface, given by

$$E_z = \frac{-2Q}{r^2+h^2}\cos\theta = \frac{-2Q}{r^2+h^2}\cdot\frac{h}{(r^2+h^2)^{\frac{1}{2}}} = \frac{-2Qh}{(r^2+h^2)^{3/2}}$$

determines the surface charge density σ.

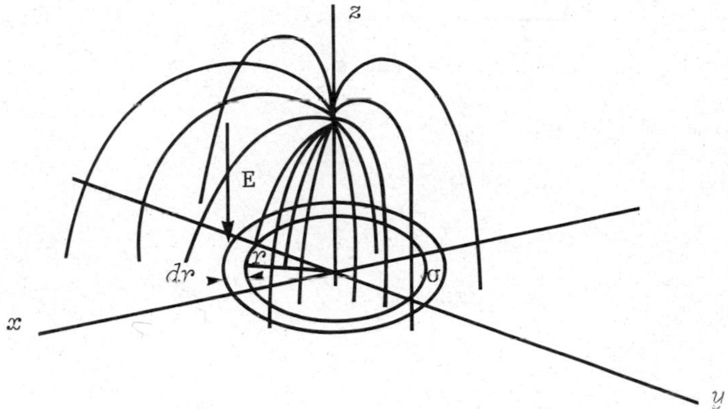

Solution: Consider a very small sphere around the charge q. If it is small enough, the field will be that of a point charge, so that one-half the flux goes through the hemisphere away from the plane and one-half through the other. Since the horizontal plane through q divides these regions, a line of flux in this plane will end at the grounded plane on a radius r dividing the plane into a region containing one-half the flux inside and one-half outside. Therefore,

$$\int_0^r \mathbf{E}\cdot d\mathbf{a} = \tfrac{1}{2}(-4\pi q)$$

$$\mathbf{E} = E_n\hat{\mathbf{z}} = \frac{-2qh}{(r^2+h^2)^{3/2}}\hat{\mathbf{z}}, \qquad d\mathbf{a} = 2\pi r\,dr\hat{\mathbf{z}}$$

$$-2\pi q = -4\pi qh\int_0^r \frac{r\,dr}{(r^2+h^2)^{3/2}}$$

$$1 = 2h\left(\frac{-1}{(r^2+h^2)^{\frac{1}{2}}}\right)_0^r = 2h\left(\frac{1}{h} - \frac{1}{(r^2+h^2)^{\frac{1}{2}}}\right)$$

$$\frac{2h}{(r^2+h^2)^{\frac{1}{2}}} = 1, \quad 4h^2 = r^2+h^2$$

$$r = \sqrt{3}\,h$$

PROBLEM 3.25 Three conducting plates are placed parallel to one another as shown. The outer plates are connected by a wire. The inner plate is isolated and carries a charge amounting to 10 esu/cm^2 of plate. In what proportion must this charge divide itself into a surface charge σ_1 on one face of the inner plate and a surface charge σ_2 on the other side of the same plate?

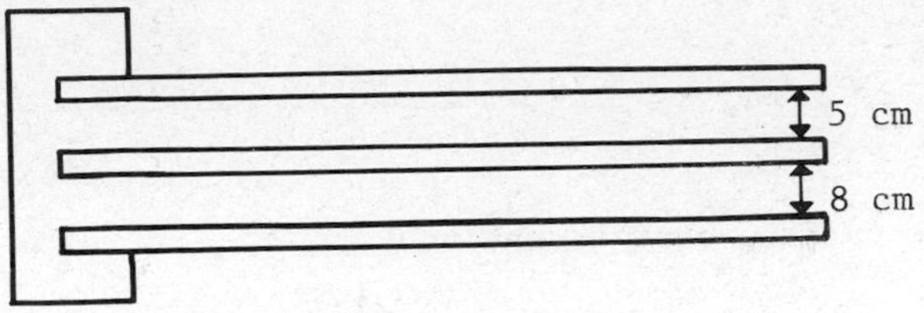

Solution: If we ignore edge effects, $E \sim -\Delta\phi/\ell$:

$$\Delta\phi \text{ (on both sides)} = \phi_2 - \phi_1$$

$$E_1 = 4\pi\sigma_1$$

$$= -\Delta\phi/8$$

$$E_2 = 4\pi\sigma_2$$

$$= -\Delta\phi/5$$

$$\frac{\sigma_2}{\sigma_1} = \frac{E_2}{E_1} = \frac{5}{8}$$

$$\sigma_1 + \sigma_2 = 10 \text{ esu/cm}^2 = \sigma_1 + \frac{5}{8}\sigma_1$$

$$\sigma_1 = 6.15 \text{ esu/cm}^2, \quad \sigma_2 = 3.85 \text{ esu/cm}^2$$

PROBLEM 3.26 Suppose that after the condition shown in the figure is attained, the body is made nonconducting again, leaving the charges frozen in place. After this, the sheets of positive and negative charge that produced the original electric field are removed. What does the residual electric field look like, both inside and outside the body?

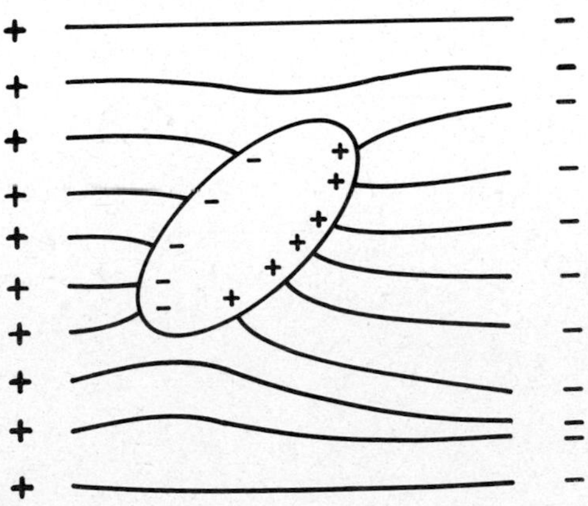

Solution: The field inside will be equal and opposite to the original field, since the two fields must add to zero. Outside, it will be, roughly, the field between equal and opposite charges as shown in the figure.

PROBLEM 3.27 Show that the average value of the potential ϕ taken over the surface of a spherical region of space that contains no charge is equal to the value of the potential at the center of the sphere, independent of the distribution of charge exterior to the sphere.

Solution: Use Green's identity

$$\int_s (\phi\nabla\psi - \psi\nabla\phi) \cdot ds = \int_v (\phi\nabla^2\psi - \psi\nabla^2\phi)\, dv$$

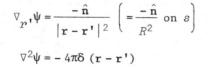

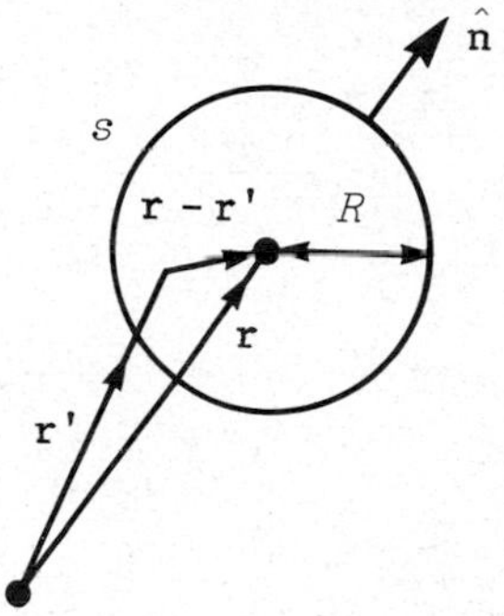

in the geometry shown, with ϕ = potential and

$$\psi = \frac{1}{|\mathbf{r} - \mathbf{r}'|} .$$

Use the results $\nabla^2\phi = -4\pi\rho$;

$$\nabla_{r'}\psi = \frac{-\hat{\mathbf{n}}}{|\mathbf{r} - \mathbf{r}'|^2} \left(= \frac{-\hat{\mathbf{n}}}{R^2} \text{ on } s \right)$$

$$\nabla^2\psi = -4\pi\delta\,(\mathbf{r} - \mathbf{r}')$$

Then,

$$-\frac{1}{R^2}\int_s \phi\, ds - \frac{1}{R}\int_s \nabla\phi \cdot ds$$

$$= \int_v \phi(r')[-4\pi\delta(\mathbf{r} - \mathbf{r}')]\, d^3\mathbf{r}' + \int_v \frac{1}{\mathbf{r} - \mathbf{r}'} 4\pi\rho(r')\, d^3\mathbf{r}'$$

where the first integrand on the right equals $-4\pi\phi(r)$ and in the second integral $\rho(r') = 0$ in v.

$$4\pi\phi(r) = \frac{1}{R^2}\int_s \phi\, ds + \frac{1}{R}\int_s \nabla\phi \cdot ds = \frac{1}{R^2}\int_s \phi\, ds + \frac{1}{R}\int_v \nabla^2\phi dv$$

where $\nabla^2\phi = -4\pi\rho(r') = 0$. Therefore,

$$4\pi\phi(r) = \frac{1}{R^2}\int_s \phi\, ds$$

$$\phi(r) = \frac{\int_s \phi\, ds}{4\pi R^2}$$

$\phi(r) = \langle\phi\rangle_{\text{average}}$ as required.

An alternate solution would be the following: for point charge and sphere, set up spherical coordinates as shown in the figure. Then, on the sphere,

$$d\phi = \frac{q}{r} = \frac{q}{(d^2 + R^2 - 2dR\cos\theta)^{\frac{1}{2}}}$$

Integrating,

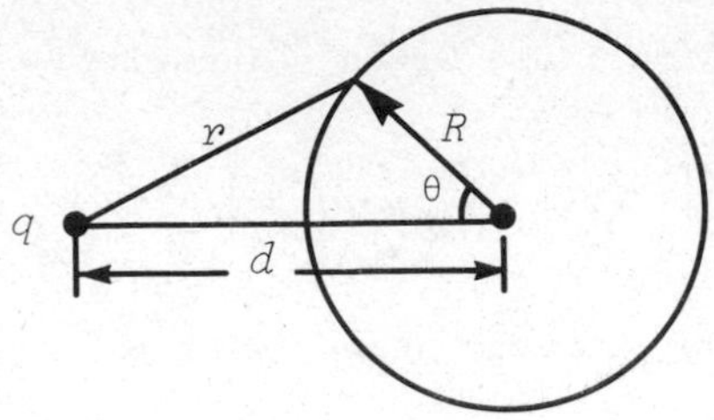

$$\phi_T = \int_0^{\pi}\int_0^{2\pi} \frac{qR^2 \sin\theta\, d\theta\, d\phi}{(d^2 + R^2 - 2dR\cos\theta)^{\frac{1}{2}}}$$

$$= 2\pi q R^2 \int_0^{\pi} \frac{\sin\theta\, d\theta}{(d^2 + R^2 - 2dR\cos\theta)^{\frac{1}{2}}}$$

$$= \frac{4\pi q R^2}{d}$$

Therefore,

$$\langle\phi\rangle = \frac{\phi_T}{4\pi R^2} = \frac{q}{d} = \text{potential due to } q \text{ at center of sphere.}$$

Therefore, for an arbitrary charge distribution outside the sphere, the result follows from the principle of superposition.

PROBLEM 3.28

(a) Find the Green's function of the first kind for the region outside a circular cylinder of radius a.

(b) Find the Green's function of the first kind for the space between two concentric circular cylinders of radii a and b $(a < b)$ and verify that this function approaches the correct limit as $b \to \infty$.

Hint: To verify the behavior in the limit, use the expansion formula

$$\ln|\mathbf{r} - \mathbf{r}'| = -\sum_{n=1}^{\infty} \frac{1}{n}\left(\frac{r_<}{r_>}\right)^n \cos n(\theta - \theta') + \ln r$$

$r_>$ means the larger of r and r' and $r_<$ is the smaller of the same two variables.

Solution:

(a)

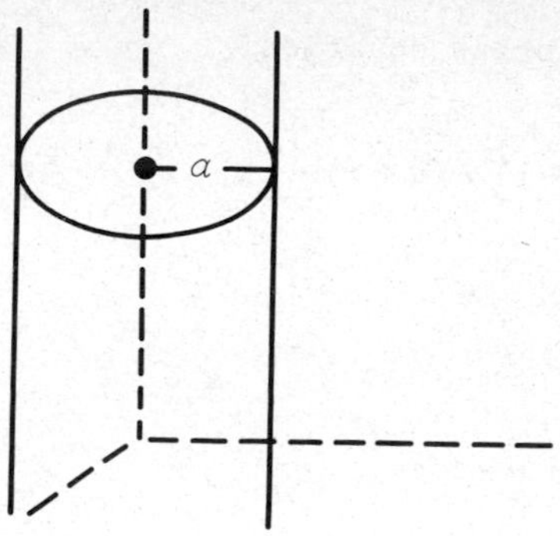

$$\nabla_r^2 G(\mathbf{r},\mathbf{r}') = -\frac{4\pi}{\rho}\delta(\rho-\rho')\delta(\theta-\theta')$$

We can write the θ delta function in terms of othonormal functions

$$\delta(\theta-\theta') = \frac{1}{2\pi}\sum_{m=-\infty}^{\infty} e^{im(\theta-\theta')}$$

Expanding the Green's function in a similar fashion,

$$G(\mathbf{r},\mathbf{r}') = \frac{1}{2\pi}\sum_{m=-\infty}^{\infty} e^{im(\theta-\theta')} g_m(\rho,\rho')$$

Substituting into the first equation,

$$\left(\frac{\partial^2}{\partial\rho^2}+\frac{1}{\rho}\frac{\partial}{\partial\rho}+\frac{1}{\rho^2}\frac{\partial^2}{\partial\theta^2}\right)\left(\frac{1}{2\pi}\sum_{m=-\infty}^{\infty} e^{im(\theta-\theta')} g_m(\rho,\rho')\right)$$

$$= -\frac{4\pi}{\rho}\delta(\rho-\rho')\frac{1}{2\pi}\sum_{m=-\infty}^{\infty} e^{im(\theta-\theta')}$$

$$\frac{1}{2\pi}\sum_{m=-\infty}^{\infty} e^{im(\theta-\theta')}\left(\frac{d^2g_m}{d\rho^2}+\frac{1}{\rho}\frac{dg_m}{d\rho}-\frac{m^2}{\rho^2}g_m\right) = -\frac{4\pi}{\rho}\delta(\rho-\rho')\frac{1}{2\pi}\sum_{m=-\infty}^{\infty} e^{im(\theta-\theta')}$$

For $\rho \neq \rho'$

$$\frac{d^2g_m}{d\rho^2}+\frac{1}{\rho}\frac{dg_m}{d\rho}-\frac{m^2}{\rho^2}g_m = 0$$

Assume g_m is of the form $g_m = \sum_{n=0}^{\infty} a_n\rho^{n+r}$, then

$$\frac{dg_m}{d\rho} = \sum_{n=0}^{\infty} a_n(n+r)\rho^{n+r-1}$$

$$\frac{d^2g_m}{d\rho^2} = \sum_{n=0}^{\infty} a_n(n+r)(n+r-1)\rho^{n+r-2}$$

Plugging into the differential equation,

$$\sum_{n=0}^{\infty} a_n[(n+r)(n-r-1)+(n+r)-m^2]\rho^{n+r-2}=0$$

For $n=0$,

$$a_0[r(r-1)+r-m^2]=0$$

$$a_0 \neq 0 \Rightarrow r^2-m^2=0 \text{ or } r=\pm m$$

for $n \neq 0$

$$a_k[(k+m)(k+m-1)+(k+m)-m^2]=0$$

Since the term in brackets never equals 0, a_k must be zero for all k except zero. Therefore, $a_0\rho^m$ and $a_0\rho^{-m}$ are solutions, and

$$g_m = A_m\rho^m + B_m\rho^{-m} \quad \text{for } \rho<\rho'$$

$$g_m = A'_m\rho^m + B'_m\rho^{-m} \quad \text{for } \rho>\rho'$$

$$g_m = 0 \quad \text{for } \rho=a \text{ and } \rho=\infty$$

$$g_m = A_m a^m + B_m a^{-m} = 0$$

$$g_m = A_m\left(\rho^m - \frac{a^{2m}}{\rho^m}\right) \quad \text{for } \rho<\rho'$$

$$g_m = B'_m\rho^{-m} \quad \text{for } \rho>\rho'$$

The symmetry in ρ and ρ' requires that A_m and B'_m be such that

$$g_m(\rho,\rho') = c_m\left(\rho_<^m - \frac{a^{2m}}{\rho_<^m}\right)\frac{1}{\rho_>^m} \tag{1}$$

If we integrate the equation

$$\frac{d^2g_m}{d\rho^2}+\frac{1}{\rho}\frac{dg_m}{d\rho}-\frac{m^2}{\rho^2}=-\frac{4\pi}{\rho}\delta(\rho-\rho')$$

from $\rho=\rho'-\varepsilon$ to $\rho=\rho'+\varepsilon$, where ε is very small, we obtain

$$\left(\frac{dg_m}{d\rho}\right)_{\rho+\varepsilon} - \left(\frac{dg_m}{d\rho}\right)_{\rho-\varepsilon} = -\frac{4\pi}{\rho'}$$

For $\rho=\rho'+\varepsilon$, $\rho_>=\rho$ and $\rho_<=\rho'$

$$\left(\frac{dg_m}{d\rho}\right)_{\rho'+\varepsilon} \simeq C_m\left(\rho'^m - \frac{a^{2m}}{\rho'^m}\right)\left(-m\,\frac{1}{\rho'^{m+1}}\right)$$

For $\rho = \rho' - \varepsilon$, $\rho_> = \rho'$, and $\rho_< = \rho$

$$\left(\frac{dg_m}{d\rho}\right)_{\rho'-\varepsilon} \simeq C_m\left(m\rho'^{m-1} + m\,\frac{a^{2m}}{\rho'^{m+1}}\right)\left(\frac{1}{\rho'^m}\right)$$

Multiply through by ρ':

$$mC_m\left[\left(-1 + \frac{a^{2m}}{\rho'^{2m}}\right) - \left(1 + \frac{a^{2m}}{\rho'^{2m}}\right)\right] = -4\pi$$

$$C_m = \frac{2\pi}{m}$$

$$G(\mathbf{r},\mathbf{r}') = \frac{1}{2\pi}\sum_{m=-\infty}^{\infty} e^{im(\theta-\theta')} g_m(\rho,\rho')$$

$$G(\mathbf{r},\mathbf{r}') = \sum_{m=-\infty}^{\infty} e^{im(\theta-\theta')}\,\frac{1}{m}\left(\rho_<^m - \frac{a^{2m}}{\rho_<^m}\right)\left(\frac{1}{\rho_>^m}\right)$$

Written in terms of real functions,

$$G(\mathbf{r},\mathbf{r}') = \frac{g_0(\rho,\rho')}{2\pi} + \frac{1}{\pi}\sum_{m=1}^{\infty} \cos m(\theta-\theta') g_m(\rho,\rho')$$

For $m=0$ since θ and θ' are not distinguishable $\rho'=0$ and since $\rho' > a$, $a \to 0$ even faster.

$$g_0(\rho,0) = \lim_{m\to 0}\left[2\pi\,\frac{d}{dm}\left(\frac{1}{\rho_>^m}\right)\right]\bigg/\left[\frac{d}{dm}(m)\right], \text{ L'Hôpital's rule}$$

$$g_0(\rho,0) = \lim_{m\to 0}\left[2\pi\left(\frac{1}{\rho_>}\right)^m \ln\left(\frac{1}{\rho_>}\right)\right]\bigg/(1)$$

$$= 2\pi\,\ln\left(\frac{1}{\rho_>}\right)$$

$$G(\mathbf{r},\mathbf{r}') = \ln\left(\frac{1}{\rho_>}\right) + 2\sum_{m=1}^{\infty} \cos m(\theta-\theta')\frac{1}{m}\left(\rho_<^m - \frac{a^{2m}}{\rho_<^m}\right)\left(\frac{1}{\rho_>^m}\right)$$

To get this in closed form, one may use the expansion for the Green's function for two-dimensional polar coordinates:

$$\ln\left(\frac{1}{\sqrt{\rho^2+\rho'^2-2\rho\rho'\cos(\phi-\phi')}}\right) = \ln\left(\frac{1}{\rho_>}\right) + \sum_{m=1}^{\infty}\frac{1}{m}\left(\frac{\rho_<}{\rho_>}\right)^m \cos[m(\phi-\phi')]$$

(b)

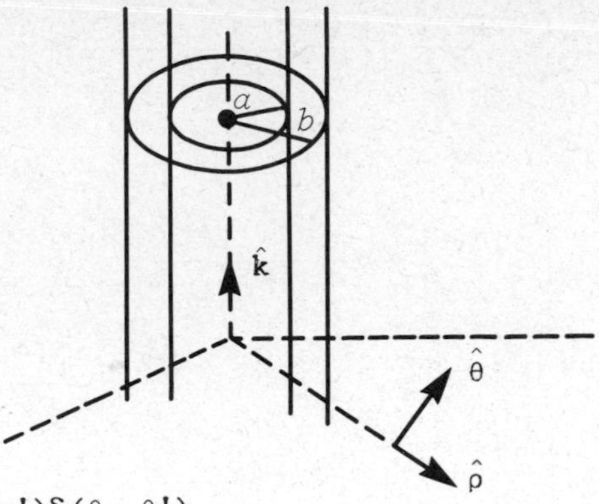

$$\nabla_r^2 G(\mathbf{r},\mathbf{r}') = -\frac{4\pi}{\rho}\delta(\rho-\rho')\delta(\theta-\theta')$$

We can write the θ delta function in terms of orthonomal functions:

$$\delta(\theta-\theta') = \frac{1}{2\pi}\sum_{m=-\infty}^{\infty} e^{im(\theta-\theta')}$$

Expand the Green's function in a similar fashion:

$$G(\mathbf{r},\mathbf{r}') = \frac{1}{2\pi}\sum_{m=-\infty}^{\infty} e^{im(\theta-\theta')} g_m(\rho,\rho') + C + d\ln\rho$$

Substituting into the first equation,

$$\left(\frac{\partial^2}{\partial\rho^2} + \frac{1}{\rho}\frac{\partial}{\partial\rho} + \frac{1}{\rho^2}\frac{\partial^2}{\partial\theta^2}\right)\left[\frac{1}{2\pi}\sum_{m=-\infty}^{\infty} e^{im(\theta-\theta')} g_m(\rho,\rho')\right]$$

$$= -\frac{4\pi}{\rho}\delta(\rho-\rho')\,\frac{1}{2\pi}\sum_{m=-\infty}^{\infty} e^{im(\theta-\theta')}$$

$$\frac{1}{2\pi}\sum_{m=-\infty}^{\infty} e^{im(\theta-\theta')}\left(\frac{\partial^2 g_m}{\partial\rho^2} + \frac{1}{\rho}\frac{\partial g_m}{\partial\rho} - \frac{m^2}{\rho^2} g_m\right) = -\frac{4\pi}{\rho}\delta(\rho-\rho')\,\frac{1}{2\pi}\sum_{m=-\infty}^{\infty} e^{im(\theta-\theta')}$$

Again,

$$g_m = A_m\rho^m + B_m\rho^{-m}, \quad \text{for } \rho < \rho'$$

$$g_m = A'_m\rho^m + B'_m\rho^{-m}, \quad \text{for } \rho > \rho'$$

$$g_m = 0, \quad \text{for } \rho = a \text{ and } \rho = b$$

$$g_m = 0 = A_m a^m + B_m a^{-m}$$

$$g_m = A_m\left(\rho^m - \frac{a^{2m}}{\rho^m}\right), \quad \text{for } \rho < \rho'$$

$$g_m = 0 = A'_m\, b^m + B'_m\, b^{-m}$$

$$g_m = B'_m \left(\frac{1}{\rho^m} - \frac{\rho^m}{b^{2m}}\right), \quad \text{for } \rho > \rho'$$

The symmetry in ρ and ρ' requires that A_m and B'_m be such that

$$g_m(\rho,\rho') = C_m\left(\rho_<^m - \frac{a^{2m}}{\rho_<^m}\right)\left(\frac{1}{\rho_>^m} - \frac{\rho_>^m}{b^{2m}}\right)$$

If we integrate the equation

$$\frac{d^2 g_m}{d\rho^2} + \frac{1}{\rho}\frac{dg_m}{d\rho} - \frac{m^2}{\rho^2}\, g_m = -\frac{4\pi}{\rho}\,\delta(\rho - \rho')$$

From $\rho = \rho' - \varepsilon$ to $\rho = \rho' + \varepsilon$, where ε is very small, we obtain

$$\left(\frac{d}{d\rho}\, g_m\right)_{\rho' + \varepsilon} - \left(\frac{d}{d\rho}\, g_m\right)_{\rho' - \varepsilon} = -\frac{4\pi}{\rho'}$$

For $\rho = \rho' + \varepsilon$, $\rho_> = \rho$, and $\rho_< = \rho'$,

$$\left(\frac{d}{d\rho}\, g_m\right)_{\rho' + \varepsilon} \simeq C_m\left(\rho'^m - \frac{a^{2m}}{\rho'^m}\right)\left(-m\,\frac{1}{\rho'^{m+1}} - m\,\frac{\rho'^{m-1}}{b^{2m}}\right)$$

For $\rho = \rho' - \varepsilon$, $\rho_> = \rho'$, and $\rho_< = \rho$,

$$\left(\frac{d}{d\rho}\, g_m\right)_{\rho' - \varepsilon} \simeq C_m\left(m\rho'^{m-1} + m\,\frac{a^{2m}}{\rho'^{m+1}}\right)\left(\frac{1}{\rho'^m} - \frac{\rho'^m}{b^{2m}}\right)$$

Multiply through by ρ':

$$mC_m\left[\left(-1 - \frac{\rho'^{2m}}{b^{2m}} + \frac{a^{2m}}{\rho'^{2m}} + \frac{a^{2m}}{b^{2m}}\right) - \left(1 - \frac{\rho'^{2m}}{b^{2m}} + \frac{a^{2m}}{\rho'^{2m}} - \frac{a^{2m}}{b^{2m}}\right)\right] = -4\pi$$

$$mC_m\,[2 - 2(a/b)^{2m}] = 4\pi$$

$$C_m = \frac{2\pi}{m}\,\frac{1}{1 - (a/b)^{2m}}$$

$$G(\mathbf{r},\mathbf{r}') = \frac{1}{2\pi}\sum_{m=-\infty}^{\infty} e^{im(\theta - \theta')}\, g_m(\rho,\rho')$$

$$G(\mathbf{r},\mathbf{r}') = \frac{g_0(\rho,\rho')}{2\pi} + \frac{1}{\pi}\sum_{m=1}^{\infty} \cos m(\theta-\theta') g_m(\rho,\rho')$$

and

$$g_0 = 2\pi \ln(1/\rho_>)$$

$$G(\mathbf{r},\mathbf{r}') = \ln\left(\frac{1}{\rho_>}\right) + 2\sum_{m=1}^{\infty} \cos m(\theta-\theta') \frac{1}{m} \frac{1}{1-(a/b)^{2m}}$$

$$\times \left(\rho_<^m - \frac{a^{2m}}{\rho_<^m}\right)\left(\frac{1}{\rho_>^m} - \frac{\rho_>^m}{b^{2m}}\right)$$

In the limit as $b \to \infty$

$$G(\mathbf{r},\mathbf{r}') = \ln\left(\frac{1}{\rho_>}\right) + 2\sum_{m=1}^{\infty} \cos m(\theta-\theta') \frac{1}{m}\left(\rho_<^m - \frac{a^{2m}}{\rho_<^m}\right)\left(\frac{1}{\rho_>^m}\right)$$

PROBLEM 3.29 Using the four-dimensional form of Green's theorem, solve the inhomogeneous wave equations

$$\Box^2 A_\mu = \frac{-4\pi}{c} j_\mu$$

(a) Show that for a localized charge-current distribution the four-vector potential is

$$A_\mu(x) = \frac{1}{\pi c}\int \frac{j_\mu(\xi)\, d^4\xi}{R^2}$$

where $R^2 = (x-\xi)\cdot(x-\xi)$, x means (x_1,x_2,x_3,x_4) and $d^4\xi = d\xi_1\, d\xi_2\, d\xi_3\, d\xi_4$.

(b) From the definitions of the field strengths $F_{\mu\upsilon}$ show that

$$F_{\mu\upsilon} = \frac{2}{\pi c}\int \frac{(j\times R)_{\mu\upsilon}}{R^4}\, d^4\xi$$

where $(j\times R)_{\mu\upsilon} = j_\mu R_\upsilon - j_\upsilon R_\mu$

(c) The three-dimensional formulation of the radiation problem leads to the retarded solution

$$A_\mu(\mathbf{x},t) = \frac{1}{c}\int \left(\frac{j_\mu(\xi,t')}{r}\right)_{t'=t-r/c} d^4\xi,$$

where $r = |x-\xi|$. Show the connection between this retarded solution and the solution of the first parts of this problem by explicitly performing the integration over $d\xi_4$.

Solution:

(a) First solving $\square^2 A_\mu = -(4\pi/c) j_\mu$, use $\square^2 G(x,x') = \delta^4(x-x')$. Let $z = x - x'$, then $G(x-x') = G(z)$ and we have $\square^2 G(z) = \delta^4(z)$. Let $G(z)$ = Fourier transform of Green's function, so

$$G(z) = \frac{1}{(2\pi)^4}\int d^4k\, G(k) e^{-i\mathbf{k}\cdot\mathbf{z}},$$

$$\delta^4(z) = \frac{1}{(2\pi)^4}\int d^4k\, e^{-i\mathbf{k}\cdot\mathbf{z}},$$

therefore,

$$G(k) = -\frac{1}{\mathbf{k}\cdot\mathbf{k}}$$

and

$$G(z) = -\frac{1}{(2\pi)^4}\int \frac{d^4k\, e^{-i\mathbf{k}\cdot\mathbf{z}}}{\mathbf{k}\cdot\mathbf{k}}$$

Now $\mathbf{k}\cdot\mathbf{z} = kz\cos\alpha$, $d^4k = k^3\,dk\,dr$, $n-d$ (of radius k) sphere has volume $v_n(k) = C_n k^n$ and surface area $nC_n k^{n-1}$, $\int d\Omega = nC_n = 2\pi^2$.

$$\int d\Omega = \int \sin^2\alpha\, d\alpha\, \frac{2\pi^2}{\pi/2} = 4\pi\int \sin^2\alpha\, d\alpha$$

and

$$G(z) = -\frac{1}{(2\pi)^4}\int \frac{d^4k\, e^{i\mathbf{k}\cdot\mathbf{z}}}{\mathbf{k}\cdot\mathbf{k}} = -\frac{4\pi}{(2\pi)^4}\int \frac{dk\, k^3 e^{ikz\cos\alpha}\sin^2\alpha\, d\alpha}{k^2}$$

$$\int dk\, k e^{-ik\alpha} = \frac{1}{(i\alpha)^2} = -\frac{1}{\alpha^2}$$

So

$$G(z) = \frac{1}{z^2 4\pi^3}\int_0^\pi \tan^2\alpha\, d\alpha = \frac{1}{2^2(4\pi^3)}(\tan\alpha - \alpha)\Big|_0^\pi = \frac{-1}{4\pi^2 z^2}$$

$$A_\mu(x) = -\frac{4\pi}{c}\int j_\mu(x') G(x,x')\, dx'$$

$$A_\mu(x) = \frac{1}{\pi c}\int \frac{j_\mu(x')\, d^4x'}{(x-x')^2}$$

(b)

$$F_{\mu\nu} = \frac{\partial A_\nu}{\partial x_\mu} - \frac{\partial A_\mu}{\partial x_\nu} = \frac{1}{\pi c}\frac{\partial}{\partial x_\mu}\int \frac{j_\nu(x')\,d^4x'}{\left|\sum_{\beta=0}^{3}(x_\beta - x_\beta')^2\right|} - \frac{1}{\pi c}\frac{\partial}{\partial x_\nu}\int \frac{j_\mu(x')\,d^4x'}{\left|\sum_\beta (x_\beta - x_\beta')^2\right|}$$

$$= \frac{1}{\pi c}\int \frac{j_\nu(x')[-2(x_\mu - x_\mu')]\,d^4x'}{\left|\sum_\beta (x_\beta - x_\beta')^2\right|^2} - \frac{1}{\pi c}\frac{j_\mu(x')[-2(x_\nu - x_\nu')]\,d^4x'}{\left|\sum_\beta (x_\beta - x_\beta')^2\right|^2}$$

$$= \frac{-2}{\pi c}\int \frac{j_\nu(x')R_\mu\,d^4x'}{R^4} + \frac{2}{\pi c}\int \frac{j_\mu R_\nu\,d^4x'}{R^4} = \frac{2}{\pi c}\int \frac{j_\mu R_\nu - j_\nu R_\mu\,d^4x'}{R^4}$$

$$F_{\mu\nu} = \frac{2}{\pi c}\int \frac{(j\times R)_{\mu\nu}\,d^4x'}{R^4}$$

(c) From above,

$$A_\mu(x) = \frac{1}{\pi c}\int \frac{j_\mu(x')\,d^4x'}{|x - x'|^2}$$

$$= \frac{1}{\pi c}\int d^3x' \int \frac{j_\mu(\mathbf{x}', x_0')\,dx_0}{(x - x_0')^2 - (\mathbf{x} - \mathbf{x}')^2} = \frac{1}{\pi c}\int d^3x' \int \frac{j_\mu(\mathbf{x}', z_0 + x_0)\,dz_0}{(z_0 - |\bar{z}|)(z_0 + |\bar{z}|)}$$

Letting $x - x_0 \to z_0$; $x_0' \to z_0 + x_0$; $\mathbf{x} - \mathbf{x}' \to \bar{z}$

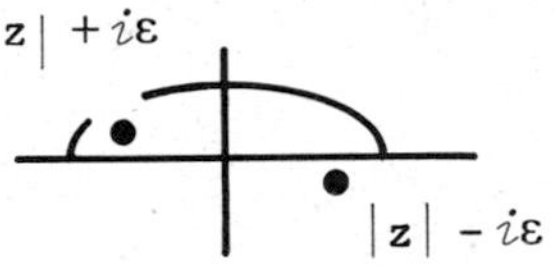

$$A_\mu(x) = \frac{1}{\pi c}\int d^3x'\; 2\pi i\left(\frac{j_\mu(\mathbf{x}', -|\mathbf{x} - \mathbf{x}'| + x_0)}{2|\mathbf{x} - \mathbf{x}'|}\right) = \frac{1}{c}\int\left(\frac{d^3x'\; i j_\mu(\mathbf{x}', x_0 - |\mathbf{x} - \mathbf{x}'|)}{|\mathbf{x} - \mathbf{x}'|}\right)$$

let $x_0 \to ict$ and $|\mathbf{x} - \mathbf{x}'| \to R$, therefore,

$$A_\mu(\mathbf{x}, t) = \frac{1}{c}\int \frac{d^3x'\; j_\mu(\mathbf{x}', t')}{R}\Bigg|_{t' = t - R/c}$$

PROBLEM 3.30 Calculate the dipole moment for the charge distribution shown in the figure:

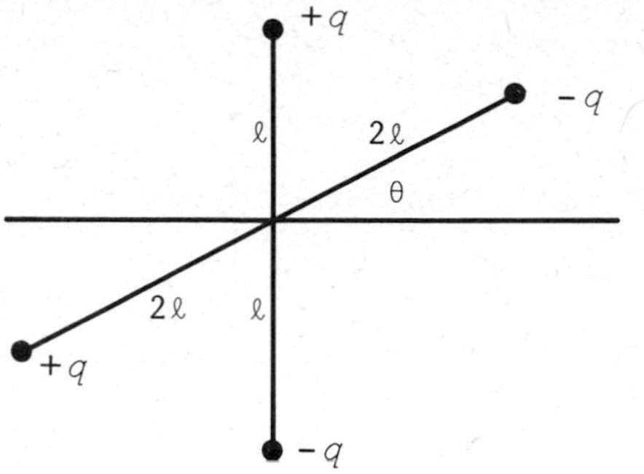

Solution:

$$\mathbf{p} = \Sigma_\alpha q_\alpha \mathbf{r}_\alpha = q(0,\ell) - q(0,-\ell) - q(2\ell\cos\theta, 2\ell\sin\theta) + q(-2\ell\cos\theta, -2\ell\sin\theta)$$

$$= q(-4\ell\cos\theta, +2\ell - 4\ell\sin\theta) = -2\ell q(2\cos\theta, 2\sin\theta - 1)$$

$$\mathbf{p} = -2\ell q[2\cos\theta\,\hat{\mathbf{x}} + (2\sin\theta - 1)\hat{\mathbf{y}}]$$

$\hat{\mathbf{n}}$ is a unit vector making an angle θ with the x axis. We could have written immediately,

$$\mathbf{p} = q(2\ell\hat{\mathbf{y}} - 4\ell\hat{\mathbf{n}}) = q[2\ell\hat{\mathbf{y}} - 4\ell(\cos\theta\,\hat{\mathbf{x}} + \sin\theta\,\hat{\mathbf{y}})]$$

$$\mathbf{p} = -2\ell q[2\cos\theta\,\hat{\mathbf{x}} + (2\sin\theta - 1)\hat{\mathbf{y}}], \quad \text{as above}$$

PROBLEM 3.31 Calculate the quadrupole tensor for the charge distribution shown in the figure for the general angle θ. Find the values of θ for which the tensor becomes diagonal. (Choose the x_3 axis to coincide with the axis of the dipole whose moment is $2\ell q$.)

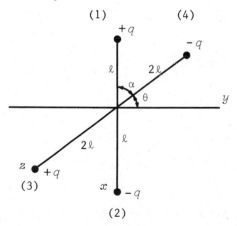

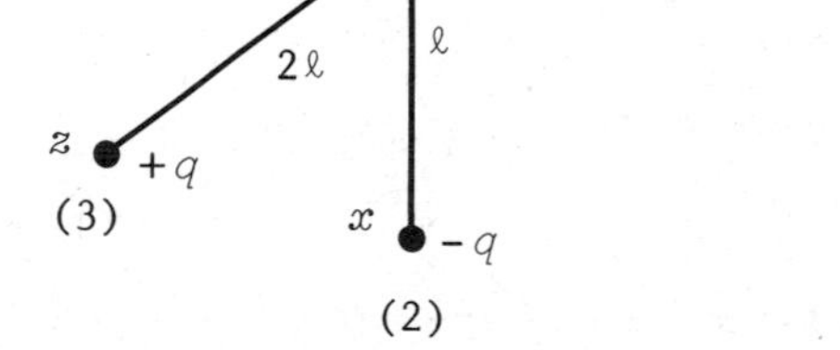

Solution:

$$q_1 = q, \quad r_1 = (0,0,\ell)$$

$$q_2 = -q, \quad r_2 = (0,0,-\ell)$$

$$q_3 = q, \quad r_3 = (0, -2\ell\sin\alpha, -2\ell\cos\alpha)$$

$$q_4 = -q, \quad r_4 = (0, 2\ell \sin \alpha, 2\ell \cos \alpha)$$

Note that we have an antisymmetric distribution of charge in the sense that for every charge q at **r**, we have a charge $-q$ at $-$**r**, therefore, the quadrupole moment tensor is identically zero, i.e., $Q_{ij} = 0$.

PROBLEM 3.32 The linear charge density on a ring of radius a is given by

$$\rho_1 = \frac{q}{a}(\cos\theta - \sin 2\theta)$$

Find the monopole, dipole, and quadrupole moments of the system and calculate the potential at an arbitrary point in space, accurate to terms in $1/r^3$.

Solution:

<u>Monopole</u>

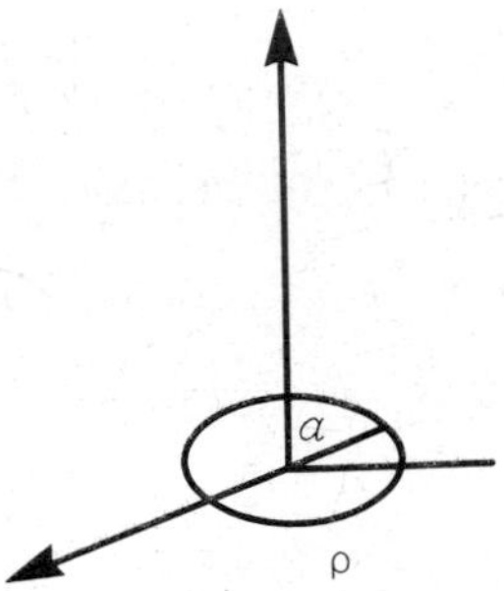

$$\rho_\ell = \frac{q}{a}(\cos\phi - \sin 2\phi)$$

$$Q = \int_0^{2\pi} a\,d\phi\,\frac{q}{a}(\cos\phi - 2\sin\phi\cos\phi)$$

$$= q\int_0^{2\pi}(\cos\phi - 2\sin\phi\cos\phi) = 0$$

($\cos\phi$ and $\sin 2\phi$ have period 2π)

$$\therefore\ Q = 0$$

<u>Dipole</u>

$$\mathbf{p} = \int_0^{2\pi} a\,d\phi(a\hat{\mathbf{e}}_r)\frac{q}{a}(\cos\phi - \sin 2\phi) = qa\int_0^{2\pi}\hat{\mathbf{e}}_r(\cos\phi - \sin 2\phi)$$

We have $\hat{\mathbf{e}}_r = \cos\phi\,\hat{\mathbf{x}} + \sin\phi\,\hat{\mathbf{y}}$. Therefore,

$$p_x = qa\int_0^{2\pi}(\cos^2\phi - \sin 2\phi\cos\phi)\,d\phi = qa\int_0^{2\pi}\left[\left(\frac{1+\cos 2\phi}{2}\right) - 2\sin\phi\cos^2\phi\right]d\phi$$

$$= qa\int_0^{2\pi}\tfrac{1}{2}\,d\phi = \pi qa$$

$$p_y = qa\int_0^{2\pi}d\phi(\cos\phi\sin\phi - 2\sin^2\phi\cos\phi) = 0$$

Therefore, dipole moment $= \mathbf{p} = \pi qa\hat{\mathbf{x}}$.

Quadrupole Moment

We clearly have $Q_{zx} = Q_{zy} = 0$

$$Q_{zz} = \int_0^{2\pi} (a\,d\phi)\,\frac{q}{a}\,(\cos\phi - 2\sin\phi\cos\phi)(-a^2)$$

$$= -qa^2 \int_0^{2\pi} d\phi(\cos\phi - 2\sin\phi\cos\phi) = 0$$

$$Q_{xy} = \int_0^{2\pi} (a\,d\phi)\,\frac{q}{a}\,(\cos\phi - \sin 2\phi)\,3xy$$

$$= \int_0^{2\pi} d\phi\, q(\cos\phi - \sin 2\phi)\,3a^2\cos\phi\sin\phi$$

$$= 3qa^2 \int_0^{2\pi} \left(\cos^2\phi\sin\phi - \frac{\sin^2 2\phi}{2}\right) d\phi$$

$$= 3qa^2 \int_0^{2\pi} \left[\cos^2\phi\sin\phi - \left(\frac{1-\cos 4\phi}{4}\right)\right]$$

$$= -3qa^2 \int_0^{2\pi} \tfrac{1}{4}\,d\phi = -\tfrac{3}{2}\pi qa^2$$

$$\therefore\ Q_{xy} = Q_{yx} = -\tfrac{3}{2}\pi qa^2$$

$$Q_{xx} = \int_0^{2\pi} (a\,d\phi)\,\frac{q}{a}\,(\cos\phi - 2\sin\phi\cos\phi)(3x^2 - a^2)$$

$$= \int_0^{2\pi} d\phi\, q(\cos\phi - 2\sin\phi\cos\phi)(3a^2\cos^2\phi - a^2)$$

$$= 3qa^2 \int_0^{2\pi} (\cos^3\phi - 2\sin\phi\cos^3\phi)\,d\phi$$

$$- qa^2 \int_0^{2\pi} (\cos\phi - 2\sin\phi\cos\phi)\,d\phi = 0$$

$$Q_{yy} = \int_0^{2\pi} d\phi\, q(\cos\phi - 2\sin\phi\cos\phi)(3a^2\sin^2\phi - a^2) = 0,$$

$$Q_{xx} = 0 \text{ and } Q_{yy} = 0$$

Therefore,

$$Q_{ij} = \begin{vmatrix} 0 & -{}^3/_2\, qa^2\pi & 0 \\ -{}^3/_2\, qa^2\pi & 0 & 0 \\ 0 & 0 & 0 \end{vmatrix}$$

Potential

Up to terms of order r^{-3}, we have

$$\Phi = \Phi^{(2)} + \Phi^{(4)}$$

$$\Phi^{(2)} = \frac{\mathbf{p}\cdot\hat{\mathbf{e}}_r}{r^2} = \frac{1}{r^2}\,(\pi qa\hat{\mathbf{x}}\cdot\hat{\mathbf{e}}_r) = \frac{\pi qa}{r^2}\sin\theta\cos\phi$$

$$\Phi^{(4)} = \frac{1}{6}\sum_{ij} Q_{ij}\frac{\partial^2}{\partial x_i\,\partial x_j}\left(\frac{1}{r}\right) = \frac{1}{3}Q_{xy}\frac{\partial^2}{\partial x\,\partial y}\left(\frac{1}{r}\right) = \frac{-1}{3}\,\frac{3}{2}\,\pi qa^2\,\frac{\partial^2}{\partial x\,\partial y}\left(\frac{1}{r}\right)$$

$$= -\tfrac{1}{2}\pi qa^2\,\frac{\partial^2}{\partial x\,\partial y}$$

Now,

$$\frac{\partial^2}{\partial x\,\partial y}\left(\frac{1}{r}\right) = \frac{\partial}{\partial x}\left[\nabla\left(\frac{1}{r}\right)\right]_y = \frac{\partial}{\partial x}\left(\frac{-1}{r^2}\,\hat{\mathbf{e}}_r\right)_y = \frac{\partial}{\partial x}\left(\frac{-\sin\theta\sin\phi}{r^2}\right)$$

$$= -\frac{\partial}{\partial x}\left(\frac{1}{r^2}\,\frac{y}{r}\right) = -y\frac{\partial}{\partial x}\left(\frac{1}{r^3}\right) = -y\left[\nabla\left(\frac{1}{r^3}\right)\right]_x$$

$$= -y\left(\frac{-3}{r^4}\,\hat{\mathbf{e}}_r\right)_x = 3r\sin\theta\sin\phi\,\frac{\sin\theta\cos\phi}{r^4} = \frac{3}{2}\,\frac{\sin^2\theta\sin 2\phi}{r^3}$$

$$\Phi^{(4)} = -\tfrac{1}{2}\pi qa^2\,\frac{3}{2}\,\frac{\sin^2\theta\sin 2\phi}{r^3} = \frac{-3}{4}\,qa^2\pi\,\frac{\sin^2\theta\sin 2\phi}{r^3}$$

$$\Phi = \frac{\pi qa}{r^2}\sin\theta\left(\cos\phi - \frac{3}{4}\,\frac{a}{r}\sin\theta\sin 2\phi\right)$$

PROBLEM 3.33 Use Gauss' law to calculate the electric field E and the potential Φ due to an infinitely long cylindrical conductor of radius a which carries a charge density per unit length ρ_ℓ. Can the potential be referred to the value at infinity in this case? How might one define the "zero" of potential?

Solution:

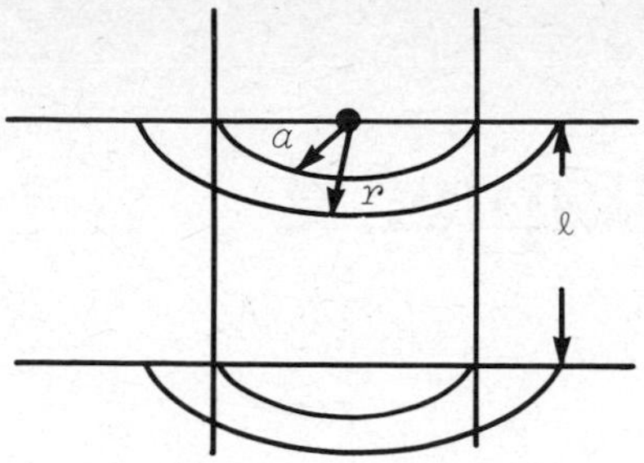

Consider length ℓ of conductor. Use Gauss' law:

$r > a, \quad E(2\pi r\ell) = 4\pi\, \rho_\ell \ell, \quad E = \frac{2\rho_\ell}{r}$

$r < a, \quad E(2\pi r\ell) = 4\pi(\ell\rho_\ell)\,\frac{\pi r^2}{\pi a^2}, \quad E = \frac{2\rho_\ell}{a^2}\, r$ } for uniform distribution of change

$r < a, \quad E = 0$ (conductor)

$E = 0$ for conductor, $r < a$

$E = \frac{2\rho_\ell r}{a^2}\,\hat{\mathbf{e}}_r, \quad r < a$

$E = \frac{2\rho_\ell}{r}\,\hat{\mathbf{e}}_r, \quad r > a$

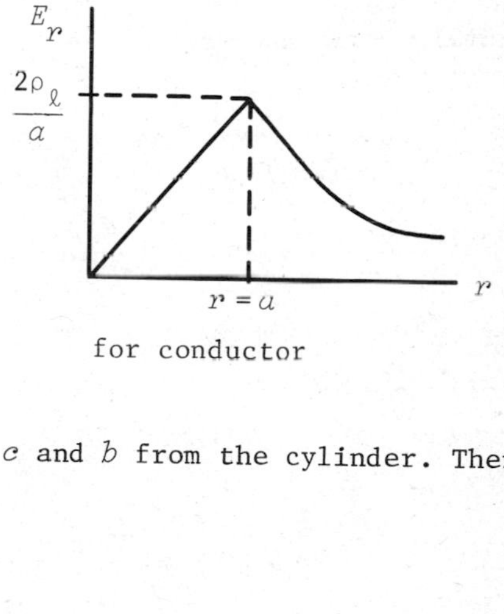

for conductor

Consider two points C and B at distances c and b from the cylinder. Then

$$\Phi = \Phi_c - \Phi_a = -\int_a^c E\,dr = -\,2\rho_\ell \ln(c/a).$$

Potential cannot meaningfully be referred to the value at infinity, for potential at c, of finite distance from cylinder, would go to infinity if we let $a \to \infty$ in above expression. Arbitrarily set $\phi(r_0) = 0$, for some finite r_0.

Chapter 4

PROBLEM 4.1 The pi meson (or pion) is a particle that is found in all three charged states, i.e., there are positive, negative, and neutral pi mesons. The mass (times c^2) of charged pions is 139.6 MeV, while the mass of neutral pi mesons is 135.0 MeV. In one model of the pion, the mass difference is assumed to be caused only by the elctrostatic energy. If one further assumes that pions are represented as spheres and that the charge of charged pions is uniformly distributed throughout this sphere, it is possible to calculate the "radius" of the pion. Under these assumptions calculate the radius of the pion. Is your result compatible with other estimates of nuclear dimensions?

Solution: The energy associated with the field is 4.6 MeV.

$$4.6\ \text{MeV} = (4.6 \times 10^6\ \text{eV})(1.6 \times 10^{-19}\ \text{J/eV}) = 7.36 \times 10^{-13}\ \text{J}$$

The total charge on the pion is 1.6×10^{-19} C.

$$\rho = \frac{q}{(4/3)\pi R^3} = \frac{3q}{4\pi R^3}$$

Inside the pion

$$\oint \mathbf{E} \cdot \hat{\mathbf{n}}\, da = \frac{1}{\varepsilon_0} \int \rho\, dv$$

$$E(4\pi r^2) = \frac{1}{\varepsilon_0} \rho (4/3) r^3$$

$$E = \frac{\rho r}{3\varepsilon_0} = \frac{3qr}{3\varepsilon_0 (4\pi R^3)}$$

for $0 < r < R$

$$E = \frac{qr}{4\pi\varepsilon_0 R^3}$$

for $r \geq R$

$$E = \frac{q}{4\pi\varepsilon_0 r^2}$$

$$U = \tfrac{1}{2}\varepsilon_0 \int_{\substack{\text{all}\\ \text{space}}} E^2\, dv = \tfrac{1}{2}\varepsilon_0 \int_0^\infty E^2\, 4\pi r^2\, dr = \frac{\varepsilon_0}{2} \int_0^R 4\pi E^2 r^2\, dr + \frac{\varepsilon_0}{2} \int_R^\infty 4\pi E^2 r^2\, dr$$

$$U = \frac{\varepsilon_0}{2} 4\pi \frac{q^2}{(4\pi)^2 \varepsilon_0^2 R^6} \int_0^R r^4 \, dr + \frac{\varepsilon_0}{2} 4\pi \frac{q^2}{(4\pi)^2 \varepsilon_0^2} \int_R^\infty \frac{r^2}{r^4} dr$$

$$U = \tfrac{1}{2} \frac{q^2}{4\pi\varepsilon_0 R^6} \frac{R^5}{5} + \frac{q^2}{(2)(4\pi\varepsilon_0)} \frac{1}{R}$$

$$U = \tfrac{1}{2} \frac{q^2}{4\pi\varepsilon_0 R} \left(\frac{1}{5} + 1\right) = \frac{3}{5} \frac{q^2}{4\pi\varepsilon_0 R}$$

But $U = 7.36 \times 10^{-13}$ J, so

$$7.36 \times 10^{-13} R = (6 \times 10^{-1})(9 \times 10^9)(1.6 \times 10^{-19})^2$$

$$R = \frac{(5.4 \times 10^9)(2.56 \times 10^{-38})}{7.36 \times 10^{-13}}$$

$$R = 1.88 \times 10^{-16} \text{ m}$$

$$R = 1.88 \times 10^{-14} \text{ cm}$$

PROBLEM 4.2 Show that when a dipole of dipole moment **p** is placed in an electric field **E**, the electrostatic energy is given by

$$U = -\mathbf{p} \cdot \mathbf{E}$$

Calculate the torque exerted on the dipole by the field **E**. Do the calculation directly as well as by using the above energy relation. Is the energy the same if the dipole were formed from two charges placed sequentially in the field? If not, calculate the difference; if so, justify your reasoning physically.

Solution:

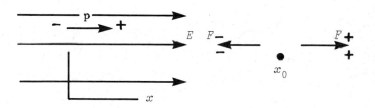

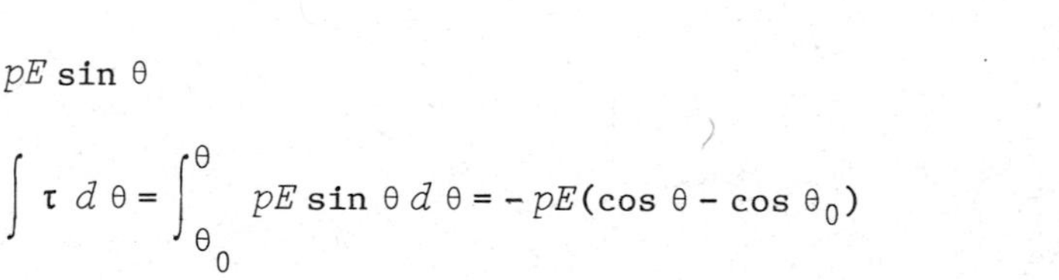

$$\tau = pE \sin\theta$$

$$\omega = \int \tau \, d\theta = \int_{\theta_0}^{\theta} pE \sin\theta \, d\theta = -pE(\cos\theta - \cos\theta_0)$$

$$\omega = U - U_0 = -pE\cos\theta + pE\cos\theta_0$$

If $U_0 = 0$ at $\theta_0 = \pi/2$,

$$U = -\mathbf{p} \cdot \mathbf{E}$$

Now suppose one starts with the charges at the same place and puts them in their final configuration.

$$\omega_+ = \int_{x_0}^{d} -\hat{\mathbf{i}}qE \cdot d\mathbf{r} + \int_{x_0}^{0} \hat{\mathbf{i}}qE \cdot d\mathbf{r} = -qE(d - x_0) + qE(0 - x_0)$$

$$\omega = -qEd = -pE$$

The reason this answer is the same is because the work done against $+q$ when $-q$ is moved is equal and opposite to the work done against $-q$ when $+q$ is moved, so this is cancelled.

PROBLEM 4.3 Do you agree with the statement that a nucleus that contains Z protons distributed more or less uniformly through the volume of a sphere of radius r has an electrostatic energy of about the value given by

$$U = \frac{3}{5} Z(Z-1) \frac{e^2}{r} \quad ?$$

Solution: One first needs to find the potential at a point inside a uniformly charged sphere.

Outside the sphere

$$r > R, \quad \mathbf{E} = \frac{1}{4\pi\varepsilon_0} \frac{Q}{r^2} \hat{\mathbf{r}}$$

Inside the sphere

$$r \leq R, \quad \mathbf{E} = \frac{1}{4\pi\varepsilon_0} \frac{Qr}{R^3} \hat{\mathbf{r}}$$

$$\phi(r) = -\int_{\infty}^{R} \mathbf{E} \cdot d\mathbf{r} - \int_{R}^{r} \mathbf{E} \cdot d\mathbf{r} = -\frac{Q}{4\pi\varepsilon_0} \left(\int_{\infty}^{R} \frac{dr}{r^2} + \int_{R}^{r} \frac{r\,dr}{R^3} \right)$$

$$\phi(r) = -\frac{Q}{4\pi\varepsilon_0} \left(-\frac{1}{r}\Big|_{\infty}^{R} + \frac{1}{R^3} \frac{r^2}{2}\Big|_{R}^{r} \right)$$

$$\phi(r) = -\frac{Q}{4\pi\varepsilon_0} \left(-\frac{1}{R} + \frac{r^2}{2R^3} - \frac{1}{2R} \right) = \frac{Q}{4\pi\varepsilon_0} \left(\frac{3}{2R} - \frac{r^2}{2R^3} \right)$$

The average potential inside is given by

$$\phi_{avg} = \frac{\int_0^R \phi(r) 4\pi r^2 dr}{4\pi R^3/3} = \frac{\frac{4\pi Q}{4\pi\varepsilon_0}\int_0^R \left(\frac{3}{2R} - \frac{r^2}{2R^3}\right) r^2\, dr}{4\pi R^3/3}$$

$$\phi_{avg} = \frac{3Q}{4\pi\varepsilon_0 R^3}\left(\frac{r^3}{2R}\bigg|_0^R - \frac{r^5}{10R^3}\bigg|_0^R\right)$$

$$\phi_{avg} = \frac{3Q}{4\pi\varepsilon_0 R^3}\left(\frac{R^2}{2} - \frac{R^2}{10}\right) = \frac{3Q}{4\pi\varepsilon_0 R}\left(\frac{2}{5}\right)$$

$$\phi_{avg} = \frac{6}{5}\frac{Q}{4\pi\varepsilon_0 R}$$

One now needs to put the z charges in the nucleus.

W	ϕ	# Proton
0	$\frac{6}{5}\frac{q_e}{4\pi\varepsilon_0 R}$	1
$\frac{6}{5}\frac{q_e^2}{4\pi\varepsilon_0 R}$	$2\left(\frac{6}{5}\frac{q_e}{4\pi\varepsilon_0 R}\right)$	2
$2\left(\frac{6}{5}\frac{q_e^2}{4\pi\varepsilon_0 R}\right)$	$3\left(\frac{6}{5}\frac{q_e}{4\pi\varepsilon_0 R}\right)$	3
$3\left(\frac{6}{5}\frac{q_e^2}{4\pi\varepsilon_0 R}\right)$	$4\left(\frac{6}{5}\frac{q_e}{4\pi\varepsilon_0 R}\right)$	4
$\vdots$	$\vdots$	$\vdots$
$(z-1)\left(\frac{6}{5}\frac{q_e^2}{4\pi\varepsilon_0 R}\right)$	$z\left(\frac{6}{5}\frac{q_e}{4\pi\varepsilon_0 R}\right)$	z

So

$$\omega = 0 + \frac{6}{5}\frac{q_e^2}{4\pi\varepsilon_0 R} + 2\left(\frac{6}{5}\frac{q_e^2}{4\pi\varepsilon_0 R}\right) + \cdots + (z-1)\left(\frac{6}{5}\frac{q_e^2}{4\pi\varepsilon_0 R}\right)$$

$$\omega = \frac{6}{5}\frac{q_e^2}{4\pi\varepsilon_0 R}\,(0 + 1 + 2 + \cdots + z - 1)$$

$$\omega = \frac{6}{5}\frac{q_e^{\,2}}{4\pi\varepsilon_0 R}\,[\tfrac{1}{2}(z)(z-1)]$$

$$\omega = \frac{3}{5}\,(z)(z-1)\,\frac{q_e^{\,2}}{4\pi\varepsilon_0 R}$$

Solving the problem another way: The average value of $1/r_{ij}$ for all pairs of points in a sphere is $1/r_{ij} = 6/5a$. If there are z charges in the nucleus,

$$U = \tfrac{1}{2}\sum_{i=1}^{z}\sum_{j=1}^{z}\frac{1}{4\pi\varepsilon_0}\frac{q_i q_j}{r_{ij}}$$

$$U = \tfrac{1}{2}\frac{1}{4\pi\varepsilon_0}\frac{1}{r_{ij}}\sum_{i=1}^{z}(q_i q_1 + q_i q_2 + q_i q_3 + \cdots + q_i q_z)$$

$$U = \tfrac{1}{2}\frac{1}{4\pi\varepsilon_0}\frac{1}{r_{ij}}\left\{q_1\sum_{\substack{i=1\\ i\neq 1}}^{z} q_i + q_2\sum_{\substack{i=1\\ i\neq 2}}^{z} q_i + q_3\sum_{\substack{i=1\\ i\neq 3}}^{z} q_i + \cdots + q_z\sum_{\substack{i=1\\ i\neq z}}^{z} q_i\right\}$$

Since all the q_i's are the same

$$U = \tfrac{1}{2}\frac{1}{4\pi\varepsilon_0}\frac{6}{5a}\,q_e^{\,2}[(z-1)+(z-1)+\cdots+(z-1)]$$

$$U = \tfrac{1}{2}\frac{1}{4\pi\varepsilon_0}\frac{6}{5a}\,q_e^{\,2}(z)(z-1)$$

$$U = \frac{3}{5}\,\frac{(z)(z-1)q_e^{\,2}}{4\pi\varepsilon_0 a}$$

PROBLEM 4.4 Let us assume that an electron is a uniformly charged, spherical particle of radius R. Assume further that the rest energy, mc^2 (where m is the mass of the electron and c is the velocity of light), is electrostatic in origin. By putting in appropriate numerical values for the charge and mass of the electron, determine is "classical radius" R.

Solution: One has, for a uniform charge distribution,

$$W = \frac{4\pi R^5 \rho_0^{\,2}}{15\varepsilon_0}, \quad \text{for } \rho_0 = \frac{e}{4\pi R^3/3}$$

$$W = \frac{4\pi R^5}{15\varepsilon_0}\;\frac{e^2}{(4\pi/3)^2 R^6} = \frac{3}{5\varepsilon_0}\frac{e^2}{4\pi R} = mc^2$$

$$R = \frac{3}{(5)\,4\pi\varepsilon_0}\,\frac{e^2}{mc^2} \simeq \frac{3}{5}\;9\times 10^9\;\frac{(1.6\times 10^{-19})^2}{(9\times 10^{-3})(3\times 10^8)^2}$$

$$= 1.69\times 10^{-19}\ \text{m}$$

By the way, the "usual" definition of the "classical electron radius" is $e^2/4\pi\varepsilon_0 mc^2 = 2.82\times 10^{-15}$ m.

PROBLEM 4.5 Given a spherical dielectric shell (inner radius a, outer radius b, dielectric constant K) and a point charge q, infinitely separated. Now let the point charge be placed at the center of the dielectric shell. Determine the change in energy of the system.

Solution: The tricky part of this problem is the infinite "self-energy" of the charge q. It must be subtracted from the total energy in this calculation. If we calculate, for a point charge at the origin (without a dielectric),

$$W_q = \tfrac{1}{2}\int \mathbf{E}\cdot\mathbf{D}\,dv = \frac{\varepsilon_0}{2}\int E^2\,dv = \frac{\varepsilon_0}{2}\iiint \left(\frac{q}{4\pi\varepsilon_0}\right)^2 \frac{1}{r^4}\,r^2\,dr\sin\theta\,d\theta\,d\phi$$

$$W_q = \text{"}\frac{q^2}{8\pi\varepsilon_0}\int_0^\infty \frac{dr}{r^2}\text{"}$$

which is infinite. We can still calculate the energy this way for the condition of this problem, but we must subtract the infinite W_q from the energy.

The fields are easily calculated (by Gauss' law, if you want);

$$\mathbf{E} = \frac{q\mathbf{r}}{4\pi\varepsilon_0 r^3},\quad \mathbf{D} = \frac{q\mathbf{r}}{4\pi r^3},\quad 0 \le r < r_a$$

$$\mathbf{E} = \frac{q\mathbf{r}}{4\pi\varepsilon_1 r^3},\quad \mathbf{D} = \frac{q\mathbf{r}}{4\pi r^3},\quad r_a \le r < r_b$$

$$\mathbf{E} = \frac{q\mathbf{r}}{4\pi\varepsilon_0 r^3},\quad \mathbf{D} = \frac{q\mathbf{r}}{4\pi r^3},\quad r_b \le r$$

Therefore,

$$W - W_q = \frac{q^2}{8\pi\varepsilon_0}\int_0^{r_a}\frac{dr}{r^2} + \frac{q^2}{8\pi\varepsilon_1}\int_{r_a}^{r_b}\frac{dr}{r^2} + \frac{q^2}{8\pi\varepsilon_0}\int_{r_b}^{\infty}\frac{dr}{r^2} - \frac{q^2}{8\pi\varepsilon_0}\int_0^{\infty}\frac{dr}{r^2}$$

$$= \left(\frac{q^2}{8\pi\varepsilon_1} - \frac{q^2}{8\pi\varepsilon_0}\right)\int_{r_a}^{r_b}\frac{dr}{r^2} = -\frac{q^2}{8\pi}\left(\frac{1}{\varepsilon_1} - \frac{1}{\varepsilon_0}\right)\left(\frac{1}{r_b} - \frac{1}{r_a}\right)$$

It is also possible to solve the potential problem for this case and use the formula

$$W - W_q = \frac{1}{2} \int_{\substack{\text{all}\\\text{space}}} \rho_F \phi(\mathbf{r})\, dv$$

with $\rho_F = q\delta(\mathbf{r})$ and $\phi(\mathbf{r})$ = (potential) - (potential due to charge itself). Finally, one can assume that the charge is distributed on a small conducting sphere or in a spherical volume and to subtract the contribution to the energy -- the energy required to assemble the charge distribution (this was, supposedly, accomplished at infinity before the charge was brought in).

PROBLEM 4.6 Determine the amount of energy required to assemble the configuration of point charges shown.

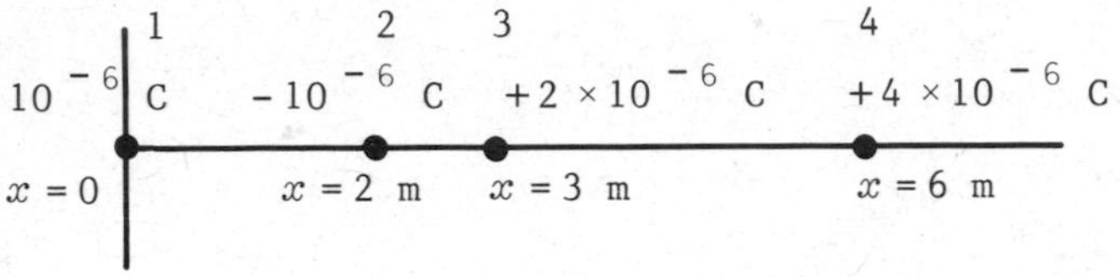

Solution:

$$U = (U_{12} + U_{13} + U_{14} + U_{21} + U_{23} + U_{24} + U_{31} + U_{32} + U_{34} + U_{41} + U_{42} + U_{43})$$

$$U = \frac{1}{2}\,\frac{1}{4\pi\varepsilon}\left(\frac{2q_1q_2}{r_{12}} + \frac{2q_1q_3}{r_{13}} + \frac{2q_1q_4}{r_{14}} + \frac{2q_2q_3}{r_{23}} + \frac{2q_2q_4}{r_{24}} + \frac{2q_3q_4}{r_{34}}\right)$$

$$U = \frac{1}{4\pi\varepsilon}\left(\frac{(10^{-6})(-10^{-6})}{2} + \frac{(10^{-6})(2\times 10^{-6})}{3} + \frac{(10^{-6})(4\times 10^{-6})}{4} + \frac{(-10^{-6})(2\times 10^{-6})}{1}\right.$$

$$\left. + \frac{(-10^{-6})(4\times 10^{-6})}{4} + \frac{(2\times 10^{-6})(4\times 10^{-6})}{6}\right)$$

The units are Joules.

$$U = \frac{1}{4\pi\varepsilon}\left(\frac{-10^{-12}}{2} + \frac{2\times 10^{-12}}{3} + \frac{4\times 10^{-12}}{6} - \frac{2\times 10^{-12}}{1} - \frac{4\times 10^{-12}}{4} + \frac{8\times 10^{-12}}{3}\right)$$

$$U = \frac{10^{-12}}{8\pi\varepsilon}\ \text{J}$$

PROBLEM 4.7 Two spherical conducting shells are charged as shown. Assume the shells are concentric and compute the amount of energy nessary to assemble this configuration neglecting the self energy of the charges themselves.

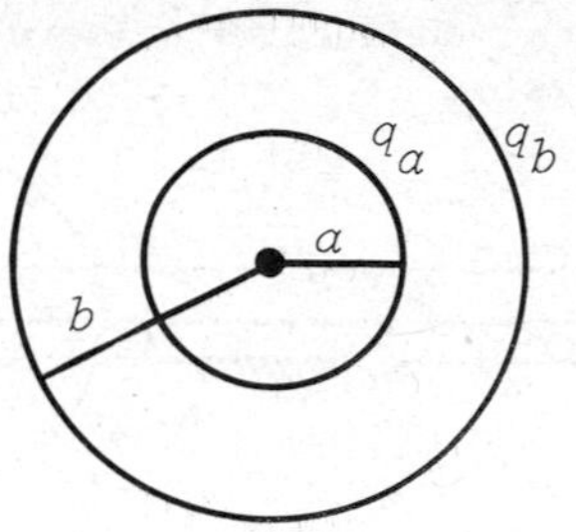

Solution:

$E = 0, \quad r < a$

$$E = \frac{q_a}{4\pi\varepsilon r^2}, \quad a < r < b$$

$$E = \frac{q_a + q_b}{4\pi\varepsilon r^2}, \quad r > b$$

$$U = \tfrac{1}{2}\varepsilon \int_{\substack{\text{all}\\ \text{space}}} \mathbf{E} \cdot \mathbf{E}\ dv = \tfrac{1}{2}\varepsilon \int_0^a 0\ dv + \tfrac{1}{2}\varepsilon \int_a^b \frac{q_a^2\ 4\pi r^2\ dr}{16\pi^2\varepsilon^2 r^4}$$

$$+ \tfrac{1}{2}\varepsilon \int_b^\infty \frac{(q_a^2 + 2q_a q_b + q_b^2)4\pi r^2\ dr}{16\pi^2\varepsilon^2 r^4}$$

$$= \tfrac{1}{2} \int_a^b \frac{q_a^2}{4\pi\varepsilon r^2}\ dr + \tfrac{1}{2} \int_b^\infty \frac{q_a^2 + 2q_a q_b + q_b^2}{4\pi\varepsilon r^2}\ dr$$

$$= \frac{-q_a^2}{8\pi\varepsilon}\left(\frac{1}{b} - \frac{1}{a}\right) + \frac{+(q_a^2 + 2q_a q_b + q_b^2)}{8\pi\varepsilon}\left(+\frac{1}{b}\right)$$

$$= \frac{q_a^2}{8\pi\varepsilon}\left(\frac{1}{a} - \frac{1}{b}\right) + \frac{q_a^2 + 2q_a q_b + q_b^2}{8\pi\varepsilon b}$$

$$= \frac{q_a^2}{8\pi\varepsilon a} - \frac{q_a^2}{8\pi\varepsilon b} + \frac{q_a^2}{8\pi\varepsilon b} + \frac{2q_a q_b}{8\pi\varepsilon b} + \frac{q_b^2}{8\pi\varepsilon b}$$

$$U = \frac{q_a^2}{8\pi\varepsilon a} + \frac{q_a q_b}{4\pi\varepsilon b} + \frac{q_b^2}{8\pi\varepsilon b}$$

PROBLEM 4.8 A point charge of 2×10^{-6} coul is located 15 cm outside the surface of an insulated uncharged conducting sphere of radius 50 cm. Find the potential energy of this system.

Solution:

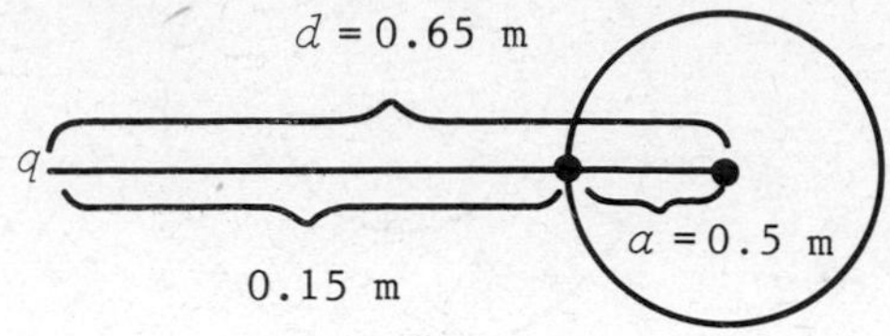

This is equal to the figure below.

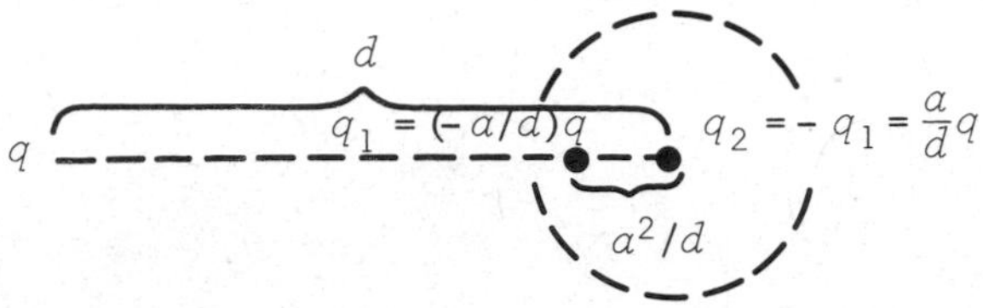

Now find the energy of configuration. The force on q is

$$F = \frac{1}{4\pi\varepsilon}\,\frac{qq_1}{(d - a^2/d)^2} + \frac{1}{4\pi\varepsilon}\,\frac{qq_2}{d^2} = \frac{q}{4\pi\varepsilon}\left(\frac{-(a/d)q}{(d - a^2/d)^2} + \frac{(a/d)q}{d^2}\right)$$

$$U = -\int_\infty^d F\,dx = -\int_\infty^d \frac{q}{4\pi\varepsilon}\left(\frac{-(a/d)q}{(d - a^2/d)^2} + \frac{(a/d)q}{d^2}\right)dd$$

$$= \frac{q}{4\pi\varepsilon}\int_\infty^d \frac{(a/d)q}{(d - a^2/d)^2}\,dd - \frac{q}{4\pi\varepsilon}\int_\infty^d \frac{(a/d)q}{d^2}\,dd$$

$$U = \frac{q^2a}{4\pi\varepsilon}\int_\infty^d \frac{1}{d(d - a^2/d)^2}\,dd - \frac{q^2a}{4\pi\varepsilon}\int_\infty^d d^{-3}\,dd$$

$$= \frac{q^2a}{4\pi\varepsilon}\int_\infty^d \frac{1}{d(d - a^2/d)^2}\,dd - \frac{q^2a}{4\pi\varepsilon}\left(-\frac{1}{2d^2}\right)_\infty^d$$

$$= \frac{q^2a}{4\pi\varepsilon}\int_\infty^d \frac{1}{d(d^2 - 2a^2 + a^4/d^2)}\,dd + \frac{q^2a}{8\pi\varepsilon d^2}$$

$$= \frac{q^2a}{4\pi\varepsilon}\int_\infty^d \frac{1}{d^3 - 2a^2d + a^4/d}\,dd + \frac{q^2a}{8\pi\varepsilon d^2}$$

$$= \frac{q^2 a}{4\pi\varepsilon} \int_{\infty}^{d} \frac{d}{d^4 - 2a^2 d^2 + a^4}\, dd + \frac{q^2 a}{8\pi\varepsilon d^2}$$

$$U = \frac{q^2 a}{4\pi\varepsilon} \int_{\infty}^{d} \frac{d}{(d^2 - a^2)^2}\, dd + \frac{q^2 a}{8\pi\varepsilon d^2}$$

$$= \frac{q^2 a}{4\pi\varepsilon} (-\tfrac{1}{2}) \left[\frac{1}{d^2 - a^2}\right]_{\infty}^{d} + \frac{q^2 a}{8\pi\varepsilon d^2}$$

$$= \frac{-q^2 a}{8\pi\varepsilon (d^2 - a^2)} + \frac{q^2 a}{8\pi\varepsilon d^2}$$

$$= \frac{q^2 a}{8\pi\varepsilon} \left(\frac{1}{d^2} - \frac{1}{d^2 - a^2}\right) = \frac{q^2 a}{8\pi\varepsilon} \left(\frac{d^2 - a^2 - d^2}{d^2 (d^2 - a^2)}\right) = \frac{q^2 a}{8\pi\varepsilon} \left(\frac{-a^2}{d^2 (d^2 - a^2)}\right)$$

$$U = \frac{-q^2 a^3}{8\pi\varepsilon d^2 (d^2 - a^2)}$$

$$U - \frac{-(2 \times 10^{-6}\ \text{C})^2 (0.5\ \text{m})^3}{8\pi (8.85 \times 10^{-12}\ \text{C}^2/\text{N} \cdot \text{m}^2)(0.65\ \text{m})^2 [(0.65\ \text{m})^2 - (0.5\ \text{m})^2]}$$

$$= -3.08 \times 10^{-2}\ \text{N} \cdot \text{m}$$

$$U = -3.08 \times 10^{-2}\ \text{J}$$

PROBLEM 4.9 A spherical distribution of charge has a radius and a charge density which varies a $\rho = Kr^2$. Determine the amount of work that is necessary to assemble this charge distribution.

Solution:

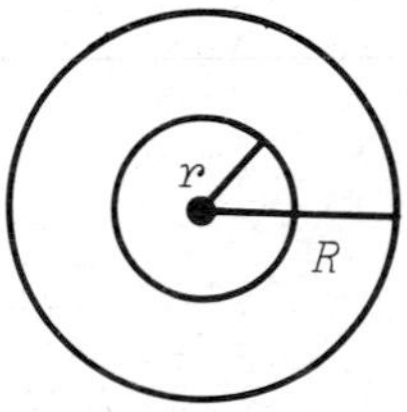

For $r < R$,

$$\oint \mathbf{D} \cdot d\mathbf{s} = \int \rho\, dv$$

$$D(4\pi r^2) = \int_0^r Kr^2 (4\pi r^2)\, dr$$

$$= \int_0^r K(4\pi r^4)\, dr$$

$$D(4\pi r^2) = \frac{4\pi K r^5}{5}$$

$$\varepsilon_0 E(4\pi) = \frac{4\pi K r^3}{5}$$

$$E = \frac{Kr^3}{5\varepsilon}$$

For $r > R$,

$$D(4\pi r^2) = \int_0^R Kr^2 (4\pi r^2)\, dr$$

$$D(4\pi r^2) = \frac{4\pi K R^5}{5}$$

$$E = \frac{KR^5}{5\varepsilon r^2}$$

$$U = \tfrac{1}{2}\varepsilon \int_{\substack{\text{all}\\ \text{space}}} \mathbf{E}\cdot\mathbf{E}\, dv, \qquad dv = 4\pi r^2\, dr$$

$$U = \tfrac{1}{2}\varepsilon \int_0^R E^2 (4\pi r^2) dr + \tfrac{1}{2}\varepsilon \int_R^\infty E^2 (4\pi r^2)\, dr$$

$$= \tfrac{1}{2}\varepsilon \int_0^R \frac{K^2 r^6}{25\varepsilon^2} (4\pi r^2) dr + \tfrac{1}{2}\varepsilon \int_R^\infty \frac{K^2 R^{10}}{25\varepsilon^2 r^4} (4\pi r^2) dr$$

$$= \int_0^R \frac{2\pi K^2}{25\varepsilon} r^8\, dr + \int_R^\infty \frac{KR^{10} 2\pi}{25\varepsilon} r^{-2}\, dr$$

$$= \frac{2\pi K^2 R^9}{25\varepsilon(9)} + \frac{-2\pi K^2 R^{10}}{25\varepsilon}\left(0 - \frac{1}{R}\right)$$

$$U = \frac{2\pi K^2 R^9}{25\varepsilon}\left(\frac{1}{9} + 1\right) \simeq \frac{0.3 K^2 R^9}{\varepsilon}$$

PROBLEM 4.10 A line of charge has a constant charge density λ. A point charge Q is placed a distance b away from the charge on the perpendicular bisector of the line. The line has a length $2L$ as shown in the figure. Find the potential energy of this configuration.

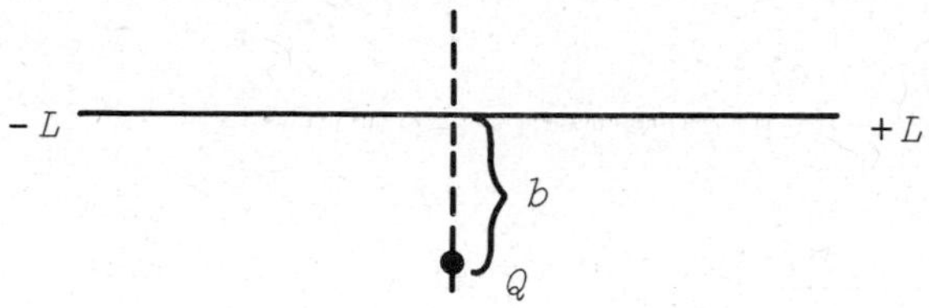

Solution:

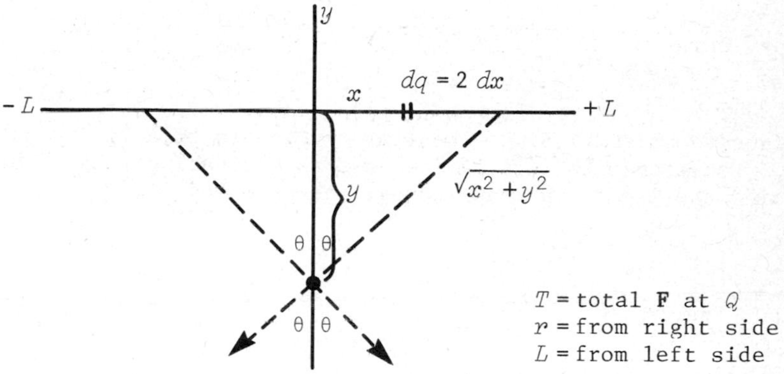

$$\mathbf{F}_T = \mathbf{F}_r + \mathbf{F}_L$$

$$d\mathbf{F}_r = \frac{1}{4\pi\varepsilon}\,\frac{Q\,dq}{x^2+y^2}\cos\theta\,(-\hat{\mathbf{j}}) + \frac{1}{4\pi\varepsilon}\,\frac{Q\,dq}{x^2+y^2}\sin\theta\,(-\hat{\mathbf{i}})$$

$$d\mathbf{F}_L = \frac{1}{4\pi\varepsilon}\,\frac{Q\,dq}{\left(\sqrt{x^2+y^2}\right)^2}\cos\theta\,(-\hat{\mathbf{j}}) + \frac{1}{4\pi\varepsilon}\,\frac{Q\,dq}{x^2+y^2}\sin\theta\,(+\hat{\mathbf{i}})$$

$$d\mathbf{F}_T = \frac{2}{4\pi\varepsilon}\,\frac{Q\,dq}{x^2+y^2}\cos\theta\,(-\hat{\mathbf{j}}), \qquad \cos\theta = \frac{y}{\sqrt{x^2+y^2}}$$

$$d\mathbf{F} = \frac{-\hat{\mathbf{j}}}{2\pi\varepsilon}\,\frac{Q\lambda y\,dx}{(x^2+y^2)^{3/2}}$$

$$\mathbf{F} = \frac{-\hat{\mathbf{j}}Q\lambda y}{2\pi\varepsilon}\int_{x=0}^{L}(x^2+y^2)^{-3/2}\,dx = \frac{-\hat{\mathbf{j}}Q\lambda y}{2\pi\varepsilon y^2}\,\frac{x}{(y^2+x^2)^{1/2}}\Bigg|_{x=0}^{L}$$

$$\mathbf{F} = \frac{-\hat{\mathbf{j}}Q\lambda}{2\pi\varepsilon y} \frac{L}{(y^2 + L^2)^{\frac{1}{2}}}$$

(Integrals used in this problem are given at the end.)

$$U = -\int_{y=\infty}^{b} F\,dy$$

$$= -\int_{y=\infty}^{b} \frac{-Q\lambda L}{2\pi\varepsilon y(y^2 + L^2)^{\frac{1}{2}}}\,dy$$

$$= \frac{Q\lambda L}{2\pi\varepsilon} \int_{y=\infty}^{b} \frac{1}{y(y^2 + L^2)^{\frac{1}{2}}}\,dy = \frac{-Q\lambda}{2\pi\varepsilon}\left[\ln\left(\frac{L + \sqrt{b^2 + L^2}}{b}\right) - \lim_{y\to\infty} \ln\left(\frac{L + \sqrt{y^2 + L^2}}{y}\right)\right]$$

(See note at end.) Since the limit of ln = ln of a limit and since the limit of a function with ln's is the same as evaluating the function and then taking the limit, one can look at the quantity

$$\lim_{y\to\infty} \ln\left(\frac{L + \sqrt{y^2 + L^2}}{y}\right)$$

as follows.

$$\lim_{y\to\infty} \ln\left(\frac{L + \sqrt{y^2 + L^2}}{y}\right) = \ln\left(\lim_{y\to\infty} \frac{L + \sqrt{y^2 + L^2}}{y}\right)$$

$$= \ln\left(\lim_{y\to\infty} \frac{L}{y} + \lim_{y\to\infty} \frac{\sqrt{y^2 + L^2}}{y}\right)$$

$$= \ln\left(\lim_{y\to\infty} \sqrt{(y^2/y^2) + (L^2/y^2)}\right)$$

$$= \ln\left(\lim_{y\to\infty} \sqrt{1 + L^2/y^2}\right) = \ln(1) = 0$$

So

$$U = \frac{-Q\lambda}{2\pi\varepsilon} \ln\left(\frac{L + \sqrt{b^2 + L^2}}{b}\right)$$

The integrals used in the problem are the following:

$$\int \frac{1}{(x^2+c^2)^{3/2}}\,dx = \frac{1}{c^2}\left(\frac{x}{(c^2+x^2)^{\frac{1}{2}}}\right), \text{ here } y = c, \quad \frac{1}{y^2}\left(\frac{x}{(x^2+y^2)^{\frac{1}{2}}}\right)$$

$$\int \frac{1}{x\sqrt{x^2+a^2}}\,dx = -\frac{1}{a}\ln\left(\frac{a+\sqrt{x^2+a^2}}{x}\right), \text{ here } y = x, \; a = L, \quad -\frac{1}{L}\ln\left(\frac{L+\sqrt{y^2+L^2}}{y}\right)$$

A note on the integral at the beginning of the problem: the reason we can replace part of the boundary equation with

$$\lim_{y\to\infty}\left(\frac{L+\sqrt{y^2+L^2}}{y}\right)$$

is because the integral in the problem

$$\frac{Q\lambda L}{2\pi\varepsilon}\int_{y=\infty}^{b}\frac{dy}{y(y^2+L^2)^{\frac{1}{2}}}$$

is the same as

$$\frac{Q\lambda L}{2\pi\varepsilon}\lim_{t\to\infty}\int_{y=t}^{b}\frac{dy}{y^2(y^2+L^2)^{\frac{1}{2}}}$$

in which case we would get

$$\lim_{t\to\infty}\frac{-Q\lambda}{2\pi\varepsilon}\left[\ln\left(\frac{L+\sqrt{b^2+L^2}}{b}\right)-\ln\left(\frac{L+\sqrt{t^2+L^2}}{t}\right)\right]$$

and since the sum of a limit is the limit of a sum, this last expression is equal to

$$\frac{-Q\lambda}{2\pi\varepsilon}\left[\ln\left(\frac{L+\sqrt{b^2+L^2}}{b}\right)-\lim_{t\to\infty}\ln\left(\frac{L+\sqrt{t^2+L^2}}{t}\right)\right]$$

which is exactly what we have derived.

PROBLEM 4.11 Find the electrostatic energy of a sphere of radius R containing a uniform volume density ρ of charge and existing in otherwise empty space. If an electron is assumed to be constructed according to this model, obtain an expression for its radius by equating the electrostatic to the self-energy.

Solution:

For

$$r > R, \quad E = \frac{e}{4\pi\varepsilon_0 r^2}$$

$$r < R, \quad E = \frac{er}{4\pi\varepsilon_0 R^3}$$

$$U = \tfrac{1}{2}\varepsilon_0 \int E^2\, dv = \tfrac{1}{2}\varepsilon_0 \int_0^R \frac{e^2 r^2}{(4\pi\varepsilon_0 R^3)^2} 4\pi r^2\, dr + \tfrac{1}{2}\varepsilon_0 \int_R^\infty \frac{e^2}{(4\pi\varepsilon_0)^2 r^4} 4\pi r^2\, dr$$

$$U = \tfrac{1}{2}\varepsilon_0 \frac{e^2}{(4\pi\varepsilon_0)^2} 4\pi \left(\frac{1}{R^6}\frac{R^5}{5} + \frac{1}{R}\right)$$

$$U = \frac{e^2}{8\pi\varepsilon_0 R}\left(\frac{1}{5} + 1\right) = \frac{3e^2}{20\pi\varepsilon_0 R}$$

$$U = mc^3 \Rightarrow \frac{3e^2}{20\pi\varepsilon_0 R} = mc^2$$

$$R = \frac{3e^2}{20\pi\varepsilon_0 mc^2}$$

PROBLEM 4.12 Calculate the polarizability for two harmonically bound, oppositely charged particles in a slowly varying external field. Use the results you obtain in calculating directly the energy of the system of two charges, by summing the potential energy of the spring and the work done in displacing the charges. Show that the energy is $-\frac{1}{2}\mathbf{p}\cdot\mathbf{E}$, where $\mathbf{p}$ is the dipole moment for the system. (Note: Neglect the effect of the two charges on each other.)

Solution:

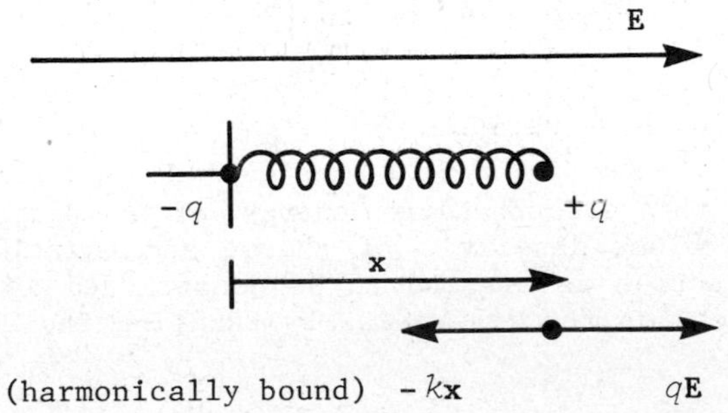

If the field is slowly varying, equilibrium will be established.

$$q\mathbf{E} - k\mathbf{x} = 0 \text{ or } \mathbf{x} = \frac{q}{k}\mathbf{E}$$

$$\rho(\mathbf{r}') = -q[\delta^3(\mathbf{r}') - \delta^3(\mathbf{r}' - \mathbf{x})]$$

$$\mathbf{p} \triangleq \int_v \mathbf{r}'\rho(\mathbf{r}')\, d\tau' = q\mathbf{x}$$

$\mathbf{p} = \alpha\mathbf{E}$, where α is the polarizability

$q\mathbf{x} = \alpha\mathbf{E}$, but $\mathbf{x} = \frac{q}{k}\mathbf{E}$ from before

$$\therefore \quad \alpha = \frac{q^2}{k}$$

Potential energy of the spring (V)

$$\mathbf{F} = -\nabla V, \quad \therefore \quad V = \frac{k|\mathbf{x}|^2}{2}$$

or

$$V = \frac{k}{2}\frac{q^2}{k^2}E^2 = \frac{q^2E^2}{2k}$$

Work done against the field in displacing the charges:

$$W = -\int_0^x \mathbf{F} \cdot d\mathbf{x}' = -\int_0^x qE\, dx' = -qEx$$

or

$$W = -qE\,\frac{q}{k}\,E = -\frac{q^2E^2}{k}$$

$$W + V = -\frac{q^2E^2}{k} + \frac{q^2E^2}{2k} = -\frac{q^2E^2}{2k}$$

Show that the energy is $-\frac{1}{2}\mathbf{p} \cdot \mathbf{E}$:

$$-\tfrac{1}{2}\mathbf{p} \cdot \mathbf{E} = -\tfrac{1}{2}\frac{q^2}{k}\mathbf{E} \cdot \mathbf{E} = -\frac{q^2E^2}{2k} \qquad \text{Q.E.D.}$$

PROBLEM 4.13 Calculate the quadruple moment of two concentric ring charges q and $-q$ having radii a and b, and show that it is the first nonvanishing moment.

Solution:

$$\rho(\mathbf{r}') = \frac{q}{2\pi}\left(\frac{1}{a^2}\delta(r' - a)\delta(\cos\,\theta') - \frac{1}{b^2}\delta(r' - b)\delta(\cos\,\theta')\right)$$

(1) Pole

$$\int_v \rho(\mathbf{r}')\,d\tau' = \frac{q}{2\pi a^2}\int_v \delta(r' - a)\delta(\cos\,\theta')d\tau' - \frac{q}{2\pi b^2}\int_v \delta(r' - b)\delta(\cos\,\theta')d\tau'$$

$$\int_v \rho(\mathbf{r}')\,d\tau' = \frac{q}{2\pi a^2}\int_v \delta(r' - a)\delta(\cos\,\theta')r'^2 \sin\theta'\,dr'\,d\theta'\,d\phi'$$

$$- \frac{q}{2\pi b^2}\int_v \delta(r' - b)\delta(\cos\,\theta')r'^2 \sin\theta'\,dr'\,d\theta'\,d\phi'$$

$$\int_v \rho(\mathbf{r}')\,d\tau' = \frac{q}{2\pi}\int_0^{2\pi}\int_0^{\pi} \delta(\cos\,\theta')\sin\theta'\,d\theta'\,d\phi'$$

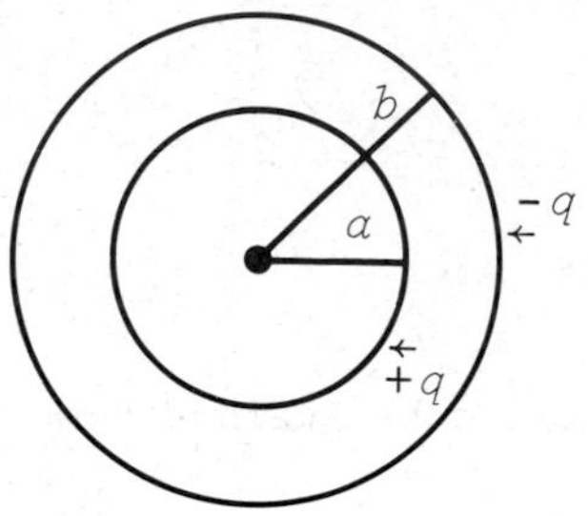

$$- \frac{q}{2\pi}\int_0^{2\pi}\int_0^{\pi} \delta(\cos\,\theta')\sin\theta'\,d\theta'\,d\phi'$$

$$= \frac{q}{2\pi}\int_0^{2\pi} d\phi' - \frac{q}{2\pi}\int_0^{2\pi} d\phi'$$

$$= q - q = 0$$ Q.E.D.

(2) Dipole

$$\mathbf{p} = \int_v \mathbf{r}'\rho(\mathbf{r}')\,d\tau'$$

$$\mathbf{p} = \frac{q}{2\pi a^2}\int_v \mathbf{r}'\delta(r' - a)\delta(\cos\,\theta')r'^2 \sin\theta'\,dr'\,d\theta'\,d\phi'$$

$$- \frac{q}{2\pi b^2}\int_v \mathbf{r}'\delta(r' - b)\delta(\cos\,\theta')r'^2 \sin\theta'\,dr'\,d\theta'\,d\phi'$$

$$\mathbf{p} = \frac{qa}{2\pi}\int_0^{2\pi}\int_0^{\pi} \hat{\mathbf{r}}'\delta(\cos\,\theta')\sin\theta'\,d\theta'\,d\phi' - \frac{qb}{2\pi}\int_0^{2\pi}\int_0^{\pi} \hat{\mathbf{r}}'\delta(\cos\,\theta')\sin\theta'\,d\theta'\,d\phi'$$

where $\mathbf{r}' = r'\hat{\mathbf{r}}'$

$$\mathbf{p} = \frac{qa}{2\pi}\int_0^{2\pi} \hat{\mathbf{r}}'\, d\phi' - \frac{qb}{2\pi}\int_0^{2\pi} \hat{\mathbf{r}}'\, d\phi' = 0$$

because $\hat{\mathbf{r}}' = \hat{\mathbf{i}}\cos\phi' + \hat{\mathbf{j}}\sin\phi'$, therefore

$$\int_0^{2\pi} \hat{\mathbf{r}}'\, d\phi = \hat{\mathbf{i}}\int_0^{2\pi} \cos\phi'\, d\phi' + \hat{\mathbf{j}}\int_0^{2\pi} \sin\phi'\, d\phi'$$

$$= 0 + 0$$

$$\underline{Q} = \frac{1}{6}\int_v \rho(\mathbf{r}')(3\mathbf{r}'\mathbf{r}' - \underline{1}\, r'^2)\, d\tau'$$

$$\underline{Q} = \frac{1}{6}\int_v \frac{q}{2\pi a^2}\,\delta(r' - a)\delta(\cos\theta')3\mathbf{r}'\mathbf{r}'\, d\tau' \tag{1}$$

$$-\frac{1}{6}\int_v \frac{q}{2\pi b^2}\,\delta(r' - b)\delta(\cos\theta')3\mathbf{r}'\mathbf{r}'\, d\tau' \tag{2}$$

$$-\frac{1}{6}\int_v \frac{q}{2\pi a^2}\,\delta(r' - a)\delta(\cos\theta')\,\underline{1}\, r'^2\, d\tau' \tag{3}$$

$$+\frac{1}{6}\int_v \frac{q}{2\pi b^2}\,\delta(r' - b)\delta(\cos 0')\,\underline{1}\, r'^2\, d\tau' \tag{4}$$

$$\text{Equation (1)} = \frac{q}{12\pi a^2}\int_v \delta(r' - a)\delta(\cos\theta')3\hat{\mathbf{r}}'\hat{\mathbf{r}}' r'^4 \sin\theta'\, dr'\, d\theta'\, d\phi'$$

$$= \frac{qa^2}{4\pi}\int_0^{2\pi}\int_0^{\pi} \delta(\cos\theta')\hat{\mathbf{r}}'\hat{\mathbf{r}}' \sin\theta'\, d\theta'\, d\phi'$$

$$= \frac{qa^2}{4\pi}\int_0^{2\pi} \hat{\mathbf{r}}'\hat{\mathbf{r}}'\, d\phi'$$

$$= \frac{qa^2}{4\pi} \int_0^{2\pi} [\hat{\mathbf{i}}\hat{\mathbf{i}} \cos^2 \phi' + \hat{\mathbf{j}}\hat{\mathbf{j}} \sin^2 \phi' + 2\hat{\mathbf{i}}\hat{\mathbf{j}} \cos \phi' \sin \phi']\, d\phi'$$

$$= \frac{qa^2}{4} (\hat{\mathbf{i}}\hat{\mathbf{i}} + \hat{\mathbf{j}}\hat{\mathbf{j}})$$

$$\text{Equation (2)} = -\frac{1}{6} \int_v \frac{q}{2\pi b^2} \delta(r' - b)\delta(\cos \theta') 3\hat{\mathbf{r}}'\hat{\mathbf{r}}' r'^4 \sin \theta'\, dr'\, d\theta'\, d\phi'$$

$$= -\frac{qb^2}{4\pi} \int_0^{2\pi} \int_0^{\pi} \delta(\cos \theta')\hat{\mathbf{r}}'\hat{\mathbf{r}}' \sin \theta'\, d\theta'\, d\phi'$$

$$= -\frac{qb^2}{4\pi} \int_0^{2\pi} \hat{\mathbf{r}}'\hat{\mathbf{r}}'\, d\phi'$$

$$= -\frac{qb^2}{4} (\hat{\mathbf{i}}\hat{\mathbf{i}} + \hat{\mathbf{j}}\hat{\mathbf{j}}) \text{ from before, Eq. (1)}$$

$$\text{Equation (3)} = -\frac{1}{6} \int_v \frac{q}{2\pi a^2} \delta(r' - a)\delta(\cos \theta')\, \underline{1}\, r'^4 \sin \theta'\, dr'\, d\theta'\, d\phi'$$

$$= -\underline{1}\, \frac{qa^2}{12\pi} \int_0^{2\pi} \int_0^{\pi} \delta(\cos \theta') \sin \theta'\, d\theta'\, d\phi'$$

$$= -\underline{1}\, \frac{qa^2}{12\pi} \int_0^{2\pi} d\phi' = -\underline{1}\, \frac{qa^2}{6}$$

$$\text{Equation (4)} = +\frac{1}{6} \int_v \frac{q}{2\pi b^2} \delta(r' - b)\delta(\cos \theta')\, \underline{1}\, r'^4 \sin \theta'\, dr'\, d\theta'\, d\phi'$$

$$= \underline{1}\, \frac{qb^2}{6}, \text{ same as Eq. (3)}$$

$$\underline{Q} = \left[(\hat{\mathbf{i}}\hat{\mathbf{i}} + \hat{\mathbf{j}}\hat{\mathbf{j}}) \left(\frac{a^2}{4} - \frac{b^2}{4} - \frac{a^2}{6} + \frac{b^2}{6} \right) + \hat{\mathbf{k}}\hat{\mathbf{k}} \left(\frac{b^2}{6} - \frac{a^2}{6} \right) \right] q$$

$$\underline{Q} = q \left[(\hat{\mathbf{i}}\hat{\mathbf{i}} + \hat{\mathbf{j}}\hat{\mathbf{j}}) \left(\frac{a^2}{12} - \frac{b^2}{12} \right) + \hat{\mathbf{k}}\hat{\mathbf{k}} \left(\frac{k^2}{6} - \frac{a^2}{6} \right) \right]$$

PROBLEM 4.14 Prove that:

(a) When we cross a simple uniform surface charge layer, the potential is continuous but its normal derivative is not. Calculate the discontinuity in the normal derivative.

(b) When we cross a uniform dipole surface layer, the potential is discontinuous but its normal derivative is not. Calculate the discontinuity in the potential.

Solution:

(a)

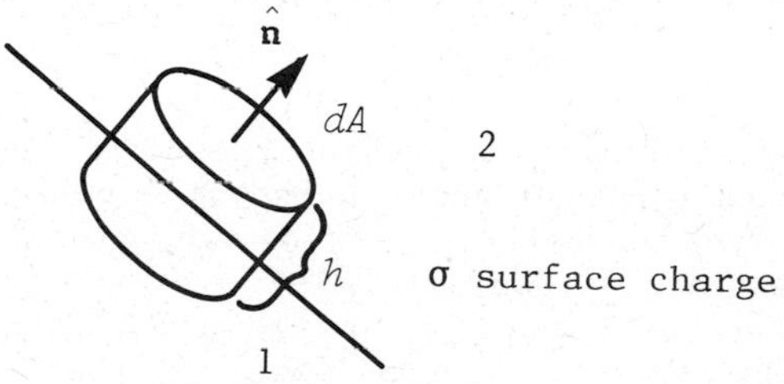

$$\nabla \cdot \mathbf{E} = \frac{\rho}{\varepsilon_0}$$

$$\int_v \nabla \cdot \mathbf{E}\, d\tau = \int_S \mathbf{n} \cdot \mathbf{E}\, da = \int_v \frac{\rho}{\varepsilon_0}\, d\tau$$

In the limit as $h \to 0$

$$\hat{\mathbf{n}} \cdot (\mathbf{E}_2 - \mathbf{E}_1) dA = \frac{\sigma}{\varepsilon_0}\, dA$$

$$\hat{\mathbf{n}} \cdot (\mathbf{E}_2 - \mathbf{E}_1) = \frac{\sigma}{\varepsilon_0}$$

Thus the normal component of E (normal derivative of the potential) is discontinuous when we cross a simple uniform charge density.

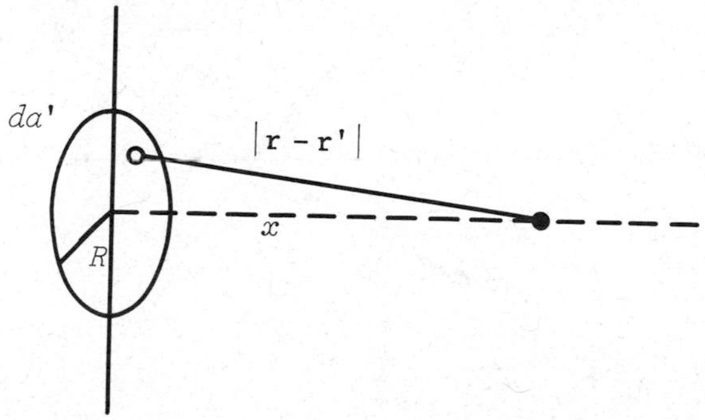

Let da' be a small, circular, flat area of the surface of radius R. Consider the point a distance x away from the surface and on the perpendicular, through the center of the circle:

$$\phi(x) = \frac{1}{4\pi\varepsilon_0}\int_0^R\int_0^{2\pi}\frac{\sigma\, da'}{(x^2+r^2)^{\frac{1}{2}}} = \frac{1}{4\pi\varepsilon_0}\int_0^R\int_0^{2\pi}\frac{\sigma r\, dr\, d\theta}{(x^2+r^2)^{\frac{1}{2}}}$$

$$\phi(x) = \frac{\sigma}{2\varepsilon_0}\int_0^R \frac{r\, dr}{(x^2+r^2)^{\frac{1}{2}}} = \frac{\sigma}{2\varepsilon_0}(x^2+r^2)^{\frac{1}{2}}\Big|_0^R$$

$$\phi(x) = \frac{\sigma}{2\varepsilon_0}\left[(x^2+r^2)^{\frac{1}{2}} - x\right]$$

If we let the observation point approach the disc, the potential due to the disc alone approaches $(\sigma/2\varepsilon_0)R$ and has the same value directly on the other side of the disc, thus the potential is continuous in crossing the disc. The potential of the surface without the disc, with a hole where the disc fits in, is continuous across the plane of the hole. Since the total potential can be obtained by linear superposition of the disc and that of the remainder, the potential is continuous across the surface charge layer.

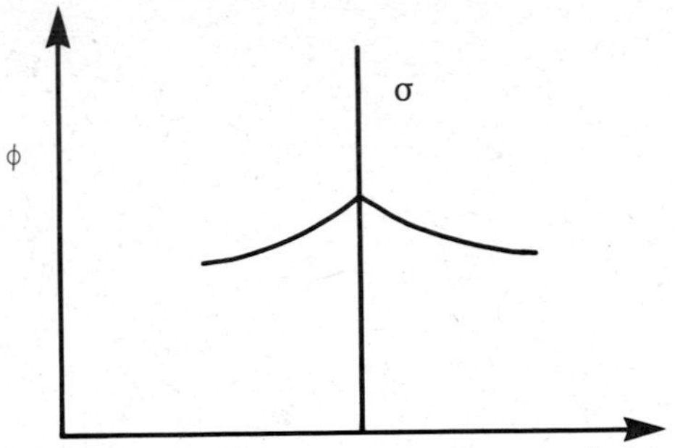

(b) Show that the potential is discontinuous when we cross a uniform dipole surface layer.

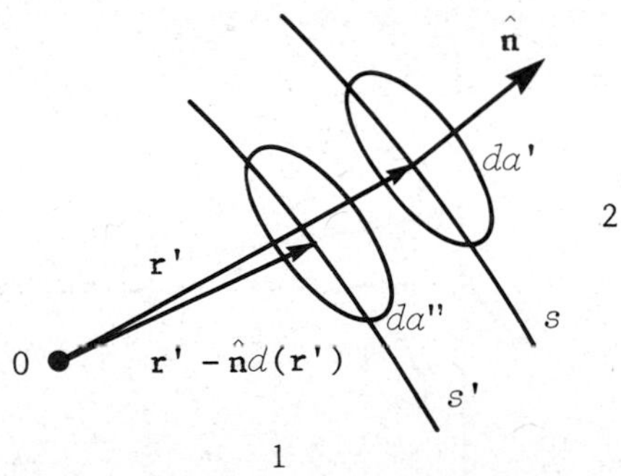

where $d(\mathbf{r}')$ is the perpendicular distance between the two surfaces at $\mathbf{r}'$:

$$\phi(\mathbf{r}) = \frac{1}{4\pi\varepsilon_0}\left(\int_s \frac{\sigma(\mathbf{r}')\,da'}{|\mathbf{r}-\mathbf{r}'|} - \int_{s'} \frac{\sigma(\mathbf{r}')\,da''}{|\mathbf{r}-\mathbf{r}'+\hat{\mathbf{n}}d|}\right)$$

Using a three-dimensional Taylor series expansion and assuming

$$|\hat{\mathbf{n}}d| << |\mathbf{r}-\mathbf{r}'|$$

$$\frac{1}{|\mathbf{r}-\mathbf{r}'+\hat{\mathbf{n}}d|} = \frac{1}{|\mathbf{r}-\mathbf{r}'|} + \hat{\mathbf{n}}d\cdot\nabla'\frac{1}{|\mathbf{r}-\mathbf{r}'|} + \cdots$$

If we let s' approach v infinitesimally close to s,

$$\phi(\mathbf{r}) = \frac{1}{4\pi\varepsilon_0}\int_s \lim_{d\to 0}\,[\sigma(\mathbf{r}')d]\,\mathbf{n}\cdot\nabla'\frac{1}{|\mathbf{r}-\mathbf{r}'|}\,da'$$

$$D(\mathbf{r}') \triangleq \lim_{d\to 0}\,\sigma(\mathbf{r}')d(\mathbf{r}')$$

which is the strength of the dipole layer

$$\hat{\mathbf{n}}\cdot\nabla'\frac{1}{|\mathbf{r}-\mathbf{r}'|}\,da' = -\frac{\cos\theta\,da'}{|\mathbf{r}-\mathbf{r}'|^2} = -d\Omega$$

where $d\Omega$ is the element of solid angle subtended at the observation point by the area element da'

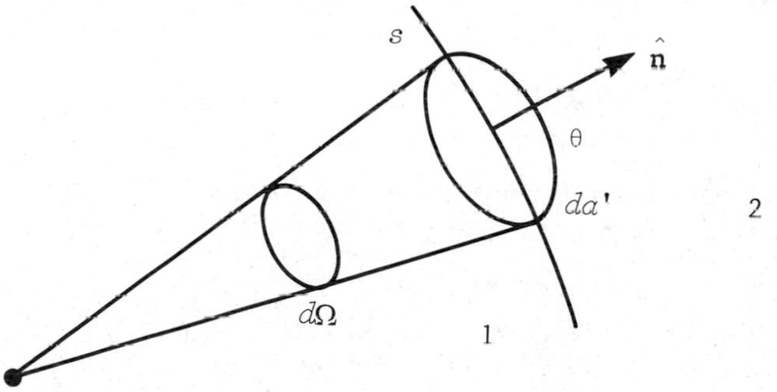

$$\therefore \quad \phi(\mathbf{r}) = \frac{-1}{4\pi\varepsilon_0}\int_s D(\mathbf{r}')\,d\Omega$$

As we let the observation point approach the disc, the potential due to the disc alone approaches $-D/2\varepsilon_0$, while directly on the other side of the disc it is $D/2\varepsilon$,* thus there is a discontinuity of D/ε_0 in crossing from the inner to the outer side of the disc. The potential of the dipole surface without the disc with a hole where the disc fits in is continuous across the plane of the hole. Since the total potential can be obtained by linear superposition of the potential of the disc and that of the remainder, the total potential jump in crossing the surface is

$$\phi_2 - \phi_1 = \frac{D}{\varepsilon_0}$$

*The reason for the change in sign is that on the other side of the dipole surface $\hat{\mathbf{n}} \cdot \nabla(da'/|\mathbf{r}-\mathbf{r}'|)$ is equal to $+d\Omega$.

Applying Gauss's law to this particular situation:

$$\nabla \cdot \mathbf{E} = \frac{\rho}{\varepsilon_0}$$

$$\int_v \nabla \cdot \mathbf{E}\, d\tau = \oint_s \hat{\mathbf{n}} \cdot \mathbf{E}\, da = \int_v \frac{\rho}{\varepsilon_0}\, d\tau$$

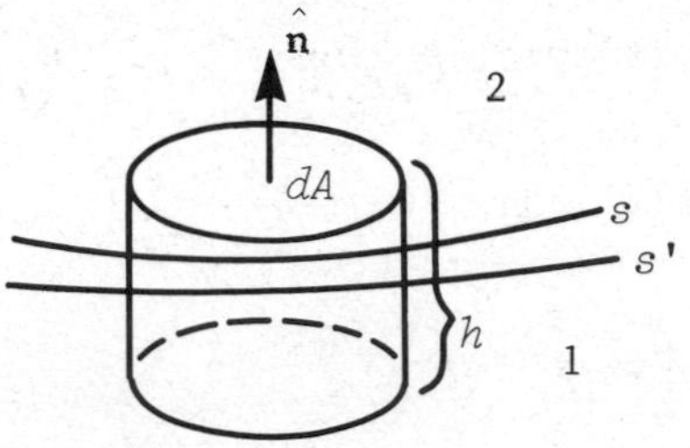

Let h go to zero as $d(\mathbf{r}')$ goes to zero.

$$\int_v \rho\, d\tau = 0$$

because of the equal and opposite charges on the two enclosed surfaces.

$$\int_s \hat{\mathbf{n}} \cdot \mathbf{E}\, da = \hat{\mathbf{n}} \cdot (\mathbf{E}_2 - \mathbf{E}_1) = 0$$

therefore, the normal component of **E** (normal derivative of the potential), is continuous across a dipole surface.

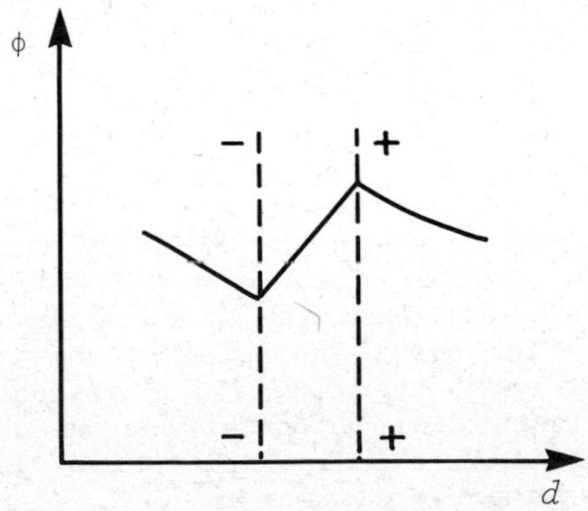

PROBLEM 4.15 Consider the field due to an electric dipole of moment **p**. What charge distribution would have to be introduced on a sphere with **p** at its center to produce zero field outside the sphere, without modifying the potential in the interior of the sphere?

Solution:

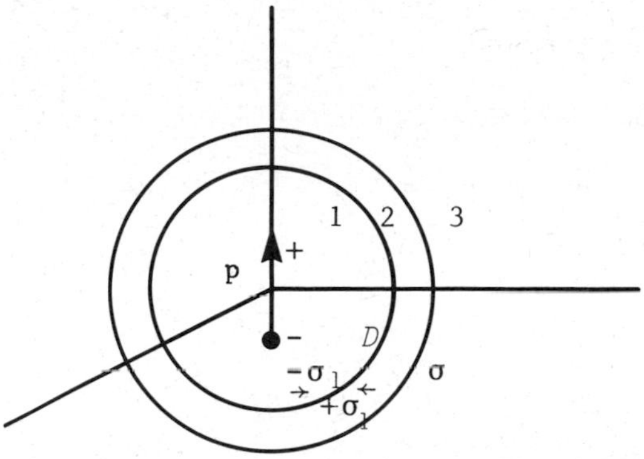

D = strength of dipole layer

$$D = \lim_{d \to 0} \sigma_1 d$$

Assume that the simple surface charge layer and the simple dipole surface layer are infinitely close together and of radius a. The boundary conditions are, for σ,

$$\phi_3 = \phi_2 \text{ at the boundary}$$

$$\hat{\mathbf{r}} \cdot (\mathbf{E}_2 - \mathbf{E}_3) = \frac{\sigma}{\varepsilon_0}$$

$$\hat{\mathbf{r}} \times (\mathbf{E}_2 - \mathbf{E}_3) = 0$$

and for D,

$$\phi_3 - \phi_1 = \frac{D}{\varepsilon_0}$$

$$\hat{\mathbf{r}} \cdot (\mathbf{E}_3 - \mathbf{E}_1) = 0$$

We are looking for D and σ such that $\mathbf{E}_2 = 0$ and

$$\mathbf{E}_1 = \frac{1}{4\pi\varepsilon_0}\left(\frac{\mathbf{p}}{r^3} + \frac{3\mathbf{r}\cdot\mathbf{p}\mathbf{r}}{r^5}\right)$$

or the field due to a dipole at the origin. Becuase of the way the problem was set up, $\mathbf{p} = p\hat{\mathbf{k}}$, so

$$\mathbf{E}_1 = \frac{p}{4\pi\varepsilon_0 r}\,(-\hat{\mathbf{k}} + 3\hat{\mathbf{r}}\cos\theta)$$

where θ is the polar angle.

$$\phi_1 = \frac{1}{4\pi\varepsilon_0}\frac{\mathbf{r}\cdot\mathbf{p}}{r^3} \text{ or } \phi_1 = \frac{p\cos\theta}{4\pi\varepsilon_0 r^2}$$

$$\hat{\mathbf{r}}\cdot(\mathbf{E}_3 - \mathbf{E}_1) = 0$$

$$E_{3r} - \hat{\mathbf{r}}\cdot\mathbf{E}_1 = 0$$

$$\hat{\mathbf{r}} = \hat{\mathbf{k}}\cos\theta + \hat{\mathbf{j}}\sin\theta\sin\phi + \hat{\mathbf{i}}\sin\theta\cos\phi$$

$\therefore$ $\hat{\mathbf{r}}\cdot\hat{\mathbf{k}} = \cos\theta$ (actually made use of this earlier)

$$E_{3r}\Big|_a - \frac{p}{4\pi\varepsilon_0 a^3}(-\cos\theta + 3\cos\theta) = 0$$

$$E_{3r}\Big|_a = +\frac{p\cos\theta}{2\pi\varepsilon_0 a^3}$$

Also

$$\hat{\mathbf{r}}\cdot(\mathbf{E}_2 - \mathbf{E}_3) = \frac{\sigma}{\varepsilon_0}$$

since $E_2 = 0$

$$-\hat{\mathbf{r}}\cdot\mathbf{E}_3 = -E_{3r} = \frac{\sigma}{\varepsilon_0}$$

$$\therefore \quad \sigma = \frac{-p\cos\theta}{2\pi a^3}$$

$$\hat{\mathbf{r}}\times(\mathbf{E}_2 - \mathbf{E}_3) = 0$$

$$-\hat{\mathbf{r}}\times\mathbf{E}_3 = 0 \text{ or } \mathbf{E}_3 = \mathbf{E}_{3r}\hat{\mathbf{r}}$$

$$\mathbf{E}_3 = \frac{p\cos\theta}{2\pi\varepsilon_0 a^3}\, r$$

$$\therefore \quad \phi_3 = -\frac{p(\cos\theta)r}{2\pi\varepsilon_0 a^3}$$

at $r = a$,

$$\phi_3 - \phi_1 = \frac{D}{\varepsilon_0}$$

$$-\frac{p\cos\theta}{2\pi\varepsilon_0 a^2} - \frac{p\cos\theta}{4\pi\varepsilon_0 a^2} = \frac{D}{\varepsilon_0} \Rightarrow D = -\frac{3p\cos\theta}{4\pi a^2}$$

PROBLEM 4.16 A nonrelativistic positron with initial velocity $v_0 \ll c$ moves "head-on" toward a nucleus of charge Ze and is eventually repelled and goes back to infinity. Calculate the total energy radiated by the positron on this trip. Treat the nucleus as infinitely heavy and ignore all terms of order v/c in your calculation.

Solution: The formula for the total power is $P = (e^2/6\pi\varepsilon_0 c^3)a^2$ $(v \ll c)$. In our case $a = Ze^2/4\pi\varepsilon_0 x^2 m$, where m is the mass of positron.

$$dW = P\,dt = \frac{P}{v}\,dx$$

$$W = \int_{\infty}^{x_{\min}} \frac{P}{v}\,dx + \int_{x_{\min}}^{\infty} \frac{P}{v}\,dx, \qquad v = \pm\sqrt{v_0^2 - 2\text{P.E.}/m}$$

where

$$\text{P.E.} = \text{potential energy} = \frac{Ze^2}{4\pi\varepsilon_0 x}$$

$$W = \int_{\infty}^{x_{\min}} \frac{P\,dx}{-\sqrt{v_0^2 - 2\text{P.E.}/m}} + \int_{x_{\min}}^{\infty} \frac{P\,dx}{\sqrt{v_0^2 - 2\text{P.E.}/m}}$$

$$= 2\int_{x_{\min}}^{\infty} \frac{P\,dx}{\sqrt{v_0^2 - 2\text{P.E.}/m}}$$

$$= 2\left(\frac{Ze^2}{4\pi\varepsilon_0 m}\right)\int_{x_{\min}}^{\infty} \frac{dx}{x\sqrt{v_0^2 - Ze^2(2)/4\pi\varepsilon_0 xm}} \times \left(\frac{e^2}{6\pi\varepsilon_0 c^3}\right)$$

Let $s = 1/x$, $ds = -dx/x^2$, $s_{\min} = 1/x_{\min}$

$$W = 2\left(\frac{Ze^2}{4\pi\varepsilon_0 m}\right)^2 \frac{e^2}{6\pi\varepsilon_0 c^3}\int_0^{s_{\min}} \frac{s^2\,dx}{\sqrt{v_0^2 - bs}}, \qquad b = \frac{2Ze^2}{4\pi\varepsilon_0 m}$$

$$s_{\min} = \frac{v_0^2}{b}$$

$$\int_0^{s_{\min}} \frac{s^2\,ds}{\sqrt{v_0^2 - bs}} = \left(-\frac{2}{b}s^2\sqrt{v_0^2 - bs} - \frac{8}{3b^2}s(v_0^2 - bs)^{3/2} - \frac{16}{15b^3}(v_0^2 - bs)^{5/2}\right)_0^{s_{\min}}$$

$$= \frac{16}{15} \frac{v_0^5}{(2Ze^2/4\pi\varepsilon_0 m)^3}$$

$$W = \frac{8}{45} \frac{mv_0^5}{zc^3}$$

PROBLEM 4.17 Assuming that the electrostatic energy of an electron is equal to its mass-energy, mc^2, where $m = 9.11 \times 10^{-31}$ kg is the electron mass and $c = 3 \times 10^8$ m/s is the velocity of light, find the radius of an electron if the electron constitutes (a) a uniformly charged sphere of total charge 1.60×10^{-19} A · s, (b) a uniformly charged spherical shell of the same charge.

Solution:

(a) $$\rho = \frac{q}{(4/3)\pi a^3} = \frac{3.82 \times 10^{-20}}{a^3} \text{ C/m}^3$$

$$r > R, \; E = \frac{8}{4\pi\varepsilon r^2}$$

$$r < R, \; E(4\pi r^2) = \frac{1}{\varepsilon} \rho(4/3)\pi r^3, \; E = \rho r/3\varepsilon$$

$$U = \tfrac{1}{2}\varepsilon \int_0^\infty E^2 \, dv = \tfrac{1}{2}\varepsilon \left(\int_0^a \frac{\rho^2 r^2}{9\varepsilon^2} 4\pi r^2 \, dr + \int_a^\infty \frac{q^2}{16\pi^2\varepsilon^2} \frac{1}{r^4} 4\pi r^2 \, dr \right)$$

$$U = \tfrac{1}{2}\varepsilon \left(\frac{4\pi\rho^2}{9\varepsilon^2} \frac{a^5}{5} + \frac{q^2}{4\pi\varepsilon^2} \frac{1}{a} \right)$$

$$amc^2 = \frac{(2\pi)(3.82 \times 10^{-20})^2}{45\varepsilon} + \frac{q^2}{8\pi\varepsilon}$$

$$a = 1.69 \times 10^{-16} \text{ m}$$

(b) $$U = \tfrac{1}{2}\varepsilon \int_a^\infty \frac{q^2}{16\pi^2\varepsilon^2} \frac{1}{r^4} 4\pi r^2 \, dr$$

$$mc^2 = \frac{q^2}{8\pi\varepsilon} \frac{1}{a}, \quad a = 1.4 \times 10^{-16} \text{ m}$$

PROBLEM 4.18 Find the electrostatic energy associated with a uniform spherical charge distribution of total charge q and radius a by using $U = \frac{1}{2} \int \phi\rho \, dv$.

Solution: $U = \frac{1}{2} \int \phi\rho \, dv$

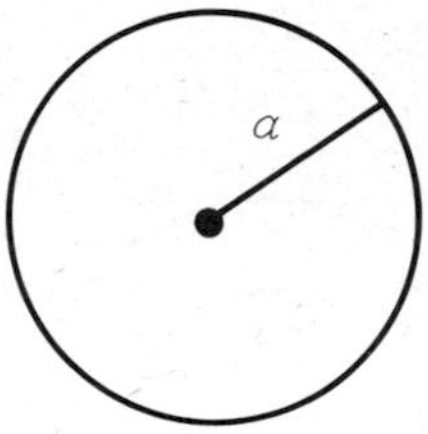

We first need to find $\phi(r)$.

$$\rho = \frac{3Q}{4\pi a^3}, \qquad E = \frac{1}{4\pi\varepsilon}\frac{Q}{r^2}, \quad r > a$$

$$E = \frac{\rho r}{3\varepsilon}, \quad r < a$$

$$\phi(r) = +\frac{Q}{4\pi\varepsilon a} - \int_a^r \frac{\rho r}{3\pi} \, dr = \frac{Q}{4\pi\varepsilon a} - \frac{\rho}{3\varepsilon}\left(\frac{r^2}{2} - \frac{a^2}{2}\right)$$

$$\phi(r) = \frac{Q}{4\pi\varepsilon a} - \frac{3Q}{12\pi a^3 \varepsilon}\left(\frac{r^2}{2} - \frac{a^2}{2}\right)$$

$$\phi(r) = \frac{Q}{4\pi\varepsilon a} + \frac{Q}{8\pi\varepsilon a} - \frac{Q}{8\pi\varepsilon}\frac{r^2}{a^3} = \frac{3Q}{8\pi\varepsilon a} - \frac{Q}{8\pi\varepsilon}\frac{r^2}{a^3}$$

Therefore,

$$U = \tfrac{1}{2}\rho \int_0^a \left(\frac{3Q}{8\pi\varepsilon a} - \frac{Q}{8\pi\varepsilon a^3} r^2\right) 4\pi r^2 \, dr$$

$$= \tfrac{1}{2}\rho \left(\frac{3Q}{2\varepsilon a}\frac{a^3}{3} - \frac{Q}{2\varepsilon a^3}\frac{a^5}{5}\right)$$

$$= \tfrac{1}{2}\frac{3Q}{4\pi a^3}\left(\frac{Qa^2}{2\varepsilon} - \frac{Qa^2}{10\varepsilon}\right)$$

$$U = \frac{3Q^2}{8\pi a\varepsilon}\left(\frac{1}{2} - \frac{1}{10}\right) = \frac{4}{10}\frac{3Q^2}{8\pi\varepsilon a}$$

$$U = \frac{3}{20}\frac{Q^2}{\pi\varepsilon a}$$

PROBLEM 4.19 Show that the total electrostatic energy of two concentric spherical shells of radii a and b $(b > a)$ formed by the uniformly distributed charges q_a and q_b, respectively, is

$$U = \frac{q_a^2}{8\pi\varepsilon_0 a} + \frac{q_b^2}{8\pi\varepsilon_0 b} + \frac{q_a q_b}{4\pi\varepsilon_0 b}$$

Solution:

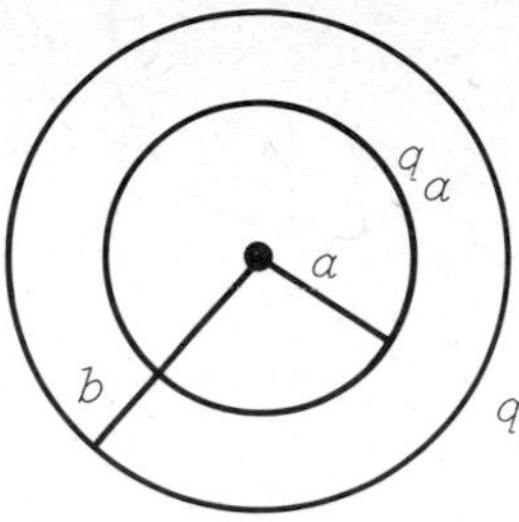

$$r > b,\ E = \frac{q_a + q_b}{4\pi\varepsilon r^2},\quad b > r > a,\ E = \frac{q_a}{4\pi\varepsilon r^2}$$

$$U = \tfrac{1}{2}\varepsilon \int_0^\infty E^2\, dv = \tfrac{1}{2}\varepsilon \int_a^b \frac{q_a^2}{16\pi^2\varepsilon^2 r^4} 4\pi r^2\, dr + \tfrac{1}{2}\varepsilon \int_b^\infty \frac{(q_a + q_b)^2}{16\pi^2\varepsilon} \frac{4\pi r^2\, dr}{r^4}$$

$$U = \tfrac{1}{2}\varepsilon \frac{q_a^2}{4\pi\varepsilon^2}\left(\frac{1}{a} - \frac{1}{b}\right) + \tfrac{1}{2}\varepsilon \frac{q_a^2 + 2q_a q_b + q_b^2}{4\pi\varepsilon^2} \frac{1}{b}$$

$$U = \frac{1}{8\pi\varepsilon}\left(\frac{q_a^2}{a} - \frac{q_a^2}{b} + \frac{q_a^2}{b} + \frac{2q_a q_b}{b} + \frac{q_b^2}{b}\right)$$

$$U = \frac{q_a^2}{8\pi\varepsilon a} + \frac{q_b^2}{8\pi\varepsilon b} + \frac{q_a q_b}{4\pi\varepsilon b}$$

PROBLEM 4.20

(a) Assuming that an atom may be regarded as a positive point charge nucleus in the center of a negative uniformly charged spherical shell, show that when the atom is excited so that the absolute value of its energy decreases n times, the radius of the shell increases n time (disregard the energy of the nucleus).

(b) The ionization energy of a hydrogen atom (the work required to excite the atom to zero energy) is 13.6 eV. Using the atomic model described in part a, find the radius of the electron shell of a hydrogen atom.

Solution:

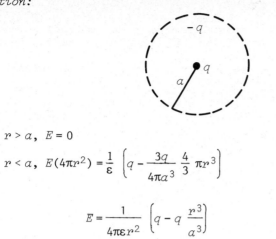

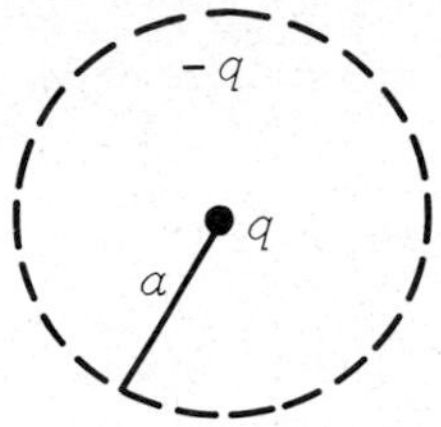

$r > a, \; E = 0$

$r < a, \; E(4\pi r^2) = \dfrac{1}{\varepsilon}\left(q - \dfrac{3q}{4\pi a^3}\,\dfrac{4}{3}\,\pi r^3\right)$

$$E = \frac{1}{4\pi\varepsilon r^2}\left(q - q\,\frac{r^3}{a^3}\right)$$

$$E = \frac{8}{4\pi\varepsilon}\left(\frac{1}{r^2} - \frac{1}{a^3}\right)$$

$$U = \tfrac{1}{2}\varepsilon \int E^2 \, dv = \tfrac{1}{2}\varepsilon\,\frac{q^2}{16\pi^2\varepsilon^2}\int_0^a \left(\frac{1}{r^4} - \frac{2}{ra^3} + \frac{r^2}{a^6}\right) 4\pi r^2 \, dr$$

$$U = \frac{q^2}{8\pi\varepsilon}\left(-\frac{1}{r} - \frac{2}{a^3}\,\frac{r^2}{2} + \frac{1}{a^6}\,\frac{r^5}{5}\right)\Bigg|_0^a$$

$$U = \frac{q^2}{8\pi\varepsilon}\left(-\frac{1}{a} - \frac{1}{a} + \frac{1}{5a}\right)$$

$$U = -\frac{1.8q^2}{8\pi\varepsilon a}$$

Let

$$U_0 = -\frac{1.8q^2}{8\pi\varepsilon a_0} \quad \text{and} \quad U_E = -\frac{1.8q^2}{8\pi\varepsilon a_1}$$

If

$$U_E = \frac{1}{n}\,U_0 \Rightarrow -\frac{1.8q^2}{8\pi\varepsilon a_1} = \frac{1}{n}\left(-\frac{1.8q^2}{8\pi\varepsilon a_0}\right)$$

$$a_1 = na_0$$

Proved.

$$E_{\text{ionization}} = U_\infty - U_0 = \frac{1.8q^2}{8\pi\varepsilon a_0}$$

$$13.6 \text{ eV} = 2.18 \times 10^{-18} \text{ J} = \frac{1.8q^2}{8\pi\varepsilon a_0}$$

$$a_0 = 9.5 \times 10^{-12} \text{ m}$$

PROBLEM 4.21

(a) What charge should be carried by a rain drop of 0.1 mm radius in order to counteract the force of gravity in a region where the earth's electric field is 130 V/m?
(b) If the breakdown field in air is 3×10^6 V/m, can the drop support this charge?

Solution:

(a) $mg = qE$

$$\frac{4}{3}\pi a^3 \rho g = qE$$

$$q = \frac{4\pi a^3 \rho g}{3E}$$

$$q = \frac{(4\pi)(10^{-4})^3(10^3)(9.8)}{(3)(130)} = 3.16 \times 10^{-10} \text{ C}$$

E from such a charge is

$$E = \frac{q}{4\pi\varepsilon r^2} = (9 \times 10^9)\left(\frac{3.16 \times 10^{-10}}{(10^{-4})^2}\right)$$

$$E = 2.84 \times 10^8 \text{ V/m}$$

(b) No.

PROBLEM 4.22 Eight equal negative charges are placed at the corners of a cube. What positive charge should be placed at the center of the cube to keep the negative charges in equilibrium?

Solution:

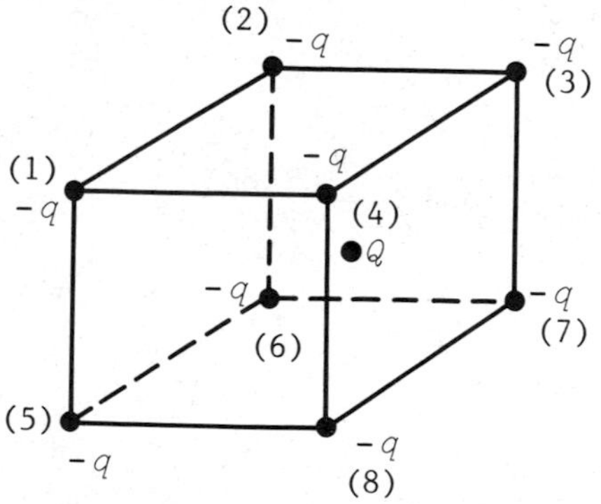

The distance of Q from a corner is

$$d^2 = \frac{a^2}{4} + \frac{a^2}{4} + \frac{a^2}{4}$$

$$d = \sqrt{3a^2/4}$$

Since there are eight interactions with the positive charge

$$U_+ = -8 \times \frac{1}{4\pi\varepsilon} \frac{Qq}{\sqrt{3a^2/4}}$$

The interaction energy of the negative charges is

$$U_- = U_{12} + U_{13} + U_{14} + U_{15} + U_{16} + U_{17} + U_{18} + U_{23} + U_{24} + U_{25} + U_{26} + U_{27} + U_{28} + U_{34}$$
$$+ U_{35} + U_{36} + U_{37} + U_{38} + U_{45} + U_{46} + U_{47} + U_{48} + U_{56} + U_{57} + U_{58} + U_{67} + U_{68} + U_{78}$$

$$U_- = \frac{q^2}{4\pi\varepsilon} \left(\frac{1}{a} + \frac{1}{a} + \frac{1}{\sqrt{2}a} + \frac{1}{a} + \frac{1}{\sqrt{2}a} + \frac{1}{\sqrt{3}a} + \frac{1}{\sqrt{2}a} + \frac{1}{a} + \frac{1}{\sqrt{2}a} + \frac{1}{\sqrt{2}a} + \frac{1}{a} + \frac{1}{\sqrt{2}a} + \frac{1}{\sqrt{3}a} + \frac{1}{a} + \frac{1}{\sqrt{3}a} \right.$$

$$\left. + \frac{1}{\sqrt{2}a} + \frac{1}{a} + \frac{1}{\sqrt{2}a} + \frac{1}{\sqrt{2}a} + \frac{1}{\sqrt{3}a} + \frac{1}{\sqrt{2}a} + \frac{1}{a} + \frac{1}{a} + \frac{1}{\sqrt{2}a} + \frac{1}{a} + \frac{1}{a} + \frac{1}{\sqrt{2}a} + \frac{1}{a} \right)$$

$$U_- = \frac{q^2}{4\pi\varepsilon a} \left(12 + \frac{12}{\sqrt{2}} + \frac{4}{\sqrt{3}} \right)$$

The total interaction energy is

$$U = \frac{q^2}{4\pi\varepsilon a} \left(12 + \frac{12}{\sqrt{2}} + \frac{4}{\sqrt{3}} \right) - \frac{Qq16}{4\pi\varepsilon\sqrt{3}a}$$

We want $\partial U/\partial a = 0$

$$\frac{\partial U}{\partial a} = -\frac{q^2}{4\pi\varepsilon a^2} \left(12 + \frac{12}{\sqrt{2}} + \frac{4}{\sqrt{3}} \right) + \frac{4Qq}{\pi\varepsilon\sqrt{3}} \left(-\frac{1}{a^2} \right) = 0$$

So

$$\frac{4Qq}{\pi\varepsilon\sqrt{3}} = \frac{q^2}{4\pi\varepsilon} \left(12 + \frac{12}{\sqrt{2}} + \frac{4}{\sqrt{3}} \right)$$

$$Q = \frac{\sqrt{3}}{16} \left(12 + \frac{12}{\sqrt{2}} + \frac{4}{\sqrt{3}} \right) q$$

$$Q = 0.108\ (12 + 8.49 + 2.3)q$$

$$Q = 2.46q$$

PROBLEM 4.23 The time-average potential of a neutral hydrogen atom is given by

$$\phi = q \frac{e^{-\alpha r}}{r} \left(1 + \frac{\alpha r}{2} \right)$$

where q is the magnitude of the electronic charge, and $\alpha^{-1} = a_0/2$, a_0 being the Bohr radius. Find the distribution of charge (both continuous and discrete) that will give this potential and interpret your result physically.

Solution:

$$\phi = q\,\frac{e^{-\alpha r}}{r}\left(1+\frac{\alpha r}{2}\right)$$

For $r \neq 0$,

$$\nabla^2\phi = -4\pi\rho$$

$$\nabla^2\phi = \frac{q}{r^2}\frac{\partial}{\partial r}\,r^2\,\frac{\partial}{\partial r}\left(\frac{e^{-\alpha r}}{r}+\frac{\alpha e^{-\alpha r}}{2}\right)$$

$$= q\,\frac{1}{r^2}\frac{\partial}{\partial r}\,r^2\left(-\frac{e^{-\alpha r}}{r^2}-\frac{\alpha e^{-\alpha r}}{r}-\alpha^2\,\frac{e^{-\alpha r}}{2}\right)$$

$$= q\,\frac{1}{r^2}\frac{\partial}{\partial r}\left(-e^{-\alpha r}-\alpha r e^{-\alpha r}-\alpha^2 r^2\,\frac{e^{-\alpha r}}{2}\right)$$

$$= -q\frac{1}{r^2}\left(-\alpha e^{-\alpha r}+\alpha e^{-\alpha r}-\alpha^2 r e^{-\alpha r}+\alpha^2 r e^{-\alpha r}-\alpha^3 r^2\,\frac{e^{-\alpha r}}{2}\right)$$

$$= q\,\alpha^3\,\frac{e^{-\alpha r}}{2}$$

$$\rho = \frac{-\nabla^2\phi}{4\pi} = \frac{-q\alpha^3 e^{-\alpha r}}{8\pi}$$

For $r = 0$, $\nabla^2\phi$ is undefined in spherical coordinates

$$\nabla^2\phi = -4\pi\rho(x)$$

$$q\,\nabla^2\,\frac{e^{-\alpha r}}{r}\left(1+\frac{\alpha r}{2}\right) = -4\pi\,\rho(x)$$

$$q\int\nabla^2\,\frac{e^{-\alpha r}}{r}\left(1+\frac{\alpha r}{2}\right)d^3r = -4\pi\int\rho(x)\,d^3r$$

Let $r \to 0$

$$q\int\nabla^2\left(\frac{1}{r}+\frac{\alpha}{2}\right)d^3r = -4\pi\rho(0)$$

$$-4\pi q\delta(\mathbf{r}) = -4\pi\rho(0)$$

since $\nabla^2\,(1/r) = -4\pi\delta(\mathbf{r})$.

$$\rho(0) = q\delta(\mathbf{r})$$

Thus

$$\rho(\vec{x}) = -q\alpha^3 \frac{e^{-\alpha r}}{8\pi} + q\delta(\mathbf{r})$$

The continuous part of the charge density is the probability density of the electron multiplied by the electron charge. The discrete part refers to the proton fixed in space at the origin.

PROBLEM 4.24 Use Gauss's theorem (and $\oint \mathbf{E} \cdot d\ell = 0$, if necessary) to prove the following:

(a) Any excess charge placed on a conductor must lie entirely on its surface. (A conductor by definition contians charges capable of moving freely under the action of applied electric fields.)

(b) A closed, hollow conductor shields its interior from fields due to charges outside, but does not shield its exterior from the fields due to charges placed inside it.

(c) The electric field at the surface of a conductor is normal to the surface and has a magnitude $4\pi\sigma$, where σ is the charge density per unit area on the surface.

Solution:

(a) Let the excess charge be positive. By Gauss's law, if $Q > 0$ then E is in the same direction as $\hat{\mathbf{n}}$. This E field will create a force on the charges toward the outer surface. As the charges leave the surface, the E field drops to zero. Then a bigger surface can be drawn to enclose the charges, and the same thing will occur. The limiting case of these processes will be when all of the charge is on the outer surface.

(b) Consider the above problem where it was shown that the E field inside a conductor is zero. A hole can be made in the conductor, which does not change the value of E. Thus, inside a region enclosed by a conductor, the E field is zero, independent of the exterior charge distribution.

If, however, charges were placed inside the conductor, by Gauss's law the E field outside the conductor, if one draws a Gaussian surface which encloses the conductor, depends only on the amount of charge inside, and is independent of the presence of conducting media. Therefore, the conductor does not shield the effect of charges inside of it.

(c) If a Gaussian surface is drawn on the surface of the conductor

$\oint \mathbf{E} \cdot \hat{\mathbf{n}}\, da = \oint \mathbf{E} \cdot \hat{\mathbf{n}}\, da$ from the top of the box

$+\oint \mathbf{E} \cdot \hat{\mathbf{n}}\, da$ for the four side faces which $= 0$

since $\mathbf{E} \cdot \hat{\mathbf{n}} = 0$ since E and $\hat{\mathbf{n}}$ are perpendicular

$+\oint \mathbf{E} \cdot \hat{\mathbf{n}}\, da$ from the bottom of the box which $= 0$

since $\mathbf{E} = 0$ inside a conductor

$$= \oint_{\text{top surface}} \mathbf{E}\cdot\hat{\mathbf{n}}\,da = EA = 4\pi q_{\text{enclosed}}\,, \quad E = 4\pi q/A, \quad q/A = \sigma, \quad \mathbf{E} = 4\pi\sigma\hat{\mathbf{n}}$$

PROBLEM 4.25 Calculate the multipole moments $q_{\ell m}$ of the charge distribution shown in the figures. Try to obtain results for the nonvanishing moments valid for all ℓ, but in each case find the first two sets of nonvanishing moments.

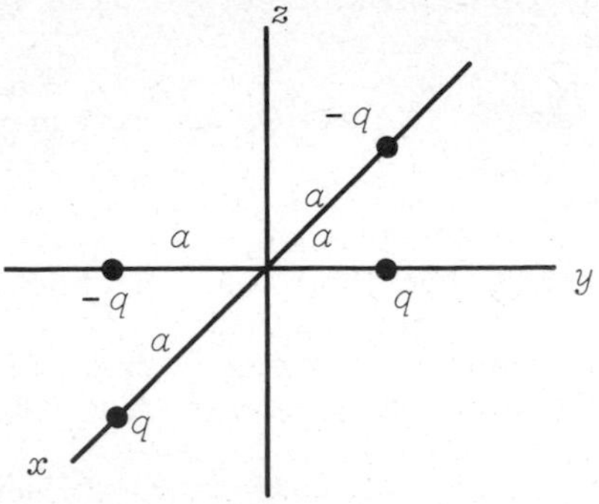

(a)

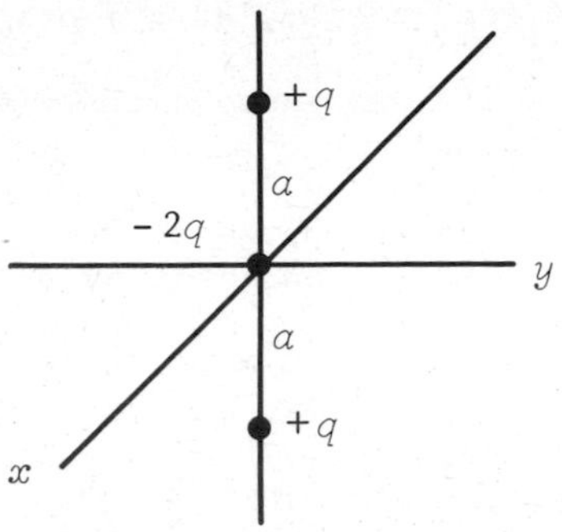

(b)

(c) For the charge distribution (b) write down the mutlipole expansion for the potential. Keeping only the lowest-order term in the expansion, plot the potential in the $x-y$ plane as a function of distance from the origin for distances greater than a.

(d) Calculate directly from Coulomb's law the exact potential for (b) in the $x-y$ plane. Plot it as a function of distance and compare with the result found in (c).

(e) Divide out the asymptotic form in parts (c) and (d) in order to see the behavior at large distances more clearly.

Solution:

(a) The charge density in spherical coordinates

$$\rho(x') = \frac{q}{r'^2}\,\delta(r'-a)\delta(\cos\theta')\left[\delta(\phi') + \delta\left(\phi' - \frac{\pi}{2}\right) - \delta(\phi'-\pi) - \delta\left(\phi' - \frac{3\pi}{2}\right)\right]$$

$$q_{\ell m} = Y^*_{\ell m}(\theta',\phi')r'^{\ell}\,\frac{q}{r'^2}\,\delta(r'-a)\delta(\cos\theta')\left[\delta(\phi') + \delta\left(\phi' - \frac{\pi}{2}\right) - \delta(\phi'-\pi)\right.$$

$$\left. - \delta\left(\phi' - \frac{3\pi}{2}\right)\right] r'^2\,dr'd(\cos\theta')d\phi'$$

$$= qa^{\ell}\left[Y^*_{\ell m}\left(\frac{\pi}{2}, 0\right) + Y^*_{\ell m}\left(\frac{\pi}{2},\frac{\pi}{2}\right) - Y^*_{\ell m}\left(\frac{\pi}{2},\pi\right) - Y^*_{\ell m}\left(\frac{\pi}{2},\frac{3\pi}{2}\right)\right]$$

$$= qa^{\ell}\sqrt{\frac{2\ell+1(\ell-m)!}{4\pi(\ell+m)!}}\,P_{\ell}^{m}(0)\left(1 + e^{-im\pi/2} - e^{-im\pi} - e^{-im\pi 3/2}\right)$$

$$(1+e^{-im\pi/2}-e^{-m\pi}-e^{-im3\pi/2}) = (1+e^{-im\pi/2})(1-e^{-im\pi})$$

$$= [1-(-1)^m][1-(-1)^m]$$

$$= \begin{cases} 0, & m = \ldots, -6, -4, -2, 0, 2, 4, \ldots \\ 2(\ell+i), & m = \ldots, -9, -5, -1, 3, 7, 11, \ldots \\ 2(\ell-i), & m = \ldots, -11, -7, -3, 1, 5, 9, \ldots \end{cases}$$

$$P_\ell^m(0) = \begin{cases} 0, & m \text{ even} \\ m+\ell, & \text{odd} \end{cases}$$

$$q_{\ell m} = qa^\ell \sqrt{\frac{2\ell+1(\ell-m)!}{4\pi(\ell+1)!}}\, P_\ell^m(0)2(1+(-i)^m), \quad m \text{ odd}, \; m+\ell \text{ even}$$

$$q_{00} = q_{10} = q_{20} = q_{30} = q_{22} = q_{32} = q_{2,-2} = q_{3,-2} = 0$$

$$q_{21} = q_{2,-1} = q_{32} = q_{3,-2} = 0$$

$$q_{11} = qa\sqrt{3/8\pi}\; P_1(0)\; 2(1-i) = 2qa\sqrt{3/2\pi}\;(1-i)$$

$$q_{1,-1} = qa\sqrt{3/8\pi}\;\; P_1^{-1}(0)2(1+i) = -qa\;\sqrt{3/2\pi}\;(1+i)$$

$$q_{3,1} = qa^3\sqrt{7/48\pi}\; P_3^1(0)2(1-i) = -3qa^3\;\sqrt{21/16}\;(1-i)$$

$$q_{3,-1} = qa^3\sqrt{21/\pi}\; P_3^{-1}(0)2(1+i) = +qa^3\;\sqrt{21/16}\;(1+i)$$

$$q_{3,3} = qa^3\sqrt{7/4\pi(720)}\; P_3^3(0)2(1+i)$$

$$q_{3,-1} = -qa^3\;\sqrt{35/8}\;(1-i)$$

$$q_{\ell m} = qa^\ell \sqrt{\frac{2\ell+1}{4\pi}}\sqrt{\frac{(\ell+m)!}{(\ell-m)!}}\,\frac{2^{m+1}}{\sqrt{\pi}}\,\frac{\Gamma(\frac{\ell+m+1}{2})}{\Gamma(\frac{\ell-m+2}{2})}\left[1-i(-)\frac{(m-1)}{2}\right](-)\left(\frac{\ell-m}{2}\right)$$

ℓ and m odd,

$q_{\ell m} = 0$ otherwise

(b) Charge density:

$$\rho(x') = \frac{q}{2\pi r'^2}[\delta(r'-a)\delta(\cos\theta'-1)+\delta(r'-a)\delta(\cos\theta'+1)$$

$$-2\delta(r')\delta(\cos\theta')](\tfrac{1}{2}\pi)$$

since we look at one point on plane, so divide by $\int_0^{2\pi} d\phi$:

$$q_{\ell m} = \int Y^*_{\ell m}(\theta',\phi')r'^{\ell}\frac{q}{2\pi r'^2}[\delta(r'-a)\delta(\cos\theta'-1)+\delta(r'-a)\delta(\cos\theta'+1)$$

$$-2\delta(r')\delta(\cos\theta')]r'^2\,dr'\,d(\cos\theta')d\phi'$$

$$= \frac{qa^{\ell}}{2\pi}\int Y^*_{\ell m}(0,\phi')\,d\phi' + \frac{qa^{\ell}}{2\pi}\int Y^*_{\ell m}(\pi,\phi')\,d\phi' + \frac{q\delta_{\ell,0}}{2\pi}\int Y^*_{0m}\left(\frac{\pi}{2},\phi'\right)d\phi'$$

$$\int Y^*_{\ell m}(0,\phi')\,d\phi' = \begin{cases}[\sqrt{(2\ell+1)/4\pi}]P_{\ell}(\cos\theta)2\pi, & m=0\\ 0, & m\neq 0\end{cases}$$

$$q_{\ell,0} = qa^{\ell}\sqrt{\frac{2\ell+1}{4\pi}}[P_{\ell}(1)+P_{\ell}(-1)] - \frac{2\delta_{\ell,0}}{\sqrt{4\pi}}$$

$$P_{\ell}(1)+P_{\ell}(-1) = \begin{cases}0, & \ell \text{ odd}\\ 2, & \ell \text{ even}\end{cases}$$

$$q_{00} = 0$$

$$q_{2n,0} = qa^{2n}\sqrt{(4n+1)/\pi}, \quad n=1,2,\ldots$$

$$q_{20} = qa^2\sqrt{5/\pi}$$

$$q_{40} = qa^4\sqrt{a/\pi}$$

(c)

$$\Phi(\vec{x}) = \sum_{\ell=0}^{\infty}\sum_{m=-\ell}^{\ell}\frac{4\pi}{2\ell+1}q_{\ell m}\frac{Y_{\ell m}(\theta,\phi)}{r^{\ell+1}}$$

$$= \sum_{n=1}^{\infty}\frac{4\pi}{4n+1}q_{2n,0}\frac{Y_{2n,0}(\theta,\phi)}{r^{2n+1}}$$

$$= \sum_{n=1}^{\infty} \frac{4\pi}{4n+1} q \frac{a^{2n}}{r^{2n+1}} \sqrt{\frac{4n+1}{\pi}} \sqrt{\frac{4n+1}{4\pi}} P_{2n}(\cos\theta)$$

$$= \sum_{n=1}^{\infty} 2q \frac{a^{2n}}{r^{2n+1}} P_{2n}(\cos\theta)$$

for lowest order,

$$\Phi(\vec{x}) = 2q \frac{a^2}{r^3} P_2(\cos\theta)$$

$$= q \frac{a^2}{r^3} (3\cos^2\theta - 1)$$

In the $x-y$ plane, $\theta = \pi/2$,

$$\Phi(r, \pi/2, \phi) = \frac{-qa^2}{r^3}$$

$$\Phi(a, \pi/2, \phi) = \frac{-q}{a}$$

(d) The potential in the $x-y$ plane is now found,

$$\Phi_c(r,\pi/2,\phi) = \frac{2q}{\sqrt{r^2+a^2}} \frac{-2q}{r}$$

$$= \frac{2q}{r}\left(\frac{1}{\sqrt{1+a^2/r^2}} - 1\right)$$

$$\Phi(a,\pi/2,\phi) = \frac{2q}{a}\left(\frac{1}{\sqrt{2}} - 1\right) = -0.59\frac{q}{a}$$

$$\rho(x') = \frac{q}{r'^2}[\delta(r'-a)\delta(\cos\theta'-1) + \delta(r'-a)\delta(\cos\theta'+1) - 2\delta(r')\delta(\cos\theta')]$$

$$q_{\ell m} = \int Y^*_{\ell m}(\theta',\phi') r'^{\ell} \frac{q}{r'^2}[\delta(r'-a)\delta(\cos\theta'-1) + \delta(r'-a)\delta(\cos\theta'+1)$$

$$- 2\delta(r')\delta(\cos\theta')] r'^2\, dr'\, d(\cos\theta') d\phi'$$

$$= qa^{\ell} \int Y^*_{\ell m}\,(0,\phi')d\phi' + qa^{\ell} \int Y^*_{\ell m}\,(\pi,\phi')d\phi' - 2q \int Y^*_{\ell=0,m}(\pi/2,\phi')d\phi'$$

for $\ell \neq 0$,

$$\int Y^*_{\ell m}\,(\theta,\phi')d\phi' = \sqrt{\frac{(2\ell+1)(\ell-m)!}{4\pi(\ell+m)!}}\, P_\ell^{\,m}(\cos\,\theta)\, \frac{e^{-im\phi}}{-im}\Bigg|_0^2$$

for $\ell = 0$,

$$\int Y^*_{00}\,(\theta,\phi')d\phi' = \frac{1}{\sqrt{4\pi}}\, P_0\,(\cos\,\theta)(2\pi) = \sqrt{\pi}$$

$$q_{00} = 0$$

$$q_{\ell m} = qa^{\ell} \sqrt{\frac{(2\ell+1)(\ell-m)!}{4\pi(\ell+m)!}}\, \frac{(2\pi)}{-im}\,[P_\ell^{\,m}(1) - P_\ell^{\,m}(-1)],\quad \ell \neq 0$$

$$P_\ell^{\,m}(+1) = 0,\ \text{for } m > 0$$

$$P_\ell^{\,m}(1) - P_\ell^{\,m}(-1) = \begin{cases} 0, & \ell+m \text{ odd} \\ 2P_\ell^{\,m}(1), & \ell+m \text{ even} \end{cases}$$

$$q_{\ell m} = \frac{2qa^{\ell}}{-im} \sqrt{\frac{\pi(2\ell+1)(\ell-m)!}{(\ell+m)!}}\, P_\ell^{\,m}(-1),\ \text{for } \ell+m \text{ even},\ m > 0$$

$$q_{00} = 0$$

$$q_{10} = 0$$

$$q_{11} = 0$$

(e)

$$\Phi(\vec{x}) = \sum_{\ell=0}^{\infty} \sum_{m=-\ell}^{+\ell} \frac{4\pi}{2\ell+1}\, q_{\ell m}\, \frac{Y_{\ell m}(\theta,\phi)}{r^{\ell+1}}$$

$$= \sum_{n=1}^{\infty} \frac{4\pi}{4n+1}\, q_{2n,0}\, \frac{Y_{2n,0}(\theta,\phi)}{r^{2n+1}}$$

$$= \sum_{n=1}^{\infty} \frac{4\pi}{4n+1} q \frac{a^{2n}}{r^{2n+1}} \sqrt{\frac{4n+1}{\pi}} \sqrt{\frac{4n+1}{4\pi}} P_{2n}(\cos\theta)$$

$$= \sum_{n=1}^{\infty} 2q \frac{a^{2n}}{r^{2n+1}} P_{2n}(\cos\theta)$$

lower order in $x-y$ plane, $\theta = \pi/2$,

$$\Phi(\vec{x}) = 2q \frac{a^2}{r^3} [3\cos^2(\pi/2) - 1] = \frac{-qa^2}{r^3}$$

Coulomb potential

$$\Phi(\vec{x}) = \frac{2q}{\sqrt{r^2+a^2}} - \frac{2q}{r} = \frac{2q}{r}\left(\frac{1}{\sqrt{1+a^2/r^2}} - 1\right)$$

$$\Phi(\vec{x}) \xrightarrow[r\to\infty]{} \frac{2q}{r}\left[\left(1 - \tfrac{1}{2}\frac{a^2}{r^2}\right) - 1\right]$$

$$\Phi(\vec{x}) \xrightarrow[r\to\infty]{} \frac{-qa^2}{r^3} \quad \text{same as multipole expansion}$$

	Multipole Expansion	Coulomb Potential	
Distance	$\Phi(r) = \frac{-qa^2}{r^3}$	$\Phi(r) = \frac{2q}{\sqrt{r^2+a^2}} - \frac{2q}{r}$	$\frac{\Phi(r)}{\Phi(a)}$
a	$\Phi(a) = \frac{-q}{a}$	$\Phi(a) = -0.586\frac{q}{a}$	1.71
$2a$	$\Phi(2a) = -0.125\frac{q}{a}$	$\Phi(2a) = -0.106\frac{q}{a}$	1.18
$\frac{3a}{2}$	$\Phi\left(\frac{3a}{2}\right) = -0.296\frac{q}{a}$	$\Phi\left(\frac{3a}{2}\right) = -0.224\frac{q}{a}$	1.32
$3a$	$\Phi(3a) = -0.037\frac{q}{a}$	$\Phi(3a) = -0.034\frac{q}{a}$	1.08
$\frac{7a}{8}$	$\Phi\left(\frac{7a}{8}\right) = -1.49\frac{q}{a}$	$\Phi\left(\frac{7a}{8}\right) = -0.780\frac{q}{a}$	1.91

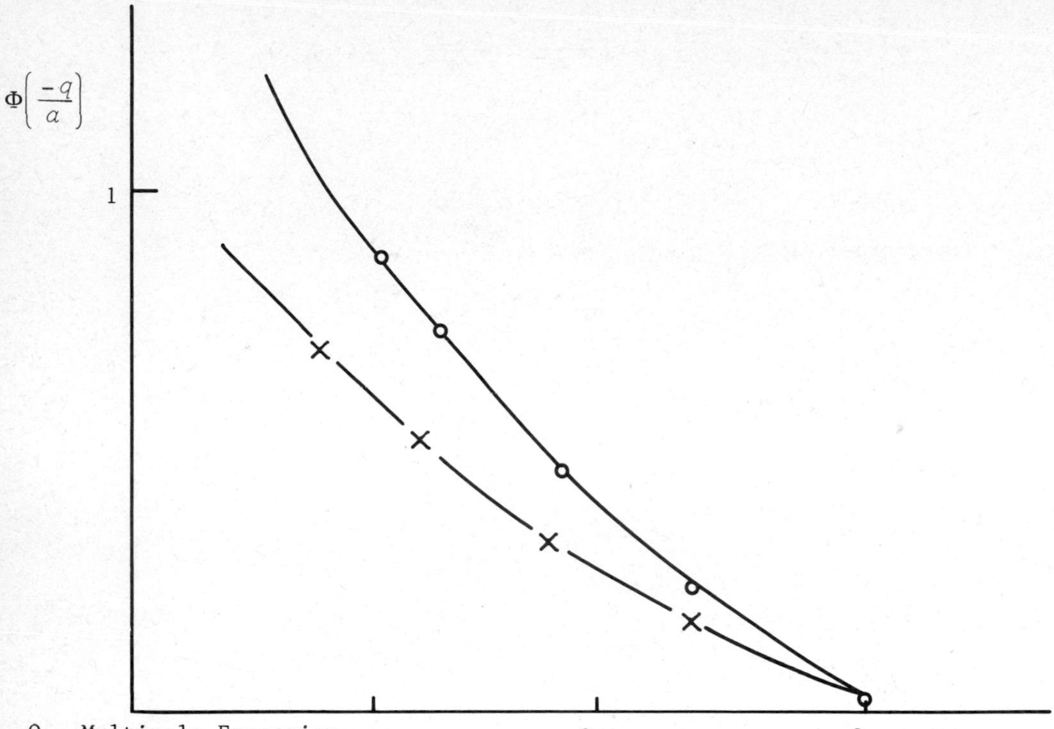

O - Multipole Expansion
X - Coulomb Potential

PROBLEM 4.26 A nucleus with quadrupole moment Q finds itself in a cylindrically symmetric electric field with a gradient $(\partial E_z/\partial z)_0$ along the z axis at the position of the nucleus.

(a) Show that the energy of quadrupole interaction is

$$W = \frac{-e}{4} Q \left(\frac{\partial E_z}{\partial z}\right)_0$$

(b) If it is known that $Q = 2 \times 10^{-24}$ cm^2 and that W/h is 10 MHz, where h is Planck's constant, calculate $(\partial E_z/\partial z)_0$ in units of $e/a_0{}^3$, where $a_0 = \hbar^2/me^2 = 0.529 \times 10^{-8}$ cm is the Bohr radius in hydrogen.

(c) Nuclear charge distributions can be approximated by a constant charge density throughout a spheroidal volume of semimajor axis a and semiminor axis b. Calculate the quadrupole moment of such a nuleus assuming that the total charge is Ze. Given that Eu^{153} $(z = 63)$ has a quadrupole moment $Q = 2.5 \times 10^{-24}$ cm^2 and a mean radius $R = (a + b)/2 = 7 \times 10^{-13}$ cm, determine the fractional difference in radius $(a - b)/R$.

Solution:

(a) $W = -\frac{1}{6} \sum_y Q_y \frac{\partial E_y}{\partial x_i}(0)$

Given: E field with $(\partial E/\partial z)_0\ \hat{\mathbf{z}}$

$$\nabla \cdot \mathbf{E} = \frac{\partial E_x}{\partial x} + \frac{\partial E_y}{\partial y} + \frac{\partial E_z}{\partial z} = 0$$

$$\left(\frac{\partial E_y}{\partial y}\right)_0 + \left(\frac{\partial E_x}{\partial x}\right)_0 = -\left(\frac{\partial E_z}{\partial z}\right)_0$$

and

$$Q_{xx} = Q_{yy} = -\tfrac{1}{2} Q_{zz}$$

$$W = -\frac{1}{6}\left(Q_{xx} \left.\frac{\partial E_x}{\partial x}\right|_0 + Q_{yy} \left.\frac{\partial E_y}{\partial y}\right|_0 + Q_{zz} \left.\frac{\partial E_z}{\partial z}\right|_0\right)$$

$$= -\frac{1}{6}\left[\frac{-Q_{zz}}{2}\left(\frac{-\partial E_z}{\partial z}\right)_0 + Q_{zz}\left(\frac{\partial E_z}{\partial z}\right)_0\right]$$

$$= -\tfrac{1}{4} Q_{zz} \left.\frac{\partial E_z}{\partial z}\right|_{z=0}$$

$Q = Q_{zz}/e$ by definition, therefore,

$$W = \frac{-e}{4} Q \left(\frac{\partial E_z}{\partial z}\right)_{z=0}$$

(b) $e = 4.8 \times 10^{-10}$ $\text{dyn}^{1/2} \cdot \text{cm}$ (statcoulomb)

$a_0 = 0.529 \times 10^{-8}$ cm

$h = 6.63 \times 10^{-27}$ dyn · cm · s

Q, (cm^2), ($a_0{}^2$)

q, ($\text{dyn}^{1/2} \cdot \text{cm}$), ($e$)

$E = \frac{F}{q}$, ($\text{dyn}^{1/2}/\text{cm}$), ($e/a_0{}^2$)

$\frac{\partial E}{\partial z}$, ($\text{dyn}^{1/2}/\text{cm}^2$), ($e/a_0{}^3$)

h, (dyn · cm · s), ($e^2/a_0 Hz$)

$$\left(\frac{\partial E_z}{\partial z}\right)_0 = \frac{h4W/h}{eQ} = \frac{(4)(6.63 \times 10^{-27}\ \text{dyn} \cdot \text{cm} \cdot \text{s})(10^7\ \text{s}^{-1})}{(4.8 \times 10^{-10}\ \text{dyn}^{1/2} \cdot \text{cm})(2 \times 10^{-24}\ \text{cm}^2)}$$

$$= 2.76 \times 10^{14}\ \text{dyn}^{\frac{1}{2}} \cdot \text{cm}^{-2}$$

$$= 2.76 \times 10^{14}\ \frac{\text{dyn}^{\frac{1}{2}}}{\text{cm}^2} \times \frac{e}{4.8 \times 10^{-10}\ \text{dyn}^{\frac{1}{2}} \cdot \text{cm}} \times \left(\frac{0.529 \times 10^{-8}\ \text{cm}}{a_0}\right)^3$$

$$= 0.085\ \frac{e}{a_0{}^3}$$

(c) Semimajor axis a, $\hat{\mathbf{z}}$
Semimajor axis b, in the $x - y$ plane
Cylindrical coordinates

$$r^2 = z^2 + \rho^2$$

$$\rho^2_{\text{max}}(z) = b^2 \left(1 - \frac{z^2}{a^2}\right)$$

$$\rho_{\text{max}}(0) = b$$

$$\rho_{\text{max}}(a) = 0$$

$$\text{Volume } (V) = \int_0^{2\pi} d\theta \int_{-a}^{a} dz \int_0^{\rho_{\text{max}}(z)} \rho\, d\rho$$

$$= 2\pi \int_{-a}^{+a} dz\ \frac{b^2(1 - z^2/a^2)}{2}$$

$$= \pi b^2 \left(2a - \frac{2a^3}{3a^2}\right) = \frac{4}{3}\pi b^2 a$$

$$\rho = \frac{ze}{(4/3)b^2 a\pi}$$

$$Q = \frac{1}{e} \int (3z^2 - r^2) e(r)\, d^3r$$

$$= \frac{z}{(4/3)\pi b^2 a} \int_0^{2\pi} d\theta \int_{-a}^{+a} dz \int_0^{\rho_{\text{max}}(z)} (2z^2 - \rho^2)\rho\, d\rho$$

$$= \frac{3z}{2b^2 a} \int_{-a}^{+a} dz \left[2z^2 \frac{b}{2} \left(1 - \frac{z^2}{a^2}\right) - \frac{b^4(1 - z^2/a^2)^2}{4} \right]$$

$$Q = \frac{3z}{2b^2 a} \int_{-a}^{a} dz \left(z^2 b^2 - \frac{z^4 b^2}{a^2} - \frac{b^4}{4} + \frac{b^4 z^2}{2a^2} - \frac{b^4 z^4}{4a^4} \right)$$

$$= \frac{3z}{2b^2 a} \left(\frac{2a^3 b^2}{3} - \frac{2a^3 b^2}{5} - \frac{ab^4}{2} + \frac{ab^4}{3} - \frac{ab^4}{10} \right)$$

$$Q = \frac{2z(a^2 - b^2)}{5}$$

Given:

$$a^2 - b^2 = 5Q/2z = 0.099 \times 10^{-24} \text{ cm}^2$$

$$a + b = 14 \times 10^{-13} \text{ cm}$$

$$a - b = \frac{9.9 \times 10^{-26} \text{ cm}^2}{14 \times 10^{-13} \text{ cm}} = 0.707 \times 10^{-13} \text{ cm}$$

$$\frac{a - b}{R} = \frac{0.707 \times 10^{-13} \text{ cm}}{7 \times 10^{-13} \text{ cm}} = 0.101$$

PROBLEM 4.27 A localized distribution of charge has a charge density

$$\rho(\mathbf{r}) = \frac{1}{64\pi} r^2 e^{-r} \sin^2 \theta$$

(a) Make a multipole expansion of the potential due to this charge density and determine all the nonvanishing multipole moments. Write down the potential at large distances as a finite expansion Legendre polynomials.

(b) Determine the potential explicitly at any point in space, and show that near the origin, correct to r^2 inclusive,

$$\Phi(\mathbf{r}) \simeq \tfrac{1}{4} - \frac{r^2}{120} P_2(\cos \theta)$$

(c) If there exists at the origin a nucleus with a quadrupole moment $Q = 10^{-24}$ cm^2, determine the magnitude of the interaction energy, assuming that the unit of charge in $\rho(\mathbf{r})$ above is the electronic charge and the unit of length is the hydrogen Bohr radius $a_0 = \hbar^2/me^2 = 0.529 \; 10^{-8}$ cm. Express your answer as a frequency by dividing by Planck's constant h.

The charge density in this problem is that for the $m = +1$ states of the $2p$ level in hydrogen, while the quadrupole interaction is of the same order as found in molecules.

Solution:

(a) $\rho(\vec{r}\,') = \dfrac{1}{64\pi} r'^2 e^{-r'} \sin^2\theta'$

$$q_{\ell m} = \int Y^*_{\ell m}(\theta',\phi') r'^{\ell} \rho(x') d^3x'$$

$$= \frac{1}{64\pi} \int Y^*_{\ell m}(\theta',\phi') r'^{\ell+4} e^{-r'} \sin^2\theta'\, d(\cos\theta') d\phi'$$

$$= \frac{1}{32}\sqrt{\frac{2\ell+1}{4\pi}} \int P_\ell(\cos\theta)\sin^2\theta'\, d\theta' \int r'^{\ell+4} e^{-r'}\, dr'$$

$$= \frac{1}{32}\sqrt{\frac{2\ell+1}{4}}\,\Gamma(\ell+r) \int P_\ell(\cos\theta)\, \frac{2}{3}[P_0(\cos\theta') - P_2(\cos\theta')] d\theta'$$

$$= \frac{\Gamma(\ell+5)}{48}\sqrt{\frac{2\ell+1}{4}}\left(2\delta_{\ell,0} - \frac{2}{5}\delta_{\ell,2}\right)$$

$$q_{00} = \frac{\Gamma(5)}{48}\frac{1}{\sqrt{4\pi}}\,2 = \frac{1}{\sqrt{4\pi}}$$

$$q_{20} = \frac{-\Gamma(7)}{48}\sqrt{\frac{5}{4\pi}}\,\frac{2}{5} = \sqrt{\frac{45}{\pi}}$$

$$\Phi = \sum Y_{\ell m}(\theta,\phi) q_{\ell m} \frac{4\pi}{2\ell+1}\frac{1}{r^{\ell+1}}$$

$$= Y_{00}\, q_{00}\frac{4\pi}{r} + Y_{20}\, q_{20}\frac{4\pi}{5r^3}$$

$$\Phi = \frac{1}{r} - \frac{6}{r^3} P_2(\cos\theta)$$

The potential at any r is

$$\Phi = \int \frac{\rho(r')dv'}{|\mathbf{x} - \mathbf{x}'|}$$

$$\Phi = \frac{1}{64\pi}\sum_{\ell,m}\frac{4\pi}{2\ell+1}\int\frac{r_<^{\ell}}{r_>^{\ell+1}}r'^4e^{-r'}\,dr'\,Y_{\ell m}(\theta,\phi)$$

$$\times\int Y^*_{\ell m}(\theta',\phi')\sin^2\theta'\,d(\cos\theta')\,d\phi'$$

$$= \frac{1}{32}\sum_{\ell,m}\frac{4\pi}{2\ell+1}\int\frac{r_<^{\ell}}{r_>^{\ell+1}}r'^4e^{-r'}\,dr'\,Y_{\ell m}(\theta,\phi)\sqrt{\frac{(2\ell+1)(\ell-m)!}{4\pi(\ell+m)!}}$$

$$\times\int P_\ell^{\,m}(\cos\theta)\sin^2\theta'\,d(\cos\theta')\delta_{m,0}$$

$$= \frac{1}{32}\sum_{\ell}\int\frac{r_<^{\ell}}{r_>^{\ell+1}}r'^4e^{-r'}\,dr'\,P_\ell(\cos\theta)$$

$$\times\int P_\ell(\cos\theta')\,\frac{2}{3}[P_0(\cos\theta')-P_2(\cos\theta')]d(\cos\theta')$$

$$=\frac{1}{48}\sum_{\ell}\int\frac{r_<^{\ell}}{r_>^{\ell+1}}r'^4e^{-r'}\,dr'\,P_\ell(\cos\theta)\left(2\delta_{\ell,0}-\frac{2}{5}\delta_{\ell,2}\right)$$

$$=\frac{1}{24}\int_0^\infty\frac{r'^4e^{-r'}\,dr'}{r_>}-\frac{P_2(\cos\theta)}{120}\int_0^\infty\frac{r_<^{\,2}r'^4e^{-r'}\,dr'}{r_>^{\,3}}$$

$$=\frac{1}{24}\left(\frac{1}{r}\int_0^r r'^4e^{-r'}\,dr'+\int_r^\infty r'^3e^{-r'}\,dr'\right)$$

$$-\frac{P_2(\cos\theta)}{120}\left(\frac{1}{r^3}\int_0^r r'^6e^{-r'}\,dr'+r^2\int_r^\infty r'e^{-r'}\,dr'\right)$$

$$=\frac{1}{24}\left(-r^3e^{-r}-4r^2e^{-r}-12re^{-r}-24e^{-r}-\frac{24e^{-r}}{r}+\frac{24}{r}\right)$$

$$+\frac{1}{24}\left(r^3e^{-r}+er^2e^{-r}+6e^{-r}r+6e^{-r}\right)$$

$$-\frac{P_2(\cos\theta)}{120r^2}\left(-r^5e^{-r}-6r^4e^{-r}-30r^3e^{-r}-120r^2e^{-r}-360re^{-r}-720e^{-r}\right.$$

$$\left.-\frac{720e^{-r}}{r}+\frac{720}{r}+\frac{P_2(\cos\theta)}{120}r^2e^{-r}(r+1)\right)$$

$$\Phi = \frac{1}{r} \frac{-e^{-r}}{24}\left(r^2 + 6r + 18 + \frac{24}{r}\right)$$

$$+\frac{P_2(\cos\theta)e^{-r}}{120r^2}\left(5r^4 + 30r^3 + 120r^2 + 360r + 720 + \frac{720}{r}\right) - \frac{P_2(\cos\theta)6}{r^3}$$

$$\Phi = \frac{1}{r}\left[1 - e^{-r}\left(1 + \frac{3}{4}r + \frac{r^2}{4} + \frac{r^3}{24}\right)\right]$$

$$-\frac{P_2(\cos\theta)6}{r^3}\left[1 - e^{-r}\left(1 + r + \frac{r^2}{2} + \frac{r^3}{6} + \frac{r^4}{24} + \frac{r^5}{144}\right)\right]$$

near the origin,

$$e^{-r} \simeq 1 - r + \frac{r^2}{2} - \frac{r^3}{6} + \frac{r^4}{24} + \ldots$$

keeping terms $\sim r^2$

$$\frac{1}{r}\left[1 - e^{-r}\left(1 + \frac{3}{4}r + \frac{r^2}{4} + \frac{r^3}{24}\right)\right]$$

$$\simeq \frac{1}{r}\left[1 - \left(1 - r + \frac{r^2}{2} - \frac{r^3}{6} + \ldots\right)\left(1 + \frac{3}{4}r + \frac{r^2}{4} + \frac{r^3}{24}\right)\right]$$

$$\simeq \frac{1}{r}\left[1 - \left(1 + \frac{3}{4}r + \frac{r^2}{4} + \frac{r^3}{24} - r - \frac{3}{4}r^2 - \frac{r^3}{4} + \frac{r^2}{2} + \frac{3r^3}{8} - \frac{r^3}{6} + \ldots\right)\right]$$

$$\simeq \frac{1}{r}\left(\frac{r}{4}\right)$$

$$\simeq \frac{1}{4}$$

$$\frac{-P_2(\cos\theta)6}{r^3}\left[1 - e^{-r}\left(1 + r + \frac{r^2}{2} + \frac{r^3}{6} + \frac{r^4}{24} + \frac{r^5}{144}\right)\right]$$

$$\simeq \frac{P_2(\cos\theta)6}{r^3}\left[1 - \left(1 - r + \frac{r^2}{2} - \frac{r^3}{6} + \frac{r^4}{24} - \frac{r^5}{120} + \ldots\right)\left(1 + r + \frac{r^2}{2} + \frac{r^3}{6} + \frac{r^4}{24} + \frac{r^5}{144}\right)\right]$$

$$\simeq \frac{P_2(\cos\theta)}{r^3}\left[6-\left(6-6r+3r^2-r^3+\frac{r^4}{4}-\frac{r^5}{20}+\ldots\right)\left(1+r+\frac{r^2}{2}+\frac{r^3}{6}+\frac{r^4}{24}+\frac{r^5}{144}\right)\right]$$

$$\simeq \frac{P_2(\cos\theta)}{r^3}\left[6-\left(6+6r+3r^2+r^3+\frac{r^4}{4}+\frac{r^5}{24}-6r-6r^2-3r^3-r^4-\frac{r^5}{4}+3r^2+3r^3\right.\right.$$
$$\left.\left.+\frac{3r^4}{2}+\frac{r^5}{2}-r^3-r^4-\frac{r^5}{2}+\frac{r^4}{4}+\frac{r^5}{4}-\frac{r^5}{20}\right)\right]$$

$$\simeq \frac{P_2(\cos\theta)}{r^3}\left(\frac{-r^5}{24}+\frac{r^5}{20}\right)$$

$$\simeq \frac{-P_2(\cos\theta)r^2}{120}$$

$$\Phi \xrightarrow[r\to 0]{} \frac{1}{4}-\frac{P_2(\cos\theta)r^2}{120}$$

(b) $\Phi = \int \frac{\rho\, dv'}{|\vec{x}-\vec{x}'|}$

$$\rho = \frac{1}{64\pi} r'^2 e^{-r'} \sin^2\theta'$$

$$|\vec{x}-\vec{x}'| = \sum \frac{r_<^{\ell}}{r_>^{\ell+1}} \frac{4\pi}{2\ell+1} Y_{\ell m}(\theta,\phi) Y^*_{\ell m}(\theta',\phi')$$

(1) $r_< = r'$, $r_> = r$, $r\to\infty$,

$$\Phi = \frac{1}{64\pi}\sum \int \frac{r'^{\ell+4}}{r^{\ell+1}} \frac{4\pi}{2\ell+1} e^{-r'} Y_{\ell m}(\theta,\phi) Y^*_{\ell m}(\theta',\phi') \sin^2\theta'\, d(\cos\theta')d\phi'$$

$$= \sum \frac{1}{32} \frac{P_\ell(\cos\theta)}{r^{\ell+1}} \int r'^{\ell+4} e^{-r'}\, dr'$$

$$\times \int P_\ell(\cos\theta')\, \frac{2}{3}[P_0(\cos\theta') - P_2(\cos\theta')]\, d(\cos\theta')$$

$$= \sum \frac{1}{32} \frac{P_\ell(\cos\theta)}{r^{\ell+1}} \Gamma(\ell+5) \frac{2}{3}\left(2\delta_{\ell,0} - \frac{2}{5}\delta_{\ell,2}\right)$$

$$= \frac{1}{32} \frac{\Gamma(5)}{r} \frac{4}{3} - \frac{1}{32} \frac{P_2(\cos\theta)}{r^3} \Gamma(7) \frac{2}{3}\frac{2}{5}$$

$$= \frac{1}{r} - \frac{6}{r^3} P_2(\cos\theta)$$

(2) $r_< = r$, $r_> = r'$, $r \to 0$,

$$\Phi = \frac{1}{64\pi} \sum \int r'^{(3-\ell)} r^\ell \frac{4\pi}{2\ell+1} e^{-r'} Y_{\ell m}(\theta,\phi) Y^*_{\ell m}(\theta',\phi') \sin^2\theta' \, d(\cos\theta') d\phi'$$

$$= \sum \frac{r^\ell}{32} P_\ell(\cos\theta)\, \Gamma(4-\ell) \frac{2}{3}\left(2\delta_{\ell,0} - \frac{2}{5}\delta_{\ell,2}\right)$$

$$= \frac{\Gamma(4)}{32} \frac{2}{3}\, 2 - \frac{r^2}{32} P_2(\cos\theta)\, \Gamma(2) \frac{2}{3}\frac{2}{5}$$

$$= \frac{1}{4} - \frac{r^2}{120} P_2(\cos\theta)$$

(c) $e = 4.8 \times 10^{-10}$ esu

$e = 4.8 \times 10^{-10}$ $\text{dyn}^{1/2} \cdot \text{cm}$

$a_0 = 0.529 \times 10^{-8}$ cm

$\frac{\partial E}{\partial z}$ $(\text{dyn}^{1/2} \cdot \text{cm}^{-2})$ measured in $\frac{e}{a_0{}^3}$

$$\Phi = \frac{1}{4} - \frac{r^2}{120} \frac{1}{2} (3\cos^2\theta - 1) = \frac{1}{4} - \frac{z^2}{120} + \frac{x^2+y^2}{240}$$

$$E_z = \frac{-\partial\Phi}{\partial z} = \frac{z}{60}$$

$$\left.\frac{\partial E}{\partial z}\right|_{z_0} = \frac{1}{60}\left(\frac{e}{a_0{}^3}\right)$$

$$\frac{W}{h} = \frac{e}{4h} Q \left(\frac{\partial E_z}{\partial z}\right)_{z=0}$$

$$= \frac{4.8 \times 10^{-10}\ \mathrm{dyn}^{\frac{1}{2}} \cdot \mathrm{cm}}{(4)(6.63 \times 10^{-27}\ \mathrm{dyn} \cdot \mathrm{cm} \cdot \mathrm{s})} 10^{-24}\ \mathrm{cm}^2 \frac{1}{60} \frac{4.8 \times 10^{-10}\ \mathrm{dyn}^{\frac{1}{2}} \cdot \mathrm{cm}}{(0.529 \times 10^{-8}\ \mathrm{cm})^3}$$

$$\frac{W}{h} = 0.98 \times 10^7\ \mathrm{Hz} \simeq 10^6\ \mathrm{Hz}$$

PROBLEM 4.28 The multipole expansion for $\mathbf{A}(\mathbf{r},t)$ has been discussed by assuming a harmonic time dependence $[\exp(-i\omega t)]$ and then examining the power series expansion of $\exp(i\mathbf{k} \cdot \mathbf{x})$. This problem provides another route to the multipole expansion, with an example to be worked out in the case of a closed current loop $(\partial\rho/\partial t = 0)$; we start with the vector potential

$$\mathbf{A}(\mathbf{r},t) = \frac{1}{c} \int d^3x \frac{\mathbf{j}(\mathbf{x}, t - |\mathbf{r} - \mathbf{x}|/c)}{|\mathbf{r} - \mathbf{x}|}$$

(a) Use a Taylor's series expansion of the integrand of the equation to obtain an approximation to $\mathbf{A}(\mathbf{r},t)$ which involves at most first derivatives and factors $(\mathbf{x} \cdot \mathbf{r}/r^2)^m$, $m = 0$ and 1.

(b) Using an analysis similar to Chapter 9 of Jackson (*Electrodynamics*, John Wiley & Sons, 1975), the closed-current-loop assumption, and the replacement $\partial/\partial r \to -\partial/\partial(ct)$, show that the result of (a) can be expressed as a sum of terms involving only the magnetic moment $\mathbf{m} = \mathbf{m}(t - r/c)$ and its time derivative.

(c) Finally, assuming $\mathbf{m} = \mathbf{m}_0 \cos\omega(t - r/c)$, find $\mathbf{E}$ and $\mathbf{B}$ and their radiation zone limits, stating your results in spherical coordinates.

Solution:

(a)

$$\frac{\mathbf{j}(\mathbf{x}, t - |\mathbf{r} - \mathbf{x}|/c)}{|\mathbf{r} - \mathbf{x}|} \cong \frac{1}{c} \frac{\mathbf{j}\left[\mathbf{x}, t - \frac{r}{c}(1 - \mathbf{x} \cdot \mathbf{r}/r^2)\right]}{\frac{r}{c}(1 - \mathbf{x} \cdot \mathbf{r}/r^2)}$$

$$\cong \frac{1}{r} \mathbf{j}(\mathbf{x}, t - r/c) - \frac{\mathbf{x} \cdot \mathbf{r}}{c^2 r} \frac{\partial}{\partial(r/c)} \left(\frac{\mathbf{j}(\mathbf{x}, t - r/c)}{r/c}\right)$$

$$= \frac{1}{r} \mathbf{j}(\mathbf{x}, t - r/c) - \frac{\mathbf{x} \cdot \mathbf{r}}{r} \frac{\partial}{\partial r} \left(\frac{\mathbf{j}(\mathbf{x}, t - r/c)}{r}\right)$$

$$\mathbf{A}(\mathbf{r},t) \cong \frac{1}{cr} \int d^3x\ \mathbf{j}(\mathbf{x}, t - r/c) - \frac{1}{cr} \int d^3x\ \mathbf{x} \cdot \mathbf{r} \frac{\partial}{\partial r} \left(\frac{\mathbf{j}(\mathbf{x}, t - r/c)}{r}\right)$$

(b) Now

$$\int d^3x\ \mathbf{j}(\mathbf{x},t-r/c) = -\int d^3x\ \mathbf{x}(\nabla_x \cdot \mathbf{j}) = \int d^3x\ \mathbf{x}\ \frac{\partial\rho}{\partial t}$$

where $\partial\rho/\partial t = 0$;

$$\int d^3x\ \mathbf{x}\cdot\mathbf{r}\ \frac{\partial}{\partial r}\left(\frac{\mathbf{j}(\mathbf{x},t-r/c)}{r}\right)$$

$$= \int d^3x\ \mathbf{x}\cdot\mathbf{r}\left[\frac{-\mathbf{j}(\mathbf{x},t-r/c)}{r^2} + \frac{1}{r}\frac{\partial}{\partial r}\ \mathbf{j}(\mathbf{x},t-r/c)\right]$$

$$= -\int d^3x\ \mathbf{x}\cdot\mathbf{r}\left(\frac{1}{r^2}\ \mathbf{j}(\mathbf{x},t-r/c) + \frac{1}{rc}\frac{\partial}{\partial t}\ \mathbf{j}(\mathbf{x},t-r/c)\right)$$

Now

$$\mathbf{m} = \frac{1}{2c}\int d^3x\ \mathbf{x}\times\mathbf{j}.$$

$(\mathbf{x}\cdot\mathbf{r})\mathbf{j}$ and $(\mathbf{x}\cdot\mathbf{r})\ \frac{\partial}{\partial t}\ \mathbf{j}$:

$(\mathbf{x}\cdot\mathbf{r})\mathbf{j} = \frac{1}{2}(\mathbf{x}\times\mathbf{j})\times\mathbf{r} + \frac{1}{2}[\mathbf{j}(\mathbf{x}\cdot\mathbf{r}) + \mathbf{x}(\mathbf{j}\times\mathbf{r})]$

$$(\mathbf{x}\cdot\mathbf{r})\ \frac{\partial}{\partial t}\ \mathbf{j} = \frac{1}{2}\left(\mathbf{x}\times\frac{\partial}{\partial t}\ \mathbf{j}\right)\times\mathbf{r} + \frac{1}{2}\left[\frac{\partial\mathbf{j}}{\partial t}(\mathbf{x}\cdot\mathbf{r}) + \mathbf{x}\left(\frac{\partial\mathbf{j}}{\partial t}\cdot\mathbf{r}\right)\right]$$

Both $(\frac{1}{2})(\ )$ terms are of the form

$[\mathbf{b}(\mathbf{x}\cdot\mathbf{r}) + \mathbf{x}(\mathbf{b}\cdot\mathbf{r})] = \mathrm{B}$

and B gives zero when integrated.

Proof: Let $\mathbf{a}$ = constant vector

$$\int d^3x\ \mathbf{a}\cdot\mathrm{B} = \int d^3x\ [(\mathbf{a}\cdot\mathbf{b})(\mathbf{x}\cdot\mathbf{r}) + (\mathbf{a}\cdot\mathbf{x})(\mathbf{b}\cdot\mathbf{r})]$$

But

$$(\ ) = \mathbf{b}\cdot\nabla_x\ \{(\mathbf{a}\cdot\mathbf{x})(\mathbf{x}\cdot\mathbf{r})\}$$

that is,

$$= \mathbf{b}\cdot\{(\mathbf{a}\cdot\mathbf{x})[(\mathbf{r}\cdot\nabla_x)\mathbf{x} + \mathbf{r}\times(\nabla_x\times\mathbf{x})] + (\mathbf{x}\cdot\mathbf{r})[(\mathbf{a}\cdot\nabla_x)\mathbf{x} + \mathbf{a}\times(\nabla_x\times\mathbf{x})]\}$$

(where $\nabla_x\times\mathbf{x} = 0$),

$$= \mathbf{b}\cdot\{(\mathbf{a}\cdot\mathbf{x})\mathbf{r} + (\mathbf{x}\cdot\mathbf{r})\mathbf{a}\} \qquad \text{Q.E.D.}$$

therefore,

$$\int d^3x\ \mathbf{a}\cdot\mathbf{B} = \int d^3x\ \mathbf{b}\cdot\nabla_x\{(\mathbf{a}\cdot\mathbf{x})(\mathbf{x}\cdot\mathbf{r})\}$$

$$= \int d^3x\ \{\nabla_x\cdot\mathbf{b}[(\mathbf{a}\cdot\mathbf{x})(\mathbf{x}\cdot\mathbf{r})] - (\mathbf{a}\cdot\mathbf{x})(\mathbf{x}\cdot\mathbf{r})\nabla_x\cdot\mathbf{b}\}$$

$$= \int_S d\mathbf{s}\cdot\mathbf{b}[(\mathbf{a}\cdot\mathbf{x})(\mathbf{x}\cdot\mathbf{r})] - \int (\mathbf{a}\cdot\mathbf{x})(\mathbf{x}\cdot\mathbf{r})\nabla_x\cdot\mathbf{b}$$

Now $\mathbf{b} = \mathbf{j}$ in one case and $\partial\mathbf{j}/\partial t$ is the other. $\mathbf{j}$ and $\partial\mathbf{j}/\partial t = 0$ on the surface, so both surface integrals are zero.

$$\nabla_x\cdot\mathbf{j} = -\frac{\partial\rho}{\partial t} = 0$$

$$\nabla_x\cdot\frac{\partial\mathbf{j}}{\partial t} = -\frac{\partial^2\rho}{\partial t^2} = 0$$

Therefore, the integrals are zero, as claimed. So,

$$\mathbf{A}(\mathbf{r},t) \simeq \frac{\mathbf{m}(t - r/c)x\mathbf{r}}{r^3} + \frac{1}{cr^2}\,\frac{\partial\mathbf{m}(t - r/c)}{\partial t}\,x\mathbf{r}$$

(c) $\mathbf{m}(t - r/c) = \mathbf{m}_0\ \cos(\omega t - \omega r/c)$

$$\frac{\partial\mathbf{m}}{\partial t} = -\,\omega\mathbf{m}_0\ \sin\omega\ (t - r/c)$$

therefore,

$$\mathbf{A}(\mathbf{r},t) \simeq \frac{\mathbf{m}_0\times\mathbf{r}}{r^3}\ \cos(\omega t - \omega r/c) - \frac{k\mathbf{m}_0\times\mathbf{r}\sin\omega(t - r/c)}{r^2}$$

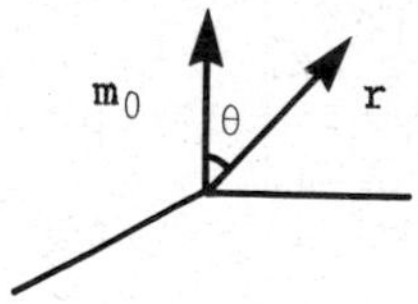

Choose $\mathbf{m}_0 = m_0\hat{\mathbf{e}}_3$ =>

$$\mathbf{m}_0\times\mathbf{r} = m_0 r\sin\theta\,\hat{\mathbf{e}}_\phi$$

$$\therefore\quad \mathbf{A} = A_\phi\hat{\mathbf{e}}_\phi,\ A_\phi = \frac{m_0\sin\theta}{r}\left(\frac{\cos\ \omega(t - r/c)}{r} - k\sin\omega(t - r/c)\right)$$

$$\mathbf{E}(r,t) = -\frac{1}{c}\frac{\partial\mathbf{A}}{\partial t} = +\frac{m_0\sin\theta}{cr}\left[\frac{\omega}{r}\sin\left(\omega t - \frac{\omega r}{c}\right) + k\omega\cos\left(\omega t - \frac{r\omega}{c}\right)\right]\hat{\mathbf{e}}_\phi$$

$$\xrightarrow[r\to\infty]{}\ \frac{+m_0 k^2}{r}\sin\theta\cos\omega\ (t - r/c)\hat{\mathbf{e}}_\phi$$

where r is in radiation zone,

$\mathbf{B} = \nabla \times \mathbf{A}$, in spherical coordinates

$$B_r = \frac{1}{r \sin\theta} \frac{\partial}{\partial\theta} \left[\sin^2\theta \frac{m_0}{r} \left(\frac{\cos\omega(t - r/c)}{r} - k \sin\omega(t - r/c) \right) \right]$$

$$= \frac{2\cos\theta}{r^2} m_0 \left(\frac{\cos\omega(t - r/c)}{r} - k \sin\omega(t - r/c) \right)$$

$$B_\theta = \frac{-1}{r} \frac{\partial}{\partial r} m_0 \sin\theta \left(\frac{\cos\omega(t - r/c)}{r} - k \sin\omega(t - r/c) \right)$$

$$= \frac{-m_0 \sin\theta}{r} \left(\frac{+k \sin\omega(t - r/c)}{r} - \frac{\cos\omega(t - r/c)}{r^2} - k^2 \cos(\omega t - \omega r/c) \right)$$

$B_\phi = 0$, in radiation zone

$$B_r \to \frac{-2km_0 \sin\omega(t - r/c)}{r^2} \to 0 \text{ ("static" behavior)}$$

$$B_\theta \to \frac{-m_0 k^2}{r} \cos\omega(t - r/c) \sin\theta \equiv E_\phi \text{ (radiation zone)}$$

So $\hat{\mathbf{r}}$, $\mathbf{B}$ and $\mathbf{E}$ are perpendicular in radiation zone.

PROBLEM 4.29 A spherical surface of radius a has a potential distribution proportional to $\sin 3\theta \cos\phi$. Find the potential at all points interior and exterior to the surface.

Solution:

$\Phi(\theta,\phi) \alpha \sin 3\theta \cos\phi \alpha (3\sin\theta - 4\sin^3\theta)(e^{i\phi} + e^{-i\phi})$

Let $\Phi(\theta,\phi) = \Upsilon(3\sin\theta - 4\sin^3\theta)(e^{i\phi} + e^{-i\phi})$

$= \Upsilon \sin\theta (3 - 4\sin^2\theta)(e^{i\phi} - e^{-i\phi})$

The solution is

$$\Phi(r,\theta,\phi) = \sum_{\ell=0}^{\infty} \sum_{m=-\ell}^{+\ell} \left(A_\ell^m r^\ell + B_\ell^m r^{-(\ell+1)} \right) Y_\ell^m(\theta,\phi)$$

Since Φ must remain finite as $r \to \infty$, we must have (within a constant factor)

$$r \geq a, \quad \Phi(r,\theta,\phi) = \sum_{\ell=0}^{\infty} \sum_{m=-\ell}^{+\ell} \frac{B_\ell^m}{r^{\ell+1}} Y_\ell^m(\theta,\phi) \tag{1}$$

and since Φ must be finite at $r = 0$,

$$r \leq a, \quad \Phi(r,\theta,\phi) = \sum_{\ell=0}^{\infty} \sum_{m=-\ell}^{+\ell} A_\ell^m r^\ell Y_\ell^m(\theta,\phi) \tag{2}$$

The boundary condition is

$$\Phi(a,\theta,\phi) = \gamma \sin\theta \ (3 - 4\sin^2\theta)(e^{i\phi} + e^{-i\phi}) \tag{3}$$

By inspection of the spherical harmonics, we immediately select the ones with $e^{\pm i\phi}$ dependence. There are

$$Y_1^{\pm 1}(\theta,\phi) = \mp\sqrt{3/8\pi} \ \sin\theta \, e^{\pm i\phi}$$

$$Y_2^{\pm 1}(\theta,\phi) = \mp\sqrt{15/8\pi} \ \cos\theta \sin\theta \, e^{\pm i\phi}$$

$$Y_3^{\pm 1} = \mp\sqrt{21/64\pi} \ (4\cos^2\theta \sin\theta - \sin^3\theta) e^{\pm i\phi}$$

We expect a suitable combination of these three harmonics to give the variation shown in Eq. (3). Actually, we do not expect to use $Y_2^{\pm 1}$. Therefore, we need to find coefficients a and b such that

$$(3 - 4\sin^2\theta) = a + b(4\cos^2\theta - \sin^2\theta)$$

that is,

$$(3 - 4\sin^2\theta) = a + b(4 - 5\sin^2 0) = a + 4b - 5b\sin^2\theta$$

that is,

$$3 = a + 4b, \quad b = 4/5$$

$$5b = 4, \quad a = -1/5$$

We can now rewrite condition (3) as follows:

$$\Phi(a,\theta,\phi) = \gamma \sin\theta (3 - 4\sin^2\theta)(e^{i\phi} + e^{-i\phi})$$

$$= \frac{\gamma}{5} \{\sin\theta \, [-1 + 4(4 - \sin^2\theta)] (e^{i\phi} + e^{-i\phi})\},$$

(let $\gamma' = \gamma/s$),

$$= \gamma' \{ -\sin\theta\,(e^{i\phi} + e^{-i\phi}) + 4(4 - \sin^2\theta)\sin\theta\,(e^{i\phi} + e^{-i\phi})\}$$

$$= \gamma' \{ (\sqrt{8\pi/3})(Y_1^1 - Y_1^{-1}) - 4(\sqrt{64\pi/21})(Y_3^1 - Y_3^{-1})\}$$

$$= \beta\{ (Y_1^1 - Y_1^{-1}) - 8(\sqrt{2/7})(Y_3^1 - Y_3^{-1})\}$$

for constant β, ($\beta = \gamma' \sqrt{8\pi/3}$)

$$\Phi(a,\theta,\phi) = \beta\,\{ (Y_1^1 - Y_1^{-1}) - 8(\sqrt{2/7})(Y_3^1 - Y_3^{-1})\} \qquad (4)$$

It is clear now that we will be able to satisfy the boundary condition at the surface of the sphere by limiting our sums to $\ell = 1(m = +1)$, $\ell = 3(m = +1)$. Therefore, by Eq. (1),

$$\Phi(a,\theta,\phi) = \frac{B_1^{+1}}{a^2} Y_1^{+1} + \frac{B_1^{-1}}{a^2} Y_1^{-1} + \frac{B_3^{+1}}{a^4} Y_1^{+1} + \frac{B_3^{-1}}{a^4} Y_1^{-1}$$

$$= \beta\,\{ (Y_1^1 - Y_1^{-1}) - 8(\sqrt{2/7})(Y_3^1 - Y_3^{-1})\}$$

Therefore,

$$\frac{B_1^{+1}}{a^2} = \beta, \quad B_1^{+1} = \beta a^2, \quad \frac{B_3^{+1}}{a^4} = -8(\sqrt{2/7})\beta, \quad B_3^{+1} = -8(\sqrt{2/7})\beta a^4$$

$$\frac{B_1^{-1}}{a^2} = -\beta, \quad B_1^{-1} = -\beta a^2, \quad \frac{B_3^{-1}}{a^4} = +8(\sqrt{2/7})\beta, \quad B_3^{-1} = +8(\sqrt{2/7})\beta a^6$$

Therefore,

$$\Phi(r,\theta,\phi) = \beta\left\{\frac{a^2}{r^2}(Y_1^1 - Y_1^{-1}) - 8(\sqrt{2/7})\frac{a^4}{r^4}(Y_3^1 - Y_3^{-1})\right\}, \quad r \geq a \qquad (5)$$

Similarly, using Eq. (2),

$$\Phi(a,\theta,\phi) = aA_1^1 Y_1^1 + aA_1^{-1} Y_1^{-1} + a^3 A_3^1 Y_3^1 + a^3 A_3^{-1} Y_3^{-1}$$

$$= \beta\,\{ (Y_1^1 - Y_1^{-1}) - 8(\sqrt{2/7})(Y_3^1 - Y_3^{-1})\}$$

So that

$$A_1^1 = \beta/a; \quad A_1^{-1} = -\beta/a; \quad A_3^1 = -\beta a^3 8(\sqrt{2/7}); \quad A_3^{-1} = \beta a^3 8(\sqrt{2/7})$$

and

$$\Phi(r,\theta,\phi) = \beta \left[\frac{r}{a} (Y_1^1 - Y_1^{-1}) - 8(\sqrt{2/7}) \frac{r^3}{a^3} (Y_3^1 - Y_3^{-1})\right], \quad r \leq a \qquad (6)$$

Equations (5) and (6) are the solutions to the problem. If you want to make them look more familiar, then it is easy to show that

$$\Phi(r,\theta,\phi) = \alpha \left[\frac{r}{a} - 4\left(\frac{r}{a}\right)^3 (4\cos^2\theta - \sin^2\theta)\right] \sin\theta\cos\phi, \quad r \leq a$$

$$\Phi(r,\theta,\phi) = \alpha \left[\frac{a^2}{r^2} - 4\frac{a^4}{r^4} (4\cos^2\theta - \sin^2\theta)\right] \sin\theta\cos\phi, \quad r \geq a$$

PROBLEM 4.30 A sphere of radius a has a surface distribution of charge that is proportional to cos 2θ. Find the potential at all points in space exterior to the sphere. Represent the charge distribution by the appropriate multipole.

Solution: We have $\sigma \sim \cos 2\theta \sim 1 + \cos^2\theta$. So, the solution is of the form

$$\phi(r,\theta) = \sum_{\ell=0}^{\infty} (A_\ell r^\ell + B_\ell r^{-(\ell+1)}) P_\ell(\cos\theta) \qquad (1)$$

with

$\phi(r,0) \to 0$ as $r \to \infty$ (because the only source of field is surface charge, on finite area) (2)

$$\left.\frac{\partial\phi}{\partial r}\right|_{r=a} = \gamma(1 + \cos^2\theta), \text{ for some constant } \gamma \qquad (3)$$

Equation (2) clearly reduces our solution to the form

$$\phi(r,\theta) = \sum_{\ell=0}^{\infty} \frac{B_\ell}{\ell+1} P_\ell(\cos\theta) \qquad (4)$$

(We could leave out A_0, but this is a constant term, and there is no loss of generality in setting it equal to zero.) Then,

$$\left.\frac{\partial\phi}{\partial r}\right|_{r=a} = \sum_{\ell=0}^{\infty} \frac{-(\ell+1)B_\ell}{r^{\ell+2}} P_\ell(\cos\theta) = \sum_{\ell=0}^{\infty} \frac{-(\ell+1)B_\ell}{a^{\ell+2}} P_\ell(\cos\theta)$$

$$= \gamma(1 + \cos^2\theta) \qquad (5)$$

Since $P_2(\cos\theta) = \frac{1}{2}(3\cos^2\theta - 1)$, this suggests we limit our sum to $\ell = 0$ and $\ell = 2$. (Do not forget that the Legendre polynomials have definite parity. A

$\ell = 1$ Legendre polynomial would be odd.) Therefore, Eq. (5) leads to

$$\frac{-B_0}{a^2} - \frac{3B_2}{a^4}\tfrac{1}{2}(3\cos^2\theta - 1) = \gamma + \gamma\cos^2\theta$$

Since this must be true for all θ, we have

$$\gamma = \frac{-B_0}{a^2} + \frac{3B_2}{2a^4}$$

$$\gamma = -\frac{9B_2}{2a^4}$$

Solving for B_0 and B_2,

$$B_0 = -\frac{4}{3}\gamma a^2, \quad B_2 = -\frac{2a^4}{9}\gamma$$

Substituting into Eq. (4), we obtain

$$\phi(r,\theta) = -\left[\left(\frac{4}{3}\gamma a^2\right)\frac{1}{r} + \frac{\gamma a^2}{9}\frac{3\cos^2\theta - 1}{r^3}\right]$$

which is the field produced by an effective point charge

$$\phi = -\frac{4}{3}\gamma a^2 = \frac{4}{3}|\gamma|a^2$$

and a quadrupole moment with

$$2p\ell = \frac{|\gamma|a^4}{9},$$

where p, ℓ are shown in the figure.

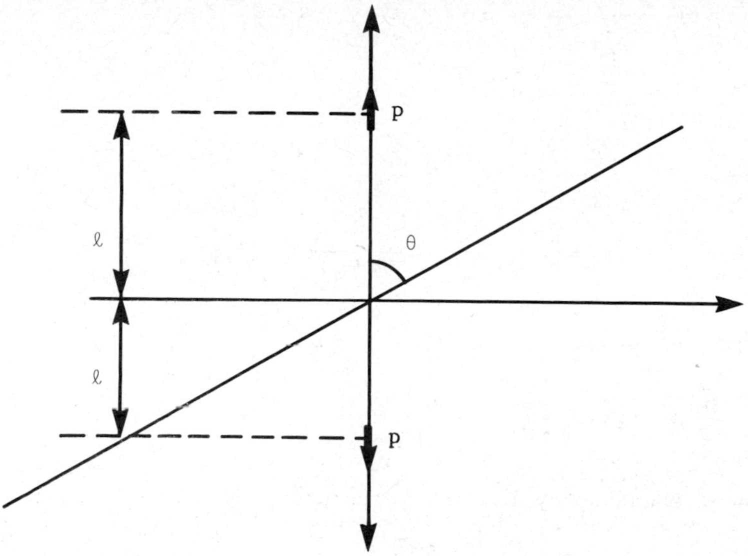

PROBLEM 4.31 A potential field is given by $\Phi_0(r,\theta,\phi) = r^2 \sin 2\theta \exp(i\phi)$. Find the potential exterior to a grounded, conducting sphere of radius a placed in this field at the origin.

Solution: The solution to the problem is

$$\Phi(r,\theta,\phi) = \sum_{\ell=0}^{\infty} \sum_{m=-\ell}^{+\ell} (A_\ell^m r^\ell + B_\ell^m r^{-(\ell+1)}) Y_\ell^m(\theta,\phi)$$

with boundary conditions

$$\Phi \to \Phi_0(r,\theta,\phi) = 2r^2 \sin\theta \cos\theta \, e^{i\phi}, \text{ as } r \to \infty \qquad (1)$$

$$\Phi = 0 \text{ at } r = a \qquad (2)$$

As $r \to \infty$,

$$\phi(r,\theta,\phi) \to \sum_{\ell,m} A_\ell^m r^\ell Y_\ell^m(\theta,\phi)$$

Since Eq. (1) depends on r^2, we should look at all spherical harmonics for $\ell = 0, 1, 2$. By simple inspection we see that we can satisfy Eq. (1), if we let all the A's $= 0$, except A_2^1, since

$$Y_2^{\pm 1} = (\sqrt{15/8\pi}) \cos\theta \sin\theta \, e^{\pm i\phi}$$

Therefore,

$$A_2^1 r^2 Y_2^1 = A_2^1 r^2 \sqrt{15/8\pi}\ \cos\theta \sin\theta\, e^{i\phi} = 2r^2 \sin\theta \cos\theta\ e^{i\phi}$$

thus,

$$A_2^1 = 2\sqrt{8\pi/15} = \sqrt{32\pi/15}$$

Our solution is now

$$\Phi(r,\theta,\phi) = \sqrt{\frac{32\pi}{15}}\, r^2\ Y_2^1(\theta,\phi) = \sum_{\ell=0}^{\infty}\ \sum_{m=-\ell}^{+\ell} \frac{B_\ell^{\ m}}{r^{(\ell+1)}}\ Y_\ell^{\ m}(\theta,\phi)$$

We must have, by Eq. (2),

$$\Phi(a,\theta,\phi) = 0 = \sqrt{\frac{32\pi}{15}}\ a^2 Y_2^1(\theta,\phi) = \sum_{\ell=0}^{\infty}\ \sum_{m=-\ell}^{+\ell} \frac{B_\ell^{\ m}}{a^{\ell+1}}\ Y_\ell^{\ m}(\theta,\phi)$$

Multiplying the above equation by $Y_{\ell'}^{m'}$ and integrating over $d\Omega$, we have

$$\sqrt{\frac{32\pi}{15}}\ a^2 \int Y_{\ell'}^{m'} Y_2^1(\theta,\phi)\,d\Omega = -\sum_{\ell=0}^{\infty}\ \sum_{m=-\ell}^{+\ell} \int \frac{B_\ell^{\ m}}{a^{\ell+1}}\ Y_\ell^{\ m}(\theta,\phi) Y_{\ell'}^{m'}(\theta,\phi)\,d\Omega$$

Using orthogonality relations, then

$$\sqrt{\frac{32\pi}{15}}\ a^5\ \delta_{m'1}\delta_{\ell'2} = -\sum_{\ell=0}^{\infty}\ \sum_{m=-\ell}^{+\ell} \frac{B_\ell^{\ m}}{a^{\ell+1}}\ \delta_{mm'}\delta_{\ell\ell'}$$

$$\therefore\ \sqrt{\frac{32\pi}{15}}\ a^2\ \delta_{m'1}\delta_{\ell'2} = -\frac{B_{\ell'}^{m'}}{a^{\ell'+1}}$$

Therefore,

$$B_2^1 = -a^5\ \sqrt{32\pi/15}$$

and all other B's are equal to zero. So, our solution is

$$\Phi(r,\theta,\phi) = \sqrt{\frac{32\pi}{15}}\, r^2\ Y_2^1(\theta,\phi) - \frac{a^5}{r^3}\sqrt{\frac{32\pi}{15}}\ Y_2^1(\theta,\phi)$$

$$= \sqrt{\frac{32\pi}{15}}\left(1 - \frac{a^5}{r^5}\right) r^2 Y_2^1(\theta,\phi)$$

or

$$\Phi(r,\theta,\phi) = 2\left(1 - \frac{a^5}{r^5}\right) r^2 \sin\theta \cos\theta\ e^{i\phi} = \left(1 - \frac{a^5}{r^5}\right) r^2 \sin 2\theta\ e^{i\phi}$$

(Do you think you could have guessed?)

PROBLEM 4.32 A conducting sphere of radius a on whose surface resides a total charge Q is placed in a uniform electric field $\mathbf{E}_0$. Find the potential at all points in space exterior to the sphere.

Solution: In spherical coordinates, if the z axis is chosen along direction of $\mathbf{E}_0$, then $\mathbf{E}_0$ is produced by potential $V_0 = -E_0 r \cos\theta$. The potential on the sphere is kept constant at $V_{sp} = +Q/a$. Therefore, solution will be of the form

$$\phi(r,\theta) = \sum_{\ell=0}^{\infty} (A_\ell r^\ell + B_\ell r^{-(\ell+1)}) P_\ell(\cos\theta) \tag{1}$$

with boundary conditions,

$$\phi(r,\theta) \to -E_0 r \cos\theta = -E_0 r\, P_1(\cos\theta), \text{ as } r \to \infty \tag{2}$$

$$\phi(a,\theta) = V_{sp} = +\frac{Q}{r} \tag{3}$$

From Eq. (2) it follows that $A_2, A_3, \ldots = 0$, that is,

$$\phi(r,\theta) = A_0 + A_1 r P_1(\cos\theta) + \sum_{\ell=0}^{\infty} \frac{B_\ell}{r^{\ell+1}} P_\ell(\cos\theta) \to A_0 + A_1 r P_1(\cos\theta)$$

as $r \to \infty$. Therefore, $A_0 = 0$ and $A_1 = -E_0$, so that

$$\phi(r,\theta) = -E_0 r P_1(\cos\theta) + \sum_{\ell=0}^{\infty} \frac{B_\ell}{r^{\ell+1}} P_\ell(\cos\theta)$$

We can satisfy boundary condition (3) only with the term $\ell = 0$ and $\ell = 1$, as follows

$$\phi(r,\theta) = -E_0 r P_1(\cos\theta) + \frac{B_0}{r} + \frac{B_1}{r^2} P_1(\cos\theta)$$

with

$$\phi(a,\theta) = +\frac{Q}{a} = -E_0 a \cos\theta + \frac{B_0}{a} + \frac{B_1}{a^2}\cos\theta$$

$$= \cos\theta\left(\frac{B_1}{a^2} - E_0 a\right) + \frac{B_0}{a}$$

Therefore,

$$B_0 = +Q$$

$$\frac{B_1}{a^2} E_0 a \Rightarrow B_1 = E_0 a^3$$

Therefore, the final solution is

$$\phi(r,\theta) = -E_0 r \cos\theta + \frac{Q}{r} + \frac{E_0 a^3}{r^2} \cos\theta$$

$$= +\frac{Q}{r} - \left(1 - \frac{a^3}{r^3}\right) r E_0 \cos\theta$$

$$\therefore \phi(r,\theta) = +\frac{Q}{r} - \left(1 - \frac{a^3}{r^3}\right) r E_0 \cos\theta$$

You may check that

$$\sigma = \left.\frac{E_r}{4\pi}\right|_{r=a} = \frac{Q}{4\pi a^2} + \frac{3}{4\pi} E_0 \cos\theta$$

and that

$$\int_{\text{sphere}} \sigma \, ds = Q$$

as it must.

PROBLEM 4.33 Obtain the potential for any point within the two-dimensional "box," subject to the boundary conditions given in the figure.

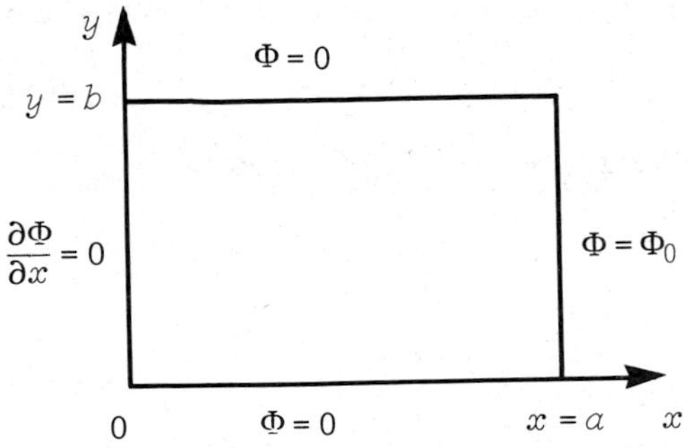

Solution: We want to solve

$$\frac{\partial^2\phi}{\partial x^2} + \frac{\partial^2\phi}{\partial y^2} = 0, \quad 0 \le x \le a, \quad 0 \le y \le b.$$

With boundary conditions

$$\phi(x,0) = 0 \tag{1}$$

$$\phi(x,b) = 0 \tag{2}$$

$$\frac{\partial\phi}{\partial x}(0,y) = 0 \tag{3}$$

$$\phi(a,y) = \Phi_0 \tag{4}$$

Boundary conditions (1) and (2) clearly suggest that we have the harmonic dependence in the y direction, that is, if $\phi(x,y) = X(x)Y(y)$, then

$$Y(y) \propto \sin ky$$

since Eq. (1) is automatically verified. To satisfy (2), we must have

$$kb = n\pi, \quad n = 1,2,\ldots \text{ i.e., } k = n\frac{\pi}{b}, \quad n = 1,2,\ldots$$

Therefore

$$\phi(x,y) \sim X(x) \sin\left(\frac{n\pi}{b}y\right), \quad n = 1,2,\ldots$$

satisfies boundary conditions (1) and (2). Boundary condition (3) suggests that we let $X(x) \sim \cosh(n\pi x/b)$, since its derivative is proportional to $\sinh(n\pi x/b)$ and this vanishes at $x = 0$, as required ($\cosh x$ is never equal to zero). Therefore, any solution

$$\sin\left(\frac{n\pi}{b}y\right) \cosh\left(\frac{n\pi}{b}x\right), \quad n = 1,2,\ldots$$

satisfies Eq. (1), (2), and (3). So, the solution will be

$$\phi(x,y) = \sum_{n=1}^{\infty} A_n \cosh\left(\frac{n\pi}{b}x\right) \sin\left(\frac{n\pi}{b}y\right)$$

where the A_n's are determined from Eq. (4), that is,

$$\phi(a,y) = \Phi_0 = \sum_{n=1}^{\infty} A_n \cosh\left(\frac{n\pi}{b}a\right) \sin\left(\frac{n\pi}{b}y\right) = \sum_{n=1}^{\infty} b_n \sin\left(\frac{n\pi}{b}y\right) \tag{5}$$

where $b_n = A_n \cosh(n\pi a/b)$. By standard Fourier methods,

$$b_n = \frac{2}{b}\int \Phi_0 \sin\left(\frac{n\pi}{b}y\right)dy = \frac{2\Phi_0}{b}\left(-\frac{b}{n\pi}\right)\cos\frac{n\pi y}{b}\bigg|_0^b = \frac{-2\Phi_0}{n\pi}(\cos n\pi - 1)$$

$$= \frac{-2\Phi_0}{n\pi}[(-1)^n - 1] = \begin{cases} 0, & \text{if } n \text{ even} \\ \dfrac{4\Phi_0}{n\pi}, & \text{if } n \text{ odd} \end{cases}$$

Therefore,

$$A_n = \begin{cases} 0, \text{ if } n \text{ even} \\ (4\Phi_0/n\pi)[1/\cosh(n\pi a/b)], \text{ if } n \text{ odd} \end{cases}$$

So that

$$\phi(x,y) = \sum_{n \text{ odd}} \frac{4\Phi_0}{4\pi \cosh(n\pi a/b)} \cosh\left(\frac{n\pi}{b} x\right) \sin\left(\frac{n\pi}{b} y\right)$$

or

$$\phi(x,y) = \frac{4\Phi_0}{\pi} \sum_{m=0}^{\infty} \frac{\cosh[(2m+1)\pi x/b] \sin[(2m+1)\pi y/b]}{(2m+1)\cosh[(2m+1)\pi a/b]}$$

PROBLEM 4.34 A potential in the $x-z$ plane is independent of z and is given by a repeating step function of magnitude $2\Phi_0$ and period $2a$.

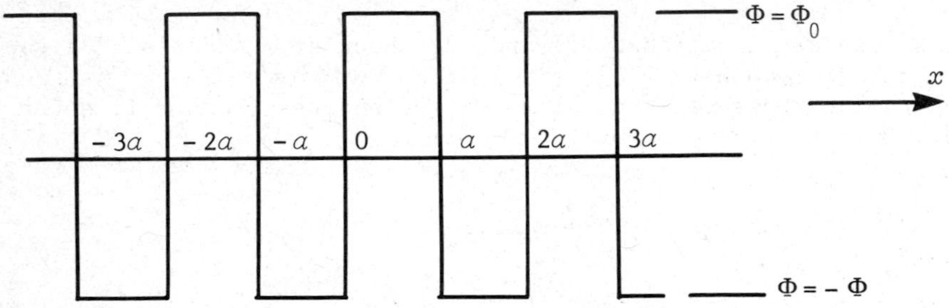

In addition, the plane defined by $y = y_0$ is held at ground potential. Find the potential for all points of space in the region $0 < y < y_0$.

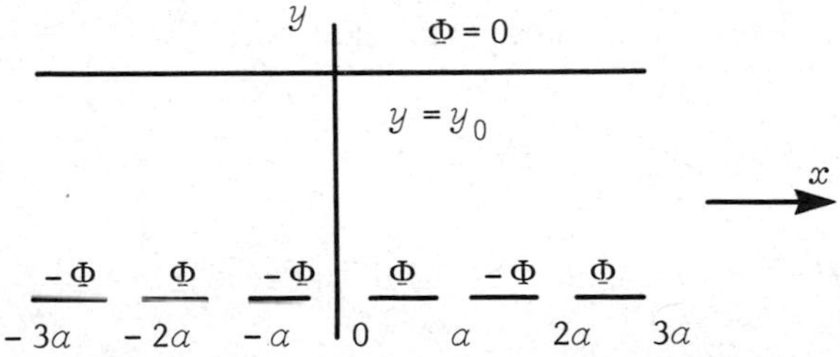

Solution: The solution is

$$\phi = \sum_n A_n \sin\left(\frac{n\pi x}{a}\right) \sinh\left(\frac{n\pi}{a}(y_0 - y)\right)$$

[You should be able to write down the solution in the above form, just from inspection of the boundary conditions. Note that we clearly need harmonic dependence along the x axis so that we can Fourier expand $\Phi(x,0)$, and $\phi(x,y_0)=0$ is guaranteed by letting $Y(y)\sim \sinh[\alpha(y_0-y)]$. Note that $\cosh x \neq 0$ for all x, so that one is really not left with much choice.]

$$\phi(x,y) = \sum_n A_n \sinh\left(\frac{n\pi}{a}(y_0 - y)\right)\sin\left(\frac{n\pi x}{a}\right)$$

We must have

$$\Phi(x,0) = \sum_n A_n \sinh\left(\frac{n\pi}{a}y_0\right)\sin\left(\frac{n\pi x}{a}\right) = \sum_n B_n \sin\left(\frac{n\pi x}{a}\right)$$

Therefore,

$$B_n = \frac{2\Phi_0}{a}\int_0^a \sin\left(\frac{n\pi x}{a}\right) = -\frac{2\Phi_0}{a}\frac{a}{n\pi}(\cos n\pi - 1)$$

$$= \begin{cases} 0, & \text{if } n \text{ even} \\ \dfrac{4\Phi_0}{n\pi}, & \text{if } n \text{ odd} \end{cases}$$

$$\therefore A_n = \frac{B_n}{\sinh\left(\frac{n\pi}{a}y_0\right)}, \quad n \text{ odd}$$

So, our final solution is

$$\phi(x,y) = \sum_{m=0}^{\infty} \frac{4\Phi_0}{(2m+1)\pi}\sin\left(\frac{(2m+1)\pi x}{a}\right)\frac{\sinh\left(\frac{2m+1}{a}\pi(y_0-y)\right)}{\sinh\left(\frac{2m+1}{a}\pi y_0\right)}$$

Chapter 5

PROBLEM 5.1

(a) If the potential difference across a cylindrical capacitor is doubled, the energy stored in the capacitor is changed by what factor?
(b) If the radii of the inner and outer cylinders are each doubled, keeping the stored charge constant, how does the stored energy change?

Solution:

(a) The energy stored in a cylindrical capacitor is $E = \frac{1}{2} CV^2$. Therefore, the energy stored in the capacitor is changed by a factor of 4.
(b) $Q = CV$

$$V^2 = \frac{Q^2}{C^2}, \quad E = \frac{1}{2} \frac{Q^2}{C}$$

If the charge is the same, there is no change in energy.

PROBLEM 5.2 If you have available a supply of 2.0 μF capacitors, each capable of withstanding 200 V without breakdown, how would you assemble a combination having an equivalent capacitance of (a) 0.40 μF or (b) 1.2 μF, each capable of withstanding 1000 V.

Solution:

(a) $C_{eq} = 0.4 \times 10^{-6}$ F, have $C = 2 \times 10^{-6}$ F

$$V = 200 \text{ V}$$

$$Q = (2 \times 10^{-6} \text{ F})(200 \text{ V}) = 4 \times 10^{-4} \text{ C}$$

$$\frac{1}{C_T} = \frac{X}{C_1}, \quad Q = CV$$

$$0.4 \times 10^{-6} \text{ F} = \frac{C_1}{X}, \quad V = \frac{4 \times 10^{-4} \text{ C}}{4 \times 10^{-7} \text{ V}} = 1000 \text{ V}$$

$$X = \frac{C_1}{0.4 \times 10^{-6} \text{ F}} = \frac{2 \times 10^{-6} \text{ F}}{0.4 \times 10^{-6} \text{ F}} = 5$$

so put five 2-μF capacitors in series.

$$\frac{1}{C_T} = \frac{1}{2 \times 10^{-6}\ \text{F}} + \frac{1}{2 \times 10^{-6}\ \text{F}} + \frac{1}{2 \times 10^{-6}\ \text{F}} + \frac{1}{2 \times 10^{-6}\ \text{F}} + \frac{1}{2 \times 10^{-6}\ \text{F}}$$

(b) Equivalent capacitance 1.2 μF = C_T, V_T = 1000 V

$$\frac{1}{C_T} = \frac{x}{C_1}, \quad C_T = \frac{C_1}{x}$$

Put three 0.4×10^{-6} F in parallel. Add capacitance $3(0.4 \times 10^{-6}$ F) $= 1.2 \times 10^{-6}$ F, voltage stays the same.

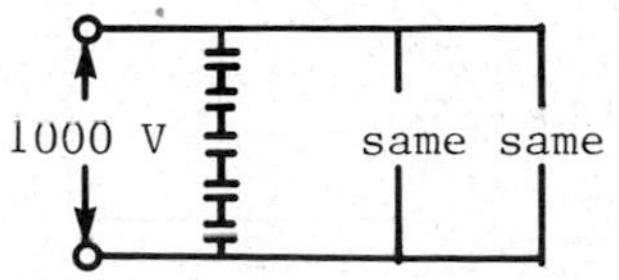

PROBLEM 5.3 In the figure the battery B supplies 12 V.

(a) Find the charge on each capacitor when switch S_1 is closed.
(b) When S_2 is also closed. Take C_1 = 1.0 μF, C_2 = 2.0 μF, C_3 = 3.0 μF, and C_4 = 4.0 μF.

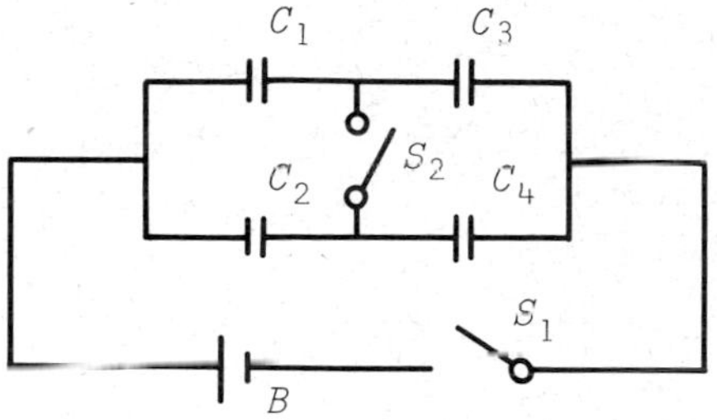

Solution:

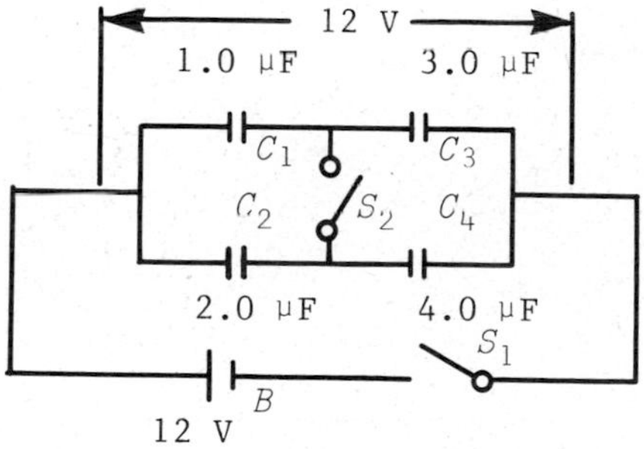

(a) When S_1 is closed

$$\frac{1}{C_a} = \frac{1}{C_1} + \frac{1}{C_3} \qquad \frac{1}{C_b} = \frac{1}{C_2} + \frac{1}{C_4}$$

$$\frac{1}{C_a} = \frac{1}{1 \times 10^{-6}\ \text{F}} + \frac{1}{3 \times 10^{-6}\ \text{F}} \qquad \frac{1}{C_b} = \frac{1}{2 \times 10^{-6}\ \text{F}} + \frac{1}{4 \times 10^{-6}\ \text{F}}$$

$$C_a = 7.5 \times 10^{-7}\ \text{F} \qquad C_b = 1.3 \times 10^{-6}\ \text{F}$$

$C_T = C_a + C_b$

$C_T = 7.5 \times 10^{-7}\ \text{F} + 1.3 \times 10^{-6}\ \text{F} = 2.08 \times 10^{-6}\ \text{F}$

$Q_T = CV = [2.08 \times 10^{-6}\ \text{F}(12\ \text{V})] = 2.5 \times 10^{-5}\ \text{C}$

$Q_T = (Q_1 + Q_3) + (Q_2 + Q_4);\quad Q_1 = Q_3,\ Q_2 = Q_4$

$Q_a = (7.5 \times 10^{-7}\ \text{F})(12\ \text{V}) = 9 \times 10^{-6}\ \text{C}$

$Q_b = (1.3 \times 10^{-6}\ \text{F})(12\ \text{V}) = 1.56 \times 10^{-5}\ \text{C}$

$Q_1 = Q_3 = 9 \times 10^{-6}\ \text{C}$

$Q_2 = Q_4 = 1.6 \times 10^{-5}\ \text{C}$

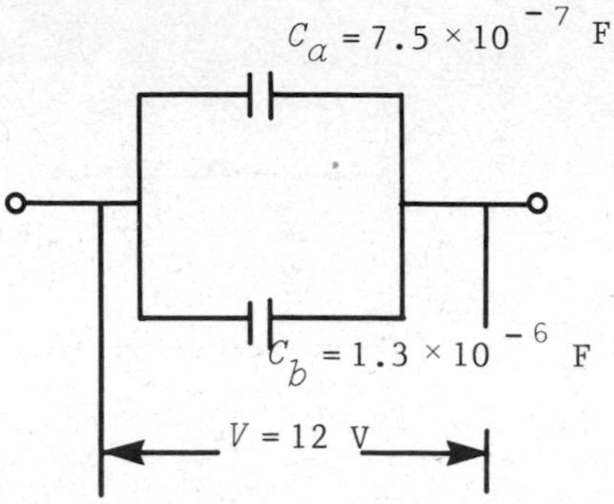

(b) When both S_1 and S_2 are closed

$C_a = 3\ \mu\text{F},\ C_b = 7\ \mu\text{F}$

$$C_T = \frac{1}{3 \times 10^{-6}\ \text{F}} + \frac{1}{7 \times 10^{-6}\ \text{F}} = 2.1 \times 10^{-6}\ \text{F}$$

$Q_T = C_T V_T = (2.1 \times 10^{-6}\ \text{F})(12\ \text{V}) = 2.52 \times 10^{-5}\ \text{C}$

$Q_a = 2.52 \times 10^{-5}\ \text{C},\ Q_b = 2.52 \times 10^{-5}\ \text{C}$

$$V_a = \frac{2.52 \times 10^{-5}\ \text{C}}{3 \times 10^{-6}\ \text{F}} = 8.4\ \text{V},\quad V_b = \frac{2.52 \times 10^{-5}\ \text{C}}{7 \times 10^{-6}\ \text{F}} = 3.6\ \text{V}$$

$Q_1 = (1 \times 10^{-6}\ \text{F})(8.4\ \text{V}) = 8.4 \times 10^{-6}\ \text{C}$

$Q_2 = (2 \times 10^{-6}\ \text{F})(8.4\ \text{V}) = 16.8 \times 10^{-6}\ \text{C} = 1.68 \times 10^{-5}\ \text{C}$

$Q_3 = (3 \times 10^{-6}\ \text{F})(3.6\ \text{V}) = 10.8 \times 10^{-6}\ \text{C}$

$Q_4 = (4 \times 10^{-6}\ \text{F})(3.6\ \text{V}) = 14.4 \times 10^{-6}\ \text{C}$

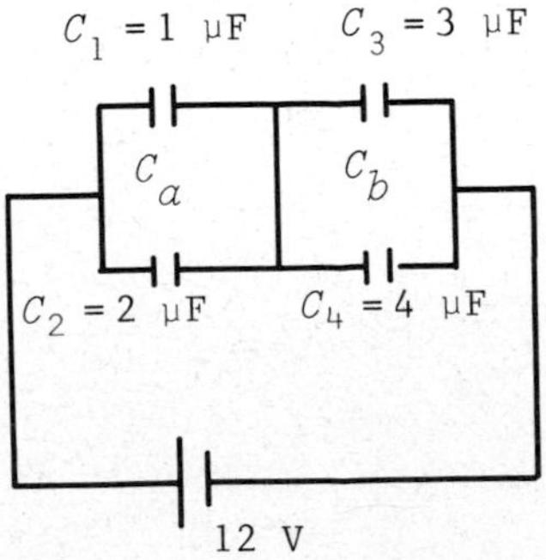

PROBLEM 5.4 The plate and cathode of a vacuum tube diode are in the form of two concentric cylinders with the cathode as the central cylinder. If the cathode diameter is 1.5 mm and the plate diameter is 18 mm with both elements having a length of 2.3 cm, what is the capacitance of the diode?

Solution:

$$\text{Cathode diameter} = 1.5 \text{ mm} \times \frac{x}{1000 \text{ mm}} = 1.5 \times 10^{-3} \text{ m}$$

$$\text{Cathode radius} = a = 7.5 \times 10^{-4} \text{ m}$$

$$\text{Plate diameter} = 18 \text{ mm} = 18 \times 10^{-3} \text{ m} = 1.8 \times 10^{-2} \text{ m}$$

$$\text{Plate radius} = b = 9 \times 10^{-3} \text{ m}$$

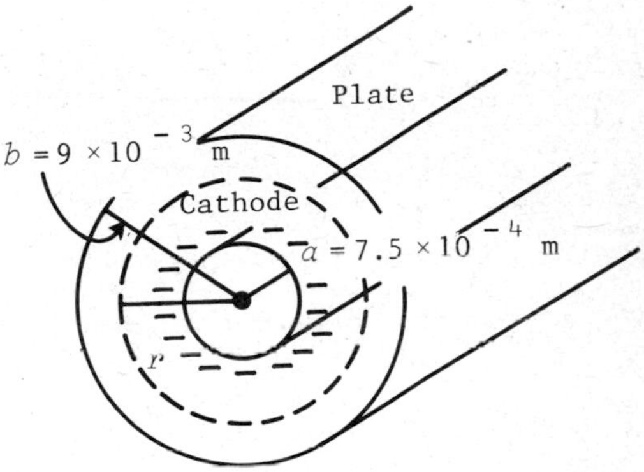

$$\varepsilon_0 \oint \mathbf{E} \cdot \hat{\mathbf{n}}\, dA = q$$

$$\varepsilon_0 EA = q$$

$$\varepsilon_0 E(2\pi r)\ell = q$$

$$\mathbf{E} = \frac{q}{2\varepsilon_0 \pi r \ell}$$

The potential difference between the two plates is

$$\mathbf{E} \cdot d\mathbf{r} = E\, dr \cos(180^\circ) = -E\, dr$$

$$V_b - V_a = -\int_a^b \mathbf{E} \cdot d\boldsymbol{\ell} = +\int_a^b E\, dr = -\int_a^b \frac{q}{2\pi\varepsilon_0 r \ell} dr$$

$$= +\frac{q}{2\pi\varepsilon_0 \ell} \int_a^b \frac{1}{r}\, dr = \frac{+q}{2\pi\varepsilon_0 \ell} (\ell\text{n}|r|)_a^b = \frac{+q}{2\pi\varepsilon_0 \ell} [\ell\text{n}(b) - \ell\text{n}(a)]$$

$$V_b - V_a = \frac{+q}{2\pi\varepsilon_0 \ell} \ell\text{n}\, \frac{b}{a}$$

$$C = \frac{q}{V} = \frac{q}{q\ell\text{n}(b/a)/(2\pi\varepsilon_0 \ell)} = \frac{2\pi\varepsilon_0 \ell}{\ell\text{n}(b/a)}$$

which depends only on geometry and ε_0

$$C = \frac{2\pi(8.85 \times 10^{-12}\ \mathrm{C^2/N \cdot m^2})(0.023\ \mathrm{m})}{\ln(9 \times 10^{-3}\ \mathrm{m}/7.5 \times 10^{-4}\ \mathrm{m})}$$

$$C = 5.15 \times 10^{-13}\ \mathrm{F}$$

PROBLEM 5.5 In the figure suppose that capacitor C_3 breaks down electrically, becoming equivalent to a conducting path. What changes in (a) the charge and (b) the potential difference occur for capacitor C_1?

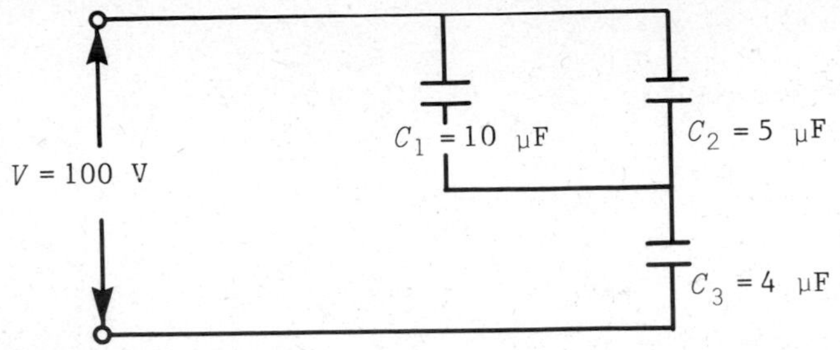

Solution:

First $q_{C_1} = 2.1 \times 10^{-4}$ C

$V_{C_1} = 21$ V

$Q = CV$

$$\frac{1}{C_T} = \frac{1}{10\ \mu\mathrm{F} + 5\ \mu\mathrm{F}} + \frac{1}{4\ \mu\mathrm{F}}$$

$$C_T = 3.16 \times 10^{-6}\ \mathrm{F}$$

$$Q_T = (3.16 \times 10^{-6}\ \mathrm{F})(100\ \mathrm{V}) = 3.16 \times 10^{-4}\ \mathrm{C}$$

$$V_3 = \frac{Q}{C} = \frac{3.16 \times 10^{-4}\ \mathrm{C}}{4 \times 10^{-6}\ \mathrm{F}} = 79\ \mathrm{V}$$

$$Q_1 = (10 \times 10^{-6}\ \mathrm{F})(21\ \mathrm{V}) = 2.1 \times 10^{-4}\ \mathrm{C}$$

After,

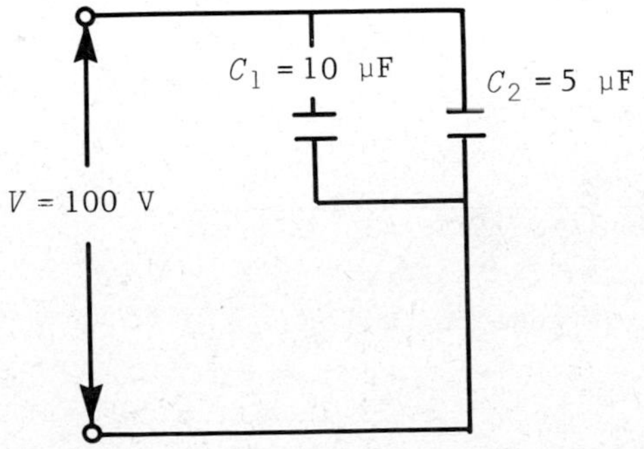

$$C_T = 10\times10^{-6}\ \text{F} + 5\times10^{-6}\ \text{F} = 1.5\times10^{-5}\ \text{F}$$

$$V_T = 100\ \text{V}$$

$$Q = (1.5\times10^{-5}\ \text{F})(100\ \text{V}) = 1.5\times10^{-3}\ \text{C}$$

$$Q_1 = (1\times10^{-5}\ \text{F})(100\ \text{V}) = 1\times10^{-3}\ \text{C}$$

$$Q_1' = 1\times10^{-3}\ \text{C}$$

$$V = 100\ \text{V}$$

Change in $Q = +7.9\times10^{-4}$ C
Change in $V = +79$ V

PROBLEM 5.6 The figure shows two capacitors in series, the rigid center section of length b being vertically movable. Show that the equivalent capacitance of the series combination is independent of the position of the center section and is given by $C = \varepsilon_0 A/(a-b)$.

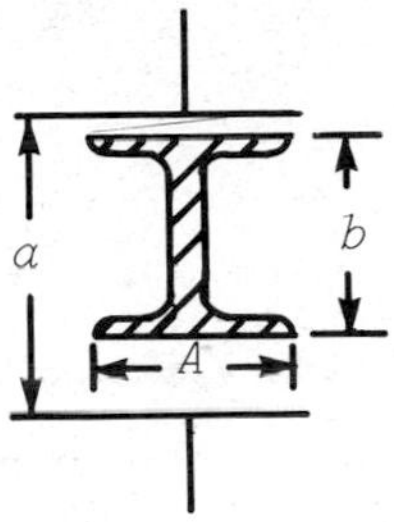

Solution:

$$\frac{1}{C_{eq}} = \frac{1}{C_1} + \frac{1}{C_2}$$

$$Q = CV$$

$$\frac{Q}{V} = C$$

$$\varepsilon_0 E_A = q$$

$$\frac{1}{C} = \frac{V}{Q}$$

$$V = \frac{W}{q_0}, \quad W = Vq$$

$$\frac{1}{C_T} = \frac{V_T}{Q_T} = \frac{V_1}{Q_1} + \frac{V_2}{Q_2}$$

$$W = \mathbf{F}\cdot\mathbf{d} = q_0 Ed$$

$$V_T = V_1 + V_2$$

$$V\varepsilon_0 = q_0 Ed$$

$$V = Ed, \quad V = E(a-b)$$

The change is the same, so E is the same between them; the total potential equals $V_1 + V_2$, $V = W/q$.

$$W = V_T q = q\,E\,D$$

$$\varepsilon_0 EA = q$$

$$V_T = ED_T, \quad V_T = E(a-b)$$

$$\varepsilon_0 EA = Q, \quad \varepsilon = \frac{Q}{A\varepsilon_0}$$

$$V = \frac{Q(a-b)}{A\varepsilon_0}$$

$$C = \frac{Q}{V} = \frac{Q}{Q(a-b)/A\varepsilon_0}, \quad C = \frac{A\varepsilon_0}{a-b}$$

$a - b$ is always the same distance, so it does not matter where the center section is.

PROBLEM 5.7 A 100-pF capacitor is charged to a potential difference of 50 V, the charging battery then being disconnected. The capacitor is then connected, as in the figure, to a second capacitor. If the measured potential difference drops to 35 V, what is the capacitance of this second capacitor?

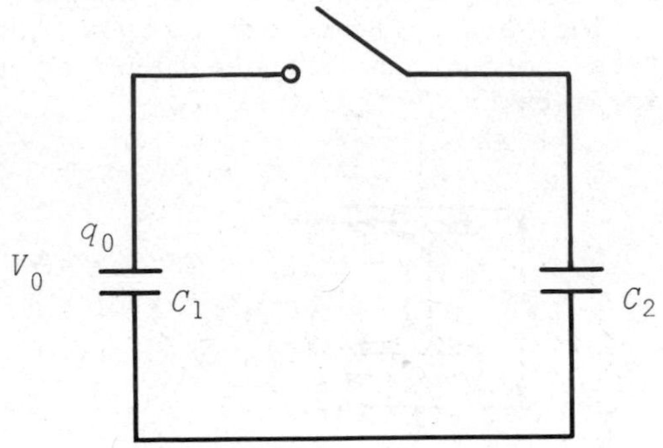

Solution:

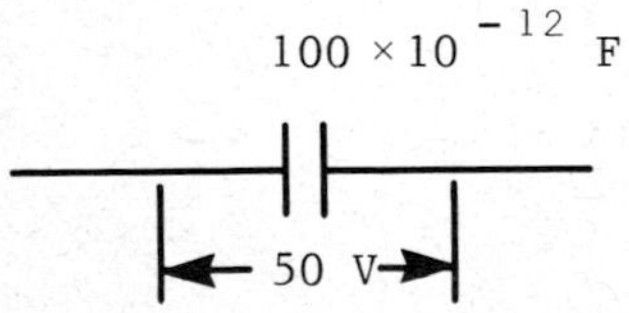

$$Q = (1 \times 10^{-10}\ \text{F})(50\ \text{V}) = 5 \times 10^{-9}\ \text{C}$$

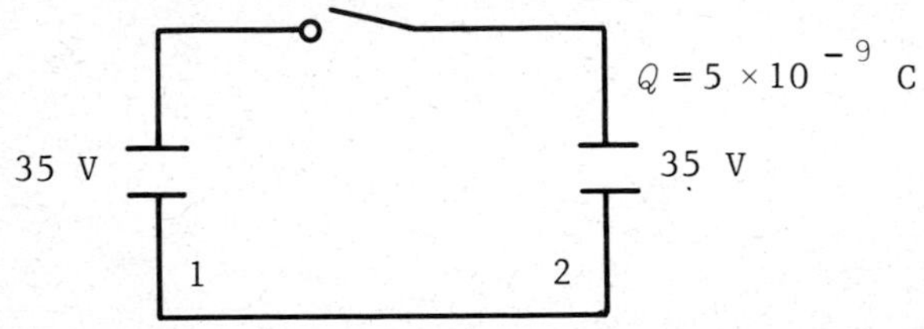

$$Q = CV$$

$$C_T = \frac{5 \times 10^{-9}\ \text{C}}{35\ \text{V}} = 1.43 \times 10^{-10}\ \text{F}$$

$$C_T - C_1 = C_2 = 1.43 \times 10^{-10}\ \text{F} = 100 \times 10^{-12}\ \text{F}$$

$$= 4.3 \times 10^{-11}\ \text{F}$$

$$= 43 \times 10^{-12}\ \text{F} = 43\ \text{pF}$$

PROBLEM 5.8 A 6.0-μF capacitor is connected in parallel with a 4.0-μF capacitor and a potential difference of 200 V is applied across the pair.

(a) What is the charge on each capacitor?
(b) What is the potential difference across each capacitor?

Solution:

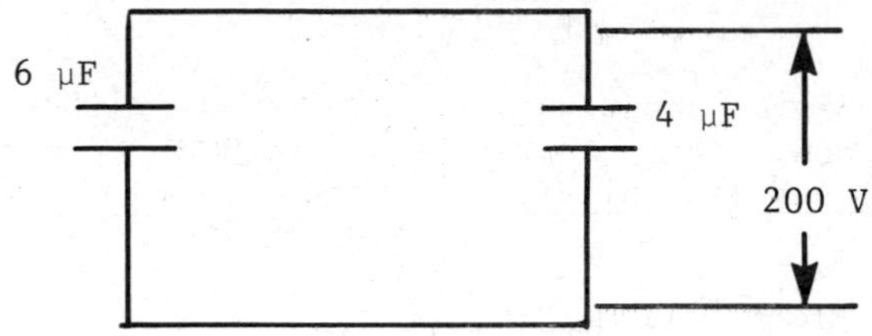

$$Q = CV$$

$$Q_T = Q_6 + Q_4$$

$$C_T V_T = C_6 V_6 + C_4 V_4$$

$$C_T = C_6 + C_4$$

$$C_T = 6 \times 10^{-6}\,\text{F} + 4 \times 10^{-6}\,\text{F} = 1 \times 10^{-5}\,\text{F}$$

(a) $Q_6 = (6 \times 10^{-6}\,\text{F})(200\text{ V}) = 1.2 \times 10^{-3}\,\text{C}$

$$Q_4 = (4 \times 10^{-6}\,\text{F})(200\text{ V}) = 8 \times 10^{-4}\,\text{C}$$

$$Q_T = (1 \times 10^{-5}\,\text{F})(200\text{ V}) = 2 \times 10^{-3}\,\text{C}$$

$$1.2 \times 10^{-3}\,\text{C} + 8 \times 10^{-4}\,\text{C} = 2 \times 10^{-3}\,\text{C}$$

(b) $V_6 = 200\text{ V}$

$$V_4 = 200\text{ V}$$

PROBLEM 5.9 Capacitors in parallel. The figure shows three capacitors connected in parallel. Show that the single capacitance C that is equivalent to this combination, is $C = C_1 + C_2 + C_3$.

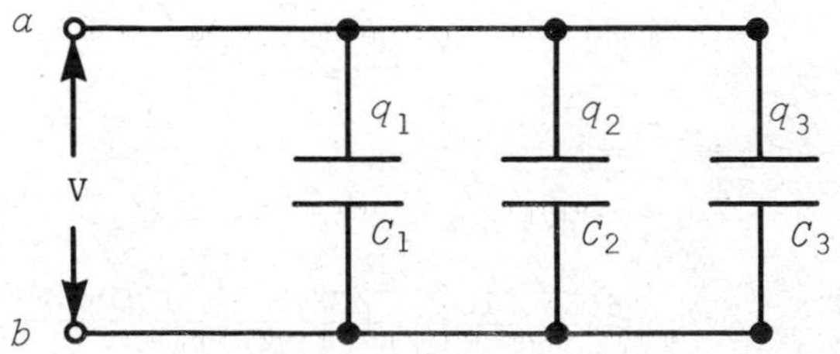

Solution: In parallel, the voltage across all three capacitors is the same.

$$Q_T = Q_1 + Q_2 + Q_3$$

$$VQ_T = VQ_1 + VQ_2 + VQ_3, \quad Q = CV$$

so

$$C_{(eq)T} = C_1 + C_2 + C_3$$

PROBLEM 5.10 A parallel-plate capacitor has circular plates of 8.0-cm radius and 1.0-mm separation. What charge will appear on the plates if a potential difference of 100 V is applied?

Solution:

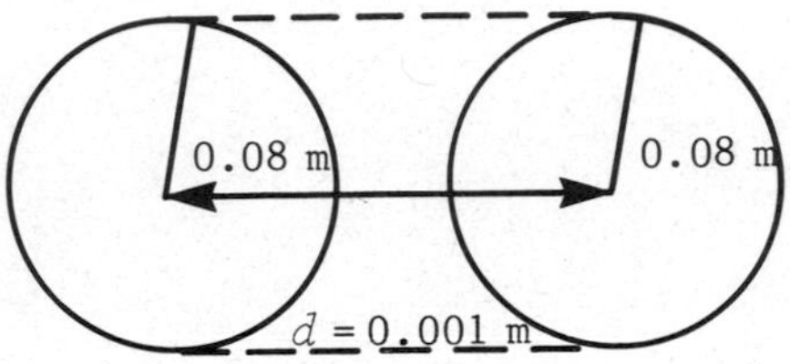

$$V = 100 \text{ V}$$

$$d = 1 \text{ mm} = 0.001 \text{ m}$$

$$A = \pi(0.08 \text{ m})^2 = 0.02 \text{ m}^2$$

$$Q = CV$$

$$\varepsilon_0 \oint \mathbf{E} \cdot \hat{\mathbf{n}}\, d\mathbf{A} = q$$

$$\varepsilon_0 EA = q$$

$$V = \frac{W}{q_0}$$

Work required to go from one plate to another $= W = Vq_0$.

$$W = \mathbf{F} \cdot \mathbf{d} = q_0\, Ed$$

$$Eq_0 = q_0\, Ed$$

$$V = Ed$$

$$C = \frac{Q}{V} = \frac{\varepsilon_0 EA}{ED} = \frac{\varepsilon_0 A}{d}$$

$$Q = CV = \left(\frac{\varepsilon_0 A}{d}\right)(V) = \frac{(8.85 \times 10^{-12} \text{ C}^2/\text{N}\cdot\text{m}^2)(100 \text{ V})(0.02 \text{ m}^2)}{0.001 \text{ m}}$$

$$Q = 1.77 \times 10^{-8} \text{ C}$$

PROBLEM 5.11 Two metallic spheres, radii a and b, are connected by a thin wire. Their separation is large compared with their dimensions. A charge Q is put onto this system.

(a) How much charge resides on each sphere?
(b) Show that the capacitance of this system is $C = 4\pi\varepsilon_0(a+b)$.

Solution:

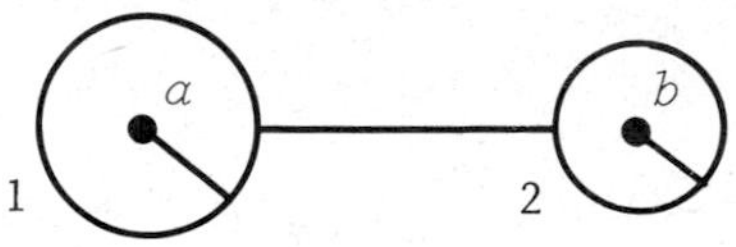

(a) $V_1C_1 = Q_a \qquad Q = CV$

$$Q_a = 4\pi\varepsilon a\left(\frac{1}{4\pi\varepsilon}\right)\left(\frac{q_1}{a}+\frac{q_2}{b}\right) = \frac{bq_1}{b}+\frac{q_2 a}{b}$$

$$Q_a = \frac{q_1 b + q_2 a}{b} = q_1 + q_2\left(\frac{a}{b}\right)$$

(b) $$C_a = \frac{q_1}{V_1} = \frac{q_1}{\frac{1}{4\pi\varepsilon}\frac{q_1}{a}} = 4\pi\varepsilon a$$

$$C_b = \frac{q_2}{V_2} = \frac{q_2}{\frac{1}{4\pi\varepsilon}\frac{q_2}{b}} = 4\pi\varepsilon b$$

$$C_1 + C_2 = 4\pi\varepsilon a + 4\pi\varepsilon b = 4\pi\varepsilon\ (a+b)$$

PROBLEM 5.12 Show that the plates of a parallel-plate capacitor attract each other with a force given by $F(q^2/2\varepsilon_0 A)$. Consider the work necessary to increase the plate separation from x to $x+dx$.

Solution:

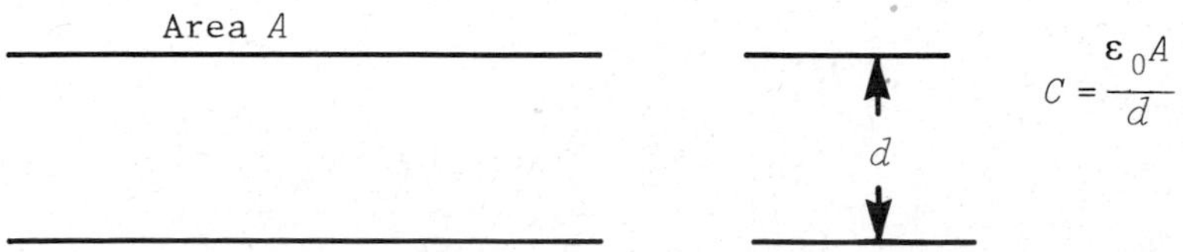

When the capacitor is charged to Q, then the stored energy is

$$U = (\tfrac{1}{2})\ CV^2 = (\tfrac{1}{2})\ Q^2/C.$$

Imagine separating the plates by an amount Δy, where Δy is small. To do this, a force F must be exerted and this force does an amount of work ΔW, where $\Delta W = F\Delta y$

But this ΔW must increase the potential energy of the system so

$$\Delta W = \Delta U = F\Delta y = \tfrac{1}{2} Q^2 \Delta\left(\frac{1}{C}\right),$$

since the charge Q remains constant. In the parallel-plate capacitor

$$\frac{1}{C} = \frac{d}{\varepsilon_0 A}, \quad \Delta\left(\frac{1}{C}\right) = \frac{\Delta y}{\varepsilon_0 A}$$

So

$$F\Delta y = \tfrac{1}{2} Q^2 \frac{\Delta y}{\varepsilon_0 A}$$

$$F = \tfrac{1}{2} \frac{Q^2}{\varepsilon_0 A}$$

PROBLEM 5.13 Two capacitors with capacitance C_1 and C_2 are initially charged with charges q_1 and q_2. Show that except in special cases the stored electrostatic energy decreases when the two capacitors are joined together in parallel. Where does the lost energy appear? Find the conditions under which they can be joined without loss of energy.

Solution: Consider two capacitors C_1 and C_2 with charges q_1 and q_2 on them.

$$U_1 = \tfrac{1}{2} \frac{q_1^2}{C_1} \qquad U_2 = \tfrac{1}{2} \frac{q_2^2}{C_2}$$

So

$$U_0 = \tfrac{1}{2} \left(\frac{q_1^2}{C_1} + \frac{q_2^2}{C_2}\right)$$

For the first case,

$$V_1 = \frac{q_1}{C_1}, \quad V_2 = \frac{q_2}{C_2}$$

When they are joined in parallel, they will reach a common potential $V = q/C$, where $q = q_1 + q_2$, but $C = C_1 + C_2$ so

$$V = \frac{q_1 + q_2}{C_1 + C_2}.$$

The final energy is

$$U = \tfrac{1}{2} CV^2 = \tfrac{1}{2} (C_1 + C_2) \frac{(q_1 + q_2)^2}{(C_1 + C_2)^2} = \tfrac{1}{2} \frac{(q_1 + q_2)^2}{C_1 + C_2}$$

$$U = \tfrac{1}{2} \left(\frac{q_1^2}{C_1 + C_2} + \frac{q_2^2}{C_1 + C_2}\right) + \frac{q_1 q_2}{C_1 + C_2}$$

It is not immediately obvious that $U < U_0$ so let us look at the difference $U - U_0$,

$$U - U_0 = \tfrac{1}{2}\left(\frac{q_1^2}{C_1 + C_2} + \frac{q_2^2}{C_1 + C_2}\right) + \frac{q_1 q_2}{C_1 + C_2} - \tfrac{1}{2}\frac{q_1^2}{C_1} - \tfrac{1}{2}\frac{q_2^2}{C_2}$$

Finding a common denominator, $C_1 C_2 (C_1 + C_2)$

$$U - U_0 = \tfrac{1}{2}\,\frac{C_1 C_2 q_1^2 + C_1 C_2 q_2^2 + 2q_1 q_2 C_1 C_2 - q_1^2 (C_1 + C_2) C_2 - q_2^2 (C_1 + C_2) C_1}{C_1 C_2 (C_1 + C_2)}$$

$$U - U_0 = \tfrac{1}{2}\,\frac{1}{C_1 C_2 (C_1 + C_2)}\,(C_1 C_2 q_1^2 + C_1 C_2 q_2^2 + 2q_1 q_2 C_1 C_2 - q_1^2 C_1 C_2 - q_1^2 C_2^2$$

$$- q_2^2 C_1^2 - q_2^2 C_1 C_2)$$

$$U - U_0 = -\tfrac{1}{2}\,\frac{1}{C_1 C_2 (C_1 + C_2)}\,(q_1^2 C_2^2 - 2q_1 q_2 C_1 C_2 + q_2^2 C_1^2)$$

$$U - U_0 = -\tfrac{1}{2}\,\frac{1}{C_1 C_2 (C_1 + C_2)}\,\{(q_1 C_2 - q_2 C_1)^2\}$$

Since the quantity in brackets is always positive, the $U - U_0$ is negative, which means that U is less than U_0 unless $q_1 C_2 = q_2 C_1$. This is the special case of interest.

PROBLEM 5.14 A radio-tuning capacitor has a maximum capacity of 100 pF (1 pF $= 10^{-12}$ F). By rotation of the moving plates, the capacity can be reduced to 10 pF. Assume the capacitor is charged to a potential difference of 300 V at maximum capacity. The tuning knob is then rotated to minimum capacity. What are the initial and final values of the potential difference? How much mechanical work is done in rotating the knob?

Solution:

$$Q = CV = (10^{-10})(3 \times 10^2) = 3 \times 10^{-8}\ \text{C}$$

$$V_f = \frac{Q}{C} = \frac{3 \times 10^{-8}}{10^{-4}} = 3000\ \text{V}$$

$$W = \tfrac{1}{2}\, C_f V_f^2 - \tfrac{1}{2}\, C_0 V_0^2$$

$$W = [(\tfrac{1}{2})(10^{-11})(9 \times 10^6) - (\tfrac{1}{2})(10^{-10})(9 \times 10^4)]\ \text{J}$$

$$W = [4.5 \times 10^{-5} - 4.5 \times 10^{-6}]\ \text{J}$$

$$W = 4.05 \times 10^{-5}\ \text{J}$$

PROBLEM 5.15 One plate of a parallel-plate capacitor is kept at a potential $\phi = 0$, the other at $\phi = v$. The capacitor contains a space charge density between the plates given by $\rho = kx^2$, where k is constant and x is measured from the plate where $\phi = 0$. Determine the expression for ϕ as a function of x.

Solution:

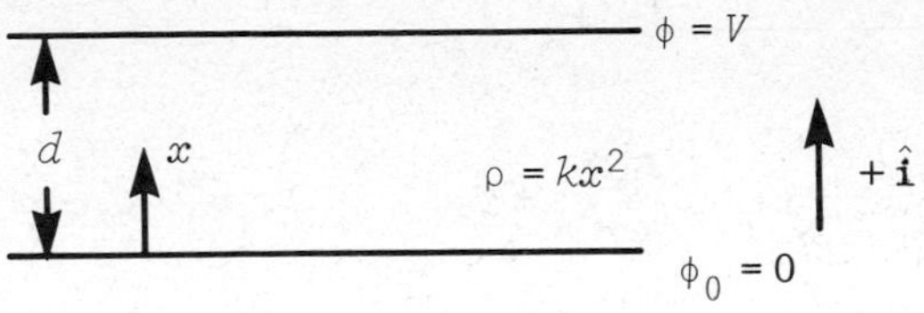

$$\nabla^2\phi = -\frac{\rho}{\varepsilon}$$

$$\frac{\partial^2\phi}{\partial x^2} = \frac{-kx^2}{\varepsilon}$$

$$\frac{\partial\phi}{\partial x} = \frac{-kx^2}{3\varepsilon} + C_1$$

$$\phi = \frac{-k}{12\varepsilon}x^4 + C_1 x + C_2$$

at $x = 0$, $\phi = 0$

$$\phi(0) = C_2 = 0$$

at $x = 0$, $\phi = V$

$$\phi(d) = \frac{-kd^4}{12\varepsilon} + C_1 d = V$$

$$C_1 = \left(V + \frac{Kd^4}{12\varepsilon}\right)\frac{1}{d}$$

$$\phi(x) = \frac{-kx^4}{12\varepsilon} + \frac{1}{d}\left(V + \frac{kd^4}{12\varepsilon}\right)x$$

PROBLEM 5.16 Two flat metal plates of area A are separated by a distance d. A third metal plate of thickness t is placed midway between the plates. Determine the capacitance of the capacitor.

Solution:

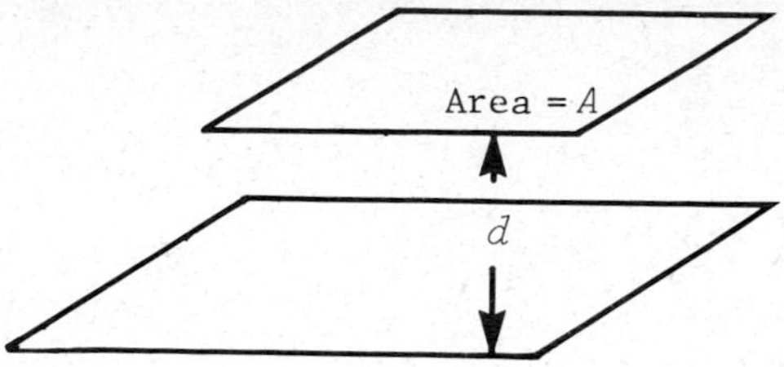

Capacitance Before

$$q = \sigma A, \quad \sigma = D = \varepsilon_0 E, \quad q = \varepsilon_0 EA, \quad E = \frac{V}{D}$$

$$C = \frac{q}{V} = \frac{\varepsilon_0 EA}{ED} = \frac{\varepsilon_0 A}{D}$$

Capacitance After

$$\sigma = D = \varepsilon_0 E$$

$$V = Ed = E(d - t)$$

$$q = \varepsilon_0 EA$$

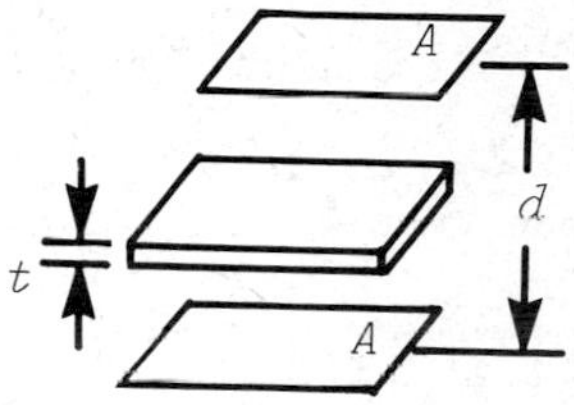

$$E = \frac{V}{d - t}$$

$$C = \frac{q}{V} = \frac{\varepsilon_0 A}{d - t}$$

The capacitance increases by an amount $C_{\text{after}} - C_{\text{before}}$:

$$\frac{d\varepsilon_0 A}{d(d-t)} - \frac{\varepsilon_0 A(d-t)}{d(d-t)} = \frac{\varepsilon_0 Ad - \varepsilon_0 Ad + \varepsilon_0 At}{d(d-t)}$$

$$\Delta C = \frac{\varepsilon_0 At}{d(d-t)}$$

PROBLEM 5.17 A parallel-plate capacitor has plate area A and separation d. A conducting plate of thickness $t < d$ is inserted between the plates, which are connected to a battery with a voltage V. How much work is done in inserting the plate if the battery is left connected during the insertion?

Solution:

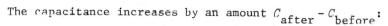

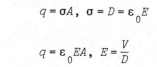

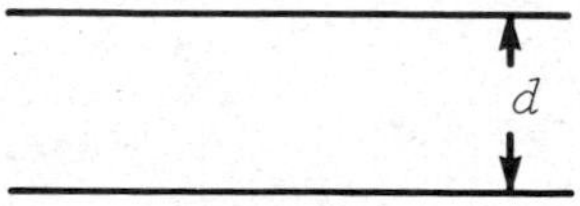

$$q = \sigma A, \quad \sigma = D = \varepsilon_0 E$$

$$q = \varepsilon_0 EA, \quad E = \frac{V}{D}$$

Capacitance Before

$$C_i = \frac{q}{V} = \frac{\varepsilon_0 EA}{ED} = \frac{\varepsilon_0 A}{D} = \frac{\varepsilon A}{d}$$

Capacitance After

$$V = E(d-t), \quad E = \frac{V}{d-t}$$

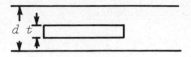

$$q = \varepsilon_0 EA$$

$$C_f = \frac{q}{V} = \frac{\varepsilon_0 A}{d-t}$$

Energy before $= U_i = \frac{1}{2} C_i V^2 = \frac{V^2}{2}\frac{\varepsilon A}{d}$

Energy after $= U_f = \frac{1}{2} C_f V^2 = \frac{V^2}{2}\frac{\varepsilon A}{d-t}$

Work done inserting plate $= U_f - U_i = \frac{V^2 \varepsilon A}{2}\left(\frac{1}{d-t} - \frac{1}{d}\right) = \frac{V^2 \varepsilon A}{2}\left(\frac{d-d+t}{d(d-t)}\right)$

Work done $= \frac{V^2 \varepsilon A t}{2d(d-t)}$

PROBLEM 5.18 The plates of a thin parallel-plate capacitor are separated by a distance d. The maximum voltage that can be applied to this capacitor before a spark occurs in the air inside the capacitor is V_0. A dielectric plate of dielectric constant ε and thickness $t < d$ is laid on the inner surface of one of the capacitor's plates. Show that the maximum voltage that can now be applied to the capacitor before a spark in the air inside the capacitor occurs is only $V = V_0\,[1 - (t/d)(1 - 1/\varepsilon)]$.

Solution:

Before

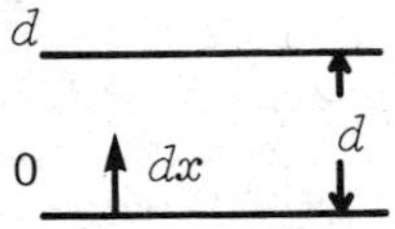

d(thin), $V_{max} = V_0$

$$DA = q$$

$$D = \frac{q}{A}$$

$$E = \frac{q}{K\varepsilon_0 A}$$

where $K = 1$ so $E = \frac{q}{\varepsilon_0 A}$

$$V = \int_0^d \frac{q}{\varepsilon_0 A}\,dx = \frac{qd}{\varepsilon_0 A} = V_0$$

<u>After</u>

For $0 < x < t$, $E = \frac{q}{K\varepsilon_0 A}$

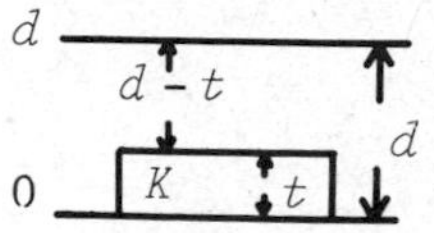

for $t < x < d$, $E = \frac{q}{\varepsilon_0 A}$

$$V = \int_0^t \frac{q}{K\varepsilon_0 A}\,dx + \int_t^d \frac{q}{\varepsilon_0 A}\,dx = \frac{qt}{K\varepsilon_0 A} + \frac{q(d-t)}{\varepsilon_0 A}$$

$$= \frac{q}{\varepsilon_0 A}\left(\frac{t}{K} + \frac{d-t}{1}\right) = \frac{q}{\varepsilon_0 A}\left(\frac{t}{K} - \frac{(d-t)K}{K}\right) = \frac{q}{\varepsilon_0 A}\left(\frac{t + Kd - Kt}{K}\right)\left(\frac{d}{d}\right)$$

$$= \frac{qd}{\varepsilon_0 A}\left(\frac{t + Kd - Kt}{dK}\right) = \frac{qd}{\varepsilon_0 A}\left(\frac{dK}{dK} - \frac{tK}{Kd} + \frac{t}{dK}\right)$$

$$V = \frac{qd}{\varepsilon_0 A}\left[1 - \left(\frac{t}{d}\right)\left(1 - \frac{1}{K}\right)\right]$$

and

$$\frac{qd}{\varepsilon_0 A} = V_0$$

so,

$$V = V_0\left[1 - \left(\frac{t}{d}\right)\left(1 - \frac{1}{K}\right)\right]$$

PROBLEM 5.19 Two identical air capacitors of 10 μF each are connected in series with a battery of constant potential difference 50 V. A dielectric sheet of dielectric constant 10 and a thickness equal to the plate separation is inserted into one of the capacitors. The battery is still connected. Compute the new voltage across this capacitor.

Solution:

<u>Before</u>

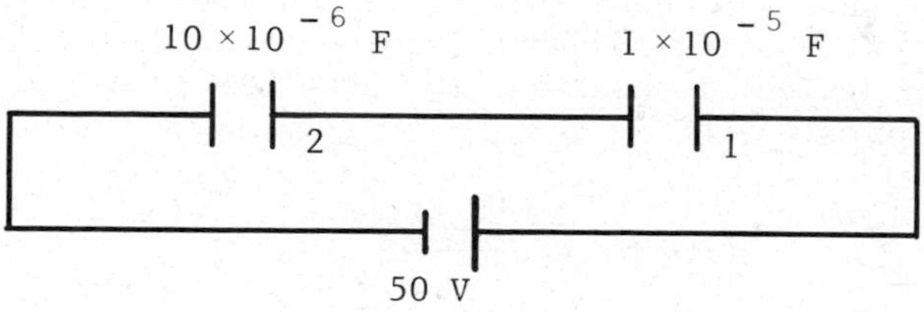

$$\frac{1}{C_T} = \frac{1}{C_1} + \frac{1}{C_2} = \frac{1}{1 \times 10^{-5}\ \text{F}} + \frac{1}{1 \times 10^{-5}\ \text{F}} = \frac{2}{1 \times 10^{-5}\ \text{F}} = 2 \times 10^5 \left(\frac{1}{\text{F}}\right)$$

$$C_T = \frac{1}{2 \times 10^{-5}\ (1/\text{F})} = 5 \times 10^{-6}\ \text{F}$$

$$V_T = 50\ \text{V}$$

The charges on the series capacitors are the same, so

$$Q_T = Q_1 = Q_2 = C_T V_T = (5 \times 10^{-6}\ \text{F})(50\ \text{V}) = 2.5 \times 10^{-4}\ \text{C}$$

$$V_1 = \frac{Q_1}{C_1} = V_2 = \frac{Q_2}{C_2} = \frac{2.5 \times 10^{-4}\ \text{C}}{1 \times 10^{-5}\ \text{F}} = 25\ \text{V}$$

<u>After</u>

C_1 goes up by a factor of K; C_2 stays the same (primes denote "after").

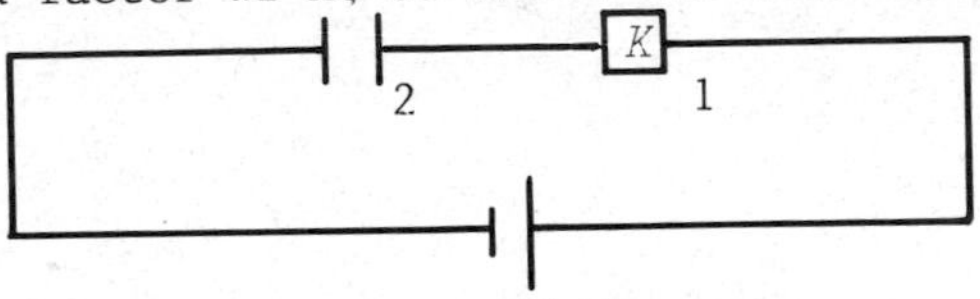

$$\frac{1}{C_T} = \frac{1}{C_1'} + \frac{1}{C_2} = \frac{1}{KC_1} + \frac{1}{C_2} = \frac{1}{(10)(1 \times 10^{-5}\ \text{F})} + \frac{1}{1 \times 10^{-5}\ \text{F}} = 1.1 \times 10^5 \left(\frac{1}{\text{F}}\right)$$

$$C_T = 9.09 \times 10^{-6}\ \text{F},\ V_T = 50\ \text{V}$$

$$Q_{each} = (9.09 \times 10^{-6}\ \text{F})(50\ \text{V}) = 4.5454 \times 10^{-4}\ \text{C}$$

so

$$V_1' = \frac{Q_1'}{C_1'} = \frac{4.5454 \times 10^{-4}\ \text{C}}{(10)(1 \times 10^{-5}\ \text{F})} = 4.5454\ \text{V}$$

The new voltage is $V_1' = 4.55$ V.

PROBLEM 5.20 A spherical capacitor is formed by two spheres whose radii are a and b, $a < b$. The inner sphere receives a uniform coat of material of thickness t and dielectric constant ε. Show that if $t \ll a$, the capacitance increases approximately by

$$\Delta C = 4\pi\varepsilon_0 t\, \frac{(\varepsilon + 1)b^2}{\varepsilon(b - a)^2}$$

Solution:

Before

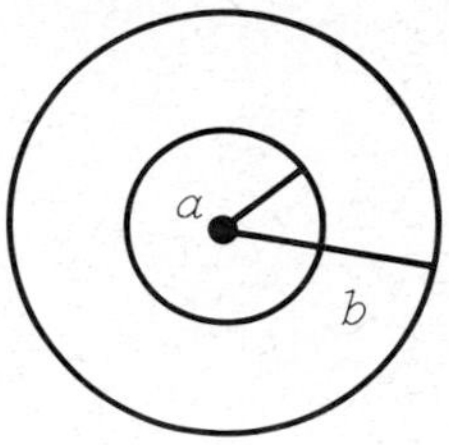

$\underline{A < r < B}$

$$\oint \mathbf{D} \cdot \hat{\mathbf{n}}\, dA = q_f$$

$$D(4\pi r^2) = q$$

$$\mathbf{D} = \frac{q\hat{\mathbf{r}}}{4\pi r^2}$$

$$\mathbf{E} = \frac{1}{\varepsilon}\,\mathbf{D} = \frac{1}{K\varepsilon_0}\,\mathbf{D}$$

where $K - 1$,

$$\mathbf{E} = \frac{q\hat{\mathbf{r}}}{4\pi\varepsilon_0 r^2}$$

$$\phi_a - \phi_b = -\int_b^a \frac{q}{\varepsilon_0(4\pi r^2)}\, dr = \frac{-q}{4\pi\varepsilon_0}\,(-)\left(\frac{1}{a} - \frac{1}{b}\right) = \frac{q}{4\pi\varepsilon_0}\left(\frac{1}{a} - \frac{1}{b}\right)$$

$$C = \frac{Q}{V} = \frac{q}{\dfrac{q}{4\pi\varepsilon_0}\left(\dfrac{1}{a} - \dfrac{1}{b}\right)} = 4\pi\varepsilon_0\left(\frac{1/(b-a)}{ab}\right)$$

$$C_{\text{before}} = 4\pi\varepsilon_0\,\frac{ab}{b-a}$$

After

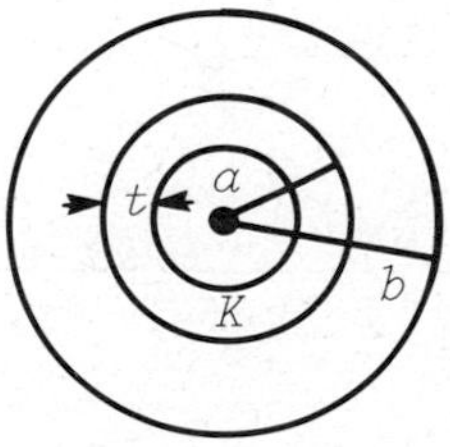

$\underline{a < r < a + t}$

$$D(4\pi r^2) = q_f$$

$$D = \frac{q_f}{4\pi r^2}$$

$$\mathbf{D} = \varepsilon \mathbf{E} = K\varepsilon_0 \mathbf{E}$$

$$E = \frac{q_f}{K\varepsilon_0 4\pi r^2}$$

$\underline{a + t < r < b}$

$$D(4\pi r^2) = q_f$$

$$D = \frac{q_f}{4\pi r^2}$$

$$E = \frac{1}{K\varepsilon_0} D, \text{ with } K = 1$$

$$E = \frac{q_f}{4\pi\varepsilon_0 r^2}$$

$$\phi_a - \phi_b = -\int_b^a E\,dr = -\int_b^{a+t} \frac{q_f\,dr}{\varepsilon_0 4\pi r^2} - \int_{a+t}^a \frac{q_f}{K\varepsilon_0 4\pi r^2}$$

$$= \frac{-q_f}{4\pi\varepsilon_0}(-)\left(\frac{1}{a+t} - \frac{1}{b}\right) - \frac{q_f}{4\pi\varepsilon_0 K}(-)\left(\frac{1}{a} - \frac{1}{a+t}\right)$$

$$= \frac{q_f}{4\pi\varepsilon_0}\left(\frac{1}{a+t} - \frac{1}{b}\right) - \frac{q_f}{4\pi\varepsilon_0 K}\left(\frac{1}{a} - \frac{1}{a+t}\right)$$

$$C_{\text{after}} = \frac{q}{V} = \frac{q}{\dfrac{q}{4\pi\varepsilon_0}\left[\left(\dfrac{1}{a+t} - \dfrac{1}{b}\right) - \dfrac{1}{K}\left(\dfrac{1}{a} - \dfrac{1}{a+t}\right)\right]}$$

$$C_{\text{after}} = 4\pi\varepsilon_0 \left[\left(\frac{1}{a+t} - \frac{1}{b}\right) - \frac{1}{K}\left(\frac{1}{a} - \frac{1}{a+t}\right)\right]^{-1}$$

$$= 4\pi\varepsilon_0 \left[\frac{b-(a+t)}{b(a+t)} - \frac{1}{K}\left(\frac{a+t-a}{a(a+t)}\right)\right]^{-1}$$

$$= 4\pi\varepsilon_0 \left(\frac{[b-(a+t)]Ka(a+t)}{b(a+t)Ka(a+t)} - \frac{tb(a+t)}{Ka(a+t)^2 b}\right)^{-1}$$

$$= 4\pi\varepsilon_0 \left(\frac{[b-(a+t)]Ka(a+t) - tb(a+t)}{Kab(a+t)^2}\right)^{-1}$$

$$= 4\pi\varepsilon_0 \left(\frac{bKa^2 + bKat - Ka^3 - 2Kta^2 - Kat^2 - tba - t^2 b}{Kab(a+t)^2}\right)^{-1}$$

$$C_{\text{after}} = 4\pi\varepsilon_0 \left(\frac{Kab(a+t)^2}{bKa^2 + bKat - Ka^3 - 2Kta^2 - Kat^2 - tba - t^2b}\right)$$

$$\Delta C = C_{\text{after}} - C_{\text{before}}$$

$$= 4\pi\varepsilon_0 \left[\left(\frac{Kab(a+t)^2}{bKa^2 + bKat - Ka^3 - 2Kta^2 - Kat^2 - tba - t^2b}\right) - \left(\frac{ab}{b-a}\right)\right]$$

$$= 4\pi\varepsilon_0 \left(\frac{Kab(a+t)^2(b-a) - (ab)(bKa^2 + bKat - Ka^3 - 2Kta^2 - Kat^2 - tba - t^2b)}{(bKa^2 + bKat - Ka^3 - 2Kta^2 - Kat^2 - tba - t^2b)(a-b)}\right)$$

$$\Delta C = 4\pi\varepsilon_0 \Big[(Ka^3b^2 - Ka^4b + 2Kta^2b^2 - 2Kta^3b + t^2Kab^2 - t^2Ka^2b - a^3b^2K - a^2b^2Kt + a^4bK + 2Ktba^3 + a^2bKt^2 + a^2b^2t + ab^2t^2)/\{(bKa^2 + bKat - Ka^3 - 2Kta^2 - Kat^2 - tba - t^2b)(b-a)\}\Big]$$

$$\Delta C = 4\pi\varepsilon_0 t \left(\frac{Ka^2b^2 + tKab^2 + a^2b^2 + tab^2}{(b-a)(bKa^2 + bKat - Ka^3 - 2Kta^2 - Kat^2 - tba - t^2b}\right)$$

[multiply by $(b-a)/(b-a)$]

$$\Delta C = \frac{4\pi\varepsilon_0 t}{(b-a)^2} \left(\frac{(Ka^2b^2 + tKab^2 + a^2b^2 + tab^2)(b-a)}{bKa^2 + bKat - Ka^3 - 2Kta^2 - Kat^2 - tba - t^2b}\right)$$

$$= \frac{4\pi\varepsilon_0 tb^2}{(b-a)^2} \left(\frac{(Ka^2 + tKa + a^2 + ta)(b-a)}{bKa^2 + bKat - Ka^3 - 2Kta^2 - Kat^2 - tba - t^2b}\right)$$

$$= \frac{4\pi\varepsilon_0 tb^2}{(b-a)^2} \left(\frac{Ka^2b + btKa + ba^2 + bta - a^3K - tKa^2 - a^3 - ta^2}{bKa^2 + bKat - Ka^3 - 2Kta^2 - Kat^2 - tba - t^2b}\right)$$

$$= \frac{4\pi\varepsilon_0 tb^2}{(b-a)^2} \left(\frac{K(a^2b + bat - a^3 - ta^2) + ba^2 + bta - a^3 - ta^2}{K(ba^2 + bat - a^3 - 2ta^2 - at^2) - tba - t^2b}\right)$$

Now since $t \ll a$, $t/a \to 0$, so one multiplies through by $1/a^2$,

$$= \frac{4\pi\varepsilon_0 tb^2}{(b-a)^2}\left(\frac{K(b-a-t)+b-a-t}{K(b-a-2t)}\right)$$

$$= \frac{4\pi\varepsilon_0 tb^2}{(b-a)^2}\left(\frac{(K+1)(b-a-t)}{K(b-a-2t)}\right)$$

Now since $t \ll a$, and since $a < b$, then also $t \ll b$ and t can be neglected in the terms $b-a-t$ and $b-a-2t$

$$\Delta C = \frac{4\pi\varepsilon_0 tb^2}{(b-a)^2}\,\frac{K+1}{K}$$

and in this problem the dielectric constant K is ε, so

$$\Delta C = \frac{4\pi\varepsilon_0 tb^2(\varepsilon+1)}{\varepsilon(b-a)^2}$$

$$\Delta C = 4\pi\varepsilon_0 t\,\frac{(\varepsilon+1)b^2}{\varepsilon(b-a)^2}$$

✓ PROBLEM 5.21 A parallel-plate capacitor of plate area A and plate separation d is filled with a dielectric whose permittivity varies linearly from ε_1 at one plate to ε_2 at the other. Neglecting edge effects, show that the capacitance of this capacitor is

$$C = \frac{A}{d}\,\frac{\varepsilon_2-\varepsilon_1}{\ln(\varepsilon_2/\varepsilon_1)}$$

Solution:

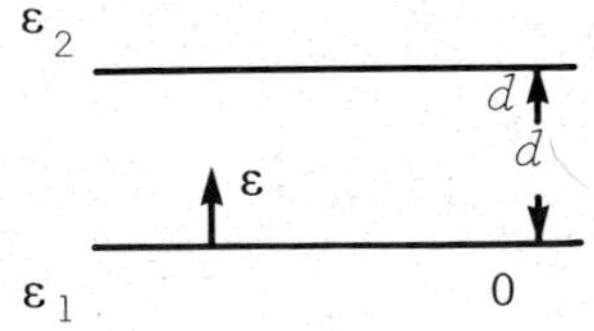

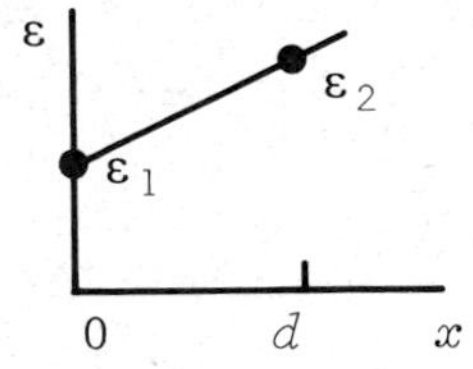

$$\text{Slope} = \frac{\varepsilon_2-\varepsilon_1}{d-0} = \frac{\varepsilon_2-\varepsilon_1}{d}$$

$$Y = mx + b$$

$$\varepsilon = \frac{\varepsilon_2-\varepsilon_1}{d}\,x + \varepsilon_1$$

$$\varepsilon_R = K = \frac{\varepsilon}{\varepsilon_0} = \frac{1}{\varepsilon_0}\left(\frac{\varepsilon_2-\varepsilon_1}{d}\,x + \varepsilon_1\right)$$

$$D = \frac{q}{A}$$

$$E = \frac{1}{\varepsilon} D = \frac{1}{K\varepsilon_0} D = \frac{1}{K\varepsilon_0} \frac{q}{A} = \frac{q}{A\varepsilon_0 \frac{1}{\varepsilon_0} \left(\frac{\varepsilon_2 - \varepsilon_1}{d} x + \varepsilon_1 \right)}$$

$$E = \frac{q}{A\left(\frac{\varepsilon_2 - \varepsilon_1}{d} x + \varepsilon_1 \right)}$$

$$\phi = \int_0^d \frac{q\, dx}{A\left(\frac{\varepsilon_2 - \varepsilon_1}{d} x + \varepsilon_1 \right)}$$

$$\phi = \frac{q}{A} \int_0^d \frac{dx}{\varepsilon_1 + \frac{\varepsilon_2 - \varepsilon_1}{d} x} = \frac{q}{A} \left\{ \frac{d}{\varepsilon_2 - \varepsilon_1} \ln \left[\varepsilon_1 + \left(\frac{\varepsilon_2 - \varepsilon_1}{d} \right) x \right] \right\}_0^d$$

$$\phi = \frac{qd}{A(\varepsilon_2 - \varepsilon_1)} [\ln (\varepsilon_1 + \varepsilon_2 - \varepsilon_1) - \ln \varepsilon_1] = \frac{qd}{A(\varepsilon_2 - \varepsilon_1)} \ln \left(\frac{\varepsilon_2}{\varepsilon_1} \right)$$

$$C = \frac{q}{V} = \frac{q}{\frac{qd}{A(\varepsilon_2 - \varepsilon_1)} \ln \left(\frac{\varepsilon_2}{\varepsilon_1} \right)}$$

$$C = \frac{A(\varepsilon_2 - \varepsilon_1)}{d \ln \left(\frac{\varepsilon_2}{\varepsilon_1} \right)}$$

PROBLEM 5.22 Two capacitors of equal capacitance C are connected in parallel, charged to a voltage V_0, and then isolated from the voltage source, as shown in the figure. A dielectric of dielectric constant K is inserted into one capacitor and completely fills the space between plates. Calculate the free charge transferred from one capacitor to the other and the final voltage V_2 across the capacitors in terms of C, V_0, and K.

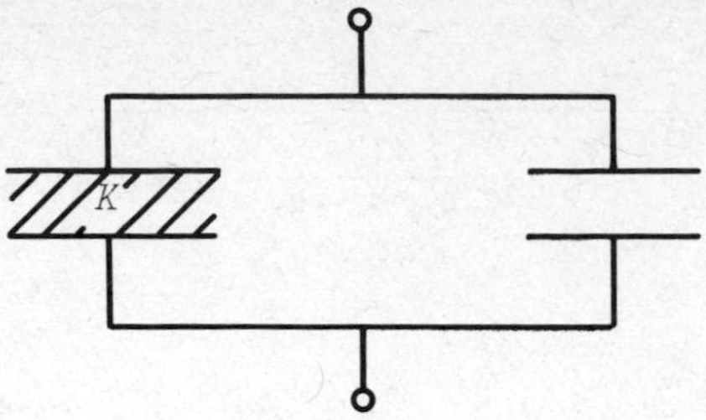

Solution:

Before

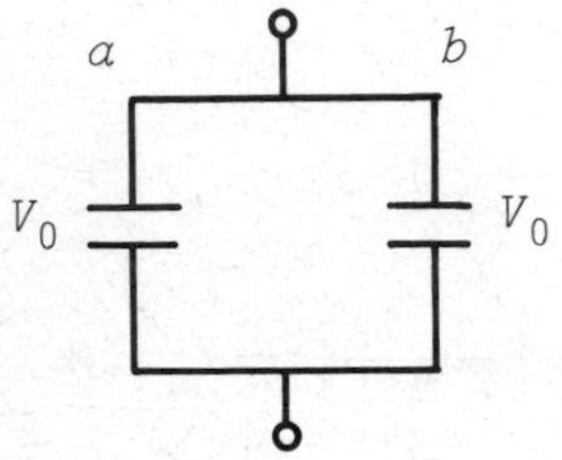

$V_a = V_0$ $\qquad$ $V_b = V_0$

$C_a = \frac{\varepsilon_0 A}{d}$ $\qquad$ $C_b = \frac{\varepsilon_0 A}{d}$

$q_a = C_a V_a = \frac{\varepsilon_0 V_0 A}{d}$ $\qquad$ $q_b = \frac{\varepsilon_0 V_0 A}{d}$

$q_{total} = q_a + q_b = \frac{2V_0 \varepsilon_0 A}{d}$

After

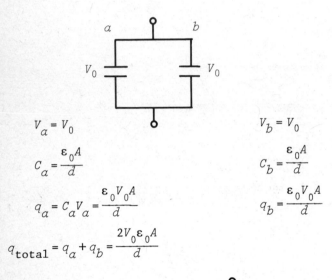

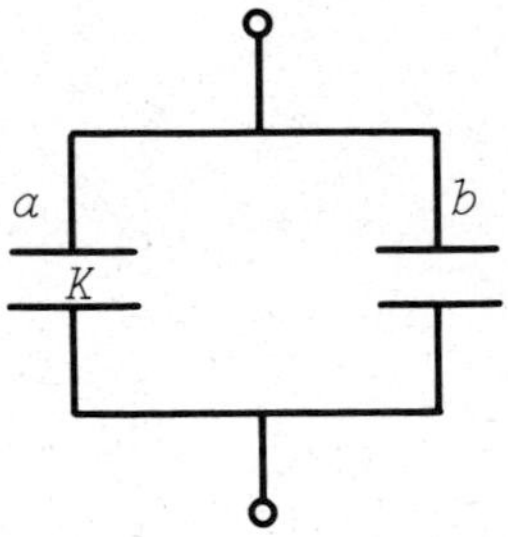

is same in parallel.

$V_a = V_2$ $\qquad$ $V_b = V_2$

$C_a = \frac{\varepsilon_0 AK}{d}$ $\qquad$ $C_b = \frac{\varepsilon_0 A}{d}$

Since $Q = CV$

$$q_{a_f} = C_{a_f} V_{a_{\text{final}}} = \frac{\varepsilon_0 AK}{d} V_2$$

$$q_{b_f} = C_{b_f} V_{b_f} = \frac{\varepsilon_0 A}{d} V_2$$

$$q_{T_f} = q_{a_f} + q_{b_f} = \frac{\varepsilon_0 AV_2}{d} (K+1)$$

Solving for V_2 by setting the total q before and after equal (by conservation of charge),

$$q_{T_i} = q_{T_f} = \frac{V_2 \varepsilon_0 A(K+1)}{d} = \frac{2V_0 \varepsilon_0 A}{d} \rightarrow V_2 = \frac{2V_0}{K+1}$$

Charge transferred $= q_{b_i} - q_{b_f} = q_{a_f} - q_{a_i}$

$$\frac{\varepsilon_0 V_0 A}{d} - \frac{\varepsilon_0 A}{d} V_2 = \frac{\varepsilon_0 AK}{d} V_2 - \frac{\varepsilon_0 A}{d} V_0 \tag{1}$$

$$V_2 = \frac{2V_0}{K+1} \tag{2}$$

$$\frac{\varepsilon_0 A}{d} (V_0 - V_2) = \frac{\varepsilon_0 A}{d} (KV_2 + V_0)$$

charge transferred = charge transferred

Check if these are equal:

$$V_0 - V_2 \stackrel{?}{=} KV_2 - V_0$$

$$V_0 - \frac{2V_0}{K+1} \stackrel{?}{=} KV_2 - V_0 = \frac{K2V_0}{K+1} - V_0$$

$$V_0\left(1 - \frac{2}{K+1}\right) \stackrel{?}{=} V_0\left(\frac{2K}{K+1} - 1\right)$$

$$1 - \frac{2}{K+1} \stackrel{?}{=} \frac{2K}{K+1} - 1$$

$$\frac{K+1-2}{K+1} \stackrel{?}{=} \frac{2K-(K+1)}{K+1}$$

$$\frac{K-1}{K+1} \stackrel{?}{=} \frac{2K-K-1}{K+1}$$

$$\frac{K-1}{K+1} \stackrel{\checkmark}{=} \frac{K-1}{K+1}$$

Free charge transferred is then

$$\frac{\varepsilon_0 A}{d}(V_0 - V_2) = \frac{\varepsilon_0 A}{d} V_0 \left(1 - \frac{2}{K+1}\right) = \frac{\varepsilon_0 A}{d} V_0 \left(\frac{K-1}{K+1}\right)$$

PROBLEM 5.23 Suppose a capacitor were connected to a battery so as to maintain constant voltage when a dielectric slab of dielectric constant K is inserted between the plates. Compare the stored energy in the capacitor before and after inserting the dielectric. On the basis of this comparison, can you argue about the direction of the force on the slab? Explain.

Solution:

Before

After

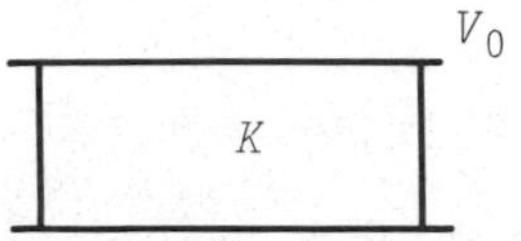

$$V_b = V_0 \qquad Q = CV$$

$$Q_b = Q \qquad V_a = V_0$$

$$C_b = \frac{\varepsilon_0 A}{d} \qquad Q_a = QK$$

$$C_a = \frac{\varepsilon_0 A}{d} K$$

$$U_a = \tfrac{1}{2} C_a V_a^2 = \tfrac{1}{2} \frac{\varepsilon_0 A}{d} K V_0^2$$

$$U_b = \tfrac{1}{2} C_b V_b^2 = \tfrac{1}{2} \frac{\varepsilon_0 A}{d} V_0^2$$

Now the capacitance goes up, so there is more energy after and

$$U_a - U_b = \tfrac{1}{2} \frac{\varepsilon_0 A}{d} V_0^2 (K-1)$$

The direction of the force on the slab is still attractive pulling the slab in, but here the battery is keeping the voltage constant and half the work done by the battery goes into increasing the stored energy in the capacitor, and half appears as work pulling the dielectric slab into the space between the plates.

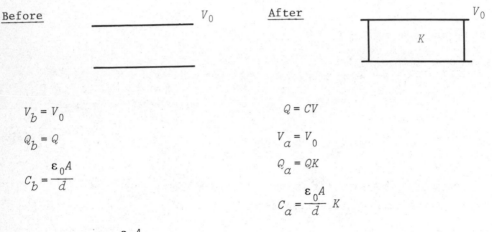

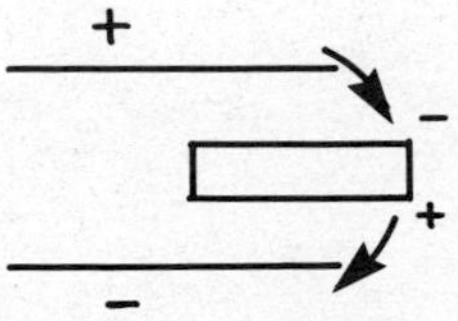

$$dU = \tfrac{1}{2}\,\frac{V_0^2\,\varepsilon_0 A}{d}\,(K-1)\,dx$$

$$F\,dx = -\,dU + V\,dQ = -\,\tfrac{1}{2}\,\frac{V_0^2\,\varepsilon_0 A}{d}\,(K-1)\,dx + \frac{V_0^2\,\varepsilon_0 A}{d}\,(K-1)\,dx$$

or

$$F = \tfrac{1}{2}\,\frac{V_0^2\,\varepsilon_0 A}{d}\,(K-1)$$

PROBLEM 5.24 The voltage between parallel plates of a capacitor is V_1. The plates are isolated electrically (i.e., the charge stays the same). A dielectric slab of dielectric constant K is inserted between the plates and completely fills the volume between them. Find the new potential V_2. Compare the stored energy before and after inserting the slab. On the basis of this comparison, present an argument as to whether electrostatic forces pull the slab into the space between plates or tend to push it away.

Solution:

Before

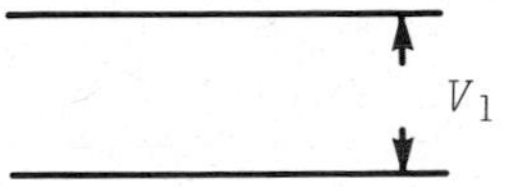

After

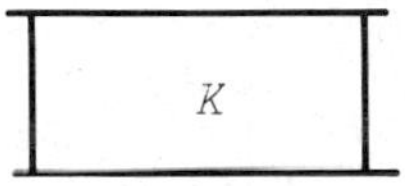

The effect of the dielectric is to decrease the original potential difference V_1 by a factor of $(1/K) \rightarrow V_2 = (V_1/K)$.

$$\text{Capacitance before} = C_b = \frac{\varepsilon_0 A}{d}$$

$$\text{Capacitance after} = C_a = \frac{\varepsilon_0 KA}{d}$$

$$F_b = U_b = \tfrac{1}{2}\,C_b V_b^2 = \tfrac{1}{2}\,\frac{\varepsilon_0 A}{d}\,V_1^2$$

$$E_a = U_a = \tfrac{1}{2}\,C_a V_a^2 = \tfrac{1}{2}\left(\frac{\varepsilon_0 KA}{d}\right) V_2^2 = \tfrac{1}{2}\,\frac{\varepsilon_0 KA}{d}\,\frac{V_1^2}{K^2}$$

$$U_a = \tfrac{1}{2}\,\frac{\varepsilon_0 A}{dK}\,V_1^2$$

$$U_b - U_a = \tfrac{1}{2}\,\frac{\varepsilon_0 A}{d}\,V_1^2 - \tfrac{1}{2}\,\frac{\varepsilon_0 A}{d}\,V_1^2\,\frac{1}{K} = \tfrac{1}{2}\,\frac{\varepsilon_0 A}{d}\,V_1^2\left(1 - \frac{1}{K}\right)$$

The stored energy goes down by a factor of $1/K$ because this is the work that the slab and plates do on anything restraining the slab from being pulled in between the plates by the forces of attraction between the surface charges that come from polarizing the molecules in the dielectric when it comes between the plates, with the free charge that is already on the plates.

PROBLEM 5.25 A certain parallel-plate capacitor of plate separation 1 mm has a capacitance of 1 μF when the space between the plates is filled with air. When a dielectric of thickness ½ mm is placed between the plates the capacitance increases to 1.8 μF. The capacitor is connected to a 100 V power source the entire time.

(a) What is the permittivity of the dielectric?
(b) How much work was done in inserting the dielectric?

Solution:

<u>Before</u> <u>After</u>

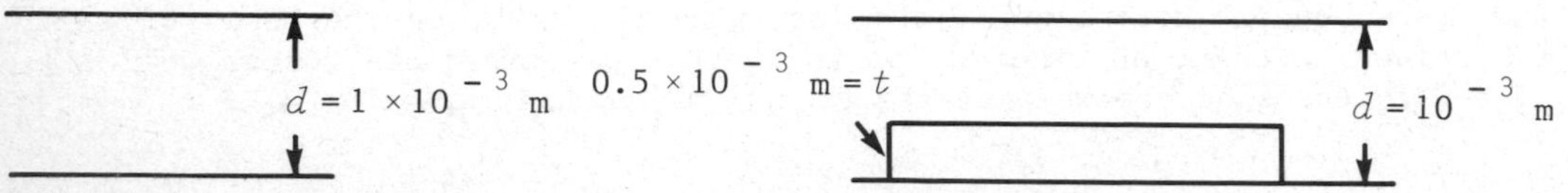

Before

$$C_b = 1 \times 10^{-6}\ \text{F}$$

$$V = 100\ \text{V}$$

(a) $$V = \frac{q_b}{C_b}$$

$$E = \frac{V}{d}$$

$$q_b = \sigma_b A$$

$$V_b = \frac{\sigma_b d}{\varepsilon_0}, \quad V_b = \int_0^d \frac{D}{\varepsilon_0}\, d\ell$$

$$DA = \sigma_b A$$

$$C_b = \frac{\sigma_b A}{\sigma_b \dfrac{d}{\varepsilon_0}}$$

$$C_b = \frac{\varepsilon_0 A}{d}$$

After

$$C_a = 1.8 \times 10^{-6}\ \text{F}$$

$$V = 100\ \text{V}$$

$$V = \frac{q_a}{C_a}, \quad q_a = \sigma_a A$$

$$V_a = \int_0^d \mathbf{E} \cdot d\ell = \int_0^t \frac{D}{\varepsilon}\, d\ell + \int_t^d \frac{D}{\varepsilon_0}\, d\ell$$

$$\int \mathbf{D} \cdot d\vec{\mathbf{s}} = q$$

$$DA = \sigma_a A, \quad D = \sigma_a$$

$$V_a = \frac{\sigma_a}{\varepsilon}\, t + \frac{\sigma_a}{\varepsilon_0}\,(d - t)$$

$$C_a = \frac{q_a}{V} = \frac{\sigma_a A}{\dfrac{\sigma_a t}{\varepsilon} + \dfrac{\sigma_a}{\varepsilon_0}\,(d - t)}$$

$$C_a = \frac{A}{\dfrac{t}{\varepsilon} + \dfrac{d - t}{\varepsilon_0}}$$

$$C_a = \frac{A}{\dfrac{t}{\varepsilon} + \dfrac{d - t}{\varepsilon_0}} = \frac{A\varepsilon\,\varepsilon_0}{t\varepsilon_0 + d\varepsilon - t\varepsilon}$$

$$= \frac{A\varepsilon_0}{t\,\dfrac{\varepsilon_0}{\varepsilon} + d - t} = \frac{dA\varepsilon_0}{d\left(\dfrac{t\varepsilon_0}{\varepsilon} + d - t\right)}$$

$$C_a = \frac{A\varepsilon_0}{d}\left(\frac{d}{t\,\dfrac{\varepsilon_0}{\varepsilon} + d - t}\right)$$

$$C_a = C_b\left(\frac{d\varepsilon}{t\varepsilon_0 + d\varepsilon - t\varepsilon}\right)$$

$$\frac{C_a}{C_b d} = \frac{\varepsilon}{t\varepsilon_0 + \varepsilon(d-t)}$$

$$[\varepsilon(d-t) + t\varepsilon_0]\,\frac{C_a}{dC_b} = \varepsilon$$

$$\varepsilon(d-t)\left(\frac{C_a}{dC_b}\right) + t\varepsilon_0\,\frac{C_a}{dC_b} = \varepsilon$$

$$\varepsilon(d-t)\left(\frac{C_a}{dC_b}\right) - \varepsilon = -t\varepsilon_0\,\frac{C_a}{dC_b}$$

$$\varepsilon\left[(d-t)\left(\frac{C_a}{dC_b}\right) - 1\right] = -t\varepsilon_0\,\frac{C_a}{dC_b}$$

$$\varepsilon\left(1 - (d-t)\,\frac{C_a}{dC_b}\right) = t\varepsilon_0\,\frac{C_a}{dC_b}$$

$$\varepsilon = \frac{t\varepsilon_0(C_a/dC_b)}{\left(1 - (d-t)\,\dfrac{C_a}{dC_b}\right)} = \frac{t\varepsilon_0 C_a}{dC_b\left(1 - (d-t)\,\dfrac{C_a}{dC_b}\right)}$$

$$\varepsilon = \frac{t\varepsilon_0 C_a}{dC_b - (d-t)C_a}$$

$$\varepsilon = \frac{(0.50\times10^{-3})(8.85\times10^{-12})(1.80\times10^{-6})}{(1\times10^{-3})(1\times10^{-6}) - (1\times10^{-3} - 0.50\times10^{-3})(1\times10^{-6})}$$

$$\varepsilon = 1.59\times10^{-11}\ \frac{\mathrm{C}^2}{\mathrm{N}\cdot\mathrm{m}^2}$$

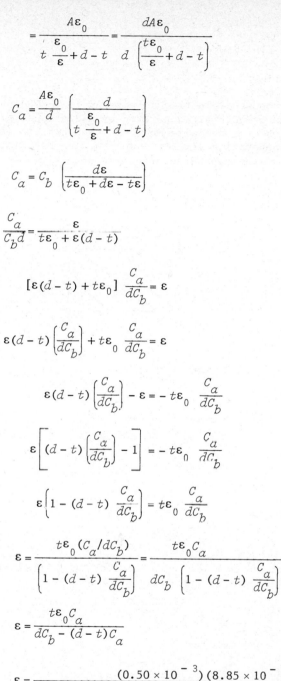

(b) $W = U_a - U_b = \frac{1}{2} C_a V^2 - \frac{1}{2} C_b V^2$

$$W = \tfrac{1}{2} (1.8 \times 10^{-6})(100)^2 - \tfrac{1}{2} (1 \times 10^{-6})(100)^2$$

$$W = 4 \times 10^{-3} \text{ J}$$

PROBLEM 5.26 Consider a condenser whose dielectric is air and is formed by two square, parallel plates of area S which are separated by a distance d. Show that the capacitance is $C = S/4\pi d$. If the alignment of the plates is disturbed so that on one edge the separation is $d + \delta$ and on the opposite edge is $d - \delta$, where $\delta << d$, show that the capacitance becomes approximately

$$C = \frac{S}{4\pi d}\left(1 + \frac{\delta^2}{3d^2}\right)$$

Solution:

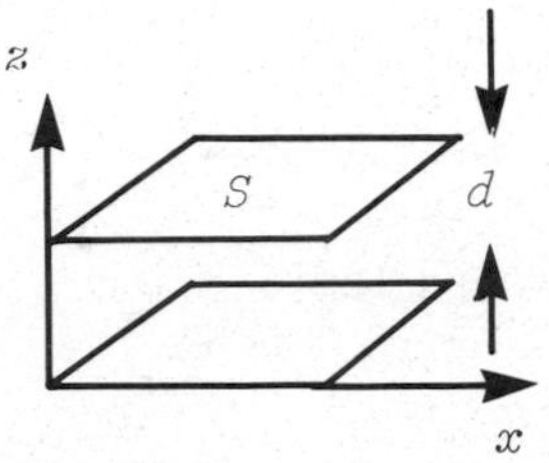

$$E = 4\pi\sigma$$

$$\phi = \int_0^d 4\pi\sigma \, dy = 4\pi\sigma d$$

$$q = s\sigma, \quad \therefore \quad C = \frac{q}{\phi} = \frac{S}{4\pi d}$$

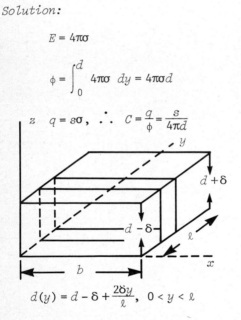

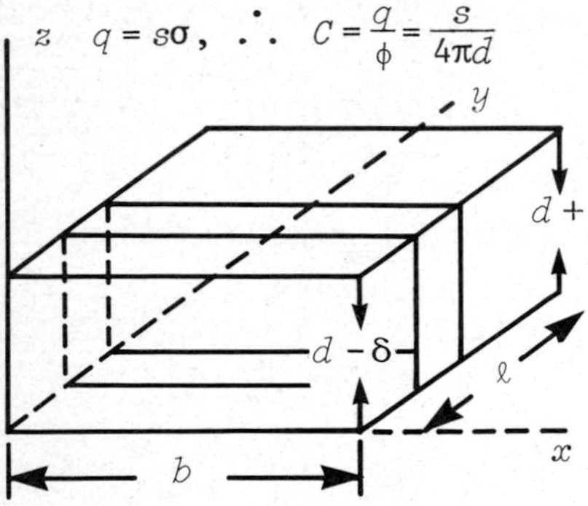

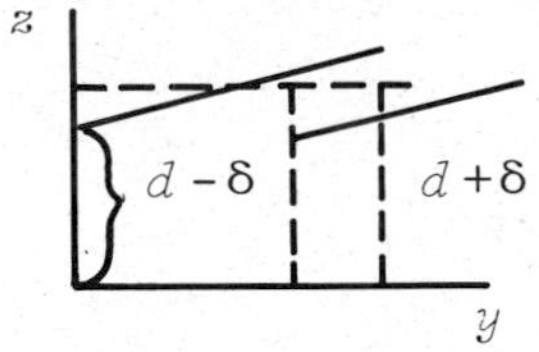

$$d(y) = d - \delta + \frac{2\delta y}{\ell}, \quad 0 < y < \ell$$

Interpret the capacitor as a system of parallel capacitors of infinitesimal width dy_1 as shown. Then, we sum (integrate) capacitances,

$$C = \int_0^{\ell} C(y)\, dy = \int_0^{\ell} \frac{b\, dy}{4\pi d(y)} = \int_0^{\ell} \frac{b\, dy}{4\pi\left[(d - \delta) - \frac{2\delta y}{\ell}\right]}$$

$$= \frac{\ell b}{8\pi\delta}\left[\ln(1 + \delta/d) - \ln(1 - \delta/d)\right] = \frac{S}{8\pi\delta}\left[\ln(1 + \delta/d) - \ln(1 - \delta/d)\right]$$

Use

$$\ln(1 + x) \simeq x - \frac{x^2}{2} + \frac{x^3}{3} \ldots \quad (x << 1)$$

$$\ln(1 - x) \simeq -\left(x + \frac{x^2}{2} + \frac{x^3}{3} \ldots\right)$$

$$C = \frac{S}{8\pi\delta}\left[\frac{\delta}{d} - \frac{1}{2}\left(\frac{\delta}{d}\right)^2 + \frac{1}{3}\left(\frac{\delta}{d}\right)^3 + \ldots + \frac{\delta}{d} + \frac{1}{2}\left(\frac{\delta}{d}\right)^2 + \frac{1}{3}\left(\frac{\delta}{d}\right)^3 + 0\left(\frac{\delta}{d}\right)^4\right]$$

$$= \frac{S}{8\pi\delta}\left[\frac{2\delta}{d} + \frac{2}{3}\left(\frac{\delta}{d}\right)^3\right] = \frac{S}{4\pi d}\left(1 + \frac{\delta^2}{3d^2}\right)$$

PROBLEM 5.27 The capacitance C of a conductor is defined to be the quantity of charge that must be placed on the conductor in order to produce a potential change of one unit, that is, $C = q/\phi$. Calculate the capacitance of the following objects:

(a) an isolated sphere of radius a;
(b) two concentric spheres of radii a and $b > a$, if the outer sphere carries a charge $+q$ and the inner one a charge $-q$;
(c) two coaxial cylinders of radii a and $b > a$, if the outer cylinder carries a charge density $+\rho_\ell$ (charge per unit length) and the inner one a charge density $-\rho_\ell$.

What are the units of capacitance calculated in case (c)?

Solution:

(a) $\phi = \frac{q}{R}$, for sphere, $\therefore C = \frac{q}{\phi} = R$

(b) For $a < r < b$,

$$E(r) = \frac{-q}{r^2}$$

(Use Gauss' law to prove it.)

$$\phi = \phi_b - \phi_a = -\int_a^b \mathbf{E} \cdot d\mathbf{r} = \int_a^b \frac{q}{r^2}\, dr = \frac{-q}{r}\Big|_a^b = q\left(\frac{b-a}{ab}\right)$$

$$\therefore C = \frac{q}{\phi} = \frac{ab}{b-a} = \frac{1}{\frac{1}{a} - \frac{1}{b}}$$ → [Note result (a) could be obtained from this by letting $b \to \infty$.]

(c) Use Gauss' law to show that, for $a \le r \le b$, then $E = -2\rho_\ell/r$, then

$$\phi = \phi_b - \phi_a = -\int_a^b \mathbf{E} \cdot d\mathbf{r} = \int_a^b \frac{2\rho_\ell}{r} = 2\rho_\ell \ln b/a$$

therefore,

$$C/\text{unit length} = \frac{\rho_\ell}{2\rho_\ell \ln b/a} = 1/(2 \ln b/a)$$

Chapter 6

PROBLEM 6.1 Prove the following relationship between the polarization, **P**, and the bound charge densities ρ_P and σ_P, for a dielectric specimen of volume v and surface S:

$$\int_v \mathbf{P}\, dv = \int_v \rho_P \mathbf{r}\, dv + \int_S \sigma_P \mathbf{r}\, da$$

Here, $\mathbf{r} = \hat{\mathbf{i}}x + \hat{\mathbf{j}}y + \hat{\mathbf{k}}z$ is the position vector from any fixed origin.

[*Hint:* Expand $\operatorname{div}(x\mathbf{P})$ according to the equation $\operatorname{div}'(f\mathbf{A}) = f \operatorname{div}'\mathbf{A} + \mathbf{A} \cdot \nabla' f$.]

Solution: Take arbitrary constant vector **a**:

$$\mathbf{a} \cdot \mathbf{r}\rho_P = -\mathbf{a} \cdot \mathbf{r}\nabla \cdot \mathbf{P} = -\nabla \cdot [(\mathbf{a} \cdot \mathbf{r})\mathbf{P}] + \mathbf{P} \cdot \nabla(\mathbf{a} \cdot \mathbf{r})$$

where the last term is $\mathbf{P} \cdot \mathbf{a}$,

$$\mathbf{a} \cdot \int \mathbf{r}\rho_P\, dv = -\int_v [\nabla \cdot (\mathbf{a} \cdot \mathbf{r})\mathbf{P}]\, dv + \int \mathbf{a} \cdot \mathbf{P}\, dv$$

$$= -\mathbf{a} \cdot \int_S \mathbf{r} P_n\, da + \mathbf{a} \cdot \int \mathbf{P}\, dv$$

$$\int_v \mathbf{P}\, dv = \int_v \mathbf{r}\rho_P\, dv + \int_S \mathbf{r}\sigma_P\, da$$

PROBLEM 6.2 A conducting sphere of radius R floats half submerged in a liquid dielectric medium of permittivity ε_1. The region above the liquid is a gas of permittivity ε_2. The total free charge on the sphere is Q. Find a radial inverse-square electric field satisfying all boundary conditions, and determine the free, bound, and total charge densities at all points on the surface of the sphere. Formulate an argument to show that this electric field is the actual one.

Solution: If the field is inverse-square radial, the potential will be given by $U = A/r$ in both media. At the interface between the liquid and the gas **E** is automatically parallel to the interface, so the boundary condition of D_n is satisfied as well as the boundary condition on E_t. At the surface

$$D_n = \begin{cases} \varepsilon_1 E_n = \dfrac{\varepsilon_1 A}{R^2} = \sigma_1 \rightarrow \text{liquid} \\ \\ \varepsilon_2 E_n = \dfrac{\varepsilon_2 A}{R^2} = \sigma_2 \rightarrow \text{gas} \end{cases} \text{"true" charge densities}$$

$$Q = 2\pi R^2\sigma_1 + 2\pi R^2\sigma_2 = 2\pi A(\varepsilon_1 + \varepsilon_2)$$

$$\sigma_{1P} = \chi_1 E_n = P_n = \frac{\chi_1 Q}{(\varepsilon_1 + \varepsilon_2)2\pi R^2}$$

$$A = \frac{Q}{2\pi(\varepsilon_1 + \varepsilon_2)}$$

$$\sigma_1 \text{ total} = \sigma_1 + \sigma_{1P}$$

$$\mathbf{E} = \frac{Q\mathbf{r}}{(\varepsilon_1 + \varepsilon_2)2\pi r^3}$$

PROBLEM 6.3 Two parallel conducting plates are separated by the distance d and maintained at the potential difference ΔU. A dielectric slab, of dielectric constant K and of uniform thickness d, is inserted snugly between the plates; however, the slab does not completely fill the volume between the plates. Find the electric field (a) in the dielectric and (b) in the vacuum region between the plates. Find the charge density σ on that part of the plate (c) in contact with the dielectric and (d) in contact with vacuum. (e) Find σ_p on the surface of the dielectric slab.

Solution: Since we are ignoring edge effects, there is no reason for **D** (and **E**) to have components in the vertical direction or in the direction out of the paper. This means $\mathbf{D} = \hat{\mathbf{i}}D_x$ and

$$\frac{\partial D_x}{\partial x} = 0 = \frac{\partial E_x}{\partial x} \quad (\rho = 0)$$

$$E_x = \frac{\Delta U}{d} \quad \text{(both in and out of the dielectric)}$$

$$D_x = \varepsilon \frac{\Delta U}{d} \quad \text{(in dielectric)}$$

$$= \varepsilon_0 \frac{\Delta U}{d} \quad \text{(outside)}$$

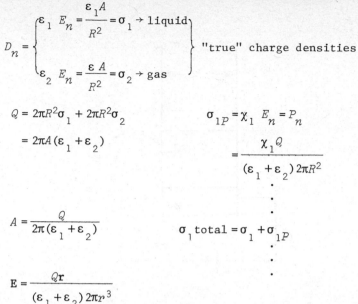

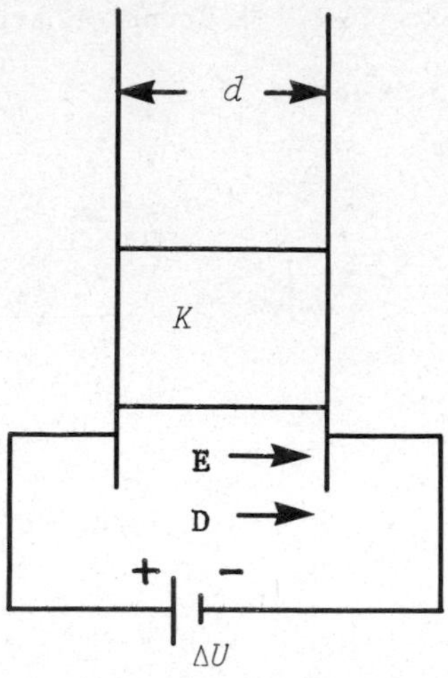

Note that E_t is automatically continuous at the vacuum-dielectric interface. This takes care of parts (a) and (b). Now **D** is zero everywhere inside the conductor, so from $D_{2n} - D_{1n} = \sigma$ we get $D = \sigma$, both at vacuum and dielectric interfaces, where σ is the "true" charge density. Therefore,

(c) $\sigma = \dfrac{\varepsilon \Delta U}{d}$ (dielectric interface)

(d) $\sigma = \dfrac{\varepsilon_0 \Delta U}{d}$ (vacuum interface)

(e) $$\sigma_p = \mathbf{n} \cdot \mathbf{P} = P_x = D_x - \varepsilon_0 E_x$$

$$= (\varepsilon - \varepsilon_0)\, \frac{\Delta U}{d}$$

PROBLEM 6.4 Two dielectric media with constant permittivities ε_1 and ε_2 are separated by a plane interface. There is no free charge on the interface. A point charge q is embedded in the medium characterized by ε_1, at a distance d from the interface. For convenience, we may take the yz plane through the origin to be the interface, and we locate q on the x axis at $x = -d$. If

$$r = \sqrt{(x+d)^2 + y^2 + z^2} \quad \text{and} \quad r' = \sqrt{(x-d)^2 + y^2 + z^2},$$

then it is easily demonstrated that $(1/4\pi\varepsilon_1)[(q/r) + (q'/r')]$ satisfies Laplace's equation at all points in medium 1 except at the position of q. Furthermore, $q''/4\pi\varepsilon_2 r$ satisfies Laplace's equation in medium 2. Show that all boundary conditions can be satisfied by these potentials, and in so doing, determine q' and q''.

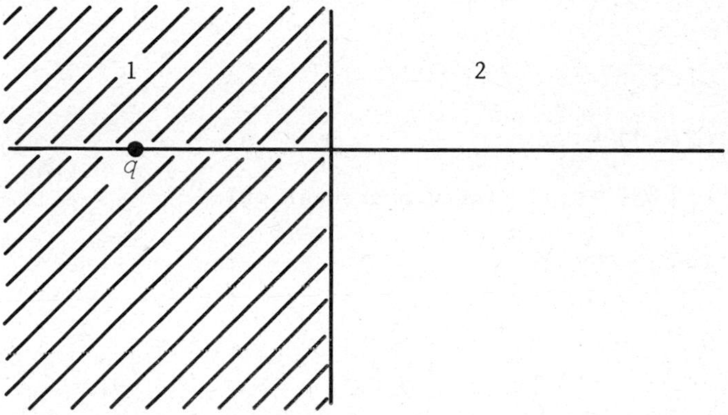

Solution:

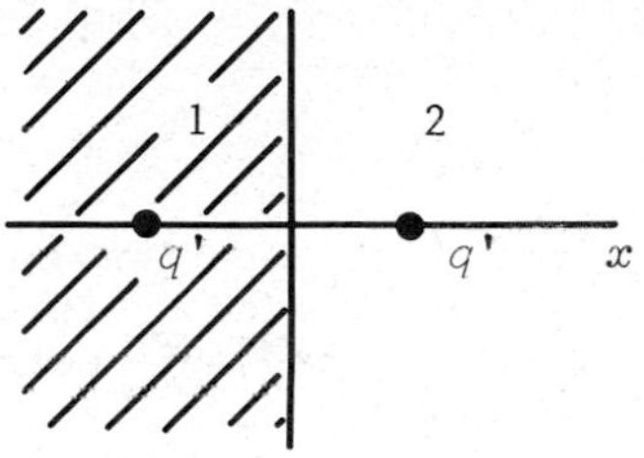

$$4\pi\varepsilon_1\phi_1 = \frac{q}{\sqrt{(x+d)^2+y^2+z^2}} + \frac{q'}{\sqrt{(x-d)^2+y^2+z^2}}$$

$$4\pi\varepsilon_2\phi_2 = \frac{q''}{\sqrt{(x+d)^2+y^2+z^2}}$$

Let us make the potential continuous at the boundary and the normal component of D(D_z) continuous at the boundary ($x = 0$):

(1) $$\frac{q/\varepsilon_1}{\sqrt{d^2+y^2+z^2}} + \frac{q'/\varepsilon_1}{\sqrt{d^2+y^2+z^2}} = \frac{q''/\varepsilon_2}{\sqrt{d^2+y^2+z^2}}\text{, all } y \text{ and } z,$$

$$\varepsilon_2(q+q') = q''\varepsilon_1$$

(2) $$-\varepsilon_1 \left.\frac{\partial\phi_1}{\partial x}\right|_{x=0} = -\varepsilon_2 \left.\frac{\partial\phi_2}{\partial x}\right|_{x=0}$$

$$\frac{dq}{(d^2+y^2+z^2)^{3/2}} - \frac{dq'}{\sqrt{d^2+y^2+z^2}} = \frac{dq''}{\sqrt{d^2+y^2+z^2}}, \text{ all } y \text{ and } z$$

$$q - q' = q''$$

$$\varepsilon_2(q - q') = \varepsilon_2 q''$$

$$\varepsilon_2(q + q') = \varepsilon_1 q'' \quad \text{(previous equation)}$$

$$2\varepsilon_2 q = (\varepsilon_2 + \varepsilon_1) q''$$

$$q'' = \frac{2\varepsilon_2}{\varepsilon_2 + \varepsilon_1} q$$

$$2\varepsilon_2 q' = (\varepsilon_1 - \varepsilon_2) q''$$

$$q' = \frac{\varepsilon_1 - \varepsilon_2}{2\varepsilon_2} q'' = \frac{\varepsilon_1 - \varepsilon_2}{\varepsilon_2 + \varepsilon_1} q$$

PROBLEM 6.5 A coaxial cable of circular cross section has a compound dielectric. The inner conductor has an outside radius a; this is surrounded by a dielectric sheath of dielectric constant K_1 and of outer radius b. Next comes another dielectric sheath of dielectric constant K_2 and outer radius c. If a potential difference U_0 is imposed between the conductors, calculate the polarization at each point in the two dielectric media.

Solution: Because of azimuthal symmetry, only the following two terms in the cylindrical harmonics expansion survive in each region: $U_1 = A_1 \ln r + B_1$ and $U_2 = A_2 \ln r + B_2$.

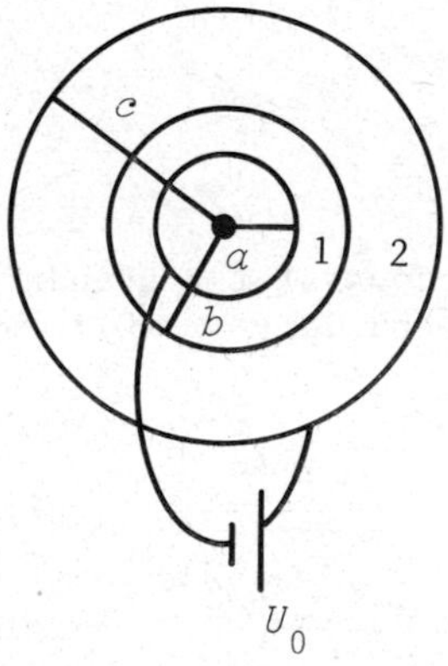

At the interface between the two dielectrics we make D_n and U continuous. In addition, we want $U_2(c) - U_1(a) = U_0$. We have one more constant than we need. So take $B_1 = 0$. Thus,

$$A_1 \ln b = A_2 \ln b + B_2, \quad B_2 = (A_1 - A_2) \ln b$$

$$-\varepsilon_1 \left.\frac{\partial U_1}{\partial r}\right|_{r=b} = -\varepsilon_2 \left.\frac{\partial U_2}{\partial r}\right|_{r=b} \rightarrow \frac{\varepsilon_1 A_1}{b} = \frac{\varepsilon_2 A_2}{b}$$

$$\frac{\varepsilon_1}{\varepsilon_2} A_1 \ln c + \left(A_1 - \frac{\varepsilon_1}{\varepsilon_2} A_1\right) \ln b - A_1 \ln a = U_0$$

$$A_1 \frac{K_1}{K_2} \ln \frac{c}{b} + \ln \frac{b}{a} = U_0, \quad A_1 = \frac{K_2 U_0}{K_1 \ln(c/b) + K_2 \ln(b/a)}$$

$$A_2 = \frac{K_1 U_0}{K_1 \ln(c/b) + K_2 \ln(b/a)}$$

$$P_r = \chi E_r = (K-1)\varepsilon_0 E_r = -\frac{(K-1)\varepsilon_0 A}{r}$$

$$P_{1r} = \chi_1 \frac{A_1}{r} = -\frac{1}{r} \frac{\varepsilon_0 K_2 (K_1 - 1) U_0}{K_1 \ln(c/b) + K_2 \ln(b/a)}, \quad P_{2r} = \frac{1}{r} \frac{\varepsilon_0 K_1 (K_2 - 1) U_0}{K_1 \ln(c/b) + K_2 \ln(b/a)}$$

PROBLEM 6.6 Two dielectric media with dielectric constants K_1 and K_2 are separated by a plane interface. There is no free charge on the interface. Find a relationship between the angles θ_1 and θ_2, where these are the angles which an arbitrary line of displacement makes with the normal to the interface: θ_1 in medium 1, θ_2 in medium 2.

Solution:

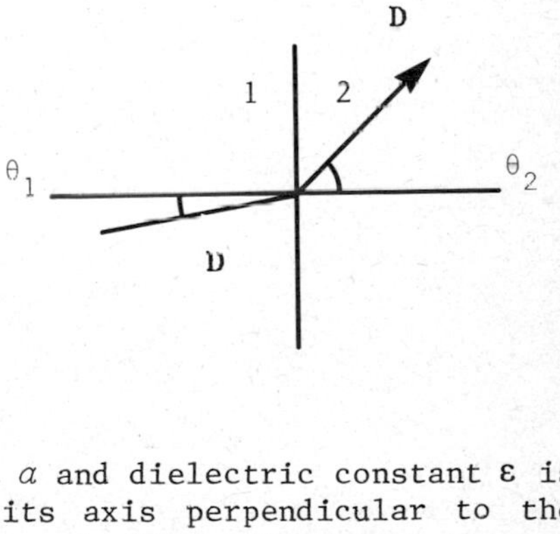

$$E_{1t} = E_{2t}, \quad D_{1n} = D_{2n}$$

$$\frac{D_{1t}}{\varepsilon_1} = \frac{D_{2t}}{\varepsilon_2}, \quad \frac{D_{1t}}{D_{1n}} = \frac{D_{2t}}{D_{2n}} \frac{\varepsilon_1}{\varepsilon_2}$$

$$\tan\theta_1 = \frac{\varepsilon_1}{\varepsilon_2} \tan\theta_2, \quad \frac{\tan\theta_1}{\tan\theta_2} = \frac{\varepsilon_1}{\varepsilon_2} = \frac{K_1}{K_2}$$

PROBLEM 6.7 A long cylindrical rod of radius a and dielectric constant ε is placed in a uniform electric field E_0, with its axis perpendicular to the field direction. Find ϕ inside and outside the rod. (Use cylindrical harmonics.)

Solution: If the rod is long, we can neglect end effects and $\phi = \phi(r, \theta)$, as $r \to \infty$, $\mathbf{E} = E_0 \hat{\mathbf{j}}$.

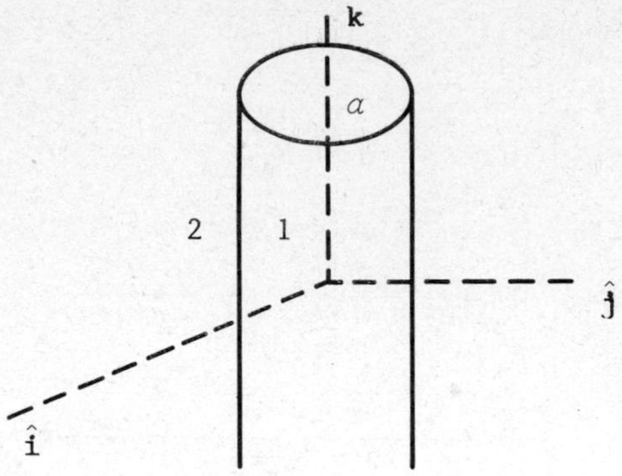

The solution to Laplace's equation,

$$\phi = \sum_{n=0}^{\infty} (A' + B' \ln r + C_n' r^n \cos n\theta + D_n' r^{-n} \cos n\theta + E_n' r^n \sin n\theta + F_n' r^{-n} \sin n\theta)$$

In region 1, ϕ_1 must be finite at $r = 0$,

$$\phi_1 = \sum_{n=0}^{\infty} (A + C_n r^n \cos n\theta + E_n r^n \sin n\theta)$$

$$\phi_2 \to -E_0 r \cos\theta \quad \text{as } r \to -\infty,$$

$$\phi_2 = \sum_{n=0}^{\infty} (D_n r^{-n} \cos n\theta + F_n r^{-n} \sin n\theta - E_0 r \sin\theta)$$

$$\phi_1(a) = \phi_2(a)$$

$$\sum_{n=0}^{\infty} (A + C_n a^n \cos n\theta + E_n a^n \sin n\theta)$$

$$= \sum_{n=0}^{\infty} (D_n a^{-n} \cos n\theta + F_n a^{-n} \sin n\theta - E_0 a \sin\theta)$$

$$A = 0, \quad C_n a^{2n} = D_n$$

For $n = 1$, $E_1 a = F_1 \dfrac{1}{a} - E_0 a$

For $n \neq 1$, $E_n a^{2n} = F_n$

$$\frac{\partial\phi_1}{\partial r} = \sum_{n=0}^{\infty} (C_n n r^{n-1} \cos n\theta + E_n n r^{n-1} \sin n\theta)$$

$$\frac{\partial\phi_2}{\partial r} = \sum_{n=0}^{\infty} (-D_n n r^{-n-1} \cos n\theta - F_n n r^{-n-1} \sin n\theta - E_0 \sin\theta)$$

$$\hat{\mathbf{r}} \cdot (\mathbf{D}_2 - \mathbf{D}_1) = 0$$

$$K \left.\frac{\partial\phi_1}{\partial r}\right|_a = \left.\frac{\partial\phi_2}{\partial r}\right|_a$$

$$K \sum_{n=0}^{\infty} (C_n n a^{n-1} \cos n\theta + E_n n a^{n-1} \sin n\theta)$$

$$= \sum_{n=0}^{\infty} (-D_n n a^{-n-1} \cos n\theta - F_n n a^{-n-1} \sin n\theta - E_0 \sin\theta)$$

$$KC_n a^{2n} = -D_n$$

$$KE_1 = -F_1 a^{-2} - E_0, \quad \text{for } n = 1$$

$$KE_n a^{2n} = -F_n, \quad \text{for } n \neq 1$$

Trying to solve the simultaneous equation for C_n and D_n and for E_n and F_n with $n \neq 1$ yields

$$C_n = D_n = E_n = F_n = 0$$

$$KE_1 = -F_1 a^{-2} - E_0$$

and

$$E_1 = F_1 a^{-2} - E_0$$

$$K(F_1 a^{-2} - E_0) = -F_1 a^{-2} - E_0$$

$$F_1 (K a^{-2} + a^{-2}) = -E_0 (1 - K)$$

$$F_1 = \frac{a^2 E_0 (K - 1)}{K + 1}$$

$$E_1 = \frac{E_0(K-1)}{K+1} - E_0$$

$$E_1 = -E_0 \left(\frac{2}{K+1}\right)$$

$$\phi_1 = -E_0 \left(\frac{2}{K+1}\right) r \sin\theta$$

$$\phi_2 = \frac{a^2 E_0 (K-1)}{(K+1)r} \sin\theta - E_0 r \sin\theta$$

PROBLEM 6.8 Consider a homogeneous dielectric ε, of infinite extent, in which there is a uniform field E_0. A spherical cavity of radius a is cut out of this dielectric. Find:

(a) ϕ in the cavity and on its surface.
(b) The polarization charge density η_p on the walls.
(c) The field outside the cavity.

(Use spherical harmonics.)

Solution:

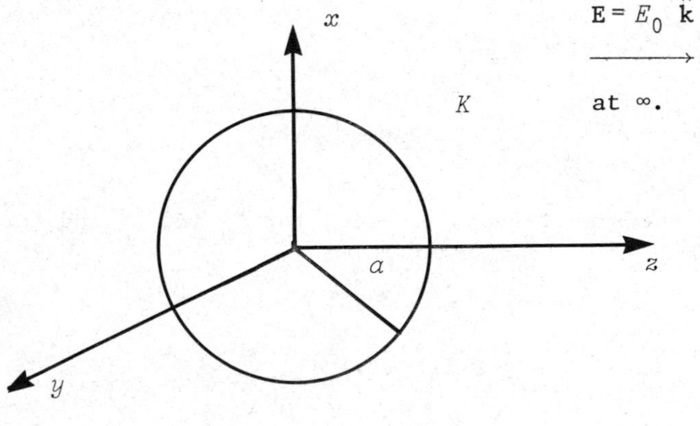

(a) Because of the azimuthal symmetry and the fact that ϕ must be finite at the origin

$$\phi_{in} = \sum_{\ell=0}^{\infty} A_\ell r^\ell P_\ell(\cos\theta)$$

$$\phi_{out} = \sum_{\ell=0}^{\infty} (B_\ell r^\ell + C_\ell r^{-(\ell+1)}) P_\ell(\cos\theta)$$

$$\phi_{out} \to -E_0 z = -E_0 r\cos\theta \quad \text{as } r \to \infty$$

$\therefore B_1 = -E_0$ and $B_n = 0$ for $n \neq 1$

$$\phi_{\text{in}}(a) = \phi_{\text{out}}(a)$$

$$\sum_{\ell=0}^{\infty} A_\ell a^\ell P_\ell(\cos\theta) = \sum_{\ell=0}^{\infty} (B_\ell a^\ell + C_\ell a^{-(\ell+1)}) P_\ell(\cos\theta)$$

$$\sum_{\ell=0}^{\infty} A_\ell a^\ell = \sum_{\ell=0}^{\infty} (B_\ell a^\ell + C_\ell a^{-(\ell+1)})$$

For $\ell = 1$,

$$A_1 a = -E_0 a + C_1 a^{-2}$$

$$A_1 = -E_0 + C_1 a^{-3}$$

For $\ell \neq 1$,

$$A_\ell = C_\ell a^{-(2\ell+1)}$$

normal D_{in} = normal D_{out}

$$-\frac{\partial\phi_{\text{in}}}{\partial r}\bigg|_{r=a} = -K\frac{\partial\phi_{\text{out}}}{\partial r}\bigg|_{r=a}$$

$$\sum_{\ell=0}^{\infty} A_\ell \ell a^{\ell-1} P_\ell(\cos\theta) = K\sum_{\ell=0}^{\infty} [B_\ell \ell a^{\ell-1} - C_\ell(\ell+1)a^{-(\ell+2)}] P_\ell(\cos\theta)$$

For $\ell = 1$,

$$A_1 = -KE_0 - KC_1 2a^{-3}$$

For $\ell \neq 1$,

$$A_\ell \ell a^{\ell-1} = -KC_\ell(\ell+1)a^{-(\ell+2)}$$

$$A_\ell = -\frac{K(\ell+1)}{\ell} a^{-(2\ell+1)} C_\ell,$$

but from before, $A_\ell = C_\ell a^{-(2\ell+1)}$. These equations can be satisfied simultaneously only if $C_\ell = A_\ell = 0$.

$$-E_0 + C_1 a^{-3} = -K(E_0 + C_1 2a^{-3})$$

$$C_1 a^{-3}(1+2K) = E_0(-K+1)$$

$$C_1 = -\frac{E_0 a^3 (K-1)}{K+1}$$

$$A_1 = -E_0 - \frac{E_0(K-1)}{2K+1} = -E_0 \left(1 + \frac{K-1}{2K+1}\right)$$

$$A_1 = -E_0 \frac{3K}{2K+1}$$

$$\phi_{\text{in}} = -\frac{E_0 3K}{2K+1} r \cos\theta$$

On the surface,

$$\phi(a) = -\frac{E_0 3Ka}{2K+1} \cos\theta$$

(b) $\mathbf{P} = \chi \mathbf{E} = \varepsilon_0 (K-1)\mathbf{E}$

$$\mathbf{P} = -\varepsilon_0 (K-1)\, \nabla \phi_{\text{out}}$$

$$\phi_{\text{out}} = -E_0 r \cos\theta - \frac{E_0 a^3 (K-1)}{2K+1} r^{-2} \cos\theta$$

$$= -E_0 r \cos\theta \left(1 + \frac{K-1}{2K+1} \frac{a^3}{r^3}\right)$$

$$\mathbf{E} = -\nabla\phi$$

$$\mathbf{E}_{\text{out}} = -\left(\hat{\mathbf{r}} \frac{\partial}{\partial r} + \hat{\theta}\, \frac{1}{r} \frac{\partial}{\partial \theta}\right) \phi_{\text{out}}$$

$$= -\hat{\mathbf{r}} \left[-E_0 \cos\theta \left(1 + \frac{(K-1)a^3}{(2K+1)r^3}\right) + E_0 \cos\theta \left(\frac{3(K-1)}{2K+1} \frac{a^3}{r^3}\right)\right]$$

$$- \hat{\theta}\; E_0 \sin\theta \left(1 + \frac{(K-1)a^3}{(2K+1)r^3}\right)$$

$$\mathbf{E}_{out}\Big|_a = \hat{\mathbf{r}}\left[+E_0 \cos\theta\left(1 - \frac{2(K-1)}{2K+1}\right)\right] - \hat{\theta}\, E_0 \sin\theta\left(1 + \frac{K-1}{2K+1}\right)$$

$$\mathbf{P} = \varepsilon_0\ (K-1)\mathbf{E}_{out}$$

$$\eta_\rho = \mathbf{P}\cdot\hat{\mathbf{n}} = -\mathbf{P}\cdot\hat{\mathbf{r}} = -\varepsilon_0\ (K-1)\ E_0 \cos\theta\left(1 - \frac{2(K-1)}{2K+1}\right)$$

$$\eta_\rho = -\varepsilon_0\ (K-1)\ E_0 \cos\theta\left(\frac{3}{2K+1}\right)$$

$$\eta_\rho = \frac{-\varepsilon_0 E_0 3(K-1)}{2K+1}\cos\theta$$

(c) From part (b),

$$\mathbf{E}_{out} = \hat{\mathbf{r}}\left[E_0 \cos\theta\left(1 - \frac{2(K-1)a^3}{(2K+1)r^3}\right)\right] - \hat{\theta}\ E_0 \sin\theta\left(1 + \frac{(K-1)a^3}{(2K+1)r^3}\right)$$

PROBLEM 6.9 A dielectric sphere of radius R has a permanent polarization **P** that is uniform in direction and magnitude. The polarized sphere gives rise to an electric field. Determine this field both inside and outside the sphere. Inside the sphere the electric field, which is in the opposite direction to the polarization, is called a depolarizing field. [*Hint:* since div **P** vanishes at all points, the electrostatic potential satisfies Laplace's equation both inside and outside the sphere. Do not assume that the dielectric is characterized by a dielectric constant.]

Solution:

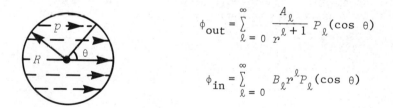

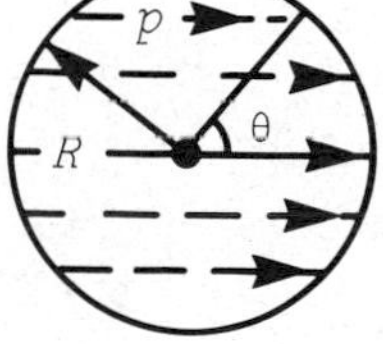

$$\phi_{out} = \sum_{\ell=0}^{\infty} \frac{A_\ell}{r^{\ell+1}} P_\ell(\cos\theta)$$

$$\phi_{in} = \sum_{\ell=0}^{\infty} B_\ell r^\ell P_\ell(\cos\theta)$$

ϕ continuous at $r = R$

$$\frac{A_\ell}{R^{\ell+1}} = B_\ell R^\ell, \text{ all } \ell$$

$$\mathbf{n}\cdot(\varepsilon_0\mathbf{E}_{in} + \mathbf{P}) = \mathbf{n}\cdot\varepsilon_0\mathbf{E}_{out,\,n}$$

$$-\varepsilon_0 \sum_{\ell=0}^{\infty} \ell B_\ell R^{\ell-1} + P\cos\theta = \varepsilon_0 \sum_{\ell=0}^{\infty} (\ell+1)\frac{A_\ell}{R^{\ell+2}} P_\ell(\cos\theta)$$

$$-\varepsilon_0 \, \ell B_\ell R^{\ell-1} = \varepsilon_0 \frac{(\ell+1)A_\ell}{R^{\ell+2}}, \quad \ell \neq 1, \quad -\varepsilon_0 B_1 + P = \varepsilon_0 \frac{2A_1}{R^3}$$

All A's and B's vanish except $\ell = 1$.

$$A_1 = B_1 R^3$$

$$P = \varepsilon_0 \left(B_1 + \frac{2A_1}{R^3} \right)$$

$$= 3\varepsilon_0 B_1, \quad B_1 = \frac{P}{3\varepsilon_0}$$

$$A_1 = \frac{P}{3\varepsilon_0} R^3$$

$$\phi_{out} = \frac{(4\pi/3)R^3 P \cos\theta}{4\pi\varepsilon_0 r^2}, \quad \phi_{in} = \frac{P}{3\varepsilon_0} r \cos\theta, \quad E_2^{(1)} = -\frac{P}{3\varepsilon_0}$$

PROBLEM 6.10 Suppose the electric polarization in a piece of sulfur has a magnitude of 10^{-5} C/m^2. Given that Avogadro's number times the quotient of the density by the atomic weight is the number of atoms per unit volume and that the charge of each sign in an atom is the atomic number times the magnitude of the electronic charge, look up necessary data in physical or chemical tables (or peek at answer).

(a) Find the mean dipole moment of a sulfur atom.
(b) How far must the centers of charge of positive and negative sign in the sulfur atom be separated to produce this moment, assuming the nucleus is a point and the electrons are arranged in a spherical cloud with the electron charge continuously distributed throughout this volume?

Solution:

(a) N = Number of molecules per unit volume.

$$N = N_0 \rho / M$$

For a sulfur atom $M = 32$, $Z = 16$, $\rho = 2\text{g/cm}^3$:

$$N_0 = 6.02 \times 10^{23} \text{ atom/g-mole}$$

$$N = (6.02 \times 10^{23}) \left(\frac{2}{3.2 \times 10^1} \right) = 3.77 \times 10^{22} \text{ atom/cm}^3 = 3.77 \times 10^{28} \text{ atom/cm}^3$$

$$P = N P_{av}$$

$$10^{-5} = 3.77 \times 10^{28} P_{av}$$

$$P_{av} = 2.65 \times 10^{-34}\ \text{C} \cdot \text{m}$$

(b) $P = qd$, where d is the separation of plus and minus charge. For sulfur, $q = Ze = (1.6 \times 10^1)(1.6 \times 10^{-19}) = 2.56 \times 10^{-18}$ C. To get a crude approximation, assume that the situation is

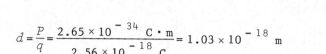

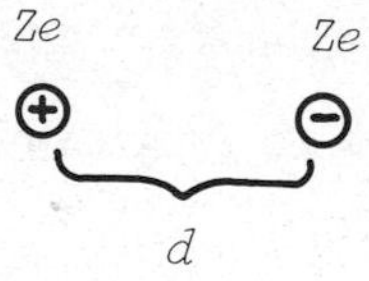

$$d = \frac{P}{q} = \frac{2.65 \times 10^{-34}\ \text{C} \cdot \text{m}}{2.56 \times 10^{-18}\ \text{C}} = 1.03 \times 10^{-18}\ \text{m}$$

This is less than the diameter of the nucleus ($\sim 10^{-14}$ m). So if we think of the electron as forming a cloud around the nucleus and that the charge distribution is symmetrical, only the negative charge contained inside a sphere of radius X contributes to the attraction. Let R_0 represent the radius of the atom. Then the charge inside X is (if the electron charge is uniform)

$$q_E = qX^3/R^3$$

The effective charge q_E is obtained from

$$P = q_E X \Rightarrow q_E\ P/X$$

so

$$\frac{P}{X} = q\frac{X^3}{R^3} \Rightarrow X^4 = \frac{PR^3}{q}$$

We could compute R exactly using the Clausius-Mossotti relation

$$\alpha = \frac{3\varepsilon_0}{N}\frac{K-1}{K+2} \text{ and } \alpha = 4\pi\varepsilon_0 R^3$$

or

$$R^3 = \frac{3}{4\pi N}\frac{K-1}{K+2}$$

where K is the dielectric constant. For sulfur $K = 4$,

$$R^3 = \frac{3}{(4\pi)(3.8 \times 10^{28})} \cdot \frac{3}{6} = 0.031 \times 10^{-28}$$

$$R^3 = 3.1 \times 10^{-30} \Rightarrow R = 1.4 \times 10^{-10}\ \text{m}$$

so

$$X^4 = \frac{(2.65 \times 10^{-34})(3.1 \times 10^{-30})}{2.56 \times 10^{-18}} = 3.2 \times 10^{-46}$$

$$X = 4.2 \times 10^{-12}\ \text{m}$$

So there is a substantial difference in this value depending on the model used.

PROBLEM 6.11 Given for HCl gas at 100°C and atmospheric pressure the specific dielectric constant is 1.00258 and the density is 1.200 g/liter. The permanent dipole moment is 3.43×10^{-30} C • m. Assuming the perfect gas law, what is the dielectric constant at 0°C and 1 atm?

Solution: Given $K = 1.00258$ and $p_0 = 3.43 \times 10^{-30}$ C • m. Since

$$K - 1 = \frac{\chi}{\varepsilon_0} \Rightarrow \chi = 2.58 \times 10^{-3}\, \varepsilon_0$$

for a gas, $\chi = (p_0^2/3KT + \alpha_I)N$.

$$N = \frac{\rho}{n} N_0 = \frac{1.2 \times 10^{-3}}{35}\, 6.02 \times 10^{23} = 2.06 \times 10^{19} \text{ molecule/cm}^3$$

$$N = 2.06 \times 10^{25} \text{ molecule/m}^3$$

So

$$2.58 \times 10^{-3}\, \varepsilon_0 = \left(\frac{(3.43 \times 10^{-30})^2}{(3)(1.38 \times 10^{-23})(3.73 \times 10^2)} + \alpha_I\right) 2.06 \times 10^{25}$$

$$1.11 \times 10^{-39} = 7.61 \times 10^{-40} + \alpha_I$$

$$\alpha_I = 3.49 \times 10^{-40}$$

At 0°C, we need N.

$$P = \eta KT$$

$$\eta_0 T_0 = \eta T$$

$$\eta_0 273 = 2.06 \times 10^{25} \cdot 373$$

$$\eta_0 = 2.81 \times 10^{25}$$

$$\chi = \left(\frac{p_0^2}{3KT} + \alpha_I\right) N = \left(\frac{(3.43 \times 10^{-30})^2}{(3)(1.38 \times 10^{-23})(273)} + 3.49 \times 10^{-40}\right) 2.81 \times 10^{25}$$

$$\chi = 3.91 \times 10^{-14}$$

$$K - 1 = \frac{\chi}{\varepsilon_0} = 4.41 \times 10^{-3}, \quad K = 1.00441$$

PROBLEM 6.12 Given a concentric cable of circular cross section, with compound dielectric. The inner conductor has an outside radius a; this is surrounded by a sheath of dielectric with specific dielectric constant K_1 and outer radius b; next comes another sheath of dielectric with constant K_2 and outer radius c; and finally the outer conductor has inside radius c.

There is a potential difference U_0 between conductors, and only the conductors bear free charge. Use Gauss' law, and find all free and bound charge densities.

Solution:

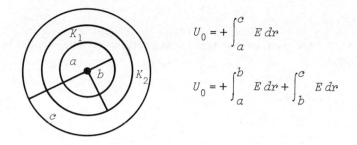

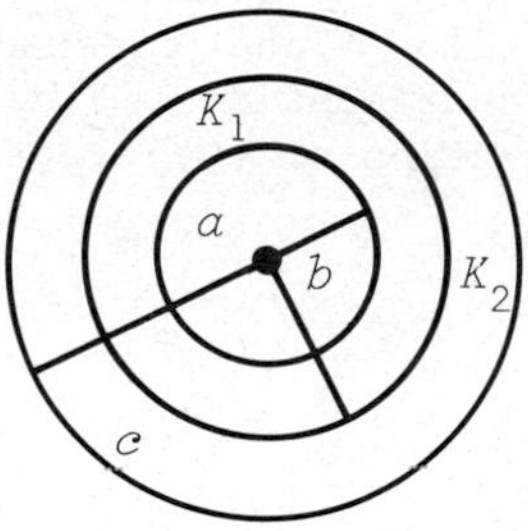

$$U_0 = +\int_a^c E\,dr$$

$$U_0 = +\int_a^b E\,dr + \int_b^c E\,dr$$

Let λ = free charge density on the surface of the conductor.

$$E = \frac{\lambda}{2\pi K\varepsilon_0 r}$$

For $a < r < b$, $E = \dfrac{\lambda}{2\pi K_1\varepsilon_0 r}$; for $b < r < c$, $E = \dfrac{\lambda}{2\pi K_2\varepsilon_0 r}$

$$U_0 = +\frac{\lambda}{2\pi\varepsilon_0}\left(\int_a^b K_1\frac{dr}{r} + \int_b^c K_2\frac{dr}{r}\right) = +\frac{\lambda}{2\pi\varepsilon_0}\left(K_1 \ln\frac{b}{a} + K_2 \ln\frac{c}{b}\right)$$

$$\lambda = \frac{2\pi\varepsilon_0 U_0}{K_1 \ln\frac{b}{a} + K_2 \ln\frac{c}{b}}$$

To get $\mathbf{P} = \mathbf{D} - \varepsilon_0\mathbf{E}$,

$$\mathbf{D} = \frac{\lambda}{2\pi r}$$

$$\mathbf{P}_1 = \hat{\mathbf{r}}\,\frac{\lambda}{2\pi r}\left(1 - \frac{1}{K_1}\right), \quad \mathbf{P}_2 = \hat{\mathbf{r}}\,\frac{\lambda}{2\pi r}\left(1 - \frac{1}{K_2}\right)$$

Surfaces,

$$r = a, \quad \sigma_a = -\frac{\lambda}{2\pi a}\left(1 - \frac{1}{K_1}\right)$$

$$r = b, \quad \sigma_b = -\frac{\lambda}{2\pi b}\left(1 - \frac{1}{K_2}\right) + \frac{\lambda}{2\pi b}\left(1 - \frac{1}{K_1}\right)$$

$$\sigma_b = \frac{\lambda}{2\pi}\left[-\frac{1}{b}\left(1 - \frac{1}{K_2}\right) + \frac{1}{a}\left(1 - \frac{1}{K_1}\right)\right]$$

$$r = c, \quad \sigma_c = \frac{\lambda}{2\pi c}\left(1 - \frac{1}{K_2}\right)$$

where λ is defined as above.

PROBLEM 6.13 Given a dielectric sphere of radius R and constant permittivity ε, in which a uniform density ρ of free charge exists. Find the electrostatic potential at the center by line integration of the electric field.

Solution:

$$r > R, \quad D = \frac{\rho(4/3\ \pi R^3)}{4\pi r^2} = \frac{\rho R^3}{3r^2}$$

$$r < R, \quad D = \frac{\rho r}{3}$$

$$r > R, \quad E = \frac{\rho R^3}{3\varepsilon_0 r^2}$$

$$r < R, \quad E = \frac{\rho r}{3K\varepsilon_0}$$

$$U(0) = -\int_\infty^0 E\,dr = -\int_\infty^R \frac{\rho R^3}{3\varepsilon_0 r^2}\,dr - \int_R^0 \frac{\rho r}{3K\varepsilon_0}$$

$$U(0) = \frac{\rho R^3}{3\varepsilon_0}\frac{1}{r}\bigg|_\infty^R - \frac{\rho r^2}{6K\varepsilon_0}\bigg|_R^0$$

$$U(0) = \frac{\rho R^2}{3\varepsilon_0} + \frac{\rho R^2}{6K\varepsilon_0} = \frac{\rho R^2}{3\varepsilon_0}\left(1 + \frac{1}{2K}\right)$$

PROBLEM 6.14 Calculate the diamagnetic susceptibility of neon at standard temperature and pressure (0°C, 1 atm) on the assumption that only the eight outer electrons in each atom contribute and that their mean radius is $R = 4.0 \times 10^{-9}$ cm.

Solution:

$$\chi = N\alpha_m\mu_0$$

where,

$$\alpha_m = \frac{Ze^2}{4m} < r >^2$$

$$P = \eta KT$$

$$1.01 \times 10^5 = \eta(1.38 \times 10^{-23})(2.73 \times 10^2)$$

$$\eta = 2.68 \times 10^{25}/\text{m}^3$$

$$\alpha_m = \frac{(8)(1.6 \times 10^{-19})^2}{(4)(9.1 \times 10^{-31})} (4 \times 10^{-11})^2 = 9 \times 10^{-29}$$

$$\chi = (2.68 \times 10^{25})(9 \times 10^{-29})(4\pi \times 10^{-7})$$

$$\chi = 3.03 \times 10^{-9}$$

PROBLEM 6.15 Calculate the relative strength of the interaction between two typical magnetic dipoles, compared with the interaction between two typical electric dipoles. To be explicit: Calculate the torque exerted on one dipole by the other when they are oriented perpendicularly to each other at a distance of 1 Å; take each magnetic dipole equal to 1 Bohr magneton and each electric dipole equal to $e \times 0.1$ Å. This calculation shows that the basic magnetic interaction is several orders of magnitude smaller than the electrical interaction in matter.

Solution: 1 Bohr magneton $= \dfrac{eh}{4\pi m}$

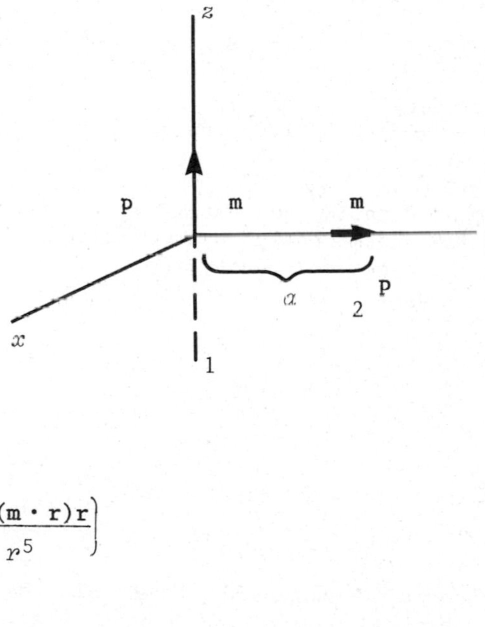

at 2, $\mathbf{B} = \dfrac{\mu_0}{4\pi}\left(-\dfrac{\mathbf{m}}{r^3} + \dfrac{3(\mathbf{m}\cdot\mathbf{r})\mathbf{r}}{r^5}\right)$

Here

$$\mathbf{m} = \hat{\mathbf{k}}\,\frac{eh}{4\pi m},\quad \mathbf{r} = \hat{\mathbf{j}}a$$

so

$$\mathbf{B} = -\hat{\mathbf{k}}\,\frac{\mu_0}{4\pi}\,\frac{eh}{4\pi m}\,\frac{1}{a^3}$$

$$\mathbf{B} = -\hat{\mathbf{k}}\frac{\mu_0 eh}{16\pi^2 ma^3}$$

The torque on 2, $\boldsymbol{\tau} = \mathbf{m}_2 \times \mathbf{B} \Rightarrow |\tau| = m_2 B$

$$\tau = \frac{eh}{4\pi m}\,\frac{\mu_0 eh}{16\pi^2 ma^3} = \frac{\mu_0 e^2h^2}{64\pi^3 m^2 a^3}$$

$$\tau = \frac{(4\pi \times 10^{-7})(1.6 \times 10^{-19})^2(6.63 \times 10^{-34})^2}{16\pi^2(9.1 \times 10^{-31})^2(10^{-10})^3} = 8.6 \times 10^{-24} \text{ N} \cdot \text{m}$$

For the electric dipole,

$$\mathbf{E} = \frac{1}{4\pi\varepsilon_0}\left(-\frac{\mathbf{P}}{r^3} + \frac{3(\mathbf{P}\cdot\mathbf{r})\mathbf{r}}{r^3}\right)$$

$$\mathbf{P} = \hat{\mathbf{k}}\ 1.6 \times 10^{-30}\ \text{C} \cdot \text{m}, \quad \hat{\mathbf{r}} = \hat{\mathbf{j}}a$$

$$\mathbf{E} = -\hat{\mathbf{k}}\frac{1}{4\pi\varepsilon_0}\,\frac{1.6 \times 10^{-30}}{a^3}$$

$$\boldsymbol{\tau} = \mathbf{P} \times \mathbf{E} \Rightarrow \tau = PE$$

$$\tau = \frac{1}{4\pi\varepsilon_0}\,\frac{(1.6 \times 10^{-30})^2}{(10^{-10})^3} = 2.3 \times 10^{-21} \text{ N} \cdot \text{m}$$

PROBLEM 6.16 Find B and H inside and outside a spherical shell of radii a and b which is magnetized permanently to a constant magnetization M. What is the effect of making the spherical cavity not concentric with the outside surface of the shell? You may find the following relation useful:

$$\mathbf{r} = (r,\ \theta,\ \phi)$$

$$\mathbf{r}' = (r',\ \theta',\ \phi')$$

$$\frac{1}{|\mathbf{r} - \mathbf{r}'|} = \frac{1}{r_>}\sum_{\ell=0}^{\infty}\left(\frac{r_<}{r_>}\right)^{\ell}\sum_{m=-\ell}^{\ell}\frac{4}{2\ell+1}Y_\ell^m(\theta,\phi)Y_\ell^{-m}(\theta',\phi')$$

where the addition theorem for Legendre polynomials has been used and the Y_ℓ^m's are normalized spherical harmonics. In particular,

$$Y_\ell^0(\theta,\phi) = \sqrt{\frac{2\ell+1}{4\pi}}\,P_\ell\ (\theta)$$

Solution:

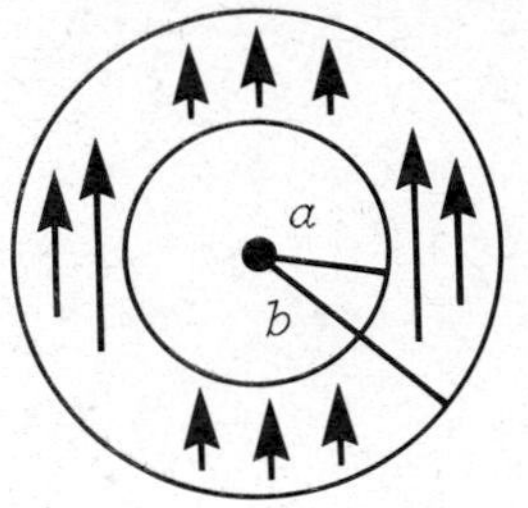

$\mathbf{M} = M_0\hat{\mathbf{k}}$

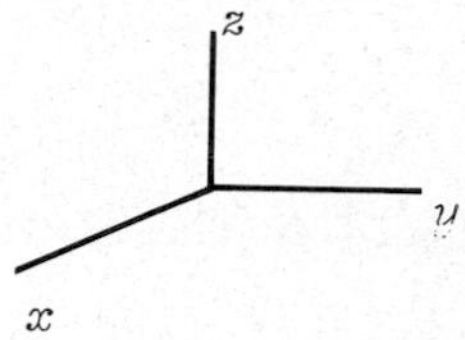

Consider two spheres, one of radius b and having a uniform magnetization $\mathbf{M} = M_0\hat{\mathbf{k}}$, the other of radius a and having a uniform magnetization $\mathbf{M} = -M_0\hat{\mathbf{k}}$. For the sphere of radius b, outside the sphere $\nabla \cdot \mathbf{B} = \nabla \times \mathbf{B} = 0$. Consequently, for $r > b$, $\mathbf{B} = \mathbf{H}$ can be written as $\mathbf{B}_{\text{out}} = -\nabla\Phi_m$ where $\nabla^2\Phi_m = 0$. With the boundary condition that $\mathbf{B} \to 0$ as $\mathbf{r} \to \infty$, the general solution for the potential is

$$\Phi_m(r, \theta) = \sum_{\ell=0}^{\infty} \alpha_\ell \frac{P_\ell(\cos\theta)}{r^{\ell+1}}$$

Inside the sphere

$$\mathbf{H} = \mathbf{B} - 4\pi\mathbf{M}$$

$$\mathbf{B}_{\text{in}} = B_0\hat{\mathbf{k}}$$

$$\mathbf{H}_{\text{in}} = (B_0 - 4\pi M_0)\hat{\mathbf{k}}$$

The boundary condition at $r = b$

$$\mathbf{H}_{\text{in}} \times \hat{\mathbf{r}} = \mathbf{H}_{\text{out}} \times \hat{\mathbf{r}}$$

$$\mathbf{B}_{\text{in}} \cdot \hat{\mathbf{r}} = \mathbf{B}_{\text{out}} \cdot \hat{\mathbf{r}}$$

$$B_0 \cos\theta = \sum_{\ell=0}^{\infty} \frac{(\ell+1)\alpha_\ell P_\ell(\cos\theta)}{b^{\ell+2}}$$

$$-(B_0 - 4\pi M_0)\sin\theta = -\sum_{\ell=0}^{\infty} \frac{\alpha_\ell}{b^{\ell+2}} \frac{dP_\ell(\cos\theta)}{d\theta}$$

solving $\alpha_1 = (4\pi/3)M_0 b^2$ (only the $\ell = 1$ term survives)

$$B_0 = \frac{8\pi}{3} M_0$$

$$\mathbf{B}_{\text{in}} = \frac{8\pi}{3} M_0\hat{\mathbf{k}} \text{ and } \mathbf{H}_{\text{in}} = \frac{-4\pi}{3} M_0\hat{\mathbf{k}}$$

Then the field inside does not depend on the radius of the sphere. Solving for the sphere of radius a with $\mathbf{M} = -M_0\hat{\mathbf{k}}$:

$$\mathbf{B}_{in} = -\frac{8\pi}{3} M_0 \hat{\mathbf{k}}$$

$$\mathbf{H}_{in} = -\left(\frac{8\pi}{3} M_0 - 4\pi M_0\right)\hat{\mathbf{k}} = \frac{4\pi}{3} M_0 \hat{\mathbf{k}}$$

Thus the field in the cavity is zero

$$\mathbf{B}_{out} = -\nabla \frac{4\pi M_0}{3} \frac{\cos\theta}{r^2}(b^2 - a^2)$$

The field outside the shell is that due to a dipole of dipole moment,

$$\mathbf{m} = \frac{4\pi}{3}(b^2 - a^2) M_0 \hat{\mathbf{k}}$$

$$\mathbf{B}_{out} = \frac{4\pi M_0 (b^2 - a^2)}{3}\left(\frac{\sin\theta}{r^3}\hat{\theta} + \frac{2\cos\theta}{r^3}\hat{\mathbf{r}}\right)$$

$$\mathbf{H}_{out} = \mathbf{B}_{out}$$

Consider the effect of displacing the sphere of radius a, $r_0\hat{\mathbf{j}}$ from the origin, where $r_0 < b - a$. Because to switch does not affect $\mathbf{B}_{in}$ and $\mathbf{H}_{in}$, the field inside the cavity is still zero:

$$\Phi_m(R, \theta, \phi) = -\frac{4\pi M_0}{3} \frac{\cos\theta'}{r'^2} a^2$$

where θ' and r' are coordinates in a coordinate system whose origin is at the center of the sphere.

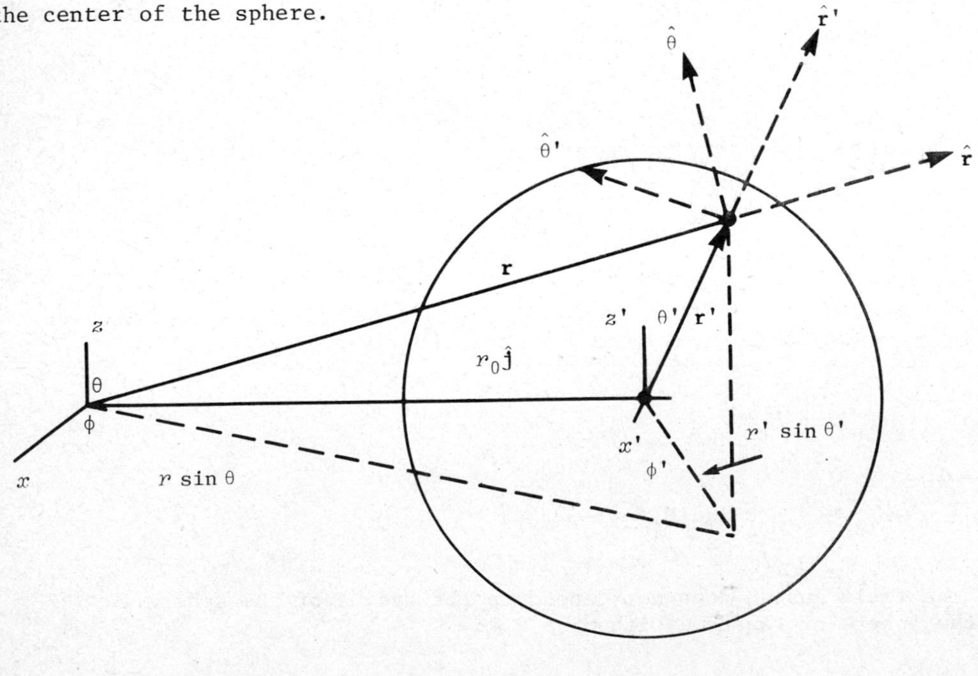

From the law of cosines, in xy plane,

$$r'^2 \sin^2 \theta' = r^2 \sin^2 \theta + r_0^2 - 2rR_0 \sin\theta \sin\phi$$

From the law of cosines, in the $(\mathbf{r}, r_0\hat{\mathbf{j}}, \mathbf{r}')$ plane,

$$r'^2 = r^2 + r_0^2 - 2rr_0 \sin\theta$$

$$\sin^2 \theta' = \frac{r^2 \sin^2\theta + r_0^2 - 2rr_0 \sin\theta \sin\phi}{r^2 + r_0^2 - 2rr_0 \sin\theta}$$

$$\cos^2 \theta' = \frac{r^2 + r_0^2 - 2rr_0 \sin\theta - r^2 \sin^2\theta - r_0^2 + 2rr_0 \sin\theta \sin\phi}{r^2 + r_0^2 - 2rr_0 \sin\theta}$$

$$\cos\theta' = \left(\frac{r^2 \cos^2\theta + 2rr_0 \sin\theta(\sin\phi - 1)}{r^2 + r_0^2 - 2rr_0 \sin\theta}\right)^{\frac{1}{2}}$$

For the displaced sphere of radius a,

$$\Phi_m(r,\theta,\phi) = -\frac{4\pi M_0 a^2}{3} \frac{[r^2 \cos^2\theta + 2rr_0 \sin\theta(\sin\phi - 1)]^{\frac{1}{2}}}{(r^2 + r_0^2 - 2rr_0 \sin\theta)^{3/2}}$$

Thus the total potential outside the sphere of radius b is

$$\Phi_m(r,\theta,\phi) = \frac{4\pi M_0}{3} \frac{b^2 \cos\theta}{r^2} - \frac{a^2[r^2\cos^2\theta + 2rr_0 \sin\theta(\sin\phi - 1)]^{\frac{1}{2}}}{(r^2 + r_0^2 - 2rr_0 \sin\theta)^{3/2}}$$

$$\mathbf{B}_{out} = -\nabla\Phi(r,\theta,\phi)$$

PROBLEM 6.17 A coil is wound on the surface of a sphere such that the field inside the sphere is uniform. What is the winding? (This form of winding is used in the Westinghouse-Goudsmit mass spectrometer.)

Solution: A uniform magnetization throughout a certain volume is equivalent to a surface current density $c(\mathbf{M} \times \hat{\mathbf{n}})$ over its surface. For a sphere whose field is in the z direction

$$\mathbf{M} \times \hat{\mathbf{n}} = M_0 \sin\theta\, \hat{\boldsymbol{\phi}}$$

The current density of a coil on the surface of a sphere would have to be proportional to $\sin\theta$ to produce a uniform field inside the sphere.

PROBLEM 6.18 A bar of dielectric material with uniform cross section A and length L has a polarization vector parallel to its length of magnitude $ax + b$, where x is a coordinate parallel to the length of the bar with zero at one face, and a and b are constants. At a point within the dielectric where

$x = L/2$:

(a) Find the value of the macroscopic electric field,

$$\mathbf{E} = -\hat{\mathbf{i}}\,\frac{aL + 2b}{\varepsilon_0}\left(1 - \frac{L}{\sqrt{L^2 + 4R^2}}\right)$$

(b) Find the value of the microscopic electric field:

$$\mathbf{E} = -\hat{\mathbf{i}}\,\frac{1}{\varepsilon_0}\left(\frac{5aL}{6} + \frac{5b}{3}\right) + \hat{\mathbf{i}}\,\frac{aL + 2b}{\varepsilon_0}\,\frac{L}{\sqrt{L^2 + 4R^2}}$$

Solution:

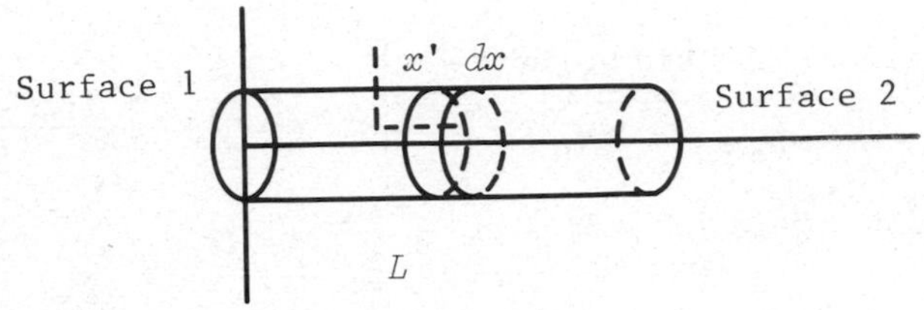

$$\mathbf{P} = (ax + b)\hat{\mathbf{i}}$$

At surface 1,

$$\sigma_1 = \mathbf{P}\cdot(-\hat{\mathbf{i}})\Big|_{x=0} = -b$$

At surface 2,

$$\sigma_2 = \mathbf{P}\cdot\hat{\mathbf{i}}\Big|_{x=L} = aL + b$$

$$\rho = -\nabla\cdot\mathbf{P} = -a$$

The macroscopic field is due to the bound surface charge and volume charge. We first need to determine E produced by a disk.

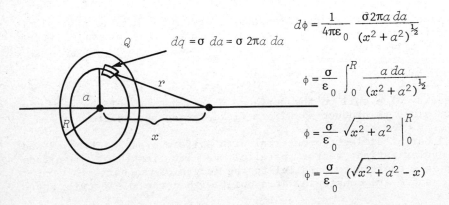

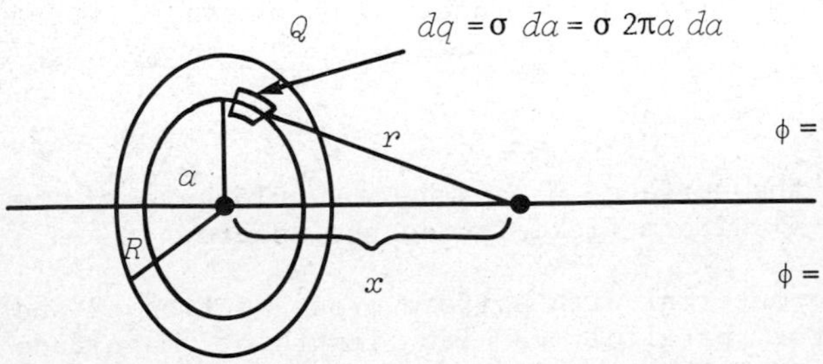

$$d\phi = \frac{1}{4\pi\varepsilon_0}\,\frac{\sigma 2\pi a\,da}{(x^2 + a^2)^{\frac{1}{2}}}$$

$$\phi = \frac{\sigma}{\varepsilon_0}\int_0^R \frac{a\,da}{(x^2 + a^2)^{\frac{1}{2}}}$$

$$\phi = \frac{\sigma}{\varepsilon_0}\sqrt{x^2 + a^2}\,\Big|_0^R$$

$$\phi = \frac{\sigma}{\varepsilon_0}\left(\sqrt{x^2 + a^2} - x\right)$$

$$E = -\hat{i}\,\frac{\partial\phi}{\partial x} = -\hat{i}\,\frac{\sigma}{\varepsilon_0}\left\{\tfrac{1}{2}\,(x^2+R^2)^{-\frac{1}{2}}\,2x - 1\right\}$$

$$E = \hat{i}\,\frac{\sigma}{\varepsilon_0}\left\{1 - \frac{X}{(x^2+R^2)^{\frac{1}{2}}}\right\}$$

From the charge on the left face,

$$E_L = -\hat{i}\,\frac{b}{\varepsilon_0}\left\{1 - \frac{L/2}{(L^2/4+R^2)^{\frac{1}{2}}}\right\}$$

From the charge on the right face,

$$E_R = -\hat{i}\,\frac{aL+b}{\varepsilon_0}\left\{1 - \frac{L/2}{(L^2/4+R^2)^{\frac{1}{2}}}\right\}$$

Since the volume charge density is constant, the electric field at P due to this charge is 0. Therefore, the macroscopic field is

$$E = -\hat{i}\left\{1 - \frac{L}{(L^2+4R^2)^{\frac{1}{2}}}\right\}\frac{aL+2b}{\varepsilon_0}$$

The microscopic field is E_L,

$$E_L = E + \frac{P}{3\varepsilon_0} = -\hat{i}\,\frac{aL+2b}{\varepsilon_0}\left\{1 - \frac{4}{(L^2+4R^2)^{\frac{1}{2}}}\right\} + \frac{i(aL/2+b)}{3\varepsilon_0}$$

$$E_L = +\hat{i}\,\frac{1}{\varepsilon_0}\left\{-aL - 2b + \frac{aL}{6} + \frac{b}{3}\right\} + i\,\frac{aL+2b}{\varepsilon_0}\,\frac{L}{\sqrt{L^2+4R^2}}$$

$$E_L = -\hat{i}\,\frac{1}{\varepsilon_0}\left\{\frac{5aL}{6} + \frac{5b}{3}\right\} + \hat{i}\,\frac{aL+2b}{\varepsilon_0}\,\frac{L}{\sqrt{L^2+4R^2}}$$

PROBLEM 6.19 Given a sphere of dielectric material of radius R with radial polarization proportional to the distance from the center.

(a) Find all bound charge densities.
(b) Show by direct integration that in this case the total bound charge is 0.
(c) What is the electric field intensity at general points outside the sphere?

Solution: Given $P = a\mathbf{r}$:

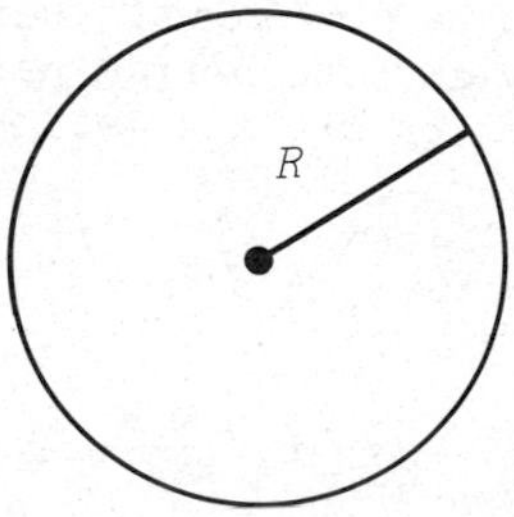

(a) $\sigma = \mathbf{P} \cdot \hat{\mathbf{n}} = aR,$

$$\rho = -\nabla \cdot \mathbf{P} = -\left(\hat{\imath}\,\frac{\partial}{\partial x} + \hat{\jmath}\,\frac{\partial}{\partial y} + \hat{k}\,\frac{\partial}{\partial z}\right) \cdot (\hat{\imath}ax + \hat{\jmath}ay + \hat{k}az)$$

$$\rho = -3a$$

(b) $q = \int \sigma\, da + \int \rho\, dV\ \sigma 4\pi R^2 + \rho(4/3)\pi R^3$

$$q = 4\pi R^3 - (3a)(4/3)(\pi R^3) = 0$$

(c) Everywhere $r > R$, $E = 0$. Inside

$$\oint \mathbf{E} \cdot d\mathbf{A} = \frac{1}{\varepsilon_0} Q_T$$

$$E(4\pi r^2) = \frac{1}{\varepsilon_0}(-3a)(4/3)\pi r^3$$

$$E = -\frac{ar}{\varepsilon_0}$$

PROBLEM 6.20 A sphere of dielectric material has radius R and uniform polarization P directed along the polar axis. For a point on the axis at distance $x > R$ from the center of the sphere, find the electrostatic potential by integration over the bound charge density: $\phi = PR^3/3\varepsilon_0 Z^2$.

Solution:

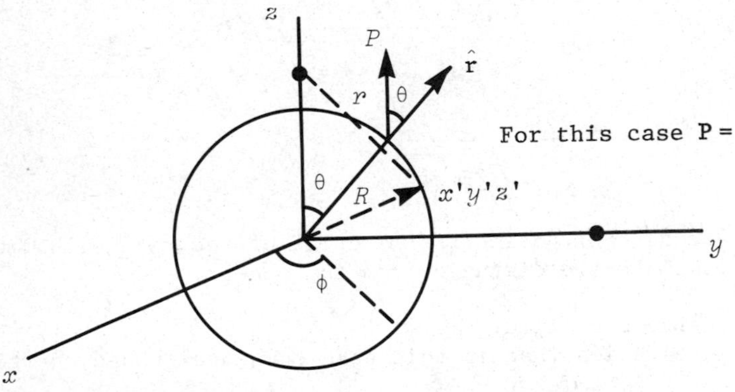

For this case $\mathbf{P} = \hat{\mathbf{k}}P_0$

Since P is constant, there is no volume charge density. However, there is a surface charge density given by $\sigma = P\cos\theta$. Therefore, the top part of the sphere is charged positively and the bottom part negatively. We will compute the potential at a point on the z axis.

$$\phi = \frac{1}{4\pi\varepsilon_0}\int \frac{\sigma\, da}{r}, \quad da = R^2 \sin\theta\, d\theta\, d\phi$$

and $\mathbf{R} + \mathbf{r} = \hat{\mathbf{k}}Z \Rightarrow \mathbf{r} = \hat{\mathbf{k}}Z - \mathbf{R}$, so $r = (\mathbf{r} \cdot \mathbf{r})^{1/2}$.

$\mathbf{r}\cdot\mathbf{r} = (\hat{\mathbf{k}}Z - \mathbf{R})\cdot(\hat{\mathbf{k}}Z - \mathbf{R}) = Z^2 - 2\mathbf{R}\cdot\hat{\mathbf{k}}Z + R^2$

$\mathbf{R}\cdot\hat{\mathbf{k}} = Z'$, where Z' is a point in the surface of the sphere. From spherical coordinates $Z' = R\cos\theta$, so

$$r = (Z^2 - 2RZ\cos\theta + R^2)^{\frac{1}{2}}$$

and

$$\phi = \frac{1}{4\pi\varepsilon_0}\iint \frac{P\cos\theta\, R^2 \sin\theta\, d\theta\, d\phi}{(Z^2 - 2RZ\cos\theta + R^2)^{\frac{1}{2}}}$$

$$\phi = -\frac{PR^2}{4\pi\varepsilon_0}\iint \frac{\cos\theta\, d(\cos\theta)\, d\phi}{(Z^2 + R^2 - 2RZ\cos\theta)^{\frac{1}{2}}}$$

Integrating over θ, this is of the form $x\,dx/(a+bx)^{\frac{1}{2}}$ and the integral $\int$ over $\phi = 2\pi$, so

$$\phi = -\frac{PR^2}{2\varepsilon_0}\left(\frac{-2[2(Z^2+R^2) + 2RZ\cos\theta]}{3(4R^2Z^2)}[Z^2 + R^2 - 2RZ\cos\theta]^{\frac{1}{2}}\right)\Bigg|_0^{\pi}$$

$$\phi = \frac{PR^2}{6\varepsilon_0}\left(\frac{(Z^2+R^2-RZ)(Z^2+R^2+2RZ)^{\frac{1}{2}} - (Z^2+R^2+RZ)(Z^2+R^2-2RZ)^{\frac{1}{2}}}{R^2Z^2}\right)$$

$$\phi = \frac{PR^2}{6\varepsilon_0}\left(\frac{\pm(Z^2+R^2-RZ)(Z+R) - (\pm)(Z^2+R^2+RZ)(Z-R)}{R^2Z^2}\right)$$

$$\phi = \frac{PR^2}{6\varepsilon_0 R^2 Z^2}[\pm(Z^3+R^3) \mp (Z^3 - R^3)]$$

$$\phi = \frac{P}{6\varepsilon_0 Z^2}2R^3 = \pm\frac{PR^3}{3\varepsilon_0 Z^2}, \quad \text{outside}$$

$$\phi = \frac{2PZ^3}{6\varepsilon_0 Z^2} = \frac{PZ}{3\varepsilon_0}, \quad \text{inside}$$

$$\mathbf{P} = \sum q_i \mathbf{r}_i = \int_L \mathbf{r}\,dq + \int_R \mathbf{r}\,dq + \int_V \mathbf{r}\,dq$$

where L, R, V refer to the left surface, the right surface, and the volume, respectively. If we are far enough away, the ends look like point charges and the bar looks like a linear charge, so the situation we work with is

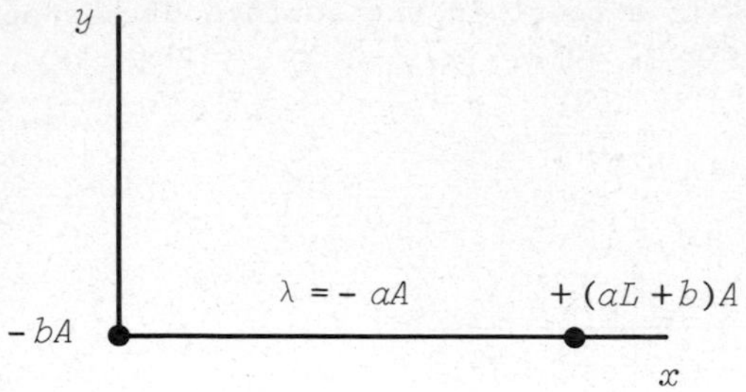

For this arrangement the dipole moment is

$$\mathbf{p} = \hat{\mathbf{i}} - bA(0) + \hat{\mathbf{i}}(aL + b)AL + \int_0^L \hat{\mathbf{i}}x\,\lambda\,dx = \hat{\mathbf{i}}(aL + b)AL + \hat{\mathbf{i}}\lambda L^2/2$$

$$\mathbf{p} = \hat{\mathbf{i}}\left[(aL + b)AL - \frac{aAL^2}{2}\right]$$

Then the potential is

$$\phi = \frac{1}{4\pi\varepsilon_0}\,\frac{(aAL^2/2 + bAL)}{r^2}$$

PROBLEM 6.21 A bar of dielectric material with uniform cross section A and length L has a polarization vector parallel to its length of magnitude $ax + b$, where x is a coordinate parallel to the length of the bar with zero at one face and a and b are constants. Find the electrostatic potential due to this bar at distances large compared with its dimensions:

$$\phi = \frac{1}{4\pi\varepsilon_0}\,\frac{(aAL^2/2 + bAL)}{R^2}$$

Solution:

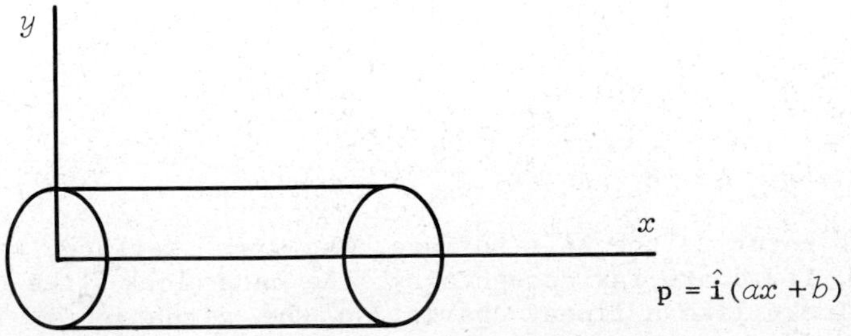

On the left face, the charge density is $\sigma_L = \mathbf{P}\cdot(-\hat{\mathbf{i}}) = -(ax + b)$ or at $x = 0$,

$\sigma_L = -b$. On the right face, at $x = L$, $\sigma_R = aL + b$. The volume charge density is $\rho = -\nabla \cdot \mathbf{P} = -a$. Therefore, the problem reduces to finding the potential to the dipole approximation from two surface charges and a volume charge. To the dipole approximation

$$\phi = \frac{1}{4\pi\varepsilon_0} \frac{\mathbf{P} \cdot \mathbf{r}_0}{r^2}$$

There is no monopole term since the total charge is 0.

PROBLEM 6.22 Given a sphere of dielectric material of radius R with a radial polarization proportional to the distance from the center.

(a) Determine the macroscopic electric field at points inside the sphere.

$$\mathbf{E} = -\frac{a}{\varepsilon_0}\mathbf{r}$$

(b) Determine the microscopic electric field at points inside the sphere.

$$\mathbf{E}_L = -\frac{2a}{3\varepsilon_0}\mathbf{r}$$

Solution: $\mathbf{P} = a\mathbf{r}$, $\nabla \cdot \mathbf{P} = 3a$, $\rho = -3a$

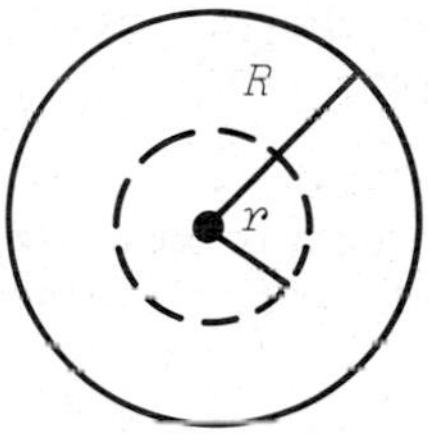

$$\oint \mathbf{E} \cdot d\mathbf{A} = \frac{1}{\varepsilon_0} Q$$

$$E(4\pi r^2) = \frac{1}{\varepsilon_0} 3a(4/3)\pi r^3$$

$$E = ar/\varepsilon_0$$

(a) $\mathbf{E} = -\dfrac{a}{\varepsilon_0}\mathbf{r}$

(b) $\mathbf{E}_L = \mathbf{E} + \dfrac{\mathbf{P}}{3\varepsilon_0} = -\dfrac{a}{\varepsilon_0}\mathbf{r} + \dfrac{a\mathbf{r}}{3\varepsilon_0} = -\dfrac{2a\mathbf{r}}{3\varepsilon_0}$

PROBLEM 6.23 An ideal dipole $\mathbf{p} = 3\hat{\mathbf{j}} + 4\hat{\mathbf{k}}$ C · m is located at the origin. Without the dipole the electric field at this point is $\mathbf{E} = \hat{\mathbf{i}}(12 + y) + \hat{\mathbf{j}}(4 + z)$ N/C.

(a) Find the torque on p.
(b) Find the potential energy of the dipole.
(c) Find the force on the dipole.

on:

(a)

$$\tau = \mathbf{p} \times \mathbf{E} = \begin{vmatrix} \hat{i} & \hat{j} & \hat{k} \\ 0 & 3 & 4 \\ 12 & 4 & 0 \end{vmatrix}$$

$$\tau = -16\hat{i} - \hat{j}(-48) + \hat{k}(-36)$$

$$\tau = -16\hat{i} + 48\hat{j} - 36\hat{k} \text{ N} \cdot \text{m}$$

(b) $U = -\mathbf{P} \cdot \mathbf{E} = -(3\hat{j} + 4\hat{k}) \cdot [\hat{i}(12 + y) + \hat{j}(4 + z)]$

$$U = -3(4 + z)$$

at (0, 0, 0) $U = -12$ J

(c) $\mathbf{F} = -\nabla U = -\left(\hat{i}\frac{\partial}{\partial x} + \hat{j}\frac{\partial}{\partial y} + \hat{k}\frac{\partial}{\partial z}\right)[-3(4 + z)]$

$$\mathbf{F} = \hat{k}3 \text{ N}$$

PROBLEM 6.24 At a point within a certain isotropic dielectric, the specific dielectric constant is 3 and the electric field intensity is 10^5 V/m. For this dielectric at this point, find

(a) The electric susceptibility.
(b) The magnitude of the electric polarization.
(c) The magnitude of the electric displacement.
(d) The permittivity.

Solution:

(a) $K = 1 + \chi/\varepsilon_0$

$$2 = \chi/\varepsilon_0$$

$$\chi = (2)(8.85 \times 10^{-12} \text{ F/m})$$

$$\chi = 1.77 \times 10^{-11} \text{ F/m}$$

(b) $\mathbf{P} = \chi\mathbf{E}$

$$\mathbf{P} = (1.77 \times 10^{-11})(10^5) = 1.77 \times 10^{-6} \text{ C/m}^2$$

(c) $\varepsilon = \varepsilon_0 + \chi$

$$\varepsilon = 3\varepsilon_0 = 2.655 \times 10^{-11} \text{ F/m}$$

$$D = \varepsilon E = 2.655 \times 10^{-6} \text{ C/m}^2$$

(d) $\varepsilon = 2.665 \times 10^{-11}$ F/m

PROBLEM 6.25 The dielectric constant of helium at 0°C and 1 atm pressure is 1.000074. Find the dipole moment induced in each helium atom when the gas is in an electric field of intensity 10^2 V/m.

Solution: $K = 1.000074$, $K = 1 + \chi$, so $\chi = 7.4 \times 10^{-5}$. Now $\mathbf{P} = \chi\varepsilon_0 E$ so

$$P = (7.4 \times 10^{-5})(8.85 \times 10^{-12})(10^2)$$

$$P = 6.55 \times 10^{-14} \text{ C/m}^2$$

Now P is the dipole moment per unit volume, so to get the dipole moment of a single helium atom, we need to find N, the number density of helium atoms. At 1 atm and 0°C, the density of helium is $\rho = 1.78 \times 10^{-4}$ g/cm^3, or

$$\rho = (1.78 \times 10^{-4} \text{ g/cm}^3)(10^{-3} \text{ kg/g})(10^6 \text{ cm}^3/\text{m}^3)$$

$$\rho = 1.78 \times 10^{-1} \text{ kg/m}^3$$

$$N = (6.02 \times 10^{23} \text{ atom/mole})\left(\frac{1}{4\text{ g}} \text{ mole}\right)(1.78 \times 10^{-4} \text{ g/cm}^3)(10^6 \text{ cm}^3/\text{m}^3)$$

$$N = 2.68 \times 10^{25} \text{ atoms/m}^3$$

Then $P = Np \Rightarrow p = P/N$, so

$$p = \frac{6.65 \times 10^{-14}}{2.68 \times 10^{25}} = 2.48 \times 10^{-39} \text{ C} \cdot \text{m}$$

PROBLEM 6.26 Given the following data for gaseous CO_2, it is required to obtain the permittivity ε at 0°C.

*Permanent dipole moment: 4.0×10^{-31} C · m

*Molecular polarizability for the induced dipole moment only: 2.22×10^{-40} C^2 · m/N

*Density: 1.250 g/liter

*Molecular weight: 28

Solution: For this case

$$N = (P/n)N_0 = \frac{1.25 \times 10^{-3}}{2.8 \times 10^1} 6.02 \times 10^{23} \text{ molecules/cm}^3$$

$$N = 2.7 \times 10^{19} \text{ molecules/cm}^3$$

$$N = 2.7 \times 10^{25} \text{ molecules/m}^3$$

Now $\varepsilon = \varepsilon_0 + \chi$

$$\chi = N\alpha, \text{ where } \alpha = \alpha_0 + \frac{P_0^2}{3KT}$$

From the data given

$$\alpha = 2.22 \times 10^{-40} + \frac{(4 \times 10^{-31})^2}{(3)(1.38 \times 10^{-23})(2.73 \times 10^2)}$$

$$\alpha = 2.22 \times 10^{-40} + 1.42 \times 10^{-41} = 2.36 \times 10^{-40}$$

so

$$\chi = (2.36 \times 10^{-40})(2.7 \times 10^{25}) = 6.4 \times 10^{-15}$$

$$\varepsilon = 8.85 \times 10^{-12} + 6.4 \times 10^{-15}$$

$$\varepsilon = 8.856 \times 10^{-12}$$

PROBLEM 6.27 Given that the polarizability of the NH_3 molecule is found by experimental data to be

$$\alpha = 2.42 \times 10^{-39} \ C^2 \cdot m/N \text{ at } 309^\circ K$$

and

$$\alpha = 1.74 \times 10^{-39} \ C^2 \cdot m/N \text{ at } 448^\circ K$$

For each temperature, find the part of the polarizability due to the permanent dipole moment and the part due to the deformation of the molecule, using the Langevin-Debye formula.

Solution: In general

$$\alpha = \alpha_0 + \frac{p_0^2}{3KT}$$

at $T_1 = 309^\circ K$,

$$\alpha_1 = \alpha_0 + \frac{p_0^2}{3KT_1}$$

at $T_2 = 448^\circ K$,

$$\alpha_2 = \alpha_0 + \frac{p_0^2}{3KT_2}$$

Subtracting, $\alpha_1 - \alpha_2 = \frac{p_0^2}{3K}\left(\frac{1}{T_1} - \frac{1}{T_2}\right)$

We can solve for p_0,

$$2.42 \times 10^{-39} - 1.74 \times 10^{-39} = \frac{p_0^2}{(3)(1.38 \times 10^{-23})}\left(\frac{1}{3.09 \times 10^2} - \frac{1}{4.48 \times 10^2}\right)$$

$$0.68 \times 10^{-39} = \frac{p_0^2}{4.14 \times 10^{-23}}(3.24 \times 10^{-3} - 2.23 \times 10^{-3})$$

$$2.8 \times 10^{-62} = p_0^2 \ (1.01 \times 10^{-3})$$

$$p_0^2 = 2.8 \times 10^{-59} = 28 \times 10^{-60}$$

$$p_0 = 5.3 \times 10^{-30} \text{ C/m}$$

Then

$$\alpha_0 = \alpha_1 - \frac{p_0^2}{3KT_1} = 2.42 \times 10^{-39} - \frac{2.8 \times 10^{-59}}{(4.14 \times 10^{-23})(3.09 \times 10^2)}$$

$$\alpha_0 = 2.42 \times 10^{-39} - 0.22 \times 10^{-38} = 0.22 \times 10^{-39} \text{ C}$$

So for this,

$$\alpha_0 = 2.2 \times 10^{-40} \text{ C}^2 \cdot \text{m/N}$$

At T_1, $\alpha_p = 2.2 \times 10^{-39}$ $\text{C}^2 \cdot \text{m/N}$

At T_2, $\alpha_p = 1.5 \times 10^{-39}$ $\text{C}^2 \cdot \text{m/N}$

PROBLEM 6.28 An electric dipole consists of two opposite charges of magnitude $q = 1 \times 10^{-6}$ C separated by $d = 2$ cm. The dipole is placed in an external field of 1×10^5 N/C.

(a) What maximum torque does the field exert on the dipole?
(b) How much work must an external agent do to turn the dipole end for end, starting from a position of alignment ($\theta = 0$)?

Solution:

(a)

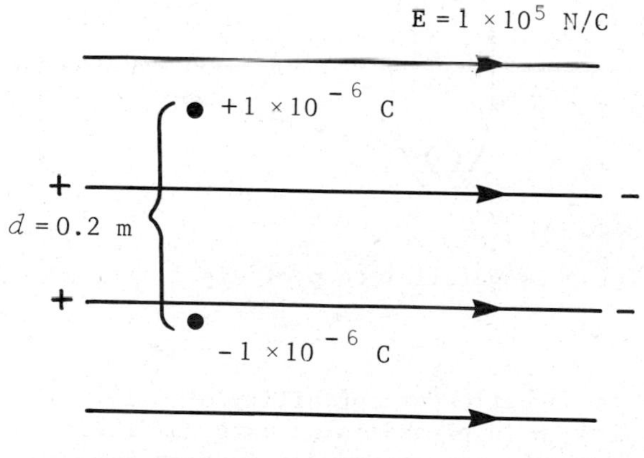

$$\tau = (E)(\sin \theta)(\text{dipole moment})$$

$$\tau = (1 \times 10^5 \text{ N/C})(1 \times 10^{-6} \text{ C})(0.02 \text{ m}) = 2 \times 10^{-3} \text{ N} \cdot \text{m}$$

(b) $w = \int \tau \, d\theta$

$$w = \int_0^{\pi} pE \sin\theta \, d\theta$$

$$w = pE \int_0^{\pi} \sin\theta \, d\theta,$$

$$w = 2pE$$

$$w = 2(1 \times 10^{-6}\ \text{C})(0.02\ \text{m})(1 \times 10^{5}\ \text{N/C})$$

$$w = 4 \times 10^{-3}\ \text{N} \cdot \text{m}$$

$$w = 4 \times 10^{-3}\ \text{J}$$

PROBLEM 6.29 A charge $q = 3 \times 10^{-6}$ C is 30 cm from a small dipole along its perpendicular bisector. The magnitude of the force on the charge is 5×10^{-6} N. Show on a diagram

(a) The direction of the force on the charge.
(b) The direction of the force on the dipole.

Solution:

(a) If the force on the charge from the dipole is 5×10^{-6} N, then the magnitude of the force on the dipole from the charge must be equal and opposite.

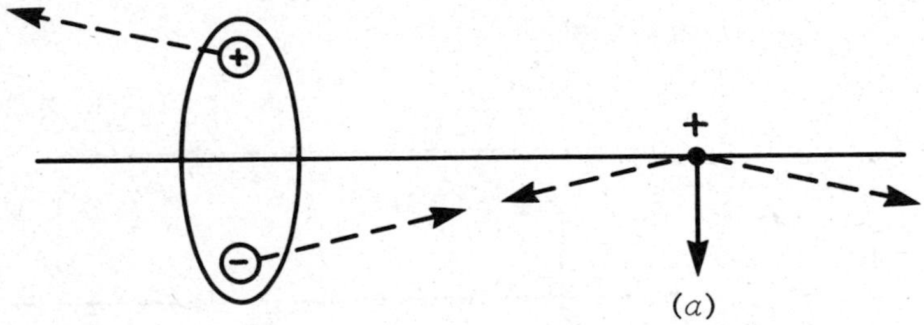

(*a*)

(b) The total force is parallel to p of dipole (from negative to positive).

PROBLEM 6.30

(a) Bismuth has a magnetic susceptibility of -1.7×10^{-6} MKS units. If a sample of bismuth is placed in an external field of 10^{-2} W/m^2, find the magnetic dipole moment induced in each bismuth atom.
(b) Using the information given in part (a), determine a value for the average radius of the electron orbit for bismuth.
(c) Repeat the above calculations for helium whose magnetic susceptibility is -5.9×10^{-7} MKS units.

Solution:

(a) $\chi = -1.7 \times 10^{-6}$

$$\mu_0 = 4\pi \times 10^{-7}$$

$$\mu = \mu_0(1+\chi) = 4\pi \times 10^{-7}(1 - 1.7 \times 10^{-6}) = 1.25 \times 10^{-6}$$

$$N = \frac{\rho}{m} N_0 = \left(\frac{9.7}{209}\right)(6.02 \times 10^{23}) = 2.79 \times 10^{28} \quad \text{(MKS units)}$$

$$\underline{m} = \frac{1}{\mu_0}\left(\frac{1}{(1/\chi)+1}\right)B = \frac{1}{1.25 \times 10^{-6}}\left(\frac{1}{1/(-1.7 \times 10^{-6})+1}\right)(10^{-2})$$

$$\underline{m} = -1.36 \times 10^{-2}$$

$$\underline{m} = Nm$$

$$m = \frac{\underline{m}}{N}\ \frac{-1.36 \times 10^{-2}}{2.79 \times 10^{28}} = -4.88 \times 10^{-31}$$

$$m = -4.88 \times 10^{-31} \quad \text{A} \cdot \text{m}^2$$

(b) $$m = \frac{-e}{2m_e} \sum_{i=1}^{z} \frac{er_i^2}{2} B$$

$$m = \frac{-e^2B}{4m_e} \sum_{i=1}^{83} r_i^2$$

$$r_i^2 = \frac{m4m_e}{e^2B(83)} = \frac{-(9.11 \times 10^{-31})(4)(-4.88 \times 10^{-31})}{(-1.6 \times 10^{-19})^2(10^{-2})(83)}$$

$$r_i^2 = 8.369 \times 10^{-23} \quad \text{(MKS units)}$$

$$r_i = 9.15 \times 10^{-12} \text{ m}$$

(c) $$m = \frac{\chi}{\mu_0 B} = \frac{5.9 \times 10^{-7}}{4\pi \times 10^{-7}}\ 10^{-2}$$

$$m = 4.7 \times 10^{-3} \text{ A/m}$$

$$N = \frac{P}{m} N_0 = \frac{1.78 \times 10^{-4}}{4}\ 6.02 \times 10^{23} = 2.68 \times 10^{19} \text{ atom/cm}^3$$

$$N = 2.68 \times 10^{25} \text{ atoms/m}^3$$

$$m = MN \Rightarrow M = \frac{4.7 \times 10^{-3}}{2.68 \times 10^{25}} = 1.75 \times 10^{-28} \text{ A} \cdot \text{m}^2$$

$$m = \frac{Ze^2}{4n} \langle r_i^2 \rangle B$$

$$1.75 \times 10^{-28} = \frac{2(1.6 \times 10^{-19})^2}{4(9.1 \times 10^{-31})} 10^{-2} r_i^2$$

$$\langle r \rangle = 1.1 \times 10^{-10} \text{ m}$$

PROBLEM 6.31 A long solenoid of 100 turns/meter carries a current of 10 A (see figure).

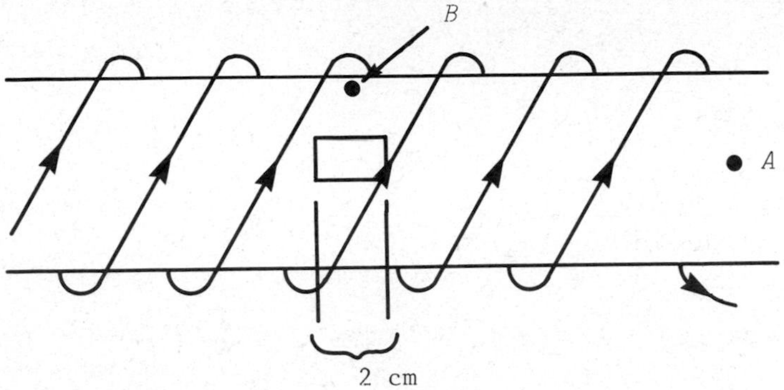

A small thin paramagnetic substance having a radius of 1 cm, length 2 cm, and susceptibility 200×10^{-4} is placed as shown. All other space is a vacuum.

(a) Find H inside the material.
(b) Find the magnetization surface current density.
(c) Find B at point A taking into account the presence of the paramagnetic material.
(d) Find B at point B taking into account the presence of the paramagnetic material.

Assume the sample acts like a dipole.

Solution:

(a) $$\oint \mathbf{H} \cdot d\ell = I_f$$

$$H\ell = nI\ell$$

$$H = nI = \left(100 \frac{\text{turns}}{\text{meter}}\right)(10 \text{ A})$$

$$H = 1 \times 10^3 \text{ A/m}$$

(b) $$M = \chi H = \frac{I_{mag}}{\ell} = (2 \times 10^{-2})(1 \times 10^3) = 20 \text{ A/m}$$

$$\frac{I_{mag}}{\ell} = 20 \text{ A/m}$$

(c) $$I_{mag} = M\ell = (20 \text{ A/m})(0.02 \text{ m}) = 0.4 \text{ A}$$

mag moment $m = IA = (0.4 \text{ A/m})(\pi)(0.01)\text{m}^2 = 1.26 \times 10^{-4} \text{ A} \cdot \text{m}^2$

$$B = \frac{\mu_0}{4\pi}\left(\frac{-m}{r^3} + \frac{3(m \cdot r)r}{r^5}\right) = \frac{\mu_0}{4\pi}\left(\frac{-1.26 \times 10^{-4}}{r^3} + \frac{3(1.26 \times 10^{-4})}{r^3}\right)$$

$$B_{mag} = \frac{\mu_0}{4\pi}\left(\frac{2.52 \times 10^{-4}}{r^3}\right)$$

$$B_{\text{free currents}} = \mu_0 nI = \mu_0 \left(100 \frac{\text{turns}}{\text{meter}}\right)(10 \text{ A}) = \mu_0(1 \times 10^3 \text{ A/m})$$

so

$$B_{total} = B_{mag} + B_{fc} = \frac{\mu_0}{4\pi}\left(\frac{2.52 \times 10^{-4}}{r^3}\right) + \mu_0(1 \times 10^3)$$

$$B_{total} = \mu_0\left(\frac{2.52 \times 10^{-4}}{4\pi r^3} + (1 \times 10^3)\right)$$

$$B = \frac{\mu_0}{4\pi}\left(\frac{-m}{r^3} + \frac{3(m \cdot r)r}{r^5}\right) = \frac{\mu_0 2m}{4\pi r^3} = \frac{\mu_0 m}{2\pi r^3}$$

$$B = \frac{\mu_0 m}{2\pi r^3}$$

$$m = \frac{0.4\pi(0.01)^2}{2} = 6.28 \times 10^{-5}$$

$$B_{mag} = \frac{\mu_0(6.28 \times 10^{-5})}{2\pi r^3} = \frac{\mu_0}{8\pi r^3}(2.52 \times 10^{-4})$$

$$B_f = \mu_0 nI = \mu_0 (100)(10) = \mu_0(1 \times 10^3)$$

$$B_{total} = \frac{\mu_0}{8\pi r^3}(2.52 \times 10^{-4}) - \mu_0(1 \times 10^3) = \mu_0\left(\frac{2.52 \times 10^{-4}}{8\pi r^3} - 10^3\right)$$

$$B_{total} = \mu_0\left(\frac{1.00268}{r^3} - (1 \times 10^3)\right)$$

PROBLEM 6.32 Why would one expect the paramagnetic susceptibility of a material to be temperature dependent and the diamagnetic susceptibility to be temperature independent?

Solution: When a diamagnetic substance is put into a magnetic field, each atom inside the sample acquires a net magnetic dipole moment that opposes the applied field. The magnetic dipole moment is related to the local field as seen by each atom by a direct proportion (with α_m as the proportionality constant). Only the field can induce a dipole moment in each atom resulting in a net magnetic dipole moment for the sample.

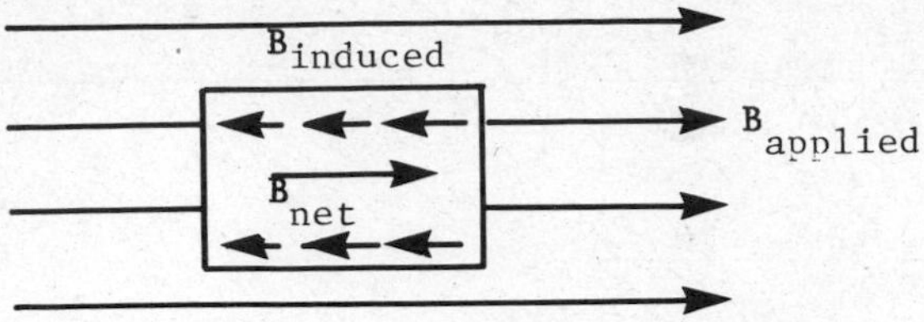

Each atom acquires a dipole moment since these (diamagnetic) substances have no unpaired spins, and how hard it is to polarize the substance depends on applied **B** fields. But in paramagnetic materials the inner-shell electrons have unpaired spins, therefore the spin angular momentum of the electrons do not all cancel out when one puts the atoms all together in a piece of material, and each atom then has a permanent dipole moment and the torques on these dipoles tend to align them with the externally (to the material) applied magnetic field (see diagram). So in materials like this (paramagnetic), where the sample has unpaired spins, the field applied outside the

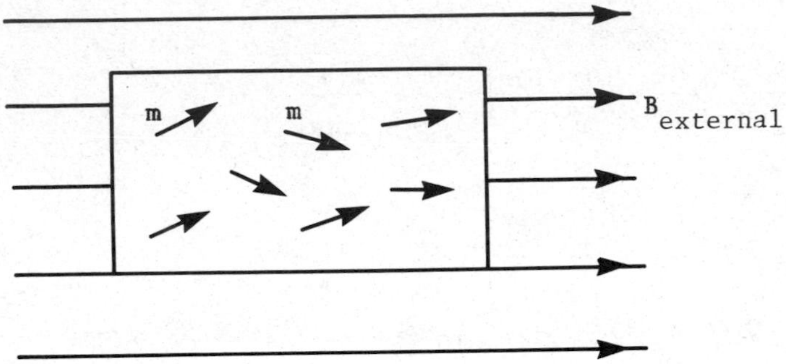

material won't be the only thing to affect the alignment, because when the sample is heated the molecules will be unaligned, but when the sample is cooled, the molecules will be less random and more aligned. Since $\mathbf{M} = \chi \mathbf{H}$, and χ is a function of H, H depends on how many molecules are aligned. In turn, the amount of molecules which become aligned (the magnetization or dipole moment/unit volume) depends on how many molecules are aligned opposing **H**, so that χ must also be a function of T.

PROBLEM 6.33 A conducting sphere of radius A has a charge $+Q$ on it and is surrounded by an insulating material whose dielectric constant varies with radius according to $K = 2\exp[-(r/a - 1)]^2$. The dielectric has a spherical outer boundary of radius B. Find the values of **D**, **E**, **P**, and ρ as a function of radius r.

Solution:

$$K = 2e^{-[(r/a)-1]^2} = 2e^{(1-c/a)^2}$$

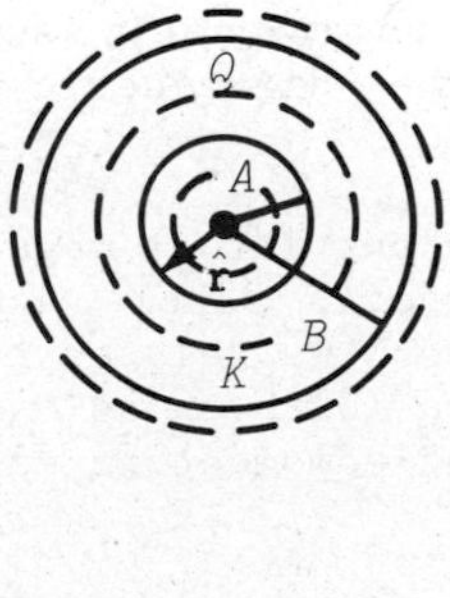

For $r < A$

$$\oint \mathbf{D} \cdot \hat{\mathbf{n}} \, dA = q_f$$

$$q_f = 0$$

$$\mathbf{D} = 0$$

$$\mathbf{D} = \varepsilon E$$

$$\mathbf{E} = \frac{1}{\varepsilon} \mathbf{D}$$

$$\mathbf{E} = 0$$

$$\mathbf{D} = \varepsilon_0 \mathbf{E} + \mathbf{P}$$

$$\mathbf{P} = \mathbf{D} - \varepsilon_0 \mathbf{E}$$

$$\mathbf{P} = 0$$

$$\rho = 0$$

For $r > B$

$$\oint \mathbf{D} \cdot \hat{\mathbf{n}} \, dA = q_f$$

$$\mathbf{D} = \frac{Q}{4\pi r^2} \, \hat{\mathbf{r}}$$

$$\mathbf{E} = \frac{1}{\varepsilon} \mathbf{D} = \frac{1}{K\varepsilon_0} \mathbf{D}, \text{ where } K = 1,$$

$$\mathbf{E} = \frac{Q}{4\pi\varepsilon \, r^2} \, \hat{\mathbf{r}}$$

$$\mathbf{P} = \mathbf{D} - \varepsilon_0 \mathbf{E} = \frac{Q\hat{\mathbf{r}}}{4\pi r^2} - \frac{\varepsilon_0 Q\hat{\mathbf{r}}}{4\pi\varepsilon_0 r^2} = 0$$

$$\mathbf{P} = 0$$

$$\rho = 0$$

For $A < r < B$

$$\oint \mathbf{D} \cdot \hat{\mathbf{n}} \, dA = q_f$$

$$D(4\pi r^2) = Q$$

$$\mathbf{D} = \frac{Q\hat{\mathbf{r}}}{4\pi r^2}$$

$$\mathbf{E} = \frac{1}{\varepsilon} \mathbf{D} = \frac{1}{K\varepsilon_0} \mathbf{D}$$

$$\mathbf{E} = \frac{Q\hat{\mathbf{r}}}{K\varepsilon_0 (4\pi r^2)}$$

$$\mathbf{P} = \mathbf{D} - \varepsilon_0 \mathbf{E} = \frac{Q\hat{\mathbf{r}}}{4\pi r^2} - \frac{\varepsilon_0 Q\hat{\mathbf{r}}}{K\varepsilon_0 4\pi r^2}$$

$$\mathbf{P} = \frac{Q\hat{\mathbf{r}}}{4\pi r^2} \left(1 - \frac{1}{K}\right)$$

$$\rho = \rho_b + \rho_f = (-\nabla \cdot \mathbf{P}) + (\nabla \cdot \mathbf{D})$$

$$= \left[-\frac{1}{r^2} \frac{\partial}{\partial r} \, r^2 \, \frac{Q}{4\pi r^2} \left(1 - \frac{1}{K}\right) \right] + \left[\frac{1}{r^2} \frac{\partial}{\partial r} \left(r^2 \, \frac{Q}{4\pi r^2} \right) \right]$$

$$= \left[-\frac{1}{r^2} \frac{\partial}{\partial r} \frac{Q}{4\pi} \left(1 - \frac{1}{2e^{(1-r/a)^2}} \right) \right] + 0$$

$$= \frac{-Q}{4\pi r^2} \left(-\frac{\partial}{\partial r} \frac{1}{2e^{(1-r/a)^2}} \right) = \frac{Q}{4\pi r^2 2} \frac{\partial}{\partial r} e^{-[(1-r/a)^2]}$$

$$= \frac{Q}{8\pi r^2} e^{-(1-r/a)^2} (-2) \left(1 - \frac{r}{a} \right) \left(-\frac{1}{a} \right)$$

$$= e^{-(1-r/a)^2} \frac{2}{a} \left(1 - \frac{r}{a} \right) \frac{Q}{8\pi r^2}$$

$$\rho_{\text{total}} = \frac{Q}{4\pi a r^2} \left(1 - \frac{r}{a} \right) e^{-(1-r/a)^2}$$

PROBLEM 6.34 A dipole of dipole moment **p** is placed at the center of a grounded conducting sphere of radius R. Calculate the electrostatic potential and electric field inside the sphere.

Solution:

$$4\pi\varepsilon_0 \Phi(\mathbf{r}) = \frac{\mathbf{p} \cdot \mathbf{r}}{r^3} + \text{term without singularities inside sphere}$$

Take **p** along the z axis and expand the solution in Legendre polynomials.

$$4\pi\varepsilon_0 \Phi(r,\theta) = \frac{p \cos\theta}{r^2} + \sum_{n=0}^{\infty} A_n P_n(\cos\theta) r^n$$

$$= \left(\frac{p}{r^2} + A_1 r \right) P_1(\cos\theta) + A_0 + \sum_{n=2}^{\infty} A_n r^n P_n(\cos\theta)$$

$$\Phi(R,\theta) = 0 \Rightarrow \frac{p}{R^2} + A_1 R = 0, \; A_0 = 0, \; A_n = 0, \; A_1 = -\frac{p}{R^3}$$

$$4\pi\varepsilon_0 \Phi(r,\theta) = p \left(\frac{1}{r^2} - \frac{r}{R^3} \right) \cos\theta$$

$$E_\phi = -\frac{1}{r \sin\theta} \frac{\partial \Phi}{\partial \phi} = 0$$

$$E_r = -\frac{\partial}{\partial r} \Phi(r,\theta) = \frac{p}{4\pi\varepsilon_0} \left(\frac{2}{r^3} - \frac{1}{R^3} \right) \cos\theta$$

$$E_\theta = -\frac{1}{r} \frac{\partial \Phi}{\partial \theta} = \frac{p}{4\pi\varepsilon_0} \left(\frac{1}{r^3} - \frac{1}{R^3} \right) \sin\theta$$

Using the method of images is also OK, but you must take the limit $q \to \infty$, $d \to 0$, $qd \to p$.

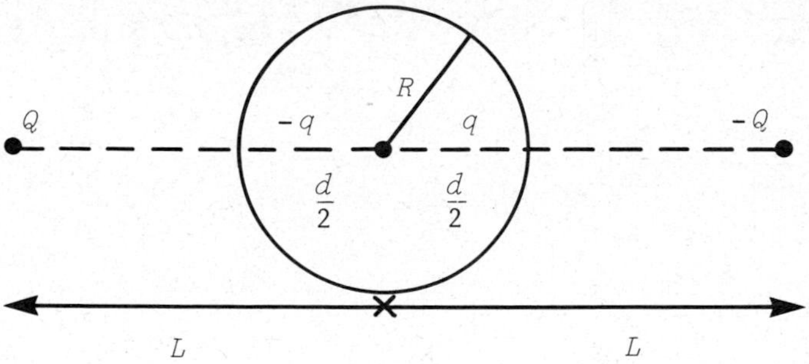

$$L = \frac{2R^2}{d}$$

$$Q = \frac{L}{R}q$$

PROBLEM 6.35 A dielectric sphere (dielectric constant K) of radius R has a free charge density ρ distributed uniformly throughout the volume.

(a) What is the electrostatic potential at the center of the sphere?
(b) How much energy is required to establish this configuration?

Solution:

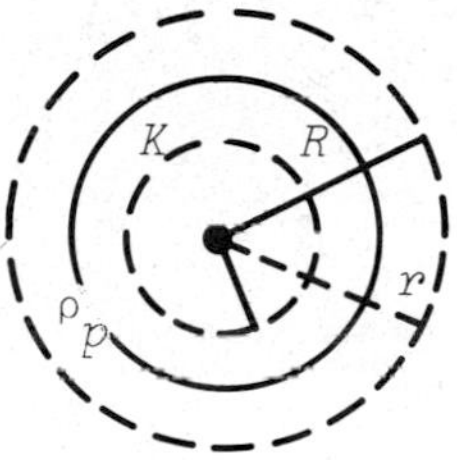

$r < R$

$$\oint \mathbf{D} \cdot \hat{\mathbf{n}}\, dA = q_f$$

$$D(4\pi r^2) = \int_0^r \rho_f 4\pi r^2\, dr$$

$$D(4\pi r^2) = \rho_f \frac{4}{3}\pi r^3$$

$$\mathbf{D} = \frac{\rho_f r}{3}\hat{\mathbf{r}}$$

$$\mathbf{D} = \varepsilon \mathbf{E} = K\varepsilon_0 \mathbf{E}$$

$$\mathbf{E} = \frac{\rho_f r}{K\varepsilon_0 3}\hat{\mathbf{r}}$$

$r > R$

$$\oint \mathbf{D} \cdot \hat{\mathbf{n}}\, dA = q_f$$

$$D(4\pi r^2) = \int_0^R \rho_f 4\pi r^2\, dr$$

$$D(4\pi r^2) = \frac{\rho_f 4\pi R^3}{3}$$

$$\mathbf{D} = \frac{\rho_f R^3}{3r^2}\hat{\mathbf{r}}$$

$$\mathbf{E} = \frac{1}{K\varepsilon_0}\mathbf{D}, \text{ where } K = 1,$$

$$\mathbf{E} = \frac{\rho_f R^3}{3\varepsilon_0 r^2}\hat{\mathbf{r}}$$

$$\phi_c - \phi_\infty = -\int_\infty^0 \mathbf{E} \cdot d\ell = -\int_\infty^R \frac{\rho_f R^3}{3\varepsilon_0 r^2}\, d\ell - \int_R^0 \frac{\rho_f r}{3K\varepsilon_0}\, d\ell$$

$$= -\frac{\rho_f R^3}{3\varepsilon_0}\left[-\left(\frac{1}{R}-\frac{1}{\infty}\right)\right] - \frac{\rho_f}{3K\varepsilon_0 2}(0^2 - R^2) = \frac{\rho_f R^2}{3\varepsilon_0 2} + \frac{\rho_f R^2}{3\varepsilon_0 K2}$$

(a) $\phi_{center} = \frac{\rho_f R^2}{3\varepsilon_0}\left(1+\frac{1}{2K}\right)$

$$U = \tfrac{1}{2}\int_0^\infty \mathbf{D}\cdot\mathbf{E}\,dv$$

$$U = \tfrac{1}{2}\left\{\int_0^R \left[\left(\frac{\rho_f r\hat{\mathbf{r}}}{3}\right)\cdot\left(\frac{\rho_f r\hat{\mathbf{r}}}{K\varepsilon_0 3}\right)\right] 4\pi r^2\,dr + \int_R^\infty \left[\left(\frac{\rho_f R^3\hat{\mathbf{r}}}{3r^2}\right)\cdot\left(\frac{\rho_f R^3\hat{\mathbf{r}}}{3\varepsilon_0 r^2}\right)\right] 4\pi r^2\,dr\right\}$$

$$= \tfrac{1}{2}\int_0^R \frac{\rho_f^2 r^2 4\pi r^2}{9K\varepsilon_0}\,dr + \tfrac{1}{2}\int_R^\infty \frac{\rho_f R^6 4\pi r^2}{9\varepsilon_0 r^4}\,dr$$

$$= \frac{\rho_f^2 4\pi}{(2)9K\varepsilon_0}\int_0^R r^4\,dr + \frac{\rho_f^2 4\pi R^6}{(2)9\varepsilon_0}\int_R^\infty \frac{1}{r^2}\,dr$$

$$= \frac{\rho_f^2 4\pi R^5}{18K\varepsilon_0(5)} + \frac{\rho_f^2 4\pi R^6}{18\varepsilon_0}\left(-\frac{1}{\infty} - -\frac{1}{R^2}\right) = \frac{\rho_f^2 4\pi R^5}{18K\varepsilon_0 5} + \frac{\rho_f^2 4\pi R^4}{18\varepsilon_0}$$

(b) $U = \frac{\rho_f^2 4\pi R^4}{18\varepsilon_0}\left(\frac{R}{5K} + 1\right)$

PROBLEM 6.36

(a) The measured value for the dielectric constant of argon is 1.000545 and its density is 1.78×10^{-3} g/cm^3. Determine the radius of the argon atom using these data.

(b) The density of liquid argon is 1.44 g/cm^3. Using this and the information in part (a) to determine the dielectric constant of the liquid argon. The experimentally measured value is 1.54. Compare your computed value with the experimental value.

Solution:

(a) $K = 1.000545$

density $= 1.78 \times 10^{-3}$ m/cm^3

$$K = \frac{\varepsilon}{\varepsilon_0} \Rightarrow \varepsilon = (1.000545)\varepsilon_0 = (1.000545)\left(8.85 \times 10^{-12}\ \frac{\text{c}^2}{\text{N}\cdot\text{m}^2}\right)$$

$$\varepsilon = 8.854823 \times 10^{-12}\ C^2/N \cdot m^2$$

The electrical permittivity is given by

$$\varepsilon = \varepsilon_0 + \chi$$

The electrical susceptibility is given by

$$\chi = \varepsilon - \varepsilon_0 = 8.854823 \times 10^{-12}\ \frac{C^2}{N \cdot m^2} - 8.85 \times 10^{-12}\ \frac{C^2}{N \cdot m^2}$$

$$\chi = 0.004823 \times 10^{-12}\ \frac{C^2}{N \cdot m^2}$$

$$p = \left(\frac{\text{average d.m.}}{\text{molecule}}\right)\left(\frac{\text{number of molecules}}{\text{volume}}\right)$$

mean dipole moment = (p)(volume)

$$\chi = \frac{N\alpha}{1 + N\alpha/3\varepsilon_0}$$

$$\chi\left(1 + \frac{N\alpha}{3\varepsilon_0}\right) = N\alpha$$

$$\chi + \frac{N\alpha\chi}{3\varepsilon_0} = N\alpha$$

$$\chi = N\alpha\left(1 - \frac{\chi}{3\varepsilon_0}\right)$$

$$N\alpha = \frac{\chi}{(1 - \chi/3\varepsilon_0)}$$

$$N\alpha = \frac{0.004823\ C^2/N \cdot m^2}{\left(1 - \frac{0.004823 \times 10^{-12}}{3(8.85 \times 10^{-12})}\right)}$$

$$N\alpha = 4.823876 \times 10^{-3}\ \frac{C^2}{N \cdot m^2}$$

$$N = \frac{\text{g/volume}}{\text{g/molecule}} = \frac{\text{molecules}}{\text{volume}} = \frac{1.78 \times 10^{-3}\ g/cm^3}{\left(\frac{39.95}{6.02 \times 10^{23}}\right)\ \text{g/molecule}}$$

$$N = 2.68 \times 10^{19}\ \frac{\text{molecules}}{cm^3}\left(\frac{1 \times 10^6 cm^3}{m^3}\right) = 2.68 \times 10^{25}\ \frac{\text{molecules}}{m^3}$$

$$\chi = \frac{N\alpha}{1 + N\alpha/3\varepsilon_0}$$

$$\chi + \frac{N\alpha\chi}{3\varepsilon_0} = N\alpha$$

$$\chi = N\alpha\left(1 - \frac{\chi}{3\varepsilon_0}\right)$$

$$\frac{\chi}{N(1 - \chi/3\varepsilon_0)} = \alpha = \frac{0.004823 \times 10^{-12}}{2.68 \times 10^{25}\left(1 - \frac{0.004823 \times 10^{-12}}{3(8.85 \times 10^{-12})}\right)}$$

$$\alpha = 1.79 \times 10^{-40}$$

$$\alpha = 4\pi\varepsilon_0 R_0^3$$

$$R_0 = \left(\frac{1.79 \times 10^{-40}}{4\pi\varepsilon_0}\right)^{1/3} = 1.17 \times 10^{-10}\ \text{m} = 1.17\ \text{Å}$$

(b) $$\rho_\ell = 1.44\ \frac{\text{g}}{\text{cm}^3}\left(\frac{10^6\text{cm}^3}{\text{m}^3}\right) = 1.44 \times 10^6\ \frac{\text{g}}{\text{m}^3}\left(\frac{\text{kg}}{10^3\ \text{g}}\right) = 1.44 \times 10^3\ \frac{\text{kg}}{\text{m}^3}$$

$$N = \frac{\text{g/volume}}{\text{g/molecule}} = \frac{1.44\ \text{g/cm}^3}{\left(\frac{39.95}{6.02 \times 10^{23}}\right)\ \text{g/molecule}} = 2.17 \times 10^{22}\ \frac{\text{molecules}}{\text{cm}^3}$$

$$N = 2.17 \times 10^{22}\ \frac{\text{molecules}}{\text{cm}^3}\left(\frac{10^6\ \text{cm}^3}{\text{m}^3}\right) = 2.17 \times 10^{28}\ \frac{\text{molecules}}{\text{m}^3}$$

$$\alpha = 4\pi\varepsilon_0 R^3 = 4\pi\varepsilon_0(1.17 \times 10^{-10}\ \text{m})^3 = 2.01 \times 10^{-29}\ \varepsilon_0$$

$$\alpha = 1.79 \times 10^{-40}$$

$$\chi = \frac{N\alpha}{(1 + N/3\varepsilon_0)} = \frac{(2.17 \times 10^{28})(1.79 \times 10^{-40})}{1 + \frac{(2.17 \times 10^{28})(1.79 \times 10^{-40})}{3(8.85 \times 10^{-12})}}$$

$$\chi = 3.388 \times 10^{-12}$$

$$\varepsilon = \chi + \varepsilon_0 = 3.388 \times 10^{-12} + 8.85 \times 10^{-12} = 1.22 \times 10^{-11}$$

Now $$K = \frac{\varepsilon}{\varepsilon_0} = \frac{1.22 \times 10^{-11}}{8.85 \times 10^{-12}} = 1.38$$

PROBLEM 6.37 A metal sphere of 2-cm radius carries a charge of 3×10^{-9} C. It is encased in a spherical shell of isotropic, homogeneous, and linear dielectric material with inner radius of 2 cm and outer radius of 4 cm, which carries no free charge. Find

(a) The electric displacement in each region of space.
(b) The electric field in each region of space.
(c) The electric polarization in the dielectric.
(d) The bound charge densities.
(e) The total bound charges on inner and outer surfaces, respectively, of the shell.

Solution:

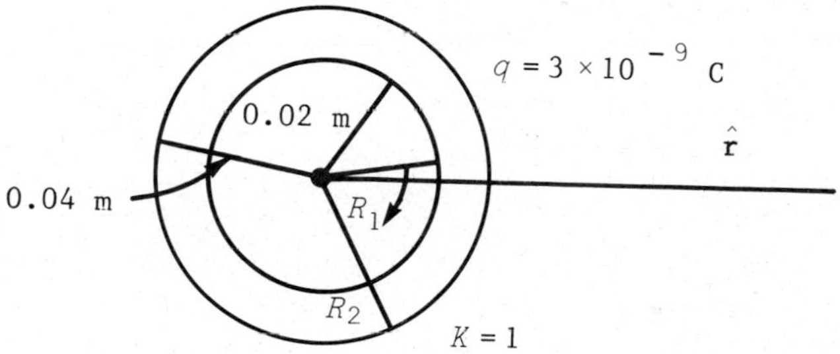

(a) $\oint \mathbf{D} \cdot \hat{\mathbf{n}}\, dA = q_f$

$0 < r < 0.02$ m, $q_f = 0$, $D = 0$

$0.02 \text{ m} < r < 0.04$ m,

$$\oint \mathbf{D} \cdot \hat{\mathbf{n}}\, dA = q_{\text{free enclosed}}$$

$$D(4\pi r^2) = 3 \times 10^{-9} \text{ C}$$

$$D = \frac{3 \times 10^{-9} \text{ C}}{4\pi r^2}$$

$r > 0.04$ m,

$$\oint \mathbf{D} \cdot \hat{\mathbf{n}}\, dA = q_f$$

$$D(4\pi r^2) = 3 \times 10^{-9} \text{ C}$$

$$D = \frac{3 \times 10^{-9} \text{ C}}{4\pi r^2}$$

(b) $\mathbf{D} = \varepsilon \mathbf{E}$

$0 < r < 0.02$ m, $E = 0$

$0.02\text{ m} < r < 0.04\text{ m},\ E = \frac{1}{4\pi\varepsilon_0 K}\frac{3\times 10^{-9}\text{ C}}{r^2},\ K = \frac{\varepsilon}{\varepsilon_0}$

$r > 0.04\text{ m},\ E = \frac{1}{4\pi\varepsilon_0 (K = 1)}\frac{3\times 10^{-9}\text{ C}}{r^2}$

(c) $\mathbf{D} = \varepsilon_0\mathbf{E} + \mathbf{P}$

$\mathbf{P} = \mathbf{D} - \varepsilon_0\mathbf{E}$

$$\mathbf{P} = \frac{3\times 10^{-9}\text{ C}}{4\pi r^2} - \varepsilon_0 \frac{1}{4\pi\varepsilon}\frac{3\times 10^{-9}\text{ C}}{r^2} = 0$$

$$\frac{Q}{4\pi r^2} - \frac{\varepsilon_0 Q}{4\pi K r^2 \varepsilon_0} = \frac{Q}{4\pi r^2}\left(1 - \frac{1}{K}\right)$$

$$\frac{Q}{4\pi r^2} - \varepsilon_0\frac{Q}{4\pi\varepsilon_0 r^2} = 0$$

(d) $\sigma_{b_{R_1}} = \mathbf{P}\cdot\hat{\mathbf{n}},\ \hat{\mathbf{n}} = -\hat{\mathbf{r}},\ \sigma_b = -P\Big|_{r = R_1}$

$$\sigma_b = \frac{Q}{4\pi r^2}\left(1 - \frac{1}{K}\right)$$

$\sigma_{b_{R_2}} = \mathbf{P}\cdot\hat{\mathbf{n}},\ \hat{\mathbf{n}} = \hat{\mathbf{r}}$

$$\sigma_{b_{R_2}} = P = \frac{Q}{4\pi r^2}\left(1 - \frac{1}{K}\right)$$

$$\rho_b = -\nabla\cdot\mathbf{P} = \frac{1}{r^2}\frac{\partial(r^2 P)}{\partial r} = \frac{1}{r^2}\frac{\partial}{\partial r}\frac{r^2 Q}{4\pi r^2}\left(1 - \frac{1}{K}\right) = 0$$

(e) $r = R_1$

inner, $\sigma_{R_1} A_{R_1} = 4\pi R_1^2(-)\dfrac{Q}{4\pi R_1^2}\left(1 - \dfrac{1}{K}\right)$

$$= -Q\left(1 - \frac{1}{K}\right) = Q\left(\frac{1}{K} - 1\right)$$

outer, $\sigma_{R_2} A_{R_2} = 4\pi R_2^2\left(\dfrac{Q(1 - 1/K)}{4\pi R_2^2}\right)$

$$= Q\left(1 - \frac{1}{K}\right)$$

When you add them the result is 0.

PROBLEM 6.38 Water vapor is a polar gas whose dielectric constant exhibits an appreciable temperature dependence. The following table gives experimental data on this effect. Assuming that water vapor obeys the ideal gas law, calculate the molecular polarizability as a function of inverse temperature and plot it. From the slope of the curve, deduce a value for the permanent dipole moment of the H_2O molecule (express the dipole moment in esu-statcoulomb-centimeters).

T (°K)	Pressure (cm Hg)	$(\varepsilon - 1) \times 10^5$
393	56.49	400.2
423	60.93	371.7
453	65.34	348.8
483	69.75	328.7

Solution:

γ = polarizability
P_m = polarization of a single molecule
E_m = molecular electric field
$p_m = \gamma E_m$

N = number of molecules in system
P = total polarization
$p = N < p_m >$

$< p_m >$ = average molecular polarization
$\mathbf{p}_0$ = molecules dipole moment
p_m = the component of $\mathbf{p}_0$ along the electric field

$$p_m = \frac{\mathbf{p}_0 \cdot \mathbf{E}_m}{|E_m|} = p_0 \cos\theta$$

$$< p_m > = < p_0 \cos\theta > = \frac{\int p_0 \cos\theta \, e^{-H/kT} \, dv}{\int e^{-H/kT} \, dv}$$

$$H = T + V = \frac{mv^2}{2} - p_0 E_m \cos\theta$$

$$< p_m > = \frac{p_0 \int \cos\theta \, e^{-p_0 E_m / kT \cos\theta} \, d(\cos\theta)}{\int e^{-p_0 E_m / kT \cos\theta} \, d(\cos\theta)}$$

$$= p_0 \left(\coth \frac{p_0 E_m}{kT} - \frac{kT}{p_0 E_m} \right)$$

Since $(p_0 E_m / kT) << 1$ for $T > 250$°K

$$\coth \frac{p_0 E_m}{kT} \simeq \frac{kT}{p_0 E_m} + \frac{1}{3} \frac{p_0 E}{kT} + \cdots$$

so

$$<p_m> = \frac{p_0^2 E_m}{kT}$$

therefore

$$\gamma = \frac{p_m}{E_m} = \frac{p_0^2}{3kT}$$

T (°K)	$(\varepsilon - 1) \times 10^5$	$\gamma = \frac{3}{4\pi}\left(\frac{\varepsilon - 1}{\varepsilon + 2}\right)$	$\frac{1}{T}$ (°K)$^{-1}$
393	400.2	3.18×10^{-4}	0.00254
423	371.7	2.95×10^{-4}	0.00236
453	348.8	2.77×10^{-4}	0.00221
483	328.7	2.60×10^{-4}	0.00207

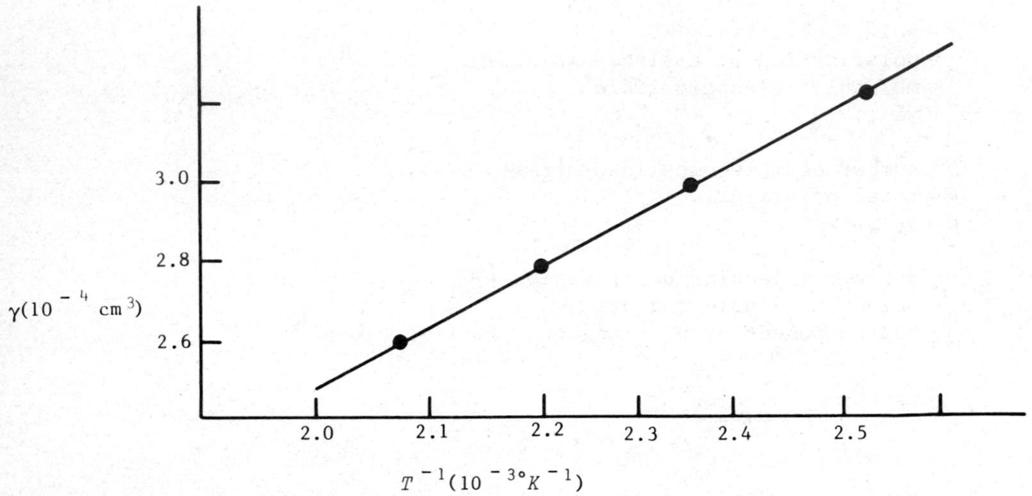

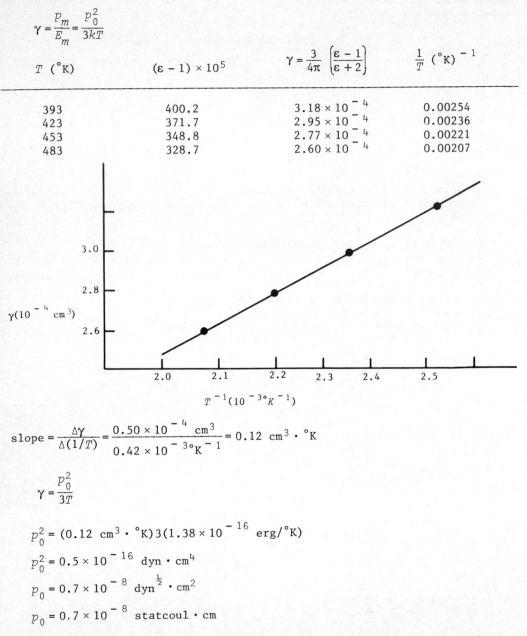

$$\text{slope} = \frac{\Delta\gamma}{\Delta(1/T)} = \frac{0.50 \times 10^{-4} \text{ cm}^3}{0.42 \times 10^{-3}\,°\text{K}^{-1}} = 0.12 \text{ cm}^3 \cdot °\text{K}$$

$$\gamma = \frac{p_0^2}{3T}$$

$$p_0^2 = (0.12 \text{ cm}^3 \cdot °\text{K})3(1.38 \times 10^{-16} \text{ erg/°K})$$

$$p_0^2 = 0.5 \times 10^{-16} \text{ dyn} \cdot \text{cm}^4$$

$$p_0 = 0.7 \times 10^{-8} \text{ dyn}^{\frac{1}{2}} \cdot \text{cm}^2$$

$$p_0 = 0.7 \times 10^{-8} \text{ statcoul} \cdot \text{cm}$$

PROBLEM 6.39 Calculate the demagnetizing factor of a long cylinder that is permanently magnetized at right angles to the cylinder axis. The magnetization is uniform.

Solution: The general solution for the potential in cylindrical coordinates (since Laplace's equation is satisfied by U^*) is

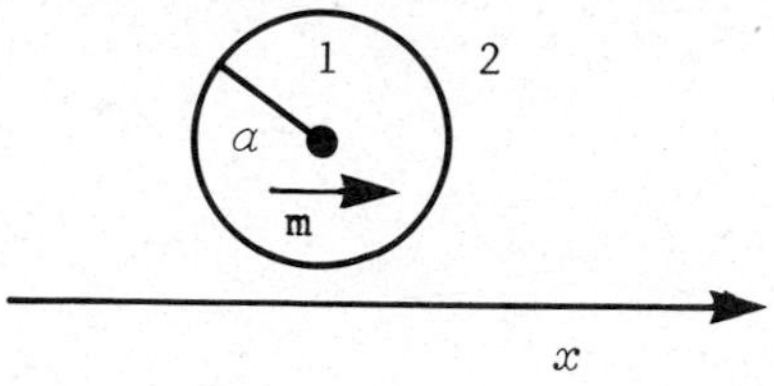

$$U^*_{out} = A_1 + B_1 \ln \rho + \sum_{n=1}^{\infty} (C_{1n}\rho^n + D_{1n}\rho^{-n}) \cos n\theta + \sum_{n=1}^{\infty} (E_{1n}\rho^n + F_{1n}\rho^{-n}) \sin n\theta$$

$$U^*_{in} = A_2 + B_2 \ln \rho + \sum_{n=1}^{\infty} (C_{2n}\rho^n + D_{2n}\rho^{-n}) \cos n\theta + \sum_{n=1}^{\infty} (E_{2n}\rho^n + F_{2n}\rho^{-n}) \sin n\theta$$

$B_1 = B_2 = 0$, however, since there are no magnetic monopoles. Making the potential continuous at $\rho = a$ leads to $A_1 = A_2$ (not physically significant) and also to $C_{1n} = 0$, for all n, for regularity at infinity and $E_{1n} = 0$, for all n, for the same reason. $D_{2n} = F_{2n} = 0$, for all n, for regularity at the origin.

Continuing with continuity at $\rho = a$,

$$\frac{D_{1n}}{a^n} = C_{2n}a^n, \quad \frac{F_{1n}}{a^n} = E_{2n}a^n$$

We need $\mathbf{B} \cdot \mathbf{n}$ continuous at the boundary.

$$B_\rho = -\mu_0 \frac{\partial U^*_{in}}{\partial \rho} + \mu_0 M_\rho, \quad \text{inside}$$

$$= -\mu_0 \frac{\partial U^*_{out}}{\partial \rho}, \quad \text{outside}$$

$M_r = \mathbf{M} \cdot \hat{\rho} = M \cos \theta$, thus

$$\frac{D_{11}}{a^2} = -C_{21} + M, \quad \text{all other equations homogeneous.}$$

$$\frac{D_{11}}{a^2} + C_{21} = M$$

So, coefficients D_{1n}, $C_{2n} = 0$, for $n > 1$ and F_{1n}, $E_{2n} = 0$, for all n. We are left with

$$\frac{D_{11}}{a} - C_{21}\, a = 0$$

$$\frac{D_{11}}{a} + C_{21}\, a = Ma$$

$$D_{11} = \tfrac{1}{2}\, Ma^2, \quad C_{21} = \frac{M}{2}$$

$$U^* = \frac{a^2 M}{2r} \cos\theta, \quad \rho \geq a$$

$$= \frac{M}{2}\, r\cos\theta, \quad \rho \leq a$$

$$= \frac{M}{2}\, x, \quad \mathbf{H} = -\hat{\mathbf{x}}\frac{M}{2} = -\frac{\mathbf{M}}{2}, \quad \text{inside}$$

Thus, there is a demagnetizing factor of ½ in this particular geometry. (Under some conventions it is $4\pi \times \frac{1}{2}$, but not in the MKS units used here.)

PROBLEM 6.40 A permanent magnet has the shape of a right circular cylinder of length L. If the magnetization $\mathbf{M}$ is uniform and has the direction of the cylinder axis, find the magnetization current densities, $\mathbf{J}_M$ and j_M. Compare the current distribution with that of a solenoid.

Solution: Since $\mathbf{M}$ is uniform, $\nabla \times \mathbf{M} = 0$ and there is no volume magnetization current. The surface magnetization current is given by $\mathbf{M} \times \mathbf{n}$. On the flat ends of the cylinder $\mathbf{n}$ and $\mathbf{M}$ are parallel, so $\mathbf{M} \times \mathbf{n} = 0$ there. On the curved surface there is a surface magnetization current. There $\mathbf{n} = \rho/\rho = \hat{\rho}$, so, $\mathbf{M} \times \mathbf{n} = M(\hat{\mathbf{z}} \times \rho)/\rho = M\hat{\theta} = \mathbf{j}_n$.

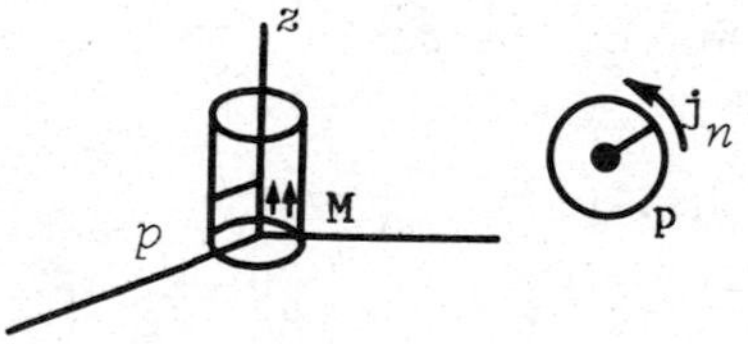

This is the same as for a tightly wound solenoid having N turns per unit length and current I where $NI = M$.

PROBLEM 6.41

(a) A permanent magnet in the shape of a right circular cylinder of length L and radius R is oriented so that its symmetry axis coincides with the z axis. The origin of coordinates is at the center of the magnet. If the cylinder has uniform axial magnetization M, (1) determine $U^*(z)$ at points on the symmetry axis, both inside and outside the magnet, and (2) use the results of part (1) to find the magnetic induction B_z at points on the symmetry axis, both inside and outside the magnet.

(b) A spherical shell of permeability $\mu = \mu_0$ and of radii a and b $(a < b)$ is placed in a field $\mathbf{B} = B_0\hat{\mathbf{k}}$. Calculate the field in the spherical cavity and show how the shell can be used as a magnetic shield.

Solution:

(a)

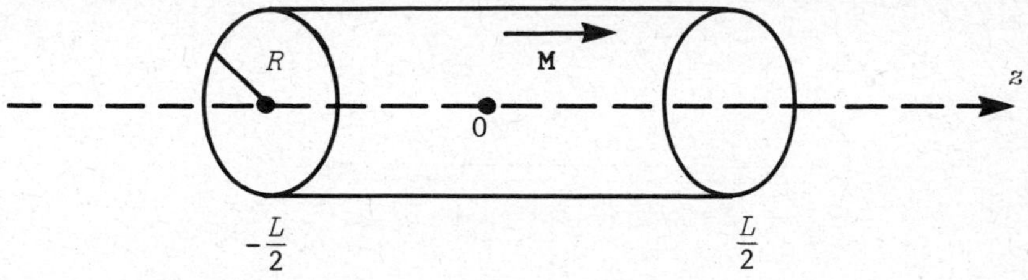

Using the concept of "magnetic pole density" and noting that $\nabla \cdot \mathbf{M}$ and $\mathbf{n} \cdot \mathbf{M} = 0$ on the curved portion of the surface, we realize that all we have is $\mathbf{M} \cdot \mathbf{n}$ contributions at the flat ends of the cylinder. Thus $\sigma_m = M$ at $Z' = L/2$ and $\sigma_m = -M$ at $Z' = -L/2$. Introduce cylindrical coordinates ρ, z, θ:

$$U^* = \frac{1}{4\pi} \iint\limits_{\text{at } \frac{L}{2}} \frac{\sigma_m(\mathbf{r}')}{|\mathbf{r} - \mathbf{r}'|} \, ds' + \frac{1}{4\pi} \iint\limits_{\text{at } -\frac{L}{2}} \frac{\sigma_m(\mathbf{r}')}{|\mathbf{r} - \mathbf{r}'|} \, ds'$$

$$U^*(Z,0,0) = \frac{\sigma_m}{4\pi} \left(\int_0^{2\pi} d\phi \int_0^R \frac{\rho' \, d\rho'}{\sqrt{\rho'^2 + (Z - L/2)}} - \int_0^{2\pi} d\phi \int_0^R \frac{\rho' \, d\rho'}{\sqrt{\rho'^2 + (Z + L/2)}} \right)$$

$$U^*(0,Z,0) = \frac{M}{2} \left[\sqrt{R^2 + (Z - L/2)^2} - \sqrt{(Z - L/2)^2} - \sqrt{R^2 + (Z + L/2)^2} + \sqrt{(Z + L/2)^2} \right]$$

$$= \frac{M}{2} \left[\sqrt{R^2 + (Z - L/2)^2} - Z + \frac{L}{2} - \sqrt{R^2 + (Z + L/2)^2} + Z + \frac{L}{2} \right]$$

$$= \frac{M}{2} \left[\sqrt{R^2 + (Z - L/2)^2} - \sqrt{R^2 + (Z + L/2)^2} \right] + \frac{ML}{2}, \quad Z \geq \frac{L}{2}$$

$$U^*(0,Z,0) = \frac{M}{2} \left[\sqrt{R^2 + (Z - L/2)^2} - \sqrt{R^2 + (Z + L/2)^2} + Z - \frac{L}{2} + Z + \frac{L}{2} \right]$$

$$= \frac{M}{2} \left[\sqrt{R^2 + (Z - L/2)^2} - \sqrt{R^2 + (Z + L/2)^2} \right] + MZ, \quad -\frac{L}{2} \leq Z \leq \frac{L}{2}$$

$$U^*(0,Z) = \frac{M}{2} \left[\sqrt{R^2 + (Z - L/2)^2} - \sqrt{R^2 + (Z + L/2)^2} \right] + \frac{M}{2} \left(-\frac{L}{2} - \frac{|Z|}{2} + \frac{|Z|}{2} - \frac{L}{2} \right)$$

$$= \frac{M}{2} \left[\sqrt{R^2 + (Z - L/2)^2} - \sqrt{R^2 + (Z + L/2)^2} \right] - \frac{ML}{2}, \quad Z \leq -\frac{L}{2}$$

$$H_z(0,Z) = -\frac{\partial U^*}{\partial Z} = +\frac{M}{2}\left\{\frac{Z+L/2}{\sqrt{R^2+(Z+L/2)^2}} - \frac{Z-L/2}{\sqrt{R^2+(Z-L/2)^2}}\right\}, \quad Z > \frac{L}{2}$$

$$H_z = \frac{M}{2}\left\{\frac{Z+L/2}{\sqrt{R^2+(Z+L/2)^2}} - \frac{Z-L/2}{\sqrt{R^2+(Z-L/2)^2}}\right\} - M, \quad -\frac{L}{2} < Z < \frac{L}{2}$$

$$H_z = \frac{M}{2}\left\{\frac{Z+L/2}{\sqrt{R^2+(Z+L/2)^2}} - \frac{Z-L/2}{\sqrt{R^2+(Z-L/2)^2}}\right\}, \quad Z < \frac{L}{2}$$

$\mathbf{B} = \mu_0(\mathbf{H}+\mathbf{M})$, ($\mathbf{M} = 0$ outside cylinder, $\mathbf{M} = \mathbf{Z}M$ inside)

$$\mathbf{B} = \mu_0\hat{\mathbf{Z}}\left\{\frac{Z+L/2}{\sqrt{R^2+(Z+L/2)^2}} - \frac{Z-L/2}{\sqrt{R^2+(Z-L/2)^2}}\right\}$$ everywhere on the Z axis

(b)

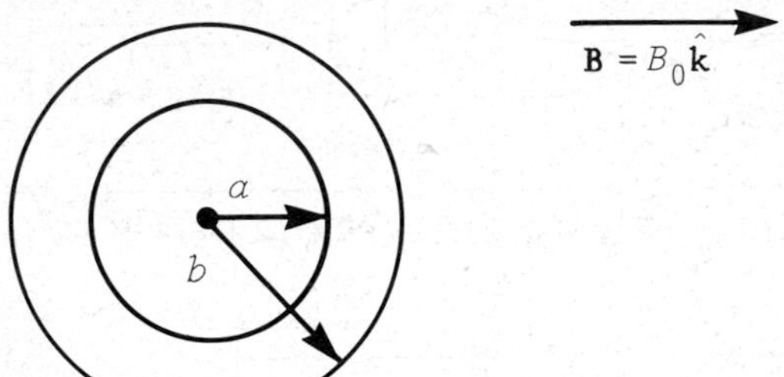

Since there are no currents present, the magnetic field is derivable from a scaler potential $\mathbf{H} = -\nabla\Phi_m$. Since $\mathbf{B} = \mu\mathbf{H}$, $\nabla\cdot\mathbf{B} = 0$ becomes $\nabla\cdot\mathbf{H} = 0$; thus $\nabla^2\Phi_m = 0$ everywhere. For $r > b$,

$$\Phi_m = -B_0\, r\cos\theta + \sum_{\ell=0}^{\infty} A_\ell r^{-(\ell+1)} P_\ell(\cos\theta)$$

from the boundary condition that $\mathbf{B} \to B_0\hat{\mathbf{k}}$ as $r \to \infty$. For $a < r < b$,

$$\Phi_m = \sum_{\ell=0}^{\infty} (B_\ell r^\ell + C_\ell r^{-(\ell+1)}) P_\ell(\cos\theta)$$

For $r < a$,

$$\Phi_m = \sum_{\ell=0}^{\infty} D_\ell r^\ell P_\ell(\cos\theta)$$

The boundary conditions at $r=a$ and $r=b$ are that H_θ and B_r be continuous. In terms of the potential Φ_m,

$$(r>b)\left(\frac{\partial\Phi_m}{\partial\theta}\right)_{r=b} = (a<r<b)\left(\frac{\partial\Phi_m}{\partial\theta}\right)_{r=b}$$

$$(r>b)\left(\frac{\partial\Phi_m}{\partial r}\right)_{r=b} = (a<r<b)\ \mu\left(\frac{\partial\Phi_m}{\partial r}\right)_{r=b}$$

$$(a<r<b)\left(\frac{\partial\Phi_m}{\partial\theta}\right)_{r=a} = (r<a)\left(\frac{\partial\Phi_m}{\partial\theta}\right)_{r=a}$$

$$(a<r<b)\mu\left(\frac{\partial\Phi_m}{\partial r}\right)_{r=a} = (r<a)\left(\frac{\partial\Phi_m}{\partial r}\right)_{r=a}$$

All coefficients with $\ell \neq 1$ vanish. For $\ell = 1$,

$$A_1 - b^3B_1 - C_1 = b^3B_0$$

$$2A_1 + \mu b^3B_1 - 2\mu C_1 = -b^3B$$

$$a^3B_1 + C_1 - a^3D_1 = 0$$

$$\mu a^3B_1 - 2\mu C_1 - a^3D_1 = 0$$

The solution for A_1 and D_1 are

$$A_1 = \left\{\frac{(2\mu+1)(\mu-1)}{(2\mu+1)(\mu+2) - 2(a^3/b^3)(\mu-1)^2}\right\}(b^3-a^3)B_0$$

$$D_1 = -\left\{\frac{9\mu}{(2\mu+1)(\mu+2) - 2(a^3/b^3)(\mu-1)^2}\right\}B_0$$

Inside the cavity,

$$\Phi_m = -\left\{\frac{9\mu}{(2\mu+1)(\mu+2) - 2(a^3/b^3)(\mu-1)^2}\right\}B_0 r\cos\theta$$

$$\mathbf{H} = \mathbf{B} = -\nabla\Phi_m = \hat{\mathbf{k}}\left\{\frac{9\mu}{(2\mu+1)(\mu+2) - 2(a^3/b^3)(\mu-1)^2}\right\}B_0$$

If $\mu \gg 1$,

$$\mathbf{B} \to \hat{\mathbf{k}}\,\frac{9\mu B_0}{2\mu^2 - 2(a^3/b^3)\mu^3} = \hat{\mathbf{k}}B_0\,\frac{9}{2\mu(1-a^3/b^3)}$$

Because of the $1/\mu$ dependence of $\mathbf{B}$, a shell made of high-permeability material causes a great reduction in the field inside of it.

PROBLEM 6.42 A dielectric spherical shell (dielectric constant ε) has inner and outer radii of a and $2a$. If the shell is placed in a uniform electric field $\mathbf{E}_0$, find the field in the interior of the shell.

Solution:

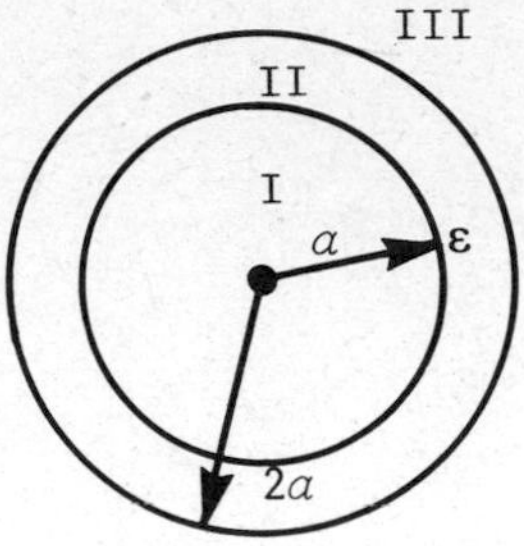

For region I

$$\Phi^{I}(r,0) = \sum_{\ell=0}^{\infty} A_\ell r^\ell P_\ell(\cos\theta) \tag{1}$$

Its derivative is given by

$$\frac{\partial\Phi}{\partial r} = \sum_{\ell=0}^{\infty} A_\ell \ell r^{\ell-1} P_\ell(\cos\theta)$$

So that

$$D_r^{I}(a) = -\frac{\partial\Phi}{\partial r}\bigg|_{r=a} = -\sum_{\ell=0}^{\infty} A_\ell \ell a^{\ell-1} P_\ell(\cos\theta) \tag{2}$$

and

$$\Phi^{I}(a,\theta) = \sum_{\ell=0}^{\infty} A_\ell a^\ell P_\ell(\cos\theta) \tag{3}$$

For region II

$$\Phi^{II}(r,0) = \sum_{\ell=0}^{\infty} \left(C_\ell r^\ell + \frac{d_\ell}{r^{\ell+1}}\right) P_\ell(\cos\theta) \tag{4}$$

therefore

$$\frac{\partial\Phi^{II}}{\partial r}(r,0) = \sum_{\ell=0}^{\infty} \left(C_\ell \ell r^{\ell-1} - \frac{(\ell+1)d_\ell}{r^{\ell+1}}\right) P_\ell(\cos\theta)$$

so that,

$$\Phi(r,a) = \sum_{\ell=0}^{\infty} \left(C_\ell a^\ell + \frac{d_\ell}{a^{\ell+1}}\right) P_\ell(\cos\theta) \tag{5}$$

$$\Phi(r,2a) = \sum_{\ell=0}^{\infty} \left(C_\ell (2a)^\ell + \frac{d_\ell}{(2a)^{\ell+1}}\right) P_\ell(\cos\,\theta) \tag{6}$$

also,

$$D_r^{II}(a) = -\varepsilon \left.\frac{\partial\Phi}{\partial r}\right|_{r=a} = -\varepsilon \sum_{\ell=0}^{\infty} \left(C_\ell \ell a^{\ell-1} - \frac{(\ell+1)d_\ell}{a^{\ell+2}}\right) P_\ell(\cos\,\theta) \tag{7}$$

$$D_r^{II}(2a) = -\varepsilon \left.\frac{\partial\Phi}{\partial r}\right|_{r=2a} = -\varepsilon \sum_{\ell=0}^{\infty} \left(C_\ell \ell (2a)^{\ell-1} - \frac{(\ell+1)^\ell d_\ell}{(2a)^{\ell+2}}\right) P_\ell(\cos\,\theta) \tag{8}$$

<u>For region III</u>

$$\Phi^{III}(r,\theta) = -E_0 r\, P_1(\cos\,\theta) + \sum_{\ell=0}^{\infty} \frac{B_\ell}{r^{\ell+1}} P_\ell(\cos\,\theta) \tag{9}$$

so that

$$\frac{\partial\Phi^{III}}{\partial r} = -E_0\, P_1(\cos\,\theta) - \sum_{\ell=0}^{\infty} \frac{(\ell+1)B_\ell}{r^{\ell+2}} P_\ell(\cos\,\theta)$$

$$\Phi^{III}(2a,\theta) = -E_0\, 2aP_1(\cos\,\theta) + \sum_{\ell=0}^{\infty} \frac{B_\ell}{(2a)^{\ell+1}} P_\ell(\cos\,\theta) \tag{10}$$

$$D_r^{III}(2a) = -\left.\frac{\partial\Phi}{\partial r}\right|_{r=2a} = E_0\, P_1(\cos\,\theta) + \sum_{\ell=0}^{\infty} \frac{(\ell+1)B_\ell}{(2a)^{\ell+2}} P_\ell(\cos\,\theta) \tag{11}$$

<u>Continuity of Φ and D_r at $r = 2a$</u>

From Eqs. (6), (8), (10), (11), we must have

$$C_\ell (2a)^\ell + \frac{d_\ell}{(2a)^{\ell+1}} = \frac{B_\ell}{(2a)^{\ell+1}}, \quad \ell \neq 1 \tag{12a}$$

$$2a\;C_1 + \frac{d_1}{4a^2} = -\,2aE_D + \frac{B_1}{4a^2} \tag{12b}$$

$$-\varepsilon\left(C_\ell \ell (2a)^{\ell-1} - \frac{(\ell+1)d_\ell}{(2a)^{\ell+2}}\right) = \frac{(\ell+1)B_\ell}{(2a)^{\ell+2}}, \quad \ell \neq 1 \tag{13a}$$

$$-\varepsilon\left(C_1 - \frac{2d_1}{8a^3}\right) = E_0 + \frac{2B_1}{8a^3} \tag{13b}$$

Continuity of Φ and D_r at $r = a$

From Eqs. (2), (3), (5), (7), we obtain

$$A_\ell a^\ell = c_\ell a^\ell + \frac{d_\ell}{a^\ell} \quad \text{for all } \ell \tag{14a}$$

for $\ell = 1$,

$$A_1 a = c_1 a + \frac{d_1}{a^2} \tag{14b}$$

$$+A_\ell \ell a^{\ell-1} = +\varepsilon\left(c_\ell \ell a^{\ell-1} - \frac{(\ell+1)d_\ell}{a^{\ell+2}}\right) \quad \text{for all } \ell \tag{15a}$$

for $\ell = 1$,

$$A_1 = \varepsilon\left(c_1 - \frac{2d_1}{a^3}\right) \tag{15b}$$

Equations (12a), (13a), (14a), and (15b) show that

$$A_\ell = c_\ell = d_\ell = B_\ell = 0, \quad \text{if } \ell \neq 1$$

Therefore, we need to solve

$$2ac_1 + \frac{d_1}{4a^2} = -2aE_0 + \frac{B_1}{4a^2}, \quad A_1 a = c_1 a + \frac{d_1}{a^2}$$

$$-\varepsilon\left(c_1 - \frac{d_1}{4a^3}\right) = E_0 + \frac{B_1}{4a^3}, \quad A_1 = \varepsilon\left(c_1 - \frac{2d_1}{a^3}\right)$$

From the first two, we obtain

$$c_1(2+\varepsilon) + \frac{d_1}{4a^3}(1-\varepsilon) = -3E_0$$

The other two yield

$$A_1 = \frac{3d_1}{a^3}\left(\frac{\varepsilon}{\varepsilon - 1}\right) \quad \text{and} \quad c_1(\varepsilon - 1) = \frac{d_1}{a^3}(1+2\varepsilon)$$

We finally obtain

$$A_1 = \frac{-36\,\varepsilon E_0}{4(2+\varepsilon)(1+2\varepsilon) - (\varepsilon - 1)^2}$$

Therefore, for $r \leq a$,

$$\Phi(r,\theta) = Ar\cos\theta$$

$$A = -\frac{36\ \varepsilon E_0}{4(2+\varepsilon)(1+2\varepsilon) - (\varepsilon - 1)^2}$$

and

$$\mathbf{E} = A(\cos\theta\ \hat{\mathbf{e}}_r - \sin\theta\ \hat{\mathbf{e}}_\theta)$$

Let us satisfy ourselves that **E** satisfies Gauss' law. We have

$$\int_S \mathbf{E}\cdot d\mathbf{s} = 2\pi A \int_0^\infty \int_0^\pi r^2 \sin\theta\cos\theta\, d\theta\, dr = 2\pi A \int_0^\infty r^2 \left\{ \frac{\sin^2\theta}{2}\bigg|_0^\pi \right\} dr = 0$$

as it must.

PROBLEM 6.43 A conducting sphere of radius a is covered with a dielectric (dielectric constant ε) in the form of a spherical shell; the radius of the entire object is b. If the covered sphere is placed in a uniform electric field $\mathbf{E}_0$ in free space, find the potential and the field components for all interior and exterior points.

Solution: In region I, $\Phi = \Phi_0$ = constant. To simplify the problem, we let $\Phi = 0$, that is, assume the conducting sphere to be grounded. If this is not the case, we may add to the potential the required additive constant, without altering our fields.

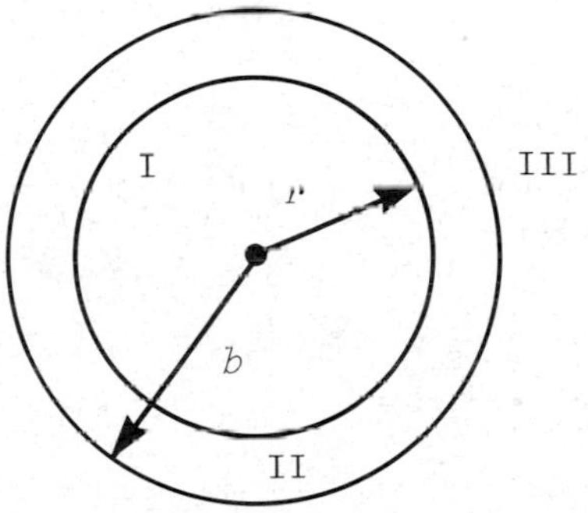

Region II

$$\Phi^{II}(r,\theta) = \sum_{\ell=0}^{\infty} \left(a_\ell r^\ell + \frac{b_\ell}{r^{\ell+1}} \right) P_\ell(\cos\theta)$$

with $\Phi(a,\theta) = 0$. Therefore,

$$a_\ell a^\ell + \frac{b_\ell}{a^{\ell+1}} = 0, \quad \text{i.e.,}\ b_\ell = -a_\ell a^{2\ell+1}$$

Also,

$$\Phi^{II}(r,\theta) = \sum_{\ell=0}^{\infty} a_\ell \left(r^\ell - \frac{a^{2\ell+1}}{r^{\ell+1}} \right) P_\ell(\cos\theta), \quad a \le r \le b \tag{1}$$

already satisfying $\Phi^{II}(a,\theta) = 0$

Region III

Since $\Phi^{III}(r,\theta) \to -E_0 r\cos\theta$ as $r\to\infty$, we can write immediately that

$$\Phi^{III}(r,\theta) = -E_0 r\cos\theta + \sum_{\ell=0}^{\infty} \frac{B_\ell}{r^{\ell+1}} P_\ell(\cos\theta) \tag{2}$$

Continuity of Φ at $r=b$

This requires using Eqs. (1) and (2), so that

$$\sum_{\ell=0}^{\infty} a_\ell\left(b^\ell - \frac{a^{2\ell+1}}{b^{\ell+1}}\right)P_\ell(\cos\theta) = -E_0 b\,P_1(\cos\theta) + \sum_{\ell=0}^{\infty}\frac{B_\ell}{b^{\ell+1}}P_\ell(\cos\theta)$$

Therefore, we must have

$$a_1\left(b - \frac{a^3}{b^2}\right) = -E_0 b + \frac{B_1}{b^2} \tag{3a}$$

$$a_\ell\left(b^\ell - \frac{a^{2\ell+1}}{b^{\ell+1}}\right) = \frac{B_\ell}{b^{\ell+1}}, \quad \ell \neq 1 \tag{3b}$$

Continuity of D_r at $r=b$

$$D^{III} = E^{III} = -\left.\frac{\partial\Phi^{III}}{\partial r}\right|_{r=b} = E_0\,P_1(\cos\theta) + \sum_{\ell=0}^{\infty}\frac{(\ell+1)B_\ell}{b^{\ell+2}}P_\ell(\cos\theta)$$

$$D^{II} = \varepsilon E^{II} = -\varepsilon\left.\frac{\partial\Phi^{II}}{\partial r}\right|_{r=b} = -\varepsilon\sum_{\ell=0}^{\infty} a_\ell\left(\ell b^{\ell-1} + \frac{(\ell+1)a^{2\ell+1}}{b^{\ell+2}}\right)P_\ell(\cos\theta)$$

Therefore, we must have

$$E_0 + \frac{2B_1}{b^3} = -\varepsilon\,a_1\left(1 + \frac{2a^3}{b^3}\right) \tag{4a}$$

$$\frac{B_0}{b^2} = -\varepsilon\,a_0\,\frac{a}{b^2} \tag{4b}$$

$$\frac{(\ell+1)B_\ell}{b^{\ell+2}} = -\varepsilon\,a_\ell\left(\ell b^{\ell-1} + \frac{(\ell+1)a^{2\ell+1}}{b^{\ell+2}}\right), \quad \ell \geq 2 \tag{4c}$$

Equations (3b), (4b), and (4c) imply that

$a_\ell = B_\ell = 0$ if $\ell \neq 1$ (homogeneous system of equations in two unknowns)

Solving for a_1 and B_1 from Eqns. (4a) and (3a), we obtain

$$a_1 = \frac{-3E_0}{2(1-a^3/b^3)+\varepsilon(1+2a^3/b^3)} = \frac{-3E_0}{(2+\varepsilon)+2(\varepsilon-1)a^3/b^3}$$

$$= \frac{-3E_0 b^3}{(2+\varepsilon)b^3+2(\varepsilon-1)a^3}$$

and

$$\frac{B_1}{b^2} = E_0 b\,\frac{\varepsilon(1+2a^3/b^3)-(1-a^3/b^3)}{\varepsilon(1+2a^3/b^3)+2(1-a^3/b^3)} = E_0 b\,\frac{(\varepsilon-1)+(2\varepsilon+1)a^3/b^3}{(\varepsilon+2)+2(\varepsilon-1)a^3/b^3}$$

$$= E_0 b\,\frac{(\varepsilon-1)b^3+(2\varepsilon+1)a^3}{(\varepsilon+2)b^3+2(\varepsilon-1)a^3}$$

Therefore, the solutions are

$$\Phi(r,\theta) = \frac{3E_0 b^3}{(\varepsilon+2)b^3+2(\varepsilon-1)a^3}\left(\frac{a^3}{r^2}-r\right)\cos\theta, \quad a \leq r \leq b$$

$$\Phi(r,\theta) = E_0\left(\frac{(\varepsilon-1)b^3+(2\varepsilon+1)a^3}{(\varepsilon+2)b^3+2(\varepsilon-1)a^3}\,\frac{b^3}{r^2}-r\right)\cos\theta, \quad r \geq b$$

To obtain the fields, rewrite these equations as

$$\Phi(r,\theta) = E_0\beta\left(\frac{a^3}{r^2}-r\right)\cos\theta; \quad \beta = 3b^3/(\varepsilon+2)b^3+2(\varepsilon-1)a^3$$

$$\Phi(r,\theta) = E_0\left(\alpha\,\frac{b^3}{r^2}-r\right)\cos\theta; \quad \alpha = \frac{(\varepsilon-1)b^3+(2\varepsilon+1)a^3}{(\varepsilon+2)b^3+2(\varepsilon-1)a^3}$$

then

$$\mathbf{E}(r,\theta) = -E_0\beta\left[\left(\frac{2a^3}{r^3}+1\right)\cos\theta\,\hat{\mathbf{e}}_r + \left(\frac{a^3}{r^3}-1\right)\sin\theta\,\hat{\mathbf{e}}_\theta\right], \quad a \leq r \leq b$$

$$\mathbf{E}(r,\theta) = -E_0\left[\left(\frac{2\alpha b^3}{r^3}+1\right)\cos\theta\,\hat{\mathbf{e}}_r + \left(\frac{\alpha b^3}{r^3}-1\right)\sin\theta\,\hat{\mathbf{e}}_\theta\right], \quad r \geq b$$

Chapter 7

PROBLEM 7.1 Given a current circuit in a prescribed magnetic field, the magnetic force on each circuit element $d\mathbf{l}$ is given by $I\,d\mathbf{l} \times \mathbf{B}$. If the circuit is allowed to move under the influences of the magnetic forces, such that a typical element is displaced $\delta\mathbf{r}$, and at the same time the current I is held constant, show by direct calculation that the mechanical work done by the circuit is $dW_m = I\,d\Phi$, where $d\Phi$ is the additional flux through the circuit.

Solution:

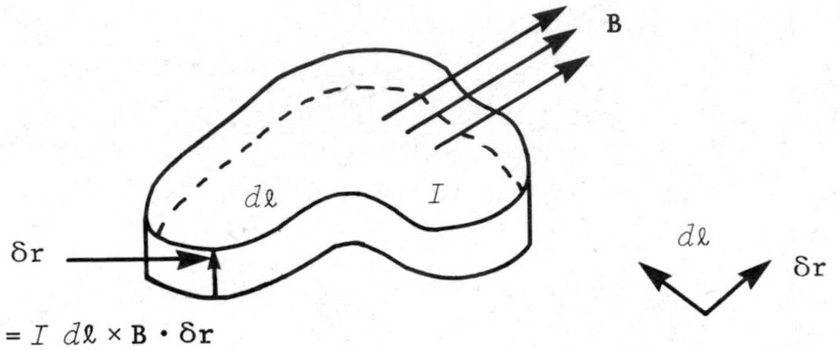

$$d\mathbf{F} = I\,d\boldsymbol{\ell} \times \mathbf{B}, \quad d\mathbf{F} \cdot \delta\mathbf{r} = I\,d\boldsymbol{\ell} \times \mathbf{B} \cdot \delta\mathbf{r}$$

$$dW = I \oint d\boldsymbol{\ell} \times \mathbf{B} \cdot \delta\mathbf{r}$$

$$= -I \oint \mathbf{B} \times d\boldsymbol{\ell} \cdot \delta\mathbf{r}$$

$$= -I \oint \mathbf{B} \cdot d\boldsymbol{\ell} \times \delta\mathbf{r}$$

$$= I \oint \mathbf{B} \cdot (\delta\mathbf{r} \times d\boldsymbol{\ell}), \quad \delta\mathbf{r} \times d\boldsymbol{\ell} = \mathbf{n}\,ds$$

so

$$dW = I \oint \mathbf{B} \cdot \mathbf{n}\,ds = I\,d\Phi$$

PROBLEM 7.2 The figure shows a cube. A uniform magnetic field is directed along the y axis. Find the magnitude and direction of the magnetic force acting on a charge q moving with a velocity $\mathbf{v}$ in each of the indicated directions.

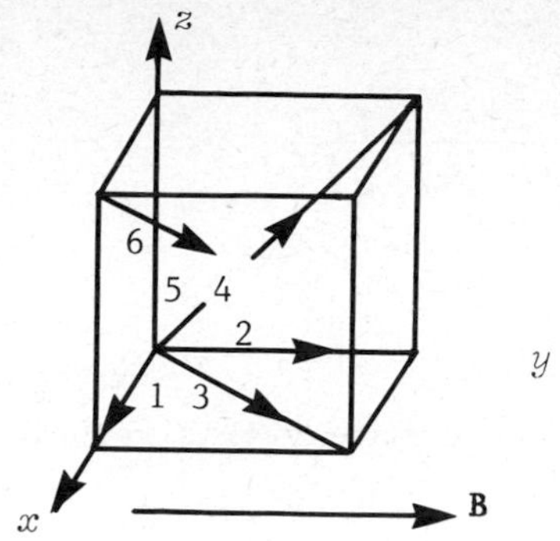

Solution:

$$\mathbf{F} = \frac{q}{c}\,\mathbf{v} \times \mathbf{B}, \quad \mathbf{B} = B\hat{\mathbf{y}}$$

Direction 1. $\mathbf{v} = v\hat{\mathbf{x}}, \quad \mathbf{F} = \frac{qvB}{c}\,(\hat{\mathbf{x}} \times \hat{\mathbf{y}}) - \frac{qvB}{c}\,\hat{\mathbf{z}}$

Direction 2. $\mathbf{v} = v\hat{\mathbf{y}}, \quad \mathbf{F} = 0$

Direction 3. $\mathbf{v} = v(\hat{\mathbf{x}} + \hat{\mathbf{y}})/\sqrt{2}, \quad \mathbf{F} = \frac{qvB}{\sqrt{2}\,c}\,\hat{\mathbf{z}}$

Direction 4. $\mathbf{v} = v(\hat{\mathbf{y}} + \hat{\mathbf{z}})/\sqrt{2}, \quad \mathbf{F} = -\frac{qvB}{\sqrt{2}\,c}\,\hat{\mathbf{x}}$

Direction 5. $\mathbf{v} = v\hat{\mathbf{z}}, \quad \mathbf{F} = -\frac{qvB}{c}\,\hat{\mathbf{x}}$

Direction 6. $\mathbf{v} = v(-\hat{\mathbf{x}} + \hat{\mathbf{y}} - \hat{\mathbf{z}})/\sqrt{3}, \quad \mathbf{F} - \frac{qvB}{\sqrt{3}\,c}\,(\hat{\mathbf{x}} - \hat{\mathbf{z}})$

PROBLEM 7.3 A magnetic field is set up in space by a circular loop L_1 of radius a carrying current I_1. A second loop L_2, whose radius b is much smaller than a, is located coaxially with L_1 so that the centers of the loops are at a distance $a/2$ apart. If L_2 carries current I_2 in the opposite clock sense to I_1, find the magnetic force between the loops.

Solution:

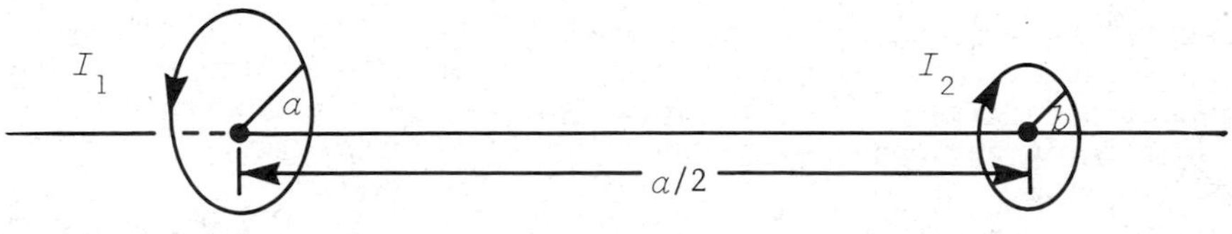

For a single loop B at a point on the axis is

$$\mathbf{B} = \frac{\mu_0 I_1 a^2}{2(a^2 + x^2)^{3/2}}\,\hat{\mathbf{i}}$$

and for this case $x = a/2$, so

$$\mathbf{B} = \frac{\mu_0 I_1 a^2}{2(5a^2/4)^{3/2}}\,\hat{\mathbf{i}} = \frac{\mu_0 I a^2}{2(5^{3/2}/8)a^3}\,\hat{\mathbf{i}}$$

$$\mathbf{B} = \frac{4\mu_0 I_1}{5^{3/2}\,a}\,\hat{\mathbf{i}}$$

Since $b \ll a$, the field at L_2 is approximately this value. The dipole moment of the loop L_2 is

$$\mathbf{m} = -I_2 \pi b^2 \hat{\mathbf{L}}.$$

The force on a dipole is

$$\mathbf{F} = \nabla(\mathbf{B} \cdot \mathbf{m})$$

$$\mathbf{B} \cdot \mathbf{m} = -\frac{\mu_0 I_1 a^2}{2(a^2+x^2)^{3/2}}\, I_2 \pi b^2 = -\frac{\mu_0 I_1 I_2 \pi a^2 b^2}{2(a^2+x^2)^{3/2}}$$

We will not consider the y and z component of $\mathbf{B}$ since $\mathbf{B} \cdot \mathbf{M}$ will involve only the x component of B because $\mathbf{M}$ only has an x component. So

$$\mathbf{F} = \hat{\mathbf{i}}\,\frac{\partial}{\partial x}\,(B_x m)$$

$$\frac{\partial}{\partial x}\,(a^2+x^2)^{-3/2} = -\frac{3}{2}\,(a^2+x^2)^{-5/2}\,2x = -\frac{3x}{(a^2+x^2)^{5/2}}$$

So

$$\mathbf{F} = \frac{\mu_0 I_1 I_2 \pi a^2 b^2}{2}\,\frac{3x}{(a^2+x^2)^{5/2}}\,\hat{\mathbf{i}}$$

At $x = a/2$,

$$\mathbf{F} = \hat{\mathbf{i}}\,\frac{\mu_0 I_1 I_2 \pi b^2 a^2}{2}\,3\,\frac{a/2}{a^5\,(5)^{5/2}/32}$$

$$\mathbf{F} = \hat{\mathbf{i}}\,\frac{8\pi\mu_0 I_1 I_2 b^2}{(5)^{5/2}\,a^2}$$

PROBLEM 7.4 Two square current loops of equal side length a and current I lie on a table, their centers being separated by a distance y which is very large compared with a. If the clock senses of the currents are opposite, find the force between them to the dipole approximation.

Solution:

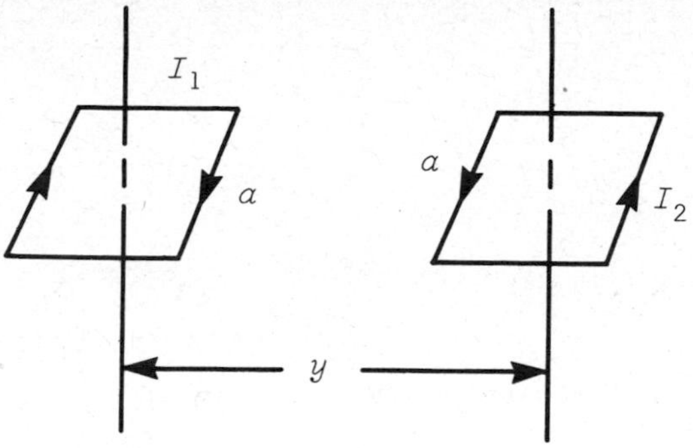

From Case 1, $\mathbf{M}_1 = -\hat{\mathbf{k}} I_1 a^2$ so to the dipole approximation at $\mathbf{r} = \hat{\mathbf{j}} y$

$$\mathbf{B} = \frac{\mu_0}{4\pi}\left(-\frac{-\hat{\mathbf{k}} I_1 a^2}{y^3} + 0\right) = \frac{\mu_0}{4\pi}\hat{\mathbf{k}}\frac{I_1 a^2}{y^3}$$

For Case 2, $\mathbf{M} = \hat{\mathbf{k}} I_2 a^2$, so

$$\mathbf{F} = \nabla(\mathbf{m}\cdot\mathbf{B}) = \nabla\left(\frac{\mu_0}{4\pi}\frac{I_1 I_2 a^4}{y^3}\right)$$

$$\mathbf{F} = \hat{\mathbf{j}}\frac{\partial}{\partial y}\left(\frac{\mu_0}{4\pi}\frac{I_1 I_2 a^4}{y^3}\right)$$

$$\mathbf{F} = -\hat{\mathbf{j}}\frac{3\mu_0}{4\pi}\frac{I_1 I_2 a^4}{y^4}$$

PROBLEM 7.5 A square loop of No. 10 copper wire is made with a length of 5 cm on a side. This wire has a resistance of 3.37 ohms and a mass of 46.8 kg/km. Suppose the loop is to be dropped between the poles of a magnet, so that its sides are, respectively, horizontal and vertical, and so that a uniform field of 1.5 Wb/m exists over its top side while the field at the bottom side is zero. When the loop falls steadily, what is

(a) The rate of fall and
(b) The current in the loop?

Solution:

(a)

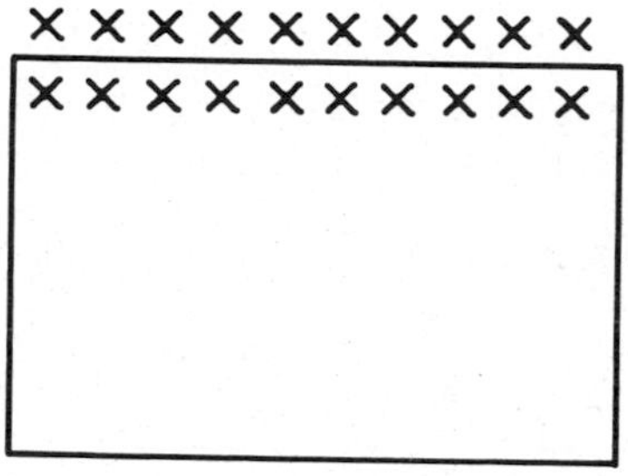

$$B = 1.5 \frac{\text{Wb}}{\text{m}^2}$$

$$\varepsilon = B\ell v = (1.5)(5 \times 10^{-2})v$$

$$I = \frac{\varepsilon}{R} = \frac{7.5 \times 10^{-2} v}{3.37} = 2.22 \times 10^{-2} \quad v$$

Now the magnetic force on the wire directed upward is

$$F = BI\ell$$

and this must balance the gravitational pull:

$$F_g = mg = (46.89/m)(0.2)g = (9.36)(9.8) \times 10^{-3} = 9.16 \times 10^{-2} \text{ N}$$

$$9.16 \times 10^{-2} = (1.5)(2.22 \times 10^{-2} \ v)(5 \times 10^{-2})$$

$$v = 0.55 \times 10^2 = 55 \text{ m/sec}$$

(b) $I = (2.22 \times 10^{-2})(55) = 1.22$ A

PROBLEM 7.6 Given two coaxial, circular loops of radius a, with a large separation x. One of them carries a steady current I and moves along the common axis with a velocity $v = dx/dt$. What is the emf induced in the other loop as a result of the changing flux from the moving loop? Use the dipole approximation.

Solution:

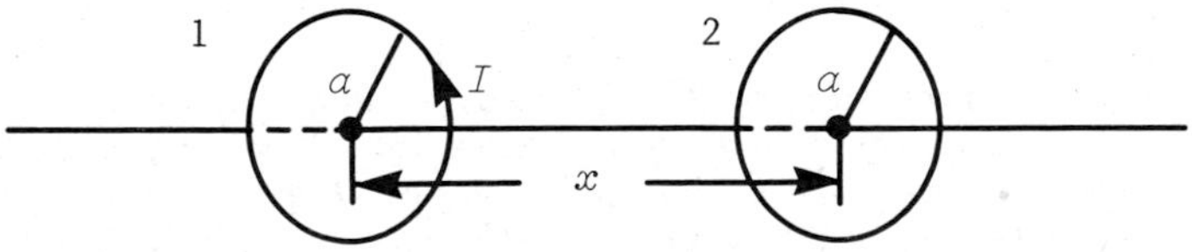

To the dipole approximation

$$\mathbf{B} = \frac{\mu_0}{4\pi}\left(-\frac{\mathbf{m}}{r^3} + \frac{3(\mathbf{m} \cdot \mathbf{r})\mathbf{r}}{r^5}\right)$$

At loop 2

$$\mathbf{B}_x = \frac{\mu_0}{4\pi}\left(-\frac{I\pi a^2}{x^3} + \frac{3I\pi a^2 x^2}{x^5}\right)$$

$$B_x = \frac{\mu_0 I}{2}\frac{a^2}{x^3}$$

Now the flux through loop 2 is

$$\phi = \frac{\mu_0 I}{2} \frac{a^2}{x^3} \pi a^2$$

and the emf is

$$\varepsilon = \frac{-d\phi}{dt} = \frac{-d\phi}{dx}\frac{dx}{dt} = -v \frac{d\phi}{dx}$$

$$\varepsilon = v \frac{\mu_0 I a^4 \pi}{2} \frac{3}{x^4}$$

$$\varepsilon = \frac{3\mu_0 \pi I a^4 v}{2x^4}$$

PROBLEM 7.7 The figure shows a uniform magnetic field **B** confined to a cylindrical volume of radius r. The field **B** is decreasing in magnitude at a constant rate of 100 G/sec. What is the instantaneous acceleration (direction and magnitude) experienced by an electron placed at P_1, at P_2, and at P_3? Assume a is 5.0 cm.

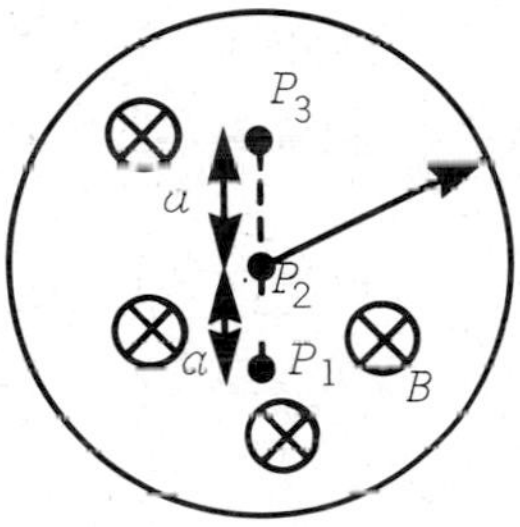

Solution:

$$\oint \mathbf{E} \cdot d\ell = -\frac{d\phi}{dt} = -\frac{dB}{dt}\pi r^2$$

$$2\pi r E = -\frac{dB}{dt}\pi r^2$$

$$E = -\tfrac{1}{2}\frac{dB}{dt} r, \quad \text{at } P_2,\ E = 0,\ F = 0$$

At $r = a$, $E = \tfrac{1}{2}(0.01)(5 \times 10^{-2}) = 2.5 \times 10^{-4}$ N/C.

At P_1, $a = \dfrac{(1.6 \times 10^{-19})(2.5 \times 10^{-4})}{9.1 \times 10^{-31}} = 4.4 \times 10^7$ m/sec^2, counterclockwise

At P_3, $a = 4.4 \times 10^7$ m/sec^2, counterclockwise

PROBLEM 7.8 A cylindrical hole of radius a is bored parallel to the axis of a long cylindrical conductor of radius b which carries a uniformly distributed current of density j. The density between the center of the conductor and the center of the hole is x_0. Find the magnetic field in the hole.

Solution: The field inside a current carrying cylinder with constant current density can be found using Ampere's Law:

$$\oint_c \mathbf{B} \cdot d\ell = \mu \int_a \mathbf{j} \cdot \hat{\mathbf{n}}\, da$$

$$2\pi r B = \mu j \pi r^2$$

$$\mathbf{B} = \tfrac{1}{2}\, \mu j r \hat{\theta}, \text{ for current out of paper}$$

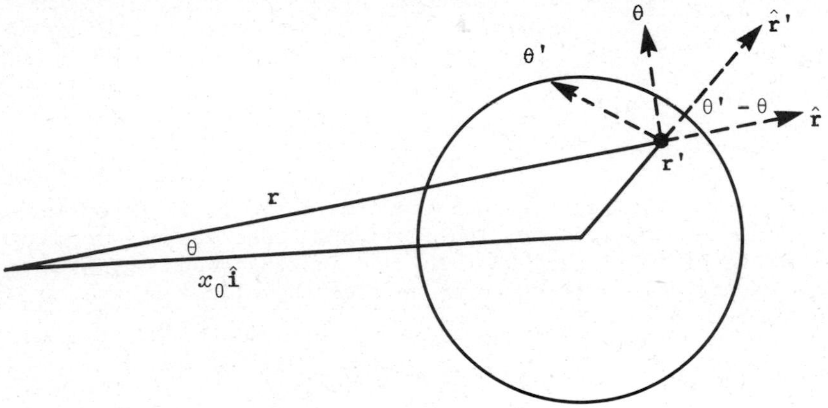

$$r' = (r^2 + x_0^2 - 2rx_0 \cos\theta)^{\frac{1}{2}}$$

$$\hat{\theta}' = \hat{\theta} \cos(\theta' - \theta) - \hat{\mathbf{r}} \sin(\theta' - \theta)$$

$$\hat{\theta}' = \hat{\theta}(\cos\theta' \cos\theta + \sin\theta' \sin\theta) - \hat{\mathbf{r}}(\sin\theta' \cos\theta - \cos\theta' \sin\theta)$$

From the law of cosines,

$$r^2 = x_0^2 + (r^2 + x_0^2 - 2rx_0 \cos\theta) - 2x_0(r^2 - x_0^2 - 2rx_0 \cos\theta)^{\frac{1}{2}} \cos(\pi - \theta')$$

Using the fact that $\cos(\pi - \theta') = -\cos\theta'$, we have

$$\cos\theta' = \frac{2rx_0 \cos\theta - 2x_0^2}{2x_0(r^2 - x_0^2 - 2rx_0 \cos\theta)^{\frac{1}{2}}}$$

$$\cos\theta' = \frac{r\cos\theta - x_0}{(r^2 - x_0^2 - 2rx_0 \cos\theta)^{\frac{1}{2}}}$$

From the law of sines,

$$\frac{\sin\theta}{(r^2 + x_0^2 - 2rx_0 \cos\theta)^{\frac{1}{2}}} = \frac{\sin(\pi - \theta')}{r}$$

$$\sin\theta' = \frac{r\sin\theta}{(r^2 + x_0^2 - 2rx_0 \cos\theta)^{\frac{1}{2}}}$$

$$\hat{\theta}' = \hat{\theta}\left(\frac{r\cos^2\theta - x_0\cos\theta}{(r^2 - x_0^2 - 2rx_0\cos\theta)^{\frac{1}{2}}} + \frac{r\sin^2\theta}{(r^2 - x_0^2 - 2rx_0\cos\theta)^{\frac{1}{2}}}\right)$$

$$-\hat{\mathbf{r}}\left(\frac{r\sin\theta\cos\theta}{(r^2 + x_0^2 - 2rx_0\cos\theta)^{\frac{1}{2}}} - \frac{r\cos\theta\sin\theta - x_0\sin\theta}{(r^2 - x_0^2 - 2rx_0\cos\theta)^{\frac{1}{2}}}\right)$$

$$\hat{\theta}' = (r^2 + x_0^2 - 2rx_0\cos\theta)^{-\frac{1}{2}}[\hat{\theta}(r - x_0\cos\theta) - \hat{\mathbf{r}}x_0\sin\theta]$$

$$r'\hat{\theta}' = \hat{\theta}(r - x_0\cos\theta) - \hat{\mathbf{r}}x_0\sin\theta$$

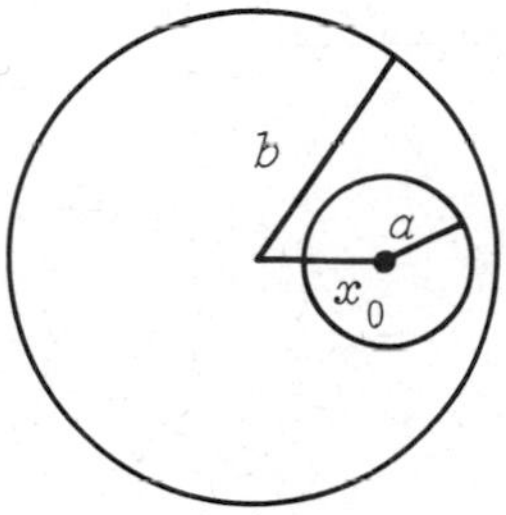

The field inside the cavity, $|\mathbf{r} - x_0\hat{\mathbf{i}}| \leq a$, is

$$\mathbf{B} = \tfrac{1}{2}\,\mu j r\hat{\theta} - \tfrac{1}{2}\,\mu j r'\hat{\theta}'$$

That is, the field is a linear superposition of the field due to the cylinder of radius b with a uniform current density $j\hat{\mathbf{k}}$ and the cylinder of radius a with its center at $x_0\hat{\mathbf{i}}$ and a uniform current density $-j\hat{\mathbf{k}}$:

$$\mathbf{B} = \tfrac{1}{2}\,\mu j r\hat{\theta} - \tfrac{1}{2}\,\mu j[\hat{\theta}(r - x_0\cos\theta) - \hat{\mathbf{r}}x_0\sin\theta]$$

$$\mathbf{B} = \tfrac{1}{2}\,\mu_0 j x_0(\hat{\theta}\cos\theta + \hat{\mathbf{r}}\sin\theta)$$

$$\mathbf{B} = \tfrac{1}{2}\,\mu_0 j x_0\hat{\mathbf{y}}$$

PROBLEM 7.9 Two spinning electrons are separated by a distance r. Assuming the spins are parallel, determine the ratio of the magnetic interaction force between the dipoles to the electric force between the electrons. Show that the magnetic force can be attractice or repulsive depending on the spin orientation. Compare the total interaction energy (magnetic plus electrostatic) of two electrons with parallel spins to that of two electrons with antiparallel spins. Can you see why electrons tend to pair off as they do so that the total spin of the system cancels?

Solution: For a spinning electron

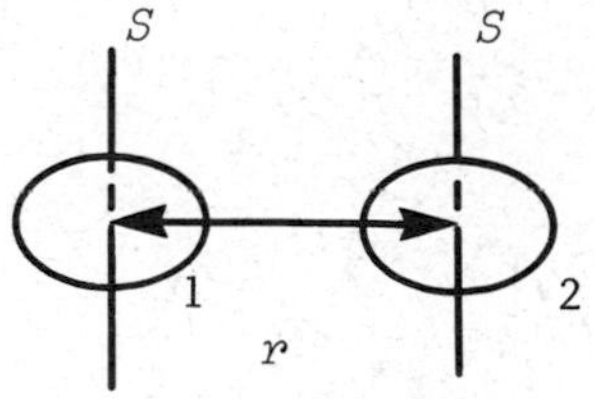

$$m = -\frac{eh}{4\pi m}$$

At 2,

$$\mathbf{B} = \frac{\mu_0}{4\pi}\left(-\frac{\mathbf{m}_1}{r^3} + \frac{3\mathbf{m}_1\cdot\mathbf{r}}{r^5}\,\mathbf{r}\right)$$

For electron 1,

$$\mathbf{m}_1 = -\hat{\mathbf{k}}\ \frac{eh}{4\pi m}$$

At 2, $\mathbf{r} = \hat{\mathbf{i}}x$, so

$$\mathbf{B} = \frac{\mu_0}{4\pi}\left(\hat{\mathbf{k}}\ \frac{eh}{4\pi m}\ \frac{1}{x^3}\right)$$

$$\mathbf{F}_m = \nabla(\mathbf{B}\cdot\mathbf{m}) = \hat{\mathbf{i}}\ \frac{\partial}{\partial x}\left(-\frac{\mu_0}{4\pi}\ \frac{e^2h^2}{(4\pi m)^2}\ \frac{1}{x^3}\right)$$

$$\mathbf{F}_m = +\frac{\mu_0 e^2h^2}{(4\pi)^3m^2}\ \frac{3}{x^4}\ \hat{\mathbf{i}}$$

$$\mathbf{F}_E = 9\times 10^9(e^2/x^2)\hat{\mathbf{i}}$$

$$\frac{F_E}{F_m} = \frac{9\times 10^9\ e^2/r^2}{[(\mu_0 e^2h^2)/((4\pi)^3m^2)](3/x^4)} = \frac{3\times 10^9x^2m^2(4\pi)^3}{\mu_0 h^2}$$

$$\frac{F_E}{F_m} = \frac{3\times 10^9(9.1\times 10^{-31})^2}{4\pi\times 10^{-7}(6.63\times 10^{-34})^2}(4\pi)^3x^2 = 8.9\times 10^{24}x^2$$

In order for $F_E = F_m$, the electrons must be on the order of 3.3×10^{-13} m apart:

$$U_m = -\mathbf{m}\cdot\mathbf{B}$$

For parallel spins,

$$U_m = \frac{\mu_0}{4\pi}\left(\frac{eh}{4\pi m}\right)^2\frac{1}{x^3}$$

For antiparallel spins,

$$U_m = \frac{\mu_0}{4\pi}\left(\frac{eh}{4\pi m}\right)^2\frac{1}{x^3}$$

For electronic interaction,

$$U_E = \frac{1}{4\pi\varepsilon_0}\ \frac{e^2}{x}$$

So

$$\frac{U_P}{U_A} = \frac{\frac{1}{4\pi\varepsilon_0}\frac{e^2}{x} + \frac{\mu_0}{4\pi}\frac{e^2h^2}{(4\pi m)^2}\frac{1}{x^3}}{\frac{1}{4\pi\varepsilon_0}\frac{e^2}{x} - \frac{\mu_0}{4\pi}\frac{e^2h^2}{(4\pi m)^2}\frac{1}{x^3}}$$

$$\frac{U_P}{U_A} = \frac{1 + \frac{\mu_0\varepsilon_0 h^2(1/x^2)}{(4\pi)^2m^2}}{1 - \frac{\mu_0\varepsilon_0 h^2}{(4\pi)^2m^2}\cdot\frac{1}{x^2}}$$

$$\frac{U_P}{U_A} = \frac{1 + \frac{h^2}{(4\pi)^2m^2c^2}\frac{1}{x^2}}{1 - \frac{h^2}{4\pi^2m^2c^2}\frac{1}{x^2}}$$

PROBLEM 7.10 Find the magnetic field at the center of square wire loop of side length 10 cm, carrying 10 A.

Solution:

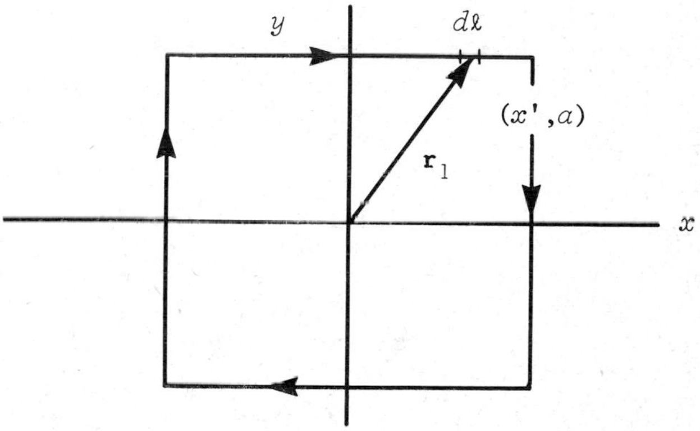

$$d\mathbf{B} = \frac{\mu_0 I}{4\pi}\frac{d\boldsymbol{\ell}\times(\mathbf{r}-\mathbf{r}_1)}{|\mathbf{r}-\mathbf{r}_1|^3}$$

For the top wire

$$\mathbf{r} = 0,\ \mathbf{r}' = \hat{\mathbf{i}}x_1' + \hat{\mathbf{j}}a$$

$$\mathbf{B} = \frac{\mu_0 I}{4\pi}\int_{-a}^{a}\frac{\hat{\mathbf{i}}dx_1\times(-\hat{\mathbf{i}}x_1 - \hat{\mathbf{j}}a)}{(x_1^2+a^2)^{3/2}}$$

$$\mathbf{B} = \frac{\mu_0 I}{4\pi}(-\hat{\mathbf{k}})\int_{-a}^{a}\frac{a\,dx_1}{(x_1^2+a^2)^{3/2}}$$

$$\mathbf{B} = -\hat{\mathbf{k}}\frac{\mu_0 I}{4\pi}\frac{ax_1}{a^2(x_1^2+a^2)^{\frac{1}{2}}}\Bigg|_{-a}^{a} = -\hat{\mathbf{k}}\frac{\mu_0 I}{4\pi}\frac{2a^2}{a^2(2a^2)^{\frac{1}{2}}} = -\hat{\mathbf{k}}\frac{\mu_0 I}{4\pi}\frac{2}{\sqrt{2}a}$$

For the four sides of the square, we have to multiply this by 4.

$$B_T = \frac{\mu_0 I}{\pi}\frac{2}{\sqrt{2}a}$$

$$B_T = \frac{(4\pi\times 10^{-7})(10^1)}{\pi}\frac{\sqrt{2}}{5\times 10^{-2}} = 1.13\times 10^{-4}\ \text{Wb/m}^2$$

PROBLEM 7.11 A long wire has a semicircular loop of radius r as shown in the figure. A current i is flowing. Find the magnetic field at the center of curvature of the loop.

Solution: The only contribution to B is from the circular loop since the cross product is zero for the straight line portion of the wire.

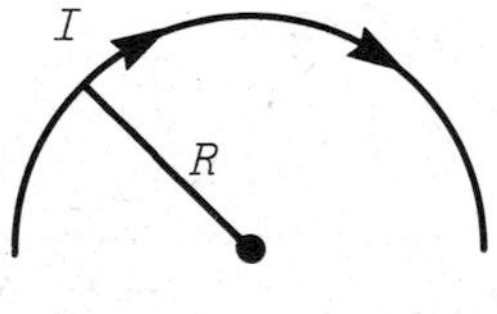

Here

$$d\mathbf{B} = \frac{\mu_0 I}{4\pi}\frac{d\ell\times\hat{\mathbf{r}}}{r^2} \Rightarrow dB = \frac{\mu_0 I}{4\pi}\frac{d\ell}{R^2}$$

$$B = \frac{\mu_0 I}{4\pi}\frac{\pi R}{R^2} = \frac{\mu_0 I}{4R}$$

PROBLEM 7.12 The Helmholtz arrangement of two coils provides a large region of uniform field. Two similar coils carrying the same current are placed on the same axis, as shown in the figure, separated by a distance equal to the coil radius. Show that at point P on the axis this arrangement gives both dB/dx and d^2B/dx^2 equal to zero.

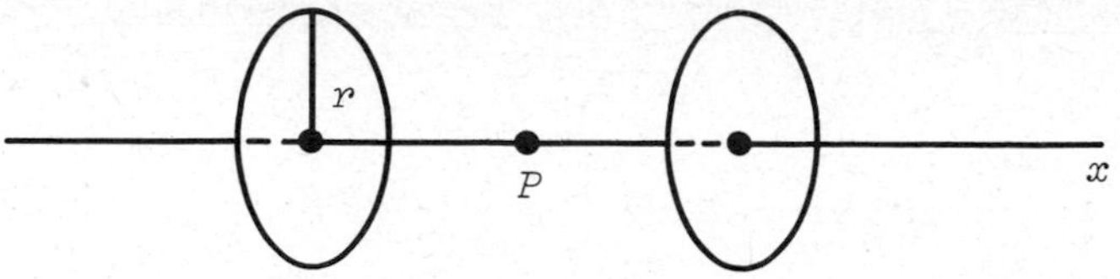

Solution: For a single turn

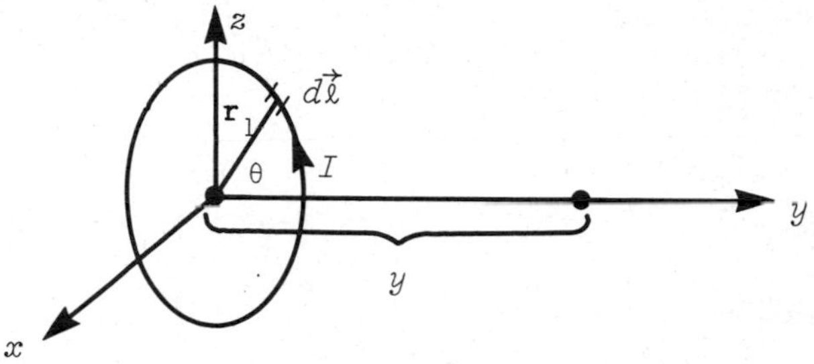

$$d\boldsymbol{\ell} = \hat{\mathbf{i}}dx + \hat{\mathbf{k}}dz$$

$$\mathbf{r} = \hat{\mathbf{j}}y$$

$$\mathbf{r}_1 = \hat{\mathbf{i}}x + \hat{\mathbf{k}}z$$

$$|r_1| = R = (x^2 + z^2)^{\frac{1}{2}}$$

$$R^2 = x^2 + z^2$$

$$\mathbf{r} - \mathbf{r}_1 = \hat{\mathbf{j}}y - \hat{\mathbf{i}}x - \hat{\mathbf{k}}z$$

$$|\mathbf{r} - \mathbf{r}_1| = (x^2 + y^2 + z^2)^{\frac{1}{2}} = (R^2 + y^2)^{\frac{1}{2}}$$

$$\mathbf{B} = \frac{\mu_0 I}{4\pi} \int_{\substack{\text{over} \\ \text{dimensions of} \\ \text{current flux}}} \frac{d\boldsymbol{\ell} \times (\mathbf{r} - \mathbf{r}_1)}{|\mathbf{r} - \mathbf{r}_1|^3} = \frac{\mu_0 I}{4\pi} \int \frac{(\hat{\mathbf{i}}dx + \hat{\mathbf{k}}dz) \times (\hat{\mathbf{j}}y - \hat{\mathbf{i}}x - \hat{\mathbf{k}}z)}{(R^2 + y^2)^{3/2}}$$

$$\mathbf{B} = \frac{\mu_0 I}{4\pi} \int \frac{\hat{\mathbf{k}}ydx + \hat{\mathbf{j}}zdx - \hat{\mathbf{i}}ydz - \hat{\mathbf{j}}xdz}{(R^2 + y^2)^{3/2}}$$

$$x = R\cos\theta \text{ and } z = R\sin\theta$$

$$dx = -R\sin\theta\, d\theta \quad \text{and } dz = R\cos\theta\, d\theta$$

$$\mathbf{B} = \frac{\mu_0 I}{4\pi} \int \frac{-\hat{\mathbf{i}}ydz + \hat{\mathbf{j}}(zdx - xdz) + \hat{\mathbf{k}}ydx}{(R^2 + y^2)^{3/2}}$$

$$\mathbf{B} = \frac{\mu_0 I}{4\pi} \int \frac{-\hat{\mathbf{i}}(yR\cos\theta\; d\theta) + \hat{\mathbf{j}}(-R^2\sin^2\theta\;\, d\theta - R^2\cos^2\theta\;\, d\theta) + \hat{\mathbf{k}}y(-R\sin\theta\;\, d\theta)}{(R^2 + y^2)^{3/2}}$$

$$\mathbf{B} = \frac{\mu_0 I}{4\pi} \int \frac{-\hat{\mathbf{i}} yR\cos\theta\, d\theta - \hat{\mathbf{j}} R^2(\sin^2\theta\, d\theta + \cos^2\theta\, d\theta) - \hat{\mathbf{k}} yR\sin\theta\, d\theta}{(R^2+y^2)^{3/2}}$$

$$= \frac{-\mu_0 I}{(R^2+y^2)^{3/2}\, 4\pi} \left(\int_0^{-2\pi} \hat{\mathbf{i}} yR\cos\theta\, d\theta + \hat{\mathbf{j}} \int_0^{-2\pi} R^2\, d\theta + \hat{\mathbf{k}} \int_0^{-2\pi} yR\sin\theta\, d\theta \right)$$

(where the first and third terms in the parentheses equal zero)

$$= \frac{-\mu_0 I^2}{4\pi (R^2+y^2)^{3/2}} [+\hat{\mathbf{j}} R^2(-2\pi)]$$

$$\mathbf{B} = \hat{\mathbf{j}} \frac{\mu_0 I R^2}{2(R^2+y^2)^{3/2}}$$

For this case, at P,

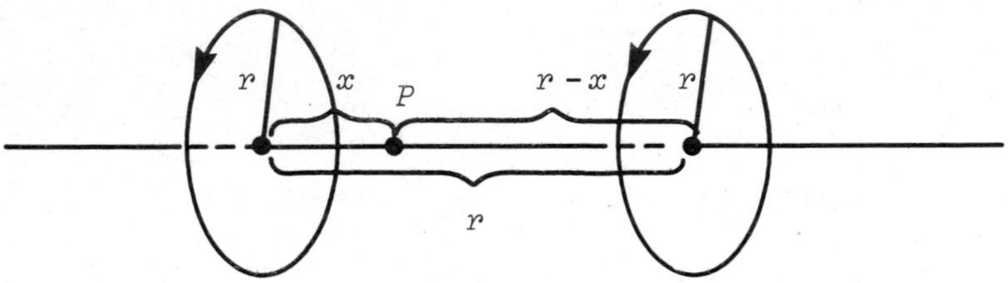

$$B = \frac{\mu_0 I r^2}{2(r^2+x^2)^{3/2}} + \frac{\mu_0 I r^2}{2[r^2+(r-x)^2]^{3/2}}$$

$$B = \frac{\mu_0 I r^2}{2} [(r^2+x^2)^{-3/2} + [r^2+(r-x)^2]^{-3/2}]$$

$$B = \frac{\mu_0 I r^2}{2} [(r^2+x^2)^{-3/2} + (2r^2-2rx+x^2)^{-3/2}]$$

$$\frac{dB}{dx} = \frac{\mu_0 I r^2}{2} \left\{ \left(-\frac{3}{2}\right)(r^2+x^2)^{-5/2}(2x) - \frac{3}{2}(2r^2-2rx+x^2)^{-5/2}(2)(-r+x) \right\}$$

$$\frac{dB}{dx} = \frac{\mu_0 I r^2}{2} \{ -3x(r^2+x^2)^{-5/2} + (-3x+3r)[r^2+(r-x)^2]^{-5/2} \}$$

$$(r^2+x^2)^{-5/2} = \left(\frac{1}{\sqrt{r^2+x^2}} \right)^5$$

At $x = \frac{r}{2}$,

$$(r^2 + x^2)^{-5/2} = \left(\frac{1}{\sqrt{r^2 + r^2/4}}\right)^5 = \left(\frac{2}{r\sqrt{5}}\right)^5$$

At $x = \frac{r}{2}$,

$$[r^2 + (r - x)^2]^{-5/2} = \left(\frac{1}{\sqrt{r^2 + r^2/4}}\right)^5 = \left(\frac{2}{r\sqrt{5}}\right)^5$$

$$\frac{dB}{dx} = \frac{\mu_0 I r^2}{2}\left[-\frac{3r}{2}\left(\frac{2}{r\sqrt{5}}\right)^5 + \left(-\frac{3r}{2} + 3r\right)\left(\frac{2}{r\sqrt{5}}\right)^5\right]$$

$$\frac{dB}{dx} = \frac{\mu_0 I r^2}{2}\left[-\frac{3r}{2}\left(\frac{2}{r\sqrt{5}}\right)^5 + \frac{3r}{2}\left(\frac{2}{r\sqrt{5}}\right)^5\right]$$

$$\frac{dB}{dx} = 0$$

Now,

$$\frac{d^2B}{dx^2} = \frac{\mu_0 I r^2}{2}\left[(-3x)(r^2 + x^2)^{-5/2} + (-3x + 3r)(2r^2 - 2rx + x^2)^{-5/2}\right]$$

$$\frac{d^2B}{dx^2} = \frac{\mu_0 I r^2}{2}\left\{\left[(-3x)\left(-\frac{5}{2}\right)(r^2 + x^2)^{-7/2}(2x) + (r^2 + x^2)^{-5/2}(-3)\right.\right.$$

$$\left. + (-3x + 3r)\left(-\frac{5}{2}\right)(2r^2 - 2rx + x^2)^{-7/2}(-r + x)(2)\right]$$

$$\left. + (2r^2 - 2rx + x^2)^{-5/2}(-3)\right\}$$

$$\frac{d^2B}{dx^2} = \frac{\mu_0 I r^2}{2}\left[15x^2(r^2 + x^2)^{-7/2} - 3(r^2 + x^2)^{-5/2} - 3(2r^2 - 2rx + x^2)^{-5/2}\right.$$

$$\left. + (15x - 15r)(x - r)(2r^2 - 2rx + x^2)^{-7/2}\right]$$

$$= \frac{\mu_0 I r^2}{2}\left[15x^2(r^2 + x^2)^{-7/2} - 3(r^2 + x^2)^{-5/2} - 3(2r^2 - 2rx + x^2)^{-5/2}\right.$$

$$+15(x-r)^2(2r^2-2rx+x^2)^{-7/2}]$$

At $x=\frac{r}{2}$,

$$\frac{d^2B}{dx^2}=\frac{\mu_0 Ir^2}{2}\left[15\,\frac{r^2}{4}\left(r^2+\frac{r^2}{4}\right)^{-7/2}-3\left(r^2+\frac{r^2}{4}\right)^{-5/2}-3\left(2r^2-\frac{2r^2}{2}+\frac{r^2}{4}\right)^{-5/2}\right.$$

$$\left.+15\left(\frac{r}{2}-r\right)^2\left(2r^2-\frac{2r^2}{2}+\frac{r^2}{4}\right)^{-7/2}\right]$$

$$\frac{d^2B}{dx^2}=\frac{\mu_0 Ir^2}{2}\left[\frac{15r^2}{4}\left(r^2+\frac{r^2}{4}\right)^{-7/2}-3\left(r^2+\frac{r^2}{4}\right)^{-5/2}-3\left(r^2+\frac{r^2}{4}\right)^{-5/2}\right.$$

$$\left.+15\left(\frac{r}{2}-r\right)^2\left(r^2+\frac{r^2}{4}\right)^{-7/2}\right]$$

$$=\frac{\mu_0 Ir^2}{2}\left[\frac{15r^2}{4}\left(\frac{5r^2}{4}\right)^{-7/2}-3\left(\frac{5r^2}{4}\right)^{-5/2}-3\left(\frac{5r^2}{4}\right)^{-5/2}+15\left(-\frac{r}{2}\right)^2\left(\frac{5r^2}{4}\right)^{-7/2}\right]$$

$$=\frac{\mu_0 Ir^2}{2}\left[\frac{15r^2}{4}\left(\frac{5r^2}{4}\right)^{-7/2}+\frac{15r^2}{4}\left(\frac{5r^2}{4}\right)^{-7/2}-6\left(\frac{5r^2}{4}\right)^{-5/2}\right]$$

$$=\frac{\mu_0 Ir^2}{2}\left[\left(\frac{15r^2}{4}+\frac{15r^2}{4}\right)\left(\frac{5r^2}{4}\right)^{-7/2}-6\left(\frac{5r^2}{4}\right)^{-5/2}\right]$$

$$=\frac{\mu_0 Ir^2}{2}\left[\frac{15r^2}{2}\left(\frac{5r^2}{4}\right)^{-7/2}-6\left(\frac{5r^2}{4}\right)^{-5/2}\right]$$

$$=\frac{\mu_0 Ir^2}{2}\left[\frac{15r^2}{2}\left(\frac{2^7}{r^7(5)^{7/2}}\right)-6\left(\frac{2^5}{r^5(5)^{5/2}}\right)\right]$$

$$=\frac{\mu_0 Ir^2}{2r^5}\left[\frac{15(2)^6}{5^{7/2}}-\frac{6(25)}{5^{5/2}}\right]=\frac{\mu_0 Ir^2}{2r^5}\quad(3.43-3.43)$$

$$\frac{d^2B}{dx^2}=0$$

What this means is that the magnetic field is very uniform near this point.

PROBLEM 7.13 Consider the current loop of width a and length b shown in the figure.

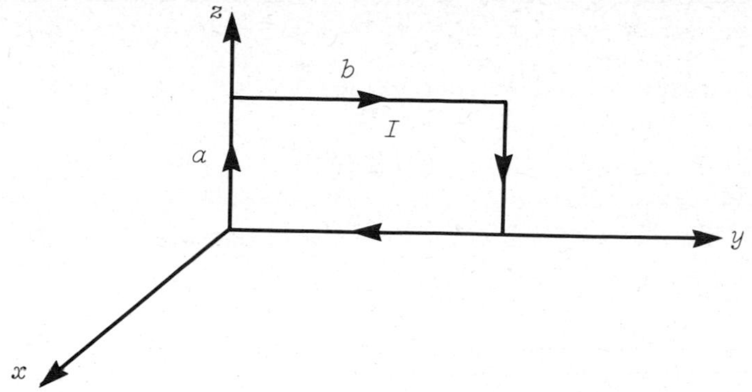

(a) Determine an expression for the vector potential that is valid anywhere.

(b) Determine an exact expression for the x component of **B** that is valid anywhere.

Solution:

(a)

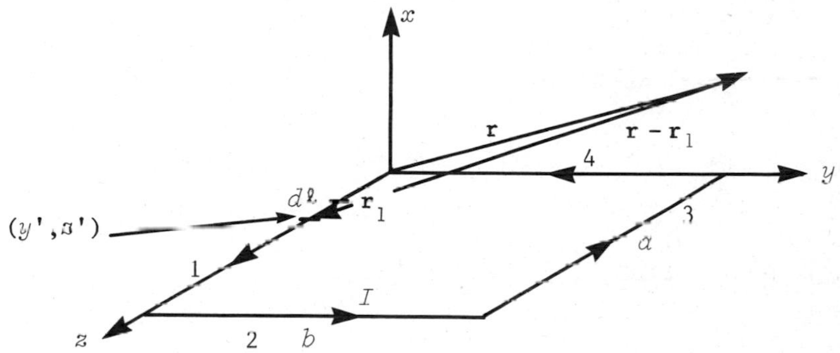

$$d\boldsymbol{\ell} = \hat{\mathbf{j}}dy + \hat{\mathbf{k}}dz$$

$$\mathbf{r}_1 = \hat{\mathbf{j}}y' + \hat{\mathbf{k}}z'$$

$$\mathbf{r} = \hat{\mathbf{i}}x + \hat{\mathbf{j}}y + \hat{\mathbf{k}}z$$

$$\mathbf{r} - \mathbf{r}_1 = \hat{\mathbf{i}}x + \hat{\mathbf{j}}(y - y') + \hat{\mathbf{k}}(z - z')$$

$$|\mathbf{r} - \mathbf{r}_1| = [(x)^2 + (y - y')^2 + (z - z')^2]^{\frac{1}{2}}$$

$$\mathbf{A} = \frac{\mu_0}{4\pi}\int \frac{I d\boldsymbol{\ell}}{r}$$

$$\mathbf{A} = \frac{\mu_0 I}{4\pi}\left(\int_{z'=0}^{z'=a} \frac{\hat{\mathbf{k}}\,dz'}{[x^2 + (y - y')^2 + (z - z')^2]^{\frac{1}{2}}} + \int_{y'=0}^{y'=b} \frac{\hat{\mathbf{j}}\,dy'}{[x^2 + (y - y')^2 + (z - z')^2]^{\frac{1}{2}}}\right.$$

$$+\int_{z'=a}^{z'=0} \frac{\hat{\mathbf{k}}\,dz'}{[x^2+(y-y')^2+(z-z')^2]^{\frac{1}{2}}} + \int_{y'=b}^{y'=0} \frac{\hat{\mathbf{j}}\,dy'}{[x^2+(y-y')^2+(z-z')^2]^{\frac{1}{2}}}\Bigg\}$$

Note that at this point one may insert values for y' and z' from the diagram into the above equations.

$$\mathbf{A} = \frac{\mu_0 I}{4\pi}\Bigg\{\int_{z'=0}^{z'=a} \frac{\hat{\mathbf{k}}\,dz'}{[x^2+y^2+(z-z')^2]^{\frac{1}{2}}} + \int_{y'=0}^{y'=b} \frac{\hat{\mathbf{j}}\,dy'}{[x^2+(y-y')^2+(z-a)^2]^{\frac{1}{2}}}$$

$$+\int_{z'=a}^{z'=0} \frac{\hat{\mathbf{k}}\,dz'}{[x^2+(y-b)^2+(z-z')^2]^{\frac{1}{2}}} + \int_{y'=b}^{y'=0} \frac{\hat{\mathbf{j}}\,dy'}{[x^2+(y-y')^2+z^2]^{\frac{1}{2}}}\Bigg\}$$

$$\mathbf{A} = \frac{\mu_0 I}{4\pi}\{-\hat{\mathbf{k}}\log[(z-z')+[(z-z')^2+y^2+z^2]^{\frac{1}{2}}]\}\Big|_0^a$$

$$-\hat{\mathbf{j}}\log\{(y-y')+[(y-y')^2+(z-a)^2+x^2]^{\frac{1}{2}}\}\Big|_0^b$$

$$-\hat{\mathbf{k}}\log\{(z-z')+[(z-z')^2+(y-b)^2+x^2]^{\frac{1}{2}}\}\Big|_a^0$$

$$-\hat{\mathbf{j}}\log\{(y-y')+[x^2+(y-y')^2+z^2]^{\frac{1}{2}}\}\Big|_b^0$$

$$\mathbf{A} = \frac{\mu_0 I}{4\pi}\Bigg[-\hat{\mathbf{k}}\log\left(\frac{(z-a)+[(z-a)^2+y^2+z^2]^{\frac{1}{2}}}{z+[z^2+y^2+x^2]^{\frac{1}{2}}}\right)$$

$$-\hat{\mathbf{j}}\log\left(\frac{(y-b)+[(y-b)^2+(z-a)^2+x^2]^{\frac{1}{2}}}{y^2+[y^2+(z-a)^2+x^2]^{\frac{1}{2}}}\right)$$

$$-\hat{\mathbf{k}}\log\left(\frac{z+[z^2+(y-b)^2+x^2]^{\frac{1}{2}}}{(z-a)+[(z-a)^2+(y-b)^2+x^2]^{\frac{1}{2}}}\right)$$

$$-\hat{\mathbf{j}}\log\left(\frac{y+(x^2+y^2+z^2)^{\frac{1}{2}}}{(y-b)+[x^2+(y-b)^2+z^2]^{\frac{1}{2}}}\right)$$

$$\mathbf{B} = \nabla\times\mathbf{A} = \begin{vmatrix} \hat{\mathbf{i}} & \hat{\mathbf{j}} & \hat{\mathbf{k}} \\ \dfrac{\partial}{\partial x} & \dfrac{\partial}{\partial y} & \dfrac{\partial}{\partial z} \\ A_x & A_y & 0 \end{vmatrix}$$

$$B_x = \hat{\mathbf{i}}\left(\frac{\partial A_y}{\partial z} - \frac{\partial A_z}{\partial y}\right)$$

$$A_y = +\frac{\mu_0 I}{4\pi}\left[\int_0^b [x^2 + (y-y')^2 + (z-a)^2]^{-\frac{1}{2}}\, dy' + \int_b^0 \frac{dy'}{[x^2 + (y-y')^2 + z^2]^{\frac{1}{2}}}\right]$$

$$B_x = +\frac{\hat{\mathbf{i}}\mu_0 I}{4\pi}\left[\int_0^b (+\tfrac{1}{2})[x^2 + (y-y')^2 + (z-a)^2]^{-3/2}\, 2z\, dy'\right.$$

$$\left. + \int_b^0 (+\tfrac{1}{2})[x^2 + (y-y')^2 + z^2]^{-3/2}\, (2z) dy'\right]$$

$$+\frac{\hat{\mathbf{k}}\mu_0 I}{4\pi}\left[\int_0^a (+\tfrac{1}{2})[x^2 + y^2 + (z-z')^2]^{-3/2}\, (2y) dz'\right.$$

$$\left. + \int_a^0 (+\tfrac{1}{2})[x^2 + (y-b)^2 + (z-z')^2]^{-3/2}\, 2y\, dz'\right]$$

$$B_x = \frac{\hat{\mathbf{i}}\mu_0 I}{4\pi}\left(\frac{-(y-y')}{[(z-a)^2 + x^2][(z-a)^2 + (y-y')^2 + z^2]^{\frac{1}{2}}}\right|_0^b$$

$$\left. - \frac{y-y_1}{[(x^2+z^2)][z^2 + (y-y')^2 + x^2]^{\frac{1}{2}}}\right|_b^0\Bigg) + \frac{\hat{\mathbf{k}}\mu_0 I}{4\pi}\left(\frac{-(z-z')}{(x^2+y^2)[x^2+y^2+(z-z')^2]^{\frac{1}{2}}}\right|_0^a$$

$$\left. - \frac{z-z'}{[x^2 + (y-b)^2][x^2 + (y-b)^2 + (z-z')^2]^{\frac{1}{2}}}\right|_a^0\Bigg)$$

$$B_x = \frac{\hat{\mathbf{i}}\mu_0 I}{4\pi}\left(\frac{-(y-b)}{[(z-a)^2 + x^2][(z-a)^2 + (y-b)^2 + z^2]^{\frac{1}{2}}}\right.$$

$$+ \frac{y}{[(z-a)^2 + x^2][(z-a)^2 + y^2 + z^2]^{\frac{1}{2}}} - \frac{y}{[(x^2+z^2)(z^2+y^2+x^2)]^{\frac{1}{2}}}$$

$$\left. + \frac{y-b}{[(x^2+z^2)[z^2 + (y-b)^2 + x^2]^{\frac{1}{2}}}\right) + \frac{\hat{\mathbf{k}}\mu_0 I}{4\pi}\left(\frac{-(z-a)}{(x^2+y^2)[x^2+y^2+(z-a)^2]^{\frac{1}{2}}}\right.$$

$$+ \frac{z}{(x^2+y^2)(x^2+y^2+z^2)^{\frac{1}{2}}} - \frac{z}{[x^2+(y-b)^2][x^2+(y-b)^2+z^2]^{\frac{1}{2}}}$$

$$\left. + \frac{z-a}{[x^2+(y-b)^2][x^2+(y-b)^2+(z-a)^2]^{\frac{1}{2}}}\right)$$

PROBLEM 7.14

(a) The electromagnet shown in the figure is made of iron with a magnetic permeability $\mu = 800\ \mu_0$. The cross section of the iron is 10 cm × 10 cm, and the length of the path around the iron and across the 2-cm gap is 200 cm. How many ampere-turns will be required in the windings to give a 5,000-G (0.5 Wb/m^2) field in the gap, assuming no bulging of magnetic field lines out of the gap?
(b) What is B inside the iron?
(c) What is H in the gap?
(d) What is H in the iron?
(e) What is the magnetization M in the iron?

Solution:

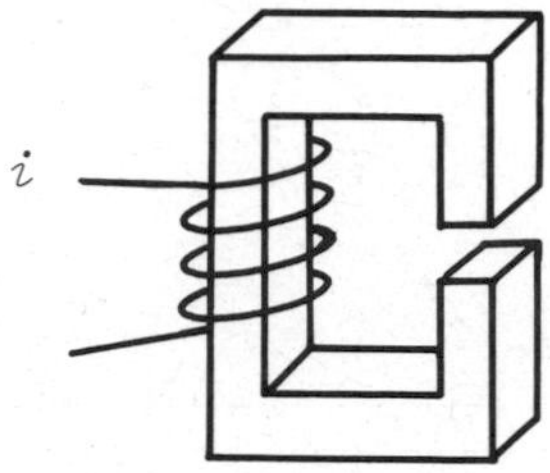

$\mu = 800\ \mu_0$

cross section = 10 cm × 10 cm

$\ell = 200$ cm (gap of 2 cm)

$B = 0.5\ Wb/m^2$

(a) $\oint \mathbf{H} \cdot d\ell = NI$

$H_{gap}\ell_{gap} + H_{iron}\ell_{iron} = NI$

(From parts c and d first find H)

$$NI = (3.98 \times 10^5)(0.02) + (4.97 \times 10^2)(1.98) = 8.9441 \times 10^3 \text{ A-turns}$$

$$NI = 8{,}944.1 \text{ A-turns}$$

(b) Lines of B are continuous so

$\phi_{iron} = \phi_{gap}$

$B_{gap} A = A\ B_{iron}$

$B_{iron} = 0.5\ Wb/m^2$

(c) $H = \frac{1}{\mu} B$

$$H_{gap} = \frac{1}{4\pi \times 10^{-7}} B_{gap} = \frac{0.5}{4\pi \times 10^{-7}}$$

$$H_{gap} = 3.98 \times 10^5 \text{ A/m}$$

(d) $$H_{iron} = \frac{1}{\mu} B_{iron} = \frac{0.5}{800(4\pi \times 10^{-7})}$$

$$H_{iron} = 4.97 \times 10^2 \text{ A/m}$$

(e) $M_{iron} = \chi\, H_{iron} = \left(\frac{\mu}{\mu_0} - 1\right) H_{iron}$ from $\mu = \mu_0(1+\chi)$

$$M_{iron} = \left(\frac{800}{1} - 1\right) H_{iron} = 799(4.97 \times 10^2)\ \text{A/m} = 3.971 \times 10^5\ \text{A/m}$$

$$M_{iron} = 3.97 \times 10^5\ \text{A/m}$$

PROBLEM 7.15 An insulating circular disk of radius a has a uniformly distributed static charge of σ C/m^2. The disk rotates about its center with an angular velocity ω. Find the magnetic field at its center.

Solution:

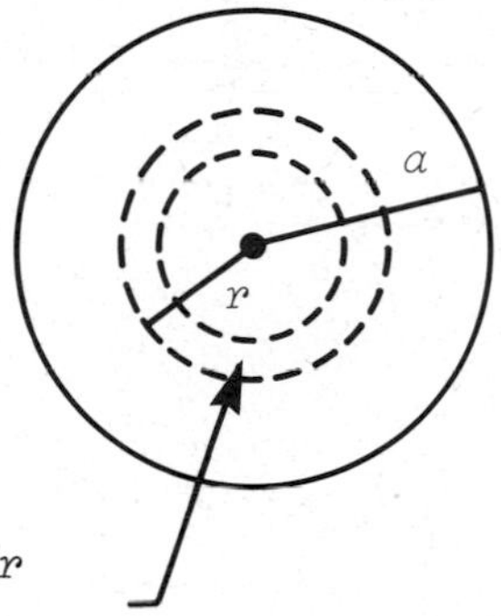

$dq = \sigma\, 2\pi r\, dr$

T = time for one revolution $= \frac{2\pi}{\omega}\frac{r}{r}$ $\qquad v = \frac{x}{t}$

$dI = \frac{dq}{T} = \frac{\sigma\, 2\pi r\, dr\, \omega}{2\pi} = \sigma r \omega\, dr$ $\qquad t = \frac{x}{v}$

Magnetic field produced by dI,

$v = \omega r$

$$dB = \frac{\mu_0 dI}{2\pi r}$$

$$B = \int_0^a \frac{\mu_0 \sigma r\, \omega\, dr}{2\pi r} = \frac{\mu_0 \sigma \omega a}{2\pi}$$

Or, using Ampere's law,

$$\int \mathbf{B} \cdot d\ell = \mu_0 i$$

$$B 2\pi r = \mu_0 i$$

$$dB = \frac{\mu_0 di}{2\pi r}$$

$$dB = \frac{\mu_0 \sigma r \omega\, dr}{2\pi r}$$

$$B = \frac{\mu_0 \sigma \omega}{2\pi}\int_0^a dr$$

$$B = \frac{\mu_0 \sigma \omega a}{2\pi}$$

PROBLEM 7.16 A single layer solenoid 20 cm long of radius 2 cm is wound uniformly with 3000 turns of wire. A current of 2 A flows through the coil.

(a) What is the value of B on the axis of the solenoid, at the middle?
(b) What is the value of B on the axis of the solenoid, at the end?
(c) What is the value of B on the axis of the solenoid, at a point 10 cm from one end?

Solution:

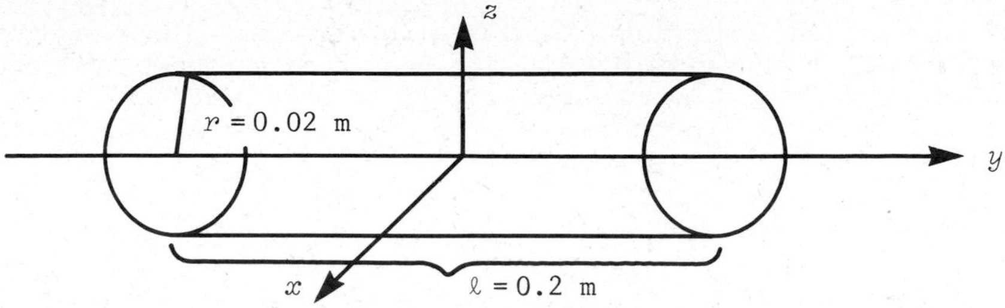

Number of turns = $\ell r = 3000(\ell)$

$$I = 2A$$

(a) For 1 loop,

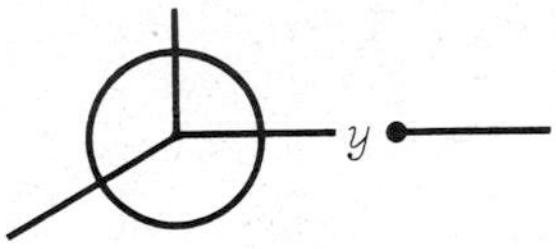

$$B = \frac{\mu_0 I r^2}{2(y^2 + r^2)^{3/2}}$$

$$n = \frac{\text{No. of turns}}{\text{length}} = \frac{3000}{0.2}$$

For a group of wires of length dx located at x (here $dx = dy$),

$$dB = \frac{\mu_0 I r^2}{2(y^2 + r^2)^{3/2}}\, n\, dy$$

For a point in the middle, y goes from $-\ell/2$ to $\ell/2$. So,

$$B = \int_{-\ell/2}^{\ell/2} \frac{\mu_0 I r^2 n\, dy}{2(y^2 + r^2)^{3/2}} = \frac{\mu_0 I r^2 n}{2}\left(\frac{y}{r^2(y^2 + r^2)^{\frac{1}{2}}}\right)\Bigg|_{-\ell/2}^{\ell/2}$$

$$B = \frac{\mu_0 I n}{2}\left(\frac{2(\ell/2)}{\sqrt{\ell^2/4 + r^2}}\right)$$

$$B = \frac{\mu_0 I n \ell}{2\sqrt{\ell^2/4 + r^2}} = \frac{(4\pi \times 10^{-7})(\ )(3000)(0.2)}{(2)\sqrt{0.2^2/4 + 0.02^2}} = 3.69 \times 10^{-2}\ \text{Wb/m}^2$$

(b) At the end, y goes from 0 to ℓ,

$$B = \frac{\mu_0 I r^2 n}{2}\left(\frac{y}{r^2(y^2+r^2)^{\frac{1}{2}}}\right)_0^{\ell}$$

$$= \frac{\mu_0 I n}{2}\left(\frac{\ell}{\sqrt{\ell^2+r^2}}\right)$$

$$= \frac{(4\pi \times 10^{-7})(2)(3000/0.2)(0.2)}{(2)\sqrt{0.2^2+0.02^2}}$$

$$B = 1.88 \times 10^{-2} \text{ Wb/m}^2$$

(c) y goes from 0.1 to 0.3,

$$B = \frac{\mu_0 I r^2 n}{2}\left[\frac{y}{r^2(y^2+r^2)^{\frac{1}{2}}}\right]_{0.1}^{0.3}$$

$$= \frac{(4\pi \times 10^{-7})(2)(3000/0.2)}{(2)}\left(\frac{0.3}{\sqrt{0.3^2+0.02^2}} - \frac{0.1}{\sqrt{0.1^2+0.02^2}}\right)$$

$$B = 3.24 \times 10^{-4} \text{ Wb/m}^2$$

PROBLEM 7.17 A surveyor is using a compass 20 ft below a power line in which there is a steady current of 100 A. Will this interfere seriously with the compass reading? The horizontal component of the earth's magnetic field at the site is 0.2 G.

Solution:

$$B = \frac{\mu_0 I}{2\pi r} = \frac{(4\pi \times 10^{-7})(10^2)}{(2\pi)(2 \times 10^1)(1.2 \times 10^1)(2.54 \times 10^{-2})} = 0.328 \times 10^{-5} \text{ Wb/m}^2$$

$$B = 0.033 \text{ G}$$

So the B field will seriously affect the reading to the point where it would not properly register the correct reading.

PROBLEM 7.18 A conductor consists of an infinite number of adjacent wires, each infinitely long and carrying a current i. Show that the lines of B will be as represented in the figure and that B for all points in front of the infinite current sheet will be given by $B = \frac{1}{2}\mu_0 n i$ where n is the number of conductors per unit length.

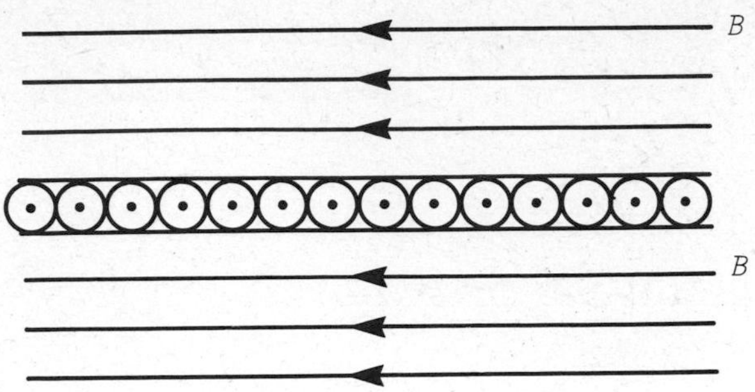

Solution: $B = \mu_0 I/2\pi r$ for each straight wire. Since there are $n\,dx$ such wires contained in a length dx, the magnetic field contribution from this is

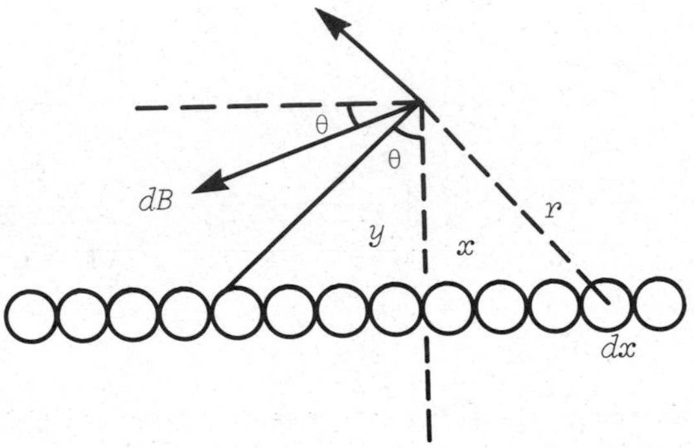

$$dB = \frac{\mu_0 In\,dx}{2\pi r}$$

where $r = (x^2 + y^2)^{\frac{1}{2}}$. Since only the horizontal components of B will survive

$$B = \int dB \cos\theta = \int \frac{\mu_0 In\,dx}{2\pi(x^2+y^2)^{\frac{1}{2}}} \frac{y}{(x^2+y^2)^{\frac{1}{2}}} \frac{\mu_0 nIy}{2\pi} \int_{-\infty}^{\infty} \frac{dx}{x^2+y^2}$$

$$B = \frac{\mu_0 nIy}{2\pi} \frac{1}{y} \tan^{-1} \frac{x}{y}\bigg|_{-\infty}^{\infty} = \frac{\mu_0 nI}{2\pi}\left(\frac{\pi}{2} + \frac{\pi}{2}\right)$$

$$B = \tfrac{1}{2}\mu_0 nI$$

PROBLEM 7.19 A long wire of circular cross section of radius A has a current density which is described by $J = J_0 r^2$. Derive an expression for B at points inside and outside the wire.

Solution: $r < A$, $B(2\pi r) = \mu_0 \int_0^r J_0 r^2\, 2\pi r\,dr$

$$B(2\pi r) = \mu_0 J_0 2\pi \frac{r^4}{4}$$

$$B = \frac{\mu_0 J_0 r^3}{4}$$

$$r > A, \quad B(2\pi r) = \mu_0 J_0 2\pi (A^4/4)$$

$$B = \frac{\mu_0 J_0 A^4}{4r}$$

PROBLEM 7.20 A toroidal sample of magnetic material of susceptibility $\chi_m = 2 \times 10^{-2}$ is wound with 1000 turns of wire carrying a current of 2 A. The toroid is 15 cm long.

(a) Find the magnetization current.
(b) Determine the magnetic field intensity H produced by the current.
(c) Calculate μ, the magnetic permeability of the material.
(d) Calculate the induced magnetization M in the material.
(e) Calculate the magnetic field B resulting from the current and the magnetization of the material.

Solution:

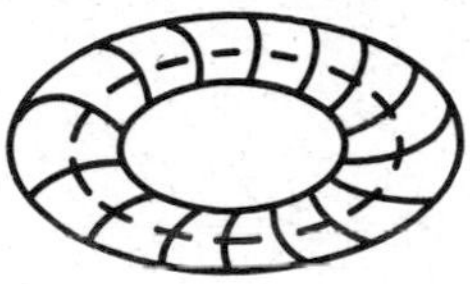

$$\oint \mathbf{H} \cdot d\boldsymbol{\ell} = I_f$$

$$H(2\pi r) = NI$$

$$H = \frac{NI}{2\pi r} = \frac{NI}{L}$$

(a) $M = \chi H = (2 \times 10^{-2})(1.33 \times 10^4) = 2.66 \times 10^2$ A/m

$I_{mag} = ML = (2.66 \times 10^2)(1.5 \times 10^{-1}) = 39.9$ A

(b) $H = \dfrac{(10^3)(2)}{0.15} = 1.33 \times 10^4$ A/m

(c) $\mu = \mu_0(1 + \chi) = (4\pi \times 10^{-7})(1 + 0.02) = 1.28 \times 10^{-6}$ Wb/A·m

(d) $M = 266$ A/m

(e) $B = \mu H = (1.28 \times 10^{-6})(1.33 \times 10^4) = 1.71 \times 10^{-2}$ Wb/m^2

PROBLEM 7.21 An iron ring of radius 5 cm has a cross section of 2 cm^2. Its permeability is $1000\,\mu_0$ (assumed constant). It is wound with 1500 turns carrying 5 A. Calculate:

(a) the magnetic field intensity,
(b) the magnetization of the iron, and
(c) the magnetization surface current per unit length of iron.

Solution:

(a) $H = \dfrac{NI}{L}$

Since the radius of the cross section is small compared to the radius of the ring.

$$H = \frac{(1.5 \times 10^3)(5)}{2(0.05)}$$

$$H = 2.39 \times 10^4 \text{ A/m}$$

(b) $M = \chi H$

To calculate χ we use

$$\mu = \mu_0(1 + \chi)$$

$$1000\ \mu_0 = \mu_0(1 + \chi)$$

$$\chi = 999$$

So,

$$M = (999)(2.39 \times 10^4)$$

$$M = 2.39 \times 10^7 \text{ A/m}$$

(c) $\frac{I}{L} = M = 2.39 \times 10^7$ A/m

PROBLEM 7.22 An infinitely long cylindrical conductor of radius a and permeability μ_1 carries a current I, which returns by means of a coaxial conducting shell of permeability μ_2 of inner radius b and outer radius c. Find the energy per unit length of the system.

Solution:

$$2\pi r H_\phi = I(r)$$

$$= \frac{r^2}{a^2} I, \quad r < a$$

$$= I, \quad a < r < b$$

$$= I - \frac{I(r^2 - b^2)}{c^2 - b^2} = \frac{I(c^2 - r^2)}{c^2 - b^2}, \quad b < r < c$$

$$= 0, \quad c < r$$

The energy per unit length is

$$\int_0^\infty 2\pi r \mu H^2\, dr = \frac{I^2}{2\pi}\left[\int_0^a \frac{\mu_1}{r}\frac{r^2}{a^2}\, dr + \int_a^b \frac{\mu_0}{r}\, dr + \int_b^c \frac{\mu_2}{r}\left(\frac{c^2 - r^2}{c^2 - b^2}\right) dr\right]$$

$$= \frac{I^2}{2\pi}\left[\frac{\mu_1}{2} + \mu_0 \ln\left(\frac{b}{a}\right) + \mu_2\left(\frac{c^2}{c^2 - b^2}\right)\ln\left(\frac{c}{b}\right) - \frac{\mu_2}{2}\right]$$

PROBLEM 7.23 An infinitely long cylinder of radius a has a constant permanent magnetization moment $M_0(\hat{\mathbf{i}}\cos\beta + \hat{\mathbf{j}}\sin\beta)$. There is a magnetic field $\hat{\mathbf{i}}H_0$ at large distances from the cylinder. Find the magnetic field at all points.

Solution: The magnetization is constant, so both inside and outside the cylinder, $\nabla^2 U = 0$. The scalar magnetic potential associated with the field $\hat{\mathbf{i}}H_0$ is

$$U = -H_0 r\cos\phi$$

The outward unit normal to the cylinder is

$$\hat{\mathbf{r}} = \hat{\mathbf{i}}\cos\phi + \hat{\mathbf{j}}\sin\phi$$

Let U^- = the potential in the region $r < a$, U^+ = the potential in the region $r > a$, $U^+ = -H_0 r\cos\phi + U_s$, where U_s = the scatter potential that goes to zero at large distances. U^- is finite for $r < a$, so

$$U_s = \frac{A}{r}\cos\phi + \frac{B}{r}\sin\phi$$

$$U^- = Cr\cos\phi + Dr\sin\phi$$

The boundary conditions at $r = a$ are the boundary conditions on U at the boundary between a material with specified magnetization and free space, for a unit normal pointing out of the material,

$$\frac{\partial U^+}{\partial n} = \frac{\partial U^-}{\partial n} - M_n \qquad (1)$$

which is from $\nabla \cdot \mathbf{B} = 0$

$$(\mathbf{B}^+ - \mathbf{B}^-)\cdot\hat{\mathbf{n}} = 0$$

and

$$U^+ = U^- \qquad (2)$$

which is from $\hat{\mathbf{n}} \times (\mathbf{H}^+ - \mathbf{H}^-) = 0$ for no current on the interface. With these boundary conditions at $r = a$,

$$M_n = M_0(\hat{\mathbf{i}}\cos\beta + \hat{\mathbf{j}}\cos\beta)\cdot(\hat{\mathbf{i}}\cos\phi + \hat{\mathbf{j}}\sin\phi)$$

$$= M_0(\cos\phi\cos\beta + \sin\phi\sin\beta)$$

The boundary conditions become

$$-H_0 a\cos\phi + \frac{A}{a}\cos\phi + \frac{B}{a}\sin\phi = C\cos\phi + D\sin\phi - M_0\cos\beta\cos\phi - M_0\sin\beta\sin\phi$$

so

$$-H_0 a + \frac{A}{a} = Ca, \qquad -H_0 - \frac{A}{a^2} = C - M_0\cos\beta, \quad B/a = Da, \quad -B/a^2 = D - M_0\sin\beta$$

The four equations are solved for the constants A, B, C, D, giving

$$A = -(M_0 a^2/2)\cos\beta, \qquad B = -(M_0 a^2/2)\sin\beta$$

$$C = -H_0 - \tfrac{1}{2}M_0\cos\beta, \qquad D = -(M_0/2)\sin\beta$$

and

$$U^+ = -H_0 r\cos\phi - \frac{M_0 a^2}{2r}\cos(\phi-\beta), \qquad U^- = -H_0 r\cos\phi - \frac{M_0}{2} r\cos(\phi-\beta)$$

The first term is not changed by the magnetic material. The second term gives the magnetic scalar potential caused by the magnetization, and

$$\mathbf{H}^- = -\nabla U^- = \hat{\mathbf{i}}H_0 + \frac{M_0}{2}(\hat{\mathbf{i}}\cos\beta + \hat{\mathbf{j}}\sin\beta)$$

$$\mathbf{B}^- = \mu_0(\mathbf{H}^- + \mathbf{M}) = \mu_0[\hat{\mathbf{i}}H_0 + \tfrac{1}{2}M_0(\hat{\mathbf{i}}\cos\beta + \hat{\mathbf{j}}\sin\beta) + M_0(\hat{\mathbf{i}}\cos\beta + \hat{\mathbf{j}}\sin\beta)]$$

$$= \mu_0[\mathbf{i}H_0 + \frac{3}{2}M_0(\hat{\mathbf{i}}\cos\beta + \hat{\mathbf{j}}\sin\beta)]$$

PROBLEM 7.24 A long thin conductor of width b carries a current of iA. Find the magnetic field in the plane of and outside the conductor at a distance from its near edge.

Solution:

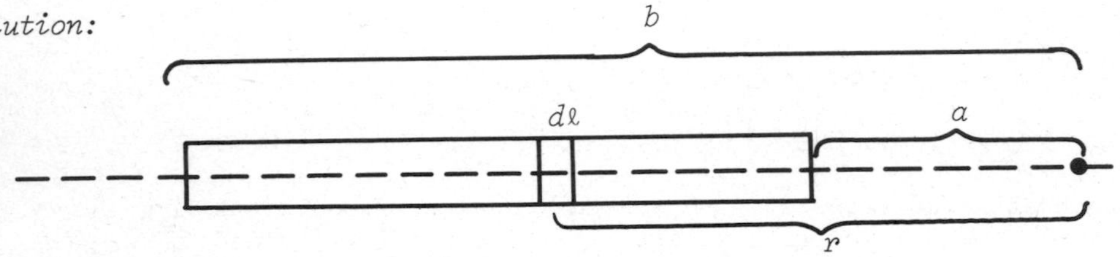

Let $I = i/b$ = current per unit length. The current in a strip $d\ell$ is $dI = I\,d\ell$. The magnetic field produced by this is

$$dB = \frac{\mu_0 I\,d\ell}{2\pi r} = \frac{(\mu_0 i/b)dr}{2\pi r}$$

$$B = \frac{\mu_0 i/b}{2\pi}\int_a^{a+b}\frac{dr}{r} = \frac{\mu_0 I}{2\pi b}\ln\frac{a+b}{a}$$

PROBLEM 7.25 A toroid coil is wound uniformly about a nonmagnetic ring whose circular axis has a radius of 10 cm, while its cross section has a radius of 1 cm. The coil has 150 turns and carries a current of 2A. Let P be a plane perpendicular to the circular axis of the ring at two diametrically opposite points.

(a) Find the component of magnetic induction normal to this plane at a general point within the material of the ring.
(b) What is the maximum percentage variation of this field in this toroidal region?

Solution:

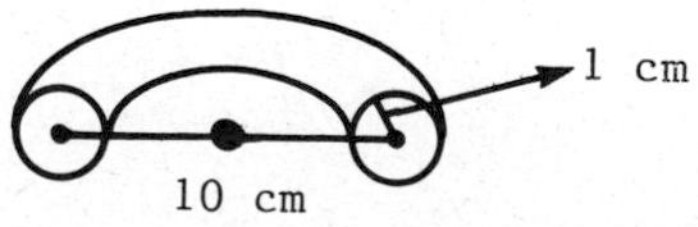

(a) From Ampere's law $\oint \mathbf{B}\cdot d\ell = \mu_0 I_T$

$$B_N(2\pi r) = \mu_0 NI$$

$$B_N = \frac{\mu_0 NI}{2\pi r}$$

$$B_N = \frac{(4\pi \times 10^{-7})(1.5 \times 10^2)(2)}{2\pi r} = \frac{6 \times 10^{-5}}{r}$$

Where r is the radius of the amperian loop.

(b) At $r = 9$ cm, $B = \dfrac{6 \times 10^{-5}}{9 \times 10^{-2}} = \dfrac{2}{3} \times 10^{-3}$ Wb/m^2

At $r = 11$ cm, $B = \dfrac{6 \times 10^{-5}}{11 \times 10^{-2}} = 5.45 \times 10^{-4}$ Wb/m^2

% variation from the maximum = 18.3%

PROBLEM 7.26 A long horizontal wire *AB* rests on the surface of a table. Another wire *CD*, vertically above the first, is 100 cm long and is free to slide up and down on the vertical metal guides. A current of 50 A is carried through the wires by means of the sliding contacts. The mass of the wire *CD* is 0.05 g/cm. To what equilibrium height will the wire *CD* rise, assuming that the magnetic force on its is due entirely to the current in the wire *AB*?

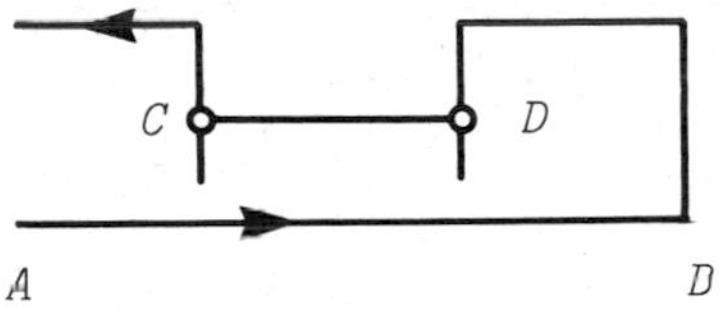

Solution:

$$F = 100 \text{ cm} \times 0.05 \frac{\text{g}}{\text{cm}} \times 980 \frac{\text{cm}}{\text{sec}^2} = 4900 \text{ dyn}$$

$$= \frac{BI\ell}{c} = \frac{(2I/hc)I\ell}{c} = \frac{2 \times 100 \text{ cm}}{h}\left(\frac{I}{10}\right)^2$$

$$h = \frac{2 \times 100 \times 5^2}{4900} = 1.02 \text{ cm}$$

PROBLEM 7.27 A thin flat conductor of great length has a uniform current density j per unit width, that is, $I = jw$.

(a) Calculate the magnetic field at a point P, at a perpendicular distance d above the center of the strip. [*Hint:* The expression for the field from a long straight strip of width dw is the same for a long straight wire.]

(b) What is the field if $d << w$, that is, if the strip becomes an infinite plane?

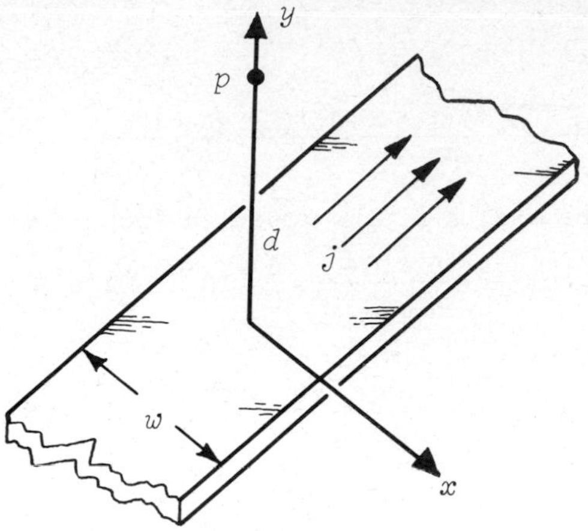

Solution:

(a) $dB = \frac{2dI}{rc} = \frac{2j\,dx}{c\sqrt{x^2+d^2}}$

$$B = \int dB \sin\theta = \frac{2j}{c}\int_{-w/2}^{w/2} \frac{dx}{\sqrt{x^2+d^2}}\,\frac{d}{\sqrt{x^2+d^2}}$$

$$= \frac{2jd}{c}\int_{-w/2}^{w/2} \frac{dx}{x^2+d^2}$$

Substitute $x = d\tan\theta$

$$dx = d\sec^2\theta\,d\theta$$

$$= \frac{2jd^2}{cd}\int_{-\tan^{-1}(w/2d)}^{\tan^{-1}(w/2d)} \frac{\sec^2\theta\,d\theta}{1+\tan^2\theta} = \frac{4j}{c}\tan^{-1}\left(\frac{w}{2d}\right)$$

(b) As $w \to \infty$, $\tan^{-1}(w/2d) \to \pi/2$ and $B = 2\pi j/c$.

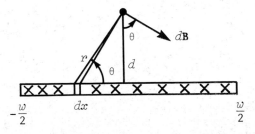

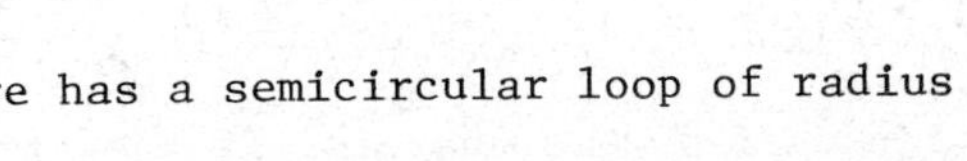

PROBLEM 7.28 A long wire has a semicircular loop of radius R as shown in the figure.

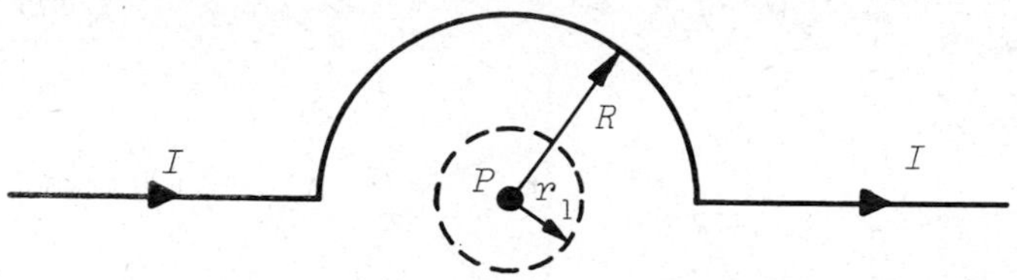

A constant I is flowing.

(a) Compute the magnetic field at P.
(b) Suppose that I begins to decrease at a constant rate $dI/dt = -K$ and that a small circular loop of radius $r \ll R$ is centered in P. Determine the magnitude of the induced emf around the small circular loop and the direction of the induced current.

Solution:

(a) $$d\mathbf{B} = \frac{\mu_0 i}{4\pi} \frac{d\ell \times \mathbf{r}}{r^2}$$

$$dB = \frac{\mu_0 i}{4\pi} \frac{d\ell}{R^2}$$

$$B = \frac{\mu_0 i}{4\pi R^2} \int_0^{2\pi R/2} d\ell$$

$$\mathbf{B} = \frac{\mu_0 I\pi(-\hat{\mathbf{k}})}{4\pi R}$$

(b) $$\frac{dI}{dt} = -K$$

$$|\Phi_R| = \frac{\mu_0 I\pi r^2}{4R}$$

$$|E| = \frac{d\Phi_B}{dt} = \left(\frac{d\Phi_B}{dI}\right)\left(\frac{dI}{dt}\right) = \frac{\mu_0 \pi r^2 K}{4R}$$

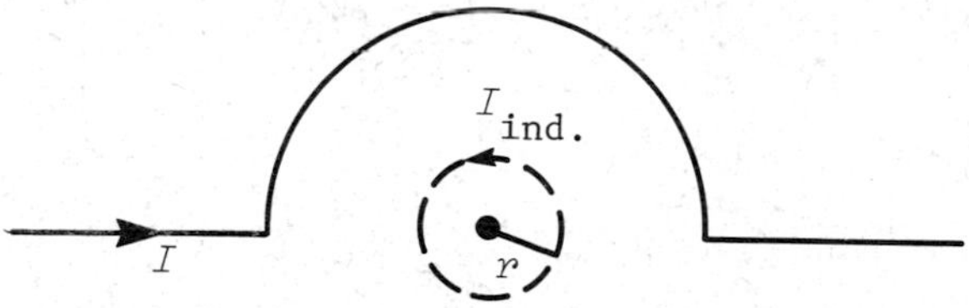

Direction of Induced Current

PROBLEM 7.29 Derive the expression for the magnetic field produced by a single turn of wire of radius R carrying a current I at a point x on the axis of the loop. Use this result to derive an expression for the magnetic field at the end of a solenoid with N turns along a length L.

Solution: First finding **B** for one loop.

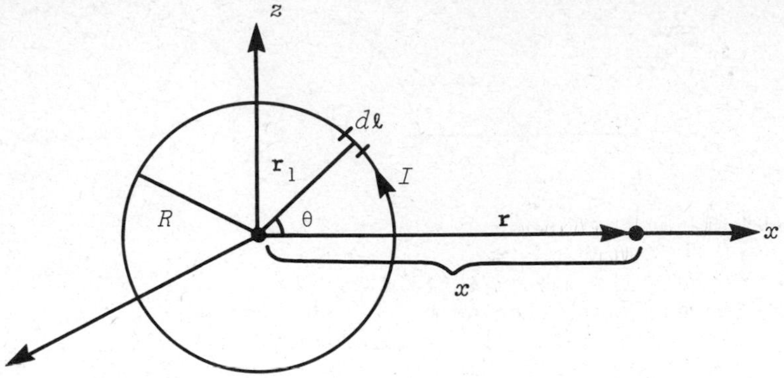

$$d\boldsymbol{\ell} = \hat{\mathbf{j}}dy + \hat{\mathbf{k}}dz$$

$$\mathbf{r} = \hat{\mathbf{i}}x$$

$$\mathbf{r}_1 = \hat{\mathbf{j}}y + \hat{\mathbf{k}}z$$

$$R = |r_1| = (y^2 + z^2)^{\frac{1}{2}}$$

$$\mathbf{r} - \mathbf{r}_1 = \hat{\mathbf{i}}x - \hat{\mathbf{j}}y - \hat{\mathbf{k}}z$$

$$|\mathbf{r} - \mathbf{r}_1| = (x^2 + y^2 + z^2)^{\frac{1}{2}} = (x^2 + R^2)^{\frac{1}{2}}$$

$$\mathbf{B} = \frac{\mu_0 I}{4\pi} \int_{\substack{\text{over}\\ \text{dimensions of}\\ \text{current flux}}} \frac{d\boldsymbol{\ell} \times (\mathbf{r} - \mathbf{r}_1)}{|\mathbf{r} - \mathbf{r}_1|^3} = \int \frac{\mu_0 I}{4\pi} \frac{(\hat{\mathbf{j}}dy + \hat{\mathbf{k}}dz) \times (\hat{\mathbf{i}}x - \hat{\mathbf{j}}y - \hat{\mathbf{k}}z)}{(R^2 + x^2)^{3/2}}$$

$$\mathbf{B} = \frac{\mu_0 I}{4\pi} \int \frac{-\hat{\mathbf{k}}x\,dy - \hat{\mathbf{i}}z\,dy + \hat{\mathbf{j}}x\,dz + \hat{\mathbf{i}}y\,dz}{(R^2 + x^2)^{3/2}}$$

$$= \frac{\mu_0 I}{4\pi(R^2 + x^2)^{3/2}} \int \hat{\mathbf{i}}(y\,dz - z\,dy) + \hat{\mathbf{j}}x\,dz - \hat{\mathbf{k}}x\,dy$$

$y = R\cos\theta$ and $z = R\sin\theta$

$dy = -R\sin\theta\,d\theta$ and $dz = +R\cos\theta\,d\theta$

$$\mathbf{B} = \frac{\mu_0 I}{4\pi(R^2 + x^2)^{3/2}} \int \hat{\mathbf{i}}(R^2\cos^2\theta\,d\theta + R^2\sin^2\theta\,d\theta) + \hat{\mathbf{j}}xR\cos\theta\,d\theta + \hat{\mathbf{k}}xR\sin\theta\,d\theta$$

$$= \frac{\mu_0 I}{4\pi(R^2 + x^2)^{3/2}} \left(\int_0^{2\pi} \hat{\mathbf{i}}R^2\,d\theta + \int_0^{2\pi} \hat{\mathbf{j}}xR\cos\theta\,d\theta + \int_0^{2\pi} \hat{\mathbf{k}}xR\sin\theta\,d\theta \right)$$

$$= \frac{\mu_0 I}{4\pi(x^2+R^2)^{3/2}} \; [\hat{\mathbf{i}}R^2 2\pi + \hat{\mathbf{j}}xR(0) + \hat{\mathbf{k}}xR(-)(1-1)]$$

$$\mathbf{B} = \frac{\mu_0 I(2\pi\hat{\mathbf{i}}R^2)}{4\pi(R^2+x^2)^{3/2}}$$

$$B = \frac{\hat{\mathbf{i}}\mu_0 IR^2}{2(x^2+R^2)^{3/2}}, \text{ for a single turn}$$

Let $n = \dfrac{\text{Number of turns}}{\text{unit length}}$

So

$$dB = \frac{\mu_0 IR^2 n\, dx}{2(x^2+R^2)^{3/2}}$$

$$B = \int_0^L \frac{\mu_0 IR^2 n\, dx}{2(x^2+R^2)^{3/2}} = \frac{\mu_0 IR^2 n}{2R^2}\left(\frac{x}{\sqrt{x^2+R^2}}\right)_0^L$$

$$B = \frac{\mu_0 In}{2}\left(\frac{L}{\sqrt{x^2+R^2}}\right)$$

$$B = \frac{\mu_0 InL}{2(x^2+R^2)^{\frac{1}{2}}}$$

PROBLEM 7.30 A current of 30 A flows in the circuit in the figure. The parallel wires are 5 cm apart. What is the force pert unit length on one of the wires?

Solution:

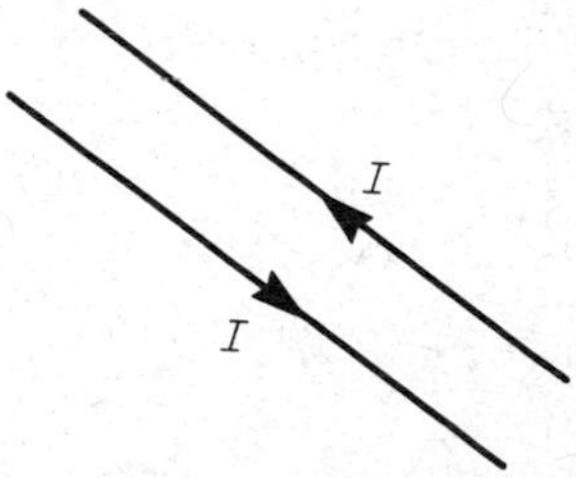

$$B = \frac{2I}{rc} = \frac{2 \times 30 \text{ A}}{5 \text{ cm} \times 10} = 1.2 \text{ G}$$

$$\frac{F}{\ell} = \frac{BI}{c} = 1.2 \times \frac{30}{10} = 3.6 \text{ dyn/cm}$$

PROBLEM 7.31 Consider the magnetic field of a circular current loop, at points on the axis of the loop, given by the equation for the magnetic field on axis:

$$B = \frac{2\pi b^2 I}{cr^3} = \frac{2\pi b^2 I}{c(b^2+z^2)^{3/2}}$$

Calculate explicitly the line integral of the field along the axis from $-\infty$ to ∞, to check the general formula $\int \mathbf{B}\cdot d\mathbf{s} = 4\pi I/c$. Why may we ignore the "return" part of the path which would be necessary to complete a closed loop?

Solution:

$$B_z = \frac{2\pi b^2 I}{c(b^2+z^2)^{3/2}}$$

Let $z = b\tan\theta$

$$d\mathbf{r} = \hat{\mathbf{z}}dz = \hat{\mathbf{z}}b\sec^2\theta\,d\theta$$

$$\oint \mathbf{B}\cdot d\mathbf{r} = \int_{-\infty}^{\infty} B_z\,dz = \frac{2\pi b^2 I}{c}\int_{-\pi/2}^{\pi/2}\frac{b\sec^2\theta\,d\theta}{(b^2+b^2\tan\theta)^{3/2}}$$

$$= \frac{2\pi I}{c}\int_{-\pi/2}^{\pi/2}\frac{\sec^2\theta\,d\theta}{\sec^3\theta}\quad\frac{2\pi I}{c}\int_{-\pi/2}^{\pi/2}\cos\theta\,d\theta = \frac{4\pi I}{c}$$

While the expression for B applies only along the axis, we see that, for $z >> b$, $B_z = 1/z^3$. We may expect, then, that in all directions $B\to 0$ as $1/r^3$ for large r. When we close the line integral at large r, we will get a contribution of order $B\ell \sim (1/r^3)(\pi r) \sim 1/r^2$, which will vanish as $r\to\infty$.

PROBLEM 7.32 Magnetic field in coaxial conductors. The leads that carry a direct current of 5000 A to a large magnet are constructed as follows: A solid aluminum rod 5 cm in diameter is surrounded by a return lead in the shape of an aluminum cylinder of 7 cm inside diameter and 9 cm outside diameter. (The annular space between the rod and the cylinder is filled with flowing oil, which serves to remove the heat.) In each of the conductors the current density will be practically constant over the cross section of the conductor. Calculate and plot on a graph the magnitude of the magentic field B, in gauss, as a function of radius in centimeters from the axis to a point outside the outer conductor. (The presence of the aluminum itself and the oil do not affect the magnetic field.)

Solution: For $r \le r_1$, $J_1 = I/\pi r_1^2$. Around a circle of radius r,

$$\oint \mathbf{B}\cdot d\mathbf{s} = B(2\pi r) = \frac{4\pi}{c}\int \mathbf{J}\cdot d\mathbf{a}$$

$$= \frac{4\pi}{c}(J_1\pi r^2) = \frac{4\pi}{c}\left(\frac{Ir^2}{r_1^2}\right)$$

$$B = \frac{2I}{r_1^2 c}\,r$$

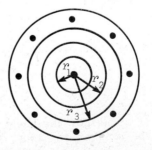

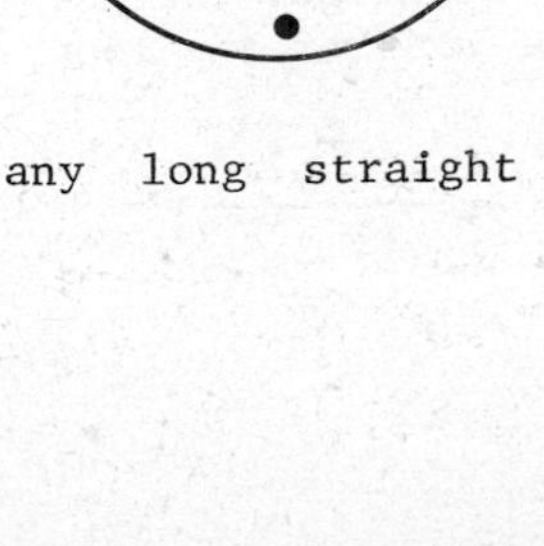

For $r_1 \le r \le r_2$, $B = 2I/rc$, the field outside any long straight wire. For $r_2 \le r \le r_3$, $J_2 = -I/\pi\,(r_3^2 - r_2^2)$,

$$2\pi r B = \frac{4\pi}{c}\left(I - \frac{I}{\pi(r_3^2 - r_2^2)}\pi(r^2 - r_2^2)\right)$$

$$B = \frac{2I}{rc}\left(1 - \frac{r^2 - r_2^2}{r_3^2 - r_2^2}\right) = \frac{2I}{rc}\left(\frac{r_3^2 - r^2}{r_3^2 - r_2^2}\right)$$

$$\frac{2I}{c} = \frac{2 \times 5000 \text{ A}}{10} = 10^3 \text{ G} \cdot \text{cm}$$

For $r_1 = 2.5$ cm, $r_2 = 3.5$ cm, and $r_3 = 4.5$ cm,

$$B = \frac{10^3}{(2.5)^2} r = 160\ r, \quad r < r_1$$

$$= \frac{10^3}{r}, \quad r_1 < r < r_2$$

$$= 125(20.25 - r^2)/r, \quad r_2 < r < r_3$$

$$= 0, \quad r_3 < r$$

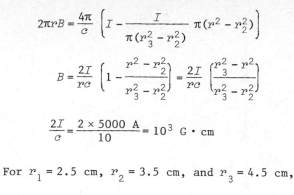

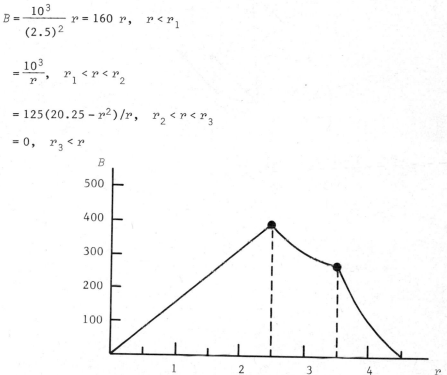

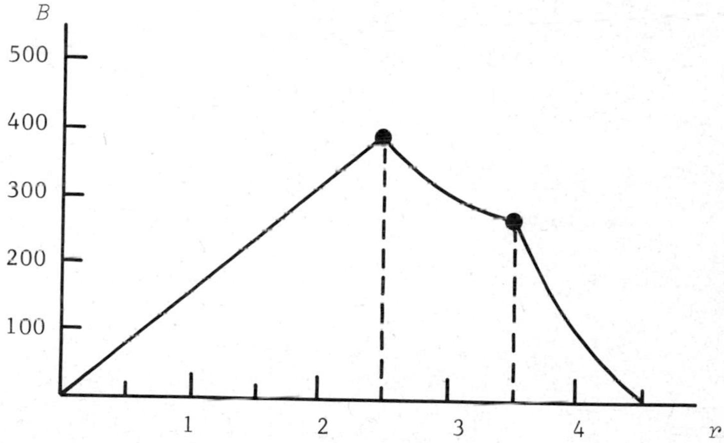

PROBLEM 7.33 The rectangular loop of wire in the figure has a mass of 0.1 g/cm and is pivoted about the z axis. The current in the loop is 20 A in the direction shown. Calculate the magnitude and direction of a magnetic field in the y direction necessary to cause the loop to swing in the direction shown to an equilibrium position 30° from the vertical.

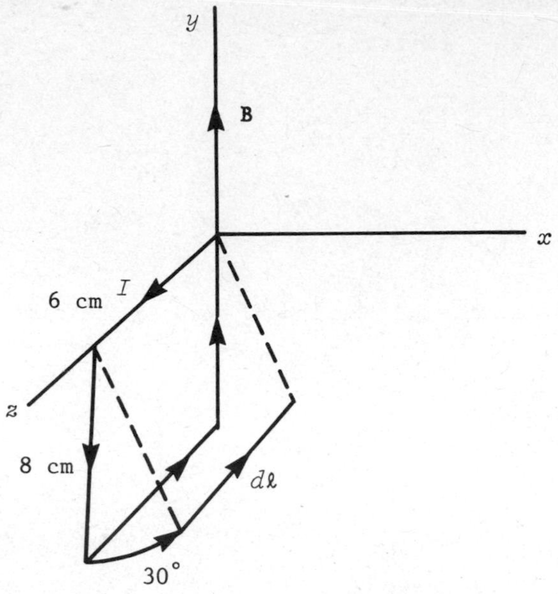

Solution:

$$F_1 = 0.1 \times 8 \times 980 = 784 \text{ dyn}$$

$$F_2 = 0.1 \times 6 \times 980 = 588 \text{ dyn}$$

$$x_1 = 4 \sin 30° = 2 \text{ cm}$$

$$x_2 = 8 \sin 30° = 4 \text{ cm}$$

$$\tau = \sum \mathbf{r} \times \mathbf{F} = 2x_1F_1 + x_2F_2 = 10976 \text{ dyn} \cdot \text{cm}$$

F_m must be to the $+x$ side to balance this torque.

$$F_m = \frac{BI\ell}{c} \text{ with } B \perp \ell = B\left(\frac{20}{10}\right)6 = 12\ B$$

$$\tau = y_2F_m = 8 \cos 30°\ F_m = 10976$$

$$= (4\sqrt{3})(12\ B)$$

$$B = \frac{10976}{48\sqrt{3}} = 132 \text{ G}$$

From $\mathbf{F} \propto \boldsymbol{\ell} \times \mathbf{B}$, B must be in $+y$ direction.

PROBLEM 7.34 A coil is wound evenly on a torus of rectangular cross section. There are N turns of wire in all. Only a few are shown in the figure. With so many turns, we shall assume that the current on the surface of the torus flows exactly radially, on the annular end faces, and exactly longitudinally on the inner and outer cylindrical surfaces. First convince yourself that on this assumption symmetry requires that the magnetic field everywhere should point in a "circumferential" direction, that is, that all field lines are circles about the axis of the torus. Second, prove that the field is zero at all points outside the torus, including the interior of the central hole. Third, find the magnitude of the field inside the torus, as a function of radius.

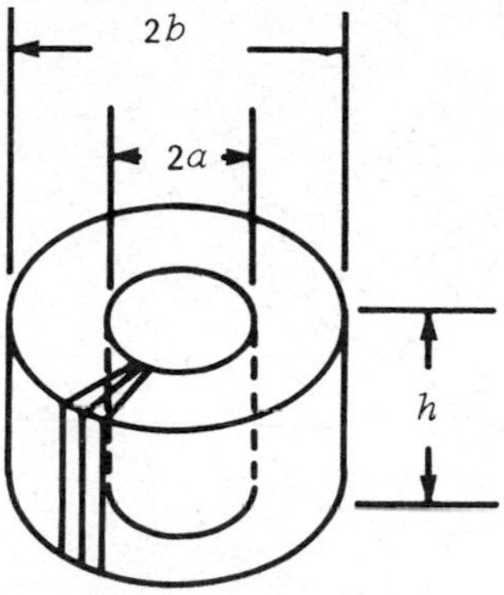

Solution: Consider any two loops symmetrically placed with respect to a point P. For corresponding line elements $d\ell_1$ and $d\ell_1'$ on the vertical part of the loop, the fields $d\mathbf{B}_1$ and $d\mathbf{B}_1'$ are equal in size and lie in the horizontal plane shown.

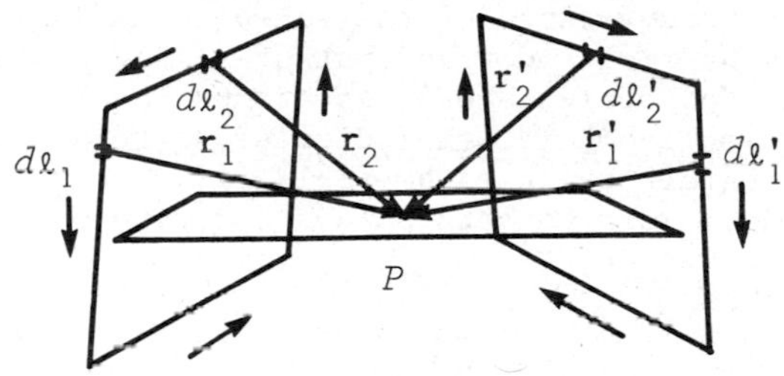

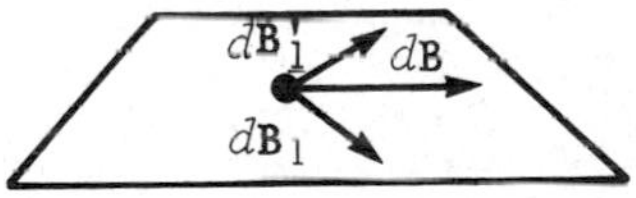

The radial components are equal and opposite, so the sum $d\mathbf{B}_1 + d\mathbf{B}_1' = d\mathbf{B}$ is a tangential field. Now consider $d\ell_2$ and $d\ell_2'$. Again, $|d\mathbf{B}_2| = |d\mathbf{B}_2'|$. The situation is more complicated since these have vertical as well as horizontal components. However, the vertical components are equal and opposite; for example, for the elements shown $dB_{2z} > 0$ and $dB_{2z}' < 0$. Also, the radial components are equal and opposite, so again only a tangential component is left. Since this is true for any symmetrically placed loops, it must be true for the sum from all the loops and the field everywhere, if not zero, must be tangential or "circumferential." Now consider the integral $\oint \mathbf{B}\cdot d\mathbf{s}$ around three circles. For $r = r_1$ inside the hole in the torus, $\oint \mathbf{B}\cdot d\mathbf{s} = 2\pi r_1 B = 0$ since no current passes through the circular surface. For $r = r_3$, outside the torus, $\oint \mathbf{B}\cdot d\mathbf{s} = 2\pi r_3 B = 0$ because each loop cuts the circular area twice, once with the current up and once down. Finally, for $r = r_2$ inside the torus, $\oint \mathbf{B}\cdot d\mathbf{s} = 2\pi r_2 B = (4\pi/c)IN$ since all N loops penetrate the circular area once, so $B_{\text{inside}} = 2IN/r_2 c$.

PROBLEM 7.35 Any ordinary solenoidal coil is really a helix. There is a certain component of longitudinal flow accompanying the main circumferential flow, because the winding has to progress from one end to the other. Suppose we regard the current as the superposition of a precisely circumferential current sheet and a precisely longitudinal "tube" of current, as shown in the figure. Describe the magnetic field of the combination, both inside and outside the cylinder. (Assume the cylinder is infinitely long.) Can you relate the ratio of the field strengths inside and outside the cylinder to the pitch of the winding of the helix this current distribution is supposed to represent?

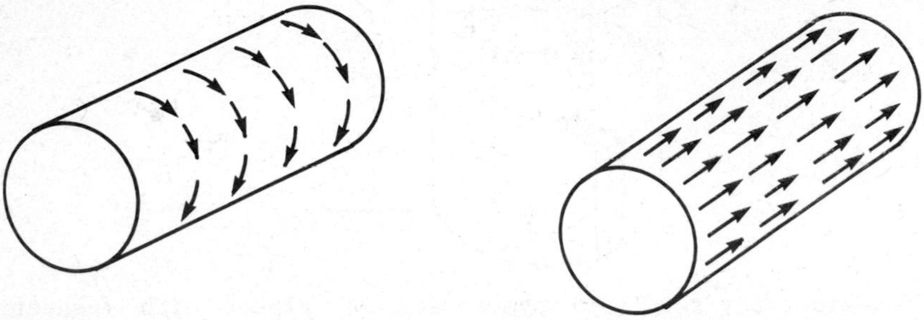

Solution: We have already seen that the longitudinal field is $B = 4\pi IN/c$ inside and 0 outside. Consider the longitudinal current (I) to be spread uniformly around the circumference of the solenoid as a cylindrical current sheet with $J = I/2\pi R$. The field must be that of a long straight wire $B = 2I/rc$ outside the solenoid. Now evaluate $\oint \mathbf{B} \cdot d\mathbf{s}$ around the path shown. Because the field B along the solenoid is perpendicular to this plane, it does not contribute to the integral. Because the field from the current sheet is tangential, the radial sides of the path do not contribute and

$$\oint \mathbf{B} \cdot d\mathbf{s} = (r_2\theta)B_{out} - (r_1\theta)B_{in} = \frac{4\pi}{c} JR\theta$$

or,

$$r_2\left(\frac{2I}{r_2c}\right) - r_1B_{in} = \frac{4\pi R}{c}\left(\frac{I}{2\pi R}\right)$$

$$\frac{2I}{c} - r_1B_{in} = \frac{2I}{c} \text{ or } B_{in} = 0$$

Thus, $B_{in} = 4\pi IN/c$ along solenoid and $B_{out} = 2I/rc$ around the solenoid. At $r = R$,

$$\frac{B_{out}}{B_{in}} = \frac{2I}{Rc}\,\frac{c}{4\pi IN} = \frac{1}{2\pi RN} = \frac{\text{pitch}}{\text{circumference}}$$

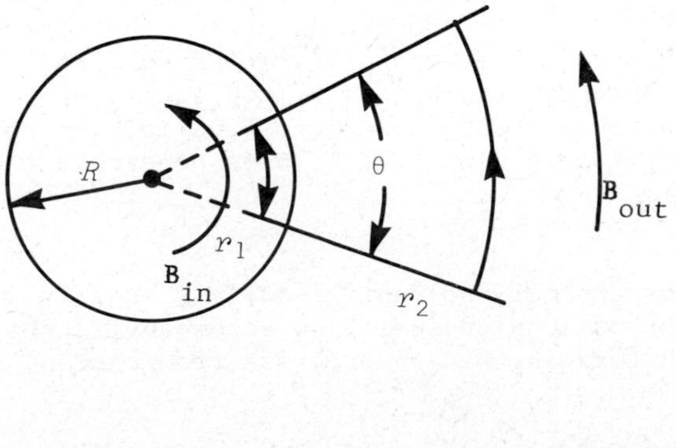

PROBLEM 7.36 Suppose we had a situation in which the component of the magnetic field parallel to the plane of a sheet had the same magnitude on both sides, but changed direction by 90° in going through the sheet. What is going on here? Would there be a force on the sheet? Should our formula for the force on a current sheet apply to cases like this?

Solution: Since $\Delta B = 4\pi j/c$ perpendicular to j, an external field of $B = 2\pi j/c$ parallel to j will lead to

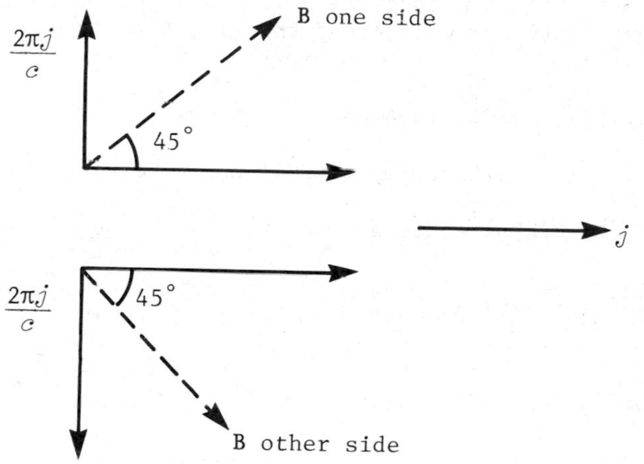

There is no force, because $\mathbf{B}_{ext} \| j$, or because $|B_z^+| = |B_z^-|$.

PROBLEM 7.37 A long wire is bent into the hairpinlike shape shown in the figure. Find an exact expression for the magnetic field at the point P which lies at the center of the half-circle.

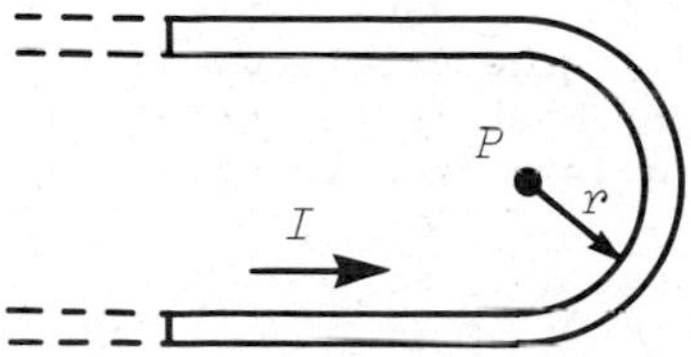

Solution: Consider instead the arrangement shown with two infinite parallel wires and a loop of radius R tangent to these. All three cause fields at the center perpendicular to the plane of the page and out of the page.

$$B = 2\left(\frac{2I}{Rc}\right) + \frac{2\pi I}{Rc}$$

But this is identical to the two circuits of the type in the problem. Hence,

$$B_p = \frac{2I}{Rc} + \frac{1}{2}\left(\frac{2\pi I}{Rc}\right) = (2+\pi)\frac{I}{Rc}$$

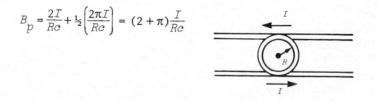

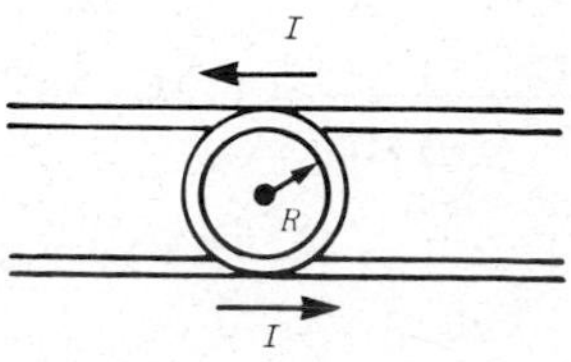

PROBLEM 7.38 A solenoid is made by winding a single layer of No. 14 copper wire on a cylindrical form 6 cm in diameter. There are five turns per centimeter and the length of the solenoid is 30 cm. Consulting wire tables, we find that No. 14 copper wire, which has a diameter of 0.163 cm, has a resistance of 0.010 ohms/m at 75°C. (We expect this coil to run hot!) If this solenoid is connected across a 24-V generator, what will be the magnetic field strength in the solenoid, in gauss, and the power dissipation, in watts?

Solution: The total length of wire in the solenoid is about $\ell = \pi d \times$ number of turns. But $d = (6.00 + 0.163)$ cm; that is, from the center of the wire on each side.

$$\ell = \pi(6.163)(5 \times 30) = 2904 \text{ cm}$$

$$R = (29.04 \text{ m})(0.010 \text{ ohm/m}) = 0.2904 \text{ ohm}$$

$$I = V/R = 24/0.2904 = 82.6 \text{ A}$$

$$B_z \simeq \frac{4\pi IN}{c} = 4\pi\left(\frac{82.6}{10}\right)(5) = 519 \text{ G}$$

with the cosine factor $[\cos(\tan^{-1} 6/30) = 0.980]$, we have, more correctly, at the center

$$B_z = 509 \text{ G}$$

$$\text{Power} = IV = 24 \times 82.6 = 1982 \text{ W}$$

PROBLEM 7.39 A hydrogen atom consists of a proton and an electron which may (for some purposes) be thought of as describing a circular orbit about the proton with radius $a_0 = \hbar^2/me^2 = 0.53 \times 10^{-8}$ cm, with speed $v = e^2/\hbar$. Here e is the electronic charge 4.8×10^{-10} esu, $\hbar \sim 10^{-27}$ erg · sec is Planck's constant divided by 2π, and m is the electron mass. What current is this circulating charge equivalent to? What is the strength in gauss of the magnetic field at the proton which arises from the electron's motion?

Solution:

$$I = \left(\frac{\text{No. of turns}}{\text{second}}\right) \times e = \left(\frac{v}{2\pi a_0}\right)e = \frac{me^5}{2\pi\hbar^3}$$

$$= \frac{(9.1 \times 10^{-28})(4.8 \times 10^{-10})^5}{2\pi(1.055 \times 10^{-27})^3} = 3.14 \times 10^6 \ \frac{\text{esu}}{\text{sec}} \ \frac{1}{3 \times 10^9 \text{ esu/C}}$$

$$= 1.05 \text{ mA}$$

$$B = \frac{2\pi I}{rc} = \frac{2\pi(1.05 \times 10^{-3})}{(0.53 \times 10^{-8} \text{ cm})(10)} = 1.24 \times 10^5 \text{ G}$$

PROBLEM 7.40 Starting with the differential expression

$$\mathbf{B} = \frac{I}{c}\, d\mathbf{l}' \times \frac{(\mathbf{x} - \mathbf{x}')}{|\mathbf{x} - \mathbf{x}'|^3}$$

for the magnetic induction at the point P with coordinate $\mathbf{x}$ produced by an increment of current $I\, d\mathbf{l}'$ at $\mathbf{x}'$, show explicitly that for a closed loop carrying a current I the magnetic induction at P is

$$\mathbf{B} = \frac{I}{c}\, \nabla\Omega$$

where Ω is the solid angle subtended by the loop at the point P. This corresponds to a magnetic scalar potential $\Phi_M = -I\Omega/c$. The sign convention for the solid angle is that Ω is positive if the point P views the "inner" side of the surface spanning the loop, that is, if a unit normal $\hat{\mathbf{n}}$ to the surface is defined by the direction of current flow via the right-hand rule, Ω is positive if $\hat{\mathbf{n}}$ points away from the point P, and negative otherwise.

Solution:

$$B = \frac{I}{C} \oint d\ell \times \frac{(\mathbf{x} - \mathbf{x}')}{|\mathbf{x} - \mathbf{x}'|^3} \tag{1}$$

Substitute

$$\frac{(\mathbf{x} - \mathbf{x}')}{|\mathbf{x} - \mathbf{x}'|^3} = -\nabla \frac{1}{|\mathbf{x} - \mathbf{x}'|} \tag{2}$$

into Eq. (1)

$$B = \frac{I}{c}\, \nabla \times \oint \frac{d\ell'}{|\mathbf{x} - \mathbf{x}'|} \tag{3}$$

We note that $\oint d\ell\phi = \int \hat{\mathbf{n}} \times \nabla\phi\, da$. This can be proven from Stoke's theorem.

$$\oint \mathbf{A} \cdot d\ell = \int \nabla \times \mathbf{A} \cdot \hat{\mathbf{n}}\, da$$

Let $\mathbf{A} = \phi\hat{\mathbf{a}}$, where $\hat{\mathbf{a}}$ is of constant magnitude and direction.

$$\oint \phi\hat{\mathbf{a}} \cdot d\ell = \int \nabla \times \phi\hat{\mathbf{a}} \cdot \hat{\mathbf{n}}\, da$$

$$\hat{\mathbf{a}} \cdot \oint \phi\, d\ell = \int \hat{\mathbf{n}} \times \nabla \cdot \phi\hat{\mathbf{a}}\, da = \int \hat{\mathbf{n}} \times \nabla\phi \cdot \hat{\mathbf{a}}\, da$$

Therefore,

$$\hat{\mathbf{a}} \cdot \oint \phi\, d\ell = \hat{\mathbf{a}} \cdot \int \hat{\mathbf{n}} \times \nabla\phi\, da \tag{4}$$

Substituting Eq. (4) into Eq. (3) gives

$$\mathbf{B} = \frac{I}{c}\,\nabla\times\int \hat{\mathbf{n}}'\times\nabla'\,\frac{1}{|\mathbf{x}-\mathbf{x}'|}\,da' \tag{5}$$

Since

$$\nabla'\,\frac{1}{|\mathbf{x}-\mathbf{x}'|} = -\nabla\frac{1}{|\mathbf{x}-\mathbf{x}'|},$$

we can bring ∇ outside the integral:

$$\mathbf{B} = \frac{I}{c}\,\nabla\times\nabla'\times\int\frac{\hat{\mathbf{n}}'}{|\mathbf{x}-\mathbf{x}'|}\,da' \tag{6}$$

Now we use the familiar identity

$$\nabla\times\nabla\times\mathbf{A} = \nabla(\nabla\cdot\mathbf{A}) - \nabla^2\mathbf{A} \tag{7}$$

$$\mathbf{B} = \frac{I}{c}\,\nabla\left\{\nabla\cdot\int\frac{\hat{\mathbf{n}}'}{|\mathbf{x}-\mathbf{x}'|}\,da'\right\} - \frac{I}{c}\,\nabla^2\int\frac{\hat{\mathbf{n}}'}{|\mathbf{x}-\mathbf{x}'|}\,da' \tag{8}$$

The second integral vanishes since

$$\nabla^2\,\frac{1}{|\mathbf{x}-\mathbf{x}'|} = -4\pi\,\delta(\mathbf{x}-\mathbf{x}')$$

and the observation at point **x'** is taken to be off the surface at **x**.

$$\mathbf{B} = \frac{I}{c}\,\nabla\int\nabla\cdot\frac{\hat{\mathbf{n}}'}{|\mathbf{x}-\mathbf{x}'|}\,da'$$

$$\mathbf{B} = \frac{I}{c}\,\nabla\int\nabla\cdot\frac{\hat{\mathbf{n}}'}{|\mathbf{x}-\mathbf{x}'|}\,da'$$

$$\mathbf{B} = \frac{I}{c}\,\nabla\int\hat{\mathbf{n}}\cdot\nabla\,\frac{1}{|\mathbf{x}-\mathbf{x}'|}\,da'$$

$$= \frac{I}{c}\,\nabla\int\hat{\mathbf{n}}\cdot\frac{(\mathbf{x}-\mathbf{x}')}{|\mathbf{x}-\mathbf{x}'|^3}\,da'$$

$$\mathbf{B} = \frac{I}{c}\,\nabla\int\frac{\cos\theta\,da'}{|\mathbf{x}-\mathbf{x}'|^2} \tag{9}$$

Substituting

$$\cos\theta\,da' = |\mathbf{x}-\mathbf{x}'|^2\,d\Omega \tag{10}$$

$$\mathbf{B} = \frac{I}{c}\,\nabla\Omega$$

PROBLEM 7.41 A circular current loop of radius a carrying a current I lies in the $x-y$ plane with its center at the origin.

(a) Show that the only nonvanishing component of the vector potential is

$$A_\phi(\rho, z) = \frac{4Ia}{c}\int_0^\infty dk \cos kz\; I_1(k\rho_<)K_1(k\rho_>)$$

where $\rho_<$ ($\rho_>$) is the smaller (larger) of a and ρ.

(b) Show that an alternative expression for A_ϕ is

$$A_\phi(\rho, z) = \frac{2\pi Ia}{c}\int_0^\infty dk\; e^{-k|z|} J_1(ka)J_1(k\rho)$$

Solution:

(a)

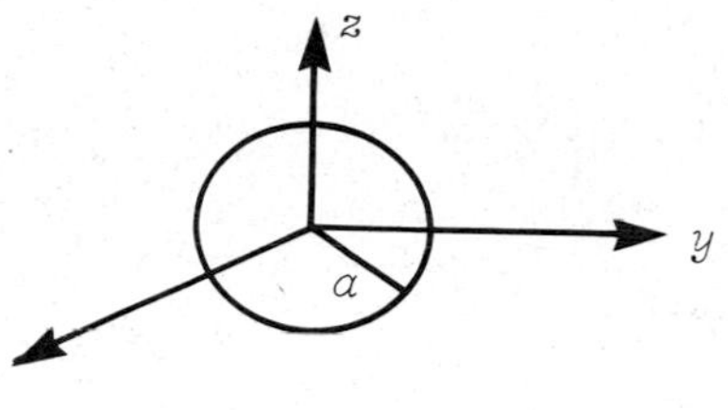

$$d\ell' = a\hat{\varphi}'\, d\phi'$$

$$\mathbf{A} = \frac{I}{c}\int \frac{d\ell'}{|\mathbf{x}-\mathbf{x}'|} = \frac{Ia}{c}\int_0^{2\pi} \frac{\hat{\varphi}'\, d\phi'}{|\mathbf{x}-\mathbf{x}'|}$$

$$= \frac{Ia}{c}\,\frac{4}{\pi}\int_0^\infty dk \cos kz \int_0^{2\pi} \hat{\varphi}'\, d\phi' \left\{\tfrac{1}{2} I_0(k\rho_<)K_0(k\rho_>)\right.$$

$$\left. + \sum_{m=1}^{\infty} \cos[m(\phi-\phi')]\, I_m(k\rho_<)K_m(k\rho_>)\right\}$$

The first integral vanishes since $\hat{\varphi}$ integrated over a circle is zero.

$$\int_0^{2\pi} \hat{\varphi}\, d\phi \cos m(\phi-\phi') = \int_0^{2\pi} (-\hat{\mathbf{i}}\sin\phi' + \hat{\mathbf{j}}\cos\phi')\cos m(\phi-\phi')\,d\phi'$$

$$= \left\{-\hat{\mathbf{i}}\int_0^{2\pi} \sin\phi' \cos m\phi \cos m\phi'\, d\phi' - \hat{\mathbf{i}}\int_0^{2\pi} \sin\phi' \sin m\phi \sin m\phi'\, d\phi'\right.$$

$$\left. + \hat{\mathbf{j}}\int_0^{2\pi} \cos\phi' \cos m\phi \cos m\phi'\, d\phi' + \hat{\mathbf{j}}\int_0^{2\pi} \cos\phi' \sin m\phi \sin m\phi'\, d\phi'\right\}$$

Using

$$\int_0^{2\pi} \cos\phi' \cos m\phi' \, d\phi' = \pi\delta_{m,1}$$

$$\int_0^{2\pi} \hat{\varphi} d\phi \cos m(\phi - \phi') = -\hat{\imath}\pi \sin\phi + \hat{\jmath}\pi \cos\phi$$

$$\mathbf{A} = \frac{Ia^4}{\pi c}\int_0^{\infty} dk \cos kz[-\hat{\imath}\pi \sin\phi + \hat{\jmath}\pi\cos\phi] I_1(k\rho_<)K_1(k\rho_>)$$

$$\hat{\varphi} = -\hat{\imath}\sin\phi + \hat{\jmath}\cos\phi$$

$$\mathbf{A} = \frac{Ia^4}{c}\hat{\varphi}\int_0^{\infty} dk \cos kz \; I_1(k\rho_<)K_1(k\rho_>)$$

(b) $$A = \frac{Ia}{c}\int_0^{2\pi} \frac{\phi \, d\phi'}{|\mathbf{x} - \mathbf{x}'|}$$

$$A = \frac{Ia}{c}\int_0^{2\pi} \hat{\varphi}' d\phi' \sum_{m=-\infty}^{+\infty} \int dk \; e^{im(\phi - \phi')} J_m(k\rho) J_m(ka) e^{-k(z_> - z_<)}$$

keeping only the real part

$$A = \frac{Ia}{c}\int_0^{2\pi} \hat{\varphi}' d\phi' \int_0^{\infty} dk \; e^{-k(z_> - z_<)}\Big\{\tfrac{1}{2}J_0(k\rho)J_0(ka)$$

$$+\sum_{m=1}^{\infty} \cos m(\phi - \phi') J_m(k\rho) \times J_m(ka)\Big\}$$

By previous arguments the first integral vanishes and

$$\int_0^{2\pi} \hat{\varphi} d\phi' \cos m(\phi - \phi') = \pi\hat{\varphi}$$

$$\mathbf{A} = \hat{\varphi}\frac{I\pi a}{c}\int_0^{\infty} e^{-k(z_> - z_<)} J_1(ka) J_1(k\rho) dk$$

$$= \frac{I\pi a\hat{\varphi}}{c}\int_0^{\infty} e^{-kz_>} J_1(ka)J_1(k\rho)dk + \frac{I\pi a}{c}\int_0^{-\infty} dk \; e^{-k|z_<|} J_1(k\rho)J_1(ka)$$

$$\mathbf{A} = \frac{2I\pi a\hat{\varphi}}{c}\int_0^{\infty} e^{-k|z|} J_1(k\rho) J_1(ka) dk$$

PROBLEM 7.42 The figure shows a rectangular coil suspended from one arm of an analytic balance. It hangs between the poles of an electromagnet with the plane of the coil parallel to the pole faces. The magnetic field is uniform in the shaded region and negligibly small in the neighborhood of the top wire.

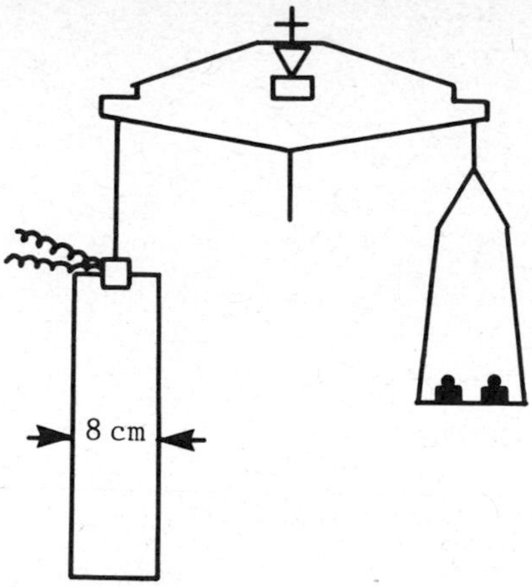

Solution: For $\ell \perp B$, $F = BI\ell/c = (I/10)B\ell$ with I in amperes, B in gauss, and ℓ in centimeters,

$$F = mg = 60.5 \text{ g} \times 980 \text{ cm/sec}^2 = 5.93 \times 10^4 \text{ dyn}$$

$$B = \frac{10F}{I\ell} = \frac{10 \times 5.93 \times 10^4}{0.5 \text{ A} \times 15 \text{ turns} \times 8 \text{ cm}} = 9883 \text{ G}$$

PROBLEM 7.43 The magnetic moment of a current distribution is defined to be

$$\mathbf{m} = \frac{1}{2c} \int \mathbf{r} \times \mathbf{J} \, d\tau$$

Show that, for the case of a filamentary current loop with current I,

$$\mathbf{m} = \frac{I}{c} \mathbf{S}$$

where $\mathbf{S}$ is the vector area of any surface spanning the loop. In this way, prove Ampere's result that the field of induction of a plane current loop at large distances is exactly like that produced by an electric dipole.

Solution: For a filament loop

$$\mathbf{m} = \frac{1}{2c} \int \mathbf{r} \times \mathbf{J} \, dz \to \frac{I}{2c} \oint_c \mathbf{r} \times d\mathbf{r}$$

Let us define a tensor $G_{ij}(\mathbf{r})$ such that

$$\mathbf{r} \times d\mathbf{r} = \overline{\overline{\mathrm{G}}} \cdot d\mathbf{r}$$

The explicit representation of $\overline{\overline{\mathrm{G}}}$ is given as follows:

$$(\mathbf{r} \times d\mathbf{r})_x = y\,dz - z\,dy, \quad \mathrm{G}_x = (0, -z, y)$$

$$(\mathbf{r} \times d\mathbf{r})_y = z\,dx - x\,dz, \quad \mathrm{G}_y = (z, 0, -x)$$

$$(\mathbf{r} \times d\mathbf{r})_z = x\,dy - y\,dx, \quad \mathrm{G}_z = (-y, x, 0)$$

then

$$\nabla \times G_x = (2,0,0)$$

$$\nabla \times G_y = (0,2,0)$$

$$\nabla \times G_z = (0,0,2)$$

Therefore, using Stoke's theorem in the three-component directions, we have

$$\frac{I}{2c}\left(\oint_c \mathbf{r}\times d\mathbf{r}\right)_x = \frac{I}{2c}\oint_c \mathbf{G}_x \cdot d\mathbf{r} = \frac{I}{2c}\int_S (\nabla\times\mathbf{G}_x)\cdot\hat{\mathbf{n}}\,da = \frac{I}{2c}\int_S 2\hat{\mathbf{n}}_x\,da = \frac{I}{c}\mathbf{S}_x$$

and, similarly, in the y,z directions, Therefore,

$$\frac{I}{2c}\oint_c \mathbf{r}\times d\mathbf{r} \equiv \frac{I}{c}\mathbf{S}$$

To prove Ampere's result, we note that if we let the plane loop be on the x,y plane, so that $\mathbf{m} = (I/c)A\hat{\mathbf{z}}$, where A is the area in the x,y plane bounded the filament, then

$$\mathbf{A} = \frac{\mathbf{m}\times\hat{\mathbf{n}}}{r^2} = \frac{IA}{c}\frac{\hat{\mathbf{z}}\times\hat{\mathbf{n}}}{r^2} \tag{1}$$

$$\mathbf{n} = \frac{\mathbf{r}}{|r|}$$

$$\hat{\mathbf{z}}\times\hat{\mathbf{n}} = \begin{vmatrix} \hat{\mathbf{e}}_r & \hat{\mathbf{e}}_\theta & \hat{\mathbf{e}}_\phi \\ \cos\theta & -\sin\theta & 0 \\ 1 & 0 & 0 \end{vmatrix} = \hat{\mathbf{e}}_\phi(\sin\theta)$$

Therefore,

$$\mathbf{A} = \frac{IA}{c}\frac{\sin\theta}{r^2}\hat{\mathbf{e}}_\phi$$

and the field of induction becomes

$$\mathbf{H} = \nabla\times\mathbf{A} = \frac{IA}{c}\left[\hat{\mathbf{e}}_r\left(\frac{1}{r\sin\theta}\right)\frac{\partial}{\partial\theta}\left(\frac{\sin\theta\sin\theta}{r^2}\right) - \hat{\mathbf{e}}_\theta\left(\frac{1}{r}\right)\frac{\partial}{\partial r}\left(\frac{\sin\theta}{r}\right)\right]$$

$$= \frac{IA}{c}\left[\hat{\mathbf{e}}_r\left(\frac{1\cdot 2\sin\theta\cos\theta}{r^3\sin\theta}\right) + \hat{\mathbf{e}}_\theta\left(\frac{1}{r}\frac{\sin\theta}{r^2}\right)\right]$$

$$\mathbf{H} = \frac{IA}{c}\left[\left(\frac{2\cos\theta}{r^3}\right)\hat{\mathbf{e}}_r + \left(\frac{\sin\theta}{r^3}\right)\hat{\mathbf{e}}_\theta\right] \quad \frac{1}{r^3\sin\theta}\ (2\sin\theta\cos\theta)$$

Now, consider a dipole $\mathbf{p} = p\mathbf{z}$, then

$$\Phi = \frac{\mathbf{p} \cdot \hat{\mathbf{n}}}{r^2} = \frac{p \cos\theta}{r^2}$$

and

$$E = -\nabla\phi = -\left(\frac{-2p\cos\theta}{r^3}\right)\hat{\mathbf{e}}_r + \left(\frac{1}{r}\,\frac{p}{r^2}\,\frac{\partial\cos\theta}{\partial\theta}\right)\hat{\mathbf{e}}_\theta$$

$$E = p\left[\left(\frac{2\cos\theta}{r^3}\right)\hat{\mathbf{e}}_r + \left(\frac{\sin\theta}{r^3}\right)\hat{\mathbf{e}}_\theta\right]$$

We see that if we identify $|\mathbf{p}| = IA/c = |\mathbf{m}|$, the field of induction of a plane current loop is like that of an electric dipole, provided we are at large distances so that Eq. (1) is a valid approximation.

Chapter 8

PROBLEM 8.1 For a homogeneous, isotropic, nonmagnetic medium of conductivity g, in which there are steady currents, show that **B** satisfies the vector Laplace equation: $\nabla^2 \mathbf{B} = 0$.

Solution:

$$\mathbf{j} = g\mathbf{E}, \quad \nabla \times \mathbf{B} = \mu_0 \mathbf{j} = \mu_0 g \mathbf{E}$$

$$\nabla \times \nabla \times \mathbf{B} = \nabla(\nabla \cdot \mathbf{B}) - \nabla^2 \mathbf{B} = \mu_0 g \ \nabla \times \mathbf{E}$$

But $\nabla \times \mathbf{E} = 0$ and $\nabla \cdot \mathbf{B} = 0$, so $\nabla^2 \mathbf{B} = 0$.

PROBLEM 8.2 Given the following set of conductors: an infinitely long, straight wire surrounded by a thin cylindrical shell of metal (at radius b) arranged coaxially with the wire. The two conductors carry equal but opposite currents, I. Find the magnetic vector potential for the system.

Solution:

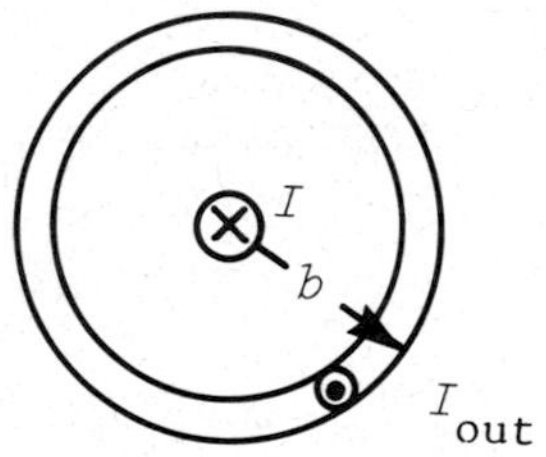

$$\nabla^2 \mathbf{A} = -\mu_0 \mathbf{j}$$

$$\nabla^2 A_z = -\mu_0 j_z(r),$$

$$\frac{1}{r}\frac{\partial}{\partial r}\left(r \frac{\partial A_z}{\partial r}\right) = -\mu_0 j_z(r)$$

$$r\frac{\partial A_z}{\partial r} = -\mu_0 \int_0^r r' j_z(r')dr' = -\frac{\mu_0}{2\pi} I, \quad \text{for } a < r < b \ (a = \text{radius of wire})$$

$$= 0, \qquad b \leq r$$

therefore,

$$\frac{\partial A_z}{\partial r} = -\frac{\mu_0 I}{2\pi} I, \quad r < b \to A_z = -\frac{\mu_0 I}{2\pi} \ln r + c$$

$$\frac{\partial A_z}{\partial r} = 0, \quad b \leq r, \ A_z = \text{const.} = -\frac{\mu_0 I}{2\pi} \ln b + c$$

Choose C such that

$$A_z = -\frac{\mu_0 I}{2\pi} \ln\left(\frac{r}{b}\right) \text{ inside and } A_z = 0 \text{ outside}$$

(This makes the argument of the log dimensionless, but is mostly irrelevant.)

PROBLEM 8.3 The vertical component of the magnetic induction between the pole faces of a particle accelerator is given by $B_z = B_z(r,z)$, where $r = (x^2+y^2)^{\frac{1}{2}}$ is the distance from the axis of the pole faces.

(a) If $|B_z|$ is a decreasing function of r, show that the lines of magnetic intensity bow outward, as shown in the figure, regardless of whether the upper pole is a north or south pole. [*Hint:* Use the fact that curl $\mathbf{B} = 0$, and that $B_r = 0$ on the median plane.]

(b) If the lines of $\mathbf{B}$ bow as shown in the figure, show that accelerated particles which drift away from the median plane experience a force tending to restore them to the median plane, regardless of whether they are positively or negatively charged.

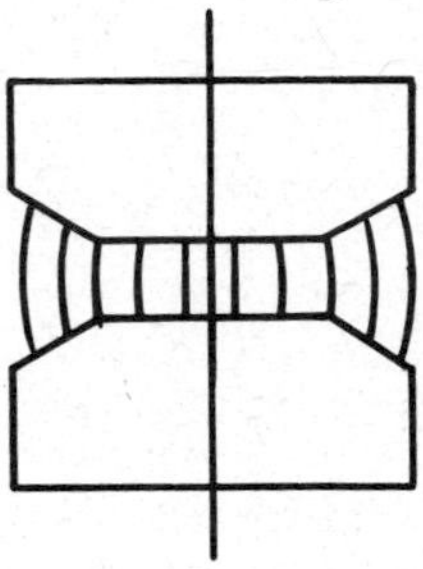

Solution:

(a) Set θ-component of $\nabla \times \mathbf{B} = 0 \rightarrow \dfrac{\partial B_r}{\partial z} - \dfrac{\partial B_z}{\partial r} = 0$

$$\frac{\partial B_r}{\partial z} = \frac{\partial B_z}{\partial r}$$

Also

$$\frac{\partial |B_z|}{\partial r} < 0, \quad \frac{\partial |B_z|}{\partial r} = \frac{\partial}{\partial r}(B_z^2)^{\frac{1}{2}} = \frac{B_z}{(B_z^2)^{\frac{1}{2}}}\frac{\partial B_z}{r}$$

$$= \frac{B_z}{|B_z|}\frac{\partial B_z}{\partial r}$$

$$\frac{\partial |B_z|}{\partial r} = \frac{B_z}{|B_z|}\frac{\partial B_r}{\partial z}$$

Thus for $B_z > 0$, $\dfrac{\partial B_r}{\partial z} < 0$

and for $B_z < 0$, $\dfrac{\partial B_r}{\partial z} > 0$

for $B_z > 0$ then, B_r decreases as z increases, it is positive below the $z = 0$ plane, is zero in the $z = 0$ plane, and becomes negative above the $z = 0$ plane thus:

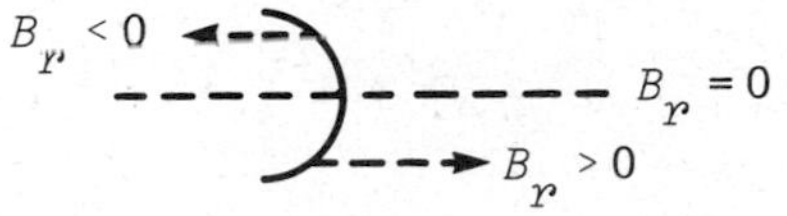

For $B_z < 0$, B_r increases as we cross the $z = 0$ plane so it starts out negative below $z = 0$ plane, goes through 0, and becomes positive, thus:

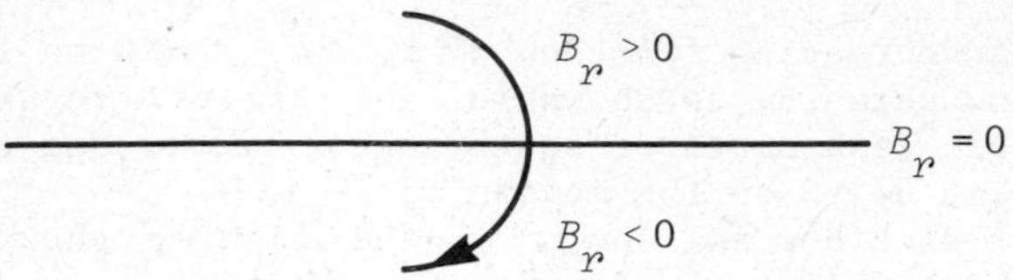

(b) For $B_z > 0$, say a positive particle comes out of paper at right, so force always restores particles toward median plane.

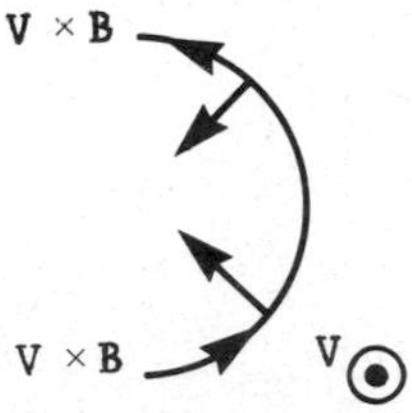

For a negative particle the direction of motion is opposite. This gives rise to two cancelling minus signs and one still has "vertical focusing" as it is called.

PROBLEM 8.4 A solenoid of square cross section (i.e., a solenoid in which the individual turns are in the shape of a square) has N turns per unit length and carries current I. The cross-sectional dimension is a. If the solenoid is very long, find the axial magnetic induction at its center.

Solution:

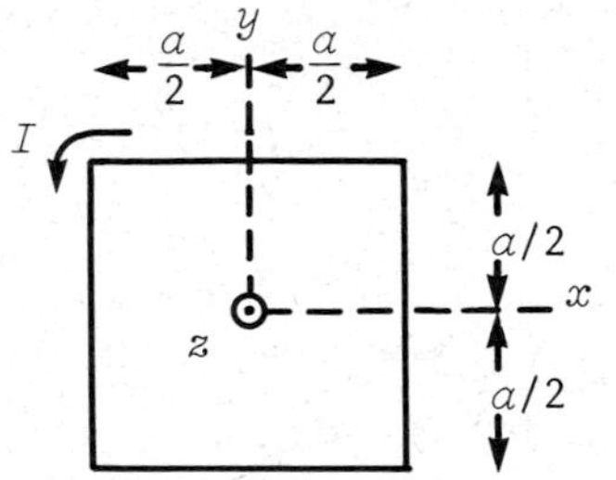

$$\mathbf{B} = \frac{\mu_0}{4\pi} \int_s \frac{\mathbf{j}_s(\mathbf{r}') \times (\mathbf{r} - \mathbf{r}')}{|\mathbf{r} - \mathbf{r}'|^3} \, ds'$$

$\mathbf{j}_s$ = (surface current density)

Consider the contribution to the field at some point on the z axis, $(0,0,z)$, due to the currents on the surface at $y = a/2$. (We should be able to obtain the contribution from the three other surfaces by inspection.) On this surface we have:

$$\mathbf{j}_s(\mathbf{r}') = -IN\hat{\mathbf{x}}, \quad \mathbf{r}' = \hat{\mathbf{x}}x' + \hat{\mathbf{y}}\frac{a}{2} + \hat{\mathbf{z}}z', \quad (\mathbf{r} = \hat{\mathbf{z}}z)$$

$$\mathbf{r} - \mathbf{r}' = -\hat{\mathbf{x}}x' - \hat{\mathbf{y}}\frac{a}{2} + \hat{\mathbf{z}}(z - z'),$$

$$\mathbf{j}_s(\mathbf{r}') \times (\mathbf{r} - \mathbf{r}') = \hat{\mathbf{z}}IN\frac{a}{2} + \hat{\mathbf{y}}IN(z - z')$$

$$\mathbf{B}(\text{partial}) = \frac{\mu_0}{4\pi} IN \left[\hat{\mathbf{z}}\frac{a}{2} \int_{-a/2}^{a/2} dx' \int_{-\infty}^{\infty} dz' \left(x'^2 + \frac{a^2}{4} + (z - z')^2 \right)^{-3/2} \right.$$

$$+\mathbf{y}\int_{-a/2}^{a/2} dx' \int_{-\infty}^{\infty} (z-z')\left(x'^2+\frac{a^2}{4}+(z-z')^2\right)^{-3/2} dz'\Bigg]$$

Let $z - z' = \eta$, $+dz' = -d\eta$ (limits of integration change sign also)

$$\mathbf{B}_p = \frac{\mu_0}{4\pi} IN \Bigg[\hat{\mathbf{z}}\,\frac{a}{2}\int_{-a/2}^{a/2} dx' \int_{-\infty}^{\infty} d\eta\left(x'^2+\frac{a^2}{4}+\eta^2\right)^{-3/2}$$

$$-\hat{\mathbf{y}}\int_{-a/2}^{a/2} dx' \int_{-\infty}^{\infty} \eta\left(x'^2+\frac{a^2}{4}+\eta^2\right)^{-3/2} d\eta\Bigg]$$

The second term is the integral of an odd function of η over an "even" interval $(-\infty,\infty)$ and vanishes for this reason. So $B_{yp} = 0$ and hence $\mathbf{B}_{\text{total}} = \hat{\mathbf{z}}B$ also. The remaining integrals are standard:

$$\int_{-\infty}^{\infty} d\eta\left(x'^2+\frac{a^2}{4}+\eta^2\right)^{-3/2} = \frac{1}{(x'^2+a^2/4)}\left(\frac{\eta}{(x'^2+a^2/4+\eta^2)^{1/2}}\right)_{-\infty}^{\infty}$$

$$= \frac{2}{x'^2+a^2/4}$$

$$\int_{-a/2}^{a/2} \frac{dx'}{x'^2+a^2/4} = \frac{2}{a}\left(\arctan\frac{2x}{a}\right)_{-a/2}^{a/2} = \frac{2}{a}\,[\arctan(1) - \arctan(-1)]$$

$$= \frac{2}{a}\,\frac{\pi}{2} = \frac{\pi}{a}$$

So

$$\mathbf{B}_p = \hat{\mathbf{z}}IN\mu_0\,\frac{\frac{a}{2}\times 2\times\frac{\pi}{a}}{4\pi}\quad\frac{\hat{\mathbf{z}}IN\mu_0}{4}$$

Therefore,

$$\mathbf{B} = 4\mathbf{B}_p = \hat{\mathbf{z}}\mu_0 NI$$

which we know to be true from much more general considerations.

PROBLEM 8.5 The rectangular loop is pivoted about the y axis and carries a current of 10 A in the indicated direction.

(a) Calculate the force on each side of the loop, and the torque in dyn · cm about the y axis, when the loop is in a uniform magnetic field of 2000 G along the x axis.
(b) Repeat the calculations of (a) when the field is along the z axis.
(c) What torque would be exerted with the field along the z axis if the loop were pivoted about an axis through its center, parallel to the y axis?

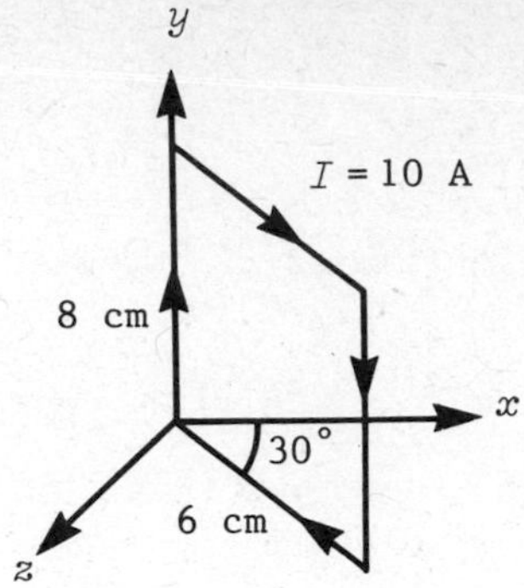

Solution:

(a) $\mathbf{F} = \frac{I}{10}\, \boldsymbol{\ell} \times \mathbf{B}$

$I/10 = 1, \quad \mathbf{B} = 2000\ \hat{\mathbf{x}}$

I: $\boldsymbol{\ell} = 8\hat{\mathbf{y}}$

$$\mathbf{F}_{\mathrm{I}} = 8(2000)(\hat{\mathbf{y}} \times \hat{\mathbf{x}}) = -1.6 \times 10^4\ \hat{\mathbf{z}}\ \text{dyn} = -\mathbf{F}_{\mathrm{III}}$$

II: $\boldsymbol{\ell} = 6\left(\frac{\sqrt{3}}{2}\ \hat{\mathbf{x}} + \tfrac{1}{2}\ \hat{\mathbf{z}}\right)$

$$\mathbf{F}_{\mathrm{II}} = 6(\tfrac{1}{2})(2000)(\hat{\mathbf{z}} \times \hat{\mathbf{x}}) = 0.6 \times 10^4\ \hat{\mathbf{y}}\ \text{dyn} = -\mathbf{F}_{\mathrm{IV}}$$

$\mathbf{F}_{\mathrm{II}}$ and $\mathbf{F}_{\mathrm{IV}}$ cause no torque about y, since $\mathbf{F} = F\hat{\mathbf{y}}$. $\mathbf{F}_{\mathrm{I}}$ causes no torque about y axis since $\mathbf{r} = 0$.

$$\boldsymbol{\tau} = \mathbf{r}_{\mathrm{III}} \times \mathbf{F}_{\mathrm{III}} = 6\left(\frac{\sqrt{3}}{2}\ \hat{\mathbf{x}} + \tfrac{1}{2}\ \hat{\mathbf{z}}\right) \times 1.6 \times 10^4\ \hat{\mathbf{z}}$$

$$= -8.31 \times 10^4\ \hat{\mathbf{y}}\ \text{dyn} \cdot \text{cm}$$

(b) $\mathbf{B} = 2000\ \hat{\mathbf{z}}$

$$\mathbf{F}_{\mathrm{I}} = 1.6 \times 10^4\ (\hat{\mathbf{y}} \times \hat{\mathbf{z}}) = 1.6 \times 10^4\ \hat{\mathbf{x}}\ \text{dyn} = -\mathbf{F}_{\mathrm{III}}$$

$$\mathbf{F}_{\mathrm{II}} = 1.2 \times 10^4\ \left(\frac{\sqrt{3}}{2}\ \hat{\mathbf{x}} + \tfrac{1}{2}\ \hat{\mathbf{z}}\right) \times \hat{\mathbf{z}} = -1.04 \times 10^4\ \hat{\mathbf{y}}\ \text{dyn} = -\mathbf{F}_{\mathrm{IV}}$$

$$\boldsymbol{\tau} = 6\left(\frac{\sqrt{3}}{2}\ \hat{\mathbf{x}} + \tfrac{1}{2}\ \hat{\mathbf{z}}\right) \times 1.6 \times 10^4\ \hat{\mathbf{x}} = 4.8 \times 10^4\ \hat{\mathbf{y}}\ \text{dyn} \cdot \text{cm}$$

(c) Since $\mathbf{F}_{\mathrm{I}} = -\mathbf{F}_{\mathrm{II}}$, the torque is the same about any axis parallel to the y axis.

PROBLEM 8.6 The square loop lies in the xy-plane and carries a current of 10 A in the direction shown. A magnetic field $B_x = 10\ x$ (B in gauss, x in centimeters) is applied. Calculate:

(a) the net force on the loop and
(b) the torque about Q.

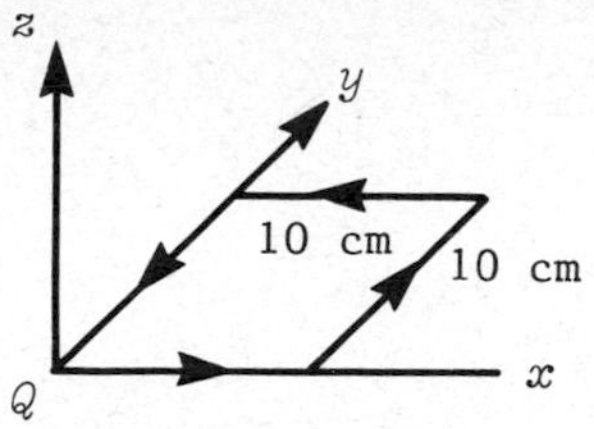

Solution:

(a) Since $\mathbf{B} = B\hat{\mathbf{x}}$ and $B(x = 0) = 0$, only the side at $x = 10$ cm contributes to $\mathbf{F}$.

$$\mathbf{F} = \frac{I}{10}\, \boldsymbol{\ell} \times \mathbf{B} = \frac{10\text{ A}}{10} (10\text{ cm})(10 \times 10\text{ G})(\hat{\mathbf{y}} \times \hat{\mathbf{x}}) = -10^3\, \hat{\mathbf{z}}\text{ dyn}$$

(b) $\tau_z = 0$ since $F_x = F_y = 0$, $\tau_y = 10\hat{\mathbf{x}} \times (-10^3\, \hat{\mathbf{z}}) = 10^4\, \hat{\mathbf{y}}$ dyn · cm

$$\tau_x = \int (\mathbf{r} \times d\mathbf{F})_x = \int \mathbf{r} \times \left(\frac{I}{c}\, d\boldsymbol{\ell} \times \mathbf{B}\right), \quad \mathbf{r} = y\hat{\mathbf{y}}$$

$$d\boldsymbol{\ell} = \hat{\mathbf{y}}\, dy$$

$$\tau_x = \int y\hat{\mathbf{y}} \times (-100\, \hat{\mathbf{z}}\, dy)$$

$$= -100\, \hat{\mathbf{x}} \int_0^{10} y\, dy = -0.5 \times 10^4\, \hat{\mathbf{x}}\text{ dyn} \cdot \text{cm}$$

PROBLEM 8.7 Two long, straight, parallel wires are 100 cm apart as shown in the figure. The upper wire carries a current of 6 A into the plane of the paper.

(a) What must the magnitude and direction of I_2 be for the resultant magnetic field at point P to be zero?
(b) What is then the resultant field at point Q?
(c) At point S?

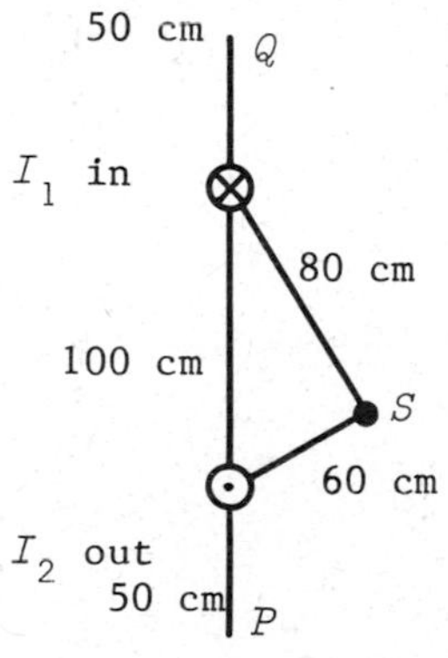

Solution:

(a) $B = 0 = \frac{2I_1}{10 \times 150 \text{ cm}} - \frac{2I_2}{10 \times 50 \text{ cm}}, \; I_2 = \frac{1}{3} I_1 = 2$ A out

(b) $B = \frac{2 \times 6}{10 \times 50} - \frac{2 \times 2}{10 \times 150} = 0.0213$ G to the right

(c) The sides 60, 80, and 100 cm are those of a 3-4-5 right triangle, so $B_1 \perp B_2$ as shown.

$$B_1 = \frac{2 \times 6}{10 \times 80} = 0.015, \quad B_2 = \frac{2 \times 2}{10 \times 60} = 0.00667$$

$$B = \sqrt{(0.015)^2 + (0.00667)^2} = 0.0164 \text{ G}$$

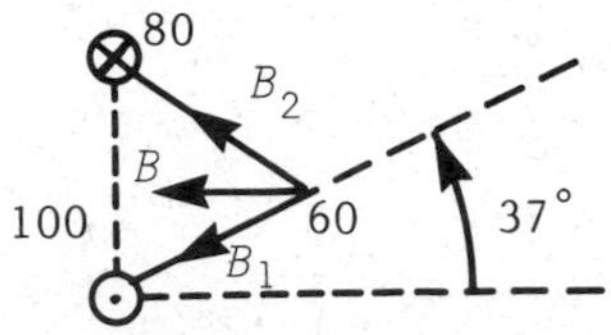

B makes an angle

$$\theta = \tan^{-1}\left(\frac{0.00667}{0.015}\right) = 24°$$

with the 60 cm side or 37° − 24° = 13° below the horizontal and to the left.

PROBLEM 8.8

(a) Show that the magnetic field produced by a linear current of finite length is $(I/cR)(\sin\alpha_1 - \sin\alpha_2)$, where R is the perpendicular distance from the point to the wire and the angles are defined by the figure. (Note the signs of the angles.)

(b) Apply this result to obtain the magnetic field at the center of a square circuit of side L.

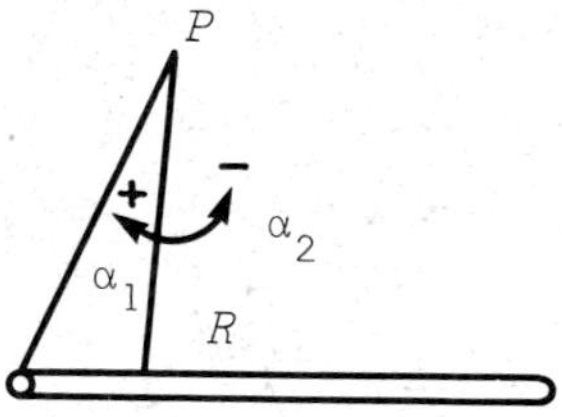

Solution:

(a) $d\mathbf{B} = \frac{I d\boldsymbol{\ell} \times \hat{\mathbf{r}}}{r^2 c}$

All $d\mathbf{B}$ are parallel and directed out of the page.

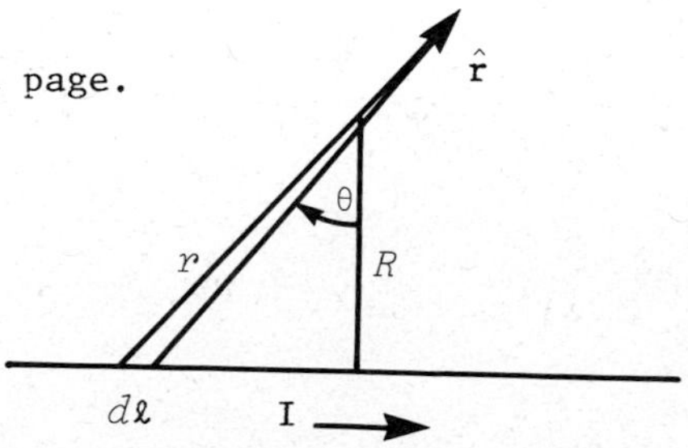

$$R = r\cos\theta$$

$$d\ell \times \mathbf{r} = d\ell \cos\theta$$

$$\ell = R\tan\theta$$

$$d\ell = R\sec^2\theta \, d\theta$$

$$B = \frac{I}{c}\int_{\alpha_2}^{\alpha_1} \frac{R\sec^2\theta\, d\theta \cos\theta}{R^2/\cos^2\theta} = \frac{I}{cR}\int_{\alpha_2}^{\alpha_1} \cos\theta\, d\theta$$

$$= \frac{I}{cR}(\sin\alpha_1 - \sin\alpha_2)$$

(b) $R = \frac{1}{2} L$, $\sin\alpha = \pm 1/\sqrt{2}$

$$B = 4 \times \frac{I}{c(\frac{1}{2} L)}\left(2 \times \frac{1}{\sqrt{2}}\right) = \frac{8\sqrt{2}\, I}{cL}$$

PROBLEM 8.9 A long straight wire carries a current of 1.5 A. An electron moves with a velocity of 5×10^6 cm/sec, parallel to the wire, 10 cm from it, and in the same direction as the current. What force acts on the electron?

Solution:

$$B = \frac{2I}{rc} = \frac{2 \times 1.5}{10 \times 10} = 3 \times 10^{-2}\ \text{G}$$

$$F = \frac{qvB}{c} = \frac{4.8 \times 10^{-10} \times 5 \times 10^6 \times 3 \times 10^{-2}}{3 \times 10^{10}} = 2.4 \times 10^{-15}\ \text{dyn away from wire.}$$

PROBLEM 8.10 Compute the strength of the magnetic field produced by an infinitely long wire carrying a current of 1 A at a distance of

(a) 0.53×10^{-8} cm and
(b) 100 cm.

Solution:

(a) Assuming the wire is finer than this (!),

$$B = \frac{2I}{rc} = \frac{2 \times 1\ \text{A}}{10 \times 0.53 \times 10^{-8}} = 3.77 \times 10^7\ \text{G}$$

(b) $$B = \frac{2I}{rc} = \frac{2 \times 1}{10 \times 100} = 2 \times 10^{-3}\ \text{G}$$

PROBLEM 8.11 A long straight wire and a rectangular loop lie on a table top. The currents are $I_1 = 10$ A and $I_2 = 20$ A.

(a) What is the force on each side of the loop?
(b) What is the torque on the loop about the indicated rotation axis, which passes through the midpoint of the loop?
(c) Find the torque after the coil has been rotated 45° about the rotation axis.

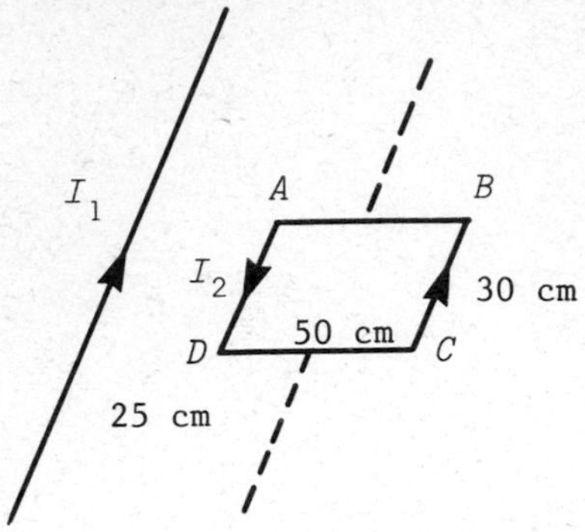

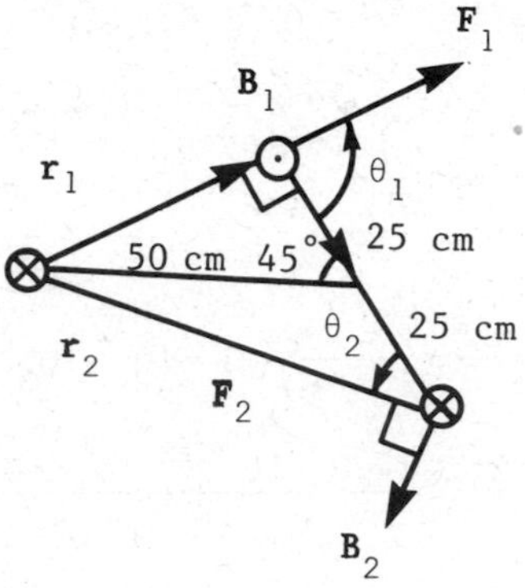

Solution:

(a) $B = \dfrac{2I_1}{rc} = \dfrac{2}{r}$ G

AD and *BC*,

$$dF = I_2 \frac{d\ell \times \mathbf{B}}{c} = \left(\frac{4}{r}\right) d\ell$$

$$F = 4 \int_B^C \frac{dr}{r} = 4 \log \left(\frac{75}{25}\right) = 4.39 \text{ dyn into loop}$$

$$AB, \; F = \frac{BI_2\ell}{c} = \left(\frac{2}{25}\right)\left(\frac{20}{10}\right) 30 = 4.8 \text{ dyn away from } I_1$$

$$CD, \; F = \left(\frac{2}{75}\right)\left(\frac{20}{10}\right) 30 = 1.6 \text{ dyn toward } I_1$$

(b) $\tau = 0$ since all forces are in the plane of the loop.

(c) $r_1^2 = (25)^2 + (50)^2 - 2(25)(50) \cos 45°$

$r_1 = 36.8$ cm

$$\cos \theta_1 = \frac{(50)^2 - (25)^2 - (36.8)^2}{2(25)(36.8)} = 0.283$$

$\theta_1 = 73.6°$

$r_2 = 69.9$ cm, $\theta_2 = 30.5°$

$$F_1 = \frac{BI\ell}{c} = \left(\frac{2}{36.8}\right)\left(\frac{20}{10}\right) 30 = 3.26 \text{ dyn}$$

$$F_2 = \left(\frac{2}{69.9}\right)\left(\frac{20}{10}\right) 30 = 1.72 \text{ dyn}$$

$$\tau = \sum \mathbf{r} \times \mathbf{F} = 25(F_1 \sin \theta_1 + F_2 \sin \theta_2) = 25(3.26 \sin 73.6° + 1.72 \sin 30.5°)$$
$$= 100 \text{ dyn} \cdot \text{cm}$$

Chapter 9

PROBLEM 9.1 Given a magnetic field of cylindrical symmetry, that is, one with a z component $B_z = B(r)$, where r is the distance from the symmetry axis. An ion of charge q and mass m revolves in a circular orbit at distance R from the symmetry axis with angular velocity $\omega = qB(R)/m$. If the magnetic field is slowly increased in magnitude, show that the emf induced around the ion's orbit is such as to accelerate the ion. Show that in order for the ion to stay in its same orbit, the average increase in $B(r)$ over the surface enclosed by the orbit must be twice as large as the increase in $B(R)$.

Solution: B increasing will generate emf around orbit in the clockwise direction by Lenz' law.

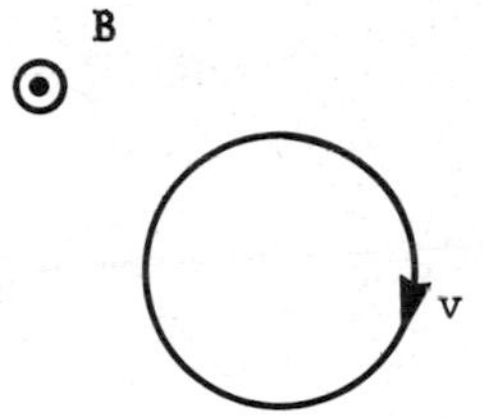

$$\varepsilon = \text{emf} = -\pi r^2 \frac{d\bar{B}}{dt} = \left(-\frac{d\Phi}{dt}\right),$$

so the energy gain per turn is

$$q\varepsilon = q\pi r^2 \frac{d\bar{B}}{dt}$$

and per unit time:

$$\frac{d}{dt}\left(\frac{mv^2}{2}\right) = \frac{qv}{2\pi r} \times \varepsilon = \frac{qvr}{2}\frac{d\bar{B}}{dt}$$

where $\bar{B}$ is the average magnetic field and v is the speed of the ion. Since the magnetic field varies slowly, we assume the speed varies slowly and use

$$\frac{mv^2}{r} = qvB, \quad v = \frac{qB}{m} r$$

(B = local field), then

$$\frac{mv^2}{2} = \tfrac{1}{2}\frac{q^2r^2}{m} B^2$$

$$\frac{d}{dt}\left(\frac{mv^2}{2}\right) = \frac{q^2r^2}{m} B \frac{dB}{dt} = qrv\frac{dB}{dt}$$

We now equate this with the above form for the rate of change of the energy:

$$qrv\,\frac{dB}{dt} = \frac{qvr}{2}\,\frac{d\bar{B}}{dt}$$

$$\frac{dB}{dt} = \tfrac{1}{2}\,\frac{d\bar{B}}{dt}$$

Relativistically: For slow acceleration (energy change per turn small). We still just balance centripetal force against magnetic force

$$\frac{\gamma m v^2}{r} = qvB, \quad \gamma = \left(1 - \frac{v^2}{c^2}\right)^{-1/2}, \quad m = \text{rest mass}$$

$$\frac{\gamma m v}{r} = qB, \quad \gamma m v = qrB$$

$$\frac{d}{dt}(\gamma m v) = mv\,\frac{d\gamma}{dt} + \gamma m\,\frac{dv}{dt} = qr\,\frac{dB}{dt} \quad \text{(constant radius!)}$$

$$\frac{d\gamma}{dt} = \frac{v}{c^2}\left(1 - \frac{v^2}{c^2}\right)^{-3/2}\frac{dv}{dt} = \gamma^3\,\frac{v}{c^2}\,\frac{dv}{dt}$$

$$v\,\frac{d}{dt}(\gamma m v) = mv^2\,\frac{d\gamma}{dt} + \gamma m v\,\frac{dv}{dt} = mc^2\,\frac{d\gamma}{dt}\left(\frac{v^2}{c^2} + \frac{1}{\gamma^2}\right) = qrv\,\frac{dB}{dt}$$

$$= mc^2\,\frac{d\gamma}{dt}\left(\frac{v^2}{c^2} + 1 - \frac{v^2}{c^2}\right)$$

$$\frac{d}{dt}(\gamma m c^2) = qrv\,\frac{dB}{dt}$$

Now, energy gain per turn $= q\,\dfrac{d\Phi}{dt} = q\pi r^2\,\dfrac{d\bar{B}}{dt}$

$$\frac{d}{dt}(\gamma m c^2) = \text{rate of energy gain} = \frac{v}{2\pi r}\,q\pi r^2\,\frac{d\bar{B}}{dt}$$

$$\frac{d}{dt}(\gamma m c^2) = \frac{vqr}{2}\,\frac{d\bar{B}}{dt}$$

$$\frac{dB}{dt} = \tfrac{1}{2}\,\frac{d\bar{B}}{dt} \quad \text{("1-2 condition" for betatron)}$$

PROBLEM 9.2 Show that the emf in a fixed circuit C is given by

$$-\frac{d}{dt}\oint_C \mathbf{A}\cdot d\mathbf{l},$$

where $\mathbf{A}$ is the vector potential.

Solution: From Faraday's law of induction we have

$$\text{emf} = -\frac{d\Phi}{dt} = -\frac{d}{dt}\int_S \mathbf{B}\cdot\mathbf{n}\,ds$$

and a surface S with "rim" C.

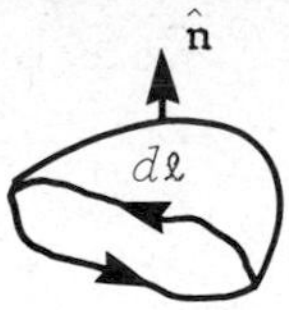

$$\text{emf} = -\frac{d}{dt}\int_S \nabla \times \mathbf{A} \cdot \mathbf{n}\, ds \quad \text{by definition of vector potential}$$

$$\varepsilon = -\frac{d}{dt}\int_C \mathbf{A} \cdot d\ell \quad \text{by Stokes' theorem}$$

PROBLEM 9.3 A metallic conductor in the shape of a wire segment of length ℓ is moved in a magnetic field $\mathbf{B}$ with velocity $\mathbf{v}$. From a consideration of the Lorentz force on the electrons in the wire, show that the ends of the wire are at the potential difference: $\mathbf{B} \cdot \ell \times \mathbf{v}$.

Solution: As the wire moves through the magnetic field magnetic forces and charge separation will take place. The latter will give rise to electrostatic fields, and once static equilibrium is achieved the net force on the charge carriers will be zero:

$$F_{\text{Lorentz}} = -e(\mathbf{E} + \mathbf{v} \times \mathbf{B}) = 0, \quad \mathbf{E} = -\mathbf{v} \times \mathbf{B}$$

$$\text{Potential difference} = -\int_0^\ell \mathbf{E} \cdot d\ell = \int_0^\ell \mathbf{v} \times \mathbf{B} \cdot d\ell = -\int_0^\ell \mathbf{B} \times \mathbf{v} \cdot d\ell$$

$$= -\mathbf{B} \times \mathbf{v} \cdot \ell \,(\text{uniform } \mathbf{B}) = -\mathbf{B} \cdot (\mathbf{v} \times \ell) = \mathbf{B} \cdot (\ell \times \mathbf{v})$$

PROBLEM 9.4 A square metal frame is located, as shown in the figure, between the poles of an electromagnet. The upper side of the frame is in a region of substantially uniform horizontal magnetic field of strength B. The lower side of the frame is outside the magnet gap where the field, though not zero, is negligible for the purposes of this problem. Show that if the frame is released, and falls under its own weight, it will acquire a downward velocity that depends only on B, for a given frame material, and is independent of the size of the frame and the cross section of the wire or rod from which it is made. What is the velocity, in cm/sec, if B is 15 kG and the frame is made of aluminum (density 2.7 g/cm^3; resistivity 2.8×10^{-6} ohm · cm)? About how far will the frame fall before it reaches its terminal velocity?

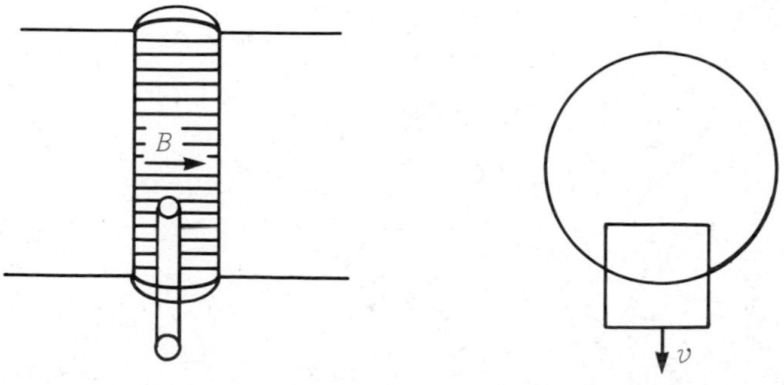

Solution:

$$F_m = \frac{Bi\ell}{c}, \quad i = \frac{\varepsilon}{R}, \quad R = 4\ell\rho/A$$

A = cross section of wire,

ℓ = side of square loop

ρ = resistivity,

$\varepsilon = B\ell v/c$

$$F_m = \frac{B\ell(\varepsilon/R)}{c} = B\ell\left(\frac{B\ell v}{c}\right)\frac{(A/4\ell\rho)}{c}$$

$$= \left(\frac{B^2\ell A}{4\rho c^2}\right)v$$

$$F_g = mg = \delta(4\ell Ag), \quad \delta = \text{mass density}$$

At equilibrium,

$$\left(\frac{B^2\ell A}{4\rho c^2}\right)v_t = 4\ell A\delta g$$

$$v_t = \left(\frac{4\rho c^2}{B^2\ell A}\right)(4\ell A\delta g) = \frac{16\rho\delta c^2 g}{B^2}$$

1 statohm = 9×10^{11} ohm = $c^2 \times 10^{-9}$ ohm

$$v_t = \frac{16 \times 2.8 \times 10^{-6}\ \text{ohm}\cdot\text{cm} \times 2.7 \times c^2 \times 980}{c^2 \times 10^{-9} \times (1.5 \times 10^4)^2} = 0.527\ \text{cm/s}$$

$$m = \frac{dv}{dt} = F_g - F_m = \alpha - \beta v, \quad v_t = \frac{\alpha}{\beta}$$

$$\int_0^v \frac{dv}{v - v_t} = -\frac{g}{v_t}\int_0^t dt$$

$$\ell n\left(\frac{v - v_t}{-v_t}\right) = -\frac{gt}{v_t}$$

$$v = v_t\left[1 - \exp\left(\frac{-gt}{v_t}\right)\right] \simeq v_t \text{ when } t = 3v_t/g$$

$$\int_0^y dy = v_t\int_0^t\left[1 - \exp\left(\frac{-gt}{v_t}\right)\right]dt$$

$$y = v_t \left[t - \frac{v_t}{g}\right]\left[1 - \exp\left(\frac{-gt}{v_t}\right)\right]$$

$$\simeq v_t \left[3\left(\frac{v_t}{g}\right) - \frac{v_t}{g}\right] \quad \text{as } v \to v_t$$

$$y \simeq \frac{2v_t^2}{g} \simeq \frac{2(0.5)^2}{980} \simeq 5 \times 10^{-4} \text{ cm}$$

PROBLEM 9.5 In the interstellar space in our Galaxy the magnetic field is believed to be generally of the order of magnitude of 10^{-6} G. The matter in this space is typically nothing but hydrogen atoms, about one per cubic centimeter, with thermal velocities of the order of 10^5 cm/sec. How does the amount of energy stored in the magnetic field, in any given volume, compare with that stored as kinetic energy of matter?

Solution:

$$\frac{B^2}{8\pi} = \frac{10^{-12}}{8\pi} \simeq 4 \times 10^{-14} \text{ erg/cm}^3$$

$$\frac{mv^2}{2} = \frac{1.6 \times 10^{-24} \times (10^5)^2}{2} \simeq 0.8 \times 10^{-14} \text{ erg/cm}^3$$

Thus of the same order of magnitude.

PROBLEM 9.6 Show that the emf induced around a closed loop in the presence of a time-dependent magnetic field is given by

$$\text{emf} = -\frac{d}{dt} \oint \mathbf{A} \cdot d\ell$$

where A is the vector potential.

Solution:

$$\nabla \times \mathbf{E} = -\frac{\partial \mathbf{B}}{\partial t}$$

$$\int \nabla \times \mathbf{E} \cdot \mathbf{n}\, da = -\int \frac{\partial \mathbf{B}}{\partial t} \cdot \hat{\mathbf{n}}\, da = -\frac{d}{dt} \int \mathbf{B} \cdot \hat{\mathbf{n}}\, da$$

$$\oint \mathbf{E} \cdot d\ell = -\frac{d}{dt} \int \nabla \times \mathbf{A} \cdot \hat{\mathbf{n}}\, da$$

$$\varepsilon = -\frac{d}{dt} \oint \mathbf{A} \cdot d\ell = \text{emf}$$

PROBLEM 9.7 Consider two closed loops carrying current I_1 and I_2. Let Φ_{ij} be the flux due to the current in loop j linking loop i. Show that

$$\Phi_{ij} = M_{ij} I_j$$

where

$$M_{ij} = \frac{\mu_0}{\mu\pi} \oint_{c_i} \oint_{c_j} \frac{d\ell_i \cdot d\ell_j}{|\mathbf{r}_j - \mathbf{r}_i|}$$

Solution:

$$\Phi_{ij} = \int da_i \, \mathbf{B}_j \, (\mathbf{r}_i) \cdot \mathbf{n}_i = \int da_i \, \nabla_j \times \mathbf{A}_j \, (\mathbf{r}_i) \cdot \mathbf{n}_i$$

$$= \oint d\ell_i \cdot \mathbf{A}_j \, (\mathbf{r}_i)$$

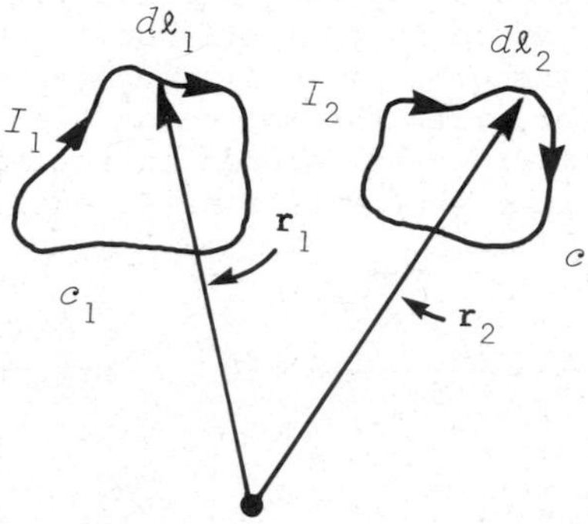

$$A_j = \frac{\mu_0 I_0}{4\pi} \oint \frac{d\ell_j}{|\mathbf{r}_i - \mathbf{r}_j|}$$

$$\Phi_{ij} = \frac{\mu_0 I_j}{4\pi} \oint \oint \frac{d\ell_i \cdot d\ell_j}{|\mathbf{r}_i - \mathbf{r}_j|} = M_{ij} I_j$$

$$M_{ij} = \frac{\mu_0}{4\pi} \oint \oint \frac{d\ell_i \cdot d\ell_j}{|\mathbf{r}_i - \mathbf{r}_j|}$$

PROBLEM 9.8 A. As shown in the diagram, a wire 6 cm long is moved with speed $v = 25$ cm/s perpendicular to a uniform field $B = 2 \times 10^3$ gauss.

(a) Calculate the magnitude and direction of the electric field induced in the wire.

(b) We may write

$$\text{emf} = -\int \mathbf{E} \cdot d\mathbf{s}$$

where **E** is the induced field. From this, calculate the emf between the ends of the wire.

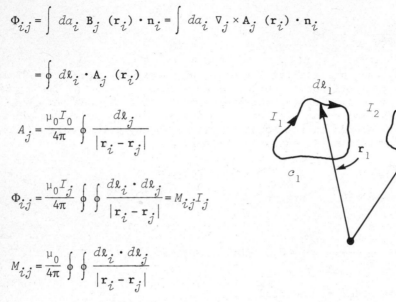

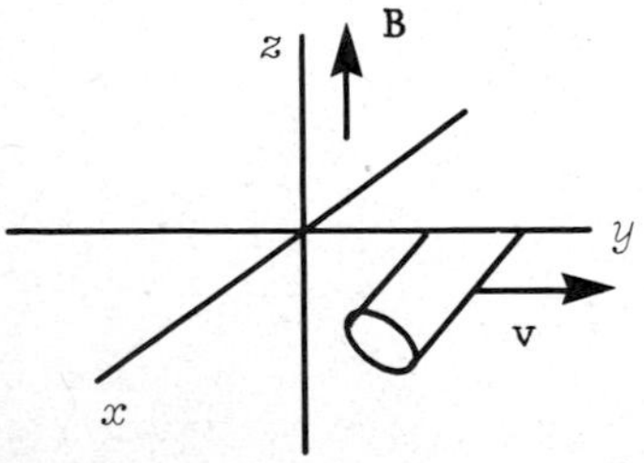

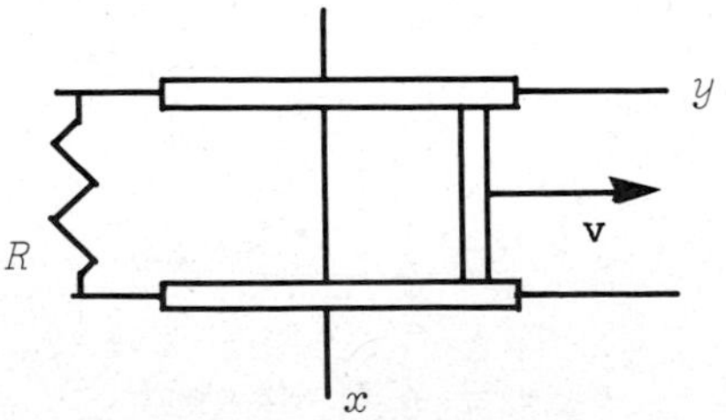

B. Suppose the ends of the wire in problem A are connected electrically to rails that are connected by a fixed resistance $R = 2$ ohms.

(a) Since $\Phi = BA$ is changing because of the change in the area A of the loop, evaluate

$$\varepsilon = -\frac{1}{c}\frac{d}{dt} BA$$

and compare to part b above.

(b) What is the magnitude and direction of the induced current in the loop?
(c) What force is necessary to keep the wire moving with constant velocity?
(d) Compare the power supplied ($\mathbf{F} \cdot \mathbf{v}$) with the heat generated in the resistor (I^2R).

Solution:

A. (a) $$\mathbf{E} = \frac{-\mathbf{v} \times \mathbf{B}}{c} = \frac{-25\hat{\mathbf{y}} \times 2 \times 10^3\hat{\mathbf{z}}}{3 \times 10^{10}}$$

$$= -1.67 \times 10^{-6}\ \hat{\mathbf{x}} \text{ statvolt/cm}$$

Hence the end of $+x$ is positive.

(b) $$\varepsilon = -\int \mathbf{E} \cdot d\mathbf{s} = -\int_0^6 E\,dx$$

$$= 1.00 \times 10^{-5} \text{ statvolts}$$

B. (a) $$\varepsilon = -\frac{1}{c}\frac{d(BA)}{dt}$$

$$= -\frac{B}{c}\frac{dA}{dt} = -\frac{B}{c}\ell v$$

$$= \frac{-2 \times 10^3 \times 6 \times 25}{3 \times 10^{10}} = -1.00 \times 10^{-5} \text{ statvolts}$$

(b) $$I = \frac{\varepsilon}{R} = \frac{10^{-5} \times 300 \text{ V/statvolt}}{2 \text{ ohms}} = 1.5 \times 10^{-3} \text{ A, clockwise}$$

(c) $$\mathbf{F}_m = \frac{I}{c}\,\boldsymbol{\ell} \times \mathbf{B} = \frac{1.5 \times 10^{-3}}{10}(6\hat{\mathbf{x}} \times 2 \cdot 10^3\hat{\mathbf{z}})$$

$$= -1.8\hat{\mathbf{y}} \text{ dyn}$$

so the outside force is

$$\mathbf{F} = -\mathbf{F}_m = 1.8\hat{\mathbf{y}} \text{ dyn}$$

(d) $$\mathbf{F} \cdot \mathbf{v} = 1.8\hat{\mathbf{y}} \cdot 25\hat{\mathbf{y}} = 45 \text{ erg/sec}$$

$$I^2R = (1.5 \times 10^{-3})^2 \times 2 = 4.5 \times 10^{-6} \text{ J/sec}$$

$$= 45 \text{ erg/sec}$$

PROBLEM 9.9 A soft-iron ring with a 1.0-cm air gap is wound with a toroidal winding such as is shown in the figure. The mean length of the iron ring is 20 cm, its cross section is 4 cm^2, and its permeability is 3000 μ_0. The 200-turn winding carries a current of 10 A.

(a) Calculate the magnetic induction in the air gap.
(b) Find B and H inside the iron ring.

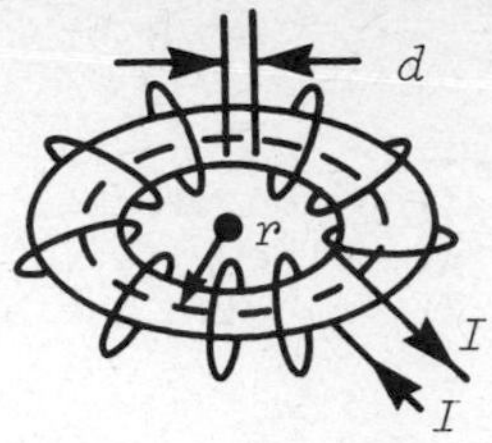

Solution: In this case, separate the gap from the rest of the material:

$$\oint \mathbf{H} \cdot d\ell = NI$$

$$H_m L_m + H_G L_G = NI$$

Assuming no divergence of B at the gap

$$\phi_m = \phi_G$$

$$B_m A_m = B_G A_G \Rightarrow B_m = B_G$$

Since $H = (1/\mu)B$, we have from above,

$$\frac{1}{\mu_m} B\, L_m + \frac{1}{\mu} B\, L_G = NI$$

$$B = \frac{NI}{\frac{1}{\mu} L_m + \frac{1}{\mu} L_G} = \frac{(200)(10)}{\frac{0.2}{3000\,\mu_0} + \frac{0.01}{\mu_0}}$$

$$B = \frac{(200)(10)\,\mu_0}{\frac{0.2}{3000} + \frac{0.01}{1}} = \frac{2000\,\mu_0}{0.01007}$$

(a) $B = 2.49 \times 10^{-1}\ \mathrm{W/m^2}$

(b) $H_m = \dfrac{B}{3000\,\mu_0} = \dfrac{2.49 \times 10^{-1}}{3 \times 10^3\,\mu_0} = 6.6 \times 10^1\ \mathrm{A/m}$

$$H_G = \frac{B}{\mu_0} = \frac{2.49 \times 10^{-1}}{\mu_0} = 1.98 \times 10^5\ \mathrm{A/m}$$

PROBLEM 9.10 Given a circular loop in vacuo of radius a and carrying current I, its axis being coincident with the x coordinate axis and its center being at the origin.

(a) Find, by use of the divergence property of the magnetic induction, the space rate of change of the Cartesian component B_y with respect to y, for a general point on the x axis.

(b) From this result, write an approximate formula for E_y, valid for small enough values of y.

Answer: $B_y = \dfrac{3\mu_0 I a^2 xy}{4(a^2+x^2)^{5/2}}$

(c) Find the magnetic force, due to the field of the loop in the preceding part, on a second circular loop coaxial with the first, having its center at $x = L$. This loop carries current I' in the same sense as the other, and has a radius b sufficiently small that the approximate field B_y of the preceding part is valid.

Solution:

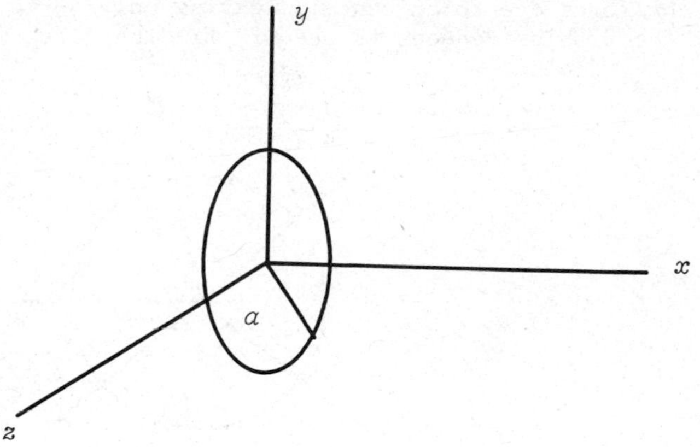

(a) For a circular loop the magnetic field intensity at a point on the x axis is

$$B_x = \frac{\mu_0 I a^2}{2(a^2+x^2)^{3/2}}$$

Now

$$\nabla \cdot \mathbf{B} = 0 \Rightarrow \frac{\partial B_x}{\partial x} + \frac{\partial B_y}{\partial y} + \frac{\partial B_z}{\partial z} = 0$$

Because of the symmetry for a point on the x axis,

$$\frac{\partial B_y}{\partial y} = \frac{\partial B_z}{\partial z}$$

so

$$\frac{\partial B_y}{\partial y} = ½ \frac{\partial B_x}{\partial x} \Rightarrow B_y = \frac{-1}{2} \int \left.\frac{\partial B_x}{\partial x}\right|_{x = \text{const}} dy$$

$$\frac{\partial B_x}{\partial x} = \frac{\mu_0 I a^2}{2} \; \frac{-3}{2} (a^2+x^2)^{-5/2}\, 2x = \frac{-3\mu_0 I a^2 x}{2(a^2+x^2)^{5/2}}$$

(b) $B_y = \dfrac{3\mu_0 I a^2 xy}{4(a^2+x^2)^{5/2}}$

(c)

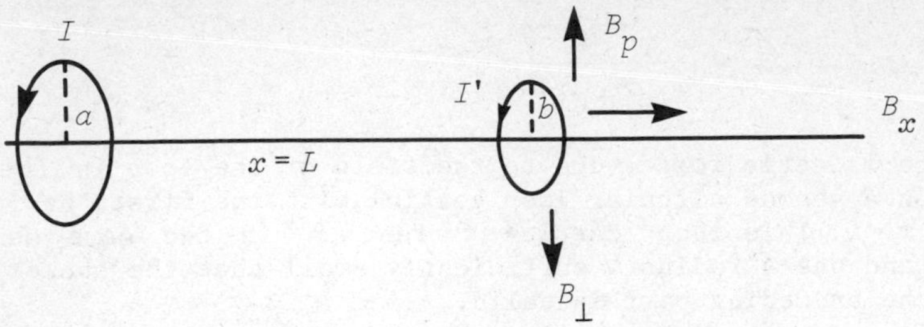

The part of the force due to B_x cancels leaving only B_p to exert a force on the loop. So the problem reduces to one of finding the force on a loop immersed in a $B_\perp$ field.

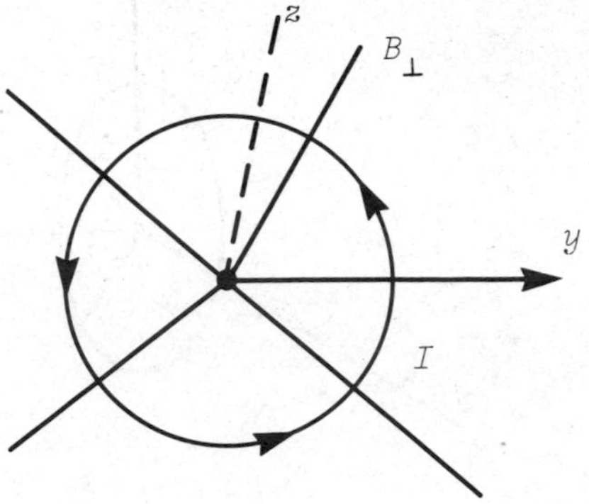

$$d\mathbf{F} = I d\ell \times \mathbf{B}_y = -\hat{\mathbf{i}}\, d\ell\, B_y \Rightarrow \mathbf{F} = -\hat{\mathbf{i}}\, B_y\, 2\pi b$$

$$\mathbf{F} = -\hat{\mathbf{i}}\, \frac{3\mu_0 I a^2 L\, b}{4(a^2 + L^2)^{5/2}}\, 2\pi b$$

PROBLEM 9.11 A rigid, straight wire of length ℓ rotates about an axis perpendicular to it and at one end, in a plane normal to a uniform magnetic field of magnitude B. Contact is made to the wire at the axle, and at the free end by a rail in the form of a circular arc. When the angular speed is w, what is the induced emf in the wire?

Solution:

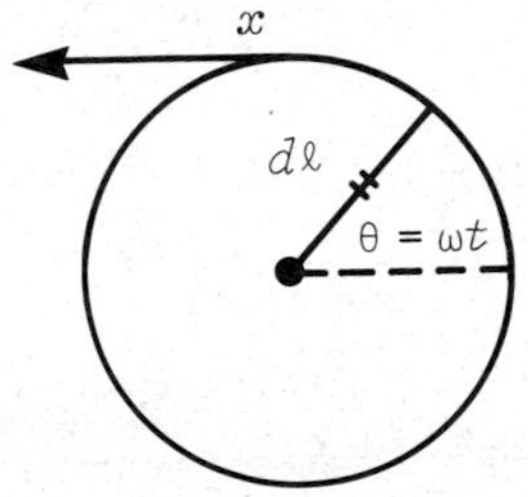

$$d\varepsilon = vB\, d\ell$$

$$v = rw, \quad d\ell = dr$$

$$d\varepsilon = Brw\, dr \Rightarrow \varepsilon = \frac{BwR^2}{2} = \tfrac{1}{2} BwR^2$$

Alternate:

$$\phi = \int B\,dA = \int B\,\tfrac{1}{2}R^2 d\theta = \tfrac{1}{2}BR^2\theta$$

$$\varepsilon = \frac{d\phi}{dt} = \tfrac{1}{2}BR^2\frac{d\theta}{dt} = \tfrac{1}{2}B\omega R^2$$

PROBLEM 9.12 A single turn, plane loop of area A rotates about an axis parallel to its plane with uniform angular speed in a uniform magnetic field. If the component of induction normal to the axis is B_n, show that the resulting induced emf in the loop is

$$\varepsilon = B_n A \left(\frac{d\theta}{dt}\right) \sin\theta$$

where θ measures the angular position of the loop.

Solution:

$$\phi = B_n A \cos\theta$$

$$\varepsilon = -\frac{d\phi}{dt} = B_n A \frac{d\theta}{dt}\sin\theta$$

PROBLEM 9.13 A pole of a bar magnet produces a magnetic induction field that is approximately radial and spherically symmetrical with magnitude $B = K/4\pi r^2$ where $K = 4 \times 10^{-5}$ W. This pole is moved along the axis of a coil of 200 turns bound together in the form of a circular loop of 5 cm radius. When the pole is just passing through the center of the coil with a velocity of 20 cm/s, what is the induced emf in the coil? (W = Weber)

Solution:

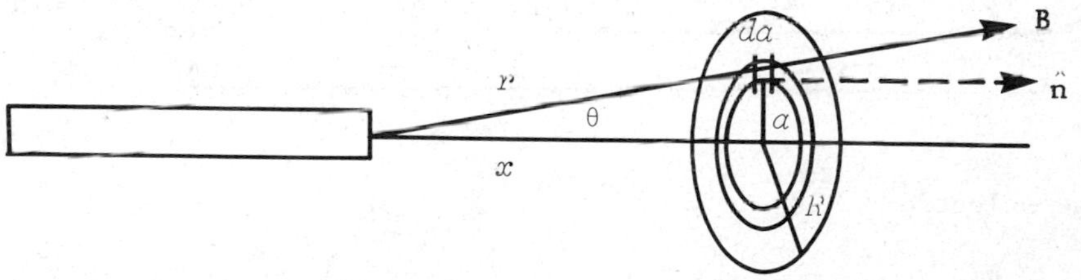

For any x,

$$\phi = \int \mathbf{B}\cdot\hat{\mathbf{n}}\,dA = \int B\cos\theta\, 2\pi a\,da = \int \frac{K}{4\pi r^2}\frac{x}{r}\,2\pi a\,da$$

$$\phi = \frac{Kx}{2}\int_0^R \frac{a\,da}{(x^2+a^2)^{3/2}} = \frac{Kx}{2}\left(\frac{-1}{(x^2+a^2)^{1/2}}\right)\Bigg|_0^R$$

$$\phi = -\frac{Kx}{2}\left(\frac{1}{\sqrt{x^2+R^2}} - \frac{1}{x}\right) = \frac{K}{2}\left(1 - \frac{x}{(x^2+R^2)^{1/2}}\right)$$

$$\varepsilon = -N\frac{d\phi}{dt} = +N\left(\frac{K}{2}\right)\left[(x)(-\tfrac{1}{2})(x^2+R^2)^{-3/2}\,2x\,\frac{dx}{dt} + (x^2+R^2)^{-\frac{1}{2}}\frac{dx}{dt}\right]$$

$$\varepsilon = \frac{NK}{2}\left(\frac{v}{(x^2+R^2)^{\frac{1}{2}}} - \frac{x^2v}{(x^2+R^2)^{3/2}}\right)$$

Putting in the numbers at $x = 0$

$$\varepsilon = \frac{(2\times10^2)(4\times10^{-5})}{2}\left(\frac{2\times10^{-1}}{5\times10^{-2}}\right) = \frac{8\times10^{-4}}{5\times10^{-2}}$$

$$\varepsilon = 16 \text{ mV}$$

PROBLEM 9.14 A very short coil of 50 turns of fine wire is wound over the center of a long, air-core solenoid of 1000 turns with a length of 1 m and a cross-section of 1 cm. Find the mutual inductance of the two coils, making a plausible approximation which should be justified by a brief argument.

Solution:

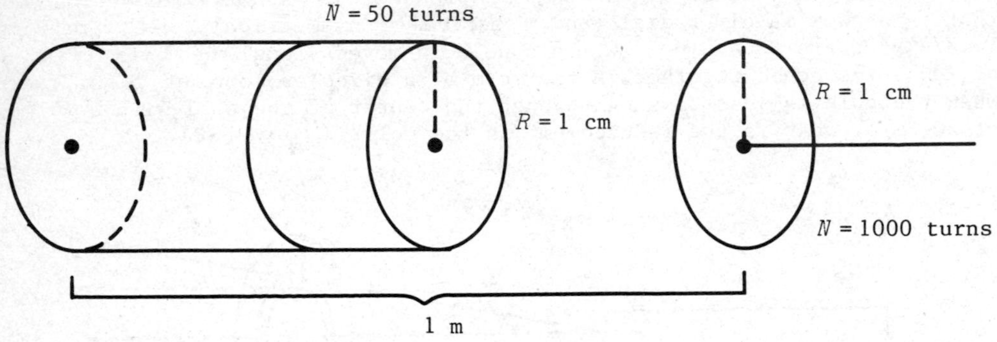

For the solenoid

$$B = \frac{\mu_0 NI}{L} = \frac{4\pi\times10^{-7}\times10^3}{1}\,I$$

$$B = 4\pi\times10^{-4}\,I$$

$$m = \frac{N_2\phi_{21}}{I}$$

$$\phi_{21} = BA = 4\pi\times10^{-4}\,I\,\pi(10^{-2})^2$$

$$\phi_{21} = 4\pi^2\times10^{-8}\,I$$

$$m = (5\times10^1)(4\pi^2\times10^{-8})$$

$$m = 2\times10^{-6}\,\pi^2$$

$$m = 1.97 \times 10^{-5} \text{ H}$$

$$m = 1.97 \ \mu\text{H}$$

PROBLEM 9.15 Two circular loops have parallel axes, and the distance r between their centers is sufficiently great compared with their radii, that dipole approximations can be used. Show how one should be placed relative to the other so that at finite distances r their mutual inductance is zero.

Solution:

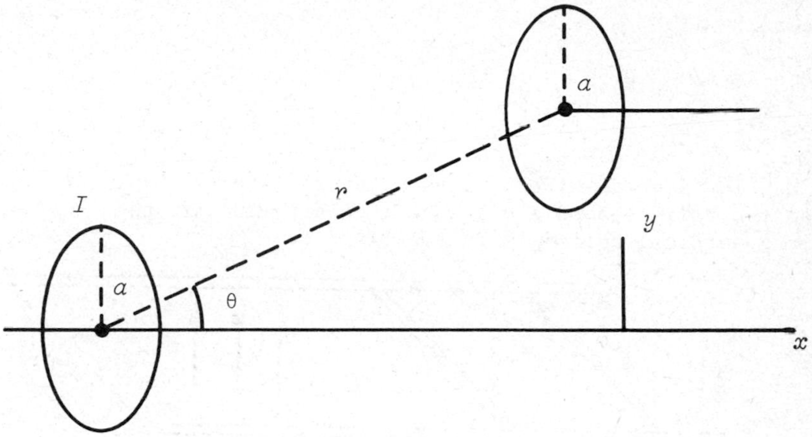

For the first loop,

$$\mathbf{m} = \hat{\mathbf{i}}\ I\pi a^2$$

$$\mathbf{r} = \hat{\mathbf{i}}x + \hat{\mathbf{j}}y + \hat{\mathbf{k}}z$$

For this loop,

$$\mathbf{B} = \frac{\mu_0}{4\pi}\left(-\frac{\mathbf{m}}{r^3} + \frac{3(\mathbf{m}\cdot\mathbf{r})\mathbf{r}}{r^5}\right)$$

Since $r >> a$, we approximate $\mathbf{r} = \hat{\mathbf{i}}x + \hat{\mathbf{j}}y$ where $x = r\cos\theta$ and $y = r\sin\theta$, so that

$$\mathbf{r} = r(\hat{\mathbf{i}}\cos\theta + \hat{\mathbf{j}}\sin\theta)$$

Thus,

$$\mathbf{B} = \frac{\mu_0}{4\pi}\left(-\hat{\mathbf{i}}\,\frac{I\pi a^2}{r^3} + \frac{(3\ I\pi a^2 r\cos\theta)(r)(\hat{\mathbf{i}}\cos\theta + \hat{\mathbf{j}}\sin\theta)}{r^5}\right)$$

$$\mathbf{B} = \frac{\mu_0}{4\pi}\,\frac{I\pi a^2}{r^3}\,(-\hat{\mathbf{i}} + \hat{\mathbf{i}}3\cos^2\theta + \hat{\mathbf{j}}3\cos\theta\sin\theta)$$

$$\mathbf{B} = \frac{\mu_0}{4}\,\frac{Ia^2}{r^3}\,[\hat{\mathbf{i}}(-1 + 3\cos^2\theta) + 3\hat{\mathbf{j}}\cos\theta\sin\theta]$$

The flux through 2 is $\phi_{21} = \int \mathbf{B} \cdot \hat{\mathbf{i}}\, dA$. In order for $m = 0$, $\phi = 0$, so, since

$$\phi = \frac{\mu_0}{4} \int \frac{Ia^2}{r^3} (-1 + 3\cos^2\theta)\, dA$$

for the flux to be 0,

$$3\cos^2\theta = 1$$

$$\cos^2\theta = 1/3$$

$$\cos\theta = 1/\sqrt{3} = 0.59$$

$$\theta = 54^\circ$$

PROBLEM 9.16 A metal wire of mass m and resistance R slides without friction on two rails spaced a distance d apart, as in the figure. The track lies in a vertical uniform magnetic field **B**.

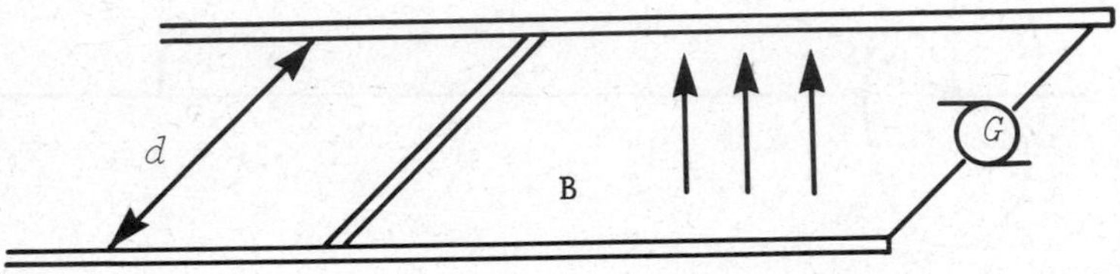

(a) A constant current I flows from generator G along one rail, across the wire, and back down the other rail. Find the velocity (speed and direction) of the wire as a function of time, assuming it to be at rest at $t = 0$.

(b) The generator is replaced by a battery with constant emf ε. The velocity of wire now approaches a constant final value. What is this terminal speed? How as a function of time does the speed approach this value?

(c) What is the current in part (b) when the terminal speed has been reached? One must assume that the field due to the current I is much smaller than the field **B**.

Solution: We assume the cross bar has resistance R at any instant $I = v/R$

$$F = BId = ma = m\frac{dv}{dt}$$

(a) $v = \frac{BId}{m} t$

(b) In this case

$$I = \frac{1}{R}\left(\varepsilon - \frac{d\phi}{dt}\right) = \frac{1}{R}(\varepsilon - B\,dv)$$

$$F = BId = Bd\frac{1}{R}(\varepsilon - B\,dv) = m\frac{dv}{dt}$$

$$\frac{B\varepsilon d}{R} - \frac{B^2d^2}{R} v = m\frac{dv}{dt}$$

$$\frac{Bd}{mR}(\varepsilon - B\,dv) = \frac{dv}{dt}$$

$$\frac{Bd}{mR}\,dt = \frac{dv}{\varepsilon - B\,dv}$$

$$\frac{Bd}{mR}\,t = -\frac{1}{Bd}\,\ell n\,\frac{\varepsilon - B\,dv}{\varepsilon}$$

$$-\frac{B^2d^2}{mR}\,t = \ell n\,\frac{\varepsilon - B\,dv}{\varepsilon}$$

$$e^{-B^2d^2t/mR} = 1 - \frac{Bd}{\varepsilon}\,v$$

$$v = \frac{\varepsilon}{Bd}\left(1 - e^{-B^2d^2t/mR}\right)$$

(c) $v_T = \varepsilon/Bd$, so the final current

$$I_f = \frac{1}{R}\left(\varepsilon - Bd\,\frac{\varepsilon}{Bd}\right) = 0$$

PROBLEM 9.17 A toroidal coil of N turns has a square cross section, each side of the square being of length a and inner radius b.

(a) Show that the self-inductance is

$$L = \frac{N^2a}{2\pi\varepsilon_0 c^2}\,\ell n\left(1 + \frac{a}{b}\right)$$

(b) Express in similar terms the mutual inductance of the system formed by the coil and a long, straight wire along the axis of symmetry of the toroidal coil. Assume the conductors closing the circuit of which the long straight wire is a part are situated far from the coil, so that their influence may be neglected.

(c) Find the ratio of the self-inductance of the coil to the mutual inductance of the system.

Solution:

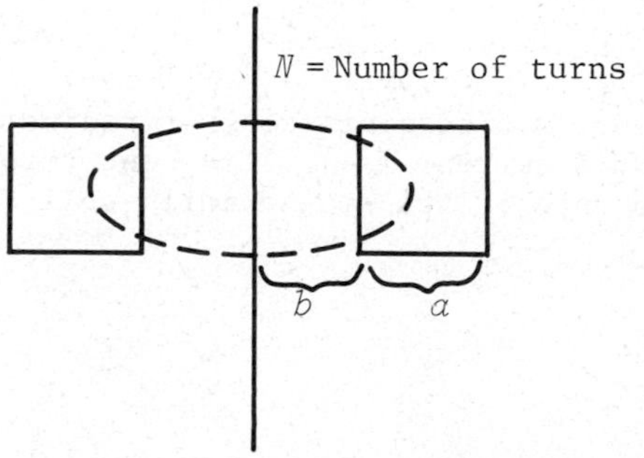

Consider the amperian loop shown. From Ampere's law

$$\oint \mathbf{B} \cdot d\boldsymbol{\ell} = \mu_0 I_T$$

$$B(2\pi r) = \mu_0 NI$$

$$B = \frac{\mu_0 NI}{2\pi r}$$

$$L = \frac{N\phi}{I}, \quad \phi = \int \mathbf{B} \cdot d\mathbf{A} = \int B\,dA = \int_b^{b+a} Ba\,dr$$

$$\phi = \int_b^{b+a} \frac{\mu_0 NI}{2\pi}\, a\, \frac{dr}{r}$$

$$\phi = \frac{\mu_0 NIa}{2\pi} \ln \frac{b+a}{b}$$

so

$$L = \frac{\mu_0 N^2 a}{2\pi} \ln \left(1 + \frac{a}{b}\right)$$

(b) Now suppose a long straight wire is as shown in the figure. For the wire,

$$B = \frac{\mu_0 I}{2\pi r} \qquad m = \frac{N\phi}{I}$$

$$\phi = \int \mathbf{B} \cdot d\mathbf{A} = \frac{\mu_0 Ia}{2\pi} \int_b^{a+b} \frac{dr}{r} = \frac{\mu_0 Ia}{2\pi} \ln \left(1 + \frac{a}{b}\right)$$

$$m = \frac{\mu_0 Na}{2\pi} \ln \left(1 + \frac{a}{b}\right)$$

$$\frac{L}{m} = \frac{(N^2 a\mu_0/2\pi) \ln (1 + a/b)}{(Na\mu_0/2\pi) \ln (1 + a/b)}$$

$$\frac{L}{m} = N$$

PROBLEM 9.18 Consider two coaxial loops or radius a which are separated by a distance d; $d \gg a$. A current $I = K_0 t^2$ is sent through one coil--coil (a) as shown and the resistance of the second coil--coil (b) is R.

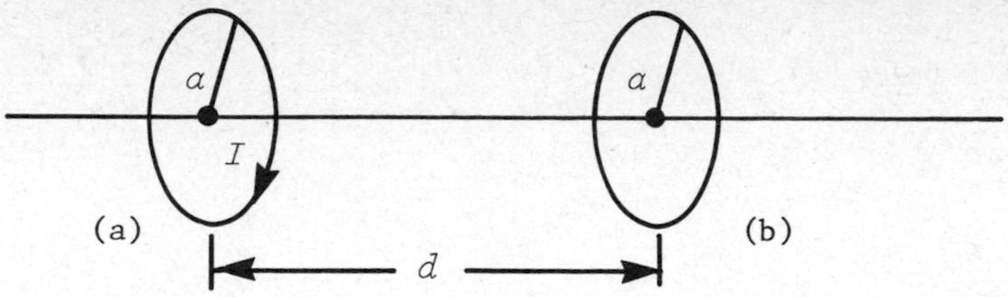

(a) Neglecting self-inductance, what is the torque on loop b?
(b) Show that, if the self-inductance is neglected, the force on the loop b is

$$\frac{24\pi^4 a^8 K_0^2 t^3}{(4\pi\varepsilon_0 c^2)^2 d^7 R}$$

In what direction is the force?
(c) In what way (qualitatively) is the true force and torque different from your estimate; that is, how does the self-inductance affect the torque and force?
(d) Explain what would happen to parts (a) and (b) if the loop b were rotated 90° about an axis normal to the common axis of the loops.

Solution:

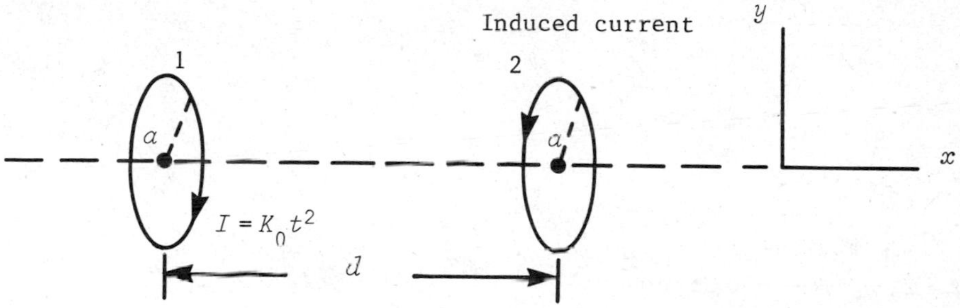

Since $d \gg a$, we can use the dipole approximation that $\boldsymbol{\tau} = \mathbf{m} \times \mathbf{B}$ where $\mathbf{m}$ is the magnetic moment of loop 2 and $\mathbf{B}$ is the field at 2 due to the current in 1. Since $\mathbf{m}$ and $\mathbf{B}$ are parallel

(a) $\tau = 0$

(b) To the dipole approximation $\mathbf{F} = \nabla(\mathbf{m}_2 \cdot \mathbf{B})$

$$\mathbf{B} = \frac{\mu_0}{4\pi}\left(-\frac{m_1}{r^3} + \frac{3(\mathbf{m}_1 \cdot \mathbf{r})(\mathbf{r})}{r^5}\right)$$

where $\mathbf{m}_2 = -\hat{\mathbf{i}} I\pi a^2$ and $\mathbf{m}_1 = -\hat{\mathbf{i}} K\pi a^2 t^2$. Since $\mathbf{r} = \hat{\mathbf{i}} x$,

$$\mathbf{B} = \frac{\mu_0}{4\pi}\left(\frac{\hat{\mathbf{i}} K\pi a^2 t^2}{x^3} - \frac{3\hat{\mathbf{i}} K\pi a^2 t^2}{x^3}\right) = -\frac{\mu_0}{2} K a^2 \frac{t^2}{x^3}\,\hat{\mathbf{i}}$$

To get $\mathbf{m}_2$, we need the current in coil 2.

$$\varepsilon_2 = -\frac{d\phi_{21}}{dt} = -\pi a^2 \frac{dB}{dt}$$

$$\varepsilon_2 = \frac{\pi a^2 \mu_0 K a^2}{2x^3} 2t$$

so the current in coil 2 is $I_2 = \varepsilon_2 / R$

$$I_2 = \frac{\pi a^4 \mu_0 K t}{x^3}$$

and is as shown. Then

$$\mathbf{m}_2 = \hat{\mathbf{i}} I_2 \pi a^2 = \frac{\hat{\mathbf{i}} \pi^2 a^6 \mu_0 K t}{x^3}$$

So that

$$\mathbf{m}_2 \cdot \mathbf{B} = -\frac{\mu_0^2 K^2 a^8 t^3 \pi^2}{4x^6}$$

and

$$\mathbf{F} = \hat{\mathbf{i}} \frac{\partial}{\partial x} (\mathbf{m}_2 \cdot \mathbf{B})$$

$$\mathbf{F} = \hat{\mathbf{i}} \frac{6\mu_0^2 \pi^2 K^2 a^8 t^3}{4d^7} \quad \text{at } x = d$$

(c) By including self-inductance all effects would be less as a result of the current flowing being less. For loop 2, the current equation is

$$\varepsilon_2 - L \frac{dI_2}{dt} = I_2 R$$

where

$$\varepsilon_2 = -m \frac{dI_1}{dt}$$

so

$$\varepsilon_2 = L \frac{dI_2}{dt} + I_2 R$$

So the addition of L serves to reduce I_2.

(d) There would be no forces or torques since there would be no flux change through the coil 2.

PROBLEM 9.19 A uniform field E_m is set up in a medium of specific inductive capacity K. Prove that the field inside a spherical cavity in the medium is given by

$$E = 3KE_m / (2K + 1).$$

[*Hints:* There are two methods. One is to use an educated guess with possible image charges. This is a short method.]

Solution:

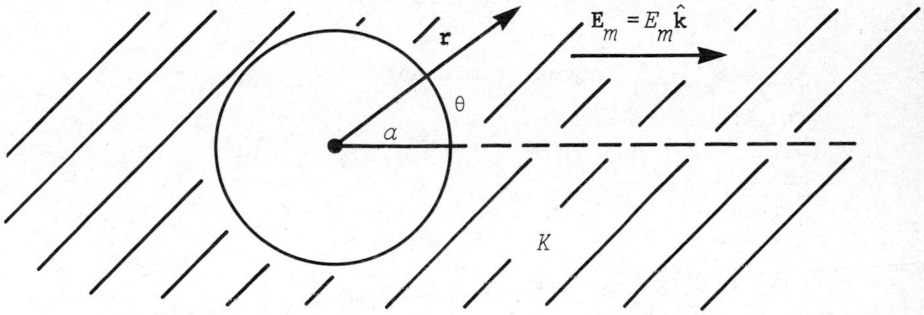

Because there are no free charges in this problem, $\nabla^2\phi(\mathbf{r}) = 0$. Because of the axial symmetry and the fact that ϕ must be finite at the origin, we have

$$\phi_{in} = \sum_{\ell=0}^{\infty} a_\ell r^\ell P_\ell(\cos\theta)$$

$$\phi_{out} = \sum_{\ell=0}^{\infty} (B_\ell r^\ell + C_\ell r^{-(\ell+1)})P_\ell(\cos\theta)$$

As $r \to \infty$, $\mathbf{E} \to E_m\hat{\mathbf{k}}$, therefore, $\phi \to -E_m z$

or

$$\phi \to -E_m r\cos\theta$$

As $r \to \infty$,

$$\phi_{out} = \sum_{\ell=0}^{\infty} B_\ell r^\ell P_\ell(\cos\theta) = -E_m r\cos\theta$$

Equating coefficients we find

$$B_\ell = 0 \text{ for } \ell \neq 1$$

and

$$B_1 = -E_m$$

Using the boundary condtions at $r = a$,

$$\phi_{in}(a) = \phi_{out}(a)$$

$$\sum_{\ell=0}^{\infty} A_\ell a^\ell P_\ell(\cos\theta) = \sum_{\ell=0}^{\infty} (B_\ell a^\ell + C_\ell a^{-(\ell+1)})P_\ell(\cos\theta)$$

For $\ell = 1$,

$$A_1 a = (-E_m a + C_1 a^{-2})$$

$$A_1 = -E_m + C_1 a^{-3}$$

For $\ell \neq 1$, $A_\ell = C_\ell a^{-(2\ell+1)}$

$$D_{\text{in}(m)} = D_{\text{out}(m)}, \text{ normal components}$$

$$-\varepsilon_0 \left.\frac{\partial \phi_{\text{in}}}{\partial r}\right|_a = -K\varepsilon_0 \left.\frac{\partial \phi_{\text{out}}}{\partial r}\right|_a$$

$$\sum_{\ell=0}^{\infty} A_\ell \ell r^{\ell-1} \frac{\partial}{\partial r} P_\ell(\cos\theta)\Big|_a = K \sum_{\ell=0}^{\infty} [B_\ell \ell r^{\ell-1} - C_\ell(\ell+1)r^{-(\ell+2)}] \frac{\partial}{\partial r} P_\ell(\cos\theta)\Big|_a$$

For $\ell = 1$,

$$A_1 = -K(E_m + 2C_1 a^{-3})$$

For $\ell \neq 1$,

$$A_\ell \ell a^{\ell-1} = -KC_\ell(\ell+1)a^{-(\ell+2)}$$

$$A_\ell = -KC_\ell \frac{\ell+1}{\ell} a^{-(2\ell+1)}$$

The equation for $\ell \neq 1$ can only be solved simultaneously with $A_\ell = C_\ell = 0$:

$$K(E_m + 2C_1 a^{-3}) = -E_m + C_1 a^{-3}$$

$$E_m(K-1) = -C_1 a^{-3}(1+2K)$$

$$C_1 = -E_m a^3 \frac{K-1}{2K+1}$$

$$A_1 = -E_m + C_1 a^{-3}$$

$$A_1 = -E_m - E_m \frac{K-1}{2K+1}$$

$$A_1 = -E_m \left(\frac{3K}{2K+1}\right)$$

$$\therefore \phi_{\text{in}} = -E_m \left(\frac{3K}{2K+1}\right) r\cos\theta$$

or

$$\phi_{in} = -E_m \left(\frac{3K}{2K+1}\right) z$$

$$\mathbf{E}_{in} = -\nabla\phi_{in} = E_m \left(\frac{3K}{2K+1}\right)\hat{\mathbf{k}}$$

PROBLEM 9.20 Calculate the current distribution and the resistance of a spherical conductor with a potential V across it, whose surface is completely covered by two hemispherical electrodes. The two electrode potentials are $\pm V/2$. Leave your answer in form of a series.

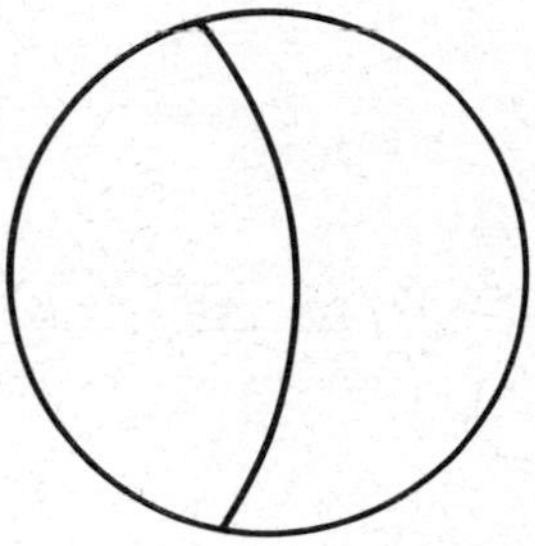

Solution:

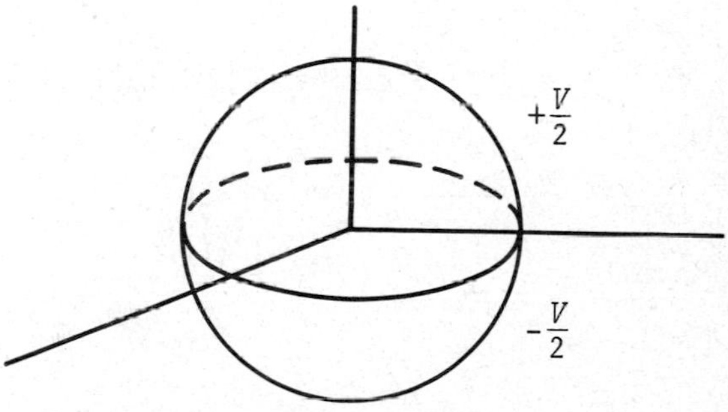

Because of the azimuthal symmetry the potential inside the sphere

$$\phi(r,\theta) = \sum_{\ell=0}^{\infty} A_\ell r^\ell P_\ell(\cos\theta)$$

$B_\ell = 0$ because ϕ must be finite at the origin. On the surface of the sphere,

$$\phi(a,\theta) = \sum_{\ell=0}^{\infty} A_\ell a^\ell P_\ell(\cos\theta)$$

$$A_\ell = \frac{2\ell+1}{2a^\ell}\int_0^\pi V(\theta)P_\ell(\cos\theta)\sin\theta\, d\theta$$

$$V(\theta) = \begin{cases} +\dfrac{V}{2}, & 0 \le \theta < \dfrac{\pi}{2} \\ \\ -\dfrac{V}{2}, & \dfrac{\pi}{2} < \theta \le \pi \end{cases}$$

Using the Rodrigues' formula,

$$P_\ell(\cos\theta) = \frac{1}{2^\ell \ell!} \frac{d^\ell (\cos^2\theta - 1)^\ell}{d(\cos\theta)^\ell}$$

we can integrate

$$\phi(r,\theta) = \frac{V}{2}\left[\frac{3}{2}\frac{r}{a} P_1(\cos\theta) - \frac{7}{8}\left(\frac{r}{a}\right)^3 P_3(\cos\theta) + \frac{11}{16}\left(\frac{r}{a}\right)^5 P_5(\cos\theta) + \cdots\right]$$

$$\phi(r,\theta) = \frac{V}{2\sqrt{\pi}} \sum_{j=1}^{\infty} (-1)^{j-1} \frac{(2j-\frac{1}{2})\Gamma(j-\frac{1}{2})}{j!} \left(\frac{r}{a}\right)^{2j-1} P_{2j-1}(\cos\theta)$$

$$\mathbf{E} = -\nabla\phi = -\left(\hat{\mathbf{r}}\frac{\partial\phi}{\partial r} + \hat{\theta}\frac{1}{r}\frac{\partial\phi}{\partial\theta}\right)$$

let $\mu = \cos\theta$, $d\mu = -\sin\theta\, d\theta$

$$\mathbf{E} = -\frac{V}{2\sqrt{\pi}} \sum_{j=1}^{\infty} (-1)^{j-1} \frac{(2j-\frac{1}{2})\Gamma(j-\frac{1}{2})}{j!} \left[\hat{\mathbf{r}}(2j-1)\left(\frac{r}{a}\right)^{2j-2} \frac{1}{a} P_{2j-1}(\mu) - \hat{\theta}\frac{1}{r}\left(\frac{r}{a}\right)^{2j-1} \sqrt{1-\mu^2}\frac{\partial}{\partial\mu} P_{2j-1}(\mu)\right]$$

$$(\mu^2 - 1)\frac{dP_\ell}{d\mu} = \ell\mu P_\ell - \ell P_{\ell-1}$$

$$\mathbf{E} = -\frac{V}{2\sqrt{\pi}} \sum_{j=1}^{\infty} (-1)^{j-1} \frac{(2j-\frac{1}{2})\Gamma(j-\frac{1}{2})}{j!} \left[\hat{\mathbf{r}}(2j-1)\left(\frac{r}{a}\right)^{2j-2} \frac{1}{a} P_{2j-1}(\mu) - \hat{\theta}\frac{1}{r}\left(\frac{r}{a}\right)^{2j-1} (1-\mu^2)^{-\frac{1}{2}}[(2j-1)\mu P_{2j-1} - (2j-1)P_{2j-2}]\right]$$

$$\mathbf{j} = \sigma\mathbf{E}$$

$$\mathbf{j} = \frac{V\sigma}{2\sqrt{\pi}} \sum_{j=1}^{\infty} (-1)^{j-1} \frac{(2j-\frac{1}{2})(2j-1)\Gamma(j-\frac{1}{2})}{j!} \left[\hat{\mathbf{r}}\frac{1}{a}\left(\frac{r}{a}\right)^{2j-2} P_{2j-1}(\mu)\right.$$

$$+\hat{\theta}\,\frac{1}{r}\left(\frac{r}{a}\right)^{2j-1}(1-\mu^2)^{-\frac{1}{2}}(\mu P_{2j-1}-P_{2j-2})\Bigg]$$

$V = IR$

$$V = R\int \hat{\mathbf{n}}\cdot\hat{\mathbf{j}}\,da = \sigma R\int \hat{\mathbf{n}}\cdot\hat{\mathbf{j}}\,da$$

$$\frac{1}{\sigma} = R\,\frac{1}{2\sqrt{\pi}}\sum_{j=1}^{\infty}(-1)^{j-1}\frac{(2j-\frac{1}{2})(2j-1)\Gamma(j-\frac{1}{2})}{j!}\int_S \frac{1}{a}P_{2j-1}(\mu)\left(\frac{a}{a}\right)^{2j-1}\hat{\mathbf{r}}\cdot\hat{\mathbf{n}}\,da$$

$$\frac{2\sqrt{\pi}}{\sigma} = R\sum_{j=1}^{\infty}(-1)^{j-1}\frac{(2j-\frac{1}{2})(2j-1)\Gamma(j-\frac{1}{2})}{j!}\left[\int_0^{2\pi}\int_0^{\pi/2} -P_{2j-1}(\mu)a\sin\theta\,d\theta\,d\phi\right.$$

$$\left.+\int_0^{2\pi}\int_{\pi/2}^{\pi} P_{2j}(\mu)a\sin\theta\,d\theta\,d\phi\right]$$

$$\frac{2\sqrt{\pi}}{\sigma} = R\sum_{j=1}^{\infty}(-1)^{j-1}\frac{(2j-\frac{1}{2})(2j-1)\Gamma(j-\frac{1}{2})}{j!}\left[2\pi a\left(\int_1^0 P_{2j-1}(\mu)d\mu\right.\right.$$

$$\left.\left.-\int_0^{-1} P_{2j-1}(\mu)d\mu\right)\right]$$

$$\frac{2\sqrt{\pi}}{\sigma} = R\sum_{j=1}^{\infty}(-1)^{j-1}\frac{(2j-\frac{1}{2})(2j-1)\Gamma(j-\frac{1}{2})}{j!}\,4\pi a\int_1^0 P_{2j-1}(\mu)d\mu$$

because P_{2j-1} is odd

$$\frac{2\sqrt{\pi}}{\sigma} = R\sum_{j=1}^{\infty}(-1)^{j-1}\frac{(2j-\frac{1}{2})(2j-1)\Gamma(j-\frac{1}{2})}{j!}\,4\pi a\left[-(-\tfrac{1}{2})^{j-1}\frac{(2j-3)!!}{2\left(\frac{\ell+1}{2}\right)!}\right]$$

$$\frac{1}{R} = \sigma\sqrt{\pi}\sum_{j=1}^{\infty}(-1)^{j}\frac{(2j-1)(2j-\frac{1}{2})\Gamma(j-\frac{1}{2})}{j!}\,a(-\tfrac{1}{2})^{j-1}\frac{(2j-3)!!}{\left(\frac{\ell+1}{2}\right)!}$$

PROBLEM 9.21

(a) Design a voltmeter that reads 10 V full scale and an ammeter that reads 100 mA full scale from a galvanometer that has an internal resistance of 35 ohms and a full scale current of 1 mA.

(b) What would the voltmeter that you designed in the previous step read across the 20,000 ohm resistor in the circuit below?

Solution:

(a) A voltmeter, 10 V full scale:

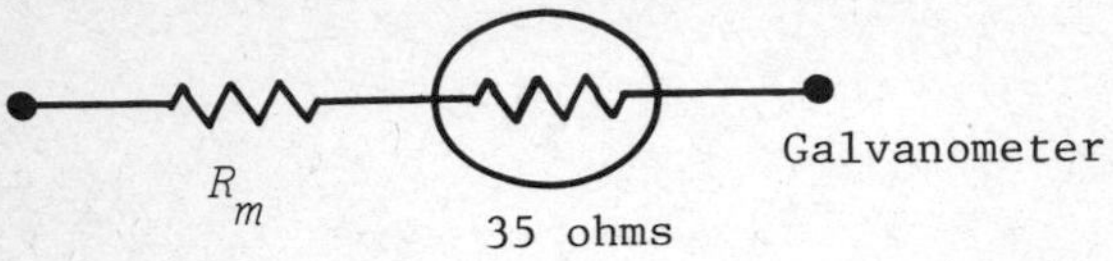

$I_g = 1 \times 10^{-3}$ A

$R_g = 35$ ohms

$I = 1 \times 10^{-3}$ A, $I_g = 1 \times 10^{-3}$ A

$$1R + 1R = 10 \text{ V}$$

$$1 \times 10^{-3} \text{ A}(R_m + 35 \text{ ohms}) = 10 \text{ V}$$

$$R_m + 35 \text{ ohms} = 1 \times 10^4 \text{ ohms}$$

$$R_m = 10{,}000 \text{ ohms} = 35 \text{ ohms} = 9965 \text{ ohms}$$

Put a $R_m = 9965$ ohms in series with galvanometer.

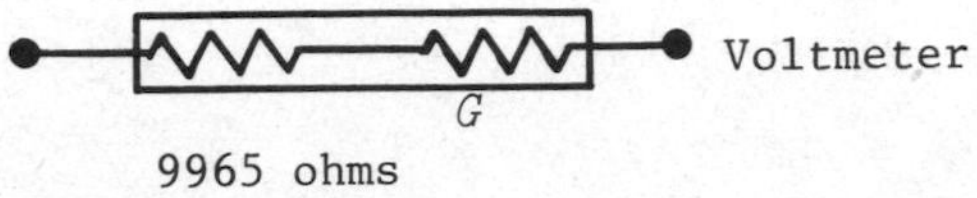

An ammeter, 100×10^{-3} A full scale = 0.1 A full scale

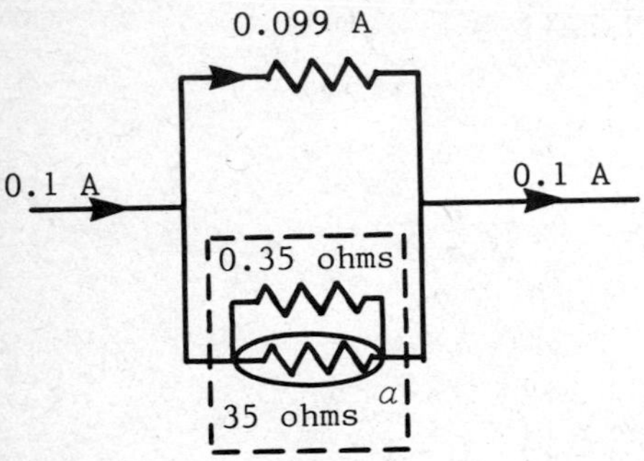

$I_g = 1 \times 10^{-3}$ A

$R_g = 35$ ohms

Put R_S in parallel with the galvanometer. Voltage is the same so $IR = IR$

$$(R_S)(0.099 \text{ A}) = (0.001 \text{ A})(35 \text{ ohms})$$

Put $R_S = 0.35$ ohms with galvanometer.

b) $R_T = 10{,}000 \text{ ohms} + 6666.67 \text{ ohms}$

$= 16{,}666.67$ ohms

$$I = \frac{V}{R} = \frac{10 \text{ V}}{1.67 \times 10^4 \text{ ohms}} = 6 \times 10^{-4} \text{ A}$$

$$V = (6 \times 10^{-4} \text{ A})(6666.67 \text{ ohms}) = 4 \text{ V}$$

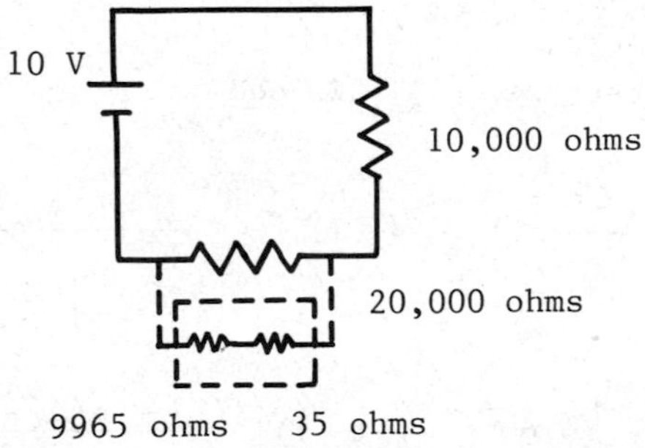

PROBLEM 9.22 Find the resistance of the following combination of 2-ohm resistors.

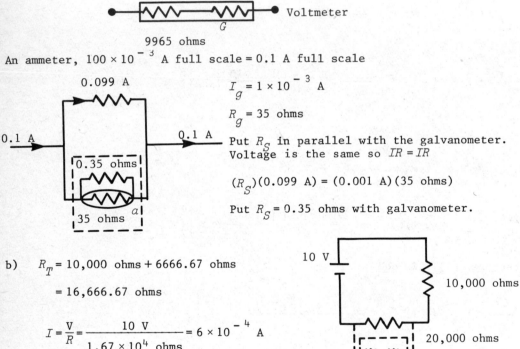

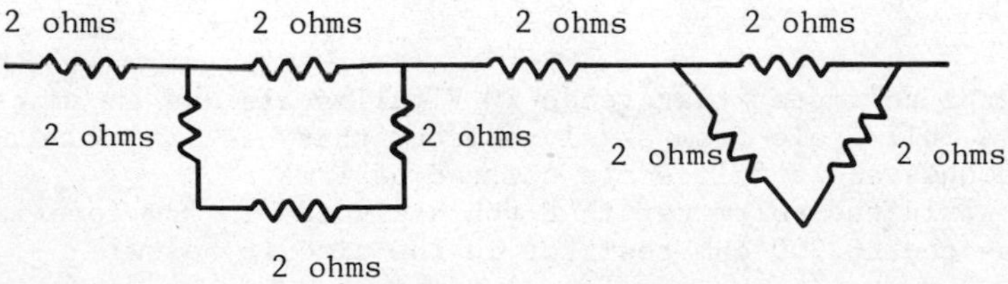

Solution:

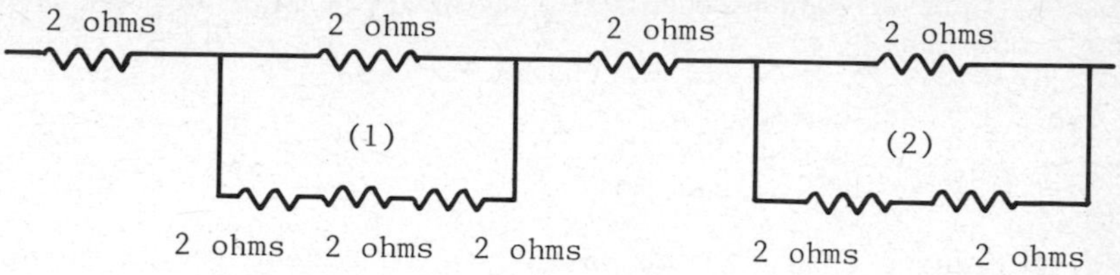

$$\frac{1}{R_1} = \frac{1}{2 \text{ ohms}} + \frac{1}{6 \text{ ohms}}, \quad R = 1.5 \text{ ohms}$$

$$\frac{1}{R_2} = \frac{1}{4 \text{ ohms}} + \frac{1}{2 \text{ ohms}}$$

$$R = 1.33 \text{ ohms}$$

$$R_T = 2 \text{ ohms} + 1.5 \text{ ohms} + 2 \text{ ohms} + 1.33 \text{ ohms} = 6.83 \text{ ohms}$$

PROBLEM 9.23 The constant c that turns up in Maxwell's equations can be determined by electrical experiments involving low-frequency fields only. Consider the arrangement shown in the figure. The force between capacitor plates is balanced against the force between parallel wires carrying current in the same direction. A voltage alternating sinusoidally at a frequency f Hz is applied to the parallel-plate capacitor C_1 and also to the capacitor C_2. The charge flowing into and out of C_2 constitutes the current in the rings. Suppose that C_2 and the various distances involved have been adjusted so that the downward force on the upper plate of C_1 exactly balances the downward force on the upper ring. (Of course, the weights of the two sides should be adjusted to balance with the voltage turned off.) Show that under these conditions the constant c can be computed from measured quantities as follows:

$$c = (2\pi)^{3/2} a \left(\frac{b}{h}\right)^{1/2} \left(\frac{C_2}{C_1}\right) f \text{ cm/sec}$$

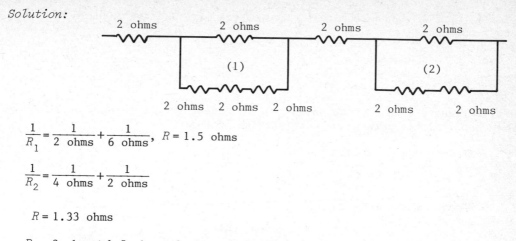

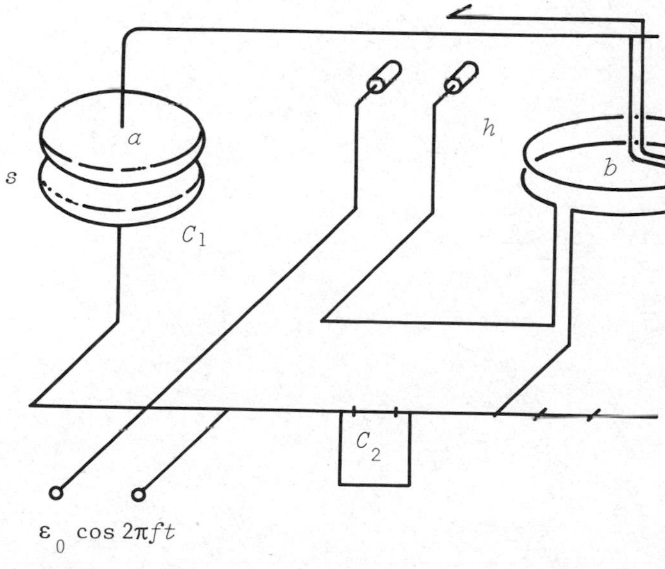

Solution: If we ignore the resistance of the wire and the inductance of the loops, the charge on C_2 is

$$Q = C_2 V = C_2 \varepsilon_0 \cos(2\pi f t)$$

so $i_2 = dQ/dt = -2\pi f C_2 \varepsilon_0 \sin(2\pi f t)$

We will require that $b \ggg h$, so that we may treat the force between the current loops as that betwen two long straight wires:

$$F_m = \frac{B i_2 \ell}{2} = \left(\frac{2i_2}{hc}\right) \frac{i_2 (2\pi b)}{c}$$

$$= \frac{4\pi b}{hc^2} [-2\pi f C_2 \varepsilon_0 \sin(2\pi f t)]^2$$

$$= \frac{16\pi^3 b f^2 C_2^2 \varepsilon_0^2}{hc^2} \sin^2(2\pi f t)$$

$$\overline{F}_m = \frac{8\pi^3 b f^2 C_2^2 \varepsilon_0^2}{hc^2}$$

The force between the plates of C_1 is

$$F_e = \frac{(E^2/8\pi)(\pi a^2 S)}{S} = E^2 a^2/8$$

$$= \frac{1}{8} a^2 \left(\frac{\varepsilon_0 \cos(2\pi f t)}{S}\right)^2 = \frac{1}{8} \frac{a^2 \varepsilon_0^2}{S^2} \cos^2(2\pi f t)$$

Since

$$C_1 = \frac{\text{Area}}{4\pi S} = \frac{\pi a^2}{4\pi S}$$

$$\frac{1}{S} = \frac{4C_1}{a^2}$$

and

$$\overline{F}_e = \frac{1}{8} a^2 \varepsilon_0^2 \left(\frac{4C_1}{a^2}\right)^2 (\tfrac{1}{2}) = \frac{C_1^2 \varepsilon_0^2}{a^2}$$

At balance, $\overline{F}_e = \overline{F}_m$ or

$$\frac{C_1^2 \varepsilon_0^2}{a^2} = \frac{8\pi^3 b f^2 C_2^2 \varepsilon_0^2}{hC^2}$$

$$c^2 = \frac{(2\pi)^3 b a^2 f^2 C_2^2}{hC_1^2}$$

$$c = (2\pi)^{3/2} \left(\frac{b}{h}\right)^{\frac{1}{2}} \left(\frac{C_2}{C_1}\right) af$$

PROBLEM 9.24 What is the maximum electromotive force induced in a coil of 4000 turns, average radius 12 cm, rotating at 30 revolutions per second in the earth's magnetic field where the field intensity is 0.5 gauss?

Solution:

$$\varepsilon = -\frac{1}{c}\frac{d\Phi}{dt} = \frac{-SB\omega}{c}\cos(\omega t + \alpha)$$

$$\varepsilon_{max} = SB\omega\, N/c$$

$$S = \pi(12)^2, \quad \omega = 30 \times 2\pi$$

$$\varepsilon_{max} = \pi(12)^2(0.5)(60\pi)(4000)/3 \times 10^{10}$$

$$= 5.68 \times 10^{-3} \text{ statvolts} = 1.71 \text{ V}$$

PROBLEM 9.25 A rectangular loop is moving through a uniform magnetic field in such a manner that the electromotive force is, and remains zero. Describe the ways in which the loop might be moving.

Solution: It can be translating in any direction or rotating about its symmetry axis. It can have no rotation about a diameter unless that diameter is parallel to B.

PROBLEM 9.26 Does the prediction of a simple sinusoidal variation of electromotive force for the rotating loop in the figure depend on the loop being rectangular, on the magnetic field being uniform, or on both? Explain. Can you suggest an arrangement of rotating loop and stationary coils which will give a definitely nonsinusoidal emf?

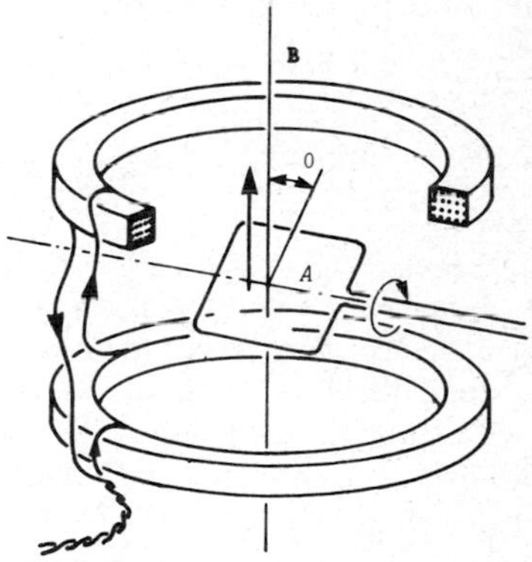

Solution: The derivation requires that

$$\Phi = \int \mathbf{B} \cdot d\mathbf{a} = \mathbf{B} \cdot \int d\mathbf{a} = \mathbf{B} \cdot \mathbf{A}$$

that is, that B be uniform. It places no requirement on the shape of the loop. Consider a coil rotating between the ends of the two long solenoids.

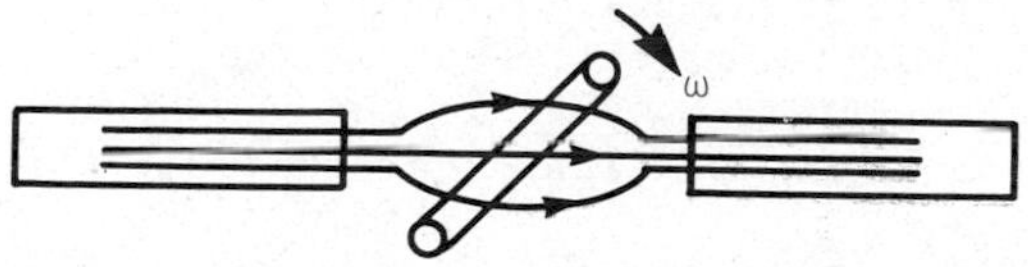

The emf will be very low until the coil approaches the axis and then will vary rapidly to maximum and back down.

PROBLEM 9.27 Calculate the electromotive force in the moving loop in the figure at the instant when it is in the position there shown. Assume the resistance of the loop is so great that the effect of the current in the loop itself is negligble. Estimate very roughly how large a resistance would be safe, in this respect. Indicate the direction in which current would flow in the loop, at the instant shown.

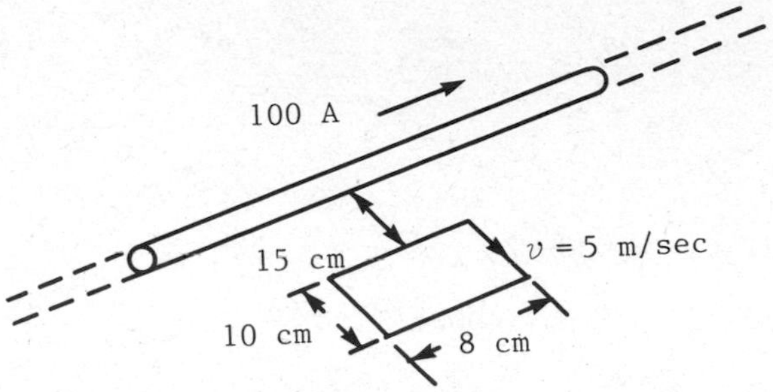

Solution:

$$B = \frac{2I}{Rc} = \frac{2 \times 100}{10r} = \frac{20}{r}, \quad B_1 = 1.333 \text{ G}, \quad B_2 = 0.800 \text{ G}$$

$$\varepsilon = \frac{(B_1 - B_2)wv}{c} = \frac{(1.333 - 0.800)(8 \text{ cm})(500 \text{ cm/sec})}{3 \times 10^{10}}$$

$$= 7.11 \times 10^{-8} \text{ statvolt} = 2.13 \times 10^{-5} \text{ V}$$

The generated field must be small compared to the field from the wire, about 1 G. For a circular loop of about these dimensions,

$$B \simeq \frac{2\pi I(\text{A})}{10r} \simeq \frac{I}{10} \simeq \frac{2 \times 10^{-5} \text{ V}}{10R} \simeq \frac{2 \times 10^{-6}}{R}$$

so $R >> 10^{-6}$, say $\gtrsim 10^{-3}$ ohm, will be safe. The flux as seen in the figure is down through the loop and is decreasing. Therefore, the induced current must be in the direction to produce flux downward. This requires a clockwise current in the loop as shown.

PROBLEM 9.28 A ring-shaped coil of N turns and area A is located in the field of a magnet. It is connected to an external circuit by a twisted pair of leads. The resistance of the circuit, including the coil itself, is R. Suppose the flux through the coil is somehow changed from its initial steady value ϕ_i to a final steady value ϕ_f. Show that the total charge Q which passes through the circuit, as a result, is independent of the rate of change of flux. A coil like this, called a "flip coil," is often used to measure the field strength in a magnet. Suppose the coil is placed with its plane perpendicular to the field **B**. What is the relation between B, NA, R, and the charge Q when the coil is flipped through 90°? through 180°?

Solution:

$$Q = i\Delta t = \frac{\varepsilon \Delta t}{R} = \frac{1}{c}\frac{\Delta\phi}{\Delta t}\frac{\Delta t}{R} = \frac{\Delta\phi}{Rc}$$

$$= \frac{(\phi_f - \phi_i)}{Rc}$$

For 90°, $\Phi_i = NBA$, $\Phi_f = 0$: $Q = \frac{NBA}{Rc}$

For 180°, $\Phi_i = NBA$, $\Phi_f = -NBA$: $Q = \frac{2NBA}{Rc}$

PROBLEM 9.29 Two electrodes consisting of metal spheres 1 ft in diameter are suspended in the deep ocean by insulated cables. The spheres hang at a depth of 200 ft; the horizontal distance between them is 1000 ft. The circuit is completed by an insulated cable supported near the surface, running back to a ship over one of the spheres. Taking the conductivity of seawater as 0.04 $(\text{ohm}\cdot\text{cm})^{-1}$, estimate the resistance to be expected in this circuit. You will have to decide first whether the resistance of the ocean path connecting the spheres is caused mainly by the region in the immediate vicinity of each sphere or by the large intervening volume of ocean. To clarify this question you might consider the resistance between two concentric spheres, one much larger than the other, with a homogeneous medium in between. Also, you might sketch approximately the lines of current flow in the ocean between the spheres. This question of the resistance of a circuit, part of which consists of a conducting probe inserted into a poorly conducting medium, is important not only in geophysics but also in many physiological investigations.

Solution: It is clear that the current density is highest close to the spheres, where all the current must be crossing the relatively small surface area of the spheres. Well away from the spheres the current can spread across the whole ocean, so

$$R = \rho \int \frac{d\ell}{A} = \frac{\rho}{4\pi}\int_{r_0}^{r} \frac{dr}{r^2} = \frac{\rho}{4\pi}\left(\frac{1}{r_0} - \frac{1}{r}\right),$$

where r_0 is the radius of the spheres and r is some much larger distance, say several hundred feet. To a very good approximation then,

$$R \simeq \frac{1}{4\pi(0.04)(\text{ohm}\cdot\text{cm})^{-1}}\left(\frac{1}{\frac{1}{2}\text{ft}\times 30\ \text{cm/ft}}\right)$$

$$= 0.133\ \text{ohm}$$

This is, of course, the resistance at each sphere, so

$$R_{\text{total}} \simeq 0.27\ \text{ohm}$$

+ the resistance of the connecting cables.

PROBLEM 9.30 A system of charges and currents is completely contained inside the fixed volume V. The dipole moment of the charge-current distribution is defined by

$$\mathbf{p} = \int_v \mathbf{r}\rho \, dv$$

where $\mathbf{r}$ is the position vector from a fixed origin. Prove that

$$\int_v \mathbf{j} \, dv = \frac{d}{dt} \mathbf{p}$$

[*Hint:* First prove the identity

$$\int_v \mathbf{j} \, dv = \oint_S \mathbf{r}\mathbf{j} \cdot n \, da - \int_v \mathbf{r} \operatorname{div} \mathbf{j} \, dv$$

and note that $\mathbf{j}$ vanishes on the surface S.]

Solution:

$$\mathbf{p} = \int_v \mathbf{r}\rho \, dv, \quad \frac{d\mathbf{p}}{dt} = \int_v \mathbf{r} \frac{\partial \rho}{\partial t} \, dv = -\int_v \mathbf{r}\nabla \cdot \mathbf{j} \, dv, \text{ from continuity.}$$

Consider

$$\mathbf{a} \cdot \frac{d\mathbf{p}}{dt} = -\int (\mathbf{a} \cdot \mathbf{r})\nabla \cdot \mathbf{j} \, dv,$$

where $\mathbf{a}$ is arbitrary.

$$(\mathbf{a} \cdot \mathbf{r})\nabla \cdot \mathbf{j} = \nabla \cdot [\mathbf{j}(\mathbf{a} \cdot \mathbf{r})] - \mathbf{j} \cdot \nabla(\mathbf{a} \cdot \mathbf{r})$$

$$= \nabla \cdot [\mathbf{j}(\mathbf{a} \cdot \mathbf{r})] - \mathbf{j} \cdot \mathbf{a}$$

$$\mathbf{a} \cdot \frac{d\mathbf{p}}{dt} = -\int_v \nabla \cdot [\mathbf{j}(\mathbf{a} \cdot \mathbf{r})] \, dv + \int_v \mathbf{j} \cdot \mathbf{a} \, dv$$

$$= -\int_S \mathbf{n} \cdot \mathbf{j}(\mathbf{a} \cdot \mathbf{r}) \, ds + \mathbf{a} \cdot \int_v \mathbf{j} \, dv$$

$$= -\mathbf{a} \cdot \int_S \mathbf{r}(\mathbf{n} \cdot \mathbf{j}) \, ds + \mathbf{a} \cdot \int \mathbf{j} \, dv$$

if $\mathbf{j} = 0$ on boundary,

$$\frac{d\mathbf{p}}{dt} = \int \mathbf{j} \, dv$$

since $\mathbf{a}$ is arbitrary.

PROBLEM 9.31 In a 6-Bev electron synchrotron, electrons travel around the machine in an approximately circular path 240 m long. It is normal to have about 10^{11} electrons circling on this path during a cycle of acceleration. The speed of the electrons is practically that of light. What is the current? Nothing in our definition of current as rate of charge transport requires the velocities of the carriers to be nonrelativistic, and there is no rule against a given charged particle getting counted many times during a second as part of the current.

Solution:

$$\frac{\text{Number of turns}}{\text{second}} = \frac{3 \times 10^8 \text{ m/sec}}{240 \text{ m}}$$

$$= 1.25 \times 10^6 \text{ Hz}$$

$$I = \frac{Q}{t} = 10^{11} \times 4.8 \times 10^{-10} \times 1.25 \times 10^6 = 6.00 \times 10^7 \text{ esu/sec}$$

$$= 2.00 \times 10^{-2} \text{ A}$$

PROBLEM 9.32 We have 5×10^{10} doubly charged positive ions per cubic centimeter, all moving west with a speed of 10^7 cm/sec. In the same region there are 10^{11} electrons per cubic centimeter moving northeast with a speed of 10^8 cm/sec. (Do not ask how we managed it!) What is the direction of j? What is its magnitude in esu per second per square centimeter?

Solution: Taking east as $\hat{\mathbf{x}}$ and north as $\hat{\mathbf{y}}$,

$$\mathbf{j} = \sum nq\mathbf{v} = \{[5 \times 10^{10} \times 2 \times (-10^7\,\hat{\mathbf{x}})] + 10^{11} \times (-1) \times \left(\frac{1}{\sqrt{2}}\,10^8\,\hat{\mathbf{x}} + \frac{1}{\sqrt{2}}\,10^8\,\hat{\mathbf{y}}\right)\}e$$

$$= (-8.07 \times 10^{18}\,\hat{\mathbf{x}} - 7.07 \times 10^{18}\,\hat{\mathbf{y}})e$$

$$\theta = \tan^{-1}(-7.07/-8.07) = 221.3^\circ$$

$$j = 10.7 \times 10^{18}\,e = 5.15 \times 10^9 \text{ esu/sec/cm}^2$$

PROBLEM 9.33 In a Van de Graaff electrostatic generator, a rubberized belt 30 cm wide travels at a velocity of 20 m/sec. The belt is given a surface charge at the lower roller, the surface charge density being high enough to cause a field of 40 statvolts/cm on each side of the belt. What is the current in milliamperes?

Solution:

$$\sigma = \frac{E}{2\pi} = \frac{40}{2\pi} = 6.37 \text{ esu/cm}^2$$

$$I = \frac{Q}{t} = \text{area} \times (\sigma/t)$$

$$= 30 \text{ cm} \times 2000 \text{ cm/sec} \times 6.37 \text{ esu/cm}^2$$

$$= 3.82 \times 10^5 \text{ esu/sec} = 1.27 \times 10^{-4} \text{ A} = 0.127 \text{ mA}$$

PROBLEM 9.34 An experimenter wants to make a layer of aluminum 50 Å thick by evaporating aluminum onto a clean glass surface under vacuum. He first lays down by evaporation a fairly thick layer, with a central strip masked so that it remains bare. Then, using another mask in the manner of a stencil, he evaporates onto the glass a strip of the same width running across the gap, meanwhile using the heavy patches as terminals for measurement of resistance. At what value of the resistance should he stop the evaporation? (Resistivity of pure aluminum at room temperature is 2.83×10^{-6} ohm · cm.)

Solution:

$$R = \rho L/A = \rho L/Lt = \rho/t$$

$$R = 2.83 \times 10^{-6} \text{ ohm} \cdot \text{cm}/50 \times 10^{-8} \text{ cm}$$

$$R = 5.66 \text{ ohms}$$

PROBLEM 9.35 A transmission line consists of two, parallel perfect conductors of arbitrary, but constant, cross section. Current flows down one conductor and returns via the other.

Show that the product of the inductance per unit length L and the capacitance per unit length C is

$$LC = \frac{\mu\varepsilon}{c^2}$$

where μ and ε are the permeability and the dielectric constant of the medium surrounding the conductors, while c is the velocity of light in vacuo.

Solution: For a perfect conductor, current flows on the surface. Assume $\mathbf{I} = I\hat{\mathbf{z}}$

$$\mathbf{A} = A\hat{\mathbf{z}}$$

$$\mathbf{B} = \nabla \times \mathbf{A} = \hat{\mathbf{x}}\frac{\partial A_z}{\partial y} - \hat{\mathbf{y}}\frac{\partial A_z}{\partial x}$$

$$\mathbf{E} = \mathbf{B} \times \hat{\mathbf{z}} = -\hat{\mathbf{x}}\frac{\partial A_z}{\partial x} - \hat{\mathbf{y}}\frac{\partial A_z}{\partial y}$$

Apply boundary conditions at surface of conductor. Since $B = E = 0$ inside conductor, the values for the field outside the conductor are given by

$$\hat{\mathbf{n}} \times \mathbf{H} = \frac{4\pi}{c} K\hat{\mathbf{z}} = \frac{1}{\mu}\left(-\frac{\partial A_z}{\partial x} - \frac{\partial A_z}{\partial y}\right)\hat{\mathbf{z}}$$

$$\mathbf{D}\cdot\hat{\mathbf{n}} = 4\pi\sigma = \varepsilon\left(-\frac{\partial A_z}{\partial x} - \frac{\partial A_z}{\partial y}\right)\hat{\mathbf{z}}$$

From these two equations, we relate σ to K,

$$\sigma = \frac{\varepsilon\mu}{c}K \quad \text{or} \quad Q = \frac{\varepsilon\mu}{c}I$$

The magnetostatic and electrostatic energy per unit length,

$$W_m = \frac{1}{8\pi}\int \mathbf{B}\cdot\mathbf{H}\,dA = \tfrac{1}{2}\,cv^2 = \tfrac{1}{2}\,\frac{Q^2}{c}$$

$$W_B = \frac{1}{8\pi}\int \mathbf{E}\cdot\mathbf{D}\,dA = \tfrac{1}{2}\,LI^2$$

where c is the capacitance per unit length and L is the inductance per unit length. Since $|B|^2 = |E|^2$,

$$\frac{W_m}{\varepsilon} = \mu W_B \quad \text{or} \quad \tfrac{1}{2}\,\frac{Q^2}{\varepsilon c} = \frac{\mu}{2}\,LI^2$$

Finally,

$$LC = \frac{Q^2}{\varepsilon\mu I^2} = \frac{\varepsilon\mu}{c^2}$$

PROBLEM 9.36

(a) Show that for a system of current-carrying elements in empty space the total energy in the magnetic field is

$$W = \frac{1}{2c^2}\int d^3x \int d^3x'\,\frac{\mathbf{J}(\mathbf{x})\cdot\mathbf{J}(\mathbf{x}')}{\mathbf{x}-\mathbf{x}'}$$

where $\mathbf{J}(\mathbf{x})$ is the current density.

(b) If the current configuration consists of n circuits carrying elements $I_1, I_2, \ldots, I_n$, show that the energy can be expressed as

$$W = \tfrac{1}{2}\sum_{i=1}^{n} L_i I_j^2 + \sum_{i=1}^{n}\sum_{j>1}^{n} M_{ij}I_iI_j$$

Exhibit integral expressions for the self-inductances (L_i) and the mutual inductances (M_{ij}).

Solution:

(a) $W = \dfrac{1}{2c}\displaystyle\int \mathbf{J}(\mathbf{x})\cdot\mathbf{A}(\mathbf{x})\,d^3x$

Substitute

$$A(\mathbf{x}) = \frac{1}{c}\int \frac{\mathbf{J}(\mathbf{x}')}{|\mathbf{x}-\mathbf{x}'|}\, d^3x'$$

$$W = \frac{1}{2c^2}\int\int d^3x\, d^3x' \,\frac{\mathbf{J}(\mathbf{x})\cdot\mathbf{J}(\mathbf{x}')}{|\mathbf{x}-\mathbf{x}'|}$$

$$\frac{dW}{dt} = Li\,\frac{dI_i}{dt} + M_{ij}\,\frac{dI_j}{dt}$$

$$W = \sum_i \int I_i L_i\, dI_i + \sum_{\substack{i \\ j>i}} \int I_i M_{ij}\, dI_j$$

$$W = \tfrac{1}{2}\sum_i L_i I_i^2 + \sum_{\substack{i \\ j>1}} M_{ij} I_i I_j$$

only sum over $j > i$ because $M_{ij} = M_{ji}$ and we only count this mutual inductance between i and j once.

$$M_{ij} = \frac{d\Phi_{ij}}{dI_j} = \frac{d}{dI_j}\int_{S_i} \mathbf{B}_j \cdot \hat{\mathbf{n}}\, dS_i$$

$$= \frac{d}{dI_j}\int_{S_i} (\nabla \times \mathbf{A}_j)\cdot\hat{\mathbf{n}}\, dS_i \quad \text{(apply Stoke's theorem)}$$

$$= \frac{d}{dI_j}\oint_i \mathbf{A}_j \cdot d\boldsymbol{\ell}_i$$

Now substitute

$$A_j = \frac{I_j}{c}\oint_j \frac{d\boldsymbol{\ell}_j}{|\mathbf{r}_i - \mathbf{r}_j|}$$

$$M_{ij} = \frac{d}{dI_j}\,\frac{I_j}{c}\oint_i\oint_j \frac{d\boldsymbol{\ell}_i \cdot d\boldsymbol{\ell}_j}{|\mathbf{r}_i - \mathbf{r}_j|}$$

$$= \frac{1}{c}\oint_i\oint_j \frac{d\boldsymbol{\ell}_i \cdot d\boldsymbol{\ell}_j}{\mathbf{r}_i - \mathbf{r}_j}$$

$$L_i = \frac{d\Phi_i}{dI_i} = \frac{d}{dI_i} \int_{S_i} \mathbf{B}_i \cdot \hat{\mathbf{n}}\, dS_i$$

$$= \frac{d}{dI_i} \int_{S_i} (\nabla \times \mathbf{A}_i) \cdot \hat{\mathbf{n}}\, dS_i = \frac{d}{dI_i} \oint_i \mathbf{A}_i \cdot d\ell_i$$

substitute

$$A_i = \frac{I_i}{c} \oint_i \frac{d\ell_i}{|\mathbf{r}_i - \mathbf{r}_i'|}$$

$$I_i = \frac{1}{c} \oint_i \oint_i \frac{d\ell_i \quad d\ell_i'}{|\mathbf{r}_i - \mathbf{r}_i'|}$$

Chapter 10

PROBLEM 10.1 Given a medium in which $\rho = 0$, $J = 0$, $\varepsilon = \varepsilon_0$, but where the magnetization $\mathbf{M}(x,y,z,t)$ is a given function. Show that the Maxwell equations are correctly obtained from a single vector function $\mathbf{Y}$, where $\mathbf{Y}$ satisfies the equation

$$\nabla^2 \mathbf{Y} - \frac{1}{c^2}\frac{\partial^2 \mathbf{Y}}{\partial t^2} = -\mu_0 \mathbf{M}$$

and where

$$\mathbf{B} = \text{curl curl } \mathbf{Y}, \quad \mathbf{E} = -\text{curl}\frac{\partial \mathbf{Y}}{\partial t}$$

Solution:

$$\nabla \cdot \mathbf{B} = 0$$

$$\nabla \cdot \mathbf{D} = \rho_f, \quad \nabla \times \mathbf{H} = \mathbf{J}_f + \frac{\partial \mathbf{D}}{\partial t}, \quad \nabla \times \mathbf{E} = -\frac{\partial \mathbf{B}}{\partial t}$$

for us:

$$\nabla \cdot \mathbf{B} = 0 \tag{1}$$

$$\nabla \cdot \mathbf{D} = 0 \tag{2}$$

$$\nabla \times \mathbf{H} = \frac{\partial \mathbf{D}}{\partial t} \tag{3}$$

$$\nabla \times \mathbf{E} = -\frac{\partial \mathbf{B}}{\partial t} \tag{4}$$

$$\nabla \cdot \mathbf{B} = \nabla \cdot (\nabla \times (\nabla \times \mathbf{Y})) = \text{div curl (of a vector called } \nabla \times \mathbf{Y}) = 0 \quad (1)$$

$$\nabla \cdot \mathbf{D} = \varepsilon_0 \nabla \cdot \mathbf{E} = -\varepsilon_0 \nabla \cdot \left(\nabla \times \frac{\partial \mathbf{Y}}{\partial t}\right)$$

$$= \text{div curl}\left(\text{of a vector called } \frac{\partial \mathbf{Y}}{\partial t}\right) = 0 \quad (2)$$

$$\nabla \times \mathbf{E} = -\nabla \times \left(\nabla \times \frac{\partial \mathbf{Y}}{\partial t}\right) = -\frac{\partial}{\partial t}\nabla \times \nabla \times \mathbf{Y} \text{ (since } \nabla\text{'s are independent of } t)$$

we are given $\mathbf{B} = \nabla \times \nabla \times \mathbf{Y}$, so

$$\nabla \times \mathbf{E} = -\frac{\partial \mathbf{B}}{\partial t} \tag{4}$$

$$\nabla \times \mathbf{H} = \nabla \times \left(\frac{\mathbf{B}}{\mu_0} - \mathbf{M}\right) = \frac{1}{\mu_0} \nabla \times \mathbf{B} - \nabla \times \mathbf{M}$$

$$= \frac{1}{\mu_0} \nabla \times (\nabla \times \nabla \times \mathbf{Y}) - \nabla \times \mathbf{M} = \frac{1}{\mu_0} \nabla \times [\nabla(\nabla \cdot \mathbf{Y}) - \nabla^2 \mathbf{Y}] - \nabla \times \mathbf{M}$$

where $\nabla \cdot \mathbf{Y} = 0$

$$= \frac{-1}{\mu_0} \nabla \times (+\nabla^2 \mathbf{Y} + \mu_0 \mathbf{M})$$

$$= -\varepsilon_0 \nabla \times \left(\frac{1}{\varepsilon_0 \mu_0} \nabla^2 \mathbf{Y} + \frac{\mathbf{M}}{\varepsilon_0}\right)$$

$$= -\varepsilon_0 \nabla \times (c^2 \nabla^2 \mathbf{Y} + \mu_0 \mathbf{M})$$

$$= -\varepsilon_0 \nabla \times \left(\frac{\partial^2 \mathbf{Y}}{\partial t^2}\right)$$

$$= \varepsilon_0 \frac{\partial}{\partial t} \left(-\nabla \times \frac{\partial \mathbf{Y}}{\partial t}\right) = \varepsilon_0 \frac{\partial}{\partial t} \mathbf{E}$$

$$\nabla \times \mathbf{H} = \frac{\partial \mathbf{D}}{\partial t} \tag{3}$$

PROBLEM 10.2 Given a medium in which $\rho = 0$, $\mathbf{J} = 0$, $\mu = \mu_0$, but where the polarization $\mathbf{P}$ is a given function of position and time: $\mathbf{P} = \mathbf{P}(x,y,z,t)$. Show that the Maxwell equations are correctly obtained from a single vector function $\mathbf{Z}$ (the Hertz vector), where $\mathbf{Z}$ satisfies the equation

$$\nabla^2 \mathbf{Z} - \frac{1}{c^2} \frac{\partial^2 \mathbf{Z}}{\partial t^2} = -\frac{\mathbf{P}}{\varepsilon_0}$$

and

$$\mathbf{E} = \text{curl curl } \mathbf{Z} - \frac{1}{\varepsilon_0} \mathbf{P}, \quad \mathbf{B} = \frac{1}{c^2} \text{curl } \frac{\partial \mathbf{Z}}{\partial t}$$

Solution: For the conditions specified, Maxwell's equations take on the following form

$$\nabla \cdot \mathbf{E} = -\frac{1}{\varepsilon_0} \nabla \cdot \mathbf{P}, \quad \nabla \cdot \mathbf{B} = 0, \quad \nabla \times \mathbf{E} = -\frac{\partial \mathbf{B}}{\partial t}, \quad \nabla \times \mathbf{B} = \frac{1}{c^2} \frac{\partial \mathbf{E}}{\partial t} + \frac{1}{c^2 \varepsilon_0} \frac{\partial \mathbf{P}}{\partial t}$$

The first two are automatically satisfied because div curl = 0. The others also follow rather easily.

PROBLEM 10.3 A parallel-plate capacitor with plates having the shape of circular disks has the region between its plates filled with a dielectric of permittivity ε. The dielectric is imperfect, having a conductivity g. The capacitance of the capacitor is C. The capacitor is charged to a potential difference ΔU and isolated.

(a) Find the charge on the capacitor as a function of time.
(b) Find the displacement current in the dielectric.
(c) Find the magnetic field in the dielectric.

Solution: Let σ be the surface charge density on positive plate. Current flow away from positive plate such that the current density will be given by $j = -\partial\sigma/\partial t$. We also have, for parallel-plate capacitors, $E = \sigma/\varepsilon$ and from Ohm's Law $j = gE$, $j = (g/\varepsilon)\sigma$. We thus obtain

$$\frac{\partial\sigma}{\partial t} = -\frac{g}{\varepsilon}\sigma \rightarrow \sigma = \sigma_0 e^{-gt/\varepsilon}$$

(a) $Q = A\sigma = A\sigma_0 e^{-gt/\varepsilon} = Q_0 e^{-gt/\varepsilon} = C\Delta U e^{-gt/\varepsilon}$

(b) Displacement current $I_D = A\dfrac{\partial D}{\partial t} = A\varepsilon\dfrac{\partial E}{\partial t} = A\dfrac{\partial\sigma}{\partial t} = \dfrac{\partial Q}{\partial t} = -\dfrac{g}{\varepsilon} C\Delta U e^{-gt/\varepsilon}$

(c) $I_{\text{total}} = I + I_D = -A\dfrac{\partial\sigma}{\partial t} + A\dfrac{\partial\sigma}{\partial t} = 0$

In fact, j_{total} is zero, so zero magnetic field.

PROBLEM 10.4 Given the electromagnetic wave

$$\mathbf{E} = \hat{\mathbf{i}}E_0 \cos\omega(\sqrt{\varepsilon\mu}\, z - t) + \hat{\mathbf{j}}E_0 \sin\omega(\sqrt{\varepsilon\mu}\, z - t)$$

where E_0 is a constant. Find the corresponding magnetic field **B** and the Poynting vector **S**.

Solution:

$$\mathbf{E} = E_0[\hat{\mathbf{i}}\cos\omega(\sqrt{\varepsilon\mu}\, z - t) + \hat{\mathbf{j}}\sin\omega(\sqrt{\varepsilon\mu}\, z - t)]$$

$$\begin{aligned}-\frac{\partial\mathbf{B}}{\partial t} = \nabla\times\mathbf{E} &= -E_0[\hat{\mathbf{i}}\times\nabla\cos\omega(\sqrt{\varepsilon\mu}\, z - t) + \hat{\mathbf{j}}\times\nabla\sin\omega(\sqrt{\varepsilon\mu}\, z - t)]\\ &= \omega\sqrt{\varepsilon\mu}\, E_0[\hat{\mathbf{i}}\times\hat{\mathbf{k}}\sin\omega(\sqrt{\varepsilon\mu}\, z - t) - \hat{\mathbf{j}}\times\hat{\mathbf{k}}\cos\omega(\sqrt{\varepsilon\mu}\, z - t)]\\ &= -\omega\sqrt{\varepsilon\mu}\, E_0[\hat{\mathbf{j}}\sin\omega(\sqrt{\varepsilon\mu}\, z - t) + \hat{\mathbf{i}}\cos\omega(\sqrt{\varepsilon\mu}\, z - t)]\end{aligned}$$

integrate:

$$\mathbf{B} = \sqrt{\varepsilon\mu}\, E_0[\hat{\mathbf{j}}\cos\omega(\sqrt{\varepsilon\mu}\, z - t) - \hat{\mathbf{i}}\sin\omega(\sqrt{\varepsilon\mu}\, z - t)]$$

$$\begin{aligned}\mathbf{S} = \mathbf{E}\times\mathbf{H} = \frac{1}{\mu}\mathbf{E}\times\mathbf{B} &= (\sqrt{\varepsilon/\mu})E_0^2[\hat{\mathbf{k}}\cos^2\omega(\sqrt{\varepsilon\mu}\, z - t) + \hat{\mathbf{k}}\sin^2\omega(\sqrt{\varepsilon\mu}\, z - t)]\\ &= \sqrt{\varepsilon/\mu}\, E_0^2\, \hat{\mathbf{k}}\end{aligned}$$

PROBLEM 10.5 Given a plane wave characterized by an E_x, B_y propagating in the positive z direction,

$$\mathbf{E} = \hat{\mathbf{i}} E_0 \sin \frac{2\pi}{\lambda} (z - ct)$$

Show that is is possible to take the scalar potential $\phi = 0$, and find a possible vector potential **A**. Be certain that the Lorentz condition ($\nabla \cdot \mathbf{A} = 0$) is satisfied.

Solution: Try

$$\mathbf{A} = -\hat{\mathbf{i}} \frac{\lambda}{2\pi c} E_0 \cos \frac{2\pi}{\lambda} (z - ct)$$

$$-\frac{\partial \mathbf{A}}{\partial t} = +\hat{\mathbf{i}} \frac{\lambda}{2\pi c} E_0 \times \left(+\frac{2\pi c}{\lambda}\right) \sin \frac{2\pi}{\lambda} (z - ct)$$

$$= \hat{\mathbf{i}} E_0 \sin \frac{2\pi}{\lambda} (z - ct)$$

which is acceptable since $\phi = 0$. Lorentz condition in this case is just $\nabla \cdot \mathbf{A} = 0$, which is satisfied since **A** has x component only and is independent of x.

$$\left(\nabla \cdot \mathbf{A} = \frac{\partial A_x}{\partial x} + 0 + 0 = 0\right)$$

PROBLEM 10.6 Show that in free space with $\rho = 0$, $J = 0$, the Maxwell equations are correctly obtained from a single vector function **A** satisfying

$$\text{div } \mathbf{A} = 0, \quad \nabla^2 \mathbf{A} - \frac{1}{c^2} \frac{\partial^2 \mathbf{A}}{\partial t^2} = 0$$

Solution: Since $\phi = 0$, we have

$$\mathbf{E} = -\frac{\partial \mathbf{A}}{\partial t}, \qquad \mathbf{B} = \nabla \times \mathbf{A}$$

$\nabla \cdot \mathbf{B} = 0$ is automatic and that $\nabla \cdot \mathbf{E} = 0$ follows from $\nabla \cdot \mathbf{A} = 0$.

$$\nabla \times \mathbf{E} = -\frac{\partial}{\partial t} \nabla \times \mathbf{A} = -\frac{\partial \mathbf{B}}{\partial t}$$

so all that is left is the curl **B** equation.

$$\nabla \times \mathbf{B} = \nabla \times \nabla \times \mathbf{A} = -\nabla^2 \mathbf{A} + \nabla(\nabla \cdot \mathbf{A})$$

with $\nabla \cdot \mathbf{A} = 0$, given. From wave equation for **A**,

$$\nabla^2 \mathbf{A} = \frac{+1}{c^2}\frac{\partial^2 \mathbf{A}}{\partial t^2} = \frac{-1}{c^2}\frac{\partial \mathbf{E}}{\partial t}$$

So

$$\nabla \times \mathbf{B} = \frac{1}{c^2}\frac{\partial \mathbf{E}}{\partial t} = \mu_0 \varepsilon_0 \frac{\partial \mathbf{E}}{\partial t} = \mu_0 \frac{\partial \mathbf{D}}{\partial t}$$

which is the remaining Maxwell's equation.

PROBLEM 10.7 Derive the wave equation for H in free space.

Solution:

$$\nabla \times \mathbf{H} = \mathbf{j} + \frac{\partial \mathbf{D}}{\partial t}$$

$$\mathbf{D} = \varepsilon \mathbf{E}$$

$$\mathbf{j} = g\mathbf{E}$$

taking the curl,

$$\nabla \times \nabla \times \mathbf{H} = \nabla \times \mathbf{j} + \varepsilon \frac{\partial}{\partial t} \nabla \times \mathbf{E}$$

$$\nabla \times \nabla \times \mathbf{H} = g\nabla \times \mathbf{E} + \varepsilon \frac{\partial}{\partial t} \nabla \times \mathbf{E}$$

$$\text{curl } \mathbf{E} = -\frac{\partial \mathbf{B}}{\partial t} = -\mu \frac{\partial \mathbf{H}}{\partial t}$$

$$\mathbf{B} = \mu \mathbf{H}$$

and

$$\nabla \times \nabla \times \mathbf{H} = \nabla(\nabla \cdot \mathbf{H}) - \nabla^2 \mathbf{H}$$

$$\nabla(\nabla \cdot \mathbf{H}) - \nabla^2 \mathbf{H} = -g\mu \left(\frac{\partial \mathbf{H}}{\partial t}\right) - \varepsilon\mu \frac{\partial}{\partial t}\left(\frac{\partial \mathbf{H}}{\partial t}\right)$$

$$\nabla\left(\nabla \cdot \frac{1}{\mu}\mathbf{B}\right) - \nabla^2 \mathbf{H} = -g\mu \left(\frac{\partial \mathbf{H}}{\partial t}\right) - \varepsilon\mu \frac{\partial}{\partial t}\left(\frac{\partial \mathbf{H}}{\partial t}\right)$$

$$\nabla \cdot \frac{1}{\mu}(\nabla \cdot \mathbf{B}) - \nabla^2 \mathbf{H} = -g\mu \left(\frac{\partial \mathbf{H}}{\partial t}\right) - \varepsilon\mu \frac{\partial}{\partial t}\left(\frac{\partial \mathbf{H}}{\partial t}\right)$$

where $\nabla \cdot \mathbf{B} = 0$

$$-\nabla^2 \mathbf{H} = -g\mu \frac{\partial \mathbf{H}}{\partial t} - \varepsilon\mu \frac{\partial^2 \mathbf{H}}{\partial t^2}$$

$$+\nabla^2 H = g\mu \frac{\partial H}{\partial t} + \varepsilon\mu \frac{\partial^2 H}{\partial t^2}$$

$$\nabla^2 H - g\mu \frac{\partial H}{\partial t} - \varepsilon\mu \frac{\partial^2 H}{\partial t^2} = 0$$

PROBLEM 10.8 Derive the wave equation for E (in free space), $\nabla \cdot D = 0$.

Solution:

$$\nabla \times E = -\frac{\partial B}{\partial t}$$

$$\nabla \times H = j + \frac{\partial D}{\partial t}$$

$$B = \mu H$$

$$D = \varepsilon E$$

$$j = gE$$

taking the curl of both sides

$$\nabla \times \nabla \times E = -\frac{\partial}{\partial t}(\nabla \times B) = -\frac{\mu\partial}{\partial t}(\nabla \times H)$$

$$\nabla(\nabla \cdot E) - \nabla^2 E = -\frac{\mu\partial}{\partial t}\left(j + \frac{\partial D}{\partial t}\right)$$

$$\nabla\frac{1}{\varepsilon}(\nabla \cdot D) - \nabla^2 E = -\mu\frac{\partial}{\partial t}\left(j + \frac{\partial D}{\partial t}\right)$$

$$-\nabla^2 E = -\mu g\frac{\partial}{\partial t}E - \mu\frac{\partial^2}{\partial t^2}\varepsilon E$$

$$\nabla^2 E - \mu g\frac{\partial E}{\partial t} - \mu\varepsilon\frac{\partial^2 E}{\partial t^2} = 0$$

PROBLEM 10.9 Starting with Maxwell's equations in linear, isotropic, conducting media, derive the boundary ("jump") conditions on the field vectors at an interface between vacuum and a perfect conductor.

Solution:

$$\nabla \cdot B = 0$$

$$\nabla \times E = -\frac{\partial B}{\partial t}$$

which is the first Boundary Conditions at the interface between two media

$$\nabla \cdot \mathbf{D} = \rho_f$$

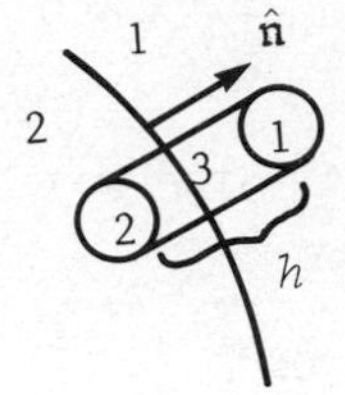

$$\nabla \times \mathbf{H} = \mathbf{j}_f + \frac{\partial \mathbf{D}}{\partial t}$$

$$\nabla \cdot \mathbf{B} = 0$$

$$\oint_s \mathbf{F} \cdot \hat{\mathbf{n}}\, da = \int_v \nabla \cdot \mathbf{F}\, dv$$

$$\int_v \nabla \cdot \mathbf{B}\, dv = \oint_s \mathbf{B} \cdot \hat{\mathbf{n}}\, da = 0$$

$$\oint_s \mathbf{B} \cdot \hat{\mathbf{n}}\, da = \oint_{s_1} \mathbf{B} \cdot \hat{\mathbf{n}}_1\, da + \oint_{s_2} \mathbf{B} \cdot \hat{\mathbf{n}}_2\, da + \oint_{s_3} \mathbf{B} \cdot \hat{\mathbf{n}}_3\, da = 0$$

now let $h \to 0$,

$$\oint_{s_1} \mathbf{B} \cdot \hat{\mathbf{n}}_1\, da - \oint_{s_2} \mathbf{B} \cdot \hat{\mathbf{n}}_2\, da = 0$$

(n's are in opposite directions)

$$B_{1n} = B_{2n} \tag{1}$$

$$\nabla \times \mathbf{E} = -\frac{\partial \mathbf{B}}{\partial t}$$

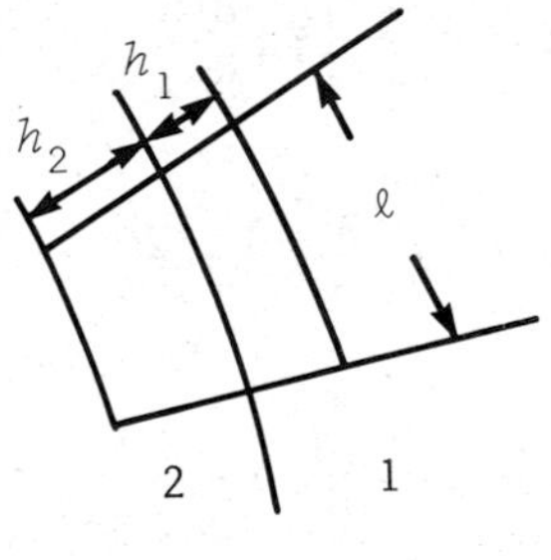

$$\int_s \nabla \times \mathbf{E} \cdot \hat{\mathbf{n}}\, da = -\int_s \frac{\partial \mathbf{B}}{\partial t} \cdot \hat{\mathbf{n}}\, da$$

$$\oint \mathbf{E} \cdot d\ell = -\int_s \frac{\partial \mathbf{B}}{\partial t} \cdot \hat{\mathbf{n}}\, da$$

$$\ell E_{1t} - \ell E_{2t} + h_1 E_{1n} + h_2 E_{2n} - h_1 E'_{1n} - h_2 E'_{2n} = -\int_s \frac{\partial \mathbf{B}}{\partial t} \cdot \hat{\mathbf{n}}\, da$$

Now shrink loop, letting h_1 and $h_2 \to 0$. So h terms vanish and so does right-hand side, provided only that $\partial \mathbf{B}/\partial t$ is bounded. So

$$\ell E_{1t} - \ell E_{2t} = 0$$

$$E_{1t} = E_{2t} \tag{2}$$

$$\nabla \cdot \mathbf{D} = \rho$$

$$\int_v \nabla \cdot \mathbf{D}\, dv = \int_v \rho\, dv$$

$$\oint_s \mathbf{D} \cdot \hat{\mathbf{n}}\, da = \int_v \rho\, dv$$

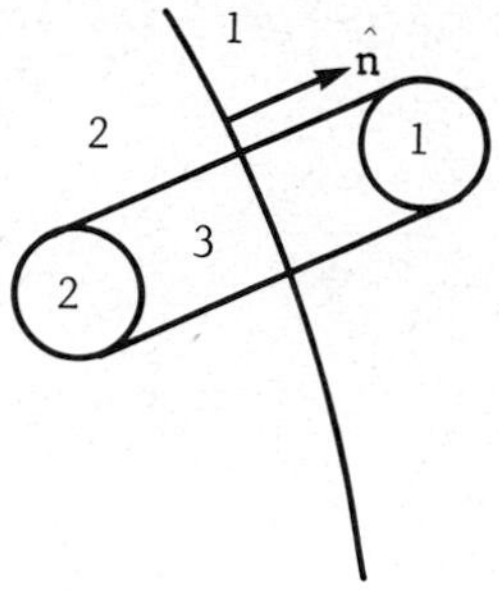

letting $h \to 0$

$$(\mathbf{D} \cdot \hat{\mathbf{n}}_1)A + (\mathbf{D} \cdot \hat{\mathbf{n}}_2)A = \sigma A$$

$$D_{1n} - D_{2n} = \sigma$$

from conservation of charge,

$$\nabla \cdot \mathbf{j} + \frac{\partial \rho}{\partial t} = 0$$

$$\nabla \cdot \mathbf{j} = -\frac{\partial \rho}{\partial t}$$

$$\int_v \nabla \cdot \mathbf{j}\, dv = -\int_v \frac{\partial \rho}{\partial t}\, dv$$

now letting $h \to 0$

$$\oint_s \mathbf{j} \cdot \hat{\mathbf{n}}\, da = -\frac{\partial}{\partial t}\, \sigma A$$

$$j_{1n} - j_{2n} = -\frac{\partial \sigma}{\partial t}$$

$$D_{1n} - D_{2n} = \sigma \qquad (3)$$

$$j_{1n} - j_{2n} = i\omega\sigma$$

if monochromatic (and σ varies as $e^{-i\omega t}$). Now

$$\mathbf{D} = \varepsilon \mathbf{E}$$

$$\mathbf{j} = g\mathbf{E}$$

$$\varepsilon_1 E_{1n} - \varepsilon_2 E_{2n} = \sigma$$

$$g_1 E_{1n} - g_2 E_{2n} = i\omega\sigma$$

Multiply by $-i\omega$

$$-i\omega\varepsilon_1 E_{1n} + i\omega\varepsilon_2 E_{2n} = -i\omega\sigma$$

$$(g_1 - i\omega\varepsilon_1)E_{1n} - (g_2 - i\omega\varepsilon_2)E_{2n} = 0$$

$$E_{2n} = \frac{g_1 - i\omega\varepsilon_1}{g_2 - i\omega\varepsilon_2} E_{1n} \tag{4a}$$

if medium 2 is a perfect conductor, $g_2 = \infty$, and

$$E_{2n} = 0 \tag{4b}$$

So

$$E_{1n} = \frac{\sigma}{\varepsilon_1} \tag{4c}$$

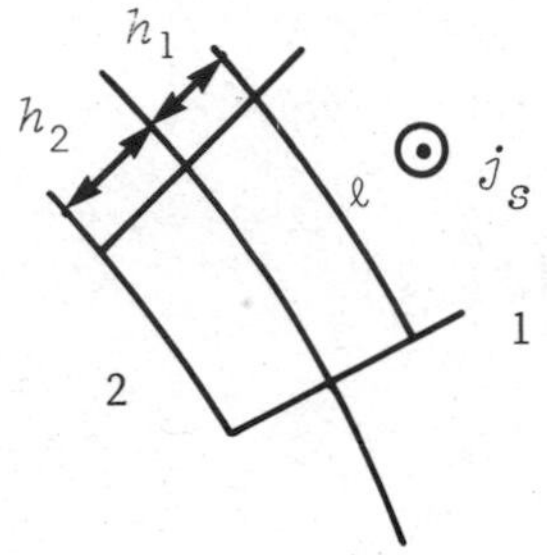

$$\nabla \times \mathbf{H} = \mathbf{j} + \frac{\partial \mathbf{D}}{\partial t}$$

$$\int_s \nabla \times \mathbf{H} \cdot \hat{\mathbf{n}}\, da = \int_s \mathbf{j} + \frac{\partial \mathbf{D}}{\partial t} \cdot \hat{\mathbf{n}}\, da$$

$$\oint \mathbf{H} \cdot d\ell = \int_s \left(\mathbf{j} + \frac{\partial \mathbf{D}}{\partial t}\right) \cdot \hat{\mathbf{n}}\, da$$

$$\ell H_{1t} - \ell H_{2t} + h_1 H_{1n} + h_2 H_{2n} - h_1 H'_{1n} - h_2 H'_{2n} = \int_s \left(\mathbf{Y} + \frac{\partial \mathbf{D}}{\partial t}\right) \cdot \hat{\mathbf{n}}\, da$$

Now shrink loop as before,

$H_{1t} - H_{2t} = j_{s\perp}$ (surface current density perpendicular to H)

$= 0$ unless conductivity is infinite.

(Unless one medium has infinite conductivity, $H_{1t} = H_{2t}$.) Apply to medium 2 (where $g = \infty$)

$$\nabla \times \mathbf{H}_2 = \mathbf{j}_2 + \frac{\partial \mathbf{D}_2}{\partial t}$$

$$\nabla \times \mathbf{H}_2 - \frac{\partial \mathbf{D}_2}{\partial t} = \mathbf{j}_2$$

$$\nabla \times \mathbf{H}_2 - \varepsilon \frac{\partial \mathbf{E}_2}{\partial t} = g_2 \mathbf{E}_2$$

$$\nabla \times \mathbf{H}_2 - \varepsilon_2(-i\omega)\mathbf{E}_2 = g_2 \mathbf{E}_2$$

$$\nabla \times \mathbf{H}_2 = (g_2 - i\omega\varepsilon_2)\mathbf{E}_2$$

$$\mathbf{E}_2 = \frac{1}{g_2 - i\omega\varepsilon_2} \nabla \times \mathbf{H}_2 \tag{5a}$$

as $g_2 \to \infty$, $\mathbf{E}_2 \to 0$

$$E_{1t} = E_{2t}$$

So

$$E_{1t} = 0 \tag{5b}$$

$$\mathbf{B}_2 = \frac{1}{i\omega} \nabla \times \mathbf{E}_2$$

therefore,

$$\mathbf{B}_2 = 0 \text{ in region 2 } (g_2 \to \infty)$$

$$B_{1n} = B_{2n} = 0$$

so

$$B_{1n} = 0 \tag{6}$$

also

$$H_{2t} = 0 \tag{7}$$

neither **E** or the time-dependent part of **B** penetrate the conductor. Now for the last boundary condition,

$$H_{1t} = J_{s\perp} \tag{8}$$

so we have

$$H_{1t} = J_{s\perp}, \quad H_{2t} = 0$$

$$E_{1t} = 0, \quad E_{1t} = E_{2t} = 0$$

$$E_{1n} = \frac{\sigma}{\varepsilon_1}, \quad E_{2n} = 0$$

$$B_{1n} = 0, \quad B_{1n} = B_{2n} = 0$$

another way is to look again at $\nabla \times \mathbf{H}$,

$$\nabla \times \mathbf{H} = \mathbf{j} + \frac{\partial \mathbf{D}}{\partial t}$$

$$\nabla \times \mathbf{H} = g\mathbf{E} + \varepsilon \frac{\partial \mathbf{E}}{\partial t} = (g - i\omega\varepsilon)\mathbf{E}$$

$$\int_s \nabla \times \mathbf{H} \cdot \hat{n} \; da = \int_s (g - i\omega\varepsilon)\mathbf{E} \cdot \hat{n} \; da$$

$$\ell H_{1t} - \ell H_{2t} + h_{\;1}H_{1n} + h_{\;2}H_{2n} - h_{\;1}H'_{1n} - h_{\;2}H'_{2n} = \frac{h}{2}\ell(g_1 - i\omega\varepsilon_1)E_1$$

these terms → 0 when loop shrinks
$$+\frac{h}{2}\ell(g_2 - i\omega\varepsilon_2)E_2$$

$$H_{1t} - H_{2t} = \frac{h}{2}(g_1 - i\omega\varepsilon_1)E_1 + \frac{h}{2}(g_2 - i\omega\varepsilon_2)E_2$$

as $\frac{h}{2} \to 0$ this term → 0

with this term, one cannot say $\frac{h}{2} \to 0$, since here $g_2 \to \infty$

Call $h/2$ the skin depth. So $(h/2)g_2E_2$ is the total current $= J_{s_\perp}$. We already said $H_{2t} = 0$, $\mathbf{B}_2 = 0$, so

$$H_{1t} = J_{s_\perp}, \text{ as before.} \tag{8}$$

PROBLEM 10.10 In a metal there are plane-wave solutions to Maxwell's equations with the form

$$E_x = E_0 \, e^{i(\omega t - kz)}$$

where k is a complex number. For low frequencies

$$k = (1 - i)\left(\frac{\sigma\omega}{2\varepsilon_0 c^2}\right)$$

(a) Write an expression for the magnetic field associated with such a wave.
(b) What is the angle between **E** and **B**?

Solution:

Given $E_x = E_0 \, e^{i(\omega t - kz)}$

$$\nabla \times \mathbf{E} = \frac{-\partial \mathbf{B}}{\partial t}$$

$$\nabla \times \mathbf{E} = \begin{vmatrix} \hat{i} & \hat{j} & \hat{k} \\ \frac{\partial}{\partial x} & \frac{\partial}{\partial y} & \frac{\partial}{\partial z} \\ E(t)e^{-ikz} & 0 & 0 \end{vmatrix} = -\hat{j}[-E(t)(-ik)\,e^{-ikz}]$$

$$\nabla \times \mathbf{E} = -\hat{\mathbf{j}}\, ik\, E_0\, e^{i(\omega t - kz)}$$

$$\mathbf{B} = -\int \nabla \times \mathbf{E}\, dt = -(-\hat{\mathbf{j}}ik)\, E_s \int e^{i\omega t}\, dt$$

$$\mathbf{B} = +\hat{\mathbf{j}}\, ikE_s \frac{1}{i\omega} e^{i\omega t}$$

(a) $\mathbf{B} = \mathbf{j}\, \frac{k}{\omega} E_0\, e^{i(\omega t - kz)}$

(b) 90°

PROBLEM 10.11 Prove a boundary condition for steady currents that must exist at the interface of two homogeneous isotropic conductors. Calculate from this and other boundary conditions the magnitude of the surface charge which must exist if the relaxation times of the two conductors are not equal.

Solution: For steady currents $\nabla \times \mathbf{j} = 0$

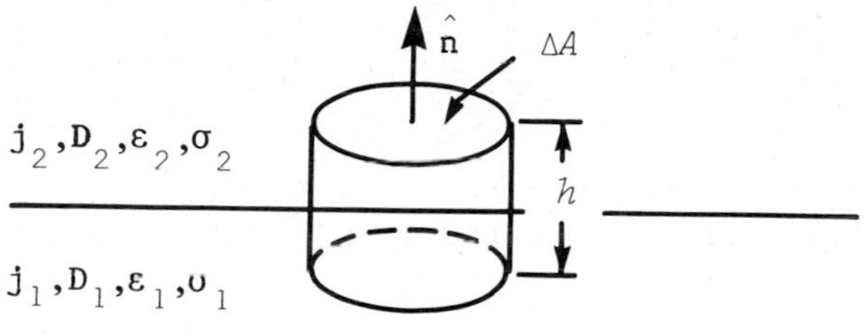

$$\int_v \nabla \cdot \mathbf{j}\, d\tau = \oint_s \mathbf{j} \cdot \hat{\mathbf{n}}\, da = 0$$

In the limit as $h \to 0$, the surface integral becomes

$$\Delta A(\mathbf{j}_2 - \mathbf{j}_1) \cdot \hat{\mathbf{n}} = 0$$

or $j_{2n} - j_{1n} = 0$, but using the supplementary condition

$$\mathbf{j} = \sigma \mathbf{E}_T$$

$$\sigma_2 E_{2n} - \sigma_1 E_{1n} = 0 \qquad (1)$$

Also using the same pill box,

$$\oint_s \hat{\mathbf{n}} \cdot \mathbf{D}\, da = \int_v \rho\, d\tau$$

$$\hat{\mathbf{n}} \cdot (\mathbf{D}_2 - \mathbf{D}_1)\Delta A = \lim_{h \to 0} \rho h\, \Delta A$$

$\eta \simeq \lim_{h \to 0} \rho h$, surface charge density

$$\hat{\mathbf{n}} \cdot (\mathbf{D}_2 - \mathbf{D}_1) = \eta$$

$$D_{2n} - D_{1n} = \eta$$

Using the supplmentary condition

$$\mathbf{D} = \varepsilon \mathbf{E}$$

$$\varepsilon_2 E_{2n} - \varepsilon_1 E_{1n} = \eta \tag{2}$$

Combining these two results we see that if $\eta = 0$, then $\varepsilon_2/\sigma_2 = \varepsilon_1/\sigma_1$ and

$$\left(\det \begin{vmatrix} \sigma_2 & -\sigma_1 \\ \varepsilon_2 & -\varepsilon_1 \end{vmatrix} = 0\right)$$

If $\eta \neq 0$, then $\varepsilon_2/\sigma_2 \neq \varepsilon_1/\sigma_1$, and

$$\left(\det \begin{vmatrix} \sigma_2 & -\sigma_1 \\ \varepsilon_2 & -\varepsilon_1 \end{vmatrix} \neq 0\right)$$

PROBLEM 10.12

(a) Write down Maxwell's equations for vacuum. Assume $\mathbf{j} = 0$, $\rho = 0$ and try solutions of the form:

$$\mathbf{E} = \mathbf{e} f(\mathbf{k} \cdot \mathbf{r} - \omega t), \quad \mathbf{B} = \mathbf{b} g(\mathbf{k} \cdot \mathbf{r} - \omega t)$$

where **e** and **b** are constant unit vectors, **k** is a constant vector, and ω is a constant.

(b) Show that **e** and **b** must be perpendicular to **k**.

(c) Show that **B** is perpendicular to both **E** and **k**.

(d) Show that $\mathbf{k}^2 = \mu_0 \varepsilon_0 \omega^2$.

(e) Show that $\mathbf{k} \times \mathbf{E}/\omega = \mathbf{B}$ (ignore possible static fields).

(f) Show that if (d) is satisfied f and g satisfy the scalar wave equation.

Solution:

(a) $\nabla \cdot \mathbf{B} = 0$, $\nabla \cdot \mathbf{E} = 0$, $\nabla \times \mathbf{E} = -\dfrac{\partial \mathbf{B}}{\partial t}$, $\nabla \times \mathbf{B} = \varepsilon_0 \mu_0 \dfrac{\partial \mathbf{E}}{\partial t}$

(b) $\nabla \cdot \mathbf{b} g = \mathbf{b} \cdot \nabla g + g \nabla \cdot \mathbf{b} = \mathbf{b} \cdot \nabla(\mathbf{k} \cdot \mathbf{r} - \omega t) \times g' = \mathbf{b} \cdot \mathbf{k} g' = 0$, where $\nabla \cdot \mathbf{b} = 0$;

therefore, $\mathbf{b} \cdot \mathbf{k} = 0$. Similarly, $\mathbf{e} \cdot \mathbf{k} = 0$ from $\nabla \cdot \mathbf{E} = 0$.

(c) $\dfrac{\partial \mathbf{B}}{\partial t} = \mathbf{b}(-\omega) g' = -\nabla \times \mathbf{E} = -\nabla \times \mathbf{e} f = -\nabla f \times \mathbf{e} - f \nabla \times \mathbf{e}$ (**e** constant and $\nabla \times \mathbf{e} = 0$)

$$= -\mathbf{k} f' \times \mathbf{e} = -\mathbf{k} \times \mathbf{e} f'$$

$\therefore$ $\mathbf{b} \perp \mathbf{e}$ and $\mathbf{k}$ so that $\mathbf{B} \perp \mathbf{E}$ and $\mathbf{k}$

(d) From above $\mathbf{b}g' = \dfrac{\mathbf{k} \times \mathbf{e}f'}{\omega}$, also

$$\nabla \times \mathbf{B} = \mu_0 \varepsilon_0 \frac{\partial \mathbf{E}}{\partial t}$$

$$\mathbf{k} \times \mathbf{b}g' = \mu_0 \varepsilon_0 (-\omega) \mathbf{e}f'$$

$$\mathbf{k} \times (\mathbf{k} \times \mathbf{b})g' = -\mu_0 \varepsilon_0 \omega \; \mathbf{k} \times \mathbf{e}f' = -\mu_0 \varepsilon_0 \omega^2 \mathbf{b}g'$$

$$[\mathbf{k}(\mathbf{k} \cdot \mathbf{b}) - \mathbf{b}k^2]g' = -\mu_0 \varepsilon_0 \omega^2 \mathbf{b}g'$$

with $\mathbf{k} \cdot \mathbf{b} = 0$ from (a), therefore,

$$k^2 = \mu_0 \varepsilon_0 \omega^2$$

(e) Integrate the first equation in (d),

$$\mathbf{b}g = \frac{\mathbf{k} \times \mathbf{e}}{\omega} f + \text{constant vector}$$

(can ignore)

$$\mathbf{B} = \frac{\mathbf{k} \times \mathbf{E}}{\omega}$$

(f) $\nabla^2 \mathbf{E} = \nabla(\nabla \cdot \mathbf{E}) - \nabla \times \nabla \times \mathbf{E}, \quad (\nabla \cdot \mathbf{E} = 0)$

$$= -\mathbf{k} \times (\mathbf{k} \times \mathbf{e})f'' = -[\mathbf{k}(\mathbf{k} \cdot \mathbf{e}) - \mathbf{e}k^2]f'' = \mathbf{e}k^2 f'', \quad (\mathbf{k} \cdot \mathbf{e} = 0)$$

$$\frac{\partial^2 \mathbf{E}}{\partial t^2} = -\omega^2 \mathbf{e}f''$$

$$\nabla^2 \mathbf{E} - \mu_0 \varepsilon_0 \omega^2 \frac{\partial^2 \mathbf{E}}{\partial t^2} = (\mathbf{e}k^2 - \mu_0 \varepsilon_0 \omega^2 \mathbf{e})f'' = [0]\,\mathbf{e}f'' = 0$$

PROBLEM 10.13 Derive the boundary condition that the normal component of **B** must satisfy at the boundary separating two materials of permeability μ_1 and μ_2.

Solution:

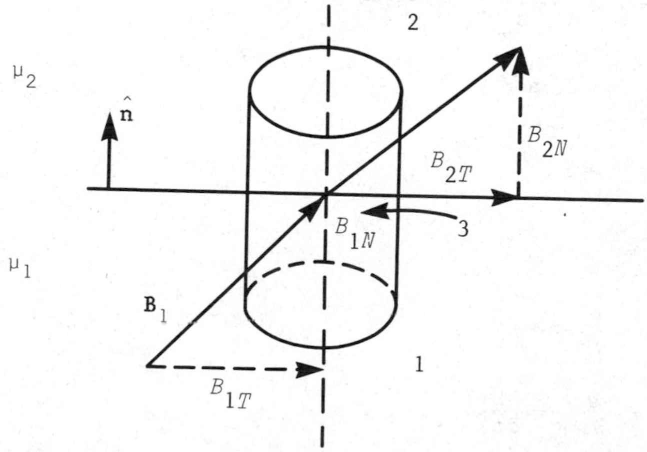

$$\nabla \cdot \mathbf{B} = 0$$

$$\oint \mathbf{B} \cdot \hat{\mathbf{n}}\, da = 0$$

$$\int_1 \mathbf{B}_1 \cdot \hat{\mathbf{n}}_1\, da + \int_2 \mathbf{B}_2 \cdot \hat{\mathbf{n}}_2\, da + \int_3 \mathbf{B}_3 \cdot \hat{\mathbf{n}}_3\, da = 0$$

if surface 3 goes to zero, the conditions for B will be going to the boundary,

$$-\mathbf{B}_1 \cdot \hat{\mathbf{n}}\, A + \mathbf{B}_2 \cdot \hat{\mathbf{n}}\, A = 0$$

$$(\mathbf{B}_2 - \mathbf{B}_1) \cdot \hat{\mathbf{n}} = 0$$

$$B_{1N} = B_{2N}$$

PROBLEM 10.14 Starting with Maxwell's equations, derive the second-order partial differential equation that the electric field must satisfy in free space. Show that

$$\mathbf{E} = \mathbf{E}_0\, e^{i(\mathbf{k} \cdot \mathbf{r} - \omega t)}$$

is one solution to this equation.

Solution:

$$\nabla \times \mathbf{E} = -\frac{\partial \mathbf{B}}{\partial t}$$

$$\nabla \times (\nabla \times \mathbf{E}) = \nabla(\nabla \cdot \mathbf{E}) - \nabla^2 \mathbf{E} = \nabla \times \left(-\frac{\partial \mathbf{B}}{\partial t}\right) = -\frac{\partial}{\partial t}(\nabla \times \mathbf{B})$$

In free space

$$\nabla \cdot \mathbf{E} = 0, \quad \mathbf{j} = 0, \quad \rho = 0$$

so

$$\nabla \times \mathbf{B} = \frac{1}{c^2}\left(\frac{\partial \mathbf{E}}{\partial t} + \frac{1}{\varepsilon_0}\mathbf{j}\right) = \frac{1}{c^2}\frac{\partial \mathbf{E}}{\partial t}$$

where $(1/\varepsilon_0)\mathbf{j} = 0$, and

$$\nabla \cdot \mathbf{E} = \frac{\rho}{\varepsilon_0} = 0$$

where $\rho/\varepsilon_0 = 0$, so

$$-\nabla^2 \mathbf{E} = -\frac{\partial}{\partial t}(\nabla \times \mathbf{B})$$

$$-\nabla^2 \mathbf{E} = -\frac{\partial}{\partial t}\left(\frac{1}{c^2}\frac{\partial \mathbf{E}}{\partial t}\right)$$

$$-\nabla^2\mathbf{E} = -\frac{1}{c^2}\frac{\partial^2\mathbf{E}}{\partial t^2}$$

$$-\nabla^2\mathbf{E} + \frac{1}{c^2}\frac{\partial^2\mathbf{E}}{\partial t^2} = 0$$

$$\nabla^2\mathbf{E} - \frac{1}{c^2}\frac{\partial^2\mathbf{E}}{\partial t^2} = 0$$

first showing that for a plane wave $\mathbf{E} = \mathbf{E}f(x - ct)$, the most general solution for a wave traveling in the x direction must satisfy

$$\frac{\partial^2\mathbf{E}}{\partial x^2} = \mu_0\varepsilon_0\frac{\partial^2\mathbf{E}}{\partial t^2}$$

E takes the form

$$\mathbf{E} = \underbrace{E_0 f(x - vt)}_{\substack{\text{wave to} \\ +x \text{ direction}}} + \mathbf{E}_1\underbrace{f(x + vt)}_{\substack{\text{wave to } -x \\ \text{direction}}}$$

So putting the $+x$ part into the wave equation, we obtain

$$\frac{\partial^2\mathbf{E}}{\partial x^2} = \varepsilon_0\mu_0\frac{\partial^2\mathbf{E}}{\partial t^2}$$

$$\frac{\partial\mathbf{E}}{\partial x} = \mathbf{E}_0 f'$$

$$\frac{\partial^2\mathbf{E}}{\partial x^2} = \mathbf{E}_0 f''$$

$$\frac{\partial\mathbf{E}}{\partial t} = \mathbf{E}_0 f'(-v)$$

$$\frac{\partial^2\mathbf{E}}{\partial t^2} = \mathbf{E}_0 v^2 f''$$

or $\quad \mathbf{E}_0 f'' = \varepsilon_0\mu_0\,\mathbf{E}_0 v^2 f''$

This can be true when $v^2\varepsilon_0\mu_0 = 1$ or $v = 1/\sqrt{\varepsilon_0\mu_0} = c$. So one can see that a solution of this form satisfies the equation, but is also has to satisfy Maxwell's equations, so for $\mathbf{E}$ just a function of x, and $\mathbf{E} = \hat{\mathbf{i}}E_x + \hat{\mathbf{j}}E_y + \hat{\mathbf{k}}E_z$.

From $\nabla \cdot \mathbf{E} = 0$, $\partial E_x/\partial x = 0$, but E_x equal to a constant can only be for a static field, so it's a solution to Maxwell's equation as well, because this is a wave we are talking about. Now that it has been shown that a solution of this form, $\mathbf{E} = \mathbf{E}_0 f(x - vt)$, satisfies the equation

$$\nabla^2 \mathbf{E} - \frac{1}{c^2}\frac{\partial^2 \mathbf{E}}{\partial t^2} = 0$$

one can now show how to get it in the form given. It does not matter what form one assumes for the solution as long as it is some function of $x - vt$, which has already been proven a solution. Defining

$$T = \frac{1}{\nu}$$

then

$$v = \frac{\lambda}{T} = \lambda\nu \text{ and } \omega = 2\pi\nu$$

and

$$\mathbf{E} = \mathbf{E} \cdot f\left[\omega\left(\frac{x}{v} - t\right)\right]$$

in general, replace x with a component of r $(\hat{\mathbf{i}}x + \hat{\mathbf{j}}y + \hat{\mathbf{k}}z)$ in v direction, for a wave in any direction instead of only the $+\hat{\mathbf{x}}$ direction.

$$\mathbf{E} = \mathbf{E}_0 f\left[\omega\left(\frac{\mathbf{r}\cdot\hat{\mathbf{v}}}{v} - t\right)\right]$$

Defining

$$\mathbf{k} = \frac{\omega\hat{\mathbf{v}}}{v}$$

then

$$\mathbf{E} = \mathbf{E}_0 f(\mathbf{k}\cdot\mathbf{r} - \omega t)$$

and choosing a function, one can choose

$$f(\mathbf{k}\cdot\mathbf{r} - \omega t) = e^{i(\mathbf{k}\cdot\mathbf{r} - \omega t)}$$

which will give

$$\mathbf{E} = \mathbf{E}_0\, e^{i(\mathbf{k}\cdot\mathbf{r} - \omega t)}$$

as one solution to this equation. Or, the given solution can be simply substituted into the second-order partial differential equation that the electric field must satisfy in free space.

PROBLEM 10.15 Show that the equation

$$\nabla \times \mathbf{H} = \frac{4\pi}{c}\mathbf{J}$$

follows directly from the equivalence between a current circuit and a suitable magnetic shell.

Solution: Consider an arbitrary current circuit C, with current I. This current distribution can be broken up into a network of elementary current loops, a typical one of which is a current following a path C_i and spanned by surface ΔS_i with normal $\hat{\mathbf{n}}_i$, as shown. The *net* current produced by this

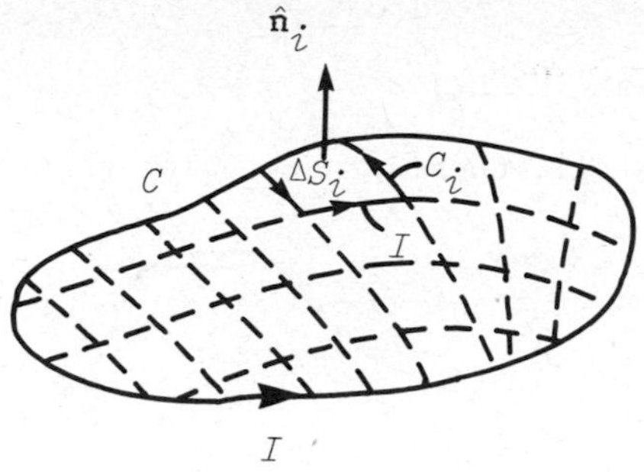

network remains invariant, and each elementary circuit can be thought of as producing a magnetic moment: $\mathbf{m}_i = (I/c)\ \Delta S_i \mathbf{n}_i$. Therefore, letting $\Delta S_i \to 0$, we see that for any *arbitrary* surface bounded by C, we may associate with each point on its surface a magnetic moment density (magnetic moment per unit area) $\mathbf{M}(\mathbf{r}) = (I/c)\hat{\mathbf{n}}(\mathbf{r})$. By construction, we expect this magnetic surface (shell) to be equivalent to the original circuit current. We now give a rigorous proof of this equivalence, simply suggested on an intuitive basis by the above argument. The magnetic potential $\mathbf{A}(\mathbf{r})$ produced by a loop with current I is given by

$$\mathbf{A}(\mathbf{r}) = \frac{1}{c}\oint_C \frac{I\,d\ell}{|\mathbf{r}-\mathbf{r}'|} = \frac{I}{c}\oint_C \frac{1}{|\mathbf{r}-\mathbf{r}'|}\,d\ell$$

using the identity

$$\oint_C f\,d\ell = \int_S (\hat{\mathbf{n}} \times \nabla f)\cdot d\mathbf{S}$$

We have

$$\mathbf{A}(\mathbf{r}) = \frac{I}{c}\oint_C \frac{1}{|\mathbf{r}-\mathbf{r}'|}\,d\ell = \frac{I}{c}\int_S \hat{\mathbf{n}} \times \nabla \frac{1}{|\mathbf{r}-\mathbf{r}'|}\,dS$$

$$= \int_S \frac{(I/c)\hat{\mathbf{n}} \times (\mathbf{r}-\mathbf{r}')}{|\mathbf{r}-\mathbf{r}'|^3}\,dS$$

which is the expression for the potential produced by a magnetic moment density

$$\mathcal{M}(\mathbf{r}) = \frac{I}{c}\,\mathbf{M}(\mathbf{r})$$

as required. Let us write with subscript 1 all quantities that refer to the circuit, and with subscript 2 all quantitites that refer to the equivalent magnetic shell. We have proved that (since $\mathbf{B} = \nabla \times \mathbf{A}$)

$$\mathbf{B}_1(\mathbf{r}) = \mathbf{B}_2(\mathbf{r}) \tag{a}$$

Since the only sources of H are free currents, we clearly have

$$H(r) \equiv H_1(r) = B_1(r) \tag{b}$$

$$H_2(r) = 0, \text{ i.e., } B_2(r) = 4\pi M \ (M \equiv M_2)$$

Let us consider the magnetic shell. Consider any point on the boundary C. We now use the *definition*,

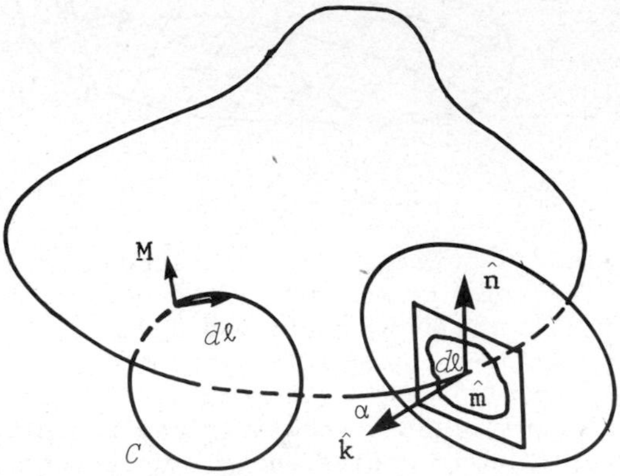

$$(\nabla \times A) \cdot \hat{k} = \lim_{\Delta S \to 0} \left(\frac{1}{\Delta S} \oint_C A \cdot d\ell \right)$$

where C is the line contour bounding the surface ΔS, contained in the plane perpendicular to $\hat{k}$. Then

$$(\nabla \times M) \cdot \hat{k} = \lim_{\Delta S \to 0} \frac{1}{\Delta S} \oint M \cdot d\ell = \lim_{\Delta S \to 0} \frac{I}{\Delta S} \frac{1}{c} \hat{n} \cdot d\ell = \lim_{\Delta S \to 0} \frac{I}{\Delta S} \frac{1}{c} \hat{m} \cdot \hat{k}$$

$$= \hat{k} \cdot \left(\lim_{\Delta S \to 0} \frac{I}{\Delta S} \hat{m} \frac{1}{c} \right) = \hat{k} \cdot \frac{J}{c}$$

($\nabla \times M = 0$ for any other point trivially.) Since $\hat{k}$ is arbitrary

$$\nabla \times M = \frac{J}{c}$$

Therefore, by (a) and (b)

$$\nabla \times H = \nabla \times B_1 = \nabla \times B_2 = \nabla \times 4\pi M = \frac{4\pi}{c} J$$

that is,

$$\nabla \times H = \frac{4\pi}{c} J$$

as required.

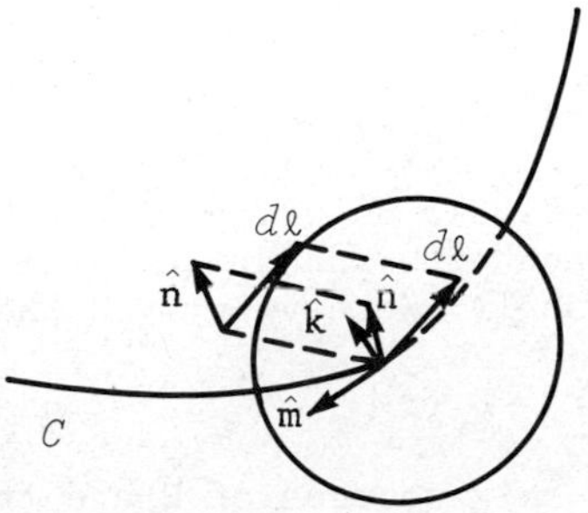

Chapter 11

PROBLEM 11.1 Show that a problem of an uncharged conducting sphere in an initially uniform electric field $\mathbf{E}_0$ may be solved by means of images. [*Hint:* a uniform electric field in the vicinity of the origin may be approximated by the field of two point charges Q and $-Q$ placed on the z axis at $z=-L$ and $z=+L$, respectively. The field becomes more nearly uniform as $L\to\infty$. It is evident that $Q/2\pi\varepsilon_0 L^2 = E_0$.]

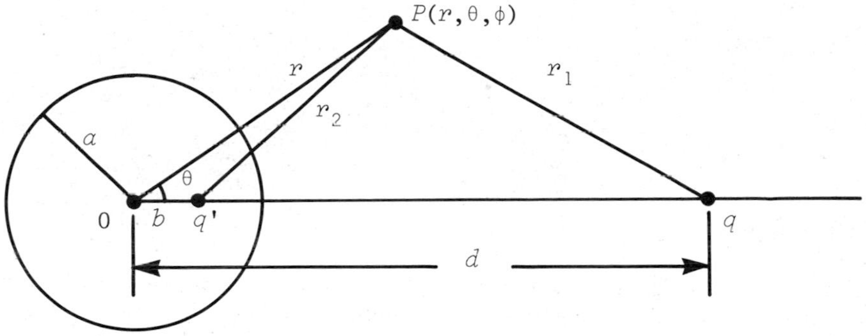

Point charge q in the vicinity of a conducting sphere; q' is the image charge, $b = a^2/d$, $q' = (-a/d)q$.

Solution: Let: $q\to Q$ and $d\to L$.

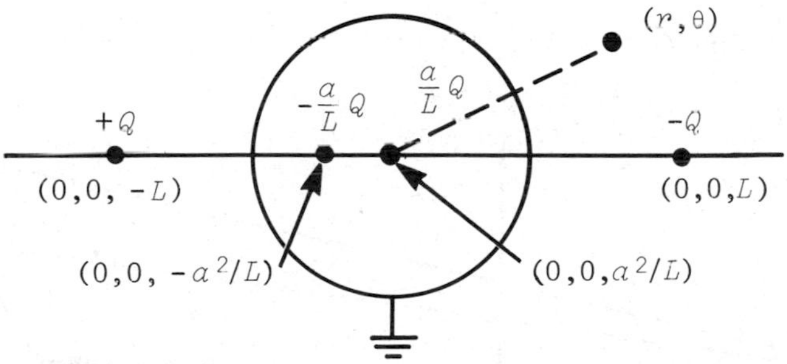

The potential at some arbitrary point (r,θ), outside the grounded sphere is the same as the potential of the four charges, Q, $-Q$ and their images.

$$4\pi\varepsilon_0 U = \left(\frac{-Q}{(r^2+L^2-2rL\cos\theta)^{\frac{1}{2}}} + \frac{Q(a/L)}{[r^2+(a^4/L^2)-2r(a^2/L)\cos\theta]^{\frac{1}{2}}}\right.$$

$$\left.+\frac{Q}{(r^2+L^2+2rL\cos\theta)^{\frac{1}{2}}} + \frac{-Q(a/L)}{[r^2+(a^4/L^2)+(2ra^2/L)\cos\theta]^{\frac{1}{2}}}\right)$$

$$4\pi\varepsilon_0 U = Q\left[-\frac{1}{L}\left(\frac{r^2}{L^2}+1-\frac{2r}{L}\cos\theta\right)^{-\frac{1}{2}}+\frac{a}{rL}\left(1+\frac{a^4}{r^2L^2}-\frac{2a^2}{rL}\cos\theta\right)^{-\frac{1}{2}}\right.$$

$$\left.+\frac{1}{L}\left(\frac{r^2}{L^2}+1+\frac{2r}{L}\cos\theta\right)^{-\frac{1}{2}}-\frac{a}{rL}\left(1+\frac{a^4}{r^2L^2}+\frac{2a^2}{rL}\cos\theta\right)^{\frac{1}{2}}\right]$$

$$=+\frac{Q}{L}\left[-1-\frac{r}{L}\cos\theta+\frac{a^3}{r^2L}\cos\theta+\frac{a}{r}+1-\frac{r}{L}\cos\theta-\frac{a}{r}+\frac{a^3}{r^2L}\cos\theta+0\left(\frac{1}{L^2}\right)\right]$$

$$=\frac{2Q}{L^2}\left[\left(\frac{a^3}{r^2}-r\right)\cos\theta+0\left(\frac{1}{L}\right)\right]$$

Let $Q,\ L\to\infty$ such that

$$\frac{Q}{L^2}\to E_0 2\pi\varepsilon_0$$

then,

$$U(r,\theta) = -E_0\left(r\cos\theta-\frac{a^3}{r^2}\cos\theta\right)$$

which is what we were asked to show.

PROBLEM 11.2 A long conducting cylinder bearing a charge λ per unit length is oriented parallel to a grounded conducting plane of infinite extent. The axis of the cylinder is at distance x_0 from the plane, and the radius of the cylinder is a. Find the location of the line image, and find also the constant M (which determines the potential of the cylinder) in terms of a and x_0.

Solution:

$$U = -\frac{\lambda}{2\pi\varepsilon_0}(\ln r_1 - \ln r_2) = -\frac{\lambda}{2\pi\varepsilon_0}\ln\frac{r_1}{r}$$

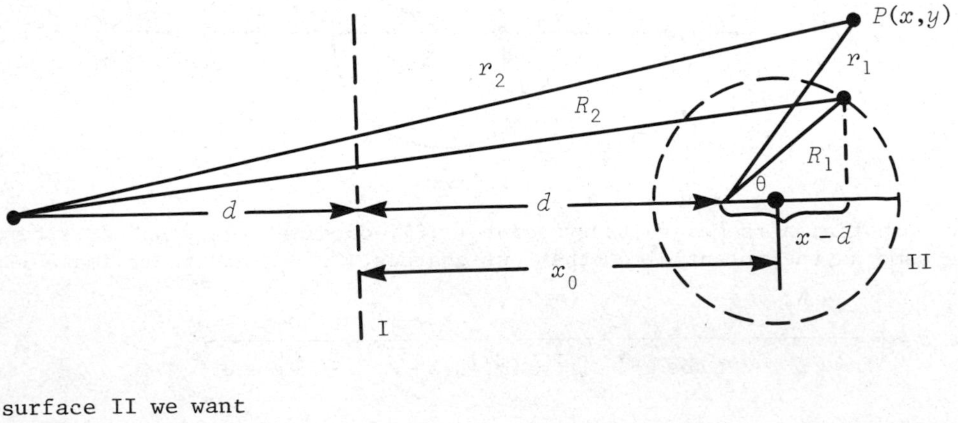

On surface II we want

$$U = -\frac{\lambda}{2\pi\varepsilon_0} \ln \frac{R_1}{R_2} = \text{constant},$$

so $R_1/R_2 = M = \text{constant}$ will do the job.

$$R_1^2 = R_2^2 M^2 = 0, \quad y^2 + (x-d)^2 - M^2[y^2 + (x+d)^2] = 0$$

$$y2(1 - M^2) + X^2(1 - M^2) - (1 + M^2)2Xd + (1 - M^2)d^2 = 0$$

$$y^2 + X^2 - \frac{(1+M^2)2Xd}{1 - M^2} + d^2 = 0$$

$$y^2 + \left(X - \frac{1+M^2}{1-M^2} d\right)^2 - \frac{(1+M^2)^2}{(1-M^2)^2} d^2 + d^2 = 0$$

So, center of circle is at $X = X_0 = \dfrac{1+M^2}{1-M^2} d$ and $d = \dfrac{1-M^2}{1+M^2} X_0$

The radius of the circle is given by

$$a^2 = d^2 \frac{1+M^2}{1-M^2} - d^2 = X_0^2 \left(1 - \frac{(1-M^2)^2}{(1+M^2)^2}\right)$$

$$\frac{(1-M^2)^2}{(1+M^2)^2} = 1 - \frac{a^2}{X_0^2} > 0, \quad \text{so the circle won't intersect plane.}$$

$$= k^2, \text{ say}$$

$$(1 - M^2) = k(1 + M^2), \quad M = \sqrt{\frac{1-K}{1+K}} = \left[\frac{1 - \sqrt{1 - (a^2/X_0^2)}}{1 + \sqrt{1 - (a^2/X_0^2)}}\right]^{1/2}$$

$$U = \frac{-\lambda}{2\pi\varepsilon_0} \tfrac{1}{2} \ln \left[\frac{1 - \sqrt{1 - (a^2/X_0^2)}}{1 + \sqrt{1 - (a^2/X_0^2)}}\right] = + \frac{\lambda}{4\pi\varepsilon_0} \ln \left[\frac{[1 + \sqrt{1 - (a^2/X_0^2)}]^2}{a^2/X_0^2}\right]$$

$$= \frac{\lambda}{2\pi\varepsilon_0} \ln \left[\frac{X_0}{a} + \sqrt{\frac{X_0^2}{a^2} - 1}\right]$$

where the term in the square brackets equals $1/M$, therefore,

$$M = \left[\frac{X_0}{a} + \sqrt{\frac{X_0^2}{a^2} - 1}\right]^{-1}$$

So once center and radius are chosen, the potential is fixed by λ.

PROBLEM 11.3 A long cylindrical conductor of radius a bearing no net charge is placed in an initially uniform electric field E_0. The direction of E_0 is perpendicular to the cylinder axis. Find the potential at points exterior to the cylinder, and find also the charge density on the cylindrical surface.

Solution: Cylindrical geometry is used, so

$$U(r) = E\ln r + \sum_{n=1}^{\infty} (A_n \cos n\theta + B_n \sin n\theta) r^n + \sum_{n=1}^{\infty} (C_n \cos n\theta + D_n \sin n\theta) r^{-n}$$

The charge density of the cylinder is given by

$$\sigma = -\varepsilon_0 \left.\frac{\partial U}{\partial r}\right|_{r=a}$$

$$= -\frac{\varepsilon_0 E}{a} - \sum_{n=1}^{\infty} na^{n-1} (A_n \cos n\theta + B_n \sin n\theta)$$

$$+ \sum_{n=1}^{\infty} (C_n \cos n\theta + D_n \sin n\theta) a_n^{-n-1}$$

$$\text{Charge per unit length} = \lambda = \int_0^{2\pi} \sigma\, a\, d\theta,$$

but

$$\int_0^{2\pi} \begin{Bmatrix} \cos n\theta \\ \sin n\theta \end{Bmatrix} d\theta = 0$$

therefore, $\lambda = -2\pi\varepsilon_0 E$, $E = -\dfrac{\lambda}{2\pi\varepsilon_0}$,

but $\lambda = 0$, therefore, $E = 0$. (In general, however, the leading term would be

$$-\frac{\lambda}{2\pi\varepsilon_0} \ln r$$

where λ is the charge per unit length. This is the two-dimensional "monopole" term in analogy with $q/4\pi\varepsilon_0 r$ in three dimensions.) Since we have a uniform field at infinity we should have $U \underset{r\to\infty}{\to} Er\cos\theta$ (giving a field in the x direction). Therefore,

$$A_1 = -E_0, \quad B_1 = 0, \quad A_n = B_n = 0, \quad n > 1$$

(We could have some constant, A_0, in the potential but it has no physical significance.) At this point, then, we have

$$U(r) = -E_0 r\cos\theta + \sum_{n=1}^{\infty} [C_n \cos n\theta + D_n \sin n\theta] r^{-n}$$

and it remains to evaluate C_n and D_n. Since we have a conductor, we know that $U(a) = \text{constant}$ (independent of θ), therefore, the coefficients of all $\cos n\theta$ and $\sin n\theta$ terms must individually vanish (orthogonality). This leads to

$$\frac{C_1}{a} = aB_1 + 0, \quad C_1 = -a^2B_1 = a^2E_0 \quad \text{and } D_n = 0, \text{ all } n \text{ and } C_n = 0, \; n \neq 1$$

(By the way, note that B_0 and C_0 do not concern us at all.) Thus, we have, finally

$$U(r) = (a^2/r - r)E_0 \cos\theta + A_0$$

$$\sigma(\theta) = -\varepsilon_0 \left.\frac{\partial U}{\partial r}\right|_{r=a} = -\varepsilon_0 \left.\left(\frac{-a^2}{r^2} - 1\right)E_0 \cos\theta\right|_{r=a}$$

$$\sigma(\theta) = 2\varepsilon_0 E_0 \cos\theta$$

PROBLEM 11.4 Two spherical conducting shells of radii r_a and r_b are arranged concentrically and are charged to the potential U_a and U_b, respectively. If $r_b > r_a$, find the potential at points between the shells, and at points $r > r_b$.

Solution: Let

$$U(r) = \sum_{n=0}^{\infty} (A_n r^n + B_n r^{-1-n})P_n(\cos\theta), \quad r_b > r \geq r_a$$

$$= \sum_{n=0}^{\infty} (C_n r^n + D_n r^{-1-n})P_n(\cos\theta), \quad r \geq r_b$$

For both regions only $n = 0$ contributes, because U is independent of θ on the boundaries and hence everywhere. Thus

$$U = C_0 + \frac{D_0}{r}, \quad r \geq r_b$$

$$C_0 + \frac{D_0}{r_b} = U_b, \text{ to match the boundary condition}$$

$$C_0 + \frac{D_0}{\infty} = 0, \text{ to match the boundary condition at } \infty$$

$$\therefore \; C_0 = 0, \quad D_0 = U_b r_b$$

and

$$U(r) = \frac{U_b r_b}{r}, \quad r > r_b$$

$$U = A_0 + \frac{B_0}{r}, \quad r_b \geq r \geq r_a$$

$$A_0 + \frac{B_0}{r_a} = U_a, \text{ to match the boundary condition at } r = r_a$$

$$A_0 + \frac{B_0}{r_b} = U_b, \text{ to match the boundary condition at } r = r_b$$

Solve above for A_0 and B_0,

$$A_0 = \frac{U_a r_a - U_b r_b}{r_a - r_b}, \quad B_0 = \frac{(U_a - U_b) r_b r_a}{r_b - r_a}$$

therefore,

$$U = \frac{U_b r_b - U_a r_a + (U_a - U_b)(r_b r_a / r)}{r_b - r_a}, \quad r_a < r < r_b$$

PROBLEM 11.5

(a) Three charges are arranged in a linear array. The charge $-2q$ is placed at the origin, and two charges, each of $+q$, are placed at $(0,0,\ell)$ and $(0,0,-\ell)$, respectively. Find a relatively simple expression for the potential $U(r)$ which is valid for distances $|\mathbf{r}| \gg \ell$. Make a plot of the equipotential surfaces in the x, z plane.

(b) What is the quadrupole moment tensor of the charge distribution discussed above?

Solution:

(a) $$4\pi\varepsilon_0 U(\mathbf{r}) = \frac{q}{|\mathbf{r} - \ell|} + \frac{q}{|\mathbf{r} + \ell|} - \frac{2q}{r} \quad (\vec{\ell} = \hat{\mathbf{z}}\ell)$$

$$|\mathbf{r} \pm \ell|^{-1} = [r^2 + (\pm 2\vec{\ell} \cdot \mathbf{r} + \ell^2)]^{-\frac{1}{2}}$$

$$= \frac{1}{r}\left[1 - \tfrac{1}{2}\left(\frac{\pm 2\vec{\ell} \cdot \mathbf{r}}{r^2} + \frac{\ell^2}{r^2}\right) + \frac{3}{8}\left(\frac{\pm 2\vec{\ell} \cdot \mathbf{r}}{r^2} + \frac{\ell^2}{r^2}\right)^2 + \ldots\right]$$

$$= \frac{1}{r}\left[1 \mp \frac{\vec{\ell} \cdot \mathbf{r}}{r^2} - \frac{\ell^2}{2r^2} + \frac{3}{2}\,\frac{(\vec{\ell} \cdot \mathbf{r})^2}{r^4} + \ldots\right]$$

$$= \frac{1}{r}\left[1 \mp \frac{\ell z}{r^2} + \tfrac{1}{2}\left(\frac{3\ell^2 z^2}{r^4} - \frac{\ell^2}{r^2}\right) + \ldots\right], \quad (\ell \cdot \mathbf{r} = \ell z)$$

Substitute this in the expression for the potential:

$$4\pi\varepsilon_0 U(\mathbf{r}) = q\left[\frac{1}{r} - \frac{z\ell}{r^3} + \tfrac{1}{2}\left(\frac{3\ell^2 z^2}{r^4} - \frac{\ell^2}{r^2}\right) + \ldots + \frac{1}{r} + \frac{z\ell}{r^3} + \tfrac{1}{2}\left(\frac{3\ell^2 z^2}{r^4} - \frac{\ell^2}{r^2}\right) + \ldots - \frac{z}{r}\right]$$

$$= \frac{q}{r}\left(\frac{3\ell^2 z^2}{r^4} - \frac{\ell^2}{r^2}\right) + \ldots$$

$$= \frac{q\ell^2}{r^3}\,(3\cos^2\theta - 1) + \ldots$$

To get the equipotentials look at $U > 0$ first. We have,

$$r^3 = C(3\cos^2\theta - 1), \quad C < 0$$

need $\cos^2\theta > 1/3$ for $U < 0$. $r = 0$ at $\cos\theta = \pm 1/3$ and $r = r_{max} = 2C$ at $\theta = 0$.

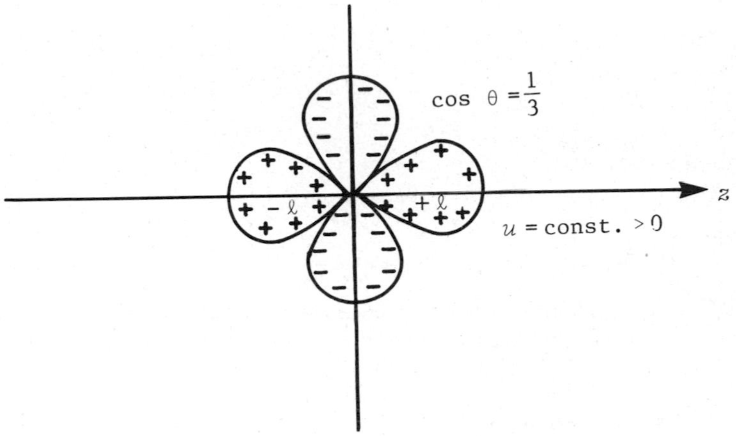

(ℓ is much smaller than what is shown.) Similar arguments hold for negative equipotentials.

(b) The general form of the quadrupole contribution to the potential is

$$4\pi\varepsilon_0 U_{quad} = \sum_{i=1}^{3}\sum_{j=1}^{3} \tfrac{1}{2}\,\frac{X_i X_j}{r^5}\,Q_{ij}, \quad (x_1 = x, \;\; y_1 = y, \;\; z_1 = z)$$

We have

$$4\pi\varepsilon_0 U = q\left(\frac{3\ell^2 z^2}{r^5} - \frac{\ell^2 r^2}{r^5}\right)$$

$$= q\left(\frac{2z^2\ell^2}{r^5} - \frac{y^2\ell^2}{r^5} - \frac{x^2\ell^2}{r^5}\right)$$

So, identifying terms in the general expression we note that all off-diagonal terms vanish and

$$Q_{11} = Q_{22} = -2q\ell^2 \quad\text{and}\quad Q_{33} = 4q\ell^2$$

PROBLEM 11.6 Given a region of space in which the electric field is everywhere directed parallel to the x axis. Prove that the electric field is independent of the y and z coordinates in this region. If there is no charge in this region, prove that the field is also independent of x.

Solution:

$$\mathbf{E} = \hat{\mathbf{i}}E_x(x, y, z), \text{ from } \nabla \times \mathbf{E} = 0 \text{ we get}$$

$$\frac{\partial E_z}{\partial y} - \frac{\partial E_y}{\partial z} = 0$$

$$\frac{\partial E_x}{\partial z} - \frac{\partial E_z}{\partial x} = 0 \rightarrow \frac{\partial E_x}{\partial z} = 0$$

$$\frac{\partial E_y}{\partial x} - \frac{\partial E_x}{\partial y} = 0 \rightarrow \frac{\partial E_x}{\partial y} = 0$$

From charge neutrality,

$$\nabla \cdot \mathbf{E} = 0 \rightarrow \frac{\partial E_x}{\partial x} = 0$$

PROBLEM 11.7 A thin, conducting, spherical shell of radius R is charged uniformly with total charge Q. By direct integration, find the potential at an arbitrary point (a) inside the shell, (b) outside the shell.

Solution: Without loss of generality, we can choose the field point to be on the z axis. We use spherical polar coordinates to integrate over $\mathbf{r}'$.

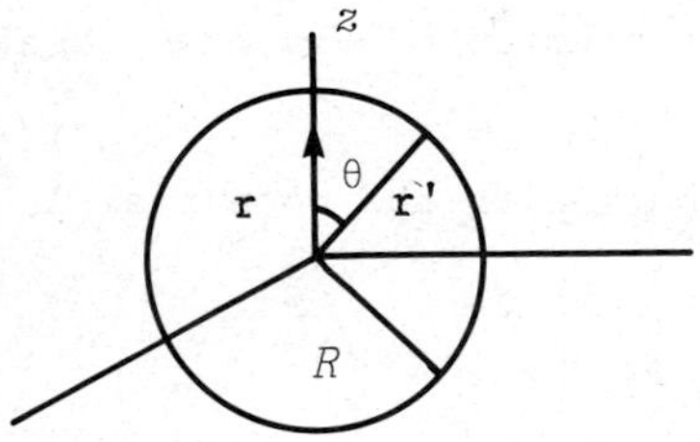

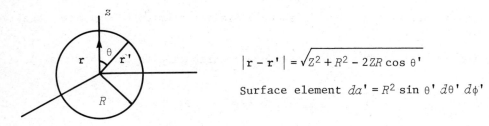

$$|\mathbf{r} - \mathbf{r}'| = \sqrt{Z^2 + R^2 - 2ZR\cos\theta'}$$

Surface element $da' = R^2 \sin\theta'\, d\theta'\, d\phi'$

$$4\pi\varepsilon_0 U(\mathbf{r}) = \int_0^{2\pi}\int_0^{\pi} \frac{\sigma R^2 \sin\theta'\, d\theta'\, d\phi'}{\sqrt{Z^2 + R^2 - 2ZR\cos\theta'}}, \quad \sigma = \frac{Q}{4\pi R^2}$$

Let $\cos\theta = \mu$, $-\sin\theta = d\mu$, take care of minus sign by changing limits of integration. Integration over ϕ is trival.

$$4\pi\varepsilon_0 U(\mathbf{r}) = \frac{Q}{4\pi} \cdot 2\pi \int_{-1}^{1} \frac{d\mu}{\sqrt{Z^2 + R^2 - 2ZR\mu}}$$

$$= \frac{Q}{Z}\left(-\frac{1}{RZ}\sqrt{Z^2+R^2-2ZR\mu}\right)\Bigg|_{-1}^{1}$$

$$= -\frac{Q}{2RZ}\left(\sqrt{Z^2+R^2-2ZR}-\sqrt{Z^2+R^2+2ZR}\right)$$

$$= \frac{Q}{2RZ}\left(|Z+R|-|R-Z|\right)$$

$$\varepsilon_0 4\pi U = \frac{Q}{R}, \quad R \geq Z$$

$$= \frac{Q}{Z}, \quad Z \geq R$$

But the choice of the direction of the z axis in space is arbitrary so

$$4\pi\varepsilon_0 U = \frac{Q}{r}, \quad r \geq R$$

$$= \frac{Q}{R}, \quad r \leq R$$

PROBLEM 11.8

(a) A circular disk of radius R has a uniform surface charge density σ. Find the electric field at a point on the axis of the disk at a distance z from the plane of the disk.
(b) A right circular cylinder of radius R and height L is oriented along the z axis. It has a nonuniform volume density of charge given by $\rho(z) = \rho_0 + \beta z$ with reference to an origin at the center of the cylinder. Find the force on a point charge q placed at the center of the cylinder.

Solution:

(a)

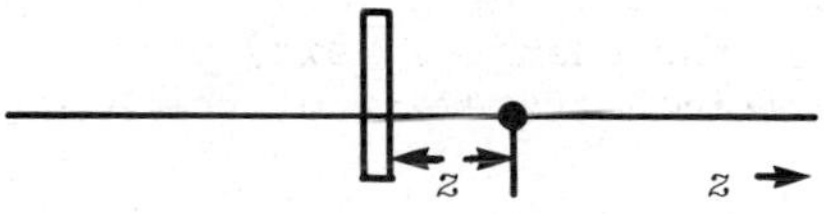

$$\mathbf{E} = \frac{1}{4\pi\varepsilon_0}\int_s \frac{\sigma(\mathbf{r}')(\mathbf{r}-\mathbf{r}')}{|\mathbf{r}-\mathbf{r}'|^3}\,dv'$$

$\sigma(\mathbf{r}')$ = surface charge density $= \sigma$ = constant. s is disk in the plane $z' = 0$. Clearly only $E_z \neq 0$,

$$E_z = \frac{\sigma z}{4\pi\varepsilon_0}\int\int dx'\,dy'\,\frac{z-0}{[(0-x')^2+(0-y')^2+(z-z')^2]^{3/2}}$$

in cylindrical coordinates: $x' = r'\cos\phi'$, $y' = r'\sin\phi'$

$$= \frac{\sigma z}{4\pi\varepsilon_0} \int_0^{2\pi} d\phi' \int_0^R r'\, dr' \frac{1}{(r'^2 + z^2)^{3/2}}$$

$$= \frac{\sigma z}{4\pi\varepsilon_0} 2\pi \left(-\frac{1}{(r'^2 + z^2)^{\frac{1}{2}}} \right)_0^R = -\frac{\sigma z}{2\varepsilon_0} \left(\frac{1}{(R^2 + z^2)^{\frac{1}{2}}} - \frac{1}{(z^2)^{\frac{1}{2}}} \right)$$

$$E_z = + \frac{\sigma}{2\varepsilon_0} \left(\frac{z}{|z|} - \frac{z}{(R^2 + z^2)^{\frac{1}{2}}} \right)$$

(b)

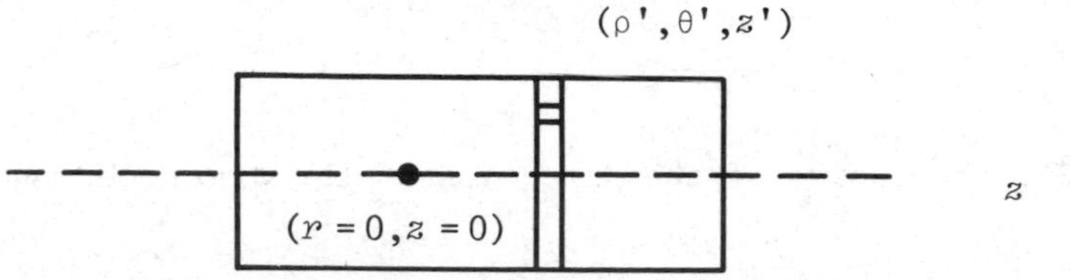

$$E_z = \int_{-\frac{1}{2}}^{\frac{1}{2}} \int_0^{2\pi} \int_0^R \frac{\rho(\mathbf{r}')(z - z')}{[(x - x')^2 + (z - x')^2 + (y - y')^2]^{3/2}} r'\, d\phi'\, dr'\, dz'$$

$$= \int_{-\frac{1}{2}}^{\frac{1}{2}} \int_0^{2\pi} \int_0^R \frac{(\rho_0 + \beta z')(-z')}{(r'^2 + z'^2)^{3/2}} d\phi'\, r'\, dr'\, dz'$$

integration over r' and ϕ' is identical to the one performed above. (Note sign, however), so,

$$E_z = -\frac{1}{2\varepsilon_0} \int_{-\frac{1}{2}}^{\frac{1}{2}} \rho(z') \left(\frac{z'}{|z'|} - \frac{z'}{(R^2 + z'^2)^{\frac{1}{2}}} \right) dz'$$

The term in parentheses in the integrand is an odd function of z', therefore only the portion $\beta z'$ of the charge density will contribute to the integral. (This is just a mathematical statement of the basic symmetry in this case.) We are left with

$$E_z = -\frac{\beta}{2\varepsilon_0} \int_{-\frac{1}{2}}^{\frac{1}{2}} \left(\frac{z'^2}{|z'|} - \frac{z'^2}{(R^2 + z'^2)^{\frac{1}{2}}} \right) dz'$$

$$= -\frac{\beta}{2\varepsilon_0} \cdot 2 \int_0^{\frac{1}{2}} \left(z' - \frac{z'^2}{(R^2 + z'^2)^{\frac{1}{2}}} \right) dz'$$

$$= -\frac{\beta}{\varepsilon_0} \left(\frac{z'^2}{2} - \frac{z'}{2} \sqrt{R^2 + z'^2} + \frac{R^2}{2} \ln(z' + \sqrt{z'^2 + R^2}) \right)_0^{\frac{1}{2}}$$

$$= -\frac{\beta}{\varepsilon_0}\left[\frac{L^2}{8} - \frac{L}{4}\sqrt{R^2 + \frac{L^2}{4}} + \frac{R^2}{2}\ \ln\left(\frac{L}{2} + \sqrt{\frac{L^2}{4} + R^2}\right) - \frac{R^2}{2}\ln R\right]$$

$$E_z = -\frac{\beta}{\varepsilon_0}\left[\frac{L}{4}\left(\frac{L}{2} - \sqrt{R^2 + \frac{L^2}{4}}\right) + \tfrac{1}{2} R^2 \ln\left(\frac{L}{2R} + \sqrt{\frac{L^2}{4R^2} + 1}\right)\right]$$

PROBLEM 11.9 An infinitely long circular cylinder is uniformly magnetized in the direction perpendicular to its axis of symmetry. Calculate **B** inside and outside the cylinder.

Solution:

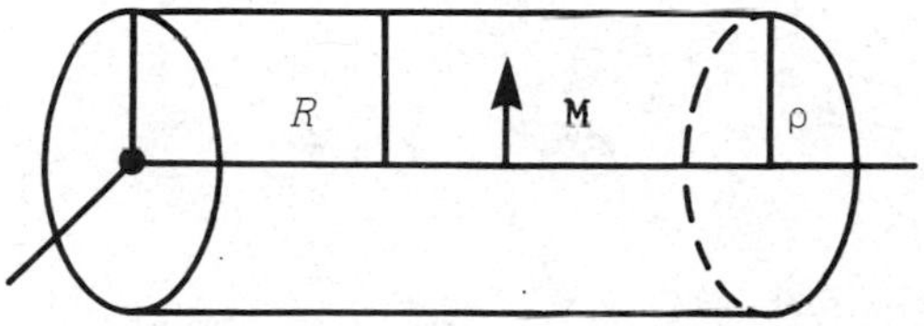

$$\mathbf{H} = -\nabla U^*, \qquad \mathbf{B} = \mu_0(\mathbf{H} + \mathbf{M})$$

$$\mu = A + B\ln r + \sum_{n=1}^{\infty} (C_n r^n + D_n r^{-n}) \cos n\theta + \sum_{n=1}^{\infty} (E_n r^n + F_n r^{-n}) \sin n\theta$$

here $\rho = r$,

$$U^*_{out} = A_1 + B_1 \ln\rho + \sum_{n=1}^{\infty} (C_{1n}\rho^n + D_{1n}\rho^{-n}) \cos n\theta + \sum_{n=1}^{\infty} (E_{1n}\rho^n + F_{1n}\rho^{-n}) \sin n\theta$$

$$U^*_{in} = A_2 + B_2 \ln\rho + \sum_{n=1}^{\infty} (C_{2n}\rho^n + D_{2n}\rho^{-n}) \cos n\theta + \sum_{n=1}^{\infty} (E_{2n}\rho^n + F_{2n}\rho^{-n}) \sin n\theta$$

The boundary conditions are

U^* regular at ∞ and origin

U^* continuous across boundary at $\rho = R$

$B_{12} = B_{2n}$

So,

$B_1 = B_2 = 0$, no magnetic monopoles

$A_1 = A_2 = 0$, no physical significance and will vanish when derivative is taken for H anyway

$D_{2n} = F_{2n} = 0$, regularity at origin

$C_{1n} = E_{1n} = 0$, regularity at infinity

so far,

$$U^*_{out} = \sum_{n=1}^{\infty} D_{1n}\rho^{-n} \cos n\theta + F_{1n}\rho^{-n} \sin n\theta$$

$$U^*_{in} = \sum_{n=1}^{\infty} C_{2n}\rho^{n} \cos n\theta + E_{2n}\rho^{n} \sin n\theta$$

now potential continuous across boundary,

$$\frac{D_{1n}}{R^n} \cos n\theta + \frac{F_{1n}}{R^n} \sin n\theta = C_{2n}R^n \cos n\theta + E_{2n}R^n \sin n\theta$$

$$\frac{D_{1n}}{R^n} = C_{2n}R^n$$

$$\frac{F_{1n}}{R^n} = E_{2n}R^n$$

now

$$B_{1n} = B_{2n}, \; \mathbf{B} = \mu_0(\mathbf{H} + \mathbf{M}), \quad F_{1n}, \; E_{2n}, \; D_{12} \cdot n, \; C_{22} \cdot n, = 0$$

by matching expressions in expansion

$$B_{\rho\ \mathrm{in}} = \mu_0 \left(\left. \frac{-\partial U^*_{in}}{\partial \rho} \right|_{\rho = R} + M_\rho \right) = -\mu_0 \left(C_{21} \cos \theta\right) + \mu_0 M \cos \theta$$

$$B_{\rho\ \mathrm{out}} = \mu_0 \left(\left. \frac{-\partial U^*_{out}}{\partial \rho} \right|_{\rho = R} \right) = (-\mu_0) \left(\frac{-D_{11}}{R^2} \cos \theta \right)$$

$$-C_2 + M = \frac{D_{11}}{R^2}$$

$$\frac{D_{11}}{R} = C_{21}R$$

$$C_2 = \frac{D_1}{R^2}$$

$$\frac{D_1}{R^2} = M - \frac{D_1}{R^2}, \quad \frac{2D_1}{R^2} = M, \quad D_1 = \frac{MR^2}{2}, \quad C_2 = \frac{M}{2}$$

$$U^*_{out} = \frac{MR^2}{2R} \cos \theta$$

$$U^*_{in} = \frac{M}{2}\, r \cos\theta = \frac{M}{2}\, z$$

$$\mathbf{H}_{in} = -\frac{M}{2}\hat{\mathbf{z}}, \quad \mathbf{B}_{out} = \mu_0 \frac{M}{2}\hat{\mathbf{z}}$$

$$\mathbf{B} = \mu_0\, \mathbf{H}_{out} = \mu_0(-)\left[\hat{\mathbf{r}}(-)\frac{MR^2}{2r^2}\cos\theta + \hat{\theta}\frac{MR^2}{2r^2}(-\sin\theta)\right]$$

$$\mathbf{B}_{out} = \mu_0 \frac{MR^2}{2}\left[\frac{\hat{\mathbf{r}}}{r^2}\cos\theta + \frac{\hat{\theta}}{r^2}\sin\theta\right]$$

PROBLEM 11.10 Consider a solenoid of circular cross section and *finite* length, carrying a current I. Express the magnetic field at an arbitrary point in space as an integral. Write this out explicitly in a coordinate system of your choice specifying limits of integration. Find $\mathbf{B}$ ($z = 0$, $\rho = 0$).

Solution:

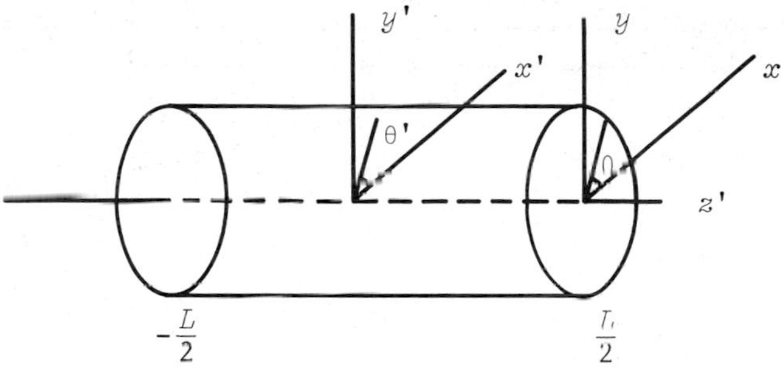

$$\mathbf{B} = \frac{\mu_0 NIR}{4\pi}\iint_{\text{surface}} \frac{\hat{\theta}'\times(\mathbf{r}-\mathbf{r}')}{|\mathbf{r}-\mathbf{r}'|^3}\, d\theta'\, dz'$$

$$\mathbf{r} = \rho\hat{\rho} + z\hat{\mathbf{z}}, \quad \mathbf{r}' = \hat{\rho}'\rho' + \hat{\mathbf{z}}'z'$$

$$\hat{\theta}'\times(\mathbf{r}-\mathbf{r}') = \hat{\theta}'\times\hat{\rho}\rho + \hat{\rho}'z + \hat{\mathbf{z}}\, R - \hat{\rho}'z'$$

$$|\mathbf{r}-\mathbf{r}'| = [z^2 + \rho^2 + z'^2 + R^2 - 2zz' - 2\rho R\cos(\theta - \theta')]^{\frac{1}{2}}$$

So

$$\mathbf{B} = \frac{\mu_0 NIR}{4\pi}\int_{-L/2}^{L/2}\int_0^{2\pi} \frac{\hat{\theta}'\times\hat{\rho}\rho + \hat{\rho}'z + \hat{\mathbf{z}}R - \hat{\rho}'z'\, dz'\, d\theta'}{(z-z')^2 + \rho^2 + R^2 - 2\rho R\cos(\theta - \theta')}$$

On axis $z = 0$, $\rho = 0$,

$$\mathbf{B} = \frac{\mu_0 NIR}{4\pi}\left[\int_{-L/2}^{L/2}\int_0^{2\pi} \frac{\rho'(z-z')\, dz'\, d\theta'}{[R^2 + (z-z')^2]^{3/2}} + \hat{\mathbf{z}}R\int_{-L/2}^{L/2}\int_0^{2\pi}\frac{dz'\, d\theta'}{[(z-z')^2 + R^2]^{3/2}}\right]$$

first term vanishes because

$$\int_0^{2\pi} \hat{\rho}'\, d\theta' = 0, \quad (\hat{\rho}' = \hat{\mathbf{x}}\cos\theta' + \hat{\mathbf{y}}\sin\theta') \text{ by symmetry}$$

second term has trivial integration over angle. So, rewrite, with $\xi = z' - z$,

$$\mathbf{B} = \frac{\mu_0 NIR^2\hat{\mathbf{z}}}{2}\int_{-L/2-z}^{L/2} \frac{d\xi}{(\xi^2+R^2)^{3/2}} = \frac{\mu_0 NI\hat{\mathbf{z}}}{2}\left(\frac{L/2-z}{[(L/2-z)^2+R^2]^{\frac{1}{2}}} + \frac{L/2+z}{[(L/2+z)^2+R^2]^{\frac{1}{2}}}\right)$$

PROBLEM 11.11 A point charge is placed at a distance d from the center of a grounded conducting sphere of radius a. Show that the ratio of the charge induced on the part of the sphere visible from q to that on the rest of the sphere is $\sqrt{(d+a)/(d-a)}$.

Solution:

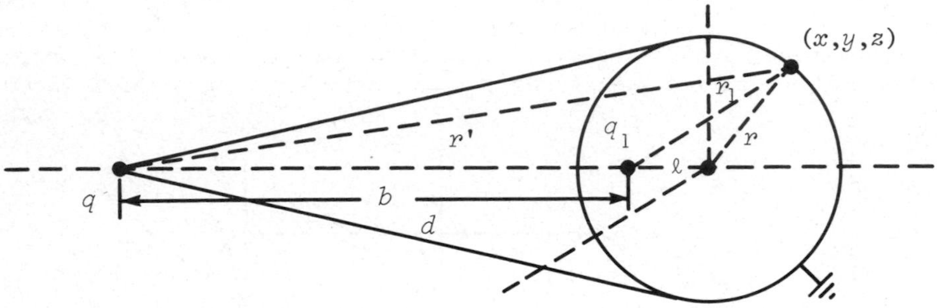

To get the value of q_1 and b,

$$a = -\frac{bqq_1}{q^2 - q_1^2}, \quad d = \frac{bq_1^2}{q^2 - q_1^2}$$

So,

$$\frac{a}{d} = -\frac{bqq_1}{bq^2} = -\frac{q_1}{q}$$

$$q_1 = -\frac{a}{d}q$$

Thus,

$$b = \frac{-(q^2 - q_1^2)a}{qq_1}$$

$$b = \frac{-[q^2 - (a^2/d^2)q^2]a}{(-a/d)q^2} = \frac{d}{a}\left(1 - \frac{a^2}{d^2}\right)a$$

$$b = d - \frac{a^2}{d}, \quad \ell = \frac{a^2}{d}$$

In order to get the surface charge density one first needs to get the potential at points on the surface of the sphere:

$$\phi = \frac{1}{4\pi\varepsilon}\left(\frac{q}{r'} + \frac{q_1}{r_1}\right)$$

From the diagram

$$-\hat{\mathbf{j}}d + \mathbf{r}' = \mathbf{r}$$

$$\mathbf{r}' = \mathbf{r} + \hat{\mathbf{j}}d$$

$$r' = (r^2 + 2yd + d^2)^{\frac{1}{2}}$$

$$\mathbf{r} = \hat{\mathbf{i}}x + \hat{\mathbf{j}}y + \hat{\mathbf{k}}z$$

$$-\hat{\mathbf{j}}\ell + \mathbf{r}_1 = \mathbf{r}$$

$$\mathbf{r}_1 = \mathbf{r} + \hat{\mathbf{j}}\ell$$

$$r_1 = (r^2 + 2y\ell + \ell^2)^{\frac{1}{2}}$$

where $\ell = a^2/d$

So

$$\phi = \frac{1}{4\pi\varepsilon}\left(\frac{q}{(r^2 + 2yd + d^2)^{\frac{1}{2}}} - \frac{(a/d)q}{(r^2 + 2y\ell + \ell^2)^{\frac{1}{2}}}\right)$$

In spherical polar coordinates $y = r \sin\theta \sin\phi$, so

$$\phi = \frac{8}{4\pi\varepsilon}\left[(r^2 + d^2 + 2rd \sin\theta \sin\phi)^{-\frac{1}{2}} - \frac{a}{d}(r^2 + \ell^2 + 2\ell r \sin\theta \sin\phi)\right]$$

To get the charge density we need E_R, the radial component of the electric field intensiy:

$$E_R = -\frac{\partial\phi}{\partial r} = -\frac{q}{4\pi\varepsilon}\left[\frac{(-1)}{2}(r^2 + d^2 + 2rd \sin\theta \sin\phi)^{-3/2}(2r + 2d \sin\theta \sin\phi)\right.$$

$$\left. - \frac{(-1)}{2}\frac{a}{d}(r^2 + \ell^2 + 2\ell r \sin\theta \sin\phi)^{-3/2}(2r + 2\ell \sin\theta \sin\phi)\right]$$

At $r = a$,

$$E_R = \frac{q}{4\pi\varepsilon}\left(\frac{a + d \sin\theta \sin\phi}{(a^2 + d^2 + 2ad \sin\theta \sin\phi)^{3/2}} - \frac{(a + \ell \sin\theta \sin\phi)(a/d)}{(a^2 + \ell^2 + 2\ell a \sin\theta \sin\phi)^{3/2}}\right)$$

The charge density $\sigma = \varepsilon_0 E_R$. Using $\ell = a^2/d$

$$\sigma = \frac{q}{4\pi}\left(\frac{a + d \sin\theta \sin\phi}{(a^2 + d^2 + 2ad \sin\theta \sin\phi)^{3/2}} - \frac{(a/d)[a + (a^2/d) \sin\theta \sin\phi]}{[a^2 + d^2 + (2a^3/d) \sin\theta \sin\phi]^{3/2}}\right)$$

$$\sigma = \frac{q}{4\pi}\left(\frac{a + d\sin\theta\sin\phi}{(a^2+d^2+2ad\sin\theta\sin\phi)^{3/2}} - \frac{(a^3/d^3)(d^2/a + d\sin\theta\sin\phi)}{(a^3/d^3)(d^2+a^2+2ad\sin\theta\sin\phi)^{3/2}}\right)$$

$$\sigma = \frac{q}{4\pi a}\,\frac{a^2-d^2}{(d^2+a^2+ad\sin\theta\sin\phi)^{3/2}}$$

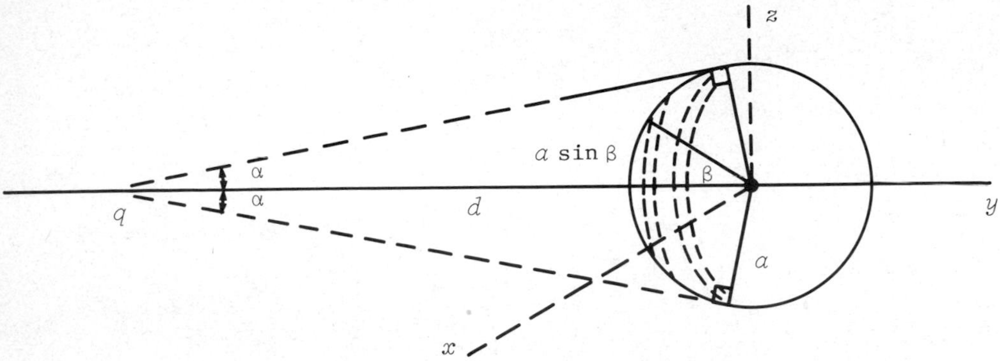

From the diagram, sin $\alpha = a/d$. We want to ge the ratio of the charge on the visible portion to the charge on the rest of the sphere. The charge on an anular ring is

$$dQ = r\ 2\pi(a^2\sin\beta)d\beta$$

The charge on the visible portion then is Q_L,

$$Q_L = \int_0^{90-\alpha} \sigma\ 2\pi a^3\sin\beta\, d\beta$$

$$\sigma = \frac{q}{4\pi a}\,\frac{a^2-d^2}{(d^2+a^2+2dy)^{3/2}}$$

But $y = -a\cos\beta$,

$$Q_L = \frac{q(a^2-d^2)}{4\pi a}\,2\pi a^2\int_0^{90-\alpha}\frac{\sin\beta\, d\beta}{(d^2+a^2-2da\cos\beta)^{3/2}}$$

Let $x = \cos\beta$, $dx = -\sin\beta\, d\beta$,

$$Q_L = qa(a^2-d^2)\int\frac{-dx}{(d^2+a^2-2dax)^{3/2}}$$

$$Q_L = -qa(a^2-d^2)\,\frac{2(d^2+a^2-2dax)^{-\frac{1}{2}}}{+2da(+1)}$$

$$Q_L = -\frac{q(a^2-d^2)}{d}\,(d^2+a^2-2da\cos\beta)^{-\frac{1}{2}}\Big|_0^{90-\alpha}$$

$$Q_L = \frac{q(a^2-d^2)}{d}\,[(d^2+a^2-2da\sin\alpha)^{-\frac{1}{2}}-(d^2+a^2-2da)^{-\frac{1}{2}}]$$

$$Q_L = -\frac{q(a^2-d^2)}{d}\left[\left(d^2+a^2-2da\,\frac{a}{d}\right)^{-\frac{1}{2}}-(d^2+a^2-2da)^{-\frac{1}{2}}\right]$$

$$Q_L = -\frac{q(a^2-d^2)}{d}\,[(d^2-a^2)^{-\frac{1}{2}}-(d-a)^{-1}]$$

$$Q_R = -\frac{q(a^2-d^2)}{d}\,(d^2+a^2-2da\cos\beta)^{-\frac{1}{2}}\Big|_{90-\alpha}^{\pi}$$

$$Q_R = -\frac{q(a^2-d^2)}{d}\,[(d^2+a^2+2da)^{-\frac{1}{2}}-(d^2+a^2-2da\sin\alpha)^{-\frac{1}{2}}]$$

$$Q_R = -\frac{q(a^2-d^2)}{d}\,[(d+a)^{-1}-(d^2-a^2)^{-\frac{1}{2}}]$$

So

$$\frac{Q_L}{Q_R} = \frac{\dfrac{1}{(d^2-a^2)^{\frac{1}{2}}}-\dfrac{1}{d-a}}{\dfrac{1}{d+a}-\dfrac{1}{(d^2-a^2)^{\frac{1}{2}}}} = \frac{\dfrac{1}{(d+a)^{\frac{1}{2}}-(d-a)^{\frac{1}{2}}}-\dfrac{1}{d-a}}{\dfrac{1}{d+a}-\dfrac{1}{(d+a)^{\frac{1}{2}}(d-a)^{\frac{1}{2}}}}$$

$$= \frac{\dfrac{1}{(d-a)^{\frac{1}{2}}}\left(\dfrac{1}{(d+a)^{\frac{1}{2}}}-\dfrac{1}{(d-a)^{\frac{1}{2}}}\right)}{\dfrac{1}{(d+a)^{\frac{1}{2}}}\left(\dfrac{1}{(d+a)^{\frac{1}{2}}}-\dfrac{1}{(d-a)^{\frac{1}{2}}}\right)} = \sqrt{\frac{d+a}{d-a}}$$

$$\frac{Q_L}{Q_R} = \sqrt{\frac{d+a}{d-a}}$$

PROBLEM 11.12 A line charge of density λ is placed inside a hollow conducting circular tube of radius a at a distance ℓ from the tube's axis. Find the voltage between the axis and the surface of the tube.

Solution: Consider two lines of charge of density λ and λ_1 separated by a distance d.

$$\phi = \frac{\lambda}{2\pi\varepsilon}\ln\frac{r_1}{b}$$

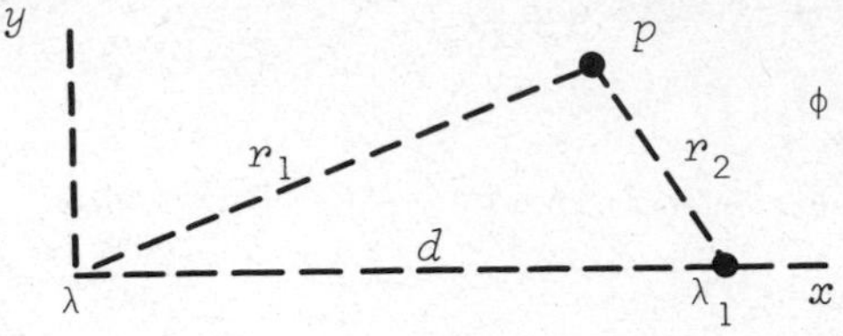

$$\phi = \frac{\lambda}{2\pi\varepsilon} \ln \frac{r_1}{b}$$

The potential at P due to these lines is

$$\phi = \frac{\lambda}{2\pi\varepsilon} \ln \frac{r_1}{b} + \frac{\lambda_1}{2\pi\varepsilon} \ln \frac{r_2}{b}$$

We will look for surfaces such that ϕ = constant, so in order for this to be true $\lambda_1 = -\lambda$

$$\phi = \frac{\lambda}{2\pi\varepsilon} \ln \frac{r_1}{r_2}$$

So for ϕ = constant, $r_1/r_2 = K$

$$r_1 = (x^2 + y^2)^{\frac{1}{2}}, \quad r_2 = [(x+d)^2 + y^2]^{\frac{1}{2}}$$

$$r_1^2 = (x^2 + y^2), \quad r_2^2 = x^2 + y^2 + 2xd + d^2 = r^2 + 2xd + d^2$$

But $r_1^2 = K^2 r_2^2$

$$x^2 + y^2 = K^2(x^2 + y^2 + 2xd + d^2)$$

$$x^2(1 - K^2) + y^2(1 - K^2) = 2K^2xd = +K^2d^2$$

$$x^2 + y^2 - \frac{2K^2xd}{1 - K^2} = \frac{K^2d^2}{1 - K^2}$$

$$\left(x - \frac{2K^2xd}{1 - K^2} + \frac{K^4d^2}{(1 - K^2)^2}\right) + y^2 = \frac{K^2d^2}{1 - K^2} + \frac{K^4d^2}{(1 - K^2)^2}$$

$$\left(x - \frac{K^2d}{1 - K^2}\right)^2 + y^2 = \frac{K^2d^2 - K^4d^2 + K^4d^2}{(1 - K^2)^2}$$

$$\left(x - \frac{K^2d}{1 - K^2}\right) + y^2 = \frac{K^2d^2}{(1 - K^2)^2}$$

This is the equation of a circle of radius $a = Kd/(1 - K^2)$ and $x_c = K^2d/(1 - K^2)$. In the particular problem,

$$\ell = x_c - d$$

$$\ell = \frac{K^2 d}{1 - K^2} - d$$

$$\ell = \frac{d(K^2 - 1 + K^2)}{1 - K^2}$$

$$\ell = \frac{d(2K^2 - 1)}{1 - K^2} \qquad \frac{x_c}{a} = K$$

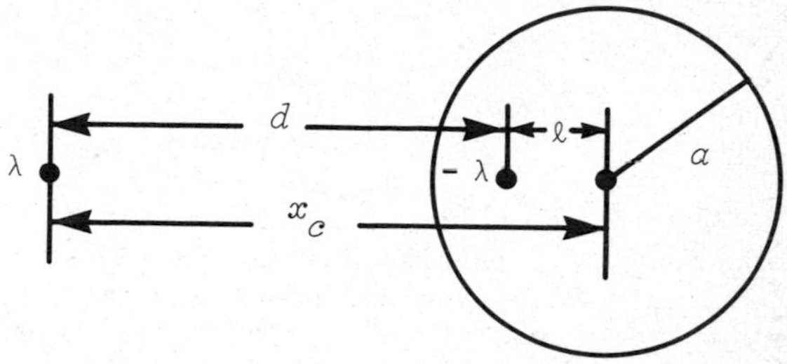

so

$$d = \frac{\ell(1 - K^2)}{2K^2 - 1}$$

therefore,

$$a = \frac{K}{1 - K^2}\,\frac{\ell(1 - K^2)}{2K^2 - 1} = \frac{K}{2K^2 - 1}$$

$$2aK^2 - a = k\ell$$

$$2aK^2 - k\ell - a = 0$$

$$K = \frac{\ell \pm \sqrt{\ell^2 + 8a^2}}{4a}$$

so

$$x_c = aK = \frac{\ell \pm \sqrt{\ell^2 + 8a^2}}{4}$$

Since x_c must be positive

$$x_c = \frac{\ell + \sqrt{\ell^2 + 8a^2}}{4} \qquad K = \frac{\ell + \sqrt{\ell^2 + 8a^2}}{4a}$$

So for these two lines of charge the potential at the center is

$$\phi_c = \frac{\lambda}{2\pi\varepsilon} \ln \frac{x_c}{b} - \frac{\lambda}{2\pi\varepsilon} \ln \frac{\ell}{b} = \frac{\lambda}{2\pi\varepsilon} \ln \frac{x_c}{\ell}$$

At the surface,

$$\phi_s = \frac{\lambda}{2\pi\varepsilon} \ln \frac{x_c - a}{b} - \frac{\lambda}{2\pi\varepsilon} \ln \frac{a}{b} = \frac{\lambda}{2\pi\varepsilon} \ln \frac{x_c - a}{a}$$

So,

$$\phi_s - \phi_c = \frac{\lambda}{2\pi\varepsilon}\left(\ln \frac{x_c - a}{a} - \ln \frac{x_c}{\ell}\right)$$

$$= \frac{\lambda}{2\pi\varepsilon}\left(\ln \frac{\ell(x_c - a)}{a x_c}\right)$$

$$\phi = V = \frac{\lambda}{2\pi\varepsilon} \ln \left[\frac{\ell}{a}\left(1 - \frac{a}{x_c}\right)\right] = \frac{\lambda}{2\pi\varepsilon} \ln \left[\frac{\ell}{a}\left(1 - \frac{1}{K}\right)\right]$$

$$V = \frac{\lambda}{2\pi\varepsilon} \ln \frac{\ell}{a}\left[1 - \frac{4a}{\ell + \sqrt{\ell^2 - 8a^2}}\right]$$

PROBLEM 11.13 A long conducting cylinder of radius a carrying a charge of line density λ is placed in an initially uniform field E in such a manner that the axis of the cylinder is normal to E. Neglecting end effects, find how the presence of the cylinder alters the field.

Solution: Consider a cylinder with a charge λ per unit length

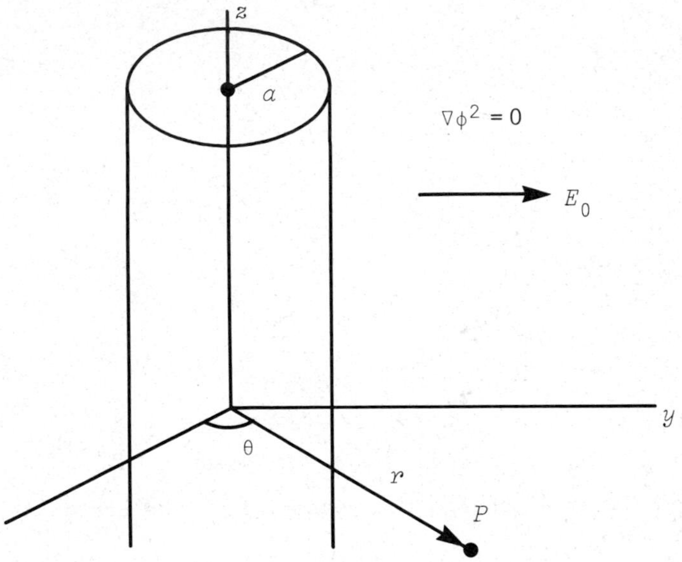

In cylindrical coordinates

$$\nabla^2\phi = \frac{1}{r}\frac{\partial}{\partial r}\left(\frac{r\,\partial\phi}{\partial r}\right) + \frac{1}{r^2}\frac{\partial^2\phi}{\partial\theta^2} = 0$$

Assume $\phi = R(r)T(\theta)$

$$\nabla^2\phi = \frac{T}{r}\frac{\partial}{\partial r}\left(\frac{r\,\partial R}{\partial r}\right) + \frac{R}{r^2}\frac{\partial^2 T}{\partial\theta^2} = 0$$

or

$$\frac{r}{R}\frac{\partial}{\partial r}\left(r\frac{\partial R}{\partial r}\right) = -\frac{1}{T}\frac{\partial^2 T}{\partial \theta^2} = k^2$$

Separating the equations,

$$r\frac{\partial}{\partial r}\left(r\frac{\partial R}{\partial r}\right) = k^2 R, \quad \frac{1}{T}\frac{\partial^2 T}{\partial \theta^2} = -k^2$$

Looking at the radial equation

$$r^2 R'' + rR' - k^2 R = 0$$

Assume solutions of the form, $R = r^n$, $R' = nr^{n-1}$, and $R'' = n(n-1)r^{n-2}$. Putting this in the radial equation, we obtain

$$n(n-1)r^n + nr - k^2 = 0$$

$$n^2 - n + n - k^2 = 0$$

$$n = \pm k$$

So the solutions are

$$R = r^k, \quad R = r^{-k}$$

The original equation is satisfied if $k = 1, 2, 3, 4$, etc., but not 0. For the angular equation

$$\frac{d^2 T}{d\theta^2} + k^2 T = 0$$

try solutions of the form $T = e^{m\theta}$, $T' = me^{m\theta}$, and $T'' = m^2 e^{m\theta}$. So after substituting,

$$m^2 e^{m\theta} + k^2 e^{m\theta} = 0, \quad m = \pm ik$$

Therefore,

$$T_1 = e^{ik\theta}, \quad T_2 = e^{-ik\theta}$$

$$T = A_1' e^{ik\theta} + A_2' e^{-ik\theta} = A_1 \cos k\theta + A_2 \sin k\theta$$

The final solution for ϕ is then

$$\phi = RT = A + A_k r^k \cos k\theta + B_k r^{-k} \cos k\theta + C_k r^k \sin k\theta + D_k r^{-k} \sin k\theta + (A_0\theta + B_0)\ln r$$

Expanding,

$$\phi = A + A_1 r \cos\theta + B_1 r^{-1} \cos\theta + C_1 r \sin\theta + D_1 r^{-1} \sin\theta + A_2 r^2 \cos 2\theta$$
$$+ B_2 r^{-2} \cos 2\theta + C_2 r^2 \sin 2\theta + D_2 r^{-2} \sin 2\theta + A_3 r^3 \cos 3\theta + B_3 r^{-3} \cos 3\theta$$
$$+ C_3 r^3 \sin 3\theta + D_3 r^{-3} \sin 3\theta + (A_0\theta + B_0)\ln r$$

Now for the boundary conditions: As $r \to \infty$,

$$\mathbf{E} = E_0 \hat{\mathbf{j}} + \frac{\lambda}{2\pi\varepsilon r} \hat{\mathbf{r}}, \quad \phi = \phi_0 - E_0 y - \frac{\lambda}{2\pi\varepsilon} \ln r$$

But since $y = r \sin\theta$,

$$\phi = \phi_0 - E_0 r \sin\theta - \frac{\lambda}{2\pi\varepsilon} \ln r$$

In order for the expansion to match this,

$$A = \phi_0, \quad A_1 = A_2 = A_3, \text{ etc.} = 0, \quad C_1 = -E_0, \quad \text{and } C_2 = C_3 = C_4, \text{ etc.} = 0$$

$$B_0 = -\frac{\lambda}{2\pi\varepsilon}$$

So the resulting expression is

$$\phi = \phi_0 - E_0 r \sin\theta + r^{-1}(B_1 \cos\theta + D_1 \sin\theta) + r^{-2}(B_2 \cos 2\theta + D_2 \sin 2\theta)$$
$$+ r^{-3}(B_3 \cos 3\theta + D_3 \sin 3\theta) - \frac{\lambda}{2\pi\varepsilon} \ln r$$

Now at $r = a$, $\phi = \text{constant}$, independent of θ, so

$$\phi = \phi_0 - E_0 a \sin\theta + a^{-1}(B_1 \cos\theta + D_1 \sin\theta) + a^{-2}(B_2 \cos 2\theta + D_2 \sin 2\theta)$$
$$+ a^{-3}(B_3 \cos 3\theta + D_3 \sin 3\theta) - \frac{\lambda}{2\pi\varepsilon} \ln r$$

Thus, $B_1 = B_2 = D_2 = B_3 = D_3$, etc. $= 0$,

$$D_1 a^{-1} = E_0 a$$

$$D_1 = E_0 a^2$$

and

$$\phi = \phi_0 - E_0 r \sin\theta + \frac{E_0 a^2}{r} \sin\theta - \frac{\lambda}{2\pi\varepsilon} \ln r$$

$$\phi = \phi_0 + E_0 \sin\theta \left(\frac{a^2}{r} - r\right) - \frac{\lambda}{2\pi\varepsilon} \ln r$$

We will now get the components of **E**:

$$E_R = -\frac{\partial\phi}{\partial r} = E_0 \sin\theta \left(-\frac{a^2}{r^2} - 1\right) + \frac{\lambda}{2\pi\varepsilon} \frac{1}{r}$$

$$E_R = -E_0 \sin\theta \left(\frac{a^2}{r^2} + 1\right) + \frac{\lambda}{2\pi\varepsilon} \frac{1}{r}$$

$$E_\theta = -\frac{1}{r}\frac{\partial\phi}{\partial\theta} = -\frac{1}{r}\,E_0\cos\theta\left(\frac{a^2}{r} - r\right)$$

$$E_\theta = -E_0\cos\theta\left(\frac{a^2}{r^2} - 1\right)$$

The surface charge density

$$\sigma = E_R\Big|_{r=a}\cdot\varepsilon$$

$$\sigma = \varepsilon\left[(-2E_0\sin\theta) + \frac{\lambda}{2\pi\varepsilon a}\right]$$

The total charge contained in a length ℓ is

$$Q = \int\sigma\, a\, d\theta\ell = \varepsilon\left[\int_0^{2\pi} -2E_0\sin\theta\, a\, d\theta\ell + \int_0^{2\pi}\frac{\lambda}{2\pi\varepsilon a}\, a\, d\theta\ell\right]$$

$$Q = \varepsilon\left(0 + \frac{2\pi a\lambda\ell}{2\pi\varepsilon a}\right) = \lambda\ell$$

PROBLEM 11.14 One plate of a thin parallel-plate capacitor of plate separation d is kept at the potential $\phi = 0$, the other at $\phi = V$. The capacitor contains a space charge of density $\rho = kx$, where k is a constant and x is the distance from the plate with $\phi = 0$. Find the potential distribution in the capacitor, the electric field in the capacitor, and the surface charge density on the inner surfaces of the plates.

Solution:

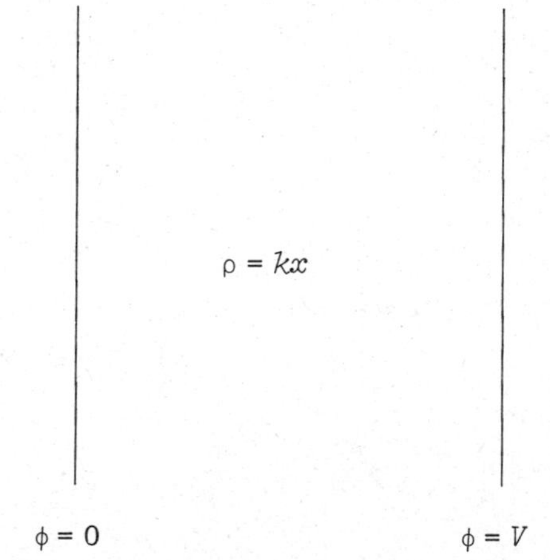

$$\nabla^2\phi = -\frac{\rho}{\varepsilon}, \qquad \frac{d^2\phi}{dx^2} = -\frac{\rho}{\varepsilon} = -\frac{K}{\varepsilon}\,x$$

$$\frac{d\phi}{dx} = -\frac{K}{\varepsilon}\frac{x^2}{2} + a$$

$$\phi = -\frac{K}{2\varepsilon}\frac{x^3}{3} + ax + b$$

at $x = 0$, $\phi = 0$, $b = 0$

at $x = d$, $\phi = V$, $V = \frac{kd^3}{6\varepsilon} + ad$

$$a = \frac{V}{d} + \frac{kd^2}{6\varepsilon}$$

So

$$\phi = -\frac{k}{2\varepsilon}\frac{x^3}{3} + \left(\frac{V}{d} + \frac{kd^2}{6\varepsilon}\right)x$$

To get the charge densities, one needs E.

$$\mathbf{E} = -\nabla\phi = +\left(\frac{k}{2\varepsilon}x^2 - \frac{V}{d} - \frac{kd^2}{6\varepsilon}\right)\hat{\mathbf{i}}$$

at $x = 0$, $\sigma_0 = \mathbf{E}\cdot\hat{\mathbf{i}} = -\frac{V}{d} - \frac{kd^2}{6\varepsilon}$

at $x = d$, $\sigma_d = -\mathbf{E}\cdot\hat{\mathbf{i}} = -\frac{kd^2}{2\varepsilon} + \frac{V}{d} + \frac{kd^2}{6\varepsilon}$

$$\sigma = -\frac{kd^2}{3\varepsilon} + \frac{V}{d}$$

PROBLEM 11.15 A spherical capacitor consists of two concentric spherical shells of radii a and b ($a < b$). The inner shell is kept at the potential $\phi = V$, the other at $\phi = 0$. The space between the shells is filled with space charge of density $\rho = kr$ where k is a constant, and r is the distance from the center. Find the potential due to this system at all points of space and find the surface charge density on the spheres.

Solution:

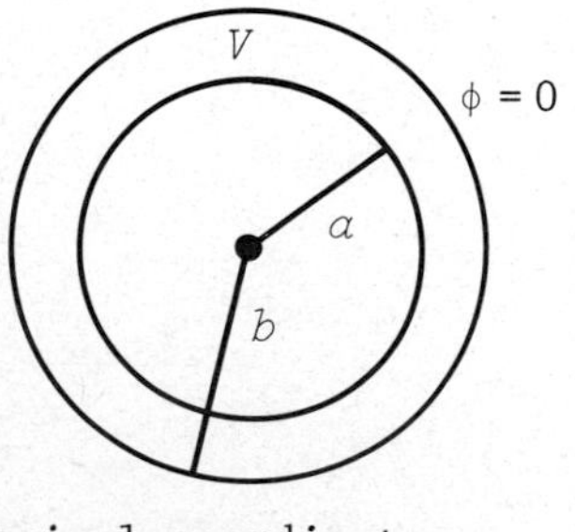

$$\nabla^2\phi = -\frac{\rho}{\varepsilon_0}$$

$$\rho = kr$$

In spherical coordinates,

$$\nabla^2 \phi = \frac{1}{r^2} \frac{\partial}{\partial r} \left(r^2 \frac{\partial \phi}{\partial r} \right) = -\frac{k}{\varepsilon} r$$

$$\frac{\partial}{\partial r} \left(r^2 - \frac{\partial \phi}{\partial r} \right) = -\frac{k}{\varepsilon} r^3$$

$$r^2 \frac{\partial \phi}{\partial r} = -\frac{k}{\varepsilon} \frac{r^4}{4} + c$$

$$\frac{\partial \phi}{\partial r} = -\frac{k}{4\varepsilon} r^2 + \frac{c}{r^2}$$

$$\phi = -\frac{k}{12\varepsilon} r^3 - \frac{ac}{r} + bd$$

at $r = a$, $\phi = V$

$$V = -\frac{k}{12\varepsilon} a^3 - \frac{c}{a} + d$$

at $r = b$, $\phi = 0$

$$0 = -\frac{k}{12\varepsilon} b^3 - \frac{c}{b} + d$$

So,

$$d = \frac{kb^3}{12\varepsilon} + \frac{c}{b}$$

Thus,

$$V = -\frac{k}{12\varepsilon} a^3 - \frac{c}{a} + \frac{kb^3}{12\varepsilon} + \frac{c}{b}$$

$$V = \frac{k}{12\varepsilon} (b^3 - a^3) + c \left(\frac{1}{b} - \frac{1}{a} \right)$$

$$V + \frac{k}{12\varepsilon} (a^3 - b^3) = \frac{a - b}{ab} c$$

$$c = \frac{ab}{a - b} V + \frac{kab}{12\varepsilon (a - b)} (a^3 - b^3)$$

So,

$$\phi = -\frac{k}{12\varepsilon} r^3 - \frac{ab}{a - b} \left(\frac{1}{r} \right) \left(V + \frac{k}{12\varepsilon} (a^3 - b^3) \right) + \frac{kb^3}{12\varepsilon} + \frac{c}{b}$$

$$\phi = -\frac{k}{12\varepsilon} r^3 + \frac{ab}{a-b} \left(V + \frac{k}{12\varepsilon} (a^3 - b^3)\right)\left(\frac{1}{b} - \frac{1}{r}\right) + \frac{kb^3}{12\varepsilon}$$

PROBLEM 11.16 Consider two very long thin-walled coaxial cylinders of radii a and b $(a < b)$. The inner cylinder is kept at the potential $\phi = V_a$, the outer at $\phi = V_b$. The space between the cylinders contains a space charge of density $\rho = kr$, where k is a constant and r is the distance from the axis. Find the potential distribution between the cylinders and find the surface charge density on the cylinders.

Solution:

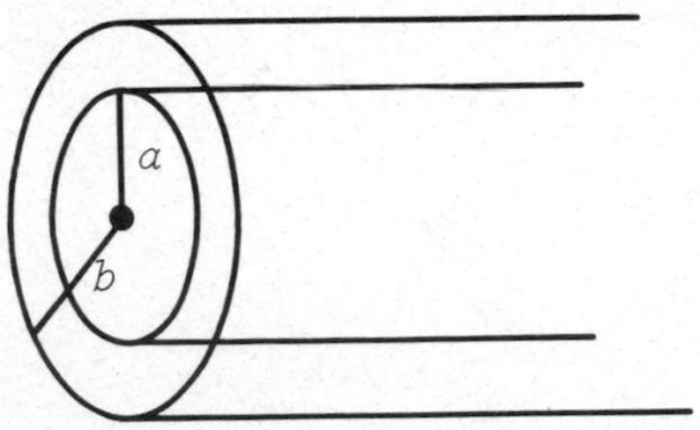

$$\nabla^2 \phi = -\frac{\rho}{\varepsilon}$$

In cylindrical coordinates,

$$\nabla^2 \phi = \frac{1}{r} \frac{\partial}{\partial r} \left(r \frac{\partial \phi}{\partial r}\right) = -\frac{kr}{\varepsilon}$$

$$\frac{\partial}{\partial r} \left(r \frac{\partial \phi}{\partial r}\right) = -\frac{kr^2}{\varepsilon}$$

$$r \frac{\partial \phi}{\partial r} = -\frac{k}{3\varepsilon} r^3 + c$$

$$\frac{\partial \phi}{\partial r} = -\frac{k}{3\varepsilon} r^2 + \frac{c}{r}$$

$$\phi = -\frac{k}{9\varepsilon} r^3 + c \ln r + d$$

at $r = a$, $\phi = V_a = -\frac{k}{8\varepsilon} a^3 + c \ln a + d$

$r = b$, $\phi = V_b = -\frac{k}{9\varepsilon} b^3 + c \ln b + d$

Subtracting,

$$V_a - V_b = \frac{k}{9\varepsilon} (b^3 - a^3) + c \ln \frac{a}{b}$$

So,

$$c = \left(\frac{1}{\ln(a/b)}\right)\left(V_a - V_b - \frac{k}{9\varepsilon} (b^3 - a^3)\right)$$

Then,

$$d = V_a + \frac{k}{9\varepsilon} a^3 - c \ln a$$

$$d = V_a + \frac{k}{9\varepsilon} a^3 - \frac{\ln a}{\ln(a/b)} \left(V_a - V_b - \frac{k}{9\varepsilon} (b^3 - a^3) \right)$$

$$\phi = -\frac{k}{9\varepsilon} r^3 + \frac{1}{\ln(a/b)} \left(V_a - V_b - \frac{k}{9\varepsilon} (b^3 - a^3) \right) \ln r$$

$$+ V_a + \frac{k}{9\varepsilon} a^3 = \frac{\ln a}{\ln(a/b)} \left(V_a - V_b - \frac{k}{9\varepsilon} (b^3 - a^3) \right)$$

$$\phi = -\frac{k}{9\varepsilon} r^3 + V_a + \frac{k}{9\varepsilon} a^3 + \frac{1}{\ln(a/b)} \left(V_a - V_b - \frac{k}{9\varepsilon} (b^3 - a^3) \right) \ln\left(\frac{r}{a}\right)$$

To get the charge densities, we use $\mathbf{E} = -\nabla\phi$:

$$\mathbf{E} = -\hat{\mathbf{r}} \frac{\partial \phi}{\partial r} = -\hat{\mathbf{r}} \left[\frac{-k}{3\varepsilon} r^2 + \frac{1/a}{r/a} \frac{1}{\ln(a/b)} \left(V_a - V_b - \frac{k}{9\varepsilon} (b^3 - a^3) \right) \right]$$

At $r = a$, $\sigma_a = \mathbf{E} \cdot \hat{\mathbf{r}}$

$$\sigma_a = \frac{k}{3\varepsilon} a^2 - \frac{1}{a} \frac{1}{\ln(a/b)} \left(V_a - V_b - \frac{k}{9\varepsilon} (b^3 - a^3) \right)$$

$$\sigma_a = \frac{k}{3\varepsilon} a^2 - \frac{1}{a} \frac{1}{\ln(a/b)} \left(V_a - V_b - \frac{k}{9\varepsilon} (b^3 - a^3) \right)$$

At $r = b$, $\sigma_b = \mathbf{E} \cdot (-\hat{\mathbf{r}})$

$$\sigma_b = -\frac{k}{3\varepsilon} b^2 + \frac{1}{b} \frac{1}{\ln(a/b)} \left(V_a - V_b - \frac{k}{9\varepsilon} (b^3 - a^3) \right)$$

PROBLEM 11.17 Two long coaxial cylindrical shells of radii a and b are kept at the potentials V_a and V_b, respectively. Show that the potential at any point between the shells is

$$\phi = V_a + (V_b - V_a) \frac{\ln(r/a)}{\ln(b/a)}$$

where r is the distance from the axis.

Solution: Given $\nabla^2\phi = 0$; in cylindrical coordinates,

$$\nabla^2\phi = \frac{1}{r} \frac{\partial}{\partial r} \left(r \frac{\partial \phi}{\partial r} \right) = 0$$

So,

$$r \frac{\partial \phi}{\partial r} = c$$

Or,

$$\frac{\partial \phi}{\partial r} = \frac{c}{r}$$

Hence,

$$\phi = c \ln r + d$$

When

$$r = a, \quad \phi = V_a = c \ln a + d$$

$$r = b, \quad \phi = V_b = d \ln b + d$$

Subtracting,

$$V_a - V_b = c \ln \frac{a}{b}, \quad c = \frac{1}{\ln(a/b)} (V_a - V_b)$$

Then,

$$d = V_a - c \ln a = V_a - \frac{\ln a}{\ln(a/b)} (V_a - V_b)$$

Hence,

$$\phi = \frac{V_a - V_b}{\ln(a/b)} \ln r + V_a - \frac{\ln a}{\ln(a/b)} (V_a - V_b)$$

$$\phi = V_a + (V_b - V_a) \frac{\ln(a/r)}{\ln(b/a)}$$

PROBLEM 11.18 A potential distribution on a spherical shell of radius a is given by

$$\phi(\theta) = \frac{k-1}{k+2} E_0 a \cos \theta$$

Prove that if all charge resides on this shell, the potential inside and outside the shell is, respectively,

$$\phi(r,\theta) = \frac{k-1}{k+2} E_0 r \cos \theta$$

$$\phi(r,\theta) = \frac{k-1}{k+2} E_0 \frac{a^3 \cos \theta}{r^2}$$

In these formulas k and E_0 are constants, and r and θ are the spherical coordinates with the origin at the center of the shell.

Solution:

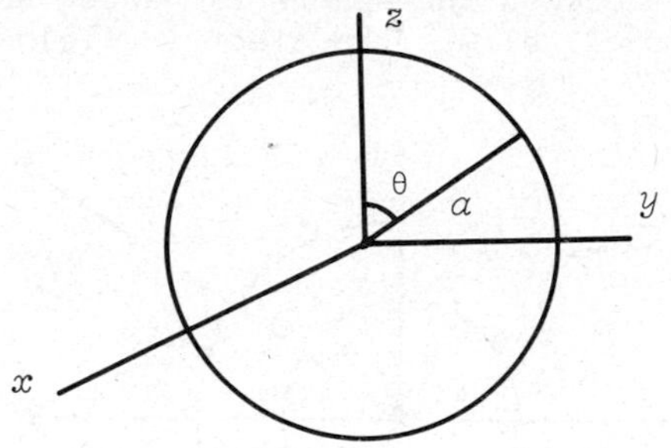

On the shell,

$$\phi = \frac{k-1}{k+2} E_0 a \cos\theta$$

In general,

$$\phi = A_0 + B_0 r^{-1} + A_1 r \cos\theta + B_1 r^{-2} \cos\theta + A_2 r^2 (3\cos^2\theta - 1)$$

$$+ B_2 r^{-3} (3\cos^2\theta - 1) + \cdots$$

For $r = a$,

$$\phi = \frac{k-1}{k+2} E_0 a \cos\theta$$

$$\phi = A_0 + B_0 a^{-1} + A_1 a \cos\theta + B_1 a^{-2} \cos\theta + A_2 a^2 (3\cos^2\theta - 1) + B_2 a^{-3} (3\cos^2\theta - 1)$$

To meet the above condition,

$$A_0 = 0, \quad B_0 = 0, \quad A_1 a + B_1 a^{-2} = \frac{k-1}{k+2} E_0 a, \quad A_2 = B_2, \text{ etc.} = 0$$

So, in general,

$$\phi = A_1 r \cos\theta + B_1 r^{-2} \cos\theta$$

For $r > a$, $\phi = 0$ at $r = \infty$, so $A_1 = 0$. Hence,

$$\phi = \frac{k-1}{k+2} E_0 \frac{a^3}{r^2} \cos\theta, \quad r > a$$

For $r < a$, the potential must be finite at $r = 0$, hence $B_1 = 0$. So,

$$\phi = \frac{k-1}{k+2} E_0 r \cos\theta, \quad r < a$$

PROBLEM 11.19 A conducting sphere carries a charge. A thin hemispherical conducting shell concentric with the sphere is placed near the sphere. Find how the presence of the shell alters the electric field of the sphere.

Solution:

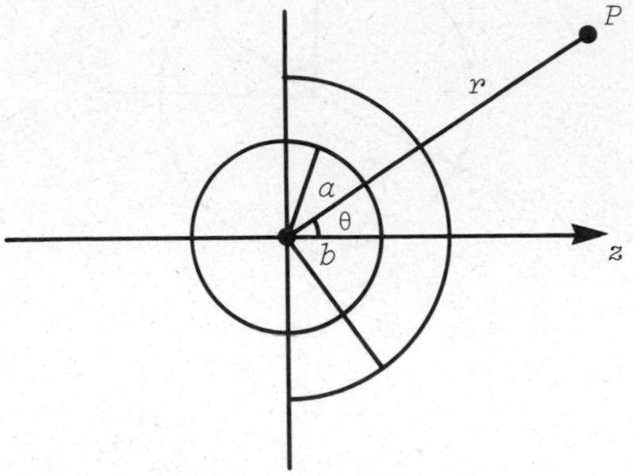

$$\nabla^2 \phi = 0$$

For $z > 0$, $\phi = f(r, \theta)$ and not ϕ. So, in spherical coordinates,

$$\nabla^2 \phi = \frac{1}{r^2} \frac{\partial}{\partial r}\left(r^2 \frac{\partial \phi}{\partial r}\right) + \frac{1}{r^2 \sin\theta} \frac{\partial}{\partial \theta}\left(\sin\theta \frac{\partial \phi}{\partial \theta}\right) = 0$$

Assume a product solution, $\phi = R(r)\Theta'(\theta)$,

$$\nabla^2 \phi = \frac{\Theta'}{r^2} \frac{\partial}{\partial r}\left(r^2 \frac{\partial R}{\partial r}\right) + \frac{R}{r^2 \sin\theta} \frac{d}{d\theta}\left(\sin\theta \frac{d\Theta'}{d\theta}\right) = 0$$

$$\frac{1}{R} \frac{1}{r^2} \frac{\partial}{\partial r}\left(r^2 \frac{\partial R}{\partial r}\right) = -\frac{1}{\Theta \sin\theta} \frac{d}{d\theta}\left(\sin\theta \frac{d\Theta'}{d\theta}\right)$$

In order for this to be true,

$$-\frac{1}{\Theta' \sin\theta} \frac{d}{d\theta}\left(\sin\theta \frac{d\Theta'}{d\theta}\right) = k$$

$$\frac{1}{R} \frac{1}{r^2} \frac{\partial}{\partial r}\left(r^2 \frac{\partial R}{\partial r}\right) = k$$

The solutions are then

$$\phi = A_n r^n P_n(\theta) + C_n r^{-(n+1)} P_n(\theta)$$

Expanding,

$$\phi = A_0 + \frac{C_0}{r} + A_1 r \cos\theta + \frac{C_1}{r^2} \cos\theta + A_2 r^2 \tfrac{1}{2}(3\cos^2\theta - 1) + \frac{C_2}{r^3} \tfrac{1}{2}(3\cos^2\theta - 1)$$

$$+ A_3 r^3 \tfrac{1}{2}(5\cos^3\theta - 3\cos\theta) + \frac{C_3}{r^4} \tfrac{1}{2}(5\cos^3\theta - 3\cos\theta)$$

As $r \to \infty$, $\phi = 0$, so all the A's must be 0. Hence,

$$\phi = \frac{C_0}{r} + \frac{C_1}{r^2} \cos\theta + \frac{C_2}{r^3} \tfrac{1}{2}(3\cos^2\theta - 1) + \frac{C_3}{r^4} \tfrac{1}{2}(5\cos^3\theta - 3\cos\theta)$$

At $r = a$, $\phi = \text{constant} = \phi_0$, independent of θ, so

$$\phi = \frac{C_0}{a} + \frac{C_1}{a^2} \cos\theta + \frac{C_2}{a^3} (\tfrac{1}{2})(3\cos^2\theta - 1) + \frac{C_3}{a^3} (\tfrac{1}{2})(5\cos^3\theta - 3\cos\theta)$$

In order for this to be true, $C_1 = C_2 = C_3 = 0$, so

$$\phi = \frac{C_0}{a}$$

which means the potential is unaffected by the presence of the hemisphere.

PROBLEM 11.20 A conducting sphere of radius a carrying a charge q is placed in an initially uniform field **E**. Find how the presence of the sphere alters the field.

Solution: In the far field,

$$\mathbf{E} = E_0\hat{\mathbf{k}} + \frac{1}{4\pi\varepsilon} \frac{Q}{r^2} \hat{\mathbf{r}}$$

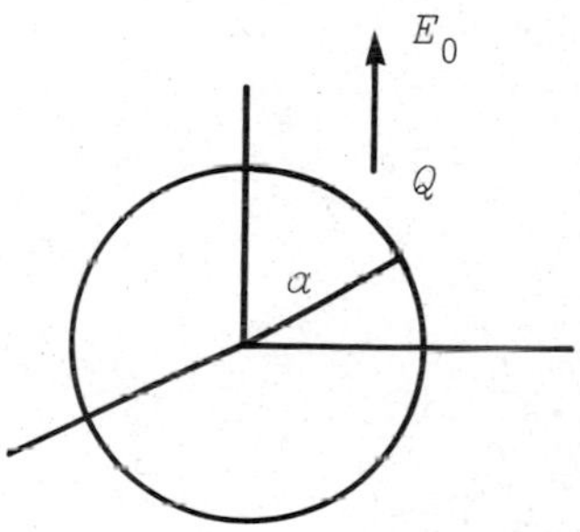

$$\phi = -E_0 z + \frac{Q}{4\pi\varepsilon r}$$

Since $z = r\cos\theta$,

$$\phi = -E_0 r \cos\theta + \frac{Q}{4\pi\varepsilon r}$$

In general,

$$\phi = A_0 + B_0 r^{-1} + A_1 r\cos\theta + B_1 r^{-2} \cos\theta + \cdots$$

As $r \to \infty$, $A_0 = 0$, $B_0 = \frac{Q}{4\pi\varepsilon r} + A_1 r\cos\theta$

So $A_1 = -E_0$. Hence,

$$\phi = \frac{Q}{4\pi\varepsilon r} - E_0 r\cos\theta + B_1 r^{-2} \cos\theta$$

At $r = a$, $\phi = \text{constant}$, so $B_1 = E_0 a^3$. Hence,

$$\phi = \frac{Q}{4\pi\varepsilon r} - E_0 r \cos\theta + \frac{E_0 a^3}{r^2}\cos\theta$$

$$\phi = \frac{Q}{4\pi\varepsilon r} + E_0\left(-r + \frac{a^3}{r^2}\right)\cos\theta$$

Chapter 12

PROBLEM 12.1

(a) Starting with the series solution

$$\Phi(\rho,\phi) = a_0 + b_0 \ln\rho + \sum_{n=1}^{\infty} a_n\rho^n \sin(n\phi + \alpha_n) + \sum_{n=1}^{\infty} b_n\rho^{-n} \sin(n\phi + \beta_n)$$

for the two-dimensional potential problem with the potential specified on the surface of a cylinder of radius b, evaluate the coefficients formally, substitute them into the series, and sum it to obtain the potential inside the cylinder in the form of Poisson's integral:

$$\Phi(\rho,\phi) = \frac{1}{2\pi}\int_0^{2\pi} \Phi(b,\phi')\,\frac{b^2-\rho^2}{b^2+\rho^2-2b\rho\cos(\phi'-\phi)}\,d\phi'$$

(b) What modification is necessary if the potential is desired in the region of space bounded by the cylinder and infinity?

(c) Solve this again by using Green's functions.

Solution:

(a)

$$\Phi(\rho,\phi) = a_0 + b_0 \ln\rho + \sum_{n=1}^{\infty} a_n\rho^n \sin(n\phi + \alpha_n) + \sum_{n=1}^{\infty} b_n\rho^{-n} \sin(n\phi + \beta_n)$$

$b_n = 0$, since origin in volume.

$$\Phi(\rho,\phi) = a_0 + \sum_{n=1}^{\infty} a_n\rho^n \sin(n\phi + \alpha_n)$$

$$\Phi(\rho,\phi) = a_0 + \sum_{n=1}^{\infty} \rho^n[a_n \sin(n\phi) + a_n \cos(n\phi)]$$

$$a_n = \frac{2b^{-n}}{2\pi}\int_0^{2\pi} \Phi(b,\phi')\sin n\phi'\,d\phi'$$

$$c_n = \frac{2b^{-n}}{2\pi}\int_0^{2\pi} \Phi(b,\phi')\cos n\phi'\,d\phi'$$

$$\Phi(\rho,\phi) = a_0 + \frac{1}{\pi}\sum_{n=1}^{\infty}\left(\frac{\rho}{b}\right)^n \left(\int_0^{2\pi} d\phi'\,\Phi(b,\phi')\sin(n\phi')\,d\phi'\sin(n\phi)\right.$$

$$+\int_0^{2\pi} d\phi(b,\phi')\cos(n\phi')d\phi\cos(n\phi)\bigg]$$

$$= a_0 + \frac{1}{\pi}\int_0^{2\pi}\Phi(b,\phi')\left[\sum_{n=1}^{\infty}\left(\frac{\rho}{b}\right)^n \sin(n\phi')\sin(n\phi)\right.$$

$$\left. + \sum_{n=1}^{\infty}\left(\frac{\rho}{b}\right)^n \cos(n\phi')\cos(n\phi)\,d\phi'\right]d\phi'$$

$$= a_0 + \frac{1}{\pi}\int_0^{2\pi}\Phi(b,\phi')\left[\sum_{n=1}^{\infty}\left(\frac{\rho}{b}\right)^n \cos[n(\phi-\phi')]\right]d\phi'$$

$$\sum_{n=1}^{\infty}\left(\frac{\rho}{b}\right)^n \cos[n(\phi-\phi')] = \text{Re}\sum_{n=1}^{\infty}\left(\frac{\rho}{b}\right)^n e^{in(\phi-\phi')}$$

$$= \text{Re}\sum_{n=1}^{\infty}\left(\frac{\rho}{b}e^{i(\phi-\phi')}\right)^n = \text{Re}\sum_{n=1}^{\infty} T^n$$

$$\left(T = \frac{\rho}{b}e^{i(\phi-\phi')}\right)$$

$$= \text{Re}\left(\frac{1}{1-T} - 1\right) = \text{Re}\left(\frac{T}{1-T}\right)$$

$$= \text{Re}\left(\frac{T(1-T^*)}{|1-T|^2}\right) = \text{Re}\left(\frac{T-|T|^2}{|1-T|^2}\right)$$

$$= \text{Re}\left(\frac{(\rho/b)e^{i(\phi-\phi')} - (\rho/b)^2}{1+(\rho/b)^2 - 2(\rho/b)\cos(\phi-\phi')}\right)$$

$$= \frac{(\rho/b)\cos(\phi-\phi') - (\rho/b)^2}{1+(\rho/b)^2 - 2(\rho/b)\cos(\phi-\phi')} = \frac{\rho b\cos(\phi-\phi') - \rho^2}{\rho^2+b^2-2\rho b\cos(\phi-\phi')}$$

$$a_0 = \frac{1}{2\pi}\int_0^{2\pi}\Phi(b,\phi')d\phi' = \frac{1}{2\pi}\int_0^{2\pi}\Phi(b,\phi')\;\frac{\rho^2+b^2-2b\rho\cos(\phi-\phi')}{\rho^2+b^2-2b\rho\cos(\phi-\phi')}\,d\phi'$$

$$\Phi(\rho,\phi) = \frac{1}{2\pi}\int_0^{2\pi}\Phi(b,\phi')\left(\frac{\rho^2+b^2-2\rho b\cos(\phi-\phi') + 2\rho b\cos(\phi-\phi') - 2\rho^2}{\rho^2+b^2-2\rho b\cos(\phi'-\phi)}\right)d\phi'$$

$$\Phi(\rho,\phi) = \frac{1}{2\pi}\int_0^{2\pi} \Phi(b,\phi') \frac{b^2 - p^2}{b^2 + \rho^2 - 2b\rho\cos(\phi' - \phi)}\, d\phi'$$

(b) $\rho \to \frac{1}{\rho}$, $b \to \frac{1}{b}$ in the Φ obtained in part a.

(c) Any image charge q is at $\mathbf{x}''$. $E_{\text{around } q} = 2q/\rho$; the potential equals $-2q'\,u(\rho/\rho_0)$. ρ_0 is the distance at which the potential equals 0.

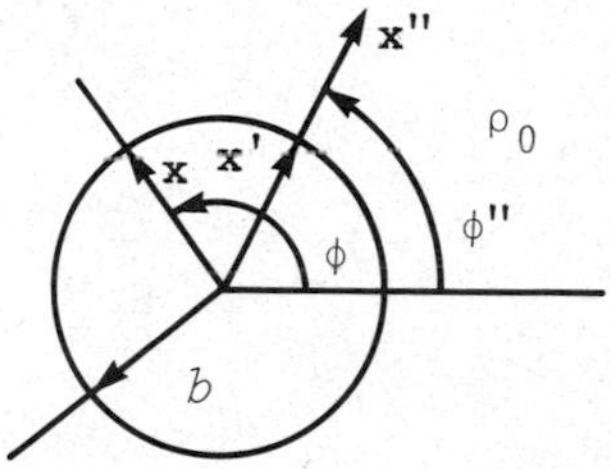

$$G(\mathbf{x},\mathbf{x}') = -2\ln\frac{|\mathbf{x}-\mathbf{x}'|}{\rho_0} - 2q\ln\frac{|\mathbf{x}-\mathbf{x}''|}{\rho_0} + C$$

$$= -2\ln|\mathbf{x}-\mathbf{x}'| + 2\ln\rho_0 - 2q\ln|\mathbf{x}-\mathbf{x}''| + 2q\ln\rho_0 + C$$

$$= -\ln|\mathbf{x}-\mathbf{x}'|^2 - 2\ln|\mathbf{x}-\mathbf{x}''|^2 + 2(1+q)\ln\rho_0 + C$$

$$= -\ln[\rho^2 + \rho'^2 - 2\rho\rho'\cos(\phi-\phi')] - q\ln[\rho^2 + \rho''^2 - 2\rho\rho''\cos(\phi-\phi')]$$

$$+ 2(1+q)\ln\rho_0 + C$$

The boundary condition at $\mathbf{x} = b\mathbf{n}$

$$G(b\mathbf{n},\mathbf{x}') = -\ln[b^2 + \rho'^2 - 2b\rho'\cos(\phi-\phi')] - q\ln[b^2 + \rho''^2 - 2b\rho'\cos(\phi-\phi')]$$

$$+ 2(1+q)\ln\rho_0 + C$$

$$= -\ln(b^2) - \ln\left(1 + \frac{\rho'^2}{b^2} - \frac{2\rho'}{b}\cos(\phi-\phi')\right) - q\ln\rho''^2$$

$$- q\ln\left(1 + \frac{b^2}{\rho''^2} - \frac{2b}{\rho''}\cos(\phi-\phi')\right) + 2(1+q)\ln\rho_0 + C = 0$$

So, to be true for all θ,

$$q = -1, \quad \frac{\rho'}{b} = \frac{b}{\rho''}, \quad \rho'' = \frac{b^2}{\rho'}$$

$$0 = -\ln b^2 + \ln\rho''^2 + 0\ln\rho_0 + C$$

$$C = \ln b^2 - \ln\rho''^2 = \ln\frac{b^2}{\rho''^2} = \ln\frac{b^2}{b^4/\rho'^2} = \ln\frac{\rho'^2}{b^2}$$

$$G(\mathbf{x},\mathbf{x}') = -\ln\left[\rho^2+\rho'^2-2\rho\rho'\cos(\phi-\phi')\right] + \ln\left(\rho^2+\frac{b^4}{\rho'^2}-\frac{2\rho b^2}{\rho'}\cos(\phi-\phi')\right)$$

$$+\ln\frac{\rho'^2}{b^2}$$

$$= -\ln\left[\rho^2+\rho'^2-2\rho\rho'\cos(\phi-\phi')\right]$$

$$+\ln\left(\frac{\rho^2\rho'^2}{b^2}+b^2-2\rho\rho'\cos(\phi-\phi')\right)$$

$$= \ln\frac{\rho^2\rho'^2/b^2+b^2-2\rho\rho'\cos(\phi-\phi')}{\rho^2+\rho'^2-2\rho\rho'\cos(\phi-\phi')}$$

$$\left.\frac{\partial G}{\partial\rho'}\right|_{\rho'=b} = \left.\left(\frac{-2\rho'+2\rho\cos(\phi-\phi')}{\rho^2+\rho'^2-2\rho\rho'\cos(\phi'-\phi)}+\frac{2\rho^2\rho'/b^2-2\rho\cos(\phi'-\phi)}{\rho^2\rho'^2/b^2+b^2-2\rho\rho'\cos(\phi-\phi')}\right)\right|_{\rho'=b}$$

$$= \frac{-2b+2\rho\cos(\phi-\phi')}{\rho^2+b^2-2\rho b\cos(\phi'-\phi)}+\frac{2\rho^2/b-2\rho\cos(\phi'-\phi)}{\rho^2+b^2-2\rho\cos(\phi'-\phi)}$$

$$= \frac{(2/b)(\rho^2-b^2)}{\rho^2+b^2-2\rho b\cos(\phi'-\phi)}$$

Use Green's theorem,

$$\Phi(\rho,\phi) = \int\int_A \tau(\rho',\phi')G(\rho,\phi,\rho',\phi')\rho'\,d\rho'\,d\phi' - \frac{1}{4\pi}\oint_S \Phi(\rho',\phi')\frac{\partial G}{\partial n'}\,dS'$$

when $\tau = 0$,

$$\Phi(\rho,\phi) = -\frac{1}{4\pi}\int_0^{2\pi} b\,d\phi'\,\Phi(b,\phi')\,\frac{(2/b)(\rho^2-b^2)}{\rho^2+b^2-2\rho b\cos(\phi'-\phi)}$$

$$\Phi(\rho,\phi) = \frac{1}{2\pi}\int_0^{2\pi}\Phi(b,\rho')\,\frac{b^2-\rho^2}{\rho^2+b^2-2\rho b\cos(\phi'-\phi)}\,d\phi'$$

PROBLEM 12.2

(a) Two halves of a long hollow conducting cylinder of inner radius b are separated by small lengthwise gaps on each side, and are kept at different potentials V_1 and V_2. Show that the potential inside is given by

$$\Phi(\rho,\phi) = \frac{V_1 + V_2}{2} + \frac{V_1 - V_2}{\pi} \tan^{-1}\left(\frac{2b\rho}{b^2 - \rho^2}\cos\phi\right)$$

where ϕ is measured from a plane perpendicular to the plane through the gap.

(b) Calculate the surface-charge density on each half of the cylinder.

Solution:

(a) $\Phi\left(b, \frac{-\pi}{2} < \phi < \frac{\pi}{2}\right) = V_1$

$\Phi\left(b, \frac{\pi}{2} < \phi < \frac{3\pi}{2}\right) = V_2$

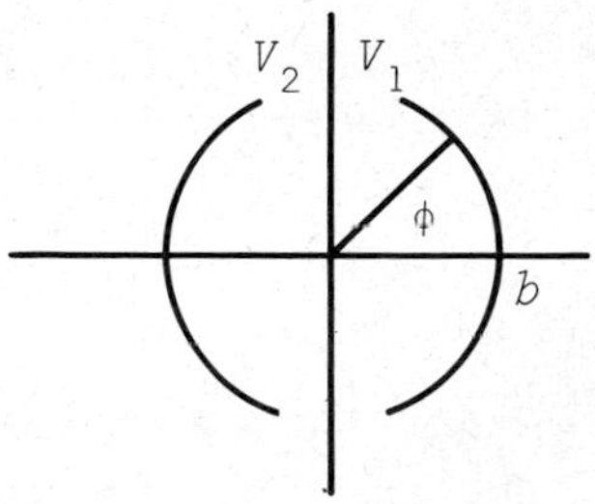

$$\Phi(\rho,\phi) = a_0 + \sum_{n=1}^{\infty} a_n \rho^n \sin(n\phi + \alpha_n)$$

$$\alpha_0 = 0, \quad \Phi(\rho,\phi) = \sum_{n=0}^{\infty} a_n \rho^n \sin(n\phi + \alpha_n)$$

$$a_n = \frac{1}{2\pi}\int_0^{2\pi} \Phi(b,\phi')b^{-n} \sin(n\phi)d\phi'$$

$$= \frac{b^{-n}}{2\pi}\left\{V_1 \int_{(-\pi/2)(0)}^{(\pi/2)(\pi)} \sin n\phi'\, d\phi' + V_2 \int_{(\pi/2)(\pi)}^{(3\pi/2)(2\pi)} \sin n\phi'\, d\phi'\right\}$$

$$\alpha_n = \frac{+\pi}{2}$$

$$a_n = \frac{b^{-n}}{2\pi}\{V_1(-1/n)[(-1)^n - 1] + V_2(-1/n)[1 - (-1)^n]\}$$

$$a_n = \frac{b^{-n}}{2\pi n}(2V_1\delta_{n\ \mathrm{odd}} - 2V_2\delta_{n\ \mathrm{odd}})$$

$$a_0 = \frac{V_1 + V_2}{2}$$

$$\Phi(\rho,\phi) = \frac{V_1 + V_2}{2} + \sum_{n=1}^{\infty} \frac{V_1 - V_2}{\pi(2n-1)}\left(\frac{\rho}{b}\right)^{2n-1} \sin[(2n-1)\phi + \pi/2]$$

$$= \frac{V_1 + V_2}{2} + \sum_{n=1}^{\infty}\left(\frac{V_1 - V_2}{2}\right)\frac{1}{2n-1}\left(\frac{\rho}{b}\right)^{2n-1}(e^{i(2n-1)\phi} + e^{-i(2n-1)\phi})$$

$$= \frac{V_1 + V_2}{2} + \left(\frac{V_1 - V_2}{2}\right)\left[\sum_{n=1}^{\infty} \frac{1}{2n-1}\left(\frac{\rho}{b} e^{i\phi}\right)^{2n-1} + \sum_{n=1}^{\infty} \frac{1}{2n-1}\left(\frac{\rho}{b} e^{-i\phi}\right)^{2n-1}\right]$$

$$\tfrac{1}{2}\ln\left(\frac{1+x}{1-x}\right) = \sum_{n=1}^{\infty} \frac{x^{2n-1}}{2n-1}$$

$$\Phi(\rho,\phi) = \frac{V_1 + V_2}{2} + \frac{V_1 - V_2}{2}\left[\tfrac{1}{2}\ln\left(\frac{1+\frac{\rho}{b} e^{i\phi}}{1-\frac{\rho}{b} e^{i\phi}}\right) + \tfrac{1}{2}\ln\left(\frac{1+\frac{\rho}{b} e^{-i\phi}}{1-\frac{\rho}{b} e^{-i\phi}}\right)\right]$$

$$\Phi(\rho,\phi) = \frac{V_1 + V_2}{2} + \frac{V_1 - V_2}{2}\left[\tfrac{1}{2}\ln\left(\frac{1+\frac{\rho^2}{b^2} + 2\frac{\rho}{b}\cos\phi}{1+\frac{\rho^2}{b^2} - 2\frac{\rho}{b}\cos\phi}\right)\right]$$

$$\Phi(\rho,\phi) = \frac{V_1 + V_2}{2} + \frac{V_1 - V_2}{\pi}\,\mathrm{Im}\sum_{n\ \mathrm{odd}} \frac{1}{n}\left(\frac{\rho}{b} e^{i\phi}\right)^{n}$$

$$= \frac{V_1 + V_2}{2} + \frac{V_1 - V_2}{\pi}\,\mathrm{Im}\left[\ln\left(\frac{1+\frac{\rho}{b} e^{i(\pi/2+\phi)}}{1-\frac{\rho}{b} e^{i(\phi+\pi/2)}}\right)\right]$$

$$\frac{1+T}{1-T} = \frac{(1+T)(1-T^*)}{|1-T|^2} = \frac{1-|T|^2 + 2i\,\mathrm{Im}T}{|1-T|^2}$$

$$\text{Phase} = \tan^{-1}\left(2\frac{\mathrm{Im}T}{1-|T|^2}\right) = \tan^{-1}\left(\frac{2\frac{\rho}{b}\sin(\phi+\pi/2)}{1-(\rho/b)^2}\right)$$

$$= \tan^{-1}\left(\frac{2b\rho\cos\phi}{b^2-\rho^2}\right)$$

$$\Phi(\rho,\phi) = \frac{V_1 + V_2}{2} + \frac{V_1 - V_2}{\pi}\tan^{-1}\left(\frac{2b\rho\cos\phi}{b^2-\rho^2}\right)$$

(b) $$\mathbf{E}\cdot\hat{\mathbf{n}} = E_\rho = \frac{-\partial\Phi}{\partial\rho} = \frac{V_1 - V_2}{\pi}\,\frac{1}{1+(2b\rho\cos\phi/b^2\rho^2)^2}\times\left(\frac{2b\cos\phi}{b^2-\rho^2} + \frac{4b\rho\cos\phi}{(b^2-\rho^2)^2}\right)$$

$$E_\rho = \frac{V_1 - V_2}{\pi}\left(\frac{2b(b^2 - \rho^2)\cos\phi + 4b\rho\cos\phi}{(b^2 - \rho^2)^2 + 4b^2\rho^2\cos^2\phi}\right)$$

$$\sigma = \left.\frac{E_\rho}{4\pi}\right|_{\rho = b} = \frac{V_1 - V_2}{4\pi^2}\,\frac{1}{b^2\cos\phi}$$

$$\sigma_1 = \sigma, \quad \text{for } -\pi/2 < \phi < \pi/2$$

$$\sigma_2 = \sigma, \quad \text{for } \pi/2 < \phi < 3\pi/2$$

PROBLEM 12.3 A spherical surface of radius R has charge uniformly distributed over its surface with a density $Q/4\pi R^2$, except for a spherical cap at the north pole, defined by the cone $\theta = \alpha$. Show that the potential inside the spherical surface can be expressed as

$$\Phi = \frac{Q}{2}\sum_{\ell=0}^{\infty}\frac{1}{2\ell+1}\left[P_{\ell+1}(\cos\alpha) - P_{\ell-1}(\cos\alpha)\right]\frac{r^\ell}{R^{\ell+1}}P_\ell(\cos\theta)$$

where, for $\ell = 0$, $P_{\ell-1}(\cos\alpha) = -1$. What is the potential outside?

Solution:

(a) $$\Phi = \int\frac{\sigma(\mathbf{x}')da'}{|\mathbf{x} - \mathbf{x}'|}$$

$$\frac{1}{|\mathbf{x} - \mathbf{x}'|} = \sum_{\ell=0}^{\infty}\frac{r^\ell}{R^{\ell+1}}P_\ell(\cos\theta)P_\ell(\cos\theta')$$

$$da' = R^2 d(\cos\theta')dr'$$

$$\sigma(\mathbf{x}') = \begin{cases}\dfrac{Q}{4\pi R^2}, & -1 < \cos\theta < \cos\alpha \\ 0 & \cos\alpha < \cos\theta < 1\end{cases}$$

$$\Phi = \sum_{\ell=0}^{\infty}\frac{r^\ell}{R^{\ell+1}}\frac{Q}{4\pi R^2}R^2\int_{-1}^{\cos\alpha}P_\ell(\cos\theta')d(\cos\theta')\int_0^{2\pi}d\phi'\,P_\ell(\cos\theta)$$

$$= \sum_{\ell=0}^{\infty}\frac{r^\ell}{R^\ell}\frac{Q}{2}\int_{-1}^{\cos\alpha}P_\ell(\cos\theta')d(\cos\theta')P_\ell(\cos\theta)$$

For $\ell \geq 1$

$$\int_{-1}^{\cos\alpha} P_\ell(\cos\theta')d(\cos\theta')$$

$$= \frac{1}{2\ell+1}\int_{-1}^{\cos\alpha}\left(\frac{dP_{\ell+1}(\cos\theta')}{d(\cos\theta')} - \frac{dP_{\ell-1}(\cos\theta')}{d(\cos\theta')}\right)d(\cos\theta')$$

$$= \frac{1}{2\ell+1}[P_{\ell+1}(\cos\theta') - P_{\ell-1}(\cos\theta')]_{-1}^{\cos\alpha}$$

$$= \frac{1}{2\ell+1}[P_{\ell+1}(\cos\alpha) - P_{\ell-1}(\cos\alpha) - (-1)^{\ell+1} + (-1)^{\ell+1}]$$

$$= \frac{1}{2\ell+1}[P_{\ell+1}(\cos\alpha) - P_{\ell-1}(\cos\alpha)]$$

For $\ell = 0$

$$\int_{-1}^{\cos\alpha} P_1(\cos\theta')d(\cos\theta') = \cos\alpha - (-1)$$

$$P_{\ell-1} = -1$$

$$\int_{-1}^{\cos\alpha} P_0(\cos\theta')d(\cos\theta') = [P_{\ell+1}(\cos\alpha) - P_{\ell-1}(\cos\alpha)]$$

$$\Phi(\mathbf{x}) = \sum_{\ell=0}^{\infty} \frac{Q}{2}\frac{1}{2\ell+1}[P_{\ell+1}(\cos\alpha) - P_{\ell-1}(\cos\alpha)]\frac{r^\ell}{r^{\ell+1}}P_\ell(\cos\theta)$$

For potential outside,

$$\frac{1}{|\mathbf{x}-\mathbf{x}'|} = \sum_{\ell=0}^{\infty}\frac{R^\ell}{r^{\ell+1}}P_\ell(\cos\theta)P_\ell(\cos\theta')$$

$$\Phi(\mathbf{x}) = \sum_{\ell=0}^{\infty}\frac{Q}{2}\frac{1}{2\ell+1}\frac{R^\ell}{r^{\ell+1}}[P_{\ell+1}(\cos\alpha) - P_{\ell-1}(\cos\alpha)]P_\ell(\cos\theta)$$

PROBLEM 12.4 A hollow sphere of inner radius a has the potential specified on its surface to be $\Phi = V(\theta,\phi)$. Prove the equivalence of the two forms of solution for the potential inside the sphere:

(a) $\Phi(\mathbf{x}) = \dfrac{a(a^2 - r^2)}{4\pi} \displaystyle\int \frac{V(\theta',\phi')}{(r^2 + a^2 - 2ar\cos\gamma)^{3/2}}\, d\Omega'$

where $\cos\gamma = \cos\theta\cos\theta' + \sin\theta\sin\theta'\cos(\phi - \phi')$.

(b) $\Phi(\mathbf{x}) = \displaystyle\sum_{\ell=0}^{\infty} \sum_{m=-\ell}^{\ell} A_{\ell m} \left(\frac{r}{a}\right)^{\ell} Y_{\ell m}(\theta,\phi)$

where $A_{\ell m} = \displaystyle\int d\Omega'\, Y^*_{\ell m}(\theta',\phi')V(\theta',\phi')$.

Solution:

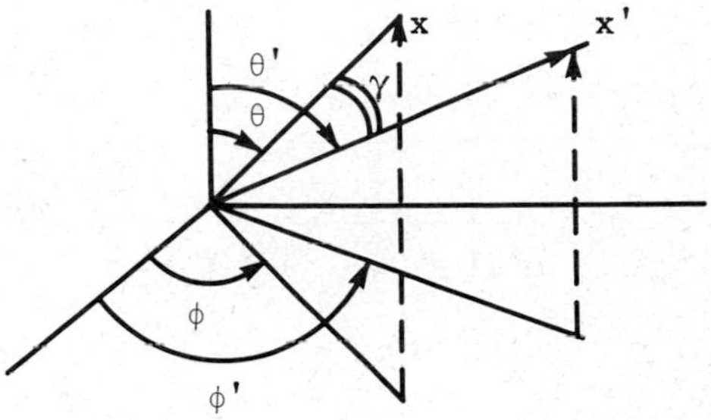

$\mathbf{x} = (r,\theta,\phi)$

$\mathbf{x}' = (a,\theta',\phi')$

$$\Phi = \frac{a(a^2 - r^2)}{4\pi} \int \frac{V(\theta',\phi')d\Omega'}{(r^2 + a^2 - 2ar\cos\gamma)^{3/2}}$$

$$= \frac{a}{4\pi}\left(r\frac{d}{dr} - a\frac{d}{da}\right) \int \frac{V(\theta',\phi')d\Omega'}{(r^2 + a^2 - 2ar\cos\gamma)^{\frac{1}{2}}}$$

$$\frac{1}{(r^2 + a^2 - 2ar\cos\gamma)^{\frac{1}{2}}} = \frac{1}{|\mathbf{x} - \mathbf{x}'|}$$

$$= 4\pi \sum_{\ell=0}^{\infty} \sum_{m=-\ell}^{+\ell} \frac{1}{2\ell+1} \frac{r^{\ell}}{a^{\ell+1}} Y^*_{\ell m}(\theta',\phi')Y_{\ell m}(\theta,\phi)$$

$$\Phi = a\left(r\frac{d}{dr} - a\frac{d}{da}\right) \sum_{\ell=0}^{\infty} \sum_{m=-\ell}^{+\ell} \frac{1}{2\ell+1} \frac{r^{\ell}}{a^{\ell+1}} \int V(\theta'\phi')Y^*_{\ell m}(\theta'\phi')d\Omega\, Y_{\ell m}(\theta,\phi)$$

$$A_{\ell m} = \int V(\theta'\phi')\, Y^*_{\ell m}(\theta'\phi')d\Omega'$$

$$\Phi = a\left(r\frac{d}{dr} - a\frac{d}{da}\right) \sum_{\ell}\sum_{m} \frac{1}{2\ell+1} \frac{r^{\ell}}{a^{\ell+1}} A_{\ell m}Y_{\ell m}(\theta,\phi)$$

$$\Phi = \sum_{\ell,m} a[\ell - (-\ell - 1)] \frac{1}{2\ell + 1} \frac{r^{\ell}}{a^{\ell+1}} A_{\ell m} Y_{\ell m}(\theta,\phi)$$

$$\Phi(\mathbf{x}) = \sum_{\ell=0}^{\infty} \sum_{m=-\ell}^{+\ell} A_{\ell m} \left(\frac{r}{a}\right)^{\ell} Y_{\ell m}(\theta,\phi)$$

PROBLEM 12.5 Show that an arbitrary function $f(x)$ can be expanded on the interval $0 \le x \le a$ in a modified Fourier-Bessel series

$$f(\rho) = \sum_{n=1}^{\infty} A_n J_v\left(k_{vn} \frac{\rho}{a}\right)$$

where y_{vn} is the nth root $dJ_v(x)/dx = 0$, and the coefficients A_n are given by

$$A_n = \frac{2}{a^2(1 - v^2/k_{vn}^2) J_v^2(y_{vn})} \int_0^a f(\rho)\rho J_v\left(k_{vn} \frac{\rho}{a}\right) d\rho$$

Solution:

$$\frac{1}{\rho}\frac{d}{d\rho}\left(\rho \frac{dJ_v\left(x_{vn} \frac{\rho}{a}\right)}{d\rho}\right) + \left(\frac{x_{vn}^2}{a^2} - \frac{v^2}{\rho^2}\right) J_v\left(x_{vn} \frac{\rho}{a}\right) = 0$$

$$\int_0^a J_v\left(x_{vn'} \frac{\rho}{a}\right)\frac{d}{d\rho}\left(\rho \frac{dJ_v\left(x_{vn} \frac{\rho}{a}\right)}{d\rho}\right) d\rho + \int_0^a \left(\frac{x_{vn}^2}{a^2} - \frac{v^2}{\rho^2}\right)\rho J_v\left(x_{vn'} \frac{\rho}{a}\right) J_v\left(x_{vn} \frac{\rho}{a}\right) d\rho = 0$$

Integrate by parts

$$\rho J_v\left(x_{vn'} \frac{\rho}{a}\right) J_v'^{1}\left(x_{vn} \frac{\rho}{a}\right)\Bigg|_0^a - \int_0^a \frac{dJ_v\left(x_{vn'} \frac{\rho}{a}\right)}{d\rho} \rho \frac{dJ\left(x_{vn} \frac{\rho}{a}\right)}{d\rho} d\rho$$

$$+ \int_0^a \left(\frac{x_{vn}^2}{a^2} - \frac{v^2}{\rho^2}\right)\rho J_v\left(x_{vn'} \frac{\rho}{a}\right) J_v\left(x_{vn} \frac{\rho}{a}\right) d\rho = 0$$

Where the first term equals zero.

$n \leftrightarrow n'$

$$\int_0^a \frac{dJ_v\left(x_{vn} \frac{\rho}{a}\right)}{d\rho} \rho \frac{dJ\left(x_{vn'} \frac{\rho}{a}\right)}{d\rho} d\rho + \int_0^a \left(\frac{x_{vn'}^2}{a^2} - \frac{v^2}{\rho^2}\right) \times \rho J_v\left(x_{vn} \frac{\rho}{a}\right) J_v\left(x_{vn'} \frac{\rho}{a}\right) d\rho = 0$$

Subtract, and then,

$$\int_0^a \left(\frac{x_{vn}^2}{a^2} - \frac{x_{vn'}^2}{a^2}\right) \rho J_v\left(x_{vn'} \frac{\rho}{a}\right) J_v\left(x_{nv} \frac{\rho}{a}\right) d\rho = 0$$

Now,

$$\frac{x_{vn}^2}{a^2} J_v\left(x_{vn} \frac{\rho}{a}\right) = \frac{v^2}{\rho^2} J_v\left(x_{nv} \frac{\rho}{a}\right) - \frac{1}{\rho} \frac{d}{d\rho} \left(\frac{\rho dJ_v\left(x_{vn} \frac{\rho}{a}\right)}{d\rho}\right)$$

and

$$\frac{v}{\rho} J_v\left(x_{nv} \frac{\rho}{a}\right) = \tfrac{1}{2} \frac{x}{a} \left[J_{v+1}\left(x_{nv} \frac{\rho}{a}\right) + J_{v-1}\left(x_{nv} \frac{\rho}{a}\right)\right]$$

$$\frac{v^2}{\rho^2} J_v\left(x_{nv} \frac{\rho}{a}\right) = \tfrac{1}{4} \frac{x^2}{a^2} \left[J_{v+2}\left(x_{nv} \frac{\rho}{a}\right) + 2J_v\left(x_{nv} \frac{\rho}{a}\right) + J_{v-2}\left(x_{nv} \frac{\rho}{a}\right)\right]$$

$$\frac{dJ_v\left(x_{vn} \frac{\rho}{a}\right)}{d\rho} = \tfrac{1}{2} \frac{x}{a} \left[J_{v-1}\left(x_{nv} \frac{\rho}{a}\right) - J_{v+1}\left(x_{nv} \frac{\rho}{a}\right)\right]$$

$$\frac{1}{\rho} \frac{d}{d\rho} \left(\rho \frac{dJ_v\left(x_{nv} \frac{\rho}{a}\right)}{d\rho}\right) = \frac{1}{\rho} \left\{\tfrac{1}{2} \frac{x}{a} \left[J_{v-1}\left(x_{nv} \frac{\rho}{a}\right) - J_{v+1}\left(x_{nv} \frac{\rho}{a}\right)\right]\right.$$

$$\left. + \rho \frac{x^2}{4a^2} \left[J_{v-2}\left(x_{nv} \frac{\rho}{a}\right) - J_v\left(x_{nv} \frac{\rho}{a}\right) - J_v\left(x_{nv} \frac{\rho}{a}\right) + J_{v+2}\left(x_{nv} \frac{\rho}{a}\right)\right]\right\}$$

$$= \frac{x}{2a\rho} \left[J_{v-1}\left(x_{nv} \frac{\rho}{a}\right) - J_{v+1}\left(x_{nv} \frac{\rho}{a}\right)\right]$$

$$+ \frac{x^2}{4a^2} \left[J_{v-2}\left(x_{nv} \frac{\rho}{a}\right) - 2J_v\left(x_{nv} \frac{\rho}{a}\right) + J_{v+2}\left(x_{nv} \frac{\rho}{a}\right)\right]$$

Now with these two results we get

$$\frac{x_{vn}^2}{a^2} J_v\left(x_{vn} \frac{\rho}{a}\right) = \frac{x}{2a\rho} \left[J_{v-1}\left(x_{nv} \frac{\rho}{a}\right) - J_{v+1}\left(x_{nv} \frac{\rho}{a}\right) + \frac{x^2}{a^2} J_v\left(x_{nv} \frac{\rho}{a}\right)\right]$$

$$\frac{x_{vn'}^2}{a^2} J_v\left(x_{vn'} \frac{\rho}{a}\right) = \frac{1}{2a\rho} \left[J_{v-1}\left(x_{n'v} \frac{\rho}{a}\right) - J_{v+1}\left(x_{n'v} \frac{\rho}{a}\right) + \frac{x^2}{a^2} J_v\left(x_{n'v} \frac{\rho}{a}\right)\right]$$

So,

$$\int_0^a \rho\left(\frac{x_{vn}^2}{a^2}-\frac{x_{n'v}^2}{a^2}\right)J_v\left(x_{vn}\,\frac{\rho}{a}\right)J_v\left(x_{nv'}\,\frac{\rho}{a}\right)d\rho$$

$$=\int_0^a \frac{x}{2a}\left[J_{v-1}\left(x_{nv}\,\frac{\rho}{a}\right)-J_{v+1}\left(x_{nv}\,\frac{\rho}{a}\right)\right]\left[J_v\left(x_{n'v}\,\frac{\rho}{a}\right)\right]$$

$$-\frac{x}{2a}\left[J_{v-1}\left(x_{n'v}\,\frac{\rho}{a}\right)-J\left(x_{n'v}\,\frac{\rho}{a}\right)\right]\left[J_v\left(x_{nv}\,\frac{\rho}{a}\right)\right]d\rho$$

$$=\int_0^a\left[\frac{dJ_v\left(x_{vn}\left(\frac{\rho}{a}\right)\right)}{d\rho}J_v\left(x_{vn'}\,\frac{\rho}{a}\right)-\frac{dJ_v\left(x_{vn'}\,\frac{\rho}{a}\right)}{d\rho}J_v\left(x_{nv}\,\frac{\rho}{a}\right)\right]d\rho=0$$

$$\frac{dJ_v\left(x_{nv}\,\frac{\rho}{a}\right)}{d\rho}=\frac{dJ_v\left(x_{n'v}\,\frac{\rho}{a}\right)}{d\rho}=0$$

Let zeros of $\dfrac{dJ_v\left(x_{nv}\,\frac{\rho}{a}\right)}{d\rho}=k_{nv}$

$$\frac{dJ_v\left(x_{n'v}\,\frac{\rho}{a}\right)}{d\rho}=k_{n'v}$$

orthogonality satisfied for $\left.\begin{array}{l} x_{nv}=k_{nv}\\ x_{n'v}=k_{n'v}\end{array}\right\}$

normalize for $x_{nv}=k_{nv}$

Bessels equation $\rightarrow \dfrac{1}{\rho}\dfrac{d}{d\rho}\left[\rho\,\dfrac{dJ_v\left(k_{nv}\,\frac{\rho}{a}\right)}{d\rho}+\left(\dfrac{k_{vn}^2}{a^2}-\dfrac{v^2}{\rho^2}\right)J_v\left(k_{nv}\,\dfrac{\rho}{a}\right)\right]=0$

$$\frac{d}{d\rho}J_v\left(k_{nv}\,\frac{\rho}{a}\right)=0$$

$$\frac{k_{vn}^2}{a^2}\left[\rho^2 J_v\left(k_{nv}\,\frac{\rho}{a}\right)\frac{d}{d\rho}J_v\left(k_{nv}\,\frac{\rho}{a}\right)\right]-v^2 J_v\left(k_{nv}\,\frac{\rho}{a}\right)\frac{d}{d\rho}J_n\left(k_{nv}\,\frac{\rho}{a}\right)=0$$

$$\tfrac{1}{2}\frac{d}{d\rho}\left[\rho J_v\left(k_{nv}\frac{\rho}{a}\right)\right]^2\frac{k_{nv}^2}{a^2}-\frac{k_{vn}^2}{a^2}\rho J_v^2\left(k_{nv}\frac{\rho}{a}\right)-\tfrac{1}{2}v^2\frac{d}{d\rho}J_v^2\left(k_{nv}\frac{\rho}{a}\right)=0$$

$$\tfrac{1}{2}\left[\rho J_v\left(k_{nv}\frac{\rho}{a}\right)\right]^2\frac{k_{nv}^2}{a^2}\Bigg|_0^a-\frac{k_v^2}{a^2}\int\rho J_v^2\left(k_{nv}\frac{\rho}{a}\right)d\rho-\tfrac{1}{2}v^2\left[J\left(k_{nv}\frac{\rho}{a}\right)\right]^2\Bigg|_0^a=0$$

$$J_v\left(k_{nv}\frac{\rho}{a}\right)\Bigg|_0^a=J_v(k_{nv})$$

$$\tfrac{1}{2}a^2J_v^2(k_{nv})\frac{k_{nv}^2}{a^2}-\frac{k_v^2}{a^2}\int\rho J_v^2\left(k_{nv}\frac{\rho}{a}\right)d\rho-\tfrac{1}{2}v^2J_v^2(k_{nv})=0$$

so

$$\int\rho J_v^2\left(k_{nv}\frac{\rho}{a}\right)d\rho=\tfrac{1}{2}a^2\left(1-\frac{v^2}{k_{nv}^2}\right)J_v^2(k_{nv})$$

on $0<\rho<a$,

$$f(\rho)=\sum_{n=1}^{\infty}A_nJ_v\left(k_{nv}\frac{\rho}{a}\right)$$

$$J_v\left(k_{nv}\frac{\rho}{a}\right)$$

form a complete set

$k_{nv}=n$th root of

$$\frac{dJ\left(k_{nv}\frac{\rho}{a}\right)}{d\rho}=0$$

$$f(\rho)=\sum_{n=1}^{\infty}A_nJ_v\left(k_{nv}\frac{\rho}{a}\right)$$

$$\int_0^a\rho f(\rho)J_v\left(k_{n'v}\frac{\rho}{a}\right)=\sum_{n=1}^{\infty}\int_0^a A_{n\rho}\,J_v\left(k_{nv}\frac{\rho}{a}\right)\times J_v\left(k_{n'v}\frac{\rho}{a}\right)d\rho$$

$$=\sum A_n\delta_{nn'}\int_0^a\rho J_v^2\left(k_{nv}\frac{\rho}{a}\right)d\rho$$

$$= \sum_n A_n \delta_{nn'} \frac{a^2}{2} \left(1 - \frac{v^2}{k^2_{nv}}\right) J^2_v(k_{vn})$$

$$A_n = \frac{2}{a^2(1 - v^2/k^2_{vn})J^2_v(k_{nv})} \int_0^a \rho f(\rho)\, J_v\left(k_{nv}\frac{\rho}{a}\right) d\rho$$

PROBLEM 12.6

(a) Two concentric spheres have radii a, b ($b > a$) and are each divided into two hemispheres by the same horizontal plane. The upper hemisphere of the inner sphere and the lower hemisphere of the outer sphere are maintained at potential V. The other hemispheres are at zero potential. Determine the potential in the region $a \leq r \leq b$ as a series in Legendre polynomials. Include terms at least up to $\ell = 4$. Check your solution against known results in the limiting cases $b \to \infty$ and $a \to 0$.

(b) Solve for the potential in part (a), using the appropriate Green function, and verify that the answer obtained in this way agrees with the direct solution from the differential equation.

Solution:

(a) $$\Phi(r,\theta) = \sum_{\ell=0}^{\infty} (A_\ell r^\ell + B_\ell r^{-(\ell+1)}) P_\ell(\cos\theta)$$

Let $\Phi'(r,\theta) = \Phi(r,\theta) - v/2$

$$\Phi'(a,\theta) = \pm\frac{v}{2} \text{ for } \cos\theta \gtrless 0$$

$$\Phi'(b,\theta) = \mp\frac{v}{2} \text{ for } \cos\theta \gtrless 0$$

$$\Phi(r,\theta) = \frac{v}{2} + \sum_{\ell=0}^{\infty} (A'_\ell r^\ell + B'_\ell r^{-(\ell+1)}) P_\ell(\cos\theta)$$

$$-\frac{v}{2}\int_{-1}^{0} P_m(x)dx + \frac{v}{2}\int_0^1 P_m(x)dx = \sum_{\ell=0}^{\infty} A_\ell' a^\ell + B_\ell' a^{-(\ell+1)} \int_{-1}^{+1} P_\ell(x)P_m(x)dx$$

$$A_\ell' a^\ell + B_\ell' a^{-(\ell+1)} = \frac{v(2\ell+1)}{2}\int_0^1 P_\ell(x)dx, \quad \text{odd } \ell$$

$$+\frac{v}{2}\int_{-1}^{0} P_m(x)dx + \frac{-v}{2}\int_0^{-1} P_m(x)dx$$

$$= \sum_{\ell=0}^{\infty} A_\ell' b^\ell + B_\ell' b^{-(\ell+1)} \int_{-1}^{+1} P_\ell(x)P_m(x)dx$$

$$A_\ell' b^\ell + B_\ell' b^{-(\ell+1)} = \frac{-v(2\ell+1)}{2} \int_0^1 P_\ell(x)dx, \quad \text{odd } \ell$$

$$A_\ell = \frac{b^{-(\ell+1)} + a^{-(\ell+1)}}{b^{-(\ell+1)}a^\ell - b^\ell a^{-(\ell+1)}} \left\{\frac{v(2\ell+1)}{2} \int_0^1 P_\ell(x)dx\right\}, \quad \text{odd } \ell$$

$$B_\ell' = \frac{b^\ell + a^\ell}{b^\ell a^{-(\ell+1)} - b^{-(\ell+1)}a^\ell} \left\{\frac{v(2\ell+1)}{2} \int_0^1 P_\ell(x)dx\right\}, \quad \text{odd } \ell$$

$$\int_0^1 P_\ell(x)dx = \frac{(-\tfrac{1}{2})^{(\ell-1)/2}(\ell-2)!!}{2\left(\frac{\ell+1}{2}\right)!}$$

Put these (A_ℓ' and B_ℓ') in Φ and get,

$$\Phi(r,\theta) = \frac{v}{2} + \frac{v}{2} \sum_{\substack{\ell \\ \text{odd}}} \frac{(2\ell+1)(-\tfrac{1}{2})^{(\ell-1)/2}(\ell-2)!!}{2\left(\frac{\ell+1}{2}\right)!} \left\{\frac{(a^\ell + b^\ell)(ab/r)^{\ell+1}}{b^{2\ell+1} - a^{2\ell+1}}\right.$$

$$\left. - \frac{(b^{\ell+1} + a^{\ell+1})r^\ell}{b^{2\ell+1} - a^{2\ell+1}}\right\} P_\ell(\cos\theta)$$

$$\Phi(r,\theta) = \frac{v}{2}\left[1 + \sum_{n=0}^{\infty} \frac{(4n+3)(2n-1)!!}{2(-2)^n(n+1)!} \left\{\frac{(a^{2n+1} + b^{2n+1})(ab/r)^{2n+2}}{b^{4n+3} - a^{4n+3}}\right.\right.$$

$$\left.\left. - \frac{(b^{2n+2} + a^{2n+2})r^{2n+1}}{b^{4n+3} - a^{4n+3}}\right\} P_{2n+1}(\cos\theta)\right]$$

$$\Phi(r,\theta) = \frac{v}{2}\left[1 + \frac{3}{2}\left\{\frac{b+a}{(b^3 - a^3)}\left(\frac{ab}{r}\right)^2 - \frac{(b^2 + a^2)r}{b^3 - a^3}\right\} P_1(\cos\theta)\right.$$

$$- \frac{7}{8}\left\{\frac{(a^3 + b^3)}{(b^7 - a^7)}\left(\frac{ab}{r}\right)^4 - \frac{(b^4 + a^4)r^3}{b^7 - a^7}\right\} P_3(\cos\theta)$$

$$\left. + \frac{11}{16}\left\{\frac{(a^5 + b^5)}{(b^{11} - a^{11})}\left(\frac{ab}{r}\right)^6 - \frac{b^6 + a^6}{b^{11} - a^{11}} r^5\right\} P_5(\cos\theta) + \cdots\right]$$

Limiting cases

$b \to \infty$

$$\frac{(a^{\ell}+b^{\ell})(ab/r)^{\ell+1}}{b^{2\ell+1}-a^{2\ell+1}} \to \left(\frac{a}{r}\right)^{\ell+1}$$

$$\frac{(b^{\ell+1}+a^{2\ell+1})r^{\ell}}{b^{2\ell+1}-a^{2\ell+1}} \to 0$$

$$\Phi = \frac{v}{2} + \sum_{n=0}^{\infty} \frac{v}{2} \frac{(4n+3)(2n-1)!!}{2(-2)^{n}(n+1)!} \left(\frac{a}{r}\right)^{2n+2} P_{2n+1}(\cos\theta)$$

$a \to 0$

$$\frac{(a^{\ell}+b^{\ell})(ab/r)^{\ell+1}}{b^{2\ell+1}-a^{2\ell+1}} \to 0$$

$$\frac{(b^{\ell+1}+a^{2\ell+1})r^{\ell}}{b^{2\ell+1}-a^{2\ell+1}} \to \left(\frac{r}{b}\right)^{\ell}$$

$$\Phi = \frac{v}{2} + \sum_{n=0}^{\infty} \frac{v}{2} \frac{(4n+3)(2n-1)!!}{2(-2)^{n}(n+1)!} \left(\frac{r}{b}\right)^{2n+1} P_{2n+1}(\cos\theta)$$

(b) Green's function for spherical shell bounded at $r = a$ and $r = b$,

$$G(\mathbf{x},\mathbf{x}') = 4\pi \sum_{\ell=0}^{\infty} \sum_{m=-\ell}^{\ell} \frac{Y^{*}_{\ell m}(\theta',\phi')Y_{\ell m}(\theta,\phi)}{2\ell+1\,[1-(a/b)^{2\ell+1}]} \times \left(r_<^{\ell} - \frac{a^{2\ell+1}}{r_<^{\ell+1}}\right)\left(\frac{1}{r_<^{\ell+1}} - \frac{r_>^{\ell}}{b^{2\ell+1}}\right)$$

The potential for $a < r < b$ is

$$\Phi(\mathbf{x}) = \int_{v} \rho(x')G(\mathbf{x},\mathbf{x}')d^{3}x - \frac{1}{4\pi} \oint \Phi(x') \frac{\partial G}{\partial n'}\, da'$$

where $\rho(x') = 0$.

$$\frac{\partial G}{\partial n'} = \frac{\partial G}{\partial r'}\bigg|_{\substack{r'=b=r_> \\ r=r_<}} = + \sum_{\ell,m} \frac{4\pi Y^{*}_{\ell m}(\theta',\phi')Y_{\ell m}(\theta,\phi)}{2\ell+1\,[1-(a/b)^{2\ell+1}]} \left(r^{\ell} - \frac{a^{2\ell+1}}{r^{\ell+1}}\right)$$

$$\times \left(\frac{-(\ell+1)}{r'^{\ell+2}} - \frac{\ell r'^{\ell-1}}{b^{2\ell+1}}\right)_{r'=b}$$

$$= - \sum_{\ell,m} \frac{4\pi Y^*_{\ell m}(\theta',\phi')Y_{\ell m}(\theta,\phi)}{[1-(a/b)^{2\ell+1}]b^{\ell+2}} \left(r^\ell - \frac{a^{2\ell+1}}{r^{\ell+1}}\right)$$

$$\frac{\partial G}{\partial n'} = \frac{-\partial G}{\partial r'}\bigg|_{\substack{r'=a=r_<\\ r=r_>}} = - \sum_{\ell,m} \frac{4\pi Y^*_{\ell m}(\theta',\phi')Y_{\ell m}(\theta,\phi)}{(2\ell+1)[1-(a/b)^{2\ell+1}]}$$

$$\times \left(\ell r'^{\ell-1} - \frac{a^{2\ell+1}(-\ell-1)}{r'^{\ell+2}}\right)_{r'=a} \left(\frac{1}{r^{\ell+1}} - \frac{r^\ell}{b^{2\ell+1}}\right)$$

$$= - \sum_{\ell,m} \frac{4\pi Y^*_{\ell m}(\theta',\phi')Y_{\ell m}(\theta,\phi)}{[1-(a/b)^{2\ell+1}]} a^{\ell-1} \left(\frac{1}{r^{\ell+1}} - \frac{r^\ell}{b^{2\ell+1}}\right)$$

$$\Phi(\mathbf{x}) = \sum_{\ell,m} \oint \frac{Y^*_{\ell m}(\theta',\phi')Y_{\ell m}(\theta,\phi)}{[1-(a/b)^{2\ell+1}]b^{\ell+2}} \left(r^\ell - \frac{a^{2\ell+1}}{r^{\ell+1}}\right) \Phi(b,\theta')b^2\, d\Omega'$$

$$+ \oint \sum_{\ell,m} \frac{Y^*_{\ell m}(\theta',\phi')Y_{\ell m}(\theta,\phi)}{[1-(a/b)^{2\ell+1}]} \Phi(a,\theta')a^{\ell-1} \left(\frac{1}{r^{\ell+1}} - \frac{r^\ell}{a^{2\ell+1}}\right) a^2\, d\Omega'$$

$$\int Y^*_{\ell m}(\theta',\phi')d\Omega' = \sqrt{\frac{(2\ell+1)(\ell-m)!}{4\pi(\ell+m)!}} \int P_\ell^m(\cos\theta')e^{-i\phi}\, d(\cos\theta')d\phi'$$

$$= \sqrt{\frac{(2\ell+1)(\ell-m)!}{4\pi(\ell+m)!}} \int P_\ell^m(\cos\theta')d(\cos\theta')\ \delta_{m,0}(2\pi)$$

$$= 2\pi\sqrt{\frac{2\ell+1}{4\pi}} \int P_\ell(\cos\theta')d(\cos\theta')$$

$$\Phi(\mathbf{x}) = \sum_\ell \int \frac{2\ell+1}{2} \frac{P_\ell(\cos\theta)P_\ell(\cos\theta')}{[1-(a/b)^{2\ell+1}]b^\ell} \left(r^\ell - \frac{a^{2\ell+1}}{r^{\ell+1}}\right) \Phi(b,\theta')d(\cos\theta')$$

$$+ \sum_\ell \int \frac{2\ell+1}{2} \frac{P_\ell(\cos\theta)P_\ell(\cos\theta')}{[1-(a/b)^{2\ell+1}]} a^{\ell+1} \left(\frac{1}{r^{\ell+1}} - \frac{r^\ell}{b^{2\ell+1}}\right) \Phi(a,\theta')d(\cos\theta')$$

as in (a),

$$\Phi' = \Phi - \frac{v}{2}$$

So,

$$\Phi'(a,\theta) = \pm\frac{v}{2}, \quad \cos \gtrless 0$$

$$\Phi'(b,\theta') = \mp\frac{v}{2}, \quad \cos \gtrless 0$$

$$\Phi(\mathbf{x}) = \frac{v}{2} + \sum_{\ell} \left(\int_{-1}^{0} + \frac{v}{2} (2\ell+1) P_\ell(\cos\theta')\, d(\cos\theta') \right.$$

$$\left. - \int_{0}^{1} \frac{v}{2} (2\ell+1) P_\ell(\cos\theta')\, d(\cos\theta') \right) \frac{P_\ell(\cos\theta)(r^\ell - a^{2\ell+1}/r^{\ell+1})}{2b^\ell[1-(a/b)^{2\ell+1}]}$$

$$+ \sum_{\ell} \left(- \int_{-1}^{0} \frac{v}{2} (2\ell+1) P_\ell(\cos\theta')\, d(\cos\theta') \right.$$

$$\left. + \int_{0}^{1} \frac{v}{2} (2\ell+1) P_\ell(\cos\theta')\, d(\cos\theta') \right) P_\ell(\cos\theta) \frac{a^{\ell+1}\left(\frac{1}{r^{\ell+1}} - \frac{r^\ell}{b^{2\ell+1}}\right)}{2[1-(a/b)^{2\ell+1}]}$$

$$\frac{2\ell+1}{2}\left(\int_0^1 P_\ell(x)\,dx - \int_{-1}^{0} P_\ell(x)\,dx \right) = (-\tfrac{1}{2})^{(\ell-1)/2} \frac{(2\ell+1)(\ell-2)!!}{2\left(\frac{\ell+1}{2}\right)!}, \quad \text{odd } \ell$$

$$\Phi(\mathbf{x}) = \frac{v}{2} + \sum_{\substack{\ell \\ \text{odd}}} \frac{v}{2} \frac{(-\frac{1}{2})^{(\ell-1)/2} (2\ell+1)(\ell-2)!!}{2\left(\frac{\ell+1}{2}\right)!} P_\ell(\cos\theta)$$

$$\left(\frac{a^{\ell+1} b^{2\ell+1}\left(\frac{1}{r^{\ell+1}} - \frac{r^\ell}{b^{2\ell+1}}\right)}{b^{2\ell+1} - a^{2\ell+1}} \right) + \sum_{\substack{\ell \\ \text{odd}}} \frac{v}{2} \frac{(-\frac{1}{2})^{(\ell-1)/2} (2\ell+1)(\ell-2)!!}{2\left(\frac{\ell+1}{2}\right)!}$$

$$P_\ell(\cos\theta) \left(\frac{-b^{\ell+1}\left(r^\ell - \frac{a^{2\ell+1}}{r^{\ell+1}}\right)}{b^{2\ell+1} - a^{2\ell+1}} \right)$$

$$\Phi(\mathbf{x}) = \frac{v}{2} + \frac{v}{2} \sum_{\substack{\ell \\ \text{odd}}} \frac{(-\frac{1}{2})^{(\ell-1)/2} (2\ell+1)(\ell-2)!!}{2\left(\frac{\ell+1}{2}\right)!} \left(\frac{(a^\ell + b^\ell)(ab/r)^{\ell+1}}{b^{2\ell+1} - a^{2\ell+1}} \right.$$

$$-\frac{(b^{\ell+1}+a^{\ell+1})r^{\ell}}{b^{2\ell+1}-a^{2\ell+1}}\Bigg)$$

PROBLEM 12.7 A hollow cube has conducting walls defined by six planes $x=0$, $y=0$, $z=0$, and $x=a$, $y=a$, $z=a$. The walls $z=0$ and $z=a$ are held at a constant potential V. The other four sides are at zero potential.

(a) Find the potential $\phi(x,y,z)$ at any point inside the cube.
(b) Evaluate the potential at the center of the cube numerically, accurate to three significant figures. How many terms in the series is it necessary to keep in order to attain this accuracy? Compare your numerical result with the average value of the potential on the walls.
(c) Find the surface-charge density on the surface $z=a$.

Solution:

Use the notation $\gamma_{nm}=\frac{\pi}{a}\sqrt{n^2+m^2}$

(a) $$\phi(x,y,z)=\sum_{n,m=1}^{\infty} A_{nm}\sin\left(\frac{n\pi x}{a}\right)\sin\left(\frac{m\pi y}{a}\right)\sinh\left(\frac{\pi}{a}\sqrt{n^2+m^2}\,z\right)$$

$$A_{nm}=\frac{4V}{a^2\sinh\gamma_{nma}}\int_0^a\sin\left(\frac{n\pi x}{a}\right)dx\int_0^a\sin\left(\frac{m\pi y}{a}\right)dy$$

$$=\frac{4V}{a^2\sinh\gamma_{nma}}\left(\frac{a}{\pi}\right)^2[1-(-1)^n]\frac{1}{n}\cdot[1-(-1)^m]\frac{1}{m}$$

$$=\frac{4Va^2(4)}{a^2\pi^2\sinh\gamma_{nma}}\frac{\delta_{n\,\mathrm{odd}}}{n}\frac{\delta_{m\,\mathrm{odd}}}{m},\quad \gamma_{nm}=\frac{\pi}{a}\sqrt{n^2+m^2}$$

$$\phi_1(x,y,z)=\frac{16V}{\pi^2}\sum_{(n,m)\mathrm{odd}}^{\infty}\frac{1}{nm\sinh(\gamma_{nma})}\sin\left(\frac{n\pi x}{a}\right)\sin\left(\frac{m\pi y}{a}\right)\sinh(\gamma_{nm}z)$$

$$\phi_2(x,y,z)=\frac{16V}{\pi^2}\sum_{(n,m)\mathrm{odd}}^{\infty}\frac{1}{nm\sinh\gamma_{nma}}\sin\left(\frac{n\pi x}{a}\right)\sin\left(\frac{m\pi y}{a}\right)\sinh[\gamma_{nm}(a-z)]$$

$$\phi=\phi_1+\phi_2$$

$$\phi(x,y,z)=\frac{16V}{\pi^2}\sum_{(n,m)\mathrm{odd}}^{\infty}\frac{\sin(n\pi x/a)\sin(m\pi y/a)}{nm\sinh(\gamma_{mna})}\{\sinh(\gamma_{nm}z)+\sinh[\gamma_{nm}(a-z)]\}$$

(b) $$\phi\left(\frac{a}{2},\frac{a}{2},\frac{a}{2}\right)=\frac{16V}{\pi^2}\sum_{(n,m)\mathrm{odd}}^{\infty}\frac{\sin(\pi n/2)\sin(\pi m/2)}{nm\sinh(\gamma_{nma})}\left[2\sinh\left(\gamma_{n,m}\frac{a}{2}\right)\right]$$

$$= \frac{16V}{\pi^2}\left(\frac{\sin(\pi/2)\sin(\pi/2)\ 2\sinh(\pi\sqrt{2}/2)}{\sinh(\sqrt{2}\pi)} + \frac{\sin(\pi/2)\sin(3\pi/2)}{3\sin\pi\sqrt{2}}\ 2\sin\frac{\pi}{2}\sqrt{2}\right)$$

$$= \frac{32V}{\pi^2}\left(\frac{\sinh(\pi\sqrt{2}/2)}{\sinh\pi\sqrt{2}} - \frac{\sinh(\pi\sqrt{\pi}/2)}{3\sinh\pi\sqrt{2}} - \frac{\sinh(\pi\sqrt{2}/2)}{3\sinh\pi\sqrt{2}} + \frac{\sinh[(3\pi/2)\sqrt{2}]}{9\sinh(3\pi\sqrt{2})} + \cdots\right)$$

$$\phi\left(\frac{a}{2},\frac{a}{2},\frac{a}{2}\right) \approx 3242\,V(0.107 - 0.002 - 0.002) \sim 0.334\,V$$

Keep three terms:

$$\phi\left(\frac{a}{2},\frac{a}{2},\frac{a}{2}\right) = \frac{2}{6}\ V = 0.333V$$

(c) $$\sigma = \left.\frac{E_{\hat{n}}}{4\pi}\right|_{z=a}$$

$$E_{\perp} = \frac{-\partial\Phi}{\partial z} = \frac{-16V}{\pi^2}\sum_{\substack{n,m\\ \text{odd}}}^{\infty} \frac{\sin(n\pi x/a)\sin(m\pi y/a)}{nm\sinh(\gamma_{nm}a)}\{\gamma_{nm}\cosh(\gamma_{nm}z)$$

$$-\gamma_{nm}\cosh[\gamma_{nm}(a-z)]\}$$

$$\left.\sigma\right|_{z=a} = \frac{4V}{\pi^3}\sum_{\substack{n,m\\ \text{odd}}}^{\infty} \frac{\gamma_{nm}\sin(n\pi x/a)\sin(m\pi y/a)}{nm\sin(\gamma_{nm}a)}[1-\cosh(\gamma_{nm}a)]$$

Chapter 13

PROBLEM 13.1 Using the definitions of the parameters involved, show how to convert the equation for a transverse sinusoidal travelling wave moving in the positive x direction,

$$z = z_0 \sin \frac{2\pi}{\lambda} (x - ut)$$

into one involving k, x, ω, and t. Complex notation may be used if you wish.

Solution:

$$z = z_0 \sin \frac{2\pi}{\lambda} (x - ut)$$

$$z = z_0 \sin \left(\frac{2\pi}{\lambda} x - \frac{2\pi}{\lambda} \frac{\omega}{2\pi} \lambda t \right)$$

$$z = z_0 \sin (kx - \omega t)$$

$$z = z_0 e^{i(kx - \omega t)}$$

taking the imaginary part.

PROBLEM 13.2 What frequency electromagnetic waves would have to be used to communicate with a submarine submerged at a depth of 100 m? The conductivity of seawater is about 4.3 mho/m.

Solution:

$$\omega = \frac{2}{\sigma \mu \delta^2} = \frac{2}{(4.3)(4\pi \times 10^{-7})(10^{-1})} = 5.86 \text{ Hz}$$

PROBLEM 13.3

(a) Given the electromagnetic wave

$$\mathbf{E} = \hat{\mathbf{i}} E_0 \cos \omega (\sqrt{\varepsilon\mu}\ z - t) + \hat{\mathbf{j}} E_0 \sin \omega (\sqrt{\varepsilon\mu}\ z - t)$$

where E_0 is constant. Find the corresponding magnetic field.

(b) Find the Poynting vector.

Solution:

(a) To get **B** knowing **E** use

$$\nabla \times \mathbf{E} = -\frac{\partial \mathbf{B}}{\partial t}$$

Now $\mathbf{E} = \hat{\mathbf{i}}E_x + \hat{\mathbf{j}}E_y$. So

$$\nabla \times \mathbf{E} = \begin{vmatrix} \hat{\mathbf{i}} & \hat{\mathbf{j}} & \hat{\mathbf{k}} \\ \dfrac{\partial}{\partial x} & \dfrac{\partial}{\partial y} & \dfrac{\partial}{\partial z} \\ E_x(z) & E_y(z) & 0 \end{vmatrix} = -\hat{\mathbf{i}}\frac{\partial E_y}{\partial z} - \hat{\mathbf{j}}\left(-\frac{\partial E_x}{\partial z}\right)$$

$$\nabla \times \mathbf{E} = -\hat{\mathbf{i}}\frac{\partial E_y}{\partial z} + \hat{\mathbf{j}}\frac{\partial E_x}{\partial z}$$

$$\mathbf{E} = \hat{\mathbf{i}}E_0 \cos \omega(\sqrt{\varepsilon\mu}\ z - t) + \hat{\mathbf{j}}E_0 \sin \omega(\sqrt{\varepsilon\mu}\ z - t)$$

$$\frac{\partial E_y}{\partial z} = E_0\omega\sqrt{\varepsilon\mu} \cos \omega(\sqrt{\varepsilon\mu}\ z - t)$$

$$\frac{\partial E_x}{\partial z} = -E_0\omega\sqrt{\varepsilon\mu} \sin \omega(\sqrt{\varepsilon\mu}\ z - t)$$

$$\mathbf{B} = -\int (\nabla \times \mathbf{E})dt = +\hat{\mathbf{i}}\int \frac{\partial E_y}{\partial z}\, dt - \hat{\mathbf{j}}\int \frac{\partial E_x}{\partial z}\, dt$$

$$\mathbf{B} = \hat{\mathbf{i}}E_0\omega\sqrt{\varepsilon\mu}\left(\frac{-1}{\omega}\right) \sin \omega(\sqrt{\varepsilon\mu}\ z - t) - \hat{\mathbf{j}}(-E_0)\omega\sqrt{\varepsilon\mu}\left(\frac{-1}{\omega}\right)(-1) \cos \omega(\sqrt{\varepsilon\mu}\ z - t)$$

$$\mathbf{B} = -\hat{\mathbf{i}}E_0\sqrt{\varepsilon\mu} \sin \omega(\sqrt{\varepsilon\mu}\ z - t) + \hat{\mathbf{j}}E_0\sqrt{\varepsilon\mu} \cos \omega(\sqrt{\varepsilon\mu}\ z - t)$$

(b) Now the Poynting vector is $\mathbf{S} = \mathbf{E} \times \mathbf{H} = \frac{1}{\mu}\ \mathbf{E} \times \mathbf{B}$

$$\mathbf{E} \times \mathbf{B} = \begin{vmatrix} \hat{\mathbf{i}} & \hat{\mathbf{j}} & \hat{\mathbf{k}} \\ E_x & E_y & 0 \\ B_x & B_y & 0 \end{vmatrix} = \hat{\mathbf{i}}(0) + \hat{\mathbf{j}}(0) + \hat{\mathbf{k}}(E_xB_y - E_yB_x)$$

$$\mathbf{S} = \hat{\mathbf{k}}\ \frac{1}{\mu}\ \{[E_0 \cos \omega(\sqrt{\varepsilon\mu}\ z - t)][E_0\sqrt{\varepsilon\mu} \cos \omega(\sqrt{\varepsilon\mu}\ z - t)]\}$$
$$+ \{[E_0 \sin \omega(\sqrt{\varepsilon\mu}\ z - t)][E_0\sqrt{\varepsilon\mu} \sin \mu(\sqrt{\varepsilon\mu}\ z - t)]\}$$

$$\mathbf{S} = \frac{1}{\mu}\ E_0^2\sqrt{\varepsilon\mu}\ \hat{\mathbf{k}}$$

$$\mathbf{S} = E_0^2\sqrt{\varepsilon/\mu}\ \hat{\mathbf{k}}$$

PROBLEM 13.4 Assume that the solutions of the wave equation in an isotropic, linear, homogeneous medium, where there is no net charge, may be written in the form of a product of functions:

$$E = E_0(z)f(t)$$

Show that this substitution leads to the ordinary differential equations

$$\frac{d^2E_0}{dz^2} + kE_0 = 0, \quad \mu\varepsilon\,\frac{d^2F}{dt^2} + kf = 0$$

where k is an arbitrary constant.

Solution: In a homogeneous medium, the wave equation is

$$\nabla^2 E - \mu\varepsilon\,\frac{\partial^2 E}{\partial t^2} - \mu\sigma\,\frac{\partial E}{\partial t} = 0$$

Assume $E = E_0(r)f(t)$

$$f(t)\,\nabla^2 E_0(r) - E_0(r)\,\mu\varepsilon\,\frac{\partial^2 f}{\partial t^2} - \mu\sigma\,E_0(r)\,\frac{\partial f}{\partial t} = 0$$

or

$$\frac{\nabla^2 E_0(r)}{E_0(r)} = \frac{\mu\varepsilon}{f(t)}\,\frac{\partial^2 f}{\partial t^2} + \frac{\mu\sigma}{f(t)}\,\frac{\partial f}{\partial t}$$

In order for two differential equations to be equal

$$\frac{\nabla^2 E_0(r)}{E_0(r)} = -k$$

$$\frac{\mu\varepsilon}{f(t)}\,\frac{\partial^2 f}{\partial t^2} + \frac{\mu\sigma}{f(t)}\,\frac{\partial f}{\partial t} = -k$$

Rewriting

$$\nabla^2 E_0(r) + kE_0(r) = 0$$

$$\mu\varepsilon\,\frac{\partial^2 f}{\partial t^2} + \mu\sigma\,\frac{\partial f}{\partial t} + kf = 0$$

PROBLEM 13.5 At what frequency is the skin depth is silver 1 mm? The conductivity of silver is 3×10^7 mho/m.

Solution:

$$\omega = \frac{2}{\sigma\mu\delta^2} = \frac{2}{(3 \times 10^7)(4\pi \times 10^{-7})(10^{-6})} = 5.3 \times 10^4 \text{ Hz}$$

PROBLEM 13.6 The radiant energy from the sun is about 8 J in 1 min on 1 cm^2 of cross section. Assuming this to be in the form of a plane wave travelling in the x direction in vacuo, plane polarized with its electric field in the zx plane, and having a wavelength of 0.555×10^{-3} mm, find expressions for the electric and magnetic fields with all constants (except phase) evaluated.

Solution:

$$\langle S\rangle = \frac{1}{2\mu_0 c} E_0^2 = \frac{8}{(60)(10^{-4})} = 1.33 \times 10^3 \text{ W/m}^2 \cdot \text{sec}$$

$$E_0^2 = 1.00 \times 10^6 \Rightarrow E_0 = 10^3 \text{ N/C}$$

$$B_0 = 3.33 \times 10^{-6}$$

$$k = \frac{2\pi}{\lambda} = \frac{6.28}{5.55 \times 10^{-7}} = 1.13 \times 10^7$$

$$\omega = 2\pi f = \frac{2\pi c}{\lambda} = 3.39 \times 10^{15}$$

$$\mathbf{E} = \hat{\mathbf{k}}\ 10^3 \sin(1.13 \times 10^7\, x - 3.39 \times 10^{15}\, t)$$

$$\mathbf{B} = \hat{\mathbf{j}}\ 3.33 \times 10^{-6} \sin(1.13 \times 10^7\, x - 3.39 \times 10^{15}\, t)$$

PROBLEM 13.7 Show that for substances of low density, like gases, which have a single resonant frequency ω_0, the index of refraction is given by

$$n \simeq 1 + \frac{Nq_e^2}{2\varepsilon_0 m_e(\omega_0^2 - \omega^2)}$$

Solution:

$$n = \frac{c}{\omega} K_R$$

$$K_R = \frac{\omega\sqrt{\mu_0 \varepsilon_R}}{\sqrt{2}} \sqrt{2} = \omega\sqrt{\mu_0\varepsilon_R}$$

$$n = c\sqrt{\mu_0\varepsilon_R} = \sqrt{\varepsilon_R/\varepsilon_0}$$

$$\varepsilon_R = \varepsilon_0 + \frac{Ne^2/m}{-\omega^2 + \omega_0^2}$$

$$n = \left(1 + \frac{Ne^2/m\varepsilon_0}{-\omega^2 + \omega_0^2}\right)^{\frac{1}{2}}$$

For a gas

$$\frac{Ne^2/m\varepsilon_0}{-\omega^2 + \omega_0^2} << 1$$

so

$$n \simeq 1 + \frac{Ne^2/2m\varepsilon_0}{-\omega^2 + \omega_0^2}$$

PROBLEM 13.8 A beam of monochromatic light (frequency ω) under vacuum is incident normally on a dielectric film of refractive index $n = \sqrt{\varepsilon/\varepsilon_0}$. The thickness of the film is d. Calculate the reflection coefficient for the reflected wave as a function of d and n. [*Hint:* Assume two waves travelling in opposite directions inside the film.]

Solution:

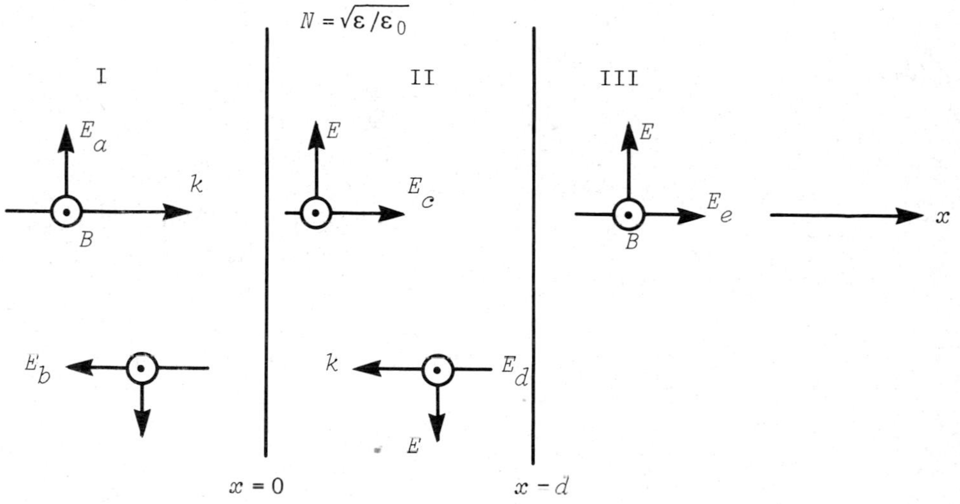

For $\mu_1 = \mu_2 = \mu_3 = \mu_0$, the boundary conditions are

(1) tangential E is continuous
(2) tangential B is continuous

For the waves

$$\mathbf{E}_1 = \mathbf{E}_{ao}\, e^{i(k_1x - \omega t)} + \mathbf{E}_{bo}\, e^{i(-k_1x - \omega t)}$$

$$\mathbf{E}_2 = \mathbf{E}_{co}\, e^{i(k_2x - \omega t)} + \mathbf{E}_{do}\, e^{i(-k_2x - \omega t)}$$

$$\mathbf{E}_3 = \mathbf{E}_{eo}\, e^{i(k_1x - \omega t)}$$

At $x = 0$ tangential E is continuous, so $E_1 = E_2$

$$E_{ao} - E_{bo} = E_{co} - E_{do} \tag{1}$$

at $x = d$

$$E_{co}e^{ik_2d} - E_{do}e^{-ik_2d} = E_{eo}e^{ik_1d} \tag{2}$$

Since the waves are plane waves, $B = kE/\omega$ in amplitude so because B is also continuous at the boundary, at $x = 0$

$$B_{ao} + B_{bo} = B_{co} + B_{do} \tag{3}$$

$$B_{co}e^{ik_2d} + B_{do}e^{-ik_2d} = B_{eo}e^{ik_1d} \tag{4}$$

Using the relation between E and B

$$k_1(E_{ao} + E_{bo}) = k_2(E_{co} + E_{do}) \tag{3'}$$

$$k_2(E_{co}e^{ik_2d} + E_{do}e^{-ik_2d}) = k_1E_{eo}e^{ik_1d} \tag{4'}$$

Now solve for the ratio of E_{bo}/E_{ao}. Rearranging the equations so they will fit the determinate form

$$-E_{bo} - E_{co} + E_{do} = -E_{ao}$$

$$e^{ik_2d}E_{co} - e^{-ik_2d}E_{do} - E_{eo}e^{ik_1d} = 0$$

$$k_1E_{bo} - k_2E_{co} - k_2E_{do} = -k_1E_{ao}$$

$$k_2e^{ik_2d}E_{co} + k_2e^{-ik_2d}E_{do} - k_1e^{ik_1d}E_{eo} = 0$$

$$E_{bo} = \frac{\begin{vmatrix} -E_{ao} & -1 & +1 & 0 \\ 0 & e^{ik_2d} & -e^{-ik_2d} & -e^{ik_1d} \\ -k_1E_{ao} & -k_2 & -k_2 & 0 \\ 0 & k_2e^{ik_2d} & k_2e^{-ik_2d} & -k_1e^{ik_1d} \end{vmatrix}}{\begin{vmatrix} -1 & -1 & 1 & 0 \\ 0 & e^{ik_2d} & -e^{-ik_2d} & -e^{ik_1d} \\ k_1 & -k_2 & -k_2 & 0 \\ 0 & k_2e^{ik_2d} & k_2e^{-ik_2d} & -k_1e^{ik_1d} \end{vmatrix}}$$

Expanding the numerator

$$N = -E_{ao}\begin{vmatrix} e^{ik_2d} & -e^{ik_2d} & -e^{ik_1d} \\ -k_2 & -k_2 & 0 \\ k_2e^{ik_2d} & k_2e^{-ik_2d} & -k_1e^{ik_1d} \end{vmatrix}$$

$$-k_1E_{ao}\begin{vmatrix} -1 & 1 & 0 \\ e^{ik_2d} & -e^{-ik_2d} & -e^{ik_1d} \\ k_2e^{ik_2d} & k_2e^{-ik_2d} & -k_1e^{ik_1d} \end{vmatrix}$$

$$N = -E_{ao}\,(k_1k_2e^{i(k_1+k_2)d} + k_2^2e^{i(k_1-k_2)d} - k_2^2e^{i(k_1+k_2)d} + k_1k_2e^{i(k_1+k_2)d})$$

$$= -k_1E_{ao}\,(-k_1e^{i(k_1-k_2)d} - k_2e^{i(k_1+k_2)d} - k_2e^{i(k_1-k_2)d} + k_1e^{i(k_1+k_2)d})$$

$$N = -E_{ao}\,[e^{i(k_1+k_2)d}(k_1k_2 - k_2^2 - k_1k_2 + k_1k_2 + k_1^2) + e^{i(k_1-k_2)d}(k_2^2 - k_1^2 - k_1k_2)]$$

Expanding the denominator

$$D = -1\begin{vmatrix} e^{ik_2d} & -e^{ik_2d} & -e^{ik_1d} \\ -k_2 & -k_2 & 0 \\ k_2e^{ik_2d} & k_2e^{-ik_2d} & -k_1e^{ik_1d} \end{vmatrix}$$

$$+k_1\begin{vmatrix} -1 & 1 & 0 \\ e^{ik_2d} & -e^{-ik_2d} & -e^{ik_1d} \\ k_2e^{ik_2d} & k_2e^{-ik_2d} & -k_1e^{ik_1d} \end{vmatrix}$$

$$D = -1\,(k_1k_2e^{i(k_1+k_2)d} + k_2^2e^{i(k_1-k_2)d} - k_2^2e^{i(k_1+k_2)d} + k_1k_2e^{i(k_1+k_2)d})$$

$$+k_1\,(-k_1e^{i(k_1-k_2)d} - k_2e^{i(k_1+k_2)d} - k_2e^{i(k_1-k_2)d} + k_1e^{i(k_1+k_2)d})$$

$$D = e^{-i(k_1+k_2)d}(-k_1k_2 + k_2^2 - k_1k_2 - k_1k_2 + k_1^2) + e^{i(k_1-k_2)d}(-k_2^2 - k_1^2 - k_1k_2)$$

$$D = e^{i(k_1+k_2)d}(k_2^2 - 3k_1k_2 + k_1^2) - e^{i(k_1-k_2)d}(k_1^2 + k_1k_2 + k_2^2)$$

Combining,

$$\frac{E_{bo}}{E_{ao}} = \frac{(k_2^2 - k_1k_2 - k_1^2)e^{i(k_1+k_2)d} + (k_1^2 + k_1k_2 - k_2^2)e^{i(k_1-k_2)d}}{(k_2^2 - 3k_1k_2 + k_1^2)e^{i(k_1+k_2)d} - (k_1^2 + k_1k_2 + k_2^2)e^{i(k_1-k_2)d}}$$

or

$$\frac{E_{bo}}{E_{ao}} = \frac{e^{ik_1d}[(k_2^2 - k_1k_2 - k_1^2)(e^{ik_2d} - e^{-ik_2d})]}{e^{ik_1d}[(k_2^2 - 3k_1k_2 + k_1^2)e^{ik_2d} - (k_1^2 + k_1k_2 + k_2^2)e^{-ik_2d}]}$$

Find the real part of this.

$$\frac{E_{bo}}{E_{ao}} = \frac{(k_2^2 - k_1k_2 - k_1^2)(e^{ik_2d} - e^{-ik_2d})[(a-b)\cos(k_2d) - i(a+b)\sin(k_2d)]}{(a-b)^2\cos^2 k_2d + (a+b)^2\sin^2 k_2d}$$

where

$$a = k_2^2 - 3k_1k_2 + k_1^2 \qquad b = k_1^2 + k_1k_2 + k_2^2$$

$$a - b = -4k_1k_2 \qquad a + b = 2k_1^2 + 2k_2^2 - 2k_1k_2 = 2(k_1^2 - k_1k_2 + k_2^2)$$

so

$$\frac{E_{bo}}{E_{ao}} = \frac{(k_2^2 - k_1k_2 - k_1^2)(2i)\sin(k_2d)[(a-b)\cos(k_2d) - i(a+b)\sin(k_2d)]}{(a-b)^2\cos^2 k_2d + (a+b)^2\sin^2 k_2d}$$

Taking the real part

$$\frac{E_{bo}}{E_{ao}} = \frac{(2)(k_2^2 - k_1k_2 - k_1^2)(2)(k_1^2 - k_1k_2 + k_2^2)\sin^2 k_2d}{16k_1^2k_2^2\cos^2 k_2d + 4(k_1^2 - k_1k_2 + k_2^2)^2\sin^2 k_2d}$$

Since

$$k = \frac{\omega n}{c}, \quad k_1 = \frac{\omega}{c}, \quad k_2 = \frac{\omega n}{c}$$

$$\frac{E_{bo}}{E_{ao}} = \frac{\left(\frac{\omega^2n^2}{c^2} - \frac{\omega^2 n}{c^2} - \frac{\omega^2}{c^2}\right)\left(\frac{\omega^2}{c^2} - \frac{\omega^2 n}{c^2} + \frac{\omega^2n^2}{c^2}\right)\sin\frac{\omega n}{c}d}{4\frac{\omega^2}{c^2}\frac{\omega^2n^2}{c^2}\cos^2\frac{\omega n}{c}d + \left(\frac{\omega^2}{c^2} - \frac{\omega^2 n}{c^2} + \frac{\omega^2n^2}{c^2}\right)^2\sin^2\frac{\omega n}{c}d}$$

$$\frac{E_{bo}}{E_{ao}} = \frac{(n^2 - n - 1)(1 - n + n^2)\sin\frac{\omega n}{c}d}{4n^2\cos^2\frac{\omega n}{c}d + (1 - n + n^2)^2\sin^2\frac{\omega n}{c}d}$$

So, if $d = N(\lambda/2)$ when $N = 0,1,2,3,4$, etc., there is no reflected wave and the transmission is 100%.

PROBLEM 13.9 At a boundary between air and a certain kind of glass, 5 1/3 % of normally incident light is reflected. For this light, find

(a) the index of refraction of the glass
(b) the amplitude ratios of the reflected and the transmitted electric fields to the incident electric field.

Solution:

(a) For reflection at normal incidence with $\mu = \mu_0$

$$R = \frac{(\sqrt{\varepsilon_1} - \sqrt{\varepsilon_2})^2}{(\sqrt{\varepsilon_1} + \sqrt{\varepsilon_2})^2}$$

Here $\varepsilon_1 = 8.85 \times 10^{-12}$ and $R = 0.053$. Now to find ε_2.

$$\pm\sqrt{R} = \frac{\sqrt{\varepsilon_1} - \sqrt{\varepsilon_2}}{\sqrt{\varepsilon_1} + \sqrt{\varepsilon_2}} \Rightarrow -\sqrt{R}\sqrt{\varepsilon_1} - \sqrt{R}\sqrt{\varepsilon_2} = \sqrt{\varepsilon_1} - \sqrt{\varepsilon_2}$$

$$-(\sqrt{R}+1)\sqrt{\varepsilon_2} = \sqrt{\varepsilon_1}(1+\sqrt{R})$$

$$\sqrt{\varepsilon_2} = \sqrt{\varepsilon_1}\,\frac{1+\sqrt{R}}{1-\sqrt{R}}$$

$$\mathrm{n} = \frac{1+\sqrt{R}}{1-\sqrt{R}}$$

Where n is the index of refraction of the glass:

$$n = \frac{1+0.232}{1-0.232} = \frac{1.232}{0.768} = 1.604$$

(b) $E_0'/E_0 = \sqrt{R} = 0.232$

$$R + T = 1, \quad T = \frac{n_2}{n_1}\left(\frac{E_0''}{E_0}\right)^2$$

$$T = 0.9467 = 1.6\left(\frac{E_0''}{E_0}\right)^2 \Rightarrow \frac{E_0''}{E_0} = \sqrt{0.592}$$

$$\frac{E_0''}{E_0} = 0.772$$

PROBLEM 13.10 A beam of light with a wavelength 4500 Å (under vacuum) is incident on a prism as shown in the figure and totally reflected through 90°. The index of refraction of the prism is 1.6. Compute the distance beyond the long side of the prism at which the electric field strength is reduced to $1/e$ of its value just at the surface. Assume the light is polarized so that E is perpendicular to the plane of incidence. Is your answer changed if E lies in the plane of incidence?

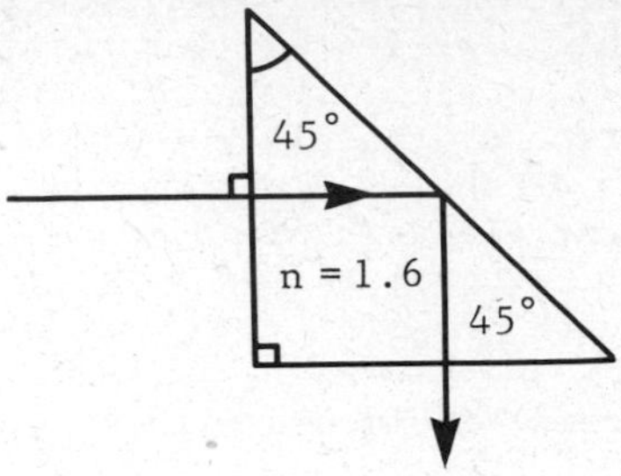

Solution:

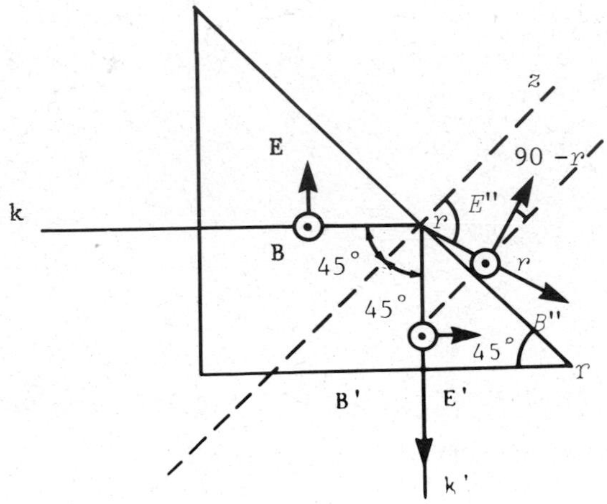

Assume

$$\mathbf{E} = \mathbf{E}_0\, e^{i(\mathbf{k}\cdot\mathbf{r} - \omega t)}, \quad \mathbf{E}'' = \mathbf{E}_0''\, e^{i(\mathbf{k}''\cdot\mathbf{r} - \omega t)}$$

$$\mathbf{E}' = \mathbf{E}_0'\, e^{i(\mathbf{k}'\cdot\mathbf{r} - \omega t)}$$

At the boundary where $z = 0$, $E_{1T} = E_{2T}$,

$$-E_0 e^{i(\mathbf{k}\cdot\mathbf{r} - \omega t)} \sin 45° + E_0' e^{i(\mathbf{k}'\cdot\mathbf{r} - \omega t)} \sin 45°$$

$$= -E_0'' e^{i(\mathbf{k}''\cdot\mathbf{r} - \omega t)} \sin(90° - r)$$

For this to be true at all points along the plane at the boundary where $z = 0$

$$\mathbf{k}\cdot\mathbf{r} = \mathbf{k}'\cdot\mathbf{r} = \mathbf{k}''\cdot\mathbf{r}, \quad \text{where } \mathbf{r} = \hat{\mathbf{i}}x + \hat{\mathbf{j}}y, \ (z = 0)$$

so from this

$$k_x x + k_y y = k_x' x + k_y' y + k_x'' x + k_y'' y$$

But $k_x = 0$ from the initial conditions which implies that k'_x and k''_x are also 0. So, we have

$$k_y = k'_y \qquad k_y = k''_y$$

Now from the initial conditions $k_y = k \sin 45°$. So $k''_y = k \sin 45°$. Also $k''^2 = k''^2_y + k''^2_z$ and since the wave is travelling in air, $k'' = 2\pi/\lambda_a$. Thus,

$$k''^2_z = \frac{4\pi^2}{\lambda_a^2} - k''^2_y = \frac{4\pi^2}{\lambda_a^2} - k^2 \sin^2 45°$$

While the wave is in the glass,

$$k = \frac{\omega n}{c} = \frac{2\pi f n}{c} = \frac{2\pi}{\lambda_a} n$$

So

$$k''^2_z = \frac{4\pi^2}{\lambda_a^2} (1 - n^2 \sin^2 45°)$$

Substituting numbers,

$$k''_y = \frac{2\pi}{\lambda_a} n \cos 45° = \frac{(6.28)(1.6)(0.707)}{4.5 \times 10^{-7}} = 1.58 \times 10^7$$

$$k''_z = \frac{2\pi}{\lambda_a} [1 - (1.6)^2(0.707)^2]^{\frac{1}{2}} = \frac{6.28}{4.5 \times 10^{-7}} (1 - 1.28)^{\frac{1}{2}}$$

$$k''_z = i\ 7.39 \times 10^{+6}$$

So for the transmitted wave

$$\mathbf{E}'' = \mathbf{E}''_0\, e^{i(k''_y y + k''_z z - \omega t)}$$

$$\mathbf{E}'' = \mathbf{E}''_0\, e^{-7.39 \times 10^6\, z}\, e^{i(k''_y y - \omega t)}$$

The transmitted field will fall to $1/e$ of its value at the surface in a distance

$$z_{1/e} = \frac{1}{7.39 \times 10^{+6}} = 1.35 \times 10^{-7} \text{ m}$$

We will now investigate the direction of the field. From the boundary condition on tangential E,

$$-E''_0 \cos r = (-E_0 + E'_0) \sin 45°$$

or

$$\frac{E''_0}{E_0} \frac{\cos r}{\sin 45°} = 1 - \frac{E'_0}{E_0} \qquad \text{(A)}$$

From boundary condition on normal E,

$$E_0 \cos 45° + E_0' \cos 45° = E_0'' \sin r$$

$$1 + \frac{E_0'}{E_0} = \frac{E_0''}{E_0} \frac{\sin r}{\cos 45°} \qquad (B)$$

Adding Eqs. (A) and (B),

$$\frac{E_0''}{E_0} \left(\frac{\cos r}{\sin 45°} + \frac{\sin r}{\cos 45°} \right) = 2$$

$$\frac{E_0''}{E_0} = \frac{2}{\dfrac{\cos r}{\sin 45°} + \dfrac{\sin r}{\cos 45°}}$$

Then

$$\frac{E_0'}{E_0} = \frac{2}{\dfrac{\cos r}{\sin 45°} + \dfrac{\sin r}{\cos 45°}} \frac{\sin r}{\cos 45°} = 1$$

So

$$R = \left(\frac{E_0'}{E_0} \right)^2 = \left(\frac{2 \sin r}{\cos r + \sin r} - 1 \right)^2$$

$$T = \frac{n_2}{n_1} \frac{E_0''}{E_0} = \frac{n_2}{n_1} \frac{4}{2(\cos r + \sin r)^2} = \frac{n_2}{n_1} \frac{2}{(\cos r + \sin r)^2}$$

Since $R + T = 1$

$$\frac{4 \sin^2 r}{(\cos r + \sin r)^2} - \frac{4 \sin r}{\cos r + \sin r} + 1 + \frac{n_2}{n_1} \frac{2}{(\cos r + \sin r)^2} = 1$$

$$\frac{1}{\cos r + \sin r} \left(\frac{2 \sin^2 r + n_2/n_1}{\cos r + \sin r} - \sin r \right) = 0$$

For this to be 0

$$\frac{2 \sin^2 r + n_2/n_1}{\cos r + \sin r} - \sin r = 0$$

or

$$2 \sin^2 r + n_2/n_1 - \cos r \sin r - \sin^2 r = 0$$

or

$$\sin^2 r - \cos r \sin r = -n_2/n_1$$

This transcendental equation should be solved for an r that will work, so there may be a direction in which the wave will go. Try $n_2 = 1$, $n_1 = 1.6$

$$\sin^2 r - \cos r \sin r = -0.625$$

$$\sin r(\sin r - \cos r) = -0.625$$

There are no angles that will work.

PROBLEM 13.11

(a) Show that the one-dimensional wave equation,

$$\frac{\partial^2\psi}{\partial x^2} - \frac{1}{c^2}\frac{\partial^2\psi}{\partial t^2} = 0$$

has the general solution,

$$\psi(x,t) = \tfrac{1}{2} f\left(t - \frac{x}{c}\right) + \tfrac{1}{2} f\left(t + \frac{x}{c}\right) + \frac{c}{2}\int_{t-x/c}^{t+x/c} F(t')dt'$$

where the boundary conditions are specified by the values of ψ and $\partial\psi/\partial x$ at $x = 0$ for all time:

$$\psi(0,t) = f(t), \quad \left.\frac{\partial\psi(x,t)}{\partial x}\right|_{x=0} = F(t)$$

(b) What is the corresponding solution if the boundary conditions are that, at $t = 0$,

$$\psi(x,0) = f(x), \quad \left.\frac{\partial\psi(x,t)}{\partial x}\right|_{t=0} = g(x) \quad ?$$

Solution:

(a) First we show that the function $\psi(x,t)$ satisfies the boundary conditions.

$$\psi(x,t) = \tfrac{1}{2} f\left(t - \frac{x}{c}\right) + \tfrac{1}{2} f\left(t + \frac{x}{c}\right) + \frac{c}{2}\int_{t-x/c}^{t+x/c} F(t')dt'$$

$$\psi(0,t) = \tfrac{1}{2} f(t) + \tfrac{1}{2} f(t) + \frac{c}{2}\int_{t}^{t} F(t')dt' = f(t)$$

$$\frac{\partial\psi(x,t)}{\partial x} = \tfrac{1}{2}\,\frac{\partial f(t-\frac{x}{c})}{\partial(t-\frac{x}{c})}\,\frac{\partial(t-\frac{x}{c})}{\partial x} + \tfrac{1}{2}\,\frac{\partial f(t+\frac{x}{c})}{\partial(t+\frac{x}{c})}\,\frac{\partial(t+\frac{x}{c})}{\partial x} + \frac{c}{2}\,\frac{\partial}{\partial x}\int_{t-x/c}^{t+x/c} F(t')dt'$$

Using the Leibnitz Rule:

$$\frac{d}{dx}\int_{u(x)}^{v(x)} f(t)dt = F(v)\,\frac{dv}{dx} - f(u)\,\frac{du}{dx}$$

$$\frac{\partial\psi(x,t)}{\partial x} = -\frac{1}{2c}\,\frac{\partial f(t-\frac{x}{c})}{\partial(t-\frac{x}{c})} + \frac{1}{2c}\,\frac{\partial f(t+\frac{x}{c})}{\partial(t+\frac{x}{c})} + \frac{c}{2}\,[F\left(t+\frac{x}{c}\right)\frac{1}{c} - F\left(t-\frac{x}{c}\right)\left(-\frac{1}{c}\right)]$$

$$\left.\frac{\partial\psi(x,t)}{\partial x}\right|_{x=0} = -\frac{1}{2c}\,\frac{\partial f}{\partial t} + \frac{1}{2c}\,\frac{\partial f}{\partial t} + \tfrac{1}{2}\,[F(t)+F(t)] = F(t)$$

Now we show that ψ is a solution to the wave equation.

$$\frac{\partial^2\psi(x,t)}{\partial x^2} = +\frac{1}{2c^2}\,\frac{\partial^2 f(t-\frac{x}{c})}{\partial(t-\frac{x}{c})^2} + \frac{1}{2c^2}\,\frac{\partial^2 f(t+\frac{x}{c})}{\partial(t+\frac{x}{c})^2} + \tfrac{1}{2}\left[\frac{\partial F(t+\frac{x}{c})}{\partial(t+\frac{x}{c})}\left(\frac{1}{c}\right) + \frac{\partial F(t-\frac{x}{c})}{\partial(t-\frac{x}{c})}\left(-\frac{1}{c}\right)\right]$$

$$\frac{\partial\psi(x,t)}{\partial t} = \tfrac{1}{2}\,\frac{\partial f(t-\frac{x}{c})}{\partial(t-\frac{x}{c})}\,\frac{\partial(t-\frac{x}{c})}{\partial t} + \tfrac{1}{2}\,\frac{\partial f(t+\frac{x}{c})}{\partial(t+\frac{x}{c})}\,\frac{\partial(t+\frac{x}{c})}{\partial t} + \frac{c}{2}\,\frac{\partial}{\partial t}\int_{t-x/c}^{t+x/c} F(t')dt'$$

$$= \tfrac{1}{2}\,\frac{\partial f(t-\frac{x}{c})}{\partial(t-\frac{x}{c})} + \tfrac{1}{2}\,\frac{\partial f(t+\frac{x}{c})}{\partial(t+\frac{x}{c})} + \frac{c}{2}\,F\left(t+\frac{x}{c}\right) - \frac{c}{2}\,F\left(t-\frac{x}{c}\right)$$

$$\frac{\partial^2\psi(x,t)}{\partial t^2} = \tfrac{1}{2}\,\frac{\partial^2 f(t-\frac{x}{c})}{\partial(t-\frac{x}{c})^2} + \tfrac{1}{2}\,\frac{\partial^2 f(t+\frac{x}{c})}{\partial(t+\frac{x}{c})^2} + \frac{c}{2}\,\frac{\partial F(t+\frac{x}{c})}{\partial(t+\frac{x}{c})} - \frac{c}{2}\,\frac{\partial F(t-\frac{x}{c})}{\partial(t-\frac{x}{c})}$$

$$\frac{\partial^2\psi}{\partial x^2} - \frac{1}{c^2}\,\frac{\partial^2\psi}{\partial t^2} = \frac{1}{2c^2}\left(\frac{\partial^2 f(t-\frac{x}{c})}{\partial(t-\frac{x}{c})^2} + \frac{\partial^2 f(t+\frac{x}{c})}{\partial(t+\frac{x}{c})^2}\right) + \frac{1}{2c}\left(\frac{\partial F(t+\frac{x}{c})}{\partial(t+\frac{x}{c})} - \frac{\partial F(t-\frac{x}{c})}{\partial(t-\frac{x}{c})}\right)$$

$$- \frac{1}{c^2}\left[\tfrac{1}{2}\left(\frac{\partial^2 f(t-\frac{x}{c})}{\partial(t-\frac{x}{c})^2} + \frac{\partial^2 f(t+\frac{x}{c})}{\partial(t+\frac{x}{c})^2}\right)\right] + \frac{c}{2}\left(\frac{\partial F(t+\frac{x}{c})}{\partial(t+\frac{x}{c})} - \frac{\partial F(t-\frac{x}{c})}{\partial(t-\frac{x}{c})}\right)$$

$$= 0$$

(b) By symmetry arguments

$$\psi(x,t) = \tfrac{1}{2} f(x-ct) + \tfrac{1}{2} f(x+ct) + \frac{1}{2c}\int_{x-ct}^{x+ct} g(x')dx'$$

The boundary conditions are satisfied:

$$\psi(x,0) = \tfrac{1}{2} f(x) + \tfrac{1}{2} f(x) + \frac{1}{2c}\int_{x}^{x} g(x')dx' = f(x)$$

$$\left.\frac{\partial\psi(x,t)}{\partial t}\right|_{t=0} = -\frac{c}{2}\left.\frac{\partial f(x-ct)}{\partial(x-ct)}\right|_{x=0} + \frac{c}{2}\frac{\partial f(x+ct)}{\partial(x+ct)}$$

$$+\frac{1}{2c}\left.[g(x+ct)c - g(x-ct)(-c)]\right|_{t=0}$$

$$= g(x)$$

$$\frac{\partial^2\psi(x,t)}{\partial x^2} = \tfrac{1}{2}\frac{\partial^2 f(x-ct)}{\partial(x-ct)^2} + \tfrac{1}{2}\frac{\partial^2 f(x-ct)}{\partial(x-ct)^2} + \frac{1}{2c}\left(\frac{\partial g(x+ct)}{\partial(x+ct)} - \frac{\partial g(x-ct)}{\partial(x-ct)}\right)$$

$$\frac{\partial^2\psi}{\partial t^2} = \frac{c^2}{2}\frac{\partial^2 f(x-ct)}{\partial(x-ct)^2} + \frac{c^2}{2}\frac{\partial^2 f(x+ct)}{\partial(x+ct)^2} + \frac{c}{2}\left(\frac{\partial g(x+ct)}{\partial(x+ct)} - \frac{\partial g(x-ct)}{\partial(x-ct)}\right)$$

$$\frac{\partial^2\psi}{\partial x^2} - \frac{1}{c^2}\frac{\partial^2\psi}{\partial t^2} = \tfrac{1}{2}\frac{\partial^2 f(x-ct)}{\partial(x-ct)^2} + \tfrac{1}{2}\frac{\partial^2 f(x-ct)}{\partial(x-ct)^2} + \frac{1}{2c}\left(\frac{\partial g(x+ct)}{\partial(x+ct)} - \frac{\partial g(x-ct)}{\partial(x-ct)}\right)$$

$$-\frac{1}{c^2}\left(\frac{c^2}{2}\frac{\partial^2 f(x-ct)}{\partial(x-ct)^2} + \frac{c^2}{2}\frac{\partial^2 f(x-ct)}{\partial(x-ct)^2}\right) + \frac{c}{2}\left(\frac{\partial g(x+ct)}{\partial(x+ct)} - \frac{\partial g(x-ct)}{\partial(x-ct)}\right)$$

$$= 0$$

PROBLEM 13.12 A plane wave is incident normally on a perfectly absorbing flat screen.

(a) From the law of conservation of linear momentum show that the pressure (called radiation pressure) exerted on the screen is equal to the field energy per unit volume in the wave.

(b) In the neighborhood of the earth the flux of electromagnetic energy from the sun is approximately 0.14 W/cm^2. If an interplanetary "sailplane" had a sail of mass 10^{-4} g/cm^2 of area and negligible other weight, what would be its maximum acceleration in centimeters per square second due to the solar radiation pressure?

Solution:

(a) From the conservation of linear momentum, we know $\Delta p/\Delta t = F$. Since pressure is defined as force per unit area, we have $\mathbb{P} = \Delta p/\Delta t A$. The energy of a photon is $E = pc$, so $\Delta p = \Delta E/c$, $\mathbb{P} = \Delta E/\Delta t c A$. Since c is the velocity normal to the area A, we can say

$$\Delta t c A = \Delta t \frac{\Delta \ell}{\Delta t} A = \Delta v$$

Finally, $\mathbb{P} = \Delta E/\Delta v$, the radiation pressure is equal to the field energy per unit volume.

(b) $I = 0.14 \dfrac{\text{W}}{\text{cm}^2}$, $\mathbb{P} = \dfrac{I}{c}$

$$a = \frac{F}{m} = \frac{F/A}{m/A} = \frac{\mathbb{P}}{m/A} = \frac{I/c}{m/A} = \frac{(0.14 \ \text{J/cm}^2)(1/3 \times 10^{-8} \ \text{sec/m})}{10^{-7} \ \text{kg/cm}^2}$$

$$= 0.0047 \ \text{m/sec}^2 = 0.47 \ \text{cm/sec}^2$$

PROBLEM 13.13 Geometrical optics is obtained as a very short wave-length approximation in electromagnetic theory. This problem develops a few basic aspects of this limit, using the eikonal approximation.

Consider a medium characterized by the scalar quantities ε and μ and in which there are no free charges or currents. In the medium **E** and **H** are assumed to take the form

$$\mathbf{E}(\mathbf{r},t) = \mathbf{E}_0(\mathbf{r})e^{ik_0 S(\mathbf{r})}e^{-i\omega t}$$

and

$$\mathbf{H}(\mathbf{r},t) = \mathbf{H}_0(\mathbf{r})e^{ik_0 S(\mathbf{r})}e^{-i\omega t}$$

where $k_0 = \omega/c$, and $S(\mathbf{r})$ is the eikonal or "optical path"; $S(r) = \text{constant}$ defines geometrical wave surfaces or wave fronts.

(a) Using Maxwell's equations for **D**, **E**, **B**, and **H** in a medium, show that

$$\sum_{i=1}^{3} \left(\frac{\partial S}{\partial x_i}\right)^2 = n^2, \quad n = \sqrt{\mu\varepsilon}$$

and discuss the approximation involved. In particular, assume that the radiation corresponds to visible light and derive a reasonable numerical limit for the validity of this "eikonal approximation."

(b) Show that under the eikonal approximation, the time average of the electric and the magnetic field energy densitites ($\langle w_e \rangle$ and $\langle w_m \rangle$) are proportional to ∇S.

(c) Show that the time average of the Poynting vector **S** can be put into the form

$$\langle \mathbf{S} \rangle = v \langle w_e + w_m \rangle \hat{\mathbf{u}}$$

and determine the speed v and the unit vector $\hat{\mathbf{u}}$.

Solution:

(a)

$$\mathbf{E}(\mathbf{r},t) = \mathbf{E}_0(\mathbf{r})e^{ik_0S(\mathbf{r})}e^{-i\omega t} = \mathbf{e}(\mathbf{r})e^{-i\omega t}, \qquad \mathbf{H}(\mathbf{r},t) = \mathbf{H}_0(\mathbf{r})e^{ik_0S(\mathbf{r})}e^{-i\omega t} = \mathbf{h}(\mathbf{r})e^{-i\omega t},$$

$$(0)\left\{\begin{aligned} \mathbf{e}(\mathbf{r}) &= \mathbf{E}_0(\mathbf{r})e^{ik_0S(\mathbf{r})} \\ \mathbf{h}(\mathbf{r}) &= \mathbf{H}_0(\mathbf{r})e^{ik_0S(\mathbf{r})} \end{aligned}\right.$$

$$\left.\begin{aligned} &\nabla \times \mathbf{H} - \frac{1}{c}\frac{\partial \mathbf{D}}{\partial t} = \frac{4\pi}{c}\mathbf{j} \\ &\nabla \times \mathbf{E} + \frac{1}{c}\frac{\partial \mathbf{B}}{\partial t} = 0 \\ &\nabla \cdot \mathbf{D} = 4\pi\rho \\ &\nabla \cdot \mathbf{B} = 0 \end{aligned}\right\} \qquad (1)$$

$$\nabla \times \mathbf{h} = \nabla \times \mathbf{H}_0 e^{ik_0S(\mathbf{r})} = [(\nabla \times \mathbf{H}_0)e^{ik_0S(\mathbf{r})} + (\nabla e^{ik_0S(\mathbf{r})}) \times \mathbf{H}_0]$$

$$= \{(\nabla \times \mathbf{H}_0)e^{ik_0S(\mathbf{r})} + ik_0e^{ik_0S(\mathbf{r})}[\nabla S(\mathbf{r})] \times \mathbf{H}_0\}$$

$$\nabla \times \mathbf{h} = [\nabla \times \mathbf{H}_0 + ik_0(\nabla S) \times \mathbf{H}_0]e^{ik_0S(\mathbf{r})}$$

$$\nabla \times \mathbf{e} = \nabla \times \mathbf{E}_0 e^{ik_0S(\mathbf{r})} = [(\nabla \times \mathbf{E}_0)e^{ik_0S(\mathbf{r})} + (\nabla e^{ik_0S(\mathbf{r})}) \times \mathbf{E}_0]$$

$$= \{(\nabla \times \mathbf{E}_0)e^{ik_0S(\mathbf{r})} + ik_0e^{ik_0S(\mathbf{r})}[\nabla S(\mathbf{r})] \times \mathbf{E}_0\}$$

$$\nabla \times \mathbf{e} = [(\nabla \times \mathbf{E}_0) + ik_0(\nabla S) \times \mathbf{E}_0]e^{ik_0S(\mathbf{r})}$$

$$\nabla \cdot \mathbf{D} = \nabla \cdot \varepsilon\mathbf{e} = \nabla \cdot \varepsilon\, \mathbf{E}_0 e^{ik_0S(\mathbf{r})} = \{\varepsilon\nabla \cdot \mathbf{E}_0 + \mathbf{E}_0 \cdot (\nabla\varepsilon) + [\nabla ik_0S] \cdot \varepsilon\mathbf{E}_0\}e^{ik_0S(\mathbf{r})}$$

$$\nabla \cdot \mathbf{D} = \nabla \cdot \varepsilon\mathbf{e} = [\varepsilon\nabla \cdot \mathbf{E}_0 + \mathbf{E}_0 \cdot (\nabla\varepsilon) + ik_0\varepsilon\, \mathbf{E}_0 \cdot (\nabla S)]e^{ik_0S(\mathbf{r})}$$

$$\nabla \cdot \mathbf{B} = \nabla \cdot \mu\mathbf{h} = \nabla \cdot \mu\, \mathbf{H}_0 e^{ik_0S(\mathbf{r})} = \{\mu\nabla \cdot \mathbf{H}_0 + \mathbf{H}_0 \cdot (\nabla\mu) + [\nabla ik_0S(\mathbf{r})] \cdot \mu\mathbf{H}_0\}e^{ik_0S(\mathbf{r})}$$

$$\nabla \cdot \mathbf{B} = \nabla \cdot \mu\mathbf{H}_0 = [\mu\nabla \cdot \mathbf{H}_0 + \mathbf{H}_0 \cdot (\nabla\mu) + ik_0\mu\, \mathbf{H}_0 \cdot (\nabla S)]e^{ik_0S(\mathbf{r})}$$

$$\left.\begin{aligned}
\nabla \times \mathbf{h} &= [(\nabla \times \mathbf{H}_0) + ik_0(\nabla S) \times \mathbf{H}_0]e^{ik_0S(\mathbf{r})} \\
\nabla \times \mathbf{e} &= [(\nabla \times \mathbf{E}_0) + ik_0(\nabla S) \times \mathbf{E}_0]e^{ik_0S(\mathbf{r})} \\
\nabla \cdot \mathbf{D} &= \nabla \cdot \varepsilon \mathbf{e} = [\varepsilon \nabla \cdot \mathbf{E}_0 + \mathbf{E}_0 \cdot (\nabla \varepsilon) + ik_0 \varepsilon\, \mathbf{E}_0 \cdot (\nabla S)]e^{ik_0S(\mathbf{r})} \\
\nabla \cdot \mathbf{B} &= \nabla \cdot \mu \mathbf{H}_0 = [\mu \nabla \cdot \mathbf{H}_0 + \mathbf{H}_0 \cdot (\nabla \mu) + ik_0 \mu\, \mathbf{H}_0 \cdot (\nabla S)]e^{ik_0S(\mathbf{r})}
\end{aligned}\right\} \quad (2)$$

$$\mathbf{E}(\mathbf{r},t) = \mathbf{e}(\mathbf{r})e^{-i\omega t}$$

$$\mathbf{H}(\mathbf{r},t) = \mathbf{h}(\mathbf{r})e^{-i\omega t}$$

Substute in Eq. (1),

$$\nabla \times \mathbf{H} - \frac{1}{c}\frac{\partial \mathbf{D}}{\partial t} = \nabla \times [\mathbf{h}(\mathbf{r})e^{-i\omega t}] = [\nabla \times \mathbf{h}(\mathbf{r})]e^{-i\omega t} + \frac{i\omega\varepsilon}{c}\,\mathbf{e}\,e^{-i\omega t}$$

$$\nabla \times \mathbf{H} - \frac{1}{c}\frac{\partial \mathbf{D}}{\partial t} = [\nabla \times \mathbf{h}(\mathbf{r}) + i(\omega/c)\,\varepsilon \mathbf{e}]e^{-i\omega t} = \frac{4\pi}{c} j,$$

$$j = 0, \quad k_0 = \frac{\omega}{c}, \quad \mathbf{D} = \varepsilon \mathbf{E}$$

$$\nabla \times \mathbf{h}(\mathbf{r}) + ik_0\varepsilon\, \mathbf{e} = 0$$

$$\nabla \times \mathbf{E} + \frac{1}{c}\frac{\partial \mathbf{B}}{\partial t} = \nabla \times [\mathbf{e}(\mathbf{r})e^{-i\omega t}] + \frac{\mu}{c}\frac{\partial}{\partial t}[\mathbf{h}e^{-i\omega t}], \quad \mathbf{B} = \mu \mathbf{h}$$

$$= [\nabla \times \mathbf{e}(\mathbf{r})]e^{-i\omega t} + \frac{\mu}{c}\,\mathbf{h}e^{-i\omega t}(-i\omega)$$

$$= [\nabla \times \mathbf{e}(\mathbf{r}) - i(\omega/c)\,\mu\, \mathbf{h}]e^{-i\omega t} = 0$$

$$\nabla \times \mathbf{e}(\mathbf{r}) - ik_0\mu\, \mathbf{h} = 0$$

$$\nabla \cdot \mathbf{D} = \nabla \cdot \varepsilon \mathbf{E} = \nabla \cdot [\varepsilon \mathbf{e}(\mathbf{r})e^{-i\omega t}] = (\nabla \cdot \varepsilon \mathbf{e})e^{-i\omega t} = 4\pi\rho = 0, \quad \rho = 0$$

$$\nabla \cdot \varepsilon \mathbf{e}(\mathbf{r}) = 0$$

$$\nabla \cdot \mathbf{B} = \nabla \cdot \mu \mathbf{H} = \nabla \cdot \mu \mathbf{h}(\mathbf{r})e^{-i\omega t} = (\nabla \cdot \mu \mathbf{h})e^{-i\omega t} = 0$$

$$\nabla \cdot \mu \mathbf{h} = 0$$

$$\left.\begin{aligned} &\nabla \times \mathbf{h}(\mathbf{r}) + ik_0\varepsilon\mathbf{e} = 0 \\ &\nabla \times \mathbf{e}(\mathbf{r}) - ik_0\mu\mathbf{h} = 0 \\ &\nabla \cdot \varepsilon\mathbf{e}(\mathbf{r}) = 0 \\ &\nabla \cdot \mu\mathbf{h}(\mathbf{r}) = 0 \end{aligned}\right\} \quad (3)$$

Substitute Eq. (2) into Eq. (3),

$$\nabla \times \mathbf{h} + ik_0\varepsilon\mathbf{e} = 0$$

$$[\nabla \times \mathbf{H}_0 + ik_0(\nabla S) \times \mathbf{H}_0]e^{ik_0S} + ik_0\varepsilon\ \mathbf{E}_0e^{ik_0S} = 0$$

$$\nabla \times \mathbf{H}_0 + ik_0(\nabla S) \times \mathbf{H}_0 + ik\ \varepsilon\ \mathbf{E}_0 = 0$$

divide both sides by ik_0,

$$(\nabla S) \times \mathbf{H}_0 + \varepsilon\mathbf{E}_0 = -\frac{1}{ik_0}\nabla \times \mathbf{H}_0$$

$$\nabla \times \mathbf{e} - ik_0\mu\mathbf{h} = 0$$

$$[\nabla \times \mathbf{E}_0 + ik_0(\nabla S) \times \mathbf{E}_0]e^{ik_0S} - ik_0\mu\mathbf{H}_0e^{-ik_0S} = 0$$

$$\nabla \times \mathbf{E}_0 + ik_0(\nabla S) \times \mathbf{E}_0 - ik_0\mu\mathbf{H}_0 = 0$$

$$(\nabla S) \times \mathbf{E}_0 - \mu\mathbf{H}_0 = -\frac{1}{ik_0}\nabla \times \mathbf{E}_0$$

$$\nabla \cdot \varepsilon\mathbf{e} = 0$$

$$[\varepsilon(\nabla \cdot \mathbf{E}_0) + \mathbf{E}_0 \cdot (\nabla\varepsilon) + ik_0\varepsilon\ \mathbf{E}_0 \cdot (\nabla S)]e^{ik_0S} = 0$$

$$\frac{\varepsilon(\nabla \cdot \mathbf{E}_0)}{\varepsilon ik_0} + \frac{\mathbf{E}_0 \cdot (\nabla\varepsilon)}{\varepsilon ik_0} + \frac{ik_0\varepsilon}{\varepsilon ik_0}\ [\mathbf{E}_0 \cdot (\nabla S)] = 0$$

$$\mathbf{E}_0 \cdot (\nabla S) = -\frac{\varepsilon(\nabla \cdot \mathbf{E}_0)}{ik_0\varepsilon} - \frac{\mathbf{E}_0 \cdot (\nabla\varepsilon)}{ik_0\varepsilon}$$

$$\mathbf{E}_0 \cdot (\nabla S) = -\frac{1}{ik_0}\left(\nabla \cdot \mathbf{E}_0 + \mathbf{E}_0 \cdot \frac{\nabla\varepsilon}{\varepsilon}\right)$$

$$\nabla \cdot \mu\mathbf{H}_0 = 0$$

$$\left[\frac{\mu(\nabla\cdot\mathbf{H}_0)}{ik_0\mu}+\frac{\mathbf{H}_0\cdot\nabla\mu}{ik_0\mu}=\frac{ik_0\mu\mathbf{H}_0}{ik_0\mu}\cdot(\nabla S)\right]=0$$

$$\mathbf{H}_0\cdot\nabla S=-\frac{1}{ik_0}\left(\nabla\cdot\mathbf{H}_0+\mathbf{H}_0\cdot\frac{\nabla\mu}{\mu}\right)$$

$$\left.\begin{aligned}
&(\nabla S)\times\mathbf{H}_0+\varepsilon\mathbf{E}_0=-\frac{1}{ik_0}\nabla\times\mathbf{H}_0\\
&(\nabla S)\times\mathbf{E}_0-\mu\mathbf{H}_0=-\frac{1}{ik_0}\nabla\times\mathbf{E}_0\\
&\mathbf{E}_0\cdot(\nabla S)=-\frac{1}{ik_0}\left(\nabla\cdot\mathbf{E}_0+\mathbf{E}_0\cdot\frac{\nabla\varepsilon}{\varepsilon}\right)\\
&\mathbf{H}_0\cdot(\nabla S)=-\frac{1}{ik_0}\left(\nabla\cdot\mathbf{H}_0+\mathbf{H}_0\cdot\frac{\nabla\mu}{\mu}\right)
\end{aligned}\right\}\qquad(4)$$

Using the short-wavelength approximation, λ_0 is very small, $k_0=2\pi/\lambda_0$ is very large, and $1/ik_0$ is very small, so in this approximation we can neglect the terms in Eq. (4) which are preceded by $1/k_0$ (or $-1/ik_0$), then,

$$(\nabla S)\times\mathbf{H}_0+\varepsilon\mathbf{E}_0=0\qquad(5a)$$

$$(\nabla S)\times\mathbf{E}_0-\mu\mathbf{H}_0=0\qquad(5b)$$

$$\mathbf{E}_0\cdot(\nabla S)=0\qquad(5c)$$

$$\mathbf{H}_0\cdot(\nabla S)=0\qquad(5d)$$

from Eq. (5b),

$$\mathbf{H}_0=\frac{1}{\mu}[(\nabla S)\cdot\mathbf{E}_0]$$

Substituting into Eq. (5a),

$$(\nabla S)\times\mathbf{H}_0+\varepsilon\mathbf{E}_0=0$$

$$(\nabla S)\times\frac{1}{\mu}[(\nabla S)\times\mathbf{E}_0]+\varepsilon\mathbf{E}_0=0$$

$$\frac{1}{\mu}[(\nabla S\cdot\mathbf{E}_0)\nabla S-[(\nabla S)\cdot(\nabla S)\mathbf{E}_0]+\varepsilon\mathbf{E}_0=0$$

$$\frac{1}{\mu}[(\nabla S\cdot\mathbf{E}_0)\nabla S-(\nabla S)^2\mathbf{E}_0]+\varepsilon\mathbf{E}_0=0$$

where $\nabla S\cdot\mathbf{E}_0=0$ by Eq. (5c).

$$\frac{-(\nabla S)^2\mathbf{E}_0}{\mu}+\varepsilon\mathbf{E}_0=0$$

$$\frac{-(\nabla S)^2}{\mu} + \varepsilon = 0$$

$$(\nabla S)^2 = \varepsilon\mu$$

letting $n = \sqrt{\varepsilon\mu}$

$$(\nabla S)^2 = n^2 = \left(\frac{\partial S}{\partial x}\right)^2 + \left(\frac{\partial S}{\partial y}\right)^2 + \left(\frac{\partial S}{\partial z}\right)^2$$

$$\sum_{i=1}^{3} \left(\frac{\partial S}{\partial x_i}\right)^2 = n^2, \quad n = \sqrt{\varepsilon\mu}$$

The approximations are valid for

$$\lambda \simeq 5000\ \text{Å}$$

$$k_0 \simeq 1.26 \times 10^7\ \text{m}^{-1}$$

since $k_0 r >> 1$,

$$r \gtrsim 10^{-5}\ \text{m}$$

(b)

$$\left.\begin{aligned} \mathbf{E}(\mathbf{r},t) &= \text{Re}[\mathbf{e}(\mathbf{r})e^{-i\omega t}] = \tfrac{1}{2}[\mathbf{e}(\mathbf{r})e^{-i\omega t} + \mathbf{e}^*(\mathbf{r})e^{i\omega t}] \\ \mathbf{H}(\mathbf{r},t) &= \text{Re}[\mathbf{h}(\mathbf{r})e^{-i\omega t}] = \tfrac{1}{2}[\mathbf{h}(\mathbf{r})e^{-i\omega t} + \mathbf{h}^*(\mathbf{r})e^{-i\omega t}] \end{aligned}\right\} \quad (6)$$

$$\langle\omega_\varepsilon\rangle = \frac{1}{2t_1}\int_{-T}^{+T} \frac{\varepsilon}{8\pi}\mathbf{E}^2 dt = \frac{\varepsilon}{8\pi 2t_1}\int_{-T}^{+T} \tfrac{1}{4}(\mathbf{e}^2 e^{-2i\omega t} + 2\mathbf{e}\cdot\mathbf{e}^* + \mathbf{e}^{*2}e^{2i\omega t})$$

$$\frac{1}{2t_1}\int_{-T}^{+T} e^{-2i\omega t} = \frac{-1}{4i\omega t_1}(e^{-2i\omega t})\Big|_{-t}^{t} = -\frac{1}{4i\omega t_1}\frac{2}{2}(e^{-2i\omega T} - e^{+2i\omega T})$$

$$= \frac{1}{2\omega t_1}\sin(2\omega T)$$

We are averaging over a time $-t_1 < t < t_1$. This interval $2t_1$ over which the electric energy density is averaged is much greater than the fundamental period T, and so this term approaches 0 and therefore, the terms with $e^{-2i\omega t}$ and $e^{-2\omega t}$ may be neglected. We are then left with,

$$\langle\omega_\varepsilon\rangle = \frac{\varepsilon}{8\pi 2t}\int_{-t_1}^{+t_1} \tfrac{1}{4}(\mathbf{e}\cdot\mathbf{e}^*)dt = \frac{\varepsilon}{16\pi}(\mathbf{e}\cdot\mathbf{e}^*)$$

$$<\omega_\varepsilon> = \frac{\varepsilon}{16\pi} (\mathbf{e} \cdot \mathbf{e}^*) \tag{7}$$

Since E and H have exactly the same term, then going through the same procedure exactly, except now substituting an **h** everywhere there is an **e**, and a μ for every ε, will yield:

$$<\omega_m> = \frac{\mu}{16\pi} \mathbf{h} \cdot \mathbf{h}^* \tag{8}$$

from Eqs. (0), (7), and (8),

$$<\omega_\varepsilon> = \frac{\varepsilon}{16\pi} (\mathbf{e} \cdot \mathbf{e}^*) = \frac{\varepsilon}{16\pi} (\mathbf{E}_0 e^{ik_0S}) \cdot (\mathbf{E}_0^* e^{-ik_0S}) = \frac{\varepsilon}{16\pi} \mathbf{E}_0 \cdot \mathbf{E}_0^*$$

$$<\omega_m> = \frac{\mu}{16\pi} \mathbf{H}_0 \cdot \mathbf{H}_0^*$$

From Eq. (5a),

$$(\nabla S) \times \mathbf{H}_0 + \varepsilon \mathbf{E}_0 = 0$$

$$\mathbf{E}_0 = \frac{1}{\varepsilon} [(\nabla S) \times \mathbf{H}_0]$$

and

$$<\omega_\varepsilon> = \frac{\varepsilon}{16\pi} \frac{1}{\varepsilon^2} [(\nabla S) \times \mathbf{H}_0] \cdot [(\nabla S) \times \mathbf{H}_0]^* = \frac{1}{16\pi\varepsilon} (\nabla S)^2 [\mathbf{H}_0 \cdot \mathbf{H}_0^*]$$

where $(\nabla S)^2 = n^2 = \varepsilon\mu$.

$$<\omega_\varepsilon> = \frac{\varepsilon\mu}{\varepsilon 16\pi} (\mathbf{H}_0 \cdot \mathbf{H}_0^*)$$

$$<\omega_\varepsilon> = \frac{\mu}{16\pi} (\mathbf{H}_0 \cdot \mathbf{H}_0^*) = \frac{1}{16\pi\varepsilon} (\mathbf{H}_0 \cdot \mathbf{H}_0^*)(\nabla S)$$

or, from Eq. (5b),

$$(\nabla S) \times \mathbf{E}_0 - \mu \mathbf{H}_0 = 0$$

$$\mathbf{H}_0 = \frac{1}{\mu} [(\nabla S) \times \mathbf{E}_0]$$

$$<\omega_m> = \frac{\mu}{16\pi\mu^2} [(\nabla S) \times \mathbf{E}_0] \cdot [(\nabla S) \times \mathbf{E}_0]^* = \frac{1}{16\pi\mu} (\nabla S)^2 [\mathbf{E}_0 \cdot \mathbf{E}_0^*]$$

where $(\nabla S)^2 = n^2 = \varepsilon\mu$.

$$\langle\omega_m\rangle = \frac{\varepsilon\mu}{16\pi\mu}\,(\mathbf{E}_0 \cdot \mathbf{E}_0^*)$$

$$\langle\omega_m\rangle = \frac{\mu}{16\pi}\,(\mathbf{E}_0 \cdot \mathbf{E}_0^*) = \frac{1}{16\pi\mu}\,(\mathbf{E}_0 \cdot \mathbf{E}_0^*)(\nabla S)^2$$

So

$$\langle\omega_e\rangle = \langle\omega_m\rangle = \frac{\mu}{16\pi}\,(\mathbf{H}_0 \cdot \mathbf{H}_0^*) = \frac{\varepsilon}{16\pi}\,(\mathbf{E}_0 \cdot \mathbf{E}_0^*) = \left(\frac{1}{16\pi\varepsilon}\,(\mathbf{H}_0 \cdot \mathbf{H}_0^*)\right)(\nabla S)^2$$

$$= \frac{1}{16\pi\mu}\,(\mathbf{E}_0 \cdot \mathbf{E}_0^*)(\nabla S)^2$$

$$\therefore \quad \langle\omega_e\rangle = \langle\omega_m\rangle \propto \nabla S$$

(c) $$\langle \mathbf{S}\rangle = \frac{1}{2t_1}\int_{-T}^{T} \frac{c}{4\pi}\,(\mathbf{e}\times\mathbf{h})\,dt$$

$$= \frac{c}{2t_1(4\pi)}\int_{-T}^{T} \tfrac{1}{4}\,(\mathbf{e}\times\mathbf{h}e^{-2i\omega t} + \mathbf{e}\times\mathbf{h}^* + \mathbf{e}^*\times\mathbf{h} + \mathbf{e}^*\times\mathbf{h}^* e^{2i\omega t})\,dt$$

Using the same approximation as in part (b), neglecting terms with $e^{-2i\omega t}$ and $e^{2i\omega t}$,

$$\langle \mathbf{S}\rangle = \frac{c}{16\pi}\,(\mathbf{e}\times\mathbf{h}^* + \mathbf{e}^*\times\mathbf{h})$$

$$\langle \mathbf{S}\rangle = \frac{c}{8\pi}\,\mathrm{Re}(\mathbf{e}\times\mathbf{h}^*) \tag{9}$$

From Eqs. (0) and (9),

$$\langle \mathbf{S}\rangle = \frac{c}{8\pi}\,\mathrm{Re}(\mathbf{e}\times\mathbf{h}^*)$$

$$\mathbf{e}(\mathbf{r}) = \mathbf{E}_0(\mathbf{r})e^{ik_0 S(\mathbf{r})}, \quad \mathbf{h}(\mathbf{r}) = \mathbf{H}_0(\mathbf{r})e^{ik_0 S(\mathbf{r})}$$

$$\langle \mathbf{S}\rangle = \frac{c}{8\pi}\,\mathrm{Re}(\mathbf{E}_0 e^{ik_0 S(\mathbf{r})} \times \mathbf{H}_0 e^{ik_0 S(\mathbf{r})})$$

$$\langle \mathbf{S}\rangle = \frac{c}{8\pi}\,\mathrm{Re}(\mathbf{E}_0 \times \mathbf{H}_0^*)$$

From Eq. (5b),

$$(\nabla S)\times\mathbf{E}_0 - \mu\mathbf{H}_0 = 0$$

$$H_0 = \frac{1}{\mu}\,[(\nabla S) \times E_0]$$

$$<S> = \frac{c}{8\pi}\,\mathrm{Re}\left\{E_0 \times \frac{1}{\mu}\,[(\nabla S) \times E_0^*]\right\} = \frac{c}{8\pi\mu}\,\mathrm{Re}\{E_0 \times [(\nabla S) \times E_0^*]\}$$

now using $\mathbf{a} \times (\mathbf{b} \times \mathbf{c}) = (\mathbf{a} \cdot \mathbf{c})\mathbf{b} - (\mathbf{a} \cdot \mathbf{b})\mathbf{c}$

$$<S> = \frac{c}{8\pi\mu}\,\mathrm{Re}\{(E_0 \cdot E_0^*)\nabla S - [E_0 \cdot (\nabla S)]E_0^*\}$$

where $E_0 \cdot (\nabla S) = 0$ from Eq. (5c).

$$<S> = \frac{c}{8\pi\mu}\,\mathrm{Re}[(E_0 \cdot E_0^*)\nabla S]$$

$$\left(<\omega_e> = \frac{\varepsilon}{16\pi}\,(E_0 \cdot E_0^*), \quad n^2 = \varepsilon\mu\right)$$

$$<S> = \frac{c}{8\pi\mu}\,\mathrm{Re}\left(\frac{<\omega_\varepsilon>}{\varepsilon/16\pi}\,\nabla S\right) = \frac{c16\pi(2)}{8\pi\mu\varepsilon}\,(<\omega_\varepsilon>\nabla S) = \frac{2c}{n^2}\,<\omega_e>\nabla S$$

now, $<\omega_e> = <\omega_m>$, $\quad <\omega> = <\omega_e> + <\omega_m> = 2<\omega_e>$

$$<\nabla S>^2 = n^2 \Rightarrow \left(\frac{\nabla S}{n}\right)^2 = 1 \Rightarrow \frac{\nabla S}{n} = \text{a unit vector } \hat{\mathbf{u}}$$

$$<S> = \frac{2c}{n^2}\,<\omega_e>\nabla S = 2<\omega_e>\,\frac{c}{n}\,\frac{\nabla S}{n} = <\omega>\,\frac{c}{n}\,\frac{\nabla S}{n}, \qquad n = |\nabla S|$$

$$\frac{c}{n} = v$$

$$<S> = <\omega_e + \omega_m>\,v\,\frac{\nabla S}{|\nabla S|}$$

where

$$\frac{\nabla S}{|\nabla S|} = \hat{\mathbf{u}} = \text{a unit vector in the direction of } <S>.$$

$$<S> = v<\omega_e + \omega_m>\hat{\mathbf{u}}$$

$$v = \frac{c}{n}, \qquad \hat{\mathbf{u}} = \frac{\nabla S}{|\nabla S|} = \frac{\nabla S}{n}$$

PROBLEM 13.14 A circularly polarized plane wave moving in the z direction has a finite extent in the x and y directions. Assuming that the amplitude modulation is slowly varying (the wave is many wavelengths broad), show that the electric and magnetic fields are given approximately by

$$\mathbf{E}(x,y,z,t) \simeq \left[E_0(x,y)(\hat{\mathbf{e}}_1 \pm i\hat{\mathbf{e}}_2) + \frac{i}{k}\left(\frac{\partial E_0}{\partial x} \pm \frac{\partial E_0}{\partial y}\right)\hat{\mathbf{e}}_3\right] e^{ikz - i\omega t}$$

$$\mathbf{B} \simeq \mp i\sqrt{\mu\varepsilon}\ \mathbf{E}$$

where $\hat{\mathbf{e}}_1$, $\hat{\mathbf{e}}_2$, $\hat{\mathbf{e}}_3$ are unit vectors in the x, y, z directions.

Solution: The unmodulated waves are given by

$$\mathbf{E}(x,y) = E_0(x,y)(\hat{\mathbf{e}}_1 \pm i\hat{\mathbf{e}}_2)e^{i(kz - \omega t)}$$

$$\mathbf{B}(x,y) = \sqrt{\varepsilon\mu}\ \hat{\mathbf{z}} \times \mathbf{B} = \sqrt{\varepsilon\mu}\ E_0(x,y)(\hat{\mathbf{e}}_2 \mp i\hat{\mathbf{e}}_1)e^{i(kz - \omega t)}$$

Since the wave packet is modulated, the varying amplitude of B and E in the $\hat{\mathbf{x}}$ and $\hat{\mathbf{y}}$ direction will give some resultant B and E field in the $\hat{\mathbf{z}}$ direction. For a slowly varying field, this can be approximated by

$$\mathbf{E}(x,y,z,t) = E_0(x,y)(\hat{\mathbf{e}}_1 \pm i\hat{\mathbf{e}}_2)e^{i(kz - \omega t)} + \int_0^t \frac{\partial \mathbf{E}(x,y)}{\partial t}\, dt$$

$$\mathbf{B}(x,y,z,t) = B_0(x,y)(\hat{\mathbf{e}}_2 \mp i\hat{\mathbf{e}}_1)e^{i(kz - \omega t)} + \int_0^t \frac{\partial \mathbf{B}(x,y)}{\partial t}\, dt$$

Substituting

$$\frac{\partial \mathbf{E}}{\partial t} = \frac{c}{\mu\varepsilon}\ (\nabla \times \mathbf{B})$$

$$= \frac{c}{\mu\varepsilon}\left(\frac{\partial B_0}{\partial x}\ \hat{\mathbf{e}}_3 \pm \frac{i\partial B_0}{\partial y}\ \hat{\mathbf{e}}_3\right)e^{i(kz - \omega t)}$$

$$\frac{\partial \mathbf{B}}{\partial t} = -c\ (\nabla \times \mathbf{E}) = -c\left(\frac{-\partial E_0}{\partial y}\ \hat{\mathbf{e}}_3 \pm \frac{i\partial E_0}{\partial x}\ \hat{\mathbf{e}}_3\right)e^{i(kz - \omega t)}$$

Now, since $B_0 = \sqrt{\varepsilon\mu}\ E_0$ and

$$c\int_0^t e^{i(kz - \omega t)} = \frac{i\sqrt{\varepsilon\mu}}{k}\ e^{i(kz - \omega t)}$$

We have

$$\mathbf{E}(x,y,z,t) = \left[E_0(\hat{\mathbf{e}}_1 \pm i\hat{\mathbf{e}}_2) + \frac{i}{k}\left(\frac{\partial E_0}{\partial x} \pm \frac{i\partial E_0}{\partial y}\right)\ \hat{\mathbf{e}}_3\right]e^{i(kz - \omega t)}$$

$$\mathbf{B}(x,y,z,t) = \left[B_0(\hat{\mathbf{e}}_2 \mp i\hat{\mathbf{e}}_1) + \frac{i}{k}\left(\frac{\partial B_0}{\partial y} \mp \frac{i\partial B_0}{\partial x}\right)\hat{\mathbf{e}}_3\right] e^{i(kz - \omega t)}$$

$$= \mp i\sqrt{\varepsilon\mu}\left[E_0(\hat{\mathbf{e}}_1 \pm \hat{\mathbf{e}}_2) + \frac{i}{k}\left(\frac{\partial E_0}{\partial x} \pm \frac{i\partial E_0}{\partial y}\right)\hat{\mathbf{e}}_3\, e^{i(kz - \omega t)}\right]$$

$$= \mp i\sqrt{\varepsilon\mu}\,\mathbf{E}(x,y,z,t)$$

PROBLEM 13.15 An approximately monochromatic plane-wave packet in one dimension has the instantaneous form, $u(x,0) = f(x)e^{ik_0 x}$, with $f(x)$ the modulation envelope. For each of the forms $f(x)$ below, calculate the wave-number spectrum $|A(k)|^2$ of the packet, evaluate explicitly the rms deviations from the means, Δx and Δk (defined in terms of the intensities $|u(x,0)|^2$ and $|A(k)|^2$), and test the inequality $\Delta x\, \Delta k \geq \frac{1}{2}$

(a) $f(x) = Ne^{-\alpha|x|/2}$

(b) $f(x) = Ne^{-\alpha^2x^2/4}$

(c) $f(x) = \begin{cases} N(1 - \alpha|x|), & \text{for } \alpha|x| < 1 \\ 0, & \text{for } \alpha|x| > 1 \end{cases}$

(d) $f(x) = \begin{cases} N, & \text{for } |x| \leq a \\ 0, & \text{for } |x| \geq a \end{cases}$

Solution:

(a) $$A(k) = \frac{N}{\sqrt{2\pi}}\int_{-\infty}^{+\infty} e^{-\alpha|x|/2}\, e^{i(k_0 - k)x}\,dx$$

$$= \frac{N}{\sqrt{2\pi}}\int_0^{\infty} e^{-\alpha x/2}\, e^{i(k_0 - k)x}\,dx + \frac{N}{\sqrt{2\pi}}\int_{-\infty}^{0} e^{\alpha x/2}\, e^{i(k_0 - k)x}\,dx$$

$$= \frac{N}{\sqrt{2\pi}}\int_0^{\infty} e^{-\alpha x/2}\, e^{i(k_0 - k)x}\,dx + \frac{N}{\sqrt{2\pi}}\int_0^{\infty} e^{-\alpha x/2}\, e^{-i(k - k_0)x}\,dx$$

$$= \frac{N}{\sqrt{2\pi}}\, 2\int_0^{\infty} e^{-\alpha x/2} \cos\,(k_0 - k)x$$

$$= \frac{2N}{\sqrt{2\pi}}\, \frac{\alpha/2}{(\alpha/2)^2 + (k_0 - k)^2}$$

$$= \frac{N}{\sqrt{2\pi}} \frac{\alpha}{(\alpha/2)^2 + (k_0 - k)^2}$$

$$|A(k)|^2 = \frac{N^2}{2\pi} \frac{\alpha^2}{[(k_0 - k)^2 + \alpha^2/4]^2}$$

To find N,

$$\int \mu^*(x)\mu(x)dx = 1$$

and

$$\int A(k)^* A(k)dk = 1$$

$$2N^2 \int_0^\infty e^{-\alpha x} dx = 2N^2 \frac{1}{\alpha} = 1, \quad N = \sqrt{\alpha/2}$$

$$\frac{N^2}{2\pi} \alpha^2 \int_{-\infty}^{+\infty} \frac{dk}{[(k_0 - k)^2 + \alpha^2/4]^2} = \frac{2\alpha^2 N^2}{\pi} \int_{-\infty}^{+\infty} \frac{dk'}{(k'^2 + a^2)^2}$$

$$= \frac{4a^2 N^2}{\pi} \int_0^\infty \frac{dk'}{(k'^2 + a^2)^2}$$

$$= \frac{2N^2}{a\pi} \int_0^\infty \frac{a\,dk'}{a^2 + k'^2} = \frac{2N^2}{a\pi}\left(\frac{\pi}{2}\right) = \frac{2N^2}{\alpha} = 1, \quad N = \sqrt{\alpha/2}$$

$$\Delta x^2 = \langle x^2 \rangle - \langle x \rangle^2$$

$$= \frac{\alpha}{2} \int_{-\infty}^{+\infty} e^{-\alpha|x|/2} x^2 e^{-\alpha|x|/2} dx - \left|\frac{\alpha}{2} \int_{-\infty}^{+\infty} e^{-\alpha|x|} x \, dx\right|^2$$

$$= \alpha \int_0^\infty e^{-\alpha x} x^2 \, dx - 0$$

$$\Delta x^2 = \alpha\left(\frac{2}{\alpha^3}\right) = \frac{3}{\alpha^2}$$

$$\Delta k^2 = \langle k^2 \rangle - \langle k \rangle^2$$

$$= \frac{\alpha}{4\pi} \int_{-\infty}^{+\infty} \frac{\alpha^2 k^2\, dk}{[(k-k_0)^2+\alpha^2/4]^2} - \left| \frac{\alpha}{4\pi} \int_{-\infty}^{+\infty} \frac{\alpha^2 k\, dk}{[(k-k_0)^2+\alpha^2/4]^2} \right|^2$$

$$= \frac{\alpha}{\pi} \int_{-\infty}^{+\infty} \frac{a^2(k'^2+2k_0k'+k_0^2)dk'}{(k'^2+a^2)^2} - \left| \frac{\alpha}{\pi} \int_{-\infty}^{+\infty} \frac{a^2(k'+k_0)dk'}{(k'^2+a^2)^2} \right|^2$$

$$= \frac{\alpha}{\pi} \int_{-\infty}^{+\infty} \frac{a^2k'^2\, dk'}{(k'^2+a^2)^2} + \frac{\alpha}{\pi} \int_{-\infty}^{\infty} \frac{a^2k_0^2\, dk'}{(k'^2+a^2)^2} - \left| \frac{\alpha}{\pi} \int_{-\infty}^{\infty} \frac{a^2k_0\, dk'}{(k'^2+a^2)^2} \right|^2$$

$$= \frac{\alpha}{\pi}\frac{1}{2} \int_{-\infty}^{+\infty} \frac{a^2\, dk'}{k'^2+a^2} + \frac{\alpha}{\pi}\frac{1}{2} \int_{-\infty}^{+\infty} \frac{k_0^2\, dk'}{k'^2+a^2} - \left| \frac{\alpha}{\pi} \int_{0}^{+\infty} \frac{k_0\, dk'}{k'^2+a^2} \right|^2$$

$$= \frac{\alpha}{\pi}\left(\frac{\alpha}{2}\frac{\pi}{2}\right) + \frac{\alpha}{\pi}\frac{1}{2}\frac{k_0^2\pi}{\alpha} - \left|\frac{2}{\pi}\frac{\pi}{2}k_0\right|^2$$

$$= \frac{\alpha^2}{4} + k_0^2 - k_0^2 = \frac{\alpha^2}{4}$$

$$\Delta x\ \Delta k = \frac{\sqrt{2}}{\alpha}\frac{\alpha}{2} = \frac{\sqrt{2}}{2}$$

(b) $$A(k) = \frac{N}{\sqrt{2\pi}} \int_{-\infty}^{\infty} e^{-\alpha^2x^2/4}\, e^{i(k_0-k)x}\, dx$$

$$= \frac{N}{\sqrt{2\pi}} \int_{0}^{\infty} e^{-\alpha^2x^2/4}\, e^{i(k_0-k)x} dx + \frac{N}{\sqrt{2\pi}} \int_{-\infty}^{\infty} e^{-\alpha^2x^2/4}\, e^{i(k_0-k)x}\, dx$$

$$= \frac{N}{\sqrt{2\pi}} \int_{0}^{\infty} e^{-\alpha^2x^2/4}\, e^{i(k_0-k)x} dx + \frac{N}{\sqrt{2\pi}} \int_{0}^{\infty} e^{-\alpha^2x^2/4}\, e^{-i(k_0-k)x}\, dx$$

$$= \frac{N\cdot 2}{\sqrt{2\pi}} \int_{0}^{\infty} e^{-\alpha^2x^2/4} \cos(k_0-k)x\, dx$$

$$= \frac{N \cdot 2}{\sqrt{2\pi}} \frac{\sqrt{\pi}\, e^{-(k_0 - k)^2/\alpha^2}}{\alpha}$$

$$= \frac{N\sqrt{2}}{\alpha} e^{-(k_0 - k)^2/\alpha^2}$$

$$|A(k)|^2 = \frac{2N^2}{\alpha^2} e^{-2(k_0 - k)^2/\alpha^2}$$

To find N,

$$\int \mu^*(x)\mu(x)dx = 2N^2 \int_0^\infty e^{-\alpha^2 x^2/2} = 2N^2 \frac{1\sqrt{\pi}}{2(\alpha/\sqrt{2})} \Rightarrow N = \sqrt{\frac{\alpha}{\sqrt{2\pi}}}$$

$$\int A^*(k)A(k)dk = \frac{4N^2}{\alpha^2} \int_0^\infty e^{-2(k_0 - k)^2/\alpha^2} dk$$

$$= \frac{4N^2}{\alpha^2} \int_0^\infty e^{-2k'^2/\alpha^2} dk$$

$$= \frac{4N^2}{\alpha^2} \left(\frac{1\sqrt{\pi}}{2(\sqrt{2}/2)}\right) = \frac{\sqrt{2\pi}\, N^2}{\alpha} = N = \sqrt{\frac{\alpha}{\sqrt{2\pi}}}$$

$$\Delta x^2 = \langle x^2\rangle - \langle x\rangle^2$$

$$= \frac{\alpha}{\sqrt{2\pi}} \int_{-\infty}^{+\infty} e^{-\alpha^2 x^2/2} x^2 dx - \left| \frac{\alpha}{\sqrt{2\pi}} \int_{-\infty}^{+\infty} e^{-\alpha^2 x^2/2} x\, dx \right|^2$$

$$= \frac{\alpha}{\sqrt{2\pi}} \left(2 \frac{1}{4(\alpha^2/2)} \sqrt{\frac{\pi}{\alpha^2/2}}\right) - 0 = \frac{1}{\alpha^2}$$

$$\Delta k^2 = \langle k^2\rangle - \langle k\rangle^2$$

$$= \frac{2}{\alpha\sqrt{2\pi}} \int_{-\infty}^{+\infty} e^{-2(k_0 - k)^2/\alpha^2} k^2 dk - \left| \frac{2}{\alpha\sqrt{2\pi}} \int_{-\infty}^{\infty} e^{-2(k_0 - k)^2/\alpha^2} k\, dk \right|^2$$

$$= \frac{2}{\alpha\sqrt{2\pi}} \int_{-\infty}^{+\infty} e^{-2k'^2/\alpha^2} k'^2 dk' + \frac{2}{\alpha\sqrt{2\pi}} \int_{-\infty}^{+\infty} e^{-2k'^2/\alpha^2} k_0^2 dk$$

$$- \left| \frac{2}{\alpha\sqrt{2\pi}} \int_{-\infty}^{+\infty} e^{-2k'^2/\alpha^2} k_0 \, dk \right|^2$$

$$= \frac{2}{\alpha\sqrt{2\pi}} \left[2 \frac{1}{4 \cdot 2/\alpha^2} \sqrt{\frac{\pi}{2/\alpha^2}} + \frac{2}{\alpha\sqrt{2\pi}} k_0^2 \left(2 \frac{1}{2 \cdot \sqrt{2}/2} \sqrt{\pi} \right) \right] - \left| \frac{2}{\alpha\sqrt{2\pi}} \frac{2k_0\sqrt{\pi}}{2\sqrt{2/\alpha^2}} \right|^2$$

$$= \frac{\alpha^2}{4} + k_0^2 - k_0^2 = \frac{\alpha^2}{4}$$

$$\Delta x \; \Delta k = \frac{1}{\alpha} \frac{\alpha}{2} = \frac{1}{2}$$

(c) $$A(k) = \frac{N}{\sqrt{2\pi}} \int_{-1/\alpha}^{+1/\alpha} (1 - \alpha|x|) e^{i(k_0 - k)x} \, dx$$

$$= \frac{N}{\sqrt{2\pi}} \int_0^{1/\alpha} (1 - \alpha x) e^{i(k_0 - k)x} \, dx + \frac{N}{\sqrt{2\pi}} \int_0^{1/\alpha} (1 - \alpha x) e^{-i(k_0 - k)x} \, dx$$

$$= \frac{2N}{\sqrt{2\pi}} \int_0^{1/\alpha} (1 - \alpha x) \cos (k_0 - k)x \, dx$$

$$= \frac{2N}{\sqrt{2\pi}} \left(\frac{\sin(k_0 - k)x}{(k_0 - k)} \Bigg|_0^{1/\alpha} - \frac{\alpha x \sin (k_0 - k)x}{(k_0 - k)} \Bigg|_0^{1/\alpha} + \frac{\alpha \cos (k_0 - k)x}{(k_0 - k)^2} \Bigg|_0^{1/\alpha} \right)$$

$$= \frac{2N}{\sqrt{2\pi}} \left(\frac{\sin[(k_0 - k)/\alpha]}{(k_0 - k)} - \frac{\sin[(k_0 - k)/\alpha]}{(k_0 - k)} + \frac{\cos[(k_0 - k)/\alpha]}{(k_0 - k)^2} - \frac{\alpha}{(k_0 - k)^2} \right)$$

Solve for N:

$$\int \mu^*(x)\mu(x)dx = N^2 \int_{-1/\alpha}^{1/\alpha} (1 - 2\alpha|x| + x^2) \, dx$$

$$= N^2 \left(\frac{2}{\alpha}\right) - 4N^2\alpha \int_0^{1/\alpha} x \, dx + 2N^2\alpha \int_0^{1/\alpha} x^2 \, dx$$

$$= N^2 \frac{2}{\alpha} - N^2 \frac{2}{\alpha} + N^2 \frac{2}{3\alpha} = 1, \quad N = \sqrt{3\alpha/2}$$

$$|A(k)|^2 = \frac{4N^2}{\pi} \frac{\alpha^2}{(k_0 - k)^4} \frac{[\cos[(k_0 - k)/\alpha] - 1]^2}{4} 4$$

$$= \frac{12\alpha^3}{\pi(k_0 - k)^4} \sin^4 \left(\frac{k_0 - k}{2\alpha}\right)$$

$$\int A^*(k_0)A(k)dk = \frac{12\alpha^3}{\pi} \int \frac{\sin^4[(k_0 - k)/2\alpha]}{(k_0 - k)^4} dk$$

$$= \frac{2 \cdot 12\alpha^2}{\pi} \int_0^\infty \frac{\sin^4 k' \, dk'(2\alpha)}{k'^4(2\alpha)^4} \quad \frac{3}{\pi}\left(\frac{\pi}{3}\right) = 1$$

$$\Delta x^2 = \langle x^2 \rangle - \langle x \rangle^2$$

$$= \frac{3\alpha}{2} \int_{-1/\alpha}^{+1/\alpha} (1 - \alpha|x|)^2 \, x^2 \, dx - \left|\frac{3}{2} \alpha \int_{-1/\alpha}^{+1/\alpha} (1 - \alpha|x|)^2 \, x \, dx\right|$$

$$= \frac{3}{2} \int_{-1/\alpha}^{+1/\alpha} x^2 \, dx - 3\alpha \int_{-1/\alpha}^{+1/\alpha} \alpha|x| \, x^2 \, dx + \frac{3\alpha^3}{2} \int_{-1/\alpha}^{+1/\alpha} x^4 \, dx - 0$$

$$= \frac{1}{\alpha^2} - \frac{3}{2\alpha^2} + \frac{3}{\alpha^2 S} = \frac{1}{10\alpha^2}$$

$$\Delta k^2 = \langle k^2 \rangle - \langle k \rangle^2$$

$$= \frac{12\alpha^3}{\pi} \int_{-\infty}^{+\infty} \frac{\sin^4[(k_0 - k)/2\alpha] \, k^2 \, dk}{(k_0 - k)^4} - \left|\frac{12}{\pi} \alpha^3 \int_{-\infty}^{+\infty} \frac{\sin^4[(k_0 - k)/2\alpha] \, k \, dk}{(k_0 - k)^4}\right|$$

$$= \frac{12\alpha^3}{\pi} \int_{-\infty}^{+\infty} \frac{\sin^4 k'[k_0^2 - 4\alpha k_0 k' + 4\alpha^2 k'^2] dk'(2\alpha)}{k'^4(2\alpha)^4} - \left|\frac{12\alpha^3}{\pi} \int_{-\infty}^{+\infty} \frac{k_0 \sin^4 k' \, dk' \, (2\alpha)}{(k'^4)(2\alpha)^4}\right|$$

$$= \frac{2 \cdot 12\alpha^3}{\pi(2\alpha)^3} \int_0^{+\infty} \frac{k_0^2 \sin^4 k' \, dk'}{k'^4} + \frac{2 \cdot 12\alpha^3}{\pi(2\alpha)} \int_0^{+\infty} \frac{\sin^4 k' \, dk'}{k'^2} - \left|\frac{2 \cdot 12}{\pi} \frac{\alpha^3 k_0}{(2\alpha)^3} \left(\frac{\pi}{3}\right)\right|$$

$$= k_0^2 - k_0^2 + \frac{12\alpha^2}{\pi} \int_0^\infty \frac{\sin^4 k' \, dk'}{k'^2}$$

$$= \frac{12\alpha^2}{\pi}\int_0^\infty \frac{\sin^2 k'\,dk'}{k'^2} - \frac{12\alpha^2}{\pi}\int_0^\infty \frac{\sin^2 k'\cos^2 k'\,dk'}{k'^2}$$

$$= \frac{12\alpha^2}{\pi}\frac{\pi}{2} - \frac{12\alpha^2}{\pi}\int_0^\infty \frac{\sin^2 2k'\,dk'}{4k'}$$

$$= 6\alpha^2 - \frac{12\alpha^2}{2\pi}\int_0^\infty \frac{\sin^2 \alpha\,d\alpha}{\alpha^2} = 6\alpha^2 - \frac{12\alpha^2}{2\pi}\frac{\pi}{2} = 3\alpha^2$$

$$\Delta x\ \Delta k = \frac{1}{\sqrt{10}\ \alpha}\sqrt{3}\ \alpha = \sqrt{3/10}$$

(d) $$A(k) = \frac{N}{\sqrt{2\pi}}\int_{-a}^{+a} e^{i(k_0 - k)x}\,dx$$

$$= \frac{2N}{\sqrt{2\pi}}\int_0^a \cos(k_0 - k)x\,dx$$

$$= \frac{2N}{\sqrt{2\pi}}\frac{\sin(k_0 - k)a}{(k_0 - k)}$$

$$|A(k)|^2 = \frac{4N^2}{2\pi}\frac{\sin^2(k_0 - k)a}{(k_0 - k)^2}$$

$$\int \mu^*(x)\mu(x)dx = \int_{-a}^{a} N^2\,dx = N^2 \cdot 2a = 1, \quad N = 1/\sqrt{2a}$$

$$\int A^*(k)A(k)dk = \frac{4N^2}{2\pi}\int_{-\infty}^{+\infty}\frac{\sin^2(k_0 - k)a\,dk}{(k_0 - k)^2}$$

$$= \frac{4N^2}{2\pi}\,2\int_0^{+\infty}\frac{\sin^2 k'}{k'^2(1/a^2)}\,dk\,\frac{1}{a}$$

$$= \frac{4N^2 a}{\pi}\left(\frac{\pi}{2}\right) = 1, \quad N = 1/\sqrt{2a}$$

$$(\Delta x)^2 = \langle x^2\rangle - \langle x\rangle^2$$

$$= \frac{1}{2a}\int_{-a}^{a} x^2\,dx - \left|\frac{1}{\pi}\,a\int_{-a}^{a} x\,dx\right|$$

$$= \frac{2}{2a}\,\frac{a^3}{3} = \frac{a^2}{3}$$

$$(\Delta k)^2 = \langle k^2\rangle - \langle k\rangle^2$$

$$= \frac{1}{\pi a}\int_{-\infty}^{+\infty}\frac{\sin^2[(k_0 - k)a]k^2\,dk}{(k_0 - k)^2} - \left|\frac{1}{\pi a}\int_{-\infty}^{+\infty}\frac{\sin^2[(k_0 - k)a]\,(k)dk}{(k_0 - k)^2}\right|^2$$

$$= \frac{1}{\pi a}\int_{-\infty}^{+\infty}\frac{\sin^2 k'[(k'/a)^2 + 2k'k_0/a + k_0^2]dk'/a}{k'^2/a} - \left|\frac{1}{\pi a}\int_{-\infty}^{+\infty}\frac{\sin^2 k'\,dk'/a\;k_0}{k'^2/a^2}\right|^2$$

$$= \frac{2}{\pi a^2}\int_0^{\infty}\sin^2 k'\,dk' + \frac{2}{\pi}\int_0^{\infty}\frac{k_0^2\sin^2 k'\,dk'}{k'^2} - \left|\frac{2k_0}{\pi}\int_0^{\infty}\frac{\sin^2 k'\,dk'}{k'^2}\right|^2$$

$$= \frac{2}{\pi a^2}\int_0^{\infty}\sin^2 k'\,dk' + \frac{2}{\pi}k_0^2\frac{\pi}{2} - k_0^2$$

$$= \frac{2}{\pi a^2}\int_0^{\infty}\sin^2 k'\,dk' = \infty, \quad \text{so } \Delta x\,\Delta k = \infty.$$

PROBLEM 13.16 Consider the nonlocal (in time) connection between **D** and **E**,

$$\mathbf{D}(\mathbf{x},t) = \mathbf{E}(\mathbf{x},t) + \int d\tau\, G(\tau)\mathbf{E}(\mathbf{x},t-\tau)$$

with the $G(\tau)$ appropriate for the single-resonance model,

$$\varepsilon(\omega) = 1 + \omega_p^2(\omega_0^2 - \omega^2 - i\gamma\omega)^{-1}$$

(a) Convert the nonlocal connection between **D** and **E** into an instantaneous relation involving derivatives of **E** with respect to time by expanding the electric field in the integral in a Taylor series in τ. Evaluate the integrals over $G(\tau)$ explicitly up to at least $\partial^2\mathbf{E}/\partial t^2$.

(b) Show that the series obtained in (a) can be obtained formally by converting the frequency-representation relation, $\mathbf{D}(\mathbf{x},\omega) = \varepsilon(\omega)\mathbf{E}(\mathbf{x},\omega)$ into a space-time relation,

$$\mathbf{D}(\mathbf{x},t) = \varepsilon\left(i\frac{\partial}{\partial t}\right)\mathbf{E}(\mathbf{x},t)$$

where the variable ω in $\varepsilon(\omega)$ is replaced by $\omega \to i(\partial/\partial t)$.

Solution:

(a) $\mathbf{D}(\mathbf{x},t) = \mathbf{E}(\mathbf{x},t) + \int d\tau\, G(\tau)\mathbf{E}(\mathbf{x},t-\tau)$

We use the single-resonance model

$$\varepsilon(\omega) = 1 + \frac{\omega_p^2}{\omega_0^2 - \omega^2 - i\gamma\omega}$$

therefore,

$$G(\tau) = \frac{1}{2\pi}\int_{-\infty}^{\infty} [\varepsilon(\omega) - 1]e^{-i\omega t}\, d\omega = \frac{\omega_p^2}{2\pi}\int_{-\infty}^{\infty} \frac{e^{-i\omega t}}{\omega_0^2 - \omega^2 - i\gamma\omega}\, d\omega$$

The poles are at

$$\omega_{1,2} = \frac{-i\gamma}{2} \pm \nu_0; \quad \nu_0^2 = \omega_0^2 - \frac{\gamma^2}{4}$$

$$G(\tau) = \omega_p^2\, e^{-\gamma\tau/2}\, \frac{\sin \nu_0\tau}{\nu_0}\, \theta(\tau)$$

$$\mathbf{D}(\mathbf{x},t) = \mathbf{E}(\mathbf{x},t) + \int_0^{\infty} \frac{\omega_p^2}{\nu_0}\, e^{-\gamma\tau/2} \sin\nu\,\tau\, \mathbf{E}(\mathbf{x},t-\tau)d\tau$$

Now, for τ small, we expand $\mathbf{E}(\mathbf{x},t-\tau)$

$$\mathbf{E}(\mathbf{x},t-\tau) = \mathbf{E}(\mathbf{x},t) - \tau\frac{\partial\mathbf{E}(\mathbf{x},t)}{\partial t} + \frac{\tau^2}{2!}\frac{\partial^2\mathbf{E}(\mathbf{x},t)}{\partial t^2} + \ldots$$

$$\mathbf{D}(\mathbf{x},t) = \mathbf{E}(\mathbf{x},t) + \frac{\omega_p^2}{\nu_0}\int_0^{\infty} \mathbf{E}(\mathbf{x},t)e^{-\gamma\tau/2} \sin\nu_0\tau\, d\tau$$

$$- \frac{\omega_p^2}{\nu_0}\frac{\partial\mathbf{E}}{\partial t}\int_0^{\infty} \tau\, e^{-\gamma\tau/2} \sin\nu_0\tau\, d\tau$$

$$+ \frac{\omega_p^2}{2\nu_0}\frac{\partial^2\mathbf{E}}{\partial t^2}\int_0^{\infty} \tau^2\, e^{-\gamma\tau/2} \sin\nu_0\tau\, d\tau + \ldots$$

now,

$$I_1 = \int_0^{\infty} e^{-\gamma\tau/2} \sin\nu_0\tau\, d\tau = \frac{1}{2i}\int_0^{\infty} [e^{(i\nu_0 - \gamma/2)\tau} - {}^{(-i\nu_0 - \gamma/2)\tau}]d\tau$$

$$= \frac{1}{2i}\left(\frac{1}{i\nu_0 - \gamma/2}\,(0-1) + \frac{1}{i\nu_0 + \gamma/2}\right)(0-1) = \frac{i}{2}\left(\frac{i\nu_0 + \gamma/2 + i\nu_0 - \gamma/2}{-\nu_0^2 - \gamma^2/4}\right)$$

$$I_1 = \frac{\nu_0}{\nu_0^2 + \gamma^2/4}$$

$$I_2 = \frac{1}{2i}\int_0^\infty \tau\,[e^{(i\nu_0 - \gamma/2)\tau} - e^{-(i\nu_0 - \gamma/2)\tau}]\,d\tau$$

$$= \frac{1}{2i}\left[\tau\left(\frac{e^{(i\nu_0 - \gamma/2)\tau}}{i\nu_0 - \gamma/2} + \frac{e^{-(i\nu_0 + \gamma/2)\tau}}{i\nu_0 + \gamma/2}\right)\Bigg|_0^\infty - \left(\frac{0-1}{(i\nu_0 - \gamma/2)^2} - \frac{0-1}{(i\nu_0 + \gamma/2)^2}\right)\right]$$

where the term in the first parentheses is zero, therefore,

$$I_2 = \frac{i}{2}\left(\frac{1}{\gamma^2/4 - \nu_0^2 + i\nu_0\gamma} - \frac{1}{\gamma^2/4 - \nu_0^2 - i\nu_0\gamma}\right)$$

$$= \frac{i}{2}\left(\frac{[(\gamma^2/4 - \nu_0^2) - i\nu_0\gamma] - [(\gamma^2/4 - \nu_0^2) + i\nu_0\gamma]}{(\gamma^2/4 - \nu_0^2)^2 + \nu_0^2\gamma^2}\right)$$

$$I_2 = \frac{\nu_0\gamma}{(\gamma^2/4 - \nu_0^2)^2 + \nu_0^2\gamma^2}$$

Finally,

$$I_3 = \frac{1}{2i}\int_0^\infty \tau^2\, e^{-\gamma\tau/2}(e^{i\nu_0\tau} - e^{-i\nu_0\tau})\,d\tau$$

$$= \frac{1}{2i}\int_0^\infty \tau^2\,(e^{(i\nu_0 - \gamma/2)\tau} - e^{-(i\nu_0 + \gamma/2)\tau})\,d\tau$$

$$= \frac{1}{2i}\left[\tau^2\left(\frac{e^{(i\nu_0 - \gamma/2)\tau}}{i\nu_0 - \gamma/2} + \frac{e^{-(i\nu_0 + \gamma/2)\tau}}{i\nu_0 + \gamma/2}\right)\Bigg|_0^\infty\right.$$

$$\left. - 2\int_0^\infty \tau\left(\frac{e^{(i\nu_0 - \gamma/2)\tau}}{i\nu_0 - \gamma/2} + \frac{e^{(i\nu_0 + \gamma/2)\tau}}{i\nu_0 + \gamma/2}\right)d\tau\right]$$

where the term in the first parentheses is zero,

$$= i\left[\tau\left(\frac{e^{(i\nu_0-\gamma/2)\tau}}{(i\nu_0-\gamma/2)^2}-\frac{e^{-(i\nu_0+\gamma/2)\tau}}{(i\nu_0+\gamma/2)^2}\right)\Bigg|_0^\infty\right.$$

$$\left.-\int_0^\infty\left(\frac{e^{(i\nu_0-\gamma/2)\tau}}{(i\nu_0-\gamma/2)^2}-\frac{e^{-(i\nu_0+\gamma/2)\tau}}{(i\nu_0+\gamma/2)^2}\right)d\tau\right]$$

where the term in the first parentheses is zero,

$$=\frac{1}{i}\left(\frac{0-1}{(i\nu_0-\gamma/2)^3}+\frac{0-1}{(i\nu_0+\gamma/2)^3}\right)$$

$$= i\left(\frac{1}{i[(\gamma^2/4)\nu_0-\nu_0^3+i\nu_0^2\gamma]-\gamma^3/8+\nu_0^2\,\gamma/2-i\nu_0\,\gamma^2/2}\right.$$

$$\left.+\frac{1}{i[(\gamma^2/4)\nu_0-\nu_0^3-i\nu_0^2\gamma]+\gamma^3/8-\nu_0^2\,\gamma/2-i\nu_0\,\gamma^2/2}\right)$$

$$= i\left(\frac{1}{(\nu_0^2\gamma/2-\gamma^3/8-\nu_0^2\gamma)+i[(\gamma^2/4)\nu_0-\nu_0^3-\nu_0\,\gamma^2/2]}\right.$$

$$\left.+\frac{1}{(\gamma^3/8-\nu_0^2\,\gamma/2+\nu_0^2\gamma)+i[(\gamma^2/4)\nu_0-\nu_0^3-\nu_0\,\gamma^2/2]}\right)$$

$$= i\left(\frac{-1}{(\nu_0^2\gamma/2-\gamma^3/8)+i(\nu_0^3+\nu_0\gamma^2/4)}+\frac{1}{(\nu_0^2\gamma/2+\gamma^3/8)-i(\nu_0^3+\nu_0\,\gamma^2/4)}\right)$$

$$= i\left(\frac{[\nu_0^2(\gamma/2)+\gamma^3/8]+i(\nu_0^3+\nu_0\gamma^2/4)-(\nu_0^2\gamma/2+\gamma^3/8)+i(\nu_0^3+\nu_0\,\gamma^2/4)}{(\gamma\nu_0^2/2+\gamma^3/8)^2+(\nu_0^3+\nu_0\gamma^2/4)^2}\right)$$

$$I_3=\frac{-2(\nu_0^3+\nu_0\gamma^2/4)}{(\gamma\nu_0^2/2+\gamma^3/8)^2+(\nu_0^3+\nu_0\gamma^2/4)^2}$$

therefore,

$$\mathrm{D}(\mathrm{x},t)=\mathrm{E}(\mathrm{x},t)\left(1+\frac{\omega_p^2}{\nu_0^2+\gamma^2/4}\right)-\frac{\partial\mathrm{E}}{\partial t}\,\frac{\omega_p^2\gamma}{(\gamma^2/4-\nu_0^2)+\nu_0^2\gamma^2}$$

$$-\frac{\partial^2 E}{\partial t^2}\ \frac{\omega_p^2(\nu_0^2+\gamma^2/4)}{(\gamma\nu_0^2/2+\gamma^3/8)^2+(\nu_0^3+\nu_0\gamma^2/4)^2}+\ldots$$

But $\nu_0^2=\omega_0^2-\gamma^2/4$, therefore,

$$\nu_0^2+\gamma^2/4=\omega_0^2$$

$$D(\mathbf{x},t)=E(\mathbf{x},t)\left(1+\frac{\omega_p^2}{\omega_0^2}\right)-\frac{\omega_p^2\gamma}{(\gamma^2/2-\omega_0^2)+\gamma^2\omega_0^2-\gamma^2/4}\ \frac{\partial E}{\partial t}$$

$$-\frac{\omega_p^2\omega_0^2}{\gamma^2\omega_0^4/4+\omega_0^4(\omega_0^2-\gamma^2/4)}\ \frac{\partial^2 E}{\partial t^2}-\ldots$$

$$D(\mathbf{x},t)=\left(1+\frac{\omega_p^2}{\omega_0^2}\ E(\mathbf{x},t)-\frac{\omega_p^2\gamma}{\gamma^2/4+(\gamma^2-1)\omega_0^2}\ \frac{\partial E}{\partial t}-\frac{\omega_p^2}{\omega_0^4}\ \frac{\partial^2 E}{\partial t^2}-\ldots\right)$$

(b) $D(\mathbf{x},\omega)=\varepsilon(\omega)E(\mathbf{x},\omega)$

Let $\omega\to i\ \partial/\partial t$ in $\varepsilon(\omega)$, then,

$$D(\mathbf{x},t)=\varepsilon\left(i\ \frac{\partial}{\partial t}\right)\ E(\mathbf{x},t)$$

where

$$\varepsilon(\omega)=1+\frac{\omega_p^2}{(\omega_0^2-\omega^2)-i\gamma\omega}=1+\frac{\omega_p^2[(\omega_0^2-\omega^2)+i\gamma\omega]}{(\omega_0^2-\omega^2)+\gamma^2\omega^2}$$

or expand:

$$\frac{\omega_p^2}{\omega_0^2\left(1-\dfrac{\omega^2-i\gamma\omega}{\omega_0^2}\right)}\simeq\frac{\omega_p^2}{\omega_0^2}\left[1-\left(\frac{\omega^2-i\gamma\omega}{\omega_0^2}\right)+\left(\frac{\omega^2-i\gamma\omega}{\omega_0^2}\right)^2+\left(\frac{\omega^2-i\gamma\omega}{\omega_0^2}\right)^3+\ldots\right]$$

therefore,

$$\varepsilon(\omega)\simeq 1+\frac{\omega_p^2}{\omega_0^2}\left(1-\frac{\omega^2}{\omega_0^2}+i\gamma\ \frac{\omega}{\omega_0^2}-\frac{\gamma^2}{\omega_0^2}\ \omega^2+o(\omega^3)\right)$$

here we have kept only terms to second order in ω.

$$\varepsilon\left(i\ \frac{\partial}{\partial t}\right)\cong 1+\frac{\omega_p^2}{\omega_0^2}+\frac{\omega_p^2}{\omega_0^4}\ \frac{\partial^2}{\partial t^2}-\frac{\gamma\omega_p^2}{\omega_0^4}\ \frac{\partial}{\partial t}+\frac{\gamma^2\omega_p^2}{\omega_0^4}\ \frac{\partial^2}{\partial t^2}+o\left(\frac{\partial^3}{\partial t^3}\right)$$

$$\cong\left(1+\frac{\omega_p^2}{\omega_0^2}\right)-\frac{\gamma\omega_p^2}{\omega_0^4}\ \frac{\partial}{\partial t}+\frac{\omega_p^2(1+\gamma^2)}{\omega_0^4}\ \frac{\partial^2}{\partial t^2}+o\left(\frac{\partial^3}{\partial t^3}\right)$$

$$D(x,t) = \left(1 + \frac{\omega_p^2}{\omega_0^2}\right) E(x,t) - \frac{\gamma\omega_p^2}{\omega_0^4} \frac{\partial E(x,t)}{\partial t} + \frac{\omega_p^2(1+\gamma^2)}{\omega_0^4} \frac{\partial^2 E(x,t)}{\partial t^2}$$

$$+ \ldots \text{ Higher order terms}$$

PROBLEM 13.17 Compute the reflection coefficient for a plane wave incident normally on

(a) a plane surface bounding two infinite dielectric media
(b) a plane slab of dielectric thickness L

Discuss the behavior of the reflection coefficient as a function of L.

Solution:

(a) Consider the figure:

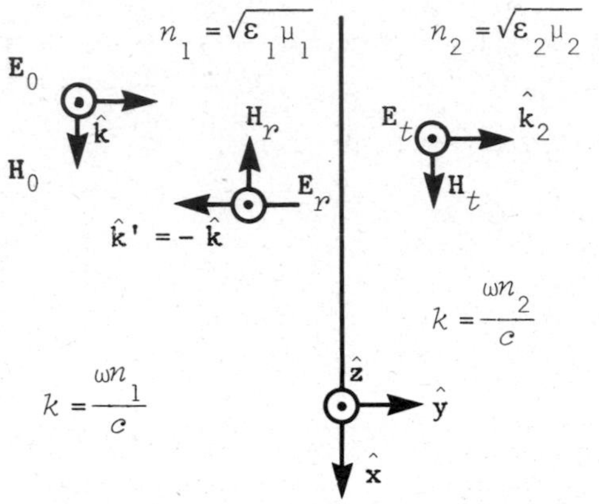

Let

$$\mathbf{E}_0 = E_0\, e^{i(\mathbf{k}\cdot\mathbf{r} - \omega t)}\, \hat{\mathbf{z}}$$

$$\mathbf{E}_r = E_r\, e^{i(\mathbf{k}\cdot\mathbf{r} - \omega t)}\, \hat{\mathbf{z}}$$

$$\mathbf{E}_t = E_t\, e^{i(\mathbf{k}_2\cdot\mathbf{r} - \omega t)}\, \hat{\mathbf{z}}$$

$$\mathbf{H}_0 = \sqrt{\varepsilon_1/\mu_1}\, E_0\, e^{i(\mathbf{k}\cdot\mathbf{r} - \omega t)}\, \hat{\mathbf{x}}$$

$$\mathbf{H}_r = -\sqrt{\varepsilon_1/\mu_1}\, E_r\, e^{i(\mathbf{k}\cdot\mathbf{r} - \omega t)}\, \hat{\mathbf{x}}$$

$$\mathbf{H}_t = \sqrt{\varepsilon_2/\mu_2}\, E_t\, e^{i(\mathbf{k}_2\cdot\mathbf{r} - \omega t)}\, \hat{\mathbf{x}}$$

The condition $\hat{\mathbf{k}}\cdot\mathbf{r} = \hat{\mathbf{k}}'\mathbf{r} = \hat{\mathbf{k}}_2\mathbf{r}$ is trivially satisfied in this case. Then from the continuity of tangential components of **E** and **H**, we have

$$E_0 + E_r = E_t$$

$$\sqrt{\varepsilon_1/\mu_1}\, E_0 - \sqrt{\varepsilon_1/\mu_1}\, E_r = \sqrt{\varepsilon_2/\mu_2}\, E_t$$

if $\mu_1 = \mu_2 = 1$, then

$$E_0 + E_r = E_t$$

$$n_1 E_0 - n_1 E_r = n_2 E_t$$

Solving

$$\begin{cases} E_0 + E_r = E_t \\ n_1 E_0 - n_1 E_r = n_2 E_0 + n_2 E_r \end{cases} \qquad \begin{cases} E_0 + E_r = E_t \\ E_0 (n_1 - n_2) = E_r (n_2 + n_1) \end{cases}$$

$$\begin{cases} E_0 + E_r = E_t \\ E_i = \left(\dfrac{n_1 - n_2}{n_1 + n_2}\right) E_0 \end{cases} \qquad E_t = \left(1 + \frac{n_1 - n_2}{n_1 + n_2}\right) E_0 = \frac{2n_1}{n_1 + n_2} E_0$$

Therefore,

$$E_r = \left(\frac{n_1 - n_2}{n_1 + n_2}\right) E_0$$

$$E_t = \left(\frac{2n_1}{n_1 + n_2}\right) E_0$$

(Notice positive direction of E's, in $+z$ direction.) So, reflection coefficient is

$$R = \frac{E_r^2}{E_0^2} - \left(\frac{n_1 - n_2}{n_1 + n_2}\right)^2$$

(b) Let

$$R_{ab} = \frac{n_a - n_b}{n_a + n_b} = -R_{ba}$$

and

$$T_{ab} = \frac{2n_a}{n_a + n_b}$$

so that, when going from medium a to b, if incident amplitude is E_0, then

$$E_x = R_{ab} E_0 \quad \text{and} \quad E_t = T_{ab} E_0$$

Consider the following figure. Then, the sum of the reflected amplitudes is ($\alpha = 2k_2 d$)

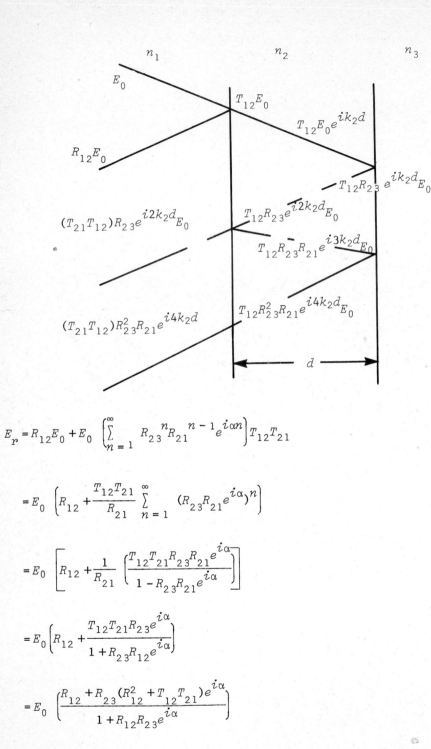

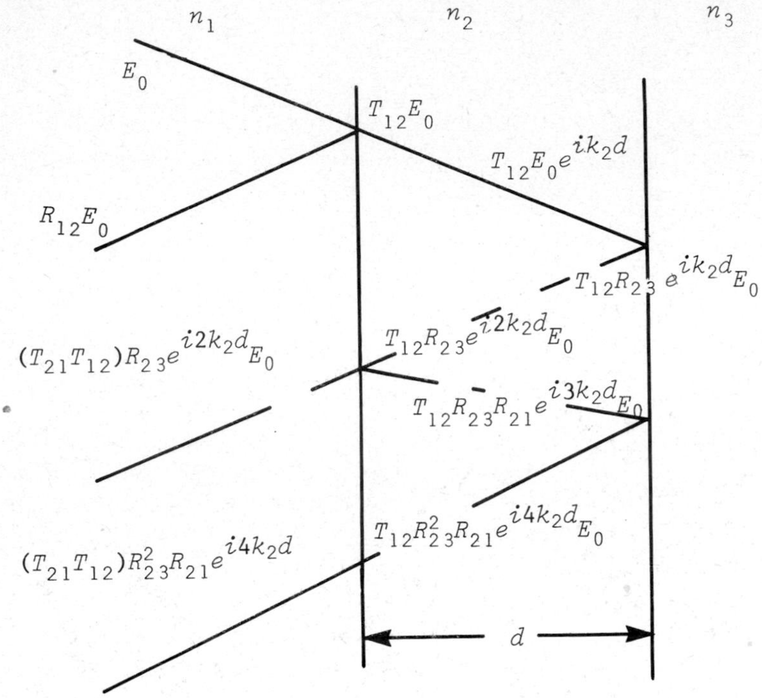

$$E_r = R_{12}E_0 + E_0\left(\sum_{n=1}^{\infty} R_{23}{}^{n}R_{21}{}^{n-1}e^{i\alpha n}\right)T_{12}T_{21}$$

$$= E_0\left(R_{12} + \frac{T_{12}T_{21}}{R_{21}}\sum_{n=1}^{\infty}(R_{23}R_{21}e^{i\alpha})^n\right)$$

$$= E_0\left[R_{12} + \frac{1}{R_{21}}\left(\frac{T_{12}T_{21}R_{23}R_{21}e^{i\alpha}}{1 - R_{23}R_{21}e^{i\alpha}}\right)\right]$$

$$= E_0\left(R_{12} + \frac{T_{12}T_{21}R_{23}e^{i\alpha}}{1 + R_{23}R_{12}e^{i\alpha}}\right)$$

$$= E_0\left(\frac{R_{12} + R_{23}(R_{12}^2 + T_{12}T_{21})e^{i\alpha}}{1 + R_{12}R_{23}e^{i\alpha}}\right)$$

but

$$R_{12}^2 + T_{12}T_{21} = \frac{(n_1 - n_2)^2 + (2n_1)(2n_2)}{(n_1 + n_2)^2} = 1$$

Therefore,

$$E_r = E_0 \frac{R_{12} + R_{23}e^{i\alpha}}{1 + R_{12}R_{23}e^{i\alpha}}$$

So,

$$R = \frac{E_r E_r^*}{E_0 E_0^*} = \left(\frac{R_{12} + R_{23}e^{i\alpha}}{1 + R_{12}R_{23}e^{i\alpha}}\right)\left(\frac{R_{12} + R_{23}e^{-i\alpha}}{1 + R_{12}R_{23}e^{-i\alpha}}\right)$$

$$= \frac{R_{12}^2 + R_{23}^2 + R_{12}R_{23}(e^{i\alpha} + e^{-i\alpha})}{1 + (R_{12}R_{23})^2 + R_{12}R_{23}(e^{i\alpha} + e^{-i\alpha})}$$

Finally,

$$\mathbb{R} = \frac{R_{12}^2 + R_{23}^2 + 2R_{12}R_{23}\cos(2k_2 d)}{1 + R_{12}^2 R_{23}^2 + 2R_{12}R_{23}\cos(2k_2 d)}$$

$$\mathbb{R} = \frac{R_{12}^2 R_{23}^2 + 2R_{12}R_{23}\cos\alpha}{1 + R_{12}^2 R_{23}^2 + 2R_{12}R_{23}\cos\alpha} = \frac{2R_{12}^2(1 - \cos\alpha)}{1 + R_{12}^4 - 2R_{12}^2\cos\alpha}$$

$$= \frac{4R_{12}^2 \sin^2\alpha/2}{1 + R_{12}^2 + 2R_{12}^2\cos\alpha}$$

(i) Let us choose d so that $\cos\alpha = -1$, that is,

$$2k_2 d = \pi + n2\pi = \pi(2n+1) = 2\pi(n + 1/2)$$

so that

$$2\frac{2\pi}{\lambda_2}d = 2\pi(n + 1/2)$$

therefore,

$$d = \frac{\lambda_2}{2}(n + 1/2), \quad n = 0,1,2,\ldots$$

Then,

$$\mathbb{R} = \frac{(R_{12} - R_{23})^2}{(1 - R_{12}R_{23})^2}$$

To obtain $\mathbb{R} = 0$, we must have $R_{12} = R_{23}$, that is,

$$\frac{n_1 - n_2}{n_1 + n_2} = \frac{n_2 - n_3}{n_2 + n_3}$$

$$(n_1 - n_2)(n_2 + n_3) = (n_2 - n_3)(n_1 + n_2)$$

$$n_1 n_2 + n_1 n_3 - n_2^2 - n_2 n_3 = n_2 n_1 + n_2^2 - n_1 n_3 - n_2 n_3$$

$$n_2^2 = n_1 n_3$$

Therefore, if we choose $d = (\lambda_2/2)(n + 1/2)$ (usually $d = \lambda_2/4$) and n_2 such that $n_2^2 = n_1 n_3$, then we get total transmission.

(ii) Suppose $n_1 = n_3$, then $R_{23} = R_{21} = -R_{12}$, so that

$$\mathbb{R} = \frac{2R_{12}^2(1 - \cos\alpha)}{1 + R_{12}^2 - \cos\alpha}$$

We see that,

$\cos\alpha = 1$, that is, $2k_2 d = 2n\pi$, $d = \frac{\lambda_2}{2} n$, $n = 0, 1, 2, \ldots$, then $\mathbb{R} = 0$

$\cos\alpha = -1$, that is, $d = \frac{\lambda_2}{2}(n + 1/2)$, $n = 0, 1, 2, \ldots$, then $\mathbb{R} = \frac{4R_{12}^2}{2 + R_{12}^4}$

$\cos\alpha = 0$, that is, $d = \frac{\lambda_2}{2}(n + 1/2)$, $n = 0, 1, 2, \ldots$, then $\mathbb{R} = \frac{2R_{12}^2}{1 + R_{12}^4}$

Chapter 14

PROBLEM 14.1 Given an oscillating electric dipole, $\mathbf{p} = \hat{\mathbf{z}}p_0e^{-i\omega t}$, located at the origin. Write down the Hertz vector and the vector potential. Obtain the forms of A, ϕ, E, and B in the radiation zone. Calculate $dp/d\Omega$, the radiated power per unit solid angle.

Solution:

$$\mathbf{p} = \hat{\mathbf{z}}p_0e^{-i\omega t}$$

for a single oscillating dipole at the origin pointing in the z direction we have

$$\mathbf{p} = \mathbf{p}(t)\delta(\mathbf{r}) = \hat{\mathbf{z}}p(t)\delta(\mathbf{r})$$

$$\mathbf{p} = \hat{\mathbf{z}}p_0e^{-i\omega t}\delta(\mathbf{r})$$

$$\mathbf{z}(\mathbf{r},t) = \frac{1}{4\pi\varepsilon_0}\int \frac{\mathbf{p}(\mathbf{r}',t - |\mathbf{r}-\mathbf{r}'|/c)}{\mathbf{r}-\mathbf{r}'}, \qquad \mathbf{A} = \frac{1}{c^2}\frac{\partial \mathbf{z}}{\partial t}$$

$$= \frac{1}{4\pi\varepsilon_0}\int \frac{\hat{\mathbf{z}}p_0(t')e^{-i\omega(t - |\mathbf{r}-\mathbf{r}'|/c)}\delta(\mathbf{r}')\,dv'}{|\mathbf{r}-\mathbf{r}'|}$$

$$= \frac{\hat{\mathbf{z}}p_0(t')}{4\pi\varepsilon_0}\int \frac{e^{-i\omega(t - |\mathbf{r}-\mathbf{r}'|/c)}}{|\mathbf{r}-\mathbf{r}'|}\delta(\mathbf{r}')\,dv'$$

here we have

$$\int f(\mathbf{r}')\delta(\mathbf{r}'-\mathbf{r})dv' = f(\mathbf{r}=0)$$

where $\mathbf{r} = 0$.

$$\mathbf{z} = \frac{\hat{\mathbf{z}}p_0(t')}{4\pi\varepsilon_0 r}e^{-i\omega t}e^{(i\omega/r)}$$

let $\omega/c = k$,

$$\mathbf{z}(\mathbf{r},t) = \frac{\hat{\mathbf{z}}p_0(t')}{4\pi\varepsilon_0 r}e^{-i\omega t}e^{ikr}$$

$$\mathbf{A} = \frac{1}{c^2}\frac{\partial \mathbf{z}}{\partial t} = \frac{\hat{\mathbf{z}} p_0 e^{ikr}(-i\omega) e^{-i\omega t}}{4\pi\varepsilon_0 r\ c^2}$$

$$\phi = -\nabla \cdot \mathbf{z}$$

Now for radiation zone differentiation (with respect to space) you treat everything as a constant except e^{ikr} terms.

$$\phi = -\hat{\mathbf{k}} \cdot \frac{\hat{\mathbf{z}} p_0 e^{-i\omega t}\ e^{ikr}\ ik}{4\pi\varepsilon_0 r}$$

$$\mathbf{B} = \nabla \times \mathbf{A} = \hat{\mathbf{k}} \times ik\mathbf{A} = i\mathbf{k} \times \mathbf{A}$$

$$\mathbf{E} = -\frac{c^2}{i\omega}\nabla \times \mathbf{B} = \frac{-c^2}{i\omega}\ \hat{\mathbf{k}} \times i\mathbf{k} \times \mathbf{A}\ ik = \left(\frac{-c^2}{i\omega}\right)(-\mathbf{k} \times \mathbf{k} \times \mathbf{A})$$

$$\mathbf{E} = \frac{c^2}{i\omega}\ \mathbf{k} \times (\mathbf{k} \times \mathbf{A})$$

$$\mathbf{S}_{av} = \tfrac{1}{2}\,\mathrm{Re}(\mathbf{E} \times \mathbf{H}^*)$$

no magnetization here, so $\mathbf{B} = \mu_0\mathbf{H}$, $\mathbf{H} = \mathbf{B}/\mu_0$.

$$\mathbf{E} \times \mathbf{H}^* = \left(\frac{c^2}{i\omega}\ \mathbf{k} \times (\mathbf{k} \times \mathbf{A})\right) \times \frac{i}{\mu_0}\ (\mathbf{k} \times \mathbf{A})$$

$$= \frac{c^2}{\omega\mu_0}\ [\mathbf{k} \times (\mathbf{k} \times \mathbf{A})] \times [\mathbf{k} \times \mathbf{A}]$$

$$= \frac{c^2}{\omega\mu_0}\ \hat{\mathbf{k}}|\mathbf{k} \times \mathbf{A}|^2 = \frac{c^2}{\omega\mu_0}\ \frac{\hat{\mathbf{k}}|\hat{\mathbf{k}} \times \mathbf{p}|^2}{16\pi^2\varepsilon_0^2 r^2 c^4}$$

$$\mathbf{S} = \tfrac{1}{2}\,\mathrm{Re}(\mathbf{E} \times \mathbf{H}^*) = \frac{c^2\hat{\mathbf{k}}|\mathbf{k} \times \mathbf{p}|^2}{\omega\mu_0\ 32\pi^2\varepsilon_0^2 r^2 c^4}$$

$$\overline{\frac{dP}{d\Omega}} = \mathbf{S} \cdot \hat{\mathbf{r}}\ r^2 = \frac{c^2\mathbf{k}|\mathbf{k} \times \mathbf{p}|^2}{\omega\mu_0 32\pi^2\varepsilon_0^2 r^2 c^4} \cdot \frac{\mathbf{k}}{k}\ r^2$$

$$\overline{\frac{dP}{d\Omega}} = \frac{c^2k^2p^2\sin^2\theta\ k^2\ r^2}{\omega\mu_0\ 32\pi^2\varepsilon_0^2 r^2 c^4 k} = \frac{k^3p^2\sin^2\theta}{32\pi^2\varepsilon_0^2 c^2\omega\mu_0}$$

$$k = \frac{\omega}{c},$$

$$\frac{dP}{d\Omega} = \frac{\omega^3\, p^2 \sin^2\theta}{c^3\; 32\pi^2 \varepsilon_0^2 c^2 \omega\mu_0} = \frac{\omega^2\, p^2 \sin^2\theta}{32\pi^2 \varepsilon_0\; c^3}$$

PROBLEM 14.2 Introduce the scalar and vector potentials and write the equations of electromagnetic theory in terms of these (assume the medium is linear). Discuss the Lorentz condition and gauge invariance.

Solution:

Two ways

(1) From Faraday's Law of induction

$$\oint_c \mathbf{E} \cdot d\mathbf{l} = -\frac{d}{dt} \int_{\text{surface}} \mathbf{B} \cdot \mathbf{n}\, da$$

$$\oint_c \mathbf{E} \cdot d\mathbf{l} = \int_s \nabla \times \mathbf{E} \cdot \hat{\mathbf{n}}\, da = -\frac{d}{dt} \int_s \mathbf{B} \cdot \hat{\mathbf{n}}\, da = -\frac{d}{dt} \int_s \nabla \times \mathbf{A} \cdot \hat{\mathbf{n}}\, da$$

$$= -\frac{d}{dt} \oint_c \mathbf{A} \cdot d\mathbf{l} = \oint_c -\frac{\partial \mathbf{A}}{\partial t} \cdot d\mathbf{l}$$

$$\oint_c \mathbf{E} \cdot d\mathbf{l} = \oint_c -\frac{\partial \mathbf{A}}{\partial t} \cdot d\mathbf{l}$$

$$\mathbf{E} = -\frac{\partial \mathbf{A}}{\partial t} + \text{a conservative quantity} \quad \text{(whose curl} = 0)$$

$$\mathbf{E} = -\frac{\partial \mathbf{A}}{\partial t} - \nabla\phi, \quad \text{general time dependent}$$

(2) Second way, since

$$\nabla \cdot \mathbf{B} = 0$$

$$\mathbf{B} = \nabla \times \mathbf{A}$$

also

$$\nabla \times \mathbf{E} = -\frac{\partial \mathbf{B}}{\partial t} = -\frac{\partial}{\partial t}(\nabla \times \mathbf{A}) = \nabla \times \left(-\frac{\partial \mathbf{A}}{\partial t}\right)$$

$$\nabla \times \left(\mathbf{E} + \frac{\partial \mathbf{A}}{\partial t}\right) = 0$$

if $\nabla \times \mathbf{V} = 0$, then $\mathbf{V} = -\nabla\phi$, so $\mathbf{E} + \frac{\partial \mathbf{A}}{\partial t} = -\nabla\phi$, and

$$\mathbf{E} = -\frac{\partial \mathbf{A}}{\partial t} - \nabla\phi, \quad \text{general time dependent}$$

(introduces the scalar and vector potentials). Now discuss gauge invariance.

(2) Since $B = \nabla \times A$ and curl $\nabla\phi = 0$, we can add some gradient of a scalar, $\nabla\psi$, to A to get a new A.

$$A' = A + \nabla\psi, \quad A = A' - \nabla\psi$$

now

$$E = -\nabla\phi - \frac{\partial A}{\partial t} = -\nabla\phi - \frac{\partial A'}{\partial t} + \nabla\frac{\partial\psi}{\partial t}$$

$$= -\nabla\left(\phi - \frac{\partial\psi}{\partial t}\right) - \frac{\partial A'}{\partial t}$$

where $\phi - \partial\psi/\partial t = \phi'$, so we have a new ϕ,

$$\phi' = \phi - \frac{\partial\psi}{\partial t}$$

These form what is known as a gauge transformation. Maxwell's equations are invariant under a gauge transformation. Now, the Lorentz condition,

$$\nabla \times H = j + \frac{\partial D}{\partial t}, \quad B = \mu H$$

$$D = \varepsilon E$$

$$\frac{1}{\mu}\nabla \times B = j + \frac{\partial E}{\partial t}$$

$$\frac{1}{\mu}(\nabla \times \nabla \times A) = j + \frac{\varepsilon\partial}{\partial t}\left(-\nabla\phi - \frac{\partial A}{\partial t}\right)$$

$$\frac{1}{\mu}[\nabla(\nabla \cdot A) - \nabla^2 A] = j - \varepsilon\nabla\frac{\partial\phi}{\partial t} - \varepsilon\frac{\partial^2 A}{\partial t^2}$$

$$\nabla(\nabla \cdot A) - \nabla^2 A = \mu j - \varepsilon\mu\frac{\partial\phi}{\partial t} - \varepsilon\mu\frac{\partial^2 A}{\partial t^2}$$

$$-\nabla^2 A + \varepsilon\mu\frac{\partial^2 A}{\partial t^2} + \nabla\left[\nabla \cdot A + \varepsilon\mu\frac{\partial\phi}{\partial t}\right] = \mu j$$

where

$$\nabla \cdot A + \varepsilon\mu\frac{\partial\phi}{\partial t} = 0 \quad \text{Lorentz condition}$$

The inhomongeneous wave equations are

$$\nabla^2 \mathbf{A} - \frac{1}{c^2}\frac{\partial^2 \mathbf{A}}{\partial t^2} = -\mu_0 \mathbf{j}$$

$$\nabla^2 \phi - \frac{1}{c^2}\frac{\partial^2 \phi}{\partial t^2} = -\frac{\rho}{\varepsilon_0}$$

$$\Box \phi = 0, \text{ homogeneous}$$

$$\Box \equiv \nabla^2 - \frac{1}{c^2}\frac{\partial^2}{\partial t^2}$$

$$\left.\begin{aligned} &\nabla \cdot \mathbf{A} + \varepsilon_0 \mu_0 \frac{\partial \phi}{\partial t} = 0 \\ &\nabla^2 \mathbf{A} - \varepsilon_0 \mu_0 \frac{\partial^2 \mathbf{A}}{\partial t^2} = -\mu_0 \mathbf{j} \\ &\nabla^2 \phi - \varepsilon_0 \mu_0 \frac{\partial^2 \phi}{\partial t^2} = -\frac{\rho}{\varepsilon_0} \end{aligned}\right\} \text{Lorentz gauge}$$

With the Lorentz condition,

$$\nabla \cdot \mathbf{A} + \varepsilon_0 \mu_0 \frac{\partial \phi}{\partial t} = 0$$

using our new $\mathbf{A}'$ and ϕ',

$$\mathbf{A}' = \mathbf{A} + \nabla\psi, \quad \mathbf{A} = \mathbf{A}' - \nabla\psi$$

$$\phi' = \phi - \frac{\partial \psi}{\partial t}, \quad \phi = \phi' + \frac{\partial \psi}{\partial t}$$

we get

$$\nabla \cdot (\mathbf{A}' - \nabla\psi) + \varepsilon_0 \mu_0 \frac{\partial}{\partial t}\left(\phi' + \frac{\partial \psi}{\partial t}\right) = 0$$

$$\nabla \cdot \mathbf{A}' - \nabla^2 \psi + \varepsilon_0 \mu_0 \frac{\partial \phi'}{\partial t} + \varepsilon_0 \mu_0 \frac{\partial^2 \psi}{\partial t^2} = 0$$

$$\nabla^2 \psi - \varepsilon_0 \mu_0 \frac{\partial^2 \psi}{\partial t^2} = 0$$

$$\nabla^2\psi - \frac{1}{c^2}\frac{\partial^2\psi}{\partial t^2} = 0$$

So that if $\Box\psi = 0$, the Lorentz condition remains satisfied. This is called a restricted gauge transformation (restricted group of gauge transformations, those which leave equations in Lorentz Gauge).

$$\nabla \cdot \mathbf{E} = \frac{\rho}{\varepsilon_0} = \nabla \cdot \left(-\nabla\phi - \frac{\partial \mathbf{A}}{\partial t}\right) = -\nabla^2\phi - \frac{\partial}{\partial t}\ \nabla \cdot \mathbf{A}$$

Poisson's equation

$$\nabla^2\phi = -\frac{\rho}{\varepsilon_0}$$

is true if $\nabla \cdot \mathbf{A} = 0$. $\nabla \cdot \mathbf{A} = 0 \rightarrow$ in Coulomb gauge where $\nabla^2\phi = -\rho/\varepsilon_0$.

PROBLEM 14.3 The current density of a full wave, end-driven, linear, antenna is given by

$$\mathbf{j} = \hat{\mathbf{z}} I_0 \sin(kd/2 - kz)\delta(x)\delta(y)$$

where

$$d = \lambda = 2\pi/k$$

Calculate the angular distribution of the power radiated from such an antenna

Solution:

$$\mathbf{j} = \hat{\mathbf{z}} I_0 \sin\left(k\frac{d}{2} - kz\right)\delta(x)\delta(y) = \hat{\mathbf{z}} I_0 \sin(kz')\delta(x)\delta(y)$$

$$d = \lambda = \frac{2\pi}{k}, \quad \pi = \frac{kd}{2}$$

$$\frac{dP}{d\Omega} = \frac{1}{32\pi^2\varepsilon_0 c}\left|\mathbf{k}\times\int \mathbf{j}_\omega(\mathbf{r}')e^{-\mathbf{k}\cdot\mathbf{r}'}\,dv'\right|^2$$

$$\int \mathbf{j}_\omega(\mathbf{r}')e^{-\mathbf{k}\cdot\mathbf{r}'}\,dv' = \int_{-d/2}^{d/2} zI_0 \sin(kz')\delta(x)\delta(y)e^{-\mathbf{k}\cdot\mathbf{r}'}\,dv'$$

(now $\mathbf{k}\cdot\mathbf{r}' = kz\cos\theta$) So,

$$= \int_{-d/2}^{d/2} \hat{\mathbf{z}} I_0 \sin kz'\ \delta(x)\delta(y)dx'\,dy'\,dz'\ e^{-kz'\cos\theta(i)}$$

since the region of integration includes the points $x = 0$, $y = 0$, the only thing left is $\int dz'$,

$$= \hat{\mathbf{z}} I_0 \int_{-d/2}^{d/2} \sin kz' \, e^{-ikz' \cos\theta} \, dz'$$

[use $\sin x$ = imaginary part of $e^{ix} = (e^{ix} - e^{-ix})/2i$.] So,

$$= \hat{\mathbf{z}} I_0 \int_{-d/2}^{d/2} \frac{e^{ikz'} - e^{-ikz'}}{2i} \, e^{-kz' \cos\theta} \, dz'$$

$$= \frac{\hat{\mathbf{z}} I_0}{2i} \int_{-d/2}^{d/2} e^{ikz'(1 - \cos 0)} - e^{-ikz'(1 + \cos\theta)}$$

$$= \frac{\hat{\mathbf{z}} I_0}{2i} \left(\frac{e^{ikz'(1 - \cos\theta)}}{ik(1 - \cos\theta)} - \frac{e^{-ikz'(1 + \cos\theta)}}{-ik(1 + \cos\theta)} \right)_{-d/2}^{d/2}$$

$$= \frac{\hat{\mathbf{z}} I_0}{2i} \left(\frac{e^{i\pi(1 - \cos\theta)}}{(1 - \cos\pi) ik} + \frac{e^{-i\pi(1 + \cos\theta)}}{(1 + \cos\theta) ik} - \frac{e^{-i\pi(1 - \cos\theta)}}{ik(1 - \cos\theta)} - \frac{e^{i\pi(1 + \cos\theta)}}{ik(1 + \cos\theta)} \right)$$

$$= \frac{\hat{\mathbf{z}} I_0}{-2k} \left(\frac{e^{i\pi(1 - \cos\theta)} - e^{-i\pi(1 - \cos\theta)}}{1 - \cos\theta} - \frac{e^{i\pi(1 + \cos\theta)} - e^{-i\pi(1 + \cos\theta)}}{1 + \cos\theta} \right)$$

$$= \frac{\hat{\mathbf{z}} I_0}{-2k} \left(\frac{e^{i\pi} e^{-i\pi\cos\theta} - e^{-i\pi} e^{i\pi\cos\theta}}{1 - \cos\theta} - \frac{e^{i\pi} e^{i\pi\cos\theta} - e^{-i\pi} e^{-i\pi\cos\theta}}{1 + \cos\theta} \right)$$

$$= \frac{\hat{\mathbf{z}} I_0 2i}{-2k(2i)} \left(\frac{(e^{i\pi\cos\theta} - e^{-i\pi\cos\theta})(1 + \cos\theta)(e^{i\pi\cos\theta} - e^{-i\pi\cos\theta})(1 - \cos\theta)}{\sin^2\theta} \right)$$

$$= \frac{\hat{\mathbf{z}} I_0 2i \sin(\pi\cos\theta)}{-k \sin^2\theta} = \int \mathbf{j}_\omega e^{-i\mathbf{k}\cdot\mathbf{r}'} \, dv'$$

Now,

$$\mathbf{k} \times \int \mathbf{j}_\omega e^{-i\mathbf{k}\cdot\mathbf{r}'} \, dv' = \mathbf{k} \times \mathbf{z} \left(\frac{I_0 2i \sin(\pi\cos\theta)}{-k \sin^2\theta} \right)$$

[and $\mathbf{k} \times \hat{\mathbf{z}} = k(1) \sin\theta(-\hat{\varphi})$]

so,

$$= \frac{k \sin\theta(+\hat{\varphi}) I_0 \sin(\pi\cos\theta) i}{k \sin^2\theta}$$

$$= \frac{\hat{\varphi} 2 i I_0 \sin(\pi\cos\theta)}{\sin\theta}$$

Now

$$\frac{dP}{d\Omega} = \frac{1}{32\pi^2\varepsilon_0 c}\left|\mathbf{k}\times\int \mathbf{j}_\omega(\mathbf{r}')e^{-i\mathbf{k}\cdot\mathbf{r}'}\,dv'\right|^2$$

$$= \frac{1}{32\pi^2\varepsilon_0 c}\left|\sqrt{\frac{\hat{\varphi}I_0 2i \sin(\pi\cos\theta)}{\sin\theta}\cdot\frac{\hat{\varphi}I_0\sin(\pi\cos\theta)2i}{\sin\theta}}\right|^2$$

$$= \frac{1}{32\pi^2\varepsilon_0 c}\,\frac{I_0^2 4\sin^2(\pi\cos\theta)}{\sin^2\theta}$$

$$\frac{dP}{d\Omega} = \frac{I_0^2}{8\pi^2\varepsilon_0 c}\,\frac{\sin^2(\pi\cos\theta)}{\sin^2\theta}$$

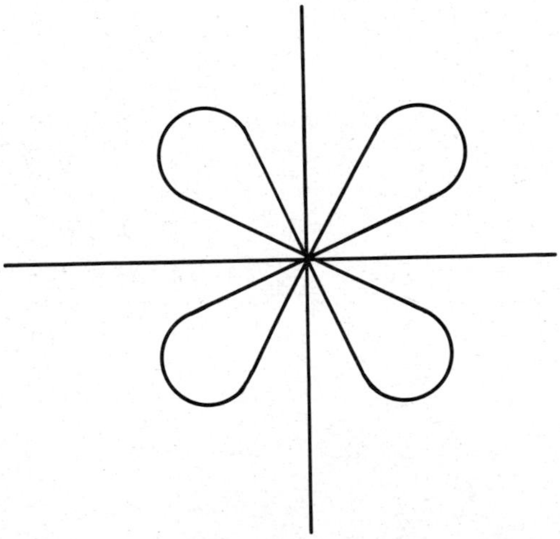

Shown here is a graph of $\dfrac{\sin^2(\pi\cos\theta)}{\sin^2\theta}$

PROBLEM 14.4

(a) What is the velocity of a beam of electrons when the simultaneous influence of an electric field of intensity 11.3 statvolt/cm and a magnetic field of 200 gauss, both fields being normal to the beam and to each other, produces no deflection of the electrons?
(b) Show in a diagram the relative orientation of the vectors **v**, **E**, and **B**. This device is called a "velocity selector" because it allows only electrons of a certain velocity to pass through without deflection.
(c) What is the radius of the electron orbit when the electric field is removed?

Solution:

(a) $\mathbf{F} = q\left(\mathbf{E} + \frac{\mathbf{v}}{c} \times \mathbf{B}\right) = 0, \quad \mathbf{v} \perp \mathbf{B}, \quad E = \frac{v}{c} B$

$$v = \frac{11.3 \times 3 \times 10^{10} \text{ cm/s}}{200 \text{ gauss}} = 1.7 \times 10^{9} \text{ cm/s}$$

(b) Take $\mathbf{E} = E\hat{\mathbf{z}}$, $\mathbf{v} = v\hat{\mathbf{y}}$, the, for $\mathbf{v} \times \mathbf{B} = -vB\hat{\mathbf{z}}$ we need $\mathbf{B} = B\hat{\mathbf{x}}$.

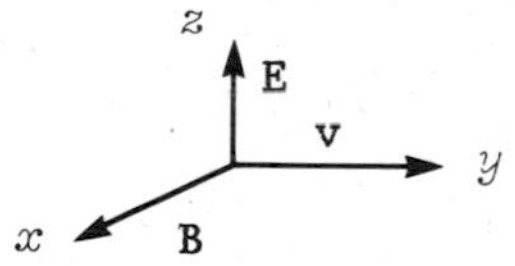

(c) For $\mathbf{v} \perp \mathbf{B}$, $M\frac{v^2}{r} = \frac{qvB}{c}$

$$r = \frac{Mvc}{qb} = \frac{9.1 \times 10^{-28} \times 1.70 \times 10^{9} \times 3 \times 10^{10}}{4.8 \times 10^{-10} \times 200}$$

$r = 0.483$ cm

Relativity has been ignored, but is small for $v/c = 0.057$.

PROBLEM 14.5 Prove that no gauge transformation will in general exist which transforms the vector potential away.

Solution:

$$\mathbf{B} = \nabla \times \mathbf{A}$$

Assume $\mathbf{A}' = \mathbf{A} + \nabla f$ such that $\mathbf{A}' = 0$. Then $\mathbf{A} = -\nabla f$, but the curl of any grad is zero, therefore it is a contradiction since **B** is not in general zero.

PROBLEM 14.6 Prove that for the case $\rho = 0$, one can, starting with arbitrary potentials, always transform to the radiation gauge.

Solution: Radiation gauge,

$$\phi' = 0, \quad \nabla \cdot \mathbf{A}' = 0$$

$$\mathbf{A}' = \mathbf{A} - \nabla f, \quad \phi' = \phi + \frac{\partial f}{\partial t}$$

In order to obtain $\nabla \cdot \mathbf{A}' = 0$ we must choose f such that $\nabla \cdot \mathbf{A} - \nabla^2 f = 0$. But this is just Poisson's equation defining f in terms of the specified function, div A. Its solution can always be obtained and is equal to

$$f = -\frac{1}{4\pi} \iiint \frac{\nabla \mathbf{A}(\mathbf{r}')dx'\,dy'\,dz'}{|\mathbf{r} - \mathbf{r}'|}$$

Thus, a gauge transformation that yields $\nabla \cdot \mathbf{A}' = 0$ can always be carried out. Now show that if $\rho = 0$ the choice of $\nabla \cdot \mathbf{A}' = 0$ also leads to $\phi' = 0$,

$$\nabla \cdot \mathbf{E} = \frac{1}{\varepsilon_0} \rho = 0$$

$$\mathbf{E} = -\frac{\partial \mathbf{A}'}{\partial t} - \nabla \phi'$$

$$\nabla \cdot \mathbf{E} = -\nabla \cdot \frac{\partial \mathbf{A}'}{\partial t} - \nabla^2 \phi' = 0$$

but $\nabla \cdot \mathbf{A}' = 0$, therefore,

$$-\nabla^2 \phi' = 0$$

This is LaPlace's equation. The only solution of this equation that is regular over all space is $\phi' = 0$.

PROBLEM 14.7 A common textbook example of a radiating system is a configuration of charges fixed relative to each other but in rotation. The charge density is obviously a function of time, but it is not in the form of

$$\rho(\mathbf{x}, t) = \rho(\mathbf{x})e^{-i\omega t}$$

(a) Show that for rotating charges one alternative is to calculate real time-dependent multiple moments using $\rho(\mathbf{x}, t)$ directly and then compute the multipole moments for a given harmonic frequency with the convention of

$$\rho(\mathbf{x}, t) = \rho(\mathbf{x})e^{-i\omega t}$$

by inspection or Fourier decomposition of the time-dependent moments. Note that care must be taken when calculating $q_{1m}(t)$ to form linear combinations that are real before making the connection.

(b) Consider a charge density $\rho(\mathbf{x}, t)$ that is periodic in time with period $T = 2\pi/\omega_0$. By making a Fourier series expansion, show that it can be written as

$$\rho(\mathbf{x},t) = \rho_0(\mathbf{x}) + \sum_{n=1}^{\infty} \text{Re}[2\rho_n(\mathbf{x})e^{-in\omega_0 t}]$$

where

$$\rho_n(\mathbf{x}) = \frac{1}{T}\int_0^T \rho(\mathbf{x},t)e^{in\omega_0 t}\, dt$$

This shows explicitly how to establish connection with

$$\rho(\mathbf{x},t) = \rho(\mathbf{x})e^{-i\omega t}$$

(c) For a single charge q rotating about the origin in the $x-y$ plane in a circle of radius R at constant angular speed ω_0, calculate the $\ell = 0$ and $\ell = 1$ multipole moments by the two methods, (a) and (b), and compare. In method (b) express the charge density $\rho_n(\mathbf{x})$ in cylindrical coordinates. Are there higher multipoles, for example, quadrupole? At what frequencies?

Solution:

(a) One form of $\rho(\mathbf{x},t)$ could be (for a distribution rotating about the z axis at a rotational frequency ω_0):

$$\rho(\mathbf{x},t) = \sum_{i=1}^{n} \frac{q_i\delta(r - r_i)\delta(\theta - \theta_i)\delta[\phi - (\phi_i + \omega_0 t)]}{r^2 \sin\theta}$$

Then the multipole moments will be given by,

$$q_{\ell,m}(t) = \int Y^*_{\ell,m}(\theta',\phi')r'^{\ell}\rho(\mathbf{x}',t)d^3x'$$

$$= \sum_{i=1}^{n} q_i$$

$$\times \int Y^*_{\ell,m}(\theta',\phi') \frac{\delta(r' - r_i)\delta(\theta' - \theta_i)\delta[\phi' - (\phi_i + \omega_0 t)]}{r'^2 \sin\theta'} r'^{\ell} r'^2 \sin\theta'\, dr'\, d\theta'\, d\phi'$$

$$q_{\ell,m}(t) = \sum_{i=1}^{n} q_i r_i Y^*_{\ell,m}(\theta_i, \phi_i + \omega_0 t)$$

Now, the $Y_{\ell,m}$'s are of the form,

$$Y_{\ell,m}(\theta,\phi) = (-1)^m \sqrt{\frac{(2\ell+1)}{4\pi}\frac{(\ell-m)!}{(\ell+m)!}}\, P_{\ell,m}(\cos\theta)e^{im\phi}$$

therefore,

$$Y_{\ell,m}(\theta_i, \phi_i + \omega_0 t) = (-1)^m \sqrt{\frac{(2\ell+1)}{4\pi}\frac{(\ell-m)!}{(\ell+m)!}}\, P_{\ell,m}(\cos\theta_i) e^{im\phi_i} e^{im\omega_0 t}$$

$$= Y_{\ell,m}(\theta_i, \phi_i) e^{im\omega_0 t}$$

$$q_{\ell,m}(t) = \sum_{i=1}^{N} q_i r_i^{\ell}\, Y^*_{\ell,m}(\theta_i, \phi_i) e^{-im\omega_0 t}$$

$$q_{\ell,m}(t) = q_{\ell,m}\, e^{-im\omega_0 t}; \quad q_{\ell,m} = \sum_{i=1}^{N} q_i r_i^{\ell}\, Y^*_{\ell,m}(\theta_i, \phi_i)$$

Therefore, the time dependence of $\rho(\mathbf{x}, t)$ can be separated out when we expand in the multipole moments. That is,

$$\rho(\mathbf{x}, t) = \rho(\mathbf{x}) e^{-i\omega t}$$

with $\omega = m\omega_0$ for the ℓ, mth multipole moment. That is, $\omega =$ the mth harmonic of ω_0.

(b) $\rho(\mathbf{x}, t)$ is periodic in time with $T = 2\pi/\omega$. Now, given a function $f(x)$ periodic with period $2L$, it can be expanded in a Fourier series:

$$f(x) = \frac{a_0}{2} + \sum_{n=1}^{\infty} \left(a_n \cos\frac{n\pi x}{L} + b_n \sin\frac{n\pi x}{L} \right)$$

where $(n = 0, 1, 2, \ldots)$

$$a_n = \frac{1}{L} \int_0^{2L} f(x) \cos\frac{n\pi x}{L}\, dx$$

$$b_n = \frac{1}{L} \int_0^{2L} f(x) \sin\frac{n\pi x}{L}\, dx$$

Thus, if we do a Fourier series expansion of $\rho(\mathbf{x}, t)$ in the time dependence, then

$$\rho(\mathbf{x}, t) = \frac{a_0}{2} + \sum_{n=1}^{\infty} a_n \cos(n\omega_0 t) + b_n \sin(n\omega_0 t)$$

where

$$a_n = \frac{\omega_0}{\pi} \int_0^{2\pi/\omega_0} \rho(\mathbf{x}, t) \cos(n\omega_0 t) dt = \frac{\omega_0}{2\pi} \int_0^{2\pi/\omega_0} \rho(\mathbf{x}, t)(e^{in\omega_0 t} + e^{-in\omega_0 t}) dt$$

$$b_n = \frac{\omega_0}{\pi} \int_0^{2\pi/\omega_0} \rho(\mathbf{x}, t) \sin n\omega_0 t\, dt = \frac{\omega_0}{2i\pi} \int_0^{2\pi/\omega_0} \rho(\mathbf{x}, t)(e^{in\omega_0 t} - e^{-in\omega_0 t}) dt$$

let

$$\rho_n = \frac{\omega_0}{\pi} \int_0^{2\pi/\omega_0} \rho(\mathbf{x},t) e^{in\omega_0 t}\, dt$$

therefore,

$$\rho(\mathbf{x},t) = \frac{a_0}{2} + \sum_{n=1}^{\infty} \left[\tfrac{1}{2}(\rho_n + \rho_{-n}) \tfrac{1}{2}(e^{in\omega_0 t} + e^{-in\omega_0 t}) \right.$$

$$\left. + \frac{1}{2i}(\rho_n - \rho_{-n}) \times \frac{1}{2i}(e^{in\omega_0 t} - e^{-in\omega_0 t}) \right]$$

$$= \frac{a_0}{2} + \tfrac{1}{4} \sum_{n=1}^{\infty} \left[\rho_n e^{in\omega_0 t} + \rho_n e^{-in\omega_0 t} + \rho_{-n} e^{in\omega_0 t} + \rho_{-n} e^{-in\omega_0 t} \right.$$

$$\left. - \rho_n e^{in\omega_0 t} + \rho_n e^{-in\omega_0 t} + \rho_{-n} e^{in\omega_0 t} - \rho_{-n} e^{-in\omega_0 t} \right]$$

$$= \frac{a_0}{2} + \tfrac{1}{2} \sum_{n=1}^{\infty} (\rho_n e^{-in\omega_0 t} + \rho_{-n} e^{in\omega_0 t})$$

$$= \frac{a_0}{2} + \tfrac{1}{2} \sum_{n=1}^{\infty} [\rho_n e^{-in\omega_0 t} + (\rho_{+n} e^{-in\omega_0 t})^*]$$

Now for

$$g = u + iv, \quad g + g^* = (u + iv) + (u - iv) = 2u$$

$$= 2\mathrm{Re}(g)$$

$$\rho(\mathbf{x},t) = \frac{a_0}{2} + \tfrac{1}{2} \sum_{n=1}^{\infty} \{\mathrm{Re}[2\rho_n(\mathbf{x},t) e^{-in\omega_0 t}]\}$$

Let $a_0/2 = p_0$ and $p_n/2 \to p_n$, then

$$\rho(\mathbf{x},t) = \rho_0(\mathbf{x}) + \sum_{n=1}^{\infty} \mathrm{Re}[2\rho_n(\mathbf{x}) e^{-in\omega_0 t}]$$

(c) From the result of part (a),

$$\rho(\mathbf{x},t) = \sum_{i=1}^{N} \frac{q_i \delta(r - r_i)\delta(\theta - \theta_i)\delta(\phi - \phi_i - \omega_0 t)}{r^2 \sin\theta}$$

we obtain for a single charge q rotating in $x-y$ plane at R with angular frequency ω_0:

$$\rho(\mathbf{x},t) = \frac{q\delta(r-R)\delta(\theta-\pi/2)\delta(\phi-\omega_0 t)}{r^2 \sin\theta}$$

choose ϕ_i = phase = 0, therefore,

$$q_{\ell,m}(t) = \sum_{i=1}^{N} q_i r_1^{\ell}\, Y^*_{\ell,m}(\theta_i,\phi_i) e^{-im\omega_0 t}$$

$$q_{\ell,m}(t) = qR^{\ell}\, Y^*_{\ell,m}(\pi/2,\ 0) e^{-im\omega_0 t}$$

$$q_{0,0}(t) = q\ R^0\ Y^*_{0,0}(\pi/2,\ 0)\cdot 1 = \frac{q}{\sqrt{4\pi}}$$

$$q_{0,0}(t) = \frac{q}{\sqrt{4\pi}}$$

$$q_{1,0}(t) = q\ R\ Y^*_{1,0}(\pi/2,\ 0)\cdot 1 = qR\ \sqrt{3/4\pi}\cos\left(\frac{\pi}{2}\right) = 0$$

$$q_{1,0}(t) = 0$$

$$q_{1,1}(t) = q\ R\left(-\sqrt{\frac{3}{8\pi}}\ \sin\frac{\pi}{2}\ e^{-i\cdot 0}\right) e^{-i\omega_0 t}$$

$$q_{1,1}(t) = q\sqrt{\frac{3}{8\pi}}\ q\ R\ e^{-i\omega_0 t}$$

$$q_{1,-1}(t) = q\ R\ \left(\sqrt{\frac{3}{8\pi}}\sin\frac{\pi}{2}\ e^{i\cdot 0}\right) e^{i\omega_0 t}$$

$$q_{1,-1}(t) = \sqrt{\frac{3}{8\pi}}\ qRe^{i\omega_0 t}$$

Now, using the result of part (b),

$$\rho(\mathbf{x},t) = \frac{q\delta(z)\delta(r-R)\delta(\phi-\omega_0 t)}{r}$$

therefore,

$$\rho_n(\mathbf{x}) = \frac{\omega_0}{2\pi}\int_0^{2\pi/\omega_0} \frac{q}{r}\ \delta(r-R)\delta(z)\delta(\phi-\omega_0 t)e^{in\omega_0 t}\ dt$$

$$\rho_n(\mathbf{x}) = \frac{q\omega_0}{2\pi r}\,\delta(r-R)\delta(z)\int_0^{2\pi}\frac{dz}{\omega_0}\,\delta(\phi - z)e^{inz}$$

$$\rho_n(\mathbf{x}) = \frac{q}{2\pi r}\,\delta(r-R)\delta(z)e^{in\phi}$$

$$\rho(\mathbf{x},t) = \rho_0(\mathbf{x}) + \sum_{n=1}^{\infty}\ \mathrm{Re}\left(\frac{q}{\pi r}\,\delta(r-R)\delta(z)e^{in\phi}\ e^{-in\omega_0 t}\right)$$

with

$$\rho_0(\mathbf{x}) = \frac{q}{2\pi r}\,\delta(r-R)\delta(z)$$

therefore,

$$q_{\ell,m}(t) = \int Y^*_{\ell,m}\ (\theta,\phi)r^{\ell}\rho(\mathbf{x},t)d^3x$$

$$= \int Y^*_{\ell,m}\ (\theta,\phi)r^{\ell}\left[\frac{q}{2\pi r}\,\frac{\delta(r-R)\delta(\phi-\pi/2)}{r\sin\theta}\right.$$

$$\left. + \sum_{n=1}^{\infty}\ \mathrm{Re}\left(\frac{q}{\pi r^2\sin\theta}\,\delta(r-R)\delta(\theta-\pi/2)e^{in\phi}e^{-in\omega_0 t}\right)\right]d^3x$$

$$= \left(\int_0^{2\pi} Y^*_{\ell,m}\ (\pi/2,\phi)d\phi\right)R^{\ell}\ \frac{q}{2\pi} + \sum_{n=1}^{\infty}\ \mathrm{Re}\left(\int_0^{2\pi} Y^*_{\ell,m}\ (\pi/2,\phi)e^{in\phi}\ d\phi\right)R^{\ell}\ \frac{q}{\pi}\ e^{-in\omega_0 t}$$

therefore,

$$q_{0,0} = \frac{q}{\sqrt{4\pi}} + \frac{q}{\sqrt{4\pi}\ \pi}\sum_{n=1}^{\infty}\ e^{-in\omega_0 t}\ \mathrm{Re}\left(\frac{1}{in}\ [1-1]\right) = \frac{q}{\sqrt{4\pi}}$$

and we get the expected result.

$$q_{1,0} = cR\ q\cos\frac{\pi}{2} + \frac{Rq}{\pi}\sqrt{\frac{3}{4\pi}}\sum_{n=1}^{\infty}\ \mathrm{Re}\ e^{-in\omega_0 t}\left(\int_0^{2\pi}\cos\frac{\pi}{2}\ e^{in\phi}\ d\phi\right)$$

therefore, $q_{1,0} = 0$ as expected.

$$q_{1,\pm 1}(t) = \mp\frac{Rq}{2\pi}\sqrt{\frac{3}{8\pi}}\int_0^{2\pi} e^{\pm i\phi}d\phi \mp \frac{Rq}{\pi}\sqrt{\frac{3}{8\pi}}\sum_{n=1}\ \mathrm{Re}\left(e^{-in\omega_0 t}\int_0^{2\pi} e^{i(n\pm 1)\phi}d\phi\right)$$

$$= \mp \frac{Rq}{\pi} \sqrt{\frac{3}{8\pi}} \sum_{n=1}^{\infty} \text{Re}(e^{-in\omega_0 t}\, 2\pi\delta_{n,\pm 1})$$

$$= \mp \sqrt{\frac{3}{8\pi}}\, 2qR\, e^{\mp i\omega_0 t} \quad \text{(neglecting the real part)}$$

$$q_{1,1}(t) = -2\sqrt{\frac{3}{8\pi}}\, qR\, e^{-i\omega_0 t}, \quad q_{1,-1}(t) = 2\sqrt{\frac{3}{8\pi}}\, qRe^{i\omega_0 t}$$

Except for the fact that there is a factor of 2 difference [due to definition of $\rho_n(\mathbf{x})$], and since we did not form real combinations in part (a), the results agree.

There will be higher multipole moments. However, the presence of the cosines in $Y_{2,m}$ will cause the quadrupole moments to vanish. The higher-order multipole moments occur at frequencies $\omega = m\omega_0$.

PROBLEM 14.8 A thin linear antenna of length d is excited in such a way that the sinusoidal current makes a full wavelength of oscillation as shown in the figure.

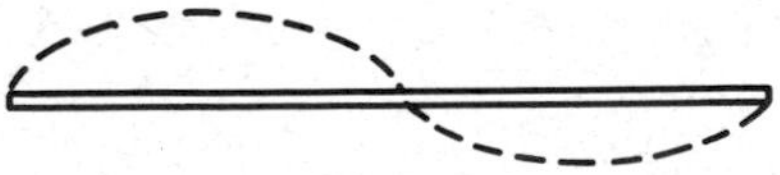

(a) Calculate exactly the power radiated per unit solid angle and plot the the angular distribution of radiation. Treat the linear antenna by the long-wavelength multipole expansion method.
(b) Determine the total power radiated and find a numerical value for the radiation resistance.
(c) Calculate the multipole moments (electric dipole, magnetic dipole, and electric quadrupole).
(d) Compare the shape of the angular distribution of radiated power for the lowest nonvanishing multipole with the exact distribution.
(e) Determine the total power radiated for the lowest multipole and the corresponding radiation resistance. Is there a paradox?

Solution:

(a)

$$\mathbf{j}(\mathbf{x}) = I_0 \sin(kz)\delta(x)\delta(y)\hat{\mathbf{e}}_3 \quad \text{for } 0 \le z \le d$$

and

$$\sin kd = 0, \quad \therefore \quad kd = n\pi,$$

choose $\sin kd/2 = 0$ also, $kd = 2\pi$

$$\therefore \quad \mathbf{j}(\mathbf{x}) = I_0 \sin\left(\frac{2\pi z}{d}\right)\delta(x)\delta(y)\hat{\mathbf{e}}_3$$

$$\therefore \quad \mathbf{A}(\mathbf{x}) = \frac{1}{c}\int \mathbf{j}(\mathbf{x}')\,\frac{e^{ik|\mathbf{x}-\mathbf{x}'|}}{|\mathbf{x}-\mathbf{x}'|}\,d^3x'$$

In the radiation zone, this becomes ($kr >> 1$):

$$\mathbf{A}(\mathbf{x}) = \frac{e^{ikr}}{cr}\int \mathbf{j}(\mathbf{x}')e^{-ik\hat{\mathbf{n}}\cdot\mathbf{x}'}\,d^3x'$$

$\hat{\mathbf{n}}$ = a unit vector of direction of $\mathbf{x}$.

$$\mathbf{A}(\mathbf{x}) = \hat{\mathbf{e}}_3\,\frac{I_0 e^{ikr}}{cr}\int_0^d \sin kz\; e^{-ikz\cos\theta}\,dz$$

where,

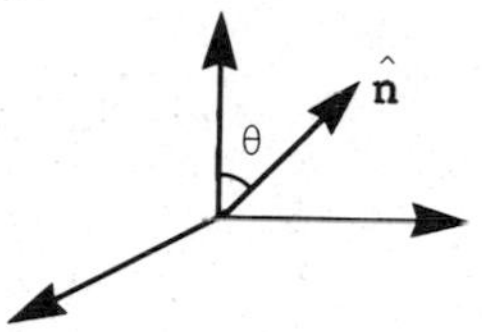

$$\therefore \quad \mathbf{A}(\mathbf{x}) = \hat{\mathbf{e}}_3\,\frac{I_0\,e^{ikr}}{cr}\,\frac{1}{2i}\int_0^d dz\left(e^{i(kz)(1-\cos\theta)} - e^{-i(kz)(1+\cos\theta)}\right)$$

$$= \hat{\mathbf{e}}_3\,\frac{I_0\,e^{ikr}}{cr}\,\frac{1}{2i}\left\{\frac{e^{ikd(1-\cos\theta)}-1}{ik(1-\cos\theta)} + \frac{e^{ikd(1+\cos 0)}-1}{ik(1+\cos\theta)}\right\}$$

$$= -\hat{\mathbf{e}}_3\,\frac{I_0 e^{ikr}}{kcr}\,\tfrac{1}{2}\left\{\frac{(e^{ikd(1-\cos\theta)}-1)(1+\cos\theta) + (e^{-ikd(1+\cos\theta)}-1)(1-\cos\theta)}{1-\cos^2\theta}\right\}$$

$$= \frac{-\hat{\mathbf{e}}_3 I_0 e^{ikr}}{2kcr\sin^2\theta}\left[(2\cos kd)e^{-ikd\cos\theta} - 2 + \cos\theta\, e^{-ikd\cos\theta}(e^{ikd} - e^{-ikd})\right]$$

$$= \frac{+\hat{\mathbf{e}}_3 I_0 e^{ikr}}{kcr\sin^2\theta}\left\{1 - e^{-ikd\cos\theta} + i\sin kd\cos\theta\; e^{-ikd\cos\theta}\right\}$$

$$= \frac{\hat{\mathbf{e}}_3 I_0 e^{ikr}}{kcr}\left\{\frac{1-e^{-i2\pi\cos\theta}}{\sin^2\theta}\right\} = \frac{\hat{\mathbf{e}}_3 I_0 e^{ikr}}{\omega r}\left\{\frac{1-\cos(kd\cos\theta)+i\sin(kd\cos\theta)}{\sin^2\theta}\right\}$$

Now, in the radiation zone $\mathbf{B} = ik\hat{\mathbf{n}}\times\mathbf{A}$ or

$$|\mathbf{B}| = k\sin\theta\;|A_3|$$

$$|A_3| = \frac{I_0}{\omega r}\frac{\sqrt{[1-\cos(kd\cos\theta)]^2+\sin^2(kd\cos\theta)}}{\sin^2\theta}$$

$$= \frac{I_0}{\omega r\sin^2\theta}\,(1+\cos^2\gamma+\sin^2\gamma-2\cos\gamma)^{\frac{1}{2}}$$

where $\gamma = kd\cos\theta$, therefore,

$$|A_3| = \frac{I_0}{\omega r\sin^2\theta}\sqrt{2}\sqrt{1-\cos\gamma} = \frac{2I_0}{\omega r\sin^2\theta}\sin\left(\frac{kd}{2}\cos\theta\right)$$

$$|A_3| = \frac{2I_0}{\omega r}\frac{\sin(\pi\cos\theta)}{\sin^2\theta}$$

$$|\mathbf{B}| = \frac{2I_0}{cr}\frac{\sin(\pi\cos\theta)}{\sin\theta}$$

since

$$\frac{dP}{d\Omega} = \frac{c}{8\pi}\,\mathrm{Re}(r^2\hat{\mathbf{n}}\cdot\mathbf{E}\times\mathbf{B}^*)$$

$$\frac{dP}{d\Omega} = \frac{I_0^2}{2\pi c}\left|\frac{\sin(\pi\cos\theta)}{\sin\theta}\right|^2$$

The plot of the angular distribution of radiation, $\theta \sim 53.9°$ is found by calculating the maximum of $f(\theta) = \sin^2(\pi\cos\theta)/\sin^2\theta$.

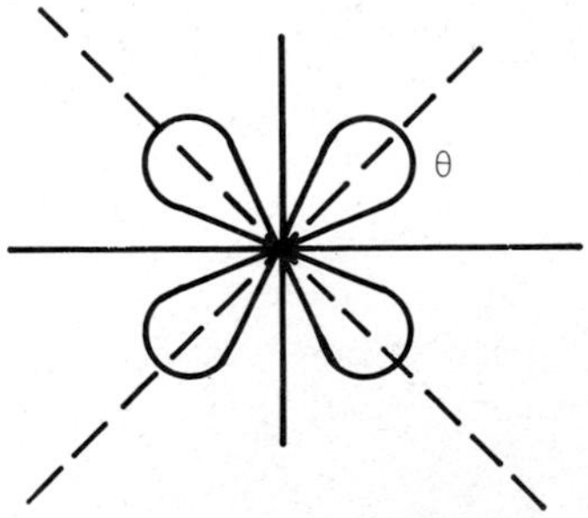

$$P = \int \frac{dP}{d\Omega}\,d\Omega = \frac{I_0^2}{2\pi c}\int_0^{2\pi} d\phi \int_0^{\pi} \frac{\sin^2(\pi\cos\theta)\sin\theta\,d\theta}{\sin^2\theta}$$

$$= \frac{I_0^2}{c} \int_{-1}^{1} \frac{\sin^2(\pi u)du}{1-u^2}$$

$$= \frac{I_0^2}{2c} \int_{-1}^{1} \left(\frac{-\sin^2(\pi u)}{1-u} - \frac{\sin^2(\pi u)}{1+u} \right) du$$

$$= \frac{-I_0^2}{2c} \int_{-1}^{1} \left(\frac{\sin^2(\pi u)}{1-u} + \frac{\sin^2(\pi u)}{1+u} \right) du$$

$$= \frac{-I_0^2}{4c} \int_{-1}^{1} \left(\frac{1-\cos 2\pi u}{1-u} + \frac{1-\cos 2\pi u}{1+u} \right) du$$

Consider

$$I_1 = \int_{-1}^{1} \frac{1-\cos 2\pi u}{1-u}\, du$$

let $z = (1-u)2\pi$, $2\pi u = 2\pi - z$, $dz = -2\pi du$, thereforc,

$$I_1 = \int_{4\pi}^{0} \frac{1-\cos(2\pi - z)}{z/2\pi} \left(\frac{-dz}{2\pi} \right) = \int_{0}^{4\pi} \frac{1-\cos z\, dz}{z}$$

and

$$I_2 = \int_{-1}^{1} \frac{1-\cos 2\pi u}{1+u}\, du = \int_{0}^{4\pi} \frac{1-\cos z}{z}\, dz$$

$$P = +\frac{I_0^2}{4c} \left(2 \int_{0}^{4\pi} \frac{1-\cos z}{z}\, dz \right)$$

Now, the cosine integral is defined by

$$\text{Ci}(x) = -\int_{x}^{\infty} \frac{\cos z}{z}\, dz = \ln \gamma\, x - \int_{0}^{x} \frac{1-\cos z}{z}\, dz, \quad \gamma = 1.781\ldots$$

$$\int_{0}^{4\pi} \frac{1-\cos z}{z}\, dz = \ln(\gamma \cdot 4\pi) + \int_{x}^{\infty} \frac{\cos z}{z}\, dz = \ln(4\pi\gamma) - \text{Ci}(4\pi)$$

$$\cong 3.11 - 0.0114 \sim 3.10$$

$$\therefore \quad P \simeq \frac{I_0^2}{2}\left(\frac{3.10}{c}\right)$$

Now

$$R_{\text{rad}} = (30c)(3.10/c) \text{ ohms}$$

$$R_{\text{rad}} \simeq 93\ \Omega$$

(c) The electric dipole moment is given by

$$\mathbf{P} = \int \mathbf{x}'\rho(x')d^3x' = \frac{i}{\omega}\int \mathbf{j}d^3x'$$

$$\mathbf{j}(\mathbf{x}) = I_0 \sin\left(\frac{2\pi z}{d}\right)\delta(x)\delta(y)\hat{\boldsymbol{\varepsilon}}_3, \quad 0 \leq z \leq d$$

$$\mathbf{P} = \hat{\boldsymbol{\varepsilon}}_3\, I_0\, \frac{i}{\omega}\int_{-\infty}^{\infty} \delta(x)dx \int_{-\infty}^{\infty} \delta(y)dy \int_0^d \sin\left(\frac{2\pi z}{d}\right)dz$$

$$= \hat{\boldsymbol{\varepsilon}}_3\, \frac{id}{2\pi\omega}\, I_0(-1)[\cos(2\pi) - \cos(0)] = 0$$

Therefore, $\mathbf{P} = 0$, thus, this distribution has zero electric dipole moment. The magnetic dipole moment is given by

$$\mathbf{m} = \frac{1}{2c}\int (\mathbf{x} \times \mathbf{j})d^3x$$

$$= \frac{1}{2c}\int d^3x(x\hat{\boldsymbol{\varepsilon}}_1 + y\hat{\mathbf{e}}_2 + z\hat{\mathbf{e}}_3) \times \hat{\boldsymbol{\varepsilon}}_3 I_0\delta(x)\delta(y)\sin\left(\frac{2\pi z}{d}\right)$$

$$= \frac{I_0}{2c}\left[-\hat{\boldsymbol{\varepsilon}}_2\int_{-\infty}^{\infty} x\delta(x)dx \int_{-\infty}^{\infty} \delta(y)dy \int_0^d \sin\left(\frac{2\pi z}{d}\right)dz\right.$$

$$\left. + \hat{\boldsymbol{\varepsilon}}_1\int_{-\infty}^{\infty} \delta(x)dx \int_{-\infty}^{\infty} y\delta(y)dy \int_0^d \sin\left(\frac{2\pi z}{d}\right)dz + 0\right]$$

Therefore, $\mathbf{m} = 0$, and the current distribution has a zero magnetic dipole moment. The quadrupole moment tensor is given by

$$Q_{\alpha\beta} = \int (3x_\alpha x_\beta - r^2\delta_{\alpha\beta})\rho(\mathbf{x})d^3x$$

where:

$$\rho(\mathbf{x}) = \frac{1}{i\omega}\nabla\cdot\mathbf{j} = \frac{1}{i\omega}\left(\frac{2\pi}{d}\right)I_0 \cos\left(\frac{2\pi z}{d}\right)\delta(x)\delta(y)$$

$$Q_{11} = \int (3x^2 - x^2 - y^2 - z^2)\rho(\mathbf{x})d^3x = \int (2x^2 - y^2 - z^2)\rho(\mathbf{x})d^3x$$

$$Q_{11} = F(y,z)\int_{-\infty}^{\infty} 2x^2\delta(x)dx - F'(x,z)\int_{-\infty}^{\infty} y^2\delta(y)dy - \frac{2\pi I_0}{i\omega d}\int_0^d z^2 \cos\left(\frac{2\pi z}{d}\right)dz$$

let

$$y = \frac{2\pi z}{d},\quad dy = \frac{2\pi}{d}\,dz$$

$$= \frac{iI_0}{\omega}\left(\frac{d}{2\pi}\right)^2 \int_0^{2\pi} y^2 \cos y\, dy$$

$$Q_{11} = \frac{iI_0 d^2}{\pi\omega}$$

$$Q_{22} = \int (2y^2 - x^2 - z^2)\rho(\mathbf{x})dx = \frac{2\pi i I_0}{\omega d}\int_0^d z^2 \cos\left(\frac{2\pi z}{d}\right)dz$$

$$Q_{22} = \frac{iI_0 d^2}{\pi\omega}$$

Now:

$$Q_{11} + Q_{22} + Q_{33} = 0$$

therefore,

$$Q_{33} = -Q_{11} - Q_{22} = 2Q_{11}$$

$$Q_{33} = \frac{2I_0 d^2}{i\pi\omega}$$

The off-diagonal elements are:

$$Q_{12} = \int 3xy\rho(\mathbf{x})d^3x = 0$$

since $\rho \propto \delta(x)\delta(y)$;

$$Q_{13} = \int 3(xz)\rho(\mathbf{x})d^3x = 0$$

since $\rho \propto \delta(x)$; also,

$$Q_{21} = \int 3yx\rho(\mathbf{x})d^3x = 0$$

$$Q_{23} = \int 3yz\rho(\mathbf{x})d^3x = 0$$

$$Q_{31} = \int 3zx\rho(\mathbf{x})d^3x = 0$$

$$Q_{32} = \int 3zy\rho(\mathbf{x})d^3x = 0$$

all since $\rho \propto \delta(x)\delta(y)$. Therefore, the quadrupole moment for this current distribution is

$$Q_{11} = Q_{22}$$

$$= -\tfrac{1}{2}\, Q_{33} = \frac{iI_0 d^2}{\pi\omega}$$

all $Q_{\alpha\beta} = 0$, for $\alpha \neq \beta$.

(d) The vector $\mathbf{Q}(\hat{\mathbf{n}})$ is then given by

$$Q_\alpha = \sum_\beta Q_{\alpha\beta}\, n_\beta$$

$\hat{\mathbf{n}}$ = unit vector in direction of $\mathbf{x}$ (source to observation point vector)

$$Q_1 = Q_{11}n_1, \quad Q_2 = Q_{22}n_2, \quad Q_3 = Q_{33}n_3$$

let

$$Q_0 = Q_{11} = Q_{22} = -\tfrac{1}{2}Q_{33} = \frac{iI_0 d^2}{\pi\omega}$$

$$\mathbf{Q}(\mathbf{n}) = Q_0\left(n_1\hat{\mathbf{e}}_1 + n_2\hat{\mathbf{e}}_2 - \frac{n_3}{2}\,\hat{\mathbf{e}}_3\right)$$

$$\frac{dP}{d\Omega} = \frac{ck^6}{288\pi} \left|[\hat{\mathbf{n}} \times \mathrm{Q}(\hat{\mathbf{n}})] \times \hat{\mathbf{n}}\right|^2$$

$$\hat{\mathbf{n}} \times \mathrm{Q}(\hat{\mathbf{n}}) = Q_0 \left(n_1 n_2 \hat{\mathbf{e}}_3 + \frac{n_1 n_3}{2} \hat{\mathbf{e}}_2 - n_1 n_2 \hat{\mathbf{e}}_3 - \frac{n_2 n_3}{2} \hat{\mathbf{e}}_1 + n_3 n_1 \hat{\mathbf{e}}_2 - n_3 n_2 \hat{\mathbf{e}}_1\right)$$

$$= - Q_0 \frac{3}{2} (n_2 \hat{\mathbf{e}}_1 - n_1 \hat{\mathbf{e}}_2) n_3$$

$$(\hat{\mathbf{n}} \times \mathrm{Q}) \times \hat{\mathbf{n}} = - Q_0 \frac{3}{2} n_3 (n_2^2 \hat{\mathbf{e}}_3 - n_2 n_3 \hat{\mathbf{e}}_2 + n_1^2 \hat{\mathbf{e}}_3 - n_1 n_3 \hat{\mathbf{e}}_1)$$

$$= \frac{3}{2} Q_0 \left[n_1 n_3^2 \hat{\mathbf{e}}_1 + n_2 n_3^2 \hat{\mathbf{e}}_2 - (n_1^2 + n_2^2) n_3 \hat{\mathbf{e}}_3\right]$$

$$|(\hat{\mathbf{n}} \times \mathrm{Q}) \times \hat{\mathbf{n}}|^2 = \left|\left(\frac{3}{2} Q_0\right)\right|^2 \left[n_1^2 n_3^4 + n_2^2 n_3^4 + (n_1^2 + n_2^2)^2 n_3^2\right]$$

Now:

$$\hat{\mathbf{n}} = n_1 \hat{\mathbf{e}}_1 + n_2 \hat{\mathbf{e}}_2 + n_3 \hat{\mathbf{e}}_3$$

$$= \sin\theta \cos\phi \, \hat{\mathbf{e}}_1 + \sin\theta \sin\phi \, \hat{\mathbf{e}}_2 + \cos\theta \, \hat{\mathbf{e}}_3$$

$$|(\hat{\mathbf{n}} \times \mathrm{Q}) \times \hat{\mathbf{n}}|^2 = \left|\frac{3}{2} Q_0\right|^2 \left[\cos^4\theta\,(\sin^2\theta \cos^2\phi + \sin^2\theta \sin^2\phi) + \cos^2\theta\,(\sin^2\theta)\right]$$

$$= \left|\frac{3}{2} Q_0\right|^2 \cos^2\theta \sin^2\theta \,(\cos^2\theta + \sin^2\theta)$$

$$= \frac{9}{16} |Q_0|^2 \sin^2(2\theta)$$

$$\frac{dP}{d\Omega} = \frac{ck^6}{288\pi} \cdot \frac{9}{16} \frac{I_0^2 d^4}{\pi^2 \omega^2} \sin^2(2\theta)$$

The maximum is at 45°. When this is compared to the exact result, the quadrupole term is nearly the same.

(e) $$P = \int d\Omega \frac{dP}{d\Omega} = \frac{I_0^2 d^4 k^5}{288\pi^3 \omega} \cdot \frac{9}{16} \cdot 2\pi \int_0^{2\pi} \sin^3\theta \cos^2\theta \, d\theta$$

$$= \frac{I_0^2 d^4 k^4}{32 \cdot 8\pi^2 c} \left(\frac{1}{5} \int_0^{\pi} \sin^3\theta \, d\theta \right)$$

$$= \frac{I_0^2 (kd)^4}{32 \cdot 8 \cdot 5 \cdot \pi^2 c} \left(-\frac{1}{3}(-1)(2) + \frac{1}{3} \cdot 2 \right)$$

$$= \frac{I_0^2 (2\pi)^4}{3 \cdot 5 \cdot 32 \cdot 2\pi^2 c} = \frac{I_0^2 \cdot 16\pi^4}{15 \cdot 2 \cdot 16 \cdot 2\pi^2 c} = \frac{I_0^2 \pi^2}{30 \cdot 2c}$$

Therefore,

$$P = \frac{I_0^2 \pi^2}{60c}$$

$$R_{\text{rad}} = 30c \left(\frac{\pi^2}{30c} \right) \text{ ohms}$$

$$R_{\text{rad}} = \pi^2 \ \Omega$$

The values for the angular distribution are close, comparing the exact solution and the values found by using the lowest multipoles, but the values for R_{rad} is much smaller, so there must be a large contribution from higher-order multipoles.

PROBLEM 14.9

(a) Show that $B^2 - E^2$ is an invariant quantity under Lorentz transformation. What is its form in four-dimensional notation?

(b) The symbol $\varepsilon_{\lambda\mu\nu\sigma}$ is defined to have the properties

$$\varepsilon_{\lambda\mu\nu\sigma} = \begin{cases} 0 \text{ if any two indices are equal} \\ \pm 1 \text{ for an even (odd) permutation of indices} \end{cases}$$

$\varepsilon_{\lambda\mu\nu\sigma}$ is a completely antisymmetric unit tensor of the fourth rank (actually a pseudotensor under spatial inversion). Prove that

$$\varepsilon_{\lambda\mu\nu\sigma} F_{\lambda\mu} F_{\nu\sigma}$$

(summation convention implied) is a Lorentz invariant, and find its form in terms of **E** and **B**.

Solution:

(a)

$$B^2 - E^2 \to \tfrac{1}{2} F_{\mu\nu} F_{\mu\nu} = -\tfrac{1}{2} F_{\mu\nu} F_{\nu\mu}$$

$$= -\tfrac{1}{2}\,\mathrm{Tr}\begin{pmatrix} 0 & -E_x & -E_y & -E_z \\ E_x & 0 & -B_z & +B_y \\ E_y & B_z & 0 & -B_x \\ B_z & -B_y & -B_x & 0 \end{pmatrix}\begin{pmatrix} 0 & E_x & E_y & E_z \\ -E_x & 0 & -B_z & B_y \\ -E_y & B_z & 0 & -B_x \\ -E_z & B_y & B_x & 0 \end{pmatrix}$$

$$= -\tfrac{1}{2}[E^2 + (-E_x^2 + B_z^2 + B_y^2) + (B_z^2 - B_x^2 - E_y^2) - E_z^2 + B_x^2 + B_y^2] = E^2 - B^2$$

(b)

$$\begin{aligned} \varepsilon_{\lambda\mu\nu\sigma} F_{\lambda\mu} F_{\nu\sigma} = & -E_xB_x - E_xB_x - E_zB_z - E_yB_y - E_yB_y - E_zB_z - E_xB_x - B_zE_z - B_zE_z - B_yE_y \\ & - B_yE_y - E_xB_x - B_zE_z - B_zE_z - B_xE_x - B_xE_x - E_yB_y - E_yB_y - B_yE_y - B_yE_y \\ & - E_zB_z - E_zB_z - E_xB_x - E_xB_x \\ = & -8\,(\mathbf{E}\cdot\mathbf{B}) \end{aligned}$$

where

$$F_{\mu\nu} = \begin{pmatrix} 0 & E_x & E_y & E_z \\ -E_x & 0 & -B_z & B_y \\ -E_y & B_z & 0 & -B_x \\ -E_z & -B_y & B_x & 0 \end{pmatrix}$$

PROBLEM 14.10

(a) Show explicitly that two successive Lorentz transformations in the same direction commute and that they are equivalent to a single Lorentz transformation with a velocity

$$\nu = \frac{\nu_1 + \nu_2}{1 + (\nu_1\nu_2/c^2)}$$

This is an alternative way to derive the parallel-velocity addition law.

(b) Show explicitly that two successive Lorentz transformations at right angles (ν_1 in the x direction, ν_2 in the y direction) do not commute.

Solution:

(a) In the first transformation $x_\mu \to x'_\mu$, in the second $x'_\mu \to x''_\mu$. Let

$$q_1 = \tanh^{-1}\frac{\nu_1}{c}, \quad q_2 = \tanh^{-1}\frac{\nu_2}{c}$$

$$\begin{pmatrix} x'_0 \\ x'_1 \\ x'_2 \\ x'_3 \end{pmatrix} = \begin{pmatrix} \cosh q_1 & -\sinh q_1 & 0 & 0 \\ -\sinh q_1 & \cosh q_1 & 0 & 0 \\ 0 & 0 & 1 & 0 \\ 0 & 0 & 0 & 1 \end{pmatrix} \begin{pmatrix} x_0 \\ x_1 \\ x_2 \\ x_3 \end{pmatrix}$$

$$\begin{pmatrix} x''_0 \\ x''_1 \\ x''_2 \\ x''_3 \end{pmatrix} = \begin{pmatrix} \cosh q_2 & -\sinh q_2 & 0 & 0 \\ -\sinh q_2 & \cosh q_2 & 0 & 0 \\ 0 & 0 & 1 & 0 \\ 0 & 0 & 0 & 1 \end{pmatrix} \begin{pmatrix} x'_0 \\ x'_1 \\ x'_2 \\ x'_3 \end{pmatrix}$$

$$= \begin{pmatrix} \cosh q_2 & -\sinh q_2 & 0 & 0 \\ -\sinh q_2 & \cosh q_2 & 0 & 0 \\ 0 & 0 & 1 & 0 \\ 0 & 0 & 0 & 1 \end{pmatrix} \begin{pmatrix} \cosh q_1 & -\sinh q_1 & 0 & 0 \\ -\sinh q_1 & \cosh q_1 & 0 & 0 \\ 0 & 0 & 1 & 0 \\ 0 & 0 & 0 & 1 \end{pmatrix} \begin{pmatrix} x_0 \\ x_1 \\ x_2 \\ x_3 \end{pmatrix}$$

$$= \begin{pmatrix} \cosh q_1 \cosh q_2 + \sinh q_2 \sinh q_1 & -\cosh q_2 \sinh q_1 - \sinh q_2 \cosh q_1 & 0 & 0 \\ -\cosh q_1 \sinh q_2 - \cosh q_2 \sinh q_1 & \sinh q_2 \sinh q_1 + \cosh q_2 \cosh q_1 & 0 & 0 \\ 0 & 0 & 1 & 0 \\ 0 & 0 & 0 & 0 \end{pmatrix} \begin{pmatrix} x_0 \\ x_1 \\ x_2 \\ x_3 \end{pmatrix}$$

$$= \begin{pmatrix} \cosh(n_1 + n_2) & -\sinh(n_1 + n_2) & 0 & 0 \\ -\sinh(n_1 + n_2) & \cosh(n_1 + n_2) & 0 & 0 \\ 0 & 0 & 1 & 0 \\ 0 & 0 & 0 & 1 \end{pmatrix} \begin{pmatrix} x_0 \\ x_1 \\ x_2 \\ x_3 \end{pmatrix}$$

The two transformations are equivalent to one with

$$\nu' = c \tanh(n_1 + n_2)$$

$$v' = \frac{c(\tanh n_1 + \tanh n_2)}{1 + \tanh n_1 \tanh n_2}$$

$$v' = \frac{v_1 + v_2}{1 + v_1 v_2/c^2}$$

$$\begin{pmatrix} x''_0 \\ x''_1 \\ x''_2 \\ x''_3 \end{pmatrix} \begin{pmatrix} \cosh q_1 & -\sinh q_1 & 0 & 0 \\ -\sinh q_1 & \cosh q_1 & 0 & 0 \\ 0 & 0 & 1 & 0 \\ 0 & 0 & 0 & 1 \end{pmatrix} \begin{pmatrix} \cosh q_2 & -\sinh q_2 & 0 & 0 \\ -\sinh q_2 & \cosh q_2 & 0 & 0 \\ 0 & 0 & 1 & 0 \\ 0 & 0 & 0 & 1 \end{pmatrix} \begin{pmatrix} x_0 \\ x_1 \\ x_2 \\ x_3 \end{pmatrix}$$

$$= \begin{pmatrix} \cosh(q_1+q_2) & -\sinh(q_1+q_2) & 0 & 0 \\ -\sinh(q_1+q_2) & \cosh(q_1+q_2) & 0 & 0 \\ 0 & 0 & 1 & 0 \\ 0 & 0 & 0 & 1 \end{pmatrix} \begin{pmatrix} x_0 \\ x_1 \\ x_2 \\ x_3 \end{pmatrix}$$

So, two successive transformations in the same direction commute. In the first transformation $x_\mu \to x'_\mu$ in $\hat{\mathbf{x}}$ direction with velocity v_1, in the second $x'_\mu \to x''_\mu$ in $\hat{\mathbf{y}}$ direction with a velocity v_2:

$$\begin{pmatrix} x'_0 \\ x'_1 \\ x'_2 \\ x'_3 \end{pmatrix} = \begin{pmatrix} \cosh q_1 & -\sinh q_1 & 0 & 0 \\ -\sinh q_1 & \cosh q_1 & 0 & 0 \\ 0 & 0 & 1 & 0 \\ 0 & 0 & 0 & 1 \end{pmatrix} \begin{pmatrix} x_0 \\ x_1 \\ x_2 \\ x_3 \end{pmatrix}$$

$$\begin{pmatrix} x''_0 \\ x''_1 \\ x''_2 \\ x''_3 \end{pmatrix} = \begin{pmatrix} \cosh q_2 & 0 & -\sinh q_2 & 0 \\ 0 & 1 & 0 & 0 \\ -\sinh q_2 & 0 & \cosh q_2 & 0 \\ 0 & 0 & 0 & 1 \end{pmatrix} \begin{pmatrix} x'_0 \\ x'_1 \\ x'_2 \\ x'_3 \end{pmatrix}$$

$$= \begin{pmatrix} \cosh q_2 & 0 & -\sinh q_2 & 0 \\ 0 & 1 & 0 & 0 \\ -\sinh q_2 & 0 & \cosh q_2 & 0 \\ 0 & 0 & 0 & 1 \end{pmatrix} \begin{pmatrix} \cosh q_1 & -\sinh q_1 & 0 & 0 \\ -\sinh q_1 & \cosh q_1 & 0 & 0 \\ 0 & 0 & 1 & 0 \\ 0 & 0 & 0 & 1 \end{pmatrix} \begin{pmatrix} x'_0 \\ x'_1 \\ x'_2 \\ x'_3 \end{pmatrix}$$

$$
= \begin{pmatrix} \cosh q_1 \cosh q_2 & -\cosh q_2 \sinh q_1 & -\sinh q_2 & 0 \\ -\sinh q_1 & \cosh q_1 & 0 & 0 \\ -\sinh q_2 \cosh q_1 & \sinh q_1 \sinh q_2 & \cosh q_2 & 0 \\ 0 & 0 & 0 & 1 \end{pmatrix} \begin{pmatrix} x'_0 \\ x'_1 \\ x'_2 \\ x'_3 \end{pmatrix}
$$

Now transform first in the $\hat{y}$ direction with a velocity v_2 and then second in the $\hat{x}$ direction with a velocity v_1:

$$
\begin{pmatrix} x'_0 \\ x'_1 \\ x'_2 \\ x'_3 \end{pmatrix} = \begin{pmatrix} \cosh q_2 & 0 & -\sinh q_2 & 0 \\ 0 & 1 & 0 & 0 \\ -\sinh q_2 & 0 & \cosh q_2 & 0 \\ 0 & 0 & 0 & 1 \end{pmatrix} \begin{pmatrix} x_0 \\ x_1 \\ x_2 \\ x_3 \end{pmatrix}
$$

$$
\begin{pmatrix} x''_0 \\ x''_1 \\ x''_2 \\ x''_3 \end{pmatrix} = \begin{pmatrix} \cosh q_1 & -\sinh q_1 & 0 & 0 \\ -\sinh q_1 & \cosh q_1 & 0 & 0 \\ 0 & 0 & 1 & 0 \\ 0 & 0 & 0 & 1 \end{pmatrix} \begin{pmatrix} x'_0 \\ x'_1 \\ x'_2 \\ x'_3 \end{pmatrix}
$$

$$
= \begin{pmatrix} \cosh q_1 & -\sinh q_1 & 0 & 0 \\ -\sinh q_1 & \cosh q_1 & 0 & 0 \\ 0 & 0 & 1 & 0 \\ 0 & 0 & 0 & 1 \end{pmatrix} \begin{pmatrix} \cosh q_2 & 0 & -\sinh q_2 & 0 \\ 0 & 1 & 0 & 0 \\ -\sinh q_2 & 0 & \cosh q_2 & 0 \\ 0 & 0 & 0 & 1 \end{pmatrix} \begin{pmatrix} x_0 \\ x_1 \\ x_2 \\ x_3 \end{pmatrix}
$$

$$
\begin{pmatrix} x''_0 \\ x''_1 \\ x''_2 \\ x''_3 \end{pmatrix} = \begin{pmatrix} \cosh q_1 \cosh q_2 & -\sinh q_1 & -\cosh q_1 \sinh q_2 & 0 \\ -\sinh q_1 \cosh q_2 & \cosh q_1 & +\sinh q_1 \sinh q_2 & 0 \\ -\sinh q_2 & 0 & \cosh q_2 & 0 \\ 0 & 0 & 0 & 1 \end{pmatrix} \begin{pmatrix} x_0 \\ x_1 \\ x_2 \\ x_3 \end{pmatrix}
$$

PROBLEM 14.11

(a) Find the form of the wave equation in system K if it has its standard form in system K' and the two coordinate systems are related by the Galilean transformation $x' = x - vt$, $t' = t$.

(b) Show explicitly that the form of the wave equation is the same in system K as in K' if the coordinates are related by the Lorentz transformation $x' = \gamma(x - vt)$, $t' = \gamma[t - (vx/c^2)]$.

Solution:

(a) In K'

$$\sum_i \left(\frac{\partial^2}{\partial x_i^2} - \frac{1}{c^2}\frac{\partial^2}{\partial t'^2}\right)\psi = 0$$

We need the form in system K. Let $x_i' = x_i - v_i t$, $t' = t$. Now,

$$\frac{\partial}{\partial x_i} = \frac{\partial x_i}{\partial x_i}\frac{\partial}{\partial x_i} + \frac{\partial t}{\partial x_i'}\frac{\partial}{\partial x_i} = \frac{\partial}{\partial x_i}, \qquad \frac{\partial^2}{\partial x_i'^2} = \frac{\partial^2}{\partial x_i^2}$$

$$\frac{\partial}{\partial t'} = \frac{\partial t}{\partial t'}\frac{\partial}{\partial t} + \frac{\partial x_i}{\partial t'}\frac{\partial}{\partial x_i} = \frac{\partial}{\partial t} + v_i\frac{\partial}{\partial x_i}$$

$$\frac{\partial^2}{\partial t'^2} = \left(\frac{\partial}{\partial t} + v_i\frac{\partial}{\partial x_i}\right)\left(\frac{\partial}{\partial t} + v_i\frac{\partial}{\partial x_i}\right) = \frac{\partial^2}{\partial t^2} + v_i\frac{\partial}{\partial t}\frac{\partial}{\partial x_i} + v_i\frac{\partial}{\partial x_i}\frac{\partial}{\partial t} + v_i\frac{\partial}{\partial x_i}v_i\frac{\partial}{\partial x_i}$$

$$\frac{\partial^2}{\partial t'^2} = \frac{\partial^2}{\partial t^2} + 2\mathbf{v}\cdot\frac{\nabla\partial}{\partial t} + \mathbf{v}\cdot\nabla\ \mathbf{v}\cdot\nabla$$

So now the wave equation becomes,

$$\left(\nabla'^2 - \frac{1}{c^2}\frac{\partial^2}{\partial t'^2}\right)\psi = \left(\nabla^2 - \frac{1}{c^2}\frac{\partial^2}{\partial t^2} - \frac{2v}{c^2}\cdot\frac{\nabla\partial}{\partial t} - \frac{1}{c^2}\mathbf{v}\cdot\nabla\ \mathbf{v}\cdot\nabla\right)\psi = 0$$

(b) In K', the wave equation is

$$\left(\sum_i \frac{\partial^2}{\partial x_i'^2} - \frac{1}{c^2}\frac{\partial^2}{\partial t'^2}\right)\psi = 0$$

Now to transform to K frame: let $x_i' = \gamma(x_i - v_i t)$

$$t' = \gamma\left(t - \frac{v_i x_i}{c^2}\right)$$

Now,

$$\frac{\partial}{\partial x_i'} = \frac{\partial x_i}{\partial x_i'}\frac{\partial}{\partial x_i} + \frac{\partial t}{\partial x_i'}\frac{\partial}{\partial t} = \gamma\frac{\partial}{\partial x_i} + \frac{v_i}{\gamma c^2}\frac{\partial}{\partial t}$$

$$\frac{\partial^2}{\partial x_i'^2} = \left(\gamma\frac{\partial}{\partial x_i} + \frac{v_i}{\gamma c^2}\frac{\partial}{\partial t}\right)\left(\gamma\frac{\partial}{\partial x_i} + \frac{v_i}{\gamma c^2}\frac{\partial}{\partial t}\right)$$

$$\frac{\partial^2}{\partial x_i'^2} = \gamma^2 \frac{\partial^2}{\partial x_i^2} + 2\frac{v_i}{c^2}\frac{\partial}{\partial x_i}\frac{\partial}{\partial t} + \frac{v_i v_i}{\gamma^2 c^2}\frac{\partial^2}{\partial t^2}$$

$$\frac{\partial^2}{\partial x_i'^2} = \gamma^2 \nabla^2 + \frac{2}{c^2}\,\mathbf{v}\cdot\nabla\frac{\partial}{\partial t} + \frac{\mathbf{v}\cdot\mathbf{v}}{\gamma^2 c^2}\frac{\partial^2}{\partial t^2}$$

$$\frac{\partial}{\partial t'} = \frac{\partial t}{\partial t'}\frac{\partial}{\partial t} + \frac{\partial x_i}{\partial t'}\frac{\partial}{\partial x_i} = \gamma\frac{\partial}{\partial t} + \frac{v_i}{\gamma}\frac{\partial}{\partial x_i}$$

$$\frac{\partial^2}{\partial t'^2} = \left(\gamma\frac{\partial}{\partial t} + \frac{v_i}{\gamma}\frac{\partial}{\partial x_i}\right)\left(\gamma\frac{\partial}{\partial t} + \frac{v_i}{\gamma}\frac{\partial}{\partial x_i}\right)$$

$$\frac{\partial^2}{\partial t'^2} = \gamma^2\frac{\partial^2}{\partial t^2} + \frac{2v_i}{\gamma}\frac{\partial}{\partial x_i}\frac{\partial}{\partial t} + \frac{v_i v_i}{\gamma^2 c^2}\frac{\partial^2}{\partial x_i^2}$$

$$\frac{\partial^2}{\partial t'^2} = \gamma^2\frac{\partial^2}{\partial t^2} + \frac{2}{\gamma}\,\mathbf{v}\cdot\nabla\frac{\partial}{\partial t} + \frac{\mathbf{v}\cdot\mathbf{v}}{\gamma^2 c^2}\nabla^2$$

Now the wave equation becomes,

$$\left(\nabla'^2 - \frac{1}{c^2}\frac{\partial^2}{\partial t'^2}\right)\psi = \gamma^2\nabla^2 + \frac{2}{c^2}\,\mathbf{v}\cdot\nabla\frac{\partial}{\partial t} + \frac{\mathbf{v}\cdot\mathbf{v}}{\gamma^2 c^2}\frac{\partial^2}{\partial t^2}$$

$$- \frac{1}{c^2}\left(\gamma^2\frac{\partial^2}{\partial t^2} + \frac{2}{\gamma}\,\mathbf{v}\cdot\nabla\frac{\partial}{\partial t} + \frac{\mathbf{v}\cdot\mathbf{v}}{\gamma^2 c^2}\nabla^2\right)$$

$$= \left(\gamma^2 - \frac{\mathbf{v}\cdot\mathbf{v}}{\gamma^2 c^2}\right)\left(\nabla^2 - \frac{1}{c^2}\frac{\partial^2}{\partial t^2}\right) = 0$$

PROBLEM 14.12 It is a well-established fact that Newton's equation of motion $m\mathbf{a}' = e\,\mathbf{E}'$ holds for a small charged body of mass m and charge e in a coordinate system K' where the body is momentarily at rest. Show that the Lorentz force equation

$$\frac{d\mathbf{p}}{dt} = e\left(\mathbf{E} + \frac{\mathbf{v}}{c}\times\mathbf{B}\right)$$

follows directly from the Lorentz transformation properties of accelerations and electromagnetic fields.

Solution: For particle at rest in system K', $ma' = dp'/dt = eE'$, $\mu' = 0$, we want a system K, $E_{\parallel}' = E_{\parallel}$, $E_{\perp}' = \gamma(E_{\perp} + \nu\times B)$, $\mu_{\parallel} = \nu$, $\mu_{\perp} = 0$,

$$\left(\frac{dp}{dt}\right)_{\shortparallel} = \left(\frac{dp'}{dt'}\right)_{\shortparallel} + \frac{\gamma}{c^2}\left[\mu \times \left(\nu \times \frac{d\mathbf{p}'}{dt'}\right)\right]_{\shortparallel} = eE_{\shortparallel}' + \frac{\gamma}{c^2}\left[\nu(\nu \cdot e\mathbf{E}') - e\mathbf{E}'(\nu \cdot \nu)\right]$$

$$= eE_{\shortparallel}' + \frac{\gamma}{c^2}\left[eE_{\shortparallel}\ \nu^2\hat{\nu} - e\mathbf{E}\nu^2\right]_{\shortparallel} = eE_{\shortparallel}' + \frac{\gamma}{c^2}\left[e\nu^2\mathbf{E}_{\perp}\right] = e\mathbf{E}_{\shortparallel}$$

$$\left(\frac{dp}{dt}\right)_{\perp} = \gamma\left(\frac{dp'}{dt'}\right)_{\perp} + \frac{\gamma}{c^2}\left[\mu \times \left(\nu \times \frac{dp'}{dt'}\right)\right]_{\perp}$$

$$= \gamma eE_{\perp}' + \frac{\gamma}{c^2}\left[\nu\gamma eE_{\shortparallel}' - e\mathbf{E}'\nu^2\right]_{\perp} = \gamma eE' - \frac{e\gamma\nu^2 E_{\perp}'}{c^2}$$

$$= (\gamma e - e\gamma\beta^2)\gamma(\mathbf{E}_{\perp} \times \nu \times \beta)$$

$$\frac{d\mathbf{p}}{dt} = e\mathbf{E}_{\shortparallel} + e\mathbf{E}_{\perp} + e\nu \times \beta = e\mathbf{E} + e\nu \times \beta$$

PROBLEM 14.13 This problem emphasizes technique as much as the answer. Consider a charge q constrained to rotate at a constant angular velocity ω_0 in a circle of radius R.

(a) Show that the vector potential $\mathbf{A}(\mathbf{r},t)$ can be expressed as

$$\mathbf{A}(\mathbf{r},t) = \sum_{n=0}^{\infty} \mathbf{A}_n(\mathbf{r})e^{-i\omega_n t} \tag{1}$$

and evaluate $\mathbf{A}_n(\mathbf{r})$ and ω_n.

(b) Determine $\mathbf{A}(\mathbf{r},t)$ of Eq. (1) in the radiation zone under the assumption that the dipole approximation is valid.
(c) From the answer of part (b), find **E** and **B** in the radiation zone.
(d) From part (c), determine the angular distribution of the power radiated, $dP/d\Omega$.

Solution:

(a) $\mathbf{j}(\mathbf{x},t) = \rho(\mathbf{x},t)\mathbf{v}(\mathbf{x},t)$

$$\rho(\mathbf{x},t) = \frac{q}{R^2}\,\delta(x - R)\delta(\phi - \omega_0 t)\delta(\theta - \pi/2)$$

$$\mathbf{v}(\mathbf{x},t) = \frac{d\mathbf{x}}{dt} = \omega_0 R\hat{\mathbf{u}},$$

$$\hat{\mathbf{u}} = -\sin\phi\ \hat{\mathbf{e}}_1 + \cos\phi\ \hat{\mathbf{e}}_2$$

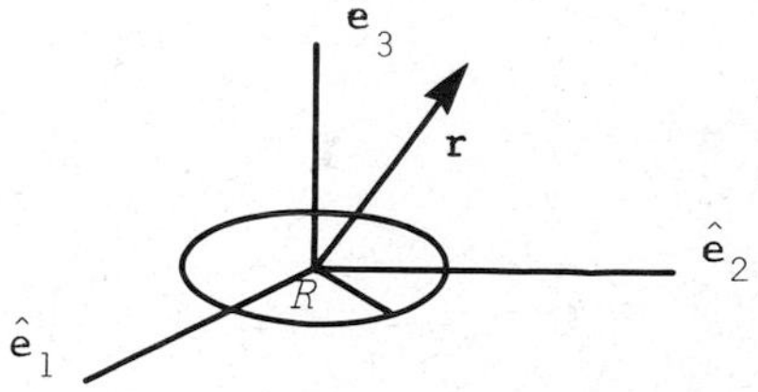

$$\mathbf{j}(\mathbf{x},t) = \frac{q\omega_0}{R}\, \hat{\mathbf{u}}\, \delta(x-R)\delta(\phi-\omega_0 t)\delta(\theta-\pi/2)$$

But,

$$\delta(\phi-\omega_0 t) = \frac{1}{2\pi}\left(1+2Re\sum_{n=1}^{\infty} e^{in\phi}e^{-i\omega_n t}\right), \quad \omega_n = n\omega_0,$$

therefore,

$$\mathbf{j}(\mathbf{x},t) = Re\sum_{n=0}^{\infty} \mathbf{j}_n(\mathbf{x})e^{-i\omega_n t}$$

$$\mathbf{j}_n(\mathbf{x}) = \frac{q\omega_0}{\alpha_n \pi R}\,\delta(x-R)\delta(\theta-\pi/2)e^{in\phi}, \quad \alpha_n = \begin{cases} 2, & n=0 \\ 1, & n>0 \end{cases}$$

(Let $\int\sum = \sum\int$)

$$\mathbf{A}(\mathbf{r},t) = \frac{Re}{c}\int d^3x\, dt' \sum_{n=0}^{\infty} \frac{\mathbf{j}_n(\mathbf{x})e^{-i\omega_n t'}}{|\mathbf{r}-\mathbf{x}|}\,\delta\left(t'-t+\frac{|\mathbf{r}-\mathbf{x}|}{c}\right)$$

$$= Re\sum_{n=0}^{\infty} \mathbf{A}_n(\mathbf{r})e^{-i\omega_n t}$$

$$\mathbf{A}_n(\mathbf{r}) = \frac{1}{c}\int d^3x\, \frac{\mathbf{j}_n(\mathbf{x})e^{ik_n|\mathbf{x}-\mathbf{r}|}}{|\mathbf{x}-\mathbf{r}|}, \quad k_n = \frac{\omega_n}{c}$$

Use

$$\sum_{n=0}^{\infty} \mathbf{A}_n(\mathbf{r})e^{-i\omega_n t}$$

and take *Re* at the end, so drop *Re*.

(b) In the radiation zone and in the dipole approximation,

$$\mathbf{A}_n(\mathbf{r}) \sim \left(\frac{1}{cr}\int d^3x\, \mathbf{j}_n(\mathbf{x})\right)e^{ik_n r}$$

$$\int d^3x\, \mathbf{j}_n(\mathbf{x}) = \frac{q\omega_0 R}{\alpha_n \pi}\int_0^{2\pi} d\phi(-\sin\phi\,\hat{\mathbf{e}}_1 + \cos\phi\,\hat{\mathbf{e}}_2)e^{in\phi}$$

$$= \frac{q\omega_0 R}{\alpha_n \pi} \int_0^{2\pi} d\phi(-\sin\phi\,\hat{\mathbf{e}}_1 + \cos\phi\,\hat{\mathbf{e}}_2)(\cos n\phi + i\sin n\phi)$$

$$= \frac{q\omega_0 R}{\alpha_n} (-i\hat{\mathbf{e}}_1 + \hat{\mathbf{e}}_2)\delta_{n1}$$

$$= \frac{-iq\omega_0 R}{\alpha_n} (\hat{\mathbf{e}}_1 + i\hat{\mathbf{e}}_2)\delta_{n1}$$

therefore, only $n = 1$ term contributes,

$$\mathbf{A}(\mathbf{r},t) \sim \frac{-iqRk_0}{r} e^{i(k_0 r - \omega_0 t)} (\hat{\mathbf{e}}_1 + i\hat{\mathbf{e}}_2)$$

$$= \frac{-ik_0 \mathbf{p}}{r} e^{i(k_0 r - \omega_0 t)}$$

$$\mathbf{p} = qR(\hat{\mathbf{e}}_1 + i\hat{\mathbf{e}}_2)$$

(c) $$\mathbf{B}_{\text{rad}}(\mathbf{r},t) = \frac{k_0^2}{r} e^{i(k_0 r - \omega_0 t)} (\hat{\mathbf{r}} \times \mathbf{p})$$

$$\mathbf{E}_{\text{rad}}(\mathbf{r},t) - \frac{k_0^2}{r} e^{i(k_0 r - \omega_0 t)} (\hat{\mathbf{r}} \times \mathbf{p}) \times \hat{\mathbf{r}}$$

$$\mathbf{S} = \frac{c}{8\pi} (\mathbf{E} \times \mathbf{B}^*)$$

only $n = 1$, no *Re* needed.

(d) $$\frac{\partial P}{\partial \Omega} = \frac{c}{8\pi} r^2 \hat{\mathbf{r}} \cdot (\mathbf{E} \times \mathbf{B}^*)$$

$$= \frac{c}{8\pi} k_0^4 \; \hat{\mathbf{r}} \cdot \{-[(\hat{\mathbf{r}} \times \mathbf{p}) \times \hat{\mathbf{r}}] \times (\hat{\mathbf{r}} \times \mathbf{p}^*)\}$$

$$[(\hat{\mathbf{r}} \times \mathbf{p}) \times \hat{\mathbf{r}}] \times [\hat{\mathbf{r}} \times \mathbf{p}^*] = \hat{\mathbf{r}}(\hat{\mathbf{r}} \times \mathbf{p}^*) \cdot (\hat{\mathbf{r}} \times \mathbf{p}) - (\hat{\mathbf{r}} \times \mathbf{p})\hat{\mathbf{r}} \cdot (\hat{\mathbf{r}} \times \mathbf{p}^*)$$

$$\frac{\partial P}{\partial \Omega} = \frac{ck_0^4}{8\pi} e^2 R^2 \; |\hat{\mathbf{r}} \times (\hat{\mathbf{e}}_1 + i\hat{\mathbf{e}}_2)|^2$$

$$\mathbf{r} \times (\hat{\mathbf{e}}_1 + i\hat{\mathbf{e}}_2) = (\sin\theta\cos\phi\,\hat{\mathbf{e}}_1 + \sin\theta\sin\phi\,\hat{\mathbf{e}}_2 + \cos\theta\,\hat{\mathbf{e}}_3) \times (\hat{\mathbf{e}}_1 + i\hat{\mathbf{e}}_2)$$

$$= i \sin\theta \cos\phi \ \hat{e}_3 - \sin\theta \sin\phi \,\hat{e}_3 + \cos\theta\,(\hat{e}_2 - i\hat{e}_1)$$

$$= i \sin\theta(\cos\phi + i\sin\phi)\hat{e}_3 + \cos\theta\,\hat{e}_2 - i\cos\theta\,\hat{e}_1$$

$$|\hat{r} \times (\hat{e}_1 + i\hat{e}_2)|^2 = \sin^2\theta + \cos^2\theta + \cos^2\theta = 1 + \cos^2\theta$$

$$\frac{\partial P}{\partial \Omega} = \frac{c}{8\pi}(eRk_0^2)^2(1 + \cos^2\theta)$$

PROBLEM 14.14 The transitional charge density for the radiative transition from the $m = 0$ $2p$ state in hydrogen to the $1s$ ground state is

$$\rho(r,\theta,\phi,t) = \frac{2e}{\sqrt{6}}\, a_0^{-4} \cdot re^{-3r/2a_0}\, Y_{00}Y_{10}e^{-i\omega_0 t}$$

where $a_0 = \hbar^2/me^2$ is the Bohr radius and $\omega = 3e^2/8\hbar a_0$ is the frequency of the levels.

(a) Evaluate all the radiation multipoles for this charge density in the long-wavelength limit.

(b) In the electric dipole approximation calculate the total time-averaged power radiated. Express your answer in units of $(\hbar\omega_0)(\alpha^4 c/a_0)$, where $\alpha = e^2/\hbar c$ is the fine-structure constant.

(c) Interpreting the classically calculated power as the photon energy $(\hbar\omega_0)$ times the transition probability, evaluate numerically the transition probability in units of s^{-1}.

Solution:

(a) $$\rho(r,\theta,\phi) = \frac{2e}{\sqrt{6}}\frac{1}{a_0^4}\, re^{-3r/2a}\, Y_{0,0}\, Y_{1,0}\, e^{-i\omega_0 t}$$

$$Y_{0,0} = \frac{1}{\sqrt{4\pi}}, \quad Y_{1,0} = \sqrt{3/4}\cos\theta$$

therefore,

$$\rho(r,\theta,\phi) = \frac{2e}{\sqrt{4\pi}\,\sqrt{6}}\sqrt{3/4}\left(\frac{1}{a_0^4}\right) \cdot re^{-3r/2a}\cos\theta\, e^{-i\omega_0 t}$$

Now, given ρ, the current density is given by: $\nabla \cdot \mathbf{j} = i\omega_0\rho$. However, we can also obtain $\mathbf{j}$ by realizing that ρ describes a physically rotating charge density:

$$\mathbf{j} = \rho v\hat{\theta} = \hat{\theta}\rho r_\perp \omega_0 = \rho r \sin\theta\, \omega_0\, \hat{\theta}$$

$$\mathbf{j}(\mathbf{x},t) = \frac{2e\omega_0}{(\sqrt{4\pi})^2\,\sqrt{2}}\frac{1}{a_0^4}\, r^2\, e^{-3r/2a_0}\cos\theta\sin\theta\, e^{-i\omega_0 t}\,\hat{\theta}$$

The radiation multipoles can then be formed from

$$\mathbf{A}(\mathbf{x}) = \frac{e^{ikr}}{cr}\int \mathbf{j}(\mathbf{x}')e^{-ik\mathbf{n}\cdot\mathbf{x}'}\,d^3x'$$

Let $e^{-ik\mathbf{n}\cdot\mathbf{x}'} = e^{+i\mathbf{k}'\cdot\hat{\mathbf{x}}'}$, that is, $\to\mathbf{k}' = -k\hat{\mathbf{n}}$

Then:

$$e^{+i\mathbf{k}'\cdot\mathbf{x}'} = 4\pi\sum_{\ell,m} i^{\ell}\, j_{\ell}(k'r)\, Y_{\ell,m}{}^{*}(\Omega_{\mathbf{k}'})Y_{\ell,m}(\Omega_{\mathbf{x}'})$$

where $j_{\ell}(k'r)$ is the spherical Bessel function of order ℓ.

$$\mathbf{A}(\mathbf{x}) = \frac{4\pi 2e\omega_0\hat{\theta}}{4\pi\sqrt{2}\,a_0^4}\,\frac{e^{ikr}}{cr}\sum_{\ell,m} i^{\ell}\, Y_{\ell,m}{}^{*}\,(\Omega_{\mathbf{x}'})\int r^2\, j_{\ell}(k'r)e^{-3r/2a_0}$$

$$\times\, Y_{\ell,m}(\theta,\phi)\cos\theta\sin\theta\, r^2\, dr\, d\Omega$$

Now let us examine

$$\int Y_{\ell,m}(\theta,\phi)\cos\theta\sin\theta\, d\Omega = I\Omega$$

Now $\sin\theta\cos\theta$ can be written as

$$\sin\theta\cos\theta = -\sqrt{\frac{8\pi}{15}}\, e^{-i\phi}\, Y_{21}(\theta,\phi)$$

$$I_{\Omega} = -\sqrt{\frac{8\pi}{15}}\int Y_{\ell,m}(\theta,\phi)e^{-i\phi}Y_{21}(\theta,\phi)d\Omega$$

(b)

$$\rho(r,\theta,\phi) = \frac{2e}{\sqrt{6}}\,\frac{1}{a_0^4}\, re^{-3r/2a_0}\, Y_{0,0}\, Y_{1,0}\, e^{-i\omega_0 t}$$

In the electronic dipole approximation, the total time-averaged power radiated is given by

$$P = \frac{ck^4}{3}\,|\mathbf{p}|^2$$

where $kc = \omega_0$ and

$$\mathbf{p} = \int \mathbf{x}'\,\rho(\mathbf{x}')d^3x'$$

$\mathbf{x}' = \hat{\mathbf{i}}x' + \hat{\mathbf{j}}y' + \hat{\mathbf{k}}z'$, $x' = r' \cos\phi' \sin\phi'$, $y' = r' \sin\phi' \sin\theta'$

$z' = r' \cos\theta'$

$$p_x = \frac{2e}{\sqrt{6}} \frac{Y_{0,0}}{a_0^4} \int_0^\infty \int_0^{2\pi} \int_0^\pi r' \cos\phi' \sin\theta' r' e^{-3r'/2a_0}$$

$$\times \sqrt{\frac{3}{4}} \cos'\theta r'^2 \, dr' \sin\theta' \, d\theta' \, d\phi'$$

Let $\alpha = 3/2a_0$

$$= \frac{2eY_{0,0}}{a_0^4 \sqrt{6}} \sqrt{\frac{3}{4\pi}} \int_0^{2\pi} \cos\phi' \, d\phi' \int_0^\pi \sin^2\theta' \cos\theta' \, d\theta' \int_0^\infty r'^4 e^{-\alpha r'} \, dr'$$

therefore,

$$p_x = 0$$

$$p_y = \frac{2eY_{0,0}}{a_0^4 \sqrt{6}} \sqrt{\frac{3}{4\pi}} \int_0^{2\pi} \sin\phi' \, d\phi' \int_0^\pi \sin^2\theta' \cos\theta' \, d\theta' \int_0^\infty r'^4 e^{-\alpha r'} dr'$$

therefore,

$$p_y = 0$$

$$p_z = \frac{2eY_{0,0}}{a_0^4 \sqrt{6}} \sqrt{\frac{3}{4\pi}} \cdot 2\pi \int_0^\infty r'^4 e^{-\alpha r'} \, dr' \int_0^\pi \cos^2\theta' \sin\theta' \, d\theta'$$

where $\Gamma = \dfrac{2eY_{0,0}}{a_0^4 \sqrt{6}} \sqrt{\dfrac{3}{4\pi}} \cdot 2\pi$

$$= \Gamma \int_0^\infty r'^4 e^{-\alpha r'} \, dr' \left(\frac{-1}{3}\right)(-1-1) = \frac{2}{3} \Gamma \int_0^\infty r'^4 e^{-\alpha r'} \, dr'$$

$$p_z = \frac{2}{3} \Gamma \left(-\frac{1}{\alpha} r^4 e^{-\alpha r} \Big|_0^\infty + \frac{4}{\alpha} \int_0^\infty r^3 e^{-\alpha r} \, dr \right) = \frac{2}{3} \Gamma \frac{4}{\alpha} \left(\frac{3}{\alpha} \int_0^\infty r^2 e^{-\alpha r} \, dr \right)$$

$$= \frac{2 \cdot 3 \cdot 4}{3} \frac{\Gamma}{\alpha^2} \left(\frac{2}{\alpha} \int_0^\infty r \, e^{-\alpha r} \, dr \right) = \frac{16\Gamma}{\alpha^3} \left(\frac{1}{\alpha} \int_0^\infty e^{-\alpha r} \, dr \right) = \frac{16\Gamma}{\alpha^5}$$

$$\mathbf{p} = 16\left(\frac{2e}{a_0^4\sqrt{24\pi}}\sqrt{\frac{3}{4\pi}}\cdot 2\pi\right)\left(\frac{2a_0}{3}\right)^5 \hat{\mathbf{k}} = \hat{\mathbf{k}}\, ea_0 \frac{2^{11}\cdot 3^{-9/2}}{2^2\sqrt{2}\sqrt{3}} = \frac{2^9}{\sqrt{2}}\frac{ea_0\hat{\mathbf{k}}}{3^5}$$

$$\mathbf{p} = \frac{16}{\sqrt{2}}\left(\frac{2}{3}\right)^5 ea_0\hat{\mathbf{k}}$$

The power is given by

$$P = \frac{\omega_0 k^3}{3}\left|\frac{16}{\sqrt{2}}\left(\frac{2}{3}\right)^5 ea_0\hat{\mathbf{k}}\right|^2$$

$$P \cong 0.74\, e^2 a_0^2 \omega_0 k^3$$

Now:

$$\alpha = \frac{e^2}{\hbar c};\quad e^2 = \alpha\hbar c;\quad \omega_0 = \frac{3}{8}\frac{e^2}{\hbar a_0}$$

$$P = 0.74(\hbar\omega_0)\alpha c a_0^2 k^3,\quad k = \frac{\omega_0}{c} = \frac{3}{8c}\frac{\alpha\hbar c}{\hbar a_0} = \frac{3}{8}\frac{a}{a_0}$$

$$= 0.74(\hbar\omega_0)\alpha c a_0^2\left(\frac{3}{8}\frac{\alpha}{a}\right)^3$$

$$P = 0.74\left(\frac{3}{8}\right)^3(\hbar\omega_0)\left(\frac{\alpha^4 c}{a_0}\right)$$

$$P \cong (3.9\times 10^{-2})(\hbar\omega_0)\left(\frac{\alpha^4 c}{a_0}\right)$$

(c) We interpret P as the photon energy $\hbar\omega_0$ times the transition probability R (s^{-1}), therefore,

$$P = \hbar\omega R = (3.9\times 10^{-2})\left(\frac{\alpha^4 c}{a_0}\right)\hbar\omega_0$$

$$R = (3.9\times 10^{-2})\left(\frac{1}{137}\right)^4\left[\frac{3\times 10^{10}\ \text{cm/s}}{0.53\ \text{Å}}\left(\frac{1\ \text{Å}}{10^{-8}\ \text{cm}}\right)\right]$$

$$R \simeq 6.3\times 10^8\ \text{s}^{-1}$$

PROBLEM 14.16 Radiation associated with a collection of charges. Consider a collection of charges q_i, with positions $\mathbf{x}_i(t)$ and masses m_i, $1 \le i \le N$.

(a) Show that **E** and **B** are given as sums over i,

$$\mathbf{E}(\mathbf{r},t) = \sum_i (\mathbf{E}_i)_{\text{ret}}$$

$$\mathbf{B}(\mathbf{r},t) = \sum_i (\mathbf{B}_i)_{\text{ret}}$$

and give expressions for $\mathbf{E}_i$ and for $\mathbf{B}_i$ in terms of $\mathbf{E}_i$.

(b) In the case of a single charged particle, the quantities $P(t')$, $dP(t')/d\Omega$, $d^2W/d\Omega\, dt$, etc., are determined by examining $\mathbf{S}\cdot\hat{\mathbf{R}}\, R^2 d\Omega$, where $\mathbf{S}$ is the Poynting vector determined from $\mathbf{E}$ and $\mathbf{B}$ in the radiation zone and $\mathbf{R} = \mathbf{r} - \mathbf{x}(t')$; $\mathbf{r}$ and $\mathbf{x}$ are the field and retarded position vectors, respectively. For the case of N charged particles, this one-particle procedure must be altered. Give a brief discussion of a procedure valid for the present case, showing that the relations

$$\frac{d^2W}{d\Omega dt} = |\mathbf{G}(t)|^2,$$

$$\frac{dW}{d\Omega} = \int_0^\infty 2|\mathbf{F}(\omega)|^2\, d\omega \equiv \int_0^\infty \frac{dI(\omega,\Omega)\, d\omega}{d\Omega}$$

and

$$\mathbf{F}(\omega) = \frac{1}{2\pi}\int_{-\infty}^{\infty} dt\; \mathbf{G}(t)\, e^{i\omega t}$$

are valid, and that only the form of $\mathbf{G}(t)$ is changed from that of the 1-particle case. Write out an expression for $\mathbf{G}(t)$ in terms of q_i, $\hat{\mathbf{n}} \equiv \hat{\mathbf{r}}$, $\boldsymbol{\beta}_i$, $\dot{\boldsymbol{\beta}}_i$ (and other quantities). Finally, convert $\int dt$ to an integral over retarded time (dt') and express $\mathbf{F}(\omega)$ as an integral involving ω, r, $\hat{\mathbf{n}}$, q_i, $\mathbf{x}_i(t')$, t', and $\boldsymbol{\beta}_i$ (but not $\dot{\boldsymbol{\beta}}_i$). Now specialize to the case $N = 2$ and nonrelativistic motion, and derive the expression

$$\frac{d^2I(\omega,\Omega)}{d\omega d\Omega} = \frac{\mu^2}{4\pi^2 c^3}\left|\int dt e^{-i\omega t}\, \ddot{\mathbf{x}}\cdot\hat{\mathbf{n}}\left(\frac{q_1}{m_1} e^{-i\mu\omega/m_1 c\; \hat{\mathbf{n}}\cdot\mathbf{x}(t)} - \frac{q_2}{m_2} e^{-i\mu\omega/m_2 c\; \hat{\mathbf{n}}\cdot\mathbf{x}(t)}\right)\right|^2$$

for $dI(\omega,\Omega)/d\omega$ (i.e., $2|\mathbf{F}(\omega)|^2$). Next, ignore the retardation factors in the preceding expression and make the following assumptions:

(i) The motion of q_1 and q_2 is in the $x-y$ plane ($\hat{\mathbf{e}}_x = \hat{\mathbf{e}}_1$, $\hat{\mathbf{e}}_y = \hat{\mathbf{e}}_2$).

(ii) One particle moves on a straight-line trajectory as viewed in a coordinate system centered on the other.

(iii) There is an initial and essentially constant relative velocity $\mathbf{v}_0 = v_0\hat{\mathbf{e}}_1$ between the charges.

(iv) The vector distance of closest approach is $\mathbf{b} = b\hat{\mathbf{e}}_2$.

(v) The only nonzero component of acceleration $\ddot{\mathbf{x}}$ is in the $\hat{\mathbf{e}}_2$ direction.

(vi) $\ddot{\mathbf{x}}$ is sufficiently small that its velocity dependence involves only v_0 for all t'.

(vii) Magnetic interactions can be neglected (nonrelativistic assumption). On this basis, show that

$$\frac{dW}{d\Omega} = \left(\frac{q_1}{m_1} - \frac{q_2}{m_2}\right)^2 A b^m [1 - f(\theta,\phi)]$$

State the value of m in b^m and write out expressions for the constant $A = A(q_1, q_2, v_0, c)$ and for the function $f(\theta,\phi)$ of the spherical angles defining $\hat{\mathbf{n}}$.

(e) Finally, now retaining the retardation factors in the result of part (c) and again using assumptions (i) - (vii) of part (d), show that $dW/d\Omega$ is still given by a power-law dependence on b, i.e., that $dW/d\Omega = A'b^{m'}$, where A' depends on θ, ϕ, q_1, q_2, m_1, m_2, v_0, and c, but not on b. Determine the value m'. The quantity A' contains, among other terms, an integral whose value involves derivatives of hypergeometric functions; as such it does not seem to be expressible in terms other than a power series.

Solution:

(a) Consider a collection of charges q_i, with positions $\mathbf{x}_i(t)$ and masses m_i, $1 \le i \le n$, observation point $\mathbf{r}$, time of observation t, t' = retarded time, $\mathbf{r}$ = distance from origin to observation point

$$R_i\hat{\mathbf{n}}_i = (\mathbf{r} - \mathbf{x}_i)$$

$$|\mathbf{r} - \mathbf{x}_i(t')| = c(t - t')$$

for one charge,

$$\left.\begin{aligned} \mathbf{E}_i(\mathbf{r}_i t) &= q_i\left[\frac{\hat{\mathbf{n}}_i - \hat{\boldsymbol{\beta}}_i}{\gamma_i^2(1 - \beta_i \cdot \hat{\mathbf{n}}_i)^3 R_i^2}\right]_{\text{ret}} + \frac{q_i}{c}\left[\frac{\hat{\mathbf{n}}_i \times (\hat{\mathbf{n}}_i - \hat{\boldsymbol{\beta}}_i) \times \dot{\boldsymbol{\beta}}_i}{(1 - \beta_i \cdot \hat{\mathbf{n}}_i)^3 R_i}\right]_{\text{ret}} \\ \beta_i &= (\hat{\mathbf{n}}_i \times \mathbf{E}_i)_{\text{ret}} \end{aligned}\right\} \qquad (1)$$

So

$$\mathbf{E}(\mathbf{r},t) = \sum_i (\mathbf{E}_i)_{\text{ret}}$$

$$\mathbf{B}(\mathbf{r},t) = \sum_i (\mathbf{B}_i)_{\text{ret}}$$

with $\mathbf{E}_i$ and $\mathbf{B}_i$ as given in Eq. (1).

(1.b) Radiating time from $(-T, T) = t'$, for $t' = 0$ origin of center of mass of particles is

$$\text{CM}_{\text{origin}} = \sum_i m_i\mathbf{x}_i$$

$$\mathbf{P}_{\text{cm}} = \sum_i \gamma_i m_i \beta_i$$

$$\text{Vol} < (cT)^3$$

$$r >> cT$$

$$\hat{\mathbf{n}}_i \to \hat{\mathbf{n}}, \quad R_i \to r$$

So,

$$\mathbf{E}(\mathbf{r},t) = \sum_i \frac{q_i}{r^2}\left(\frac{\hat{\mathbf{n}} - \boldsymbol{\beta}_i}{\gamma_i^2(1 - \boldsymbol{\beta}_i \cdot \hat{\mathbf{n}})^3}\right)_{\text{ret}} + \sum_i \frac{q_i}{c}\frac{\hat{\mathbf{n}}}{r} \times \left(\frac{(\hat{\mathbf{n}} - \boldsymbol{\beta}_i) \times \dot{\boldsymbol{\beta}}_i}{(1 - \boldsymbol{\beta}_i \cdot \hat{\mathbf{n}})^3}\right)_{\text{ret}}$$

$$\mathbf{B}(\mathbf{r},t) = \hat{\mathbf{n}} \times \mathbf{E}(\mathbf{r},t)$$

$$\mathbf{S} = \frac{c}{4\pi}\, \mathbf{E} \times \mathbf{B} = \frac{c}{4\pi}\, [\hat{\mathbf{n}}(\mathbf{E}^2) - \mathbf{E}(\hat{\mathbf{n}} \cdot \mathbf{E})]$$

$$\frac{dP}{d\Omega} = \lim_{r \to \infty} r^2\hat{\mathbf{n}} \cdot \mathbf{S}$$

if $dP/d\Omega = |\mathbf{A}(t)|^2$, so $\mathbf{A}(t) = (\sqrt{c/4\pi})r\ \mathbf{E}(t)$

$$\mathbf{A}(t) = \sum_i \mathbf{A}_i(t) = \sum_i \sqrt{\frac{c}{4\pi}}\, (R_i\mathbf{E}_i)_{\text{ret}}$$

$$\mathbf{A}(t) = \sum_i \frac{q_i}{\sqrt{4\pi c}}\, \hat{\mathbf{n}} \times \left(\frac{(\hat{\mathbf{n}} \cdot \boldsymbol{\beta}_i) \times \dot{\boldsymbol{\beta}}_i}{(1 - \boldsymbol{\beta}_i \cdot \hat{\mathbf{n}})^3}\right)_{\text{ret}}$$

$$\frac{dP}{d\Omega} = \frac{\partial^2 \omega}{\partial\Omega\, \partial t} = |\mathbf{A}(t)|^2$$

$$\frac{d\omega}{d\Omega} = \int_{-\infty}^{\infty} dt\ |\mathbf{A}(t)|^2$$

$$\left.\begin{aligned} \mathbf{A}(\omega) &= \frac{1}{2\pi}\int_{-\infty}^{\infty} dt\ \mathbf{A}(t)e^{i\omega t} \\ \mathbf{A}(t) &= \int_{-\infty}^{\infty} d\omega\ \mathbf{A}(\omega)e^{-i\omega t} \end{aligned}\right\} \qquad (2)$$

$$\frac{d\omega}{d\Omega} = \int_{-\infty}^{\infty} dt\ |\mathbf{A}(t)|^2 = \int_{-\infty}^{\infty} d\omega \int_{-\infty}^{\infty} d\omega' \int_{-\infty}^{\infty} dt\ \mathbf{A}(\omega)\mathbf{A}^*(\omega')e^{i(\omega' - \omega)t}$$

$$\frac{d\omega}{d\Omega} = 2\pi \int_{-\infty}^{\infty} d\omega\ [\mathbf{A}(\omega)\mathbf{A}^*(\omega)]$$

$$\frac{dw}{d\Omega} = 4\pi \int_0^{\infty} d\omega \; |\mathbf{A}(\omega)|^2$$

$$\frac{dw}{d\Omega} = 4\pi \int_0^{\infty} d\omega \; |\mathbf{A}(\omega)|^2$$

where the factor of $1/2\pi$ difference is included in the $\mathbf{A}(\omega)$ as in Eq. (2):

$$\mathbf{A}(\omega) = \frac{1}{2\pi} \int_{-\infty}^{\infty} dt \; \mathbf{A}(t) e^{i\omega t}$$

$$\mathbf{A}(\omega) = \frac{1}{2\pi} \int_{-\infty}^{\infty} dt \sum_i \frac{q_i}{\sqrt{4\pi c}} \; \hat{\mathbf{n}} \times \left\{ \frac{(\hat{\mathbf{n}} - \beta_i) \times \dot{\beta}_i}{(1 - \beta_i \cdot \hat{\mathbf{n}})^3} \right\}_{\text{ret}} e^{i\omega t}$$

$$t = t_i' + \left| \frac{\mathbf{r} - \mathbf{x}_i(t_i')}{c} \right|$$

$$\frac{dt}{dt_i'} = 1 + \dot{\mathbf{x}}_i(t_i') + \nabla_{x_i} \left| \frac{\mathbf{r} - \mathbf{x}_i(t_i')}{c} \right|$$

$$\frac{dt}{dt_i'} = 1 - c\beta_i(t_i') \cdot \nabla_r \left| \frac{\mathbf{r} - \mathbf{x}_i(t_i')}{c} \right|$$

$$\frac{dt}{dt_i'} = 1 - \beta_i(t_i') \cdot \hat{\mathbf{n}}_i = (1 - \beta_i \cdot \hat{\mathbf{n}})_{\text{ret}}$$

$$\mathbf{A}(\omega) = \frac{1}{2\pi\sqrt{4\pi c}} \sum_i q_i \; \hat{\mathbf{n}} \times \int_{-\infty}^{\infty} dt \left\{ \frac{[(\hat{\mathbf{n}} - \beta_i) \times \dot{\beta}_i]}{(1 - \beta_i \cdot \hat{\mathbf{n}})^3} \right\}_{\text{ret}} e^{i\omega t}$$

$$\mathbf{A}(\omega) = \frac{1}{2\pi\sqrt{4\pi c}} \sum_i q_i \; \hat{\mathbf{n}} \times \int_{-\infty}^{\infty} dt_i' \frac{dt}{dt_i'} \left. \frac{(\hat{\mathbf{n}} - \beta_i) \times \dot{\beta}_i}{(1 - \beta_i \cdot \hat{\mathbf{n}})^3} \right|_{t_i'} \exp\left[i\omega \left(\frac{t_i' + |\mathbf{r} - \mathbf{x}_i(t_i')|}{c} \right) \right]$$

$$\mathbf{A}(\omega) = \frac{1}{2\pi\sqrt{4\pi c}} \sum_i q_i \; \hat{\mathbf{n}} \times \int_{-\infty}^{\infty} dt_i' \left. \frac{(\hat{\mathbf{n}} - \beta_i) \times \dot{\beta}_i}{(1 - \beta_i \cdot \hat{\mathbf{n}})^2} \right|_{t_i'} e^{i\omega t_i} \exp\left[\frac{i\omega t}{c} \left(1 - \frac{\hat{\mathbf{n}} \cdot \mathbf{x}_i(t_i')}{r} \right) \right]$$

$$\mathbf{A}(\omega) = \frac{1}{2\pi\sqrt{4\pi c}} \sum_i q_i\, \hat{\mathbf{n}} \times \int_{-\infty}^{\infty} dt\, \frac{(\hat{\mathbf{n}} - \beta_i) \times \dot{\beta}_i}{(1 - \beta_i \cdot \hat{\mathbf{n}})^2}\bigg|_t\, e^{i\omega t} \exp\left[\frac{i\omega r}{c}\left(1 - \frac{\hat{\mathbf{n}} \cdot \mathbf{x}_i(t)}{r}\right)\right]$$

$$\mathbf{A}(\omega) = \frac{1}{2\pi\sqrt{4\pi c}} \sum_i q_i \int_{-\infty}^{\infty} dt\, \frac{d}{dt}\left(\frac{\hat{\mathbf{n}} \times (\hat{\mathbf{n}} \times \beta_i)}{(1 - \beta_i \cdot \hat{\mathbf{n}})}\right) e^{i\omega(t+r)/c} \exp\left(\frac{-i\omega}{c}\, \hat{\mathbf{n}} \times \mathbf{x}_i(t)\right)$$

$$= \frac{1}{4\pi\sqrt{\pi c}} \sum_i q_i \int_{-\infty}^{\infty} dt\, \frac{\hat{\mathbf{n}} \times (\hat{\mathbf{n}} \times \beta_i)}{(1 - \beta_i \cdot \hat{\mathbf{n}})}\, e^{i\omega r/c} \frac{d}{dt} \exp[i\omega(t - \hat{\mathbf{n}} \cdot \hat{\mathbf{x}}_i(t)/c)]$$

$$= \frac{1}{4\pi\sqrt{\pi c}} \sum_i q_i \int_{-\infty}^{\infty} dt\, \frac{\hat{\mathbf{n}} \times (\hat{\mathbf{n}} \times \beta_i)}{(1 - \beta_i \cdot \hat{\mathbf{n}})}\, e^{i\omega r/c}\, i\omega[1 - \hat{\mathbf{n}} \cdot \beta_i(t)] \exp[i\omega(t - \hat{\mathbf{n}} \cdot \mathbf{x}_i(t)/c)]$$

$$\mathbf{A}(\omega) = \frac{-e^{i\omega r}}{4\pi\sqrt{\pi c}} \sum_i \omega q_i \int_{-\infty}^{\infty} dt\, \hat{\mathbf{n}} \times (\hat{\mathbf{n}} \times \beta_i) \exp[i\omega(t - \hat{\mathbf{n}} \cdot \mathbf{x}_i(t)/c)]$$

(1.c)

$$\mathbf{A}(\omega) = \frac{e^{i\omega r/c}}{4\pi\sqrt{\pi c}} \sum_i q_i\, \hat{\mathbf{n}} + \int_{-\infty}^{\infty} dt\, \frac{(\hat{\mathbf{n}} - \beta_i) \times \dot{\beta}_i}{(1 - \beta_i \cdot \hat{\mathbf{n}})^2} \exp[i\omega(t - \hat{\mathbf{n}} \cdot \mathbf{x}_i(t)/c)]$$

$$\mathbf{A}(\omega) = \frac{e^{i\omega r/c}}{4\pi\sqrt{\pi c}}\, \hat{\mathbf{n}} \times \int_{-\infty}^{\infty} dt \left[\frac{q_1(\hat{\mathbf{n}} - \beta_1) \times \dot{\beta}_1}{(1 - \beta_1 \cdot \hat{\mathbf{n}})^2} \exp(-i\omega\, \hat{\mathbf{n}} \cdot \mathbf{x}_1(t)/c) + \frac{q_2(\hat{\mathbf{n}} - \beta_2) \times \dot{\beta}_2}{(1 - \beta_2 \cdot \hat{\mathbf{n}})^2} \exp[-i\omega \hat{\mathbf{n}} \cdot \mathbf{x}_2(t)]\right] e^{i\omega t}$$

for observation frame at center of mass of particles,

$$\mu = \frac{m_1 m_2}{m_1 + m_2}$$

$$\beta_1 \simeq 0$$

$$\beta_2 \simeq 0$$

$$\beta_{cm} = 0, \quad \dot{\beta}_{cm} = 0, \quad \mathbf{x}_{cm} = 0$$

$$\mathbf{x}_1(t) = \frac{\mu}{m_1}\, \mathbf{x}(t)$$

$$\mathbf{x}_2(t) = \frac{-\mu}{m_2}\,\mathbf{x}(t)$$

$$\beta_1 << 1$$

$$\beta_2 << 1$$

$$\beta_1 = \frac{m\dot{\mathbf{x}}}{m_1 c}$$

$$\beta_2 = -\frac{\mu\dot{\mathbf{x}}}{m_2 c}$$

$$\dot{\beta}_1 = \frac{\mu\ddot{\mathbf{x}}}{m_1 c}$$

$$\dot{\beta}_2 = -\frac{\mu\ddot{\mathbf{x}}}{m_2 c}$$

$$\mathbf{A}(\omega) = \frac{e^{i\omega r/c}}{4\pi\sqrt{\pi c}}\int_{-\infty}^{\infty} dt\; e^{i\omega t}\;\hat{\mathbf{n}}\times\left[q_1\hat{\mathbf{n}}\times\frac{\mu\ddot{\mathbf{x}}}{m_1 c}\exp[(-i\omega\mu/m_1 c)\hat{\mathbf{n}}\cdot\mathbf{x}(t)]\right.$$

$$\left.+\,q_2\hat{\mathbf{n}}\times\frac{(-)\mu\ddot{\mathbf{x}}}{m_2 c}\exp[(i\omega\mu/m_2 c)\hat{\mathbf{n}}\cdot\mathbf{x}(t)]\right]$$

$$\mathbf{A}(\omega) = \frac{e^{i\omega r/c}}{4\pi\sqrt{\pi c}}\int_{-\infty}^{\infty} dt\; e^{i\omega t}\;\hat{\mathbf{n}}\times\left(\hat{\mathbf{n}}\times\frac{\mu\ddot{\mathbf{x}}}{c}\right)\left[\frac{q_1}{m_1}\exp[-i\omega\mu/m_1 c\;\hat{\mathbf{n}}\cdot\mathbf{x}(t)]\right.$$

$$\left.-\frac{q_2}{m_2}\exp[i\omega\mu/m_2 c\;\hat{\mathbf{n}}\cdot\mathbf{x}(t)]\right]$$

$$\frac{d^2I}{d\omega d\Omega} = 4\pi\,|\mathbf{A}(\omega)|^2$$

$$\frac{d^2I(\omega,\Omega)}{d\omega d\Omega} = \frac{\mu^2}{4\pi^2 c^3}\left|\int_{-\infty}^{\infty} dt\; e^{i\omega t}\;(\hat{\mathbf{n}}\cdot\ddot{\mathbf{x}})\left[\frac{q_1}{m_1}\exp\left(\frac{-\mu\omega}{m_1 c}\,\hat{\mathbf{n}}\cdot\mathbf{x}(t)\right)\right.\right.$$

$$\left.\left.-\frac{q_2}{m_2}\times\exp\left(\frac{i\mu\omega}{m_2 c}\,\hat{\mathbf{n}}\cdot\mathbf{x}(t)\right)\right]\right|^2$$

where the constant factors are again contained in the Eq. (2).

(1.d)

$$\mathbf{v} = \frac{d\mathbf{x}}{dt} = \dot{\mathbf{x}} = \mathbf{v}_1 - \mathbf{v}_2 = v_0\hat{\mathbf{e}}_1, \quad v \sim v_0$$

$$\mathbf{x}(t) = b\hat{\mathbf{e}}_2 + v_0 t\hat{\mathbf{e}}_1, \quad \ddot{\mathbf{x}} = a_1\hat{\mathbf{e}}_1 + a_2\hat{\mathbf{e}}_2$$

$$\tfrac{1}{2}\,\mu v_0^2 = \tfrac{1}{2}\,\mu v^2 + \frac{q_1 q_2}{x}, \quad \tfrac{1}{2}\,\mu v_0^2 > \frac{q_1 q_2}{b}$$

$$m\ddot{\mathbf{x}} = \frac{q_1 q_2}{x^2}\,\hat{\mathbf{x}} = ma\hat{\mathbf{x}} = ma_1\hat{\mathbf{e}}_1 + ma_2\hat{\mathbf{e}}_2$$

$$m\ddot{\mathbf{x}} = \frac{m\hat{\mathbf{e}}_1 q_1 q_2 v_0 t}{\mu[b^2 + (v_0 t)^2]^{3/2}} + m\hat{\mathbf{e}}_2\,\frac{q_1 q_2 b}{\mu[b^2 + (v_0 t)^2]^{3/2}}$$

the only nonzero component of $a = \ddot{\mathbf{x}}$ is in the $\hat{\mathbf{e}}_2$ direction, so

$$\ddot{\mathbf{x}} = \hat{\mathbf{e}}_2\,\frac{q_1 q_2 b}{\mu[b^2 + (v_0 t)^2]^{3/2}}$$

So then,

$$\hat{\mathbf{n}} \times \ddot{\mathbf{x}} = \frac{q_1 q_2 b}{\mu[b^2 + (v_0 t)^2]^{3/2}}\,[-\hat{\mathbf{e}}_1 \cos\theta + \hat{\mathbf{e}}_3 \sin\theta\cos\phi]$$

as it was given to ignore retardation factors.

$$\frac{d^2 I}{d\omega d\Omega} = \frac{\mu^2}{4\pi^2 c^3}\,\frac{(q_1 q_2)^2}{(\mu)^2}\left(\frac{q_1}{m_1} - \frac{q_2}{m_2}\right)^2$$

$$\times \left| \int_{-\infty}^{\infty} dt\, e^{i\omega t}\,\frac{b}{[b^2 + (v_0 t)^2]^{3/2}} \right|^2 \times [-\hat{\mathbf{e}}_1 \cos\theta + \hat{\mathbf{e}}_3 \sin\theta\cos\phi]^2$$

look at the term

$$\int_{-\infty}^{\infty} dt\, e^{i\omega t}\,\frac{b}{[b^2 + (v_0 t)^2]^{3/2}} = I$$

$$I = 2\int_0^{\infty} dt\,\frac{b\cos\omega t}{[b^2 + (v_0 t)^2]^{3/2}}$$

given:

$$\int_0^{\infty} \frac{\cos \alpha x \, dx}{(x^2+\beta^2)^{3/2}} = \frac{\alpha}{\beta} k_1(\alpha \beta)$$

Let

$$x = v_0 t$$

$$\beta = b$$

$$dx = v_0 \, dt$$

$$\alpha = \frac{\omega}{v_0}$$

$$I = \frac{2b}{v_0} \frac{\omega}{v_0 b} k_1\left(\frac{\omega}{v_0} b\right) = \frac{2\omega}{v_0^2} k_1 \left(\frac{\omega b}{v_0}\right)$$

$$\frac{d^2 I}{d\omega d\Omega} = \left(\frac{q_1}{m_1} - \frac{q_2}{m_2}\right)^2 \left(\frac{\omega q_1 q_2}{\pi c v_0^2}\right)^2 \frac{1}{c} \left[k_1\left(\frac{\omega b}{v_0}\right)\right]^2 (\cos^2 \theta + \sin^2 \theta \cos^2 \phi)$$

$$\frac{d\omega}{d\Omega} = 4\pi \int_0^{\infty} d\omega \; |\mathbf{A}(\omega)|^2 = \int_0^{\infty} d\omega \; \frac{d^2 I}{d\omega d\Omega}$$

$$\frac{d\omega}{d\Omega} = \frac{1}{c} \left(\frac{q_1 q_2 \cdot 2}{2\pi c v_0^2}\right)^2 \left(\frac{q_1}{m_1} - \frac{q_2}{m_2}\right)^2 (1 - \sin^2 \theta \sin^2 \phi) \int_0^{\infty} d\omega \; \omega^2 \left[k_1\left(\frac{\omega b}{v_0}\right)\right]^2$$

$$\frac{d\omega}{d\Omega} = \frac{1}{c} \left(\frac{q_1 q_2}{\pi c v_0^2}\right)^2 \left(\frac{q_1}{m_1} - \frac{q_2}{m_2}\right)^2 (1 - \sin^2 \theta \sin^2 \phi) \left(\frac{v_0}{b}\right)^3 \int_0^{\infty} dx \; x^2 k_1^2(x)$$

given:

$$\int_0^{\infty} x^n k_\mu(x) k_\nu(x) dx$$

so,

$$\int_0^{\infty} x^n k_\mu(x) k_\nu(x) dx$$

$$= \frac{2^{n-2}}{\Gamma(n+2)} \left\{\Gamma \frac{(n+1+\mu+\nu)}{2} \Gamma \frac{(n+1-\mu+\nu)}{2} \Gamma \frac{(n+1+\mu-\nu)}{2} \Gamma \frac{(n+1-\mu-\nu)}{2}\right\}$$

So,

$$\int_0^{\infty} dx \; x^2 k_1(x) k_1(x) = \frac{1}{\Gamma(3)} \left[\Gamma \left(\frac{5}{2}\right) \Gamma \left(\frac{3}{2}\right) \Gamma \left(\frac{3}{2}\right) \Gamma \left(\frac{1}{2}\right)\right]$$

$$= \frac{(1)}{2}\,\frac{(3\sqrt{\pi})}{4}\,\frac{(\sqrt{\pi})}{2}\,\frac{(\sqrt{\pi})}{2}\,(\sqrt{\pi}) = \frac{3\pi^2}{32}$$

so

$$\frac{d\omega}{d\Omega} = \frac{1}{c}\left(\frac{q_1 q_2}{\pi c v_0^2}\right)^2 \left(\frac{q_1}{m_1} - \frac{q_2}{m_2}\right)(1 - \sin^2\theta \sin^2\phi)\left(\frac{v_0}{b}\right)^3 \frac{3\pi^2}{32}$$

$$\frac{d\omega}{d\Omega} = \frac{3v_0}{2bc}\left(\frac{q_1 q_2}{4cv_0 b}\right)^2 \left(\frac{q_1}{m_1} - \frac{q_2}{m_2}\right)^2 (1 - \sin^2\theta \sin^2\phi)$$

$$\frac{d\omega}{d\Omega} = \left(\frac{q_1}{m_1} - \frac{q_2}{m_2}\right)^2 \left[\frac{3v_0}{2c}\left(\frac{q_1 q_2}{4cv_0}\right)\right] b^{-3}\ (1 - \sin^2\theta \sin^2\phi)$$

where

$$A = \frac{3v_0}{2c}\left(\frac{q_1 q_2}{4cv_0}\right)^2$$

$f(\theta, \phi) = \sin^2\theta \sin^2\phi$

the value of min. b^m is $m = -3$

$$A = A(q_1, q_2, v_0, c) = \frac{3v_0}{2c}\left(\frac{q_1 q_2}{4cv_0}\right)^2$$

$f(\theta, \phi) = \sin^2\theta \sin^2\phi$

$$\frac{d^2 I}{d\omega d\Omega} = \frac{\mu^2}{4\pi c^3}\left|\int_{-\infty}^{\infty} dt\, e^{i\omega t}(\hat{\mathbf{n}} \times \ddot{\mathbf{x}})\left[\frac{q_1}{m_1}\exp(-(i\omega\mu/m_1 c)\hat{\mathbf{n}} \cdot \mathbf{x}(t))\right.\right.$$

$$\left.\left. - \frac{q_2}{m_2}\exp((i\mu\omega/m_2 c)\hat{\mathbf{n}} \cdot \ddot{\mathbf{x}}(t))\right]\right|^2$$

$$\hat{\mathbf{n}} \times \ddot{\mathbf{x}} = \frac{q_1 q_2 b}{\mu[b^2 + (v_0 t)^2]^{3/2}}\,[-\hat{\mathbf{e}}_1 \cos\theta + \hat{\mathbf{e}}_3 \sin\theta\cos\phi]$$

$$\frac{d^2 I}{d\omega d\Omega} = \frac{\mu^2}{4\pi c^3}\,\frac{(q_1 q_2)^2}{\mu^2}\,(\cos^2\theta + \sin^2\theta\cos^2\phi)$$

$$\left|\int_{-\infty}^{\infty} dt\, e^{i\omega t} \frac{b}{[b^2+(v_0t)^2]^{3/2}} \times \left[\frac{q_1}{m_1} \exp(-i\mu\omega/m_1c\ \hat{\mathbf{n}}\cdot\mathbf{x}(t))\right.\right.$$

$$\left.\left. - \frac{q_2}{m_2} \exp(i\mu\omega/m_2c\ \hat{\mathbf{n}}\cdot\mathbf{x}(t))\right]\right|^2$$

Look at term $|\xi|^2$, where

$$\xi = \int_{-\infty}^{\infty} dt\, e^{i\omega t} \frac{b}{[b^2+(v_0t)^2]^{3/2}}$$

$$\times \left\{\left[\frac{q_1}{m_1} \exp(-i\mu\omega/m_1c\ (\sin\theta\cos\phi\, v_0t + \sin\theta\sin\phi\, b))\right]\right.$$

$$\left. - \left[\frac{q_2}{m_2} \exp(i\mu\omega/m_2c(\sin\theta\cos\phi\, v_0t + \sin\theta\sin\phi\, b))\right]\right\}$$

$$= \frac{q_1b}{m_1} \exp(-\mu\omega/m_1c\ \sin\theta\sin\phi\, b) \int_{-\infty}^{\infty} dt \frac{\exp\{i\omega t[1-(\mu v_0/m_1c)(\sin\theta\cos\phi)]\}}{[b^2+(v_0t)^2]^{3/2}}$$

$$- \frac{q_2b}{m_2} \exp(i\mu\omega/m_2c\ \sin\theta\sin\phi\, b) \int_{-\infty}^{\infty} dt \frac{\exp\{i\omega t[1+(\mu v_0/m_2c)(\sin\theta\cos\phi)]\}}{[b^2+(v_0t)^2]^{3/2}}$$

let

$$\alpha' = \left(1 - \frac{\mu v_0}{m_1c}\sin\theta\cos\phi\right)t = \gamma t$$

$$\beta' = \left(1 + \frac{\mu v_0}{m_2c}\sin\theta\cos\phi\right)t = \delta t$$

$$\gamma = 1 - \frac{\mu v_0}{m_1c}\sin\theta\cos\phi$$

$$\delta = 1 + \frac{\mu v_0}{m_2c}\sin\theta\cos\phi$$

so,

$$\xi = \frac{q_1}{m_1\gamma} \exp(-(i\mu\omega/m_1c)(\sin\theta\sin\phi)b) \int_{-\infty}^{\infty} \frac{d\alpha' b e^{i\omega\alpha'}}{[b^2 + (v_0\alpha'/\gamma)^2]^{3/2}}$$

$$-\frac{q_2}{m_2\delta} \exp((i\mu\omega/m_2c)(\sin\theta\sin\phi)b) \int_{-\infty}^{\infty} \frac{d\beta' b e^{i\omega\beta'}}{[b^2 + (v_0\beta'/\delta)^2]^{3/2}}$$

given:

$$\int_0^{\infty} \frac{\cos\alpha x\, dx}{(x^2+\beta^2)^{3/2}} = \frac{\alpha}{\beta} k_1(\alpha\ \beta)$$

let,

$$\beta = b$$

$$\alpha = \frac{\omega}{v_0}\gamma, \quad \frac{\omega}{v_0}\delta$$

$$x = \frac{v_0\alpha'}{\gamma}, \quad \frac{v_0\beta'}{\delta}$$

$$\xi = \frac{2q_1}{m_1} \frac{\exp(-(i\mu\omega/m_1c)(\sin\theta\sin\phi)b)}{\gamma} \frac{\gamma}{v_0} \frac{b\omega\gamma}{v_0 b} k_1\left(\frac{\omega\gamma b}{v_0}\right)$$

$$-\frac{2q_2}{m_2} \frac{\exp(-(i\omega\mu/m_2c)(\sin\theta\sin\phi)b)}{\delta} \frac{\delta}{v_0} \frac{b\omega\delta}{v_0 b} k_1\left(\frac{\omega\delta b}{v_0}\right)$$

$$\xi = \frac{2q_1}{m_1} \exp(-(i\mu\omega/m_1c)(\sin\theta\sin\phi)b) \frac{\omega\gamma}{v_0^2} k_1\left(\frac{\omega b\gamma}{v_0}\right)$$

$$-\frac{2q_2}{m_2} \exp((i\omega\mu/m_2c)(\sin\theta\sin\phi)b) \frac{\omega\delta}{v_0^2} k_1\left(\frac{\omega b\delta}{v_0}\right)$$

$$|\xi|^2 = \frac{4\omega^2}{v_0^4}\left\{\left[\frac{q_1\gamma}{m_1} k_1\left(\frac{\omega b\gamma}{v_0}\right)\right]^2 + \left[\frac{q_2\delta}{m_2} k_1\left(\frac{\omega b\delta}{v_0}\right)\right]^2\right.$$

$$\left. - 2\cos\left[\frac{\omega}{c}(\sin\theta\sin\phi)b\left(\frac{\mu}{m_1}+\frac{\mu}{m_2}\right)\right] \frac{q_1q_2\gamma\delta}{m_1m_2} k_1\left(\frac{\omega b\delta}{v_0}\right) k_1\left(\frac{\omega b\gamma}{v_0}\right)\right\}$$

$$\int_0^\infty d\omega \ |\xi|^2 = \left(\frac{2}{v_0^2}\frac{q_1\gamma}{m_1}\right)^2 \int_0^\infty d\omega \ \omega^2 \left[k_1 \left(\frac{\omega b\gamma}{v_0}\right)\right]^2 + \left(\frac{2}{v_0^2}\frac{q_2\delta}{m_2}\right)^2 \int_0^\infty d\omega \ \omega^2 \left[k_1 \left(\frac{\omega b\delta}{v_0}\right)\right]^2$$

$$- 2\left(\frac{2}{v_0^2}\right)^2 \frac{q_1q_2\delta\gamma}{m_1m_2} \int_0^\infty d\omega \, \omega^2 \cos\left(\frac{\omega b}{c}\sin\theta\sin\phi\right) k_1\left(\frac{\omega b\gamma}{v_0}\right) k_1\left(\frac{\omega b\delta}{v_0}\right)$$

$$\int_0^\infty d\omega \ |\xi|^2 = \frac{1}{b^3}\left[\left(\frac{2q_1}{m_1}\right)^2 \frac{3\pi^2}{\gamma v_0 32} + \left(\frac{2q_2}{m_2}\right)^2 \frac{3\pi^2}{v_0\delta 32}\right.$$

$$\left. - \frac{8q_1q_2\delta\gamma}{v_0 m_1 m_2}\int_0^\infty dx \ x^2 k_1^2(\gamma x) \ \cos\frac{v_0}{c} \sin\theta\sin\phi\right]$$

$$\frac{dw}{d\Omega} = \frac{(q_1q_2)^2}{4\pi c^3} (1 - \sin^2\theta\sin^2\phi)\left(\frac{1}{b^3}\right)\left[\left(\frac{q_1\pi}{m_1\cdot 2}\right)^2 \frac{3}{\gamma v_0\cdot 2} + \left(\frac{2q_2\pi}{m_2 4}\right)^2 \frac{3}{v_0\delta 2}\right.$$

$$\left. - \frac{8q_1q_2\delta\gamma}{v_0(m_1m_2)} \times \int_0^\infty dx \ x^2 k_1^2(\gamma x)\cos\frac{v_0}{c} \sin\theta\sin\phi\right]$$

$$\frac{dw}{d\Omega} = \left\{\frac{(q_1q_2)^2}{4\pi c^3} (1 - \sin^2\theta\sin^2\phi)\left[\left(\frac{\pi q_1}{2m_1}\right)^2 \frac{3}{\gamma v_0 2} + \left(\frac{\pi q_2}{2m_2}\right)^2 \frac{3}{\delta v_0 2}\right.\right.$$

$$\left.\left. - \frac{8q_1q_2\delta\gamma}{v_0 m_1 m_2}\int_0^\infty dx \ x^2 k_1^2(\gamma x)\cos\frac{v_0}{c} \sin\theta\sin\phi\right]\right\}\left(\frac{1}{b^3}\right)$$

where

$$A' = -\frac{8q_1q_2\delta\gamma}{v_0 m_1 m_2}\int_0^\infty dx \ x^2 k_1^2(\gamma x)\cos\frac{v_0}{c} \sin\theta\sin\phi$$

$$b^{m'} = \frac{1}{b^3}$$

So

$$\frac{dw}{d\Omega} = A' b^{m'}$$

where the value of $m' = -3$.

PROBLEM 14.16 An infinitely long, thin cylinder contains a static charge density τ per unit length and a current I flowing parallel to its axis. Find the values of $\beta = v/c$ for those moving coordinate systems in which (a) the electric and (b) the magnetic field is zero. In each case, determine the value of the nonzero field. What is the physical interpretation of $I = \tau c$?

Solution:

Transverse area same

$$I_1' = I_1$$

$$I_2' = I_2$$

$$I_3' = \gamma(I_3 - \beta c\tau)$$

$$c\tau' = \gamma(c\tau - \beta I_3)$$

for $E = 0$, $c\tau' = 0$ or $c\tau - \beta I_3 = 0$, so $\beta = c\tau/I$ for $I > c\tau$

$$I_3' = \frac{\gamma[I_3 - (c\tau)^2]}{I_3}$$

from Ampere's Law,

$$\oint_c \mathbf{B} \cdot d\ell = \frac{4\pi I}{c}$$

$$2\pi r B = \frac{4\pi I}{c}$$

$$B = \frac{2I}{rc}$$

$$\mathbf{B} = \frac{2I}{rc}\,\hat{\theta}$$

so,

$$\mathbf{B}' = \frac{2I'}{cr}\,\hat{\theta}' = \hat{\theta}\,\frac{2}{cr'}\,\gamma\left(I - \frac{(c\tau)^2}{I}\right), \quad \mathbf{E}' = 0$$

(b) $B = 0$

$$I_3' = 0 \text{ or } I_3 - \beta c\tau = 0$$

$$B = \frac{I}{c\tau} \text{ for } I < c < \tau$$

from Gauss' law,

$$\oint_S \mathbf{E} \cdot d\mathbf{s} = 4\pi q$$

$$2\pi E r = 4\pi\,\tau$$

$$E = \frac{2\tau}{r}$$

$$\mathbf{E} = \frac{2\tau\,\hat{\mathbf{r}}}{r}$$

$$\tau' = \frac{\gamma}{c}\left(c\tau - \frac{I^2}{c\tau}\right)$$

so,

$$\mathbf{E}' = \frac{2\hat{\mathbf{r}}}{r'}\frac{\gamma}{c}\left(c\tau - \frac{I^2}{c\tau}\right)\hat{\mathbf{r}}, \quad \mathbf{B} = 0$$

When $c\tau < I$, the reference frame moves with a velocity $v = c^2\tau/I$ parallel to the axis of the cylinder; when $\tau c > I$, the reference moves with a velocity I/τ parallel to the axis of the cylinder; when $I = \tau c$, then there is no reference frame in which there is only an electric or magnetic field, since the velocity of a reference frame where this would happen would move at $v \sim c$ and the E and B fields would be ~ 0.

PROBLEM 14.17 An infinitesimal Lorentz transformation and its inverse can be written as

$$x'^{\alpha} = (g^{\alpha\beta} + \varepsilon^{\alpha\beta})x_{\beta}$$

$$x^{\alpha} = (g^{\alpha\beta} + \varepsilon'^{\alpha\beta})x'_{\beta}$$

where $\varepsilon^{\alpha\beta}$ and $\varepsilon'^{\alpha\beta}$ are infinitesimal.

(a) Show from the definition of the inverse that $\varepsilon'^{\alpha\beta} = -\varepsilon^{\alpha\beta}$.
(b) Show that from the preservation of the norm that $\varepsilon^{\alpha\beta} = -\varepsilon^{\beta\alpha}$.
(c) By writing the transformation in terms of contravariant components on both sides of the equation show that $\varepsilon^{\alpha\beta}$ is equivalent to the matrix $L = -\boldsymbol{\omega} \cdot \mathbf{s} - \boldsymbol{\xi} \cdot \mathbf{k}$, where the matrices s_i generate rotations in three dimensions, while the matrices k_i produce boosts. L is a 4×4, real matrix whose general form is

$$L - \left(\begin{array}{c|ccc} 0 & L_{01} & L_{02} & L_{03} \\ \hline L_{01} & 0 & L_{12} & L_{13} \\ L_{02} & -L_{12} & 0 & L_{23} \\ L_{03} & -L_{13} & -L_{23} & 0 \end{array}\right)$$

and

$$s_1 = \left(\begin{array}{c|ccc} 0 & 0 & 0 & 0 \\ \hline 0 & 0 & 0 & 0 \\ 0 & 0 & 0 & -1 \\ 0 & 0 & 1 & 0 \end{array}\right), \quad s_2 = \left(\begin{array}{c|ccc} 0 & 0 & 0 & 0 \\ \hline 0 & 0 & 0 & 1 \\ 0 & 0 & 0 & 0 \\ 0 & -1 & 0 & 0 \end{array}\right), \quad s_3 = \left(\begin{array}{c|ccc} 0 & 0 & 0 & 0 \\ \hline 0 & 0 & -1 & 0 \\ 0 & 1 & 0 & 0 \\ 0 & 0 & 0 & 0 \end{array}\right)$$

$$k_1 = \left(\begin{array}{c|ccc} 0 & 1 & 0 & 0 \\ \hline 1 & & & \\ 0 & & 0 & \\ 0 & & & \end{array}\right), \quad k_2 = \left(\begin{array}{c|cc} 0 & 1 & 0 \\ \hline 0 & & \\ 1 & 0 & \\ 0 & & \end{array}\right), \quad k_3 = \left(\begin{array}{c|ccc} 0 & 0 & 0 & 1 \\ \hline 0 & & & \\ 0 & & 0 & \\ 1 & & & \end{array}\right)$$

where ω and ξ are constant three-vectors.

Solution:

(a) $x'^{\alpha} = (g^{\alpha\beta} + \varepsilon^{\alpha\beta})x_{\beta}$

$$x^{\alpha} = (g^{\alpha\beta} + \varepsilon'^{\alpha\beta})x'_{\beta}$$

$$x'^{\alpha} = (g^{\alpha\beta} + \varepsilon^{\alpha\beta})g_{\beta\alpha}x^{\gamma}, \quad x'^{\alpha} = (g^{\alpha\beta} + \varepsilon^{\alpha\beta})g_{\beta\gamma}(g^{\gamma\delta} + \varepsilon'^{\gamma\delta})x'_{\delta}$$

$$(g^{\alpha\beta} + \varepsilon^{\alpha\beta})g_{\beta\gamma}(g^{\gamma\delta} + \varepsilon'^{\gamma\delta}) = g^{\alpha\delta}$$

$$g^{\alpha\beta}g_{\beta\gamma}g^{\gamma\delta} + g^{\alpha\beta}g_{\beta\gamma}\varepsilon'^{\gamma\delta} + \varepsilon^{\alpha\beta}g_{\beta\gamma}g^{\gamma\delta} = g^{\alpha\delta}$$

$$\delta^{\alpha}_{\gamma}g^{\gamma\delta} + \delta^{\alpha}_{\gamma}\varepsilon'^{\gamma\delta} + \varepsilon^{\alpha\beta}\delta^{\delta}_{\beta} = g^{\alpha\delta}$$

$$g^{\alpha\delta} + \varepsilon'^{\alpha\delta} + \varepsilon^{\alpha\delta} = g^{\alpha\delta}, \quad \varepsilon'^{\alpha\delta} = -\varepsilon^{\alpha\delta}$$

(b) $x'^{\alpha}x'_{\alpha} = x_{\beta}x^{\beta}, \quad x'^{\alpha}g_{\alpha\gamma}x'^{\gamma} = x_{\beta}g^{\beta\nu}x_{\nu}$

$$x_{\beta}(g^{\alpha\beta} + \varepsilon^{\alpha\beta})g_{\alpha\gamma}(g^{\gamma\nu} + \varepsilon^{\gamma\nu})x_{\nu} = x_{\beta}g^{\beta\nu}x_{\nu}$$

$$x_{\beta}(g^{\alpha\beta}g_{\alpha\gamma}g^{\gamma\nu} + \varepsilon^{\alpha\beta}g_{\alpha\gamma}g^{\alpha\nu} + g^{\alpha\beta}g_{\alpha\gamma}\varepsilon^{\gamma\nu}) = x_{\beta}g^{\beta}x_{\nu}$$

so,

$$g^{\alpha\beta}g_{\alpha\gamma}g^{\gamma\nu} + \varepsilon^{\alpha\beta}g_{\alpha\gamma}g^{\gamma\nu} + g^{\alpha\beta}g_{\alpha\gamma}\varepsilon^{\gamma\nu} = g^{\beta\nu}$$

$$g^{\alpha\beta}\delta_{\alpha}^{\nu} + \varepsilon^{\alpha\beta}\delta_{\alpha}^{\nu} + \delta_{\gamma}^{\beta}\varepsilon^{\gamma\nu} = g^{\beta\nu}$$

$$g^{\beta\nu} + \varepsilon^{\nu\beta} + \varepsilon^{\beta\nu} = g^{\beta\nu}$$

therefore,

$$\varepsilon^{\nu\beta} = -\varepsilon^{\beta\nu}$$

(c) $x'^{\alpha} + (g^{\alpha\beta} + \varepsilon^{\alpha\beta})g_{\beta\alpha}x^{\gamma} = (\delta_{\gamma}^{\alpha} + \varepsilon_{\gamma}^{\alpha})x^{\gamma} = e^{\varepsilon^{\alpha}{}_{\gamma}}x^{\gamma}$

$$x' = Ae = e^{L}x$$

so

$$e^{\alpha}{}_{\gamma} = L$$

PROBLEM 14.18 Considering the dipole term only, calculate the mean lifetime of the state μ_n when there is only one state μ_m below it, where

$$\mu_n = \frac{1}{\sqrt{2\pi a^3}} \frac{r}{4a} e^{-r/2a} \cos\theta$$

$$\mu_m = \frac{1}{\sqrt{\pi a^3}} e^{-r/a}, \quad a = \frac{\hbar^2}{me^2}$$

Show also that the transition from $\mu_{n'}$ to $\mu_{m'}$ with

$$\mu_{n'} = \frac{1}{\sqrt{2\pi a^3}} \left(\frac{1}{2} - \frac{r}{4a}\right) e^{-r/2a}$$

is completely forbidden.

Solution:

$$\mu_n = \frac{1}{\sqrt{2\pi a^3}} \frac{r}{4a} e^{-r/2a} \cos\theta$$

$$\mu_m = \frac{1}{\sqrt{\pi a^3}} e^{-r/a}$$

$$a = \frac{\hbar^2}{me^2}$$

We have,

$$dp_{n\to m} = \frac{d\Omega}{2\pi}\left(\frac{E_n - E_m}{\hbar}\right)\frac{e^2}{\hbar c}\left(\frac{\hbar}{mc}\right)^2\left|\int d\mathbf{r}\cdot\mu_m[\hat{\mathbf{n}}\times\nabla'\mu_n]\ \exp^{(-(i/\hbar c)(E_n - E_m)\hat{\mathbf{n}}\cdot\mathbf{r}')}\right|^2$$

Then, considering only the dipole term, we have,

$$\int d\mathbf{r}'\ \mu_m(\hat{\mathbf{n}}\times\nabla'\mu_n)\ \exp^{(-(i/\hbar c)(E_n - E_m)\hat{\mathbf{n}}\cdot\mathbf{r}')}$$

$$\simeq \exp^{(-(i/\hbar c)(E_n - E_m)\hat{\mathbf{n}}\cdot\mathbf{r}_0)}\int d\mathbf{r}\ \mu_m(\hat{\mathbf{n}}\times\nabla\mu_n)$$

Therefore, in the dipole approximation, the total transition probability is:

$$p_{n\to m} = \int_\Omega \frac{d\Omega}{2\pi}\left(\frac{E_n - E_m}{\hbar}\right)\frac{e^2}{\hbar c}\left(\frac{\hbar}{mc}\right)^2\left|\int d\mathbf{r}\ \mu_m(\hat{\mathbf{n}}\times\nabla\mu_n)\right|^2$$

We first consider $\nabla\mu_n$. We have,

$$\mu_n = \frac{1}{\sqrt{2\pi a^3}}\frac{1}{4a}\nabla(re^{-r/2a}\cos\theta)$$

$$= \frac{1}{4a\sqrt{2\pi a^3}}\left(\hat{\mathbf{e}}_r\frac{\partial}{\partial r}(re^{-r/2a}\cos\theta) + \frac{1}{r}\frac{\partial}{\partial\theta}(re^{-r/2a}\cos\theta)\hat{\mathbf{e}}_\theta\right) = 0$$

$$= \frac{1}{4a\sqrt{2\pi a^3}}\left\{\left[e^{-r/2a}\left(1 - \frac{r}{2a}\right)\cos\theta\right]\hat{\mathbf{e}}_r - (e^{-r/2a}\sin\theta)\hat{\mathbf{e}}_\theta\right\}$$

therefore,

$$\mu_m\nabla\mu_n = \frac{1}{4\sqrt{2}\ \pi a^4}\left\{\left[e^{-3\pi/2a}\left(1 - \frac{r}{2a}\right)\cos\theta\right]\hat{\mathbf{e}}_r - (e^{-3\pi/2a}\sin\theta)\hat{\mathbf{e}}_\theta\right\}$$

Let us now consider $\int d\mathbf{r}\ \mu_m\ \nabla\mu_n$. It is easy to see from the symmetry of the solution that this integral has only a z-direction component. Nevertheless, we show this result. Since,

$$\hat{\mathbf{e}}_r = (\sin\theta\cos\phi,\quad \sin\theta\sin\phi,\quad \cos\theta)$$

$$\hat{\mathbf{e}}_\theta = (\cos\theta\cos\phi,\quad \cos\theta\sin\phi,\quad -\sin\theta)$$

we have,

$$\left(\int d\mathbf{r}\ \mu_m\nabla\mu_n\right)_x \propto \int^\infty dr\,d\theta\,f(r,\theta)\int_0^{2\pi}\cos\phi\,d\phi - \int dr\,d\theta\,g(r,\theta)\int_0^{2\pi}\cos\phi\,d\phi = 0$$

$$\left(\int d\mathbf{r}\mu_m \nabla \mu_n\right)_y \alpha \int dr\, d\theta\, h(r,\theta) \int_0^{2\pi} \sin\phi\, d\phi - \int dr\, d\theta\, \ell(r,\theta) \int_0^{2\pi} \sin\phi\, d\phi = 0$$

Therefore,

$$\left(\int d\mathbf{r}\mu_m \nabla \mu_n\right)_z = \frac{1}{4\sqrt{2}\,\pi a^3} \left[\int d\mathbf{r}\, e^{-3r/2a}\left(1 - \frac{r}{2a}\right) \cos^2\theta + \int d\mathbf{r}\, e^{-3r/2a} \sin^2\theta\right]$$

$$= \frac{2\pi}{4\sqrt{2}\,\pi a^4} \left[\int_0^\infty r^2\left(1 - \frac{r}{2a}\right) e^{-3r/2a} \int_0^\pi \cos^2\theta \sin\theta\, d\theta\right.$$

$$\left. + \int_0^\infty r^2 e^{-3r/2a} \int_0^\pi \sin^3\theta\, d\theta\right]$$

$$= \frac{1}{2\sqrt{2}\, a^4} \left[\frac{2}{3} \int_0^\infty r^2\left(1 - \frac{r}{2a}\right) e^{-3r/2a} + \frac{4}{3} \int_0^\infty r^2 e^{-3r/2a}\right]$$

$$= \frac{1}{3\sqrt{2}\, a^4} \left(3 \int_0^\infty r^2 e^{-3r/2a} - \int_0^\infty \frac{r^3}{2a} e^{-3r/2a}\right)$$

$$= \frac{1}{3\sqrt{2}\, a^4} \left[3 \int_0^\infty r^2 e^{-3r/2a} - \frac{1}{2a}\left(\frac{e^{-3r/2a} r^3}{-3/2a}\right|_0^\infty + \int_0^\infty \frac{2a}{3} e^{-3r/2a}\, 3r^2\right)\right]$$

$$= \frac{1}{3\sqrt{2}\, a^4} \left(3 \int_0^\infty r^2 e^{-3r/2a} - \int_0^\infty r^2 e^{-3r/2a}\right)$$

$$= \frac{2}{3\sqrt{2}\, a^4} \int_0^\infty r^2 e^{-3r/2a}$$

Since,

$$\int_0^\infty r^2 e^{-3r/2a}\, dr = 2\left(\frac{2a}{3}\right)^3 = \frac{2^4 a^3}{3^3}$$

we have,

$$\int d\mathbf{r}\; \mu_m(\nabla \mu_n) = \frac{2}{3\sqrt{2}\, a^4} \frac{2^4 a^3}{3^3}\, \hat{\mathbf{z}} = \frac{2^5}{3^4 a\sqrt{2}}\, \hat{\mathbf{z}}$$

Therefore,

$$\left|\int d\mathbf{r}\ \mu_m(\hat{\mathbf{n}}\times\nabla\mu_n)\right|^2 = \left|\hat{\mathbf{n}}\times\int d\mathbf{r}\ \mu_m(\nabla\mu_n)\right|^2 = \left(\frac{2^4\sqrt{2}}{3^4\ a}\right)^2 |\hat{\mathbf{n}}\times\hat{\mathbf{z}}|^2$$

Using the same system of coordinates, we obtain

$$p_{n\to m} = \int \frac{d\Omega}{2\pi}\left(\frac{E_n - E_m}{\hbar}\right)\frac{e^2}{\hbar c}\left(\frac{\hbar}{mc}\right)^2 \left|d\mathbf{r}\ \mu_m(\hat{\mathbf{n}}\times\nabla\mu_n)\right|^2$$

$$= \left(\frac{E_n - E_m}{\hbar}\right)\frac{e^2}{\hbar c}\left(\frac{\hbar}{mc}\right)^2 \int \frac{d\Omega}{2\pi}\ |\hat{\mathbf{n}}\times\hat{\mathbf{z}}|^2\left(\frac{2^4\sqrt{2}}{3^4\ a}\right)^2$$

$$= \left(\frac{E_n - E_m}{\hbar}\right)\frac{e^2}{\hbar c}\left(\frac{\hbar}{mc}\right)^2\left(\frac{2}{3}\right)^8 \frac{2}{a^2}\int \frac{d\Omega}{2\pi}\sin^2\theta$$

$$= \left(\frac{2}{3}\right)^8\left(\frac{E_n - E_m}{\hbar}\right)\frac{e^2}{\hbar c}\left(\frac{\hbar}{mc}\right)^2 \frac{2}{a^2}\int_0^{\pi} d\theta\ \sin^3\theta$$

$$= \left(\frac{2}{3}\right)^8\left(\frac{E_n - E_m}{\hbar}\right)\frac{e^2}{\hbar c}\left(\frac{\hbar}{mc}\right)^2 \frac{2}{a^2}\,\frac{4}{3}$$

$$p_{n\to m} = \frac{8}{3a^2}\left(\frac{2}{3}\right)^8 \frac{e^2}{\hbar c}\left(\frac{\hbar}{mc}\right)^2\left(\frac{E_n - E_m}{h}\right) \quad \text{in sec}^{-1} \qquad (1)$$

So that the mean lifetime (on the assumption that above decay is only one) is

$$G = \frac{1}{p} = \frac{3a^2}{8}\left(\frac{3}{2}\right)^8\left(\frac{\hbar c}{e^2}\right)\left(\frac{mc}{\hbar}\right)^2\left(\frac{\hbar}{E_n - E_m}\right) \quad \text{in sec.} \qquad (2)$$

Let

$$\mu_{n'} = \frac{1}{\sqrt{2\pi a^3}}\left(\frac{1}{2} - \frac{r}{4a}\right) e^{-r/2a}$$

$$\mu_m = \frac{1}{\sqrt{\pi a^3}}\, e^{-r/a}$$

We want to show $p_{n'\to m} = 0$. Consider arbitrary direction $\hat{\mathbf{n}}$. Then,

$$dp(\hat{\mathbf{n}}) \propto \left| \int d\mathbf{r}\ \mu_m (\hat{\mathbf{n}} \times \nabla \mu_{n'}) \exp(-(iV/c)\hat{\mathbf{n}} \cdot \mathbf{r}) \right|^2$$

This is a consequence of the fact that both $\mu_{n'}$ and μ_m are functions of r only. We may write $\nabla\mu_{n'} = f(r)\hat{\mathbf{e}}_r$. Choosing spherical coordinates with the z axis along $\hat{\mathbf{n}}$ direction, we have

$$dp(\hat{\mathbf{n}}) \propto \left| \int d\mathbf{r}\ g(r)\ (\hat{\mathbf{z}} \times \hat{\mathbf{e}}_r) \exp(-(iV/c) r \cos\theta) \right|^2$$

but $\hat{\mathbf{z}} \times \hat{\mathbf{e}}_r = \sin\theta \cos\theta\, \hat{\mathbf{y}} - \sin\theta \sin\phi\, \hat{\mathbf{x}}$

Therefore,

$$\left(\int d\mathbf{r} (\hat{\mathbf{z}} \times \hat{\mathbf{e}}_r)\ g(r)\ e^{-iV(r\cos\theta)/c} \right)_z = 0$$

$$\left(\int d\mathbf{r} (\hat{\mathbf{z}} \times \hat{\mathbf{e}}_r)\ g(r)\ e^{-iV(r\cos\theta)/c} \right)_x$$

$$= - \int_0^\infty \int_0^\pi r^2 g(r)\ e^{-iV(r\cos\theta)/c} \sin^2\theta \int_0^{2\pi} \cos\phi\, d\phi = 0$$

$$\left(\int d\mathbf{r} (\hat{\mathbf{z}} \times \hat{\mathbf{e}}_r)\ g(r)\ e^{-iV(r\cos\theta)/c} \right)_y$$

$$= + \int_0^\infty \int_0^\pi r^2 g(r)\ e^{-iV(r\cos\theta)/c} \sin^2\theta \int_0^{2\pi} \sin\phi\, d\phi = 0$$

Therefore, for any $\hat{\mathbf{n}}$, $dp_{n\to m} = 0$, and our result follows. (Actually, this might be considered simpler in Cartesian coordinates, but it was so much fun this way I could not resist the temptation.)

PROBLEM 14.19 Calculate the velocity of recoil of a ^{8}Li nucleus when it emits an 18-MeV γ ray. Also calculate the recoil velocity of an electron if it emitted such a γ ray.

Solution: Use four-vector notation.

$$p = \left(\mathbf{p}, \frac{iE}{c} \right)$$

Let $c = 1$, $p^2 = -m^2$ (invariant)

$$(p^2 = \mathbf{p}^2 - \mathbf{E}^2)$$

For decay process,

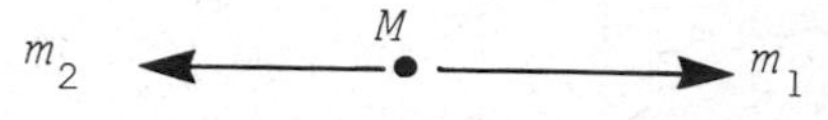

$$p = p_1 + p_2, \quad p_2 = p - p_1 \Rightarrow p_2^2 = (p - p_1) \cdot (p - p_1)$$

Therefore, in the lab system

$$\begin{aligned} -m_2^2 &= p^2 + p_1^2 - 2p \cdot p_1 \\ &= -M^2 - m_1^2 - 2(\mathbf{p} \cdot \mathbf{p}_1 - E_m E_1) \\ &= -M^2 - m_1^2 + 2ME_1 \end{aligned}$$

Therefore,

$$E_1 = \frac{M^2 + m_1^2 - m_2^2}{2M}$$

In our case,

$$E_1 = \tfrac{1}{2}\left(M^* + \frac{M^2}{M^*}\right)$$

where M^* and M are the rest masses of ^{8}Li before and after emission of the γ ray. Rewriting $p = p_1 + p_2$, $p_1 = p - p_2$ we obtain

$$p_1^2 = -m_1^2 = p^2 + p_2^2 - 2pp_2 = -M^2 - m_2^2 + 2ME_2$$

In our case, $M \to M^*$, $m_1 \to M$, we have

$$-M^2 = -M^{*2} + 2M^*E_\gamma$$

$$M^{*2} - 2M^*E_\gamma - M^2 = 0$$

therefore,

$$M^* = E_\gamma + \sqrt{E_\gamma^2 + M^2}$$

Using for M the ^{8}Li rest mass $= 8 \times 931.5 + 20.9$ MeV

$= 7472.90$ MeV

and

$$E_\gamma = 18 \text{ MeV}$$

We obtain $M^* = 7490.92$ MeV ($M^* - M = 0.02$ MeV). Therefore, $M/E_1 = 0.9999970$ (which shows that a nonrelativistic treatment would be sufficiently accurate). Now,

$$E_1 = \frac{M}{\gamma}, \quad \gamma = \sqrt{1 - (v/c)^2}$$

therefore,

$$\sqrt{1-(v/c)^2} = M/E_1 \Rightarrow (v/c)^2 = 1-(M/E_1)^2 \Rightarrow v = c\sqrt{1-(M/E_1)^2}$$

$V = 7.2\times 10^5$ m/sec

For the electron

For an electron, the same procedure can be used, with $M = 0.511$ MeV. One then obtains $v/c \simeq 0.99959$! We note how unrealistic this is, as follows: The threshold of the reaction $e^{-*} \to e^- + \gamma$ is clearly,

$$m_{e-*} \geq m_{e^-} + 18 \text{ MeV} \simeq 18.511 \text{ MeV}$$

Therefore, the rest mass of this "excited electron state" would have to be at least 18.5 MeV, as compared to "a ground-state rest mass" of $m_e = 0.5$ MeV!

PROBLEM 14.20 Verify that for a pure radiation field with

$$\mathbf{E} = \sum_{\gamma=1}^{2} \int d\mathbf{k}\ \varepsilon_k^\gamma (a_k^\gamma e^{i(\mathbf{k}\cdot\mathbf{r}-\omega t)} + c.c.)$$

one has for the momentum

$$\frac{\mathbf{P}}{\mathbf{p}} = \frac{4\pi^2}{c} \sum_{\gamma=1}^{2} \int d\mathbf{k}\ \frac{\mathbf{k}}{|k|}\ a_k^{-\gamma} a_k^\gamma$$

Solution:

Let

$$\mathbf{E} = \sum_{\gamma=1}^{2} \int d\mathbf{k}\ \varepsilon_k^\gamma (a_k^\gamma\ e^{i(kr-\omega t)} + a_k^{-\gamma}\ e^{-i(\mathbf{k}\cdot\mathbf{r}-\omega t)})$$

with $a_k^1 = b_k^2$ and $a_k^2 = -b_k^1$ and

$$\mathbf{H} = \sum_{\gamma=1}^{2} \int d\mathbf{k}\ \varepsilon_k^\gamma (b_k^\gamma e^{i(kr-\omega t)} - b_k^{-\gamma}\ e^{-i(kr-\omega t)})$$

let

$$\alpha_k^\gamma = a_k^\gamma\ e^{-i\omega t}$$

$$\beta_k^\gamma = b_k^\gamma\ e^{-i\omega t}$$

then

$$\alpha_k^{\prime} = \beta_k^2$$

$$\alpha_k^2 = \beta_k^1$$

(a)

and

$$\Xi = \sum_{\gamma=1}^{2} \int d\mathbf{k}\; \varepsilon_k^\gamma (\alpha_k^\gamma\, e^{ikr} + \bar{\alpha}_k^\gamma\, e^{-ikr})$$

$$\mathrm{H} = \sum_{\gamma=1}^{2} \int d\mathbf{k}\; \varepsilon_k^\gamma (\beta_k^\gamma\, e^{ikr} + \bar{\beta}_k^\gamma\, e^{-ikr})$$

then,

$$\mathbf{p} = \frac{1}{4\pi c}\; \mathbf{E} \times \mathbf{H}\; d\mathbf{r}$$

$$\mathbf{p} = \frac{1}{4\pi c} \sum_{\gamma=1}^{2} \sum_{\gamma=1}^{2} \int d\mathbf{k} \int d\mathbf{k}' (\varepsilon_k^\gamma \times \varepsilon_{k'}^\gamma) \int d^3\mathbf{r} [\alpha_k^\gamma \beta_{k'}^{\gamma'}\, e^{i(k+k')r}$$

$$+ \alpha_k^\gamma\, \bar{\beta}_{k'}^{\gamma'}\, e^{i(k-k')r} + \alpha_k^{-\gamma}\, \beta_{k'}^{\gamma'}\, e^{i(k'-k)r} + \bar{\alpha}_k^\gamma\, \bar{\beta}_{k'}^{\gamma'}\, e^{-i(k+k')r}]$$

$$\mathbf{p} = \frac{(2\pi)^3}{4\pi c} \sum_{\gamma,\gamma'} \int d\mathbf{k} \int d\mathbf{k}' (\varepsilon_k^\gamma \times \varepsilon_{k'}^\gamma) [\delta(k+k')(\alpha_k^\gamma\, \beta_k^{\gamma'} + \bar{\alpha}_k^\gamma\, \bar{\beta}_{k'}^{\gamma'})$$

$$+ \delta(k-k')(\alpha_k^\gamma\, \bar{\beta}_{k'}^{\gamma'} + \bar{\alpha}_k^\gamma\, \beta_{k'}^{\gamma'})]$$

$$= \frac{(2\pi^2)}{c} \sum_{\gamma,\gamma'} d\mathbf{k} [(\varepsilon_k^\gamma \times \varepsilon^{\gamma'}{}_{-k})(\alpha_k^\gamma\, \beta^{\gamma'}{}_{-k} + \bar{\alpha}_k^\gamma\, \bar{\beta}^{\gamma'}{}_{-k})$$

(b)

$$+ (\varepsilon_k^\gamma \times \varepsilon_k^{\gamma'})(\alpha_k^\gamma\, \bar{\beta}_k^{\gamma'} + \bar{\alpha}_k^\gamma\, \beta_k^{\gamma'})]$$

since

$$\varepsilon_k^1 \times \varepsilon_k^2 = \varepsilon_k^3 = -(\varepsilon_k^2 \times \varepsilon_k^1) \quad \text{and} \quad \varepsilon_k^1 \times \varepsilon_k^1 - \varepsilon_k^2 \times \varepsilon_k^2 = 0$$

we have,

$$\frac{(2\pi^2)}{c} \sum_{\gamma,\gamma'} \int d\mathbf{r} (\varepsilon_k^\gamma \times \varepsilon_k^{\gamma'})(\alpha_k^\gamma\, \bar{\beta}_k^{\gamma'} + \bar{\alpha}_k^\gamma\, \beta_k^{\gamma'})$$

$$= \frac{(2\pi^2)}{c} \int d\mathbf{k}\ \varepsilon_k^3 (\alpha_k^1 \overline{\beta_k^2} + \overline{\alpha_k^1} \beta_k^2 - \alpha_k^2 \overline{\beta_k^1} - \overline{\alpha_k^2} \beta_k^1)$$

$$= [\text{by (a)}] \rightarrow \frac{(2\pi^2)}{c} \int d\mathbf{k}\ \varepsilon_k^3 (\alpha_k^1 \overline{\alpha_k^1} + \overline{\alpha_k^1} \alpha_k^1 + \alpha_k^2 \overline{\alpha_k^2} + \overline{\alpha_k^2} \alpha_k^2)$$

$$= \frac{4\pi^2}{c} \int d\mathbf{k}\ \varepsilon_k^3 (a_k^1 \overline{a_k^1} + a_k^2 \overline{a_k^2})$$

$$= \frac{4\pi^2}{c} \sum_{\gamma=1}^{2} \int d\mathbf{k}\ \varepsilon_k^3 (\overline{a_k^\gamma}\, a_k^\gamma)$$

choosing

$$\varepsilon^1_{-k} = \varepsilon_k^1 \text{ and } \varepsilon^2_{-k} = -\varepsilon_k^2$$

then,

$$\varepsilon_k^1 \times \varepsilon^1_{-k} = \varepsilon_k^2 \times \varepsilon^2_{-k} = 0$$

but

$$\varepsilon_k^1 \times \varepsilon^2_{-k} = -(\varepsilon_k^1 \times \varepsilon_k^2) = -\varepsilon_k^3 = (\varepsilon_k^2 \times \varepsilon_k^1) = \varepsilon_k^2 \times \varepsilon^1_{-k}$$

therefore,

$$\frac{(2\pi)^2}{c} \sum_{\gamma,\gamma'} \int d\mathbf{k}\ (\varepsilon_k^\gamma \times \varepsilon^{\gamma'}_{-k})(\alpha_k^\gamma \beta^{\gamma'}_{-k} + \overline{\alpha_k^\gamma}\, \overline{\beta^{\gamma'}_{-k}})$$

$$= -\frac{(2\pi)^2}{c} \int d\mathbf{k}\ \varepsilon_k^3 (\alpha_k^1 \beta^2_{-k} + \overline{\alpha_k^1}\, \overline{\beta^2_{-k}} + \alpha_k^2 \beta^1_{-k} + \overline{\alpha_k^2}\, \overline{\beta^1_{-k}})$$

$$= [\text{by (a)}] \quad \frac{-(2\pi)^2}{c} \int d\mathbf{k}\ (\alpha_k^1 \alpha^2_{-k} - \alpha_k^2 \alpha^1_{-k} + \overline{\alpha_k^1}\, \overline{\alpha^2_{-k}} - \overline{\alpha_k^2}\, \overline{\alpha^1_{-k}}) = 0$$

(by changing $k \rightarrow -k$ in the appropriate integrals). Therefore, (b) reduces to

$$p = \frac{4\pi^2}{c} \sum_{\gamma=1}^{2} \int d\mathbf{k}\ \varepsilon_k^3 (\overline{a_k^\gamma}\, a_k^\gamma) = \frac{4\pi^2}{c} \sum_{\gamma=1}^{2} \int d\mathbf{k}\ \frac{\mathbf{k}}{|k|} (\overline{a_k^\gamma}\, a_k^\gamma)$$

PROBLEM 14.21 A beam of plane waves strikes a plane perfect conductor with the wave normal making an angle α with the normal to the surface. If the vector E is perpendicular to the plane of incidence, derive an expression for the radiation pressure in terms of the incident energy density by considering:

(a) The net transfer of electromagnetic momentum to the conductor.
(b) The Maxwell stresses at the conductor surface.
(c) The forces acting on the surface current.

Solution:

Perfect conductor => $R = 1$. Consider:

$$\mathbf{E}_0 = E_0 e^{i(\mathbf{k}\cdot\mathbf{r} - \omega t)}\hat{\mathbf{z}}$$

$$\mathbf{E}_r = -E_r e^{i(\mathbf{k}'\cdot\mathbf{r} - \omega t)}\hat{\mathbf{z}}$$

$$\mathbf{H}_0 = E_0 e^{i(\mathbf{k}\cdot\mathbf{r} - \omega t)}\hat{\mathbf{e}}_k$$

$$\mathbf{H}_r = E_r e^{i(\mathbf{k}'\cdot\mathbf{r} - \omega t)}\hat{\mathbf{e}}_{k'}$$

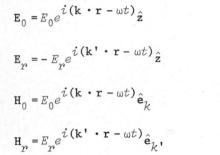

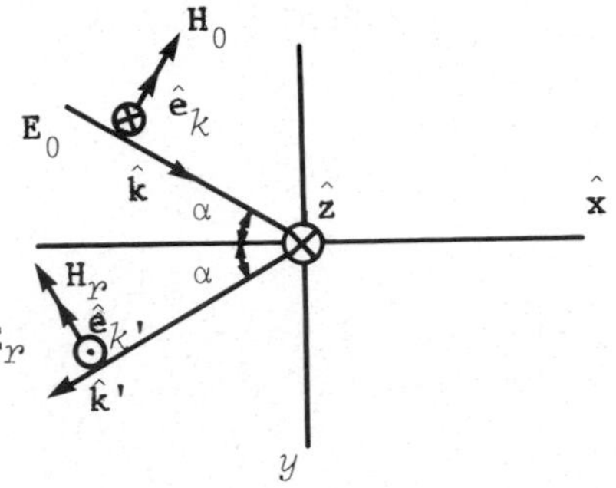

Since $(\mathbf{k}\cdot\mathbf{r})_{x=0} = (\mathbf{k}'\cdot\mathbf{r})_{x=0}$ and $|\mathbf{k}| = |\mathbf{k}'| = k$, we must have $\sin\alpha = \sin\alpha'$, that is, $\alpha = \alpha'$, as expected.

$E_{\text{tangential, continuous}} \Rightarrow E_0 = E_r$

$H_{\text{perpendicular, continuous}} \Rightarrow H_0 \sin\alpha = H_r \sin\alpha$

that is, $E_0 = E_r$. Therefore, we obtain consistent solutions with,

$$\left\{\begin{array}{ll} \mathbf{E}_0 = E_0 e^{i(\mathbf{k}\cdot\mathbf{r} - \omega t)}\hat{\mathbf{z}}, & \mathbf{H}_0 = \mathrm{E}_0 e^{i(\mathbf{k}\cdot\mathbf{r} - \omega t)}\hat{\mathbf{e}}_k \\ \mathbf{E}_r = -E_0 e^{i(\mathbf{k}\cdot\mathbf{r} - \omega t)}\hat{\mathbf{z}}, & \mathbf{H}_r = \mathrm{E}_0 e^{i(\mathbf{k}\cdot\mathbf{r} - \omega t)}\hat{\mathbf{e}}_{k'} \\ \hat{\mathbf{e}}_k = \sin\alpha\,\hat{\mathbf{x}} - \cos\alpha\,\hat{\mathbf{y}}, & \hat{\mathbf{e}}_{k'} = -\sin\alpha\,\hat{\mathbf{x}} - \cos\alpha\,\hat{\mathbf{y}} \end{array}\right\} \quad (1)$$

(a) Let G be electromagnetic momentum density. Since for an electromagnetic wave moving with speed c, G points in the same direction as $\hat{\mathbf{k}}$, the momentum flow per unit time across unit area in bounding surface is $Gc(\hat{\mathbf{x}}\cdot\hat{\mathbf{k}}) = Gc\cos\alpha$. Therefore, the force per unit area exerted on the surface of the conductor is simply the negative of the electromagnetic momentum change per unit time per unit area, that is,

$$\mathrm{p} = -c(\mathrm{G}_{\text{final}} - \mathrm{G}_{\text{initial}})\cos\alpha$$

$$= \frac{-c\cos\alpha}{4\pi c}\left[\tfrac{1}{2}(\mathbf{E}_r \times \mathbf{H}_r^*) - \tfrac{1}{2}(\mathbf{E}_0 \times \mathbf{H}_0^*)\right]_{x=0}$$

$$= -\frac{1}{8\pi}\cos\alpha\ (E_0^2\hat{\mathbf{k}}' - E_0^2\hat{\mathbf{k}}) - \frac{E_0^2}{8\pi}\cos\alpha(\hat{\mathbf{k}}' - \hat{\mathbf{k}})$$

$$= \frac{E_0^2\cos\alpha}{8\pi}\ (\hat{\mathbf{k}} - \hat{\mathbf{k}}')$$

$$= \frac{E_0^2\cos\alpha}{8\pi}\ (2\cos\alpha\ \hat{\mathbf{x}})$$

$$= \frac{E_0^2\cos^2\alpha}{4\pi}\ \hat{\mathbf{x}}$$

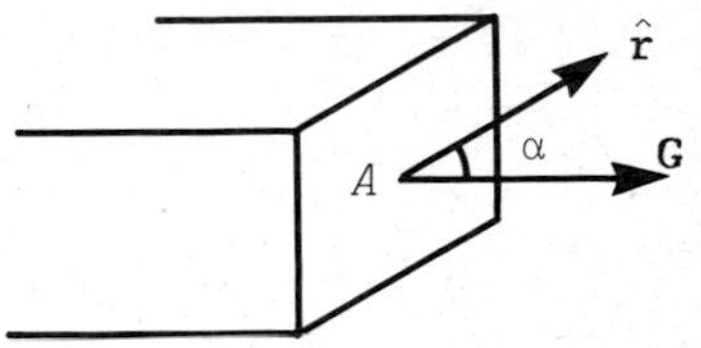

Total momentum $= \mathbf{G}\ c\ dt\cos\alpha\ A$

Since

$$\mu = \frac{1}{8\pi}\left[\tfrac{1}{2}(\mathbf{E}_0 \cdot \mathbf{E}_0^*) + \tfrac{1}{2}(\mathbf{H}\cdot\mathbf{H}_0^*)\right] = \frac{1}{16\pi}\ (2E_0^2)$$

$$= \frac{1}{8\pi}\ E_0^2\ \text{(incident wave)}$$

We obtain,

$$\mathbf{p} = 2\mu_{\text{inc.}}\ \cos^2\alpha\,\hat{\mathbf{x}} \qquad (2)$$

(b) In component form, we have

$$\mathbf{E}_0 = E_0(0,0,1)e^{i(\mathbf{k}\cdot\mathbf{r} - \omega t)}$$

$$\mathbf{H}_0 = E_0(\sin\alpha,\ -\cos\alpha,\ 0)e^{i(\mathbf{k}\cdot\mathbf{r} - \omega t)}$$

$$\mathbf{E}_r = -E_0(0,0,1)e^{i(\mathbf{k}\cdot\mathbf{r} - \omega t)}$$

$$\mathbf{H}_r = E_0(-\sin\alpha,\ -\cos\alpha,\ 0)e^{i(\mathbf{k}\cdot\mathbf{r} - \omega t)}$$

For the bounding surface, with normal $n_\alpha = (-1,0,0)$, the force per unit area transmitted across it is,

$$\mathbf{p} = \sum_\beta T_{\alpha\beta}\, n_\beta = -(T_{11}, T_{21}, T_{31})$$

$$T_{ij} = \frac{1}{8\pi}\left[E_iE_j^* + H_iH_j^* - \tfrac{1}{2}\,(\mathbf{E}\cdot\mathbf{E}^* + \mathbf{H}\cdot\mathbf{H}^*)\right]$$

For the incident wave, we have,

$$T_{11} = \frac{1}{8\pi}\left[H_1H_1^* - \tfrac{1}{2}\,(2E_0^2)\right] = \frac{1}{8\pi}\,(E_0^2 \sin^2\alpha - E_0^2)$$

$$= \frac{-E_0^2}{8\pi}\,(1 - \sin^2\alpha) = \frac{-E_0^2}{8\pi}\cos^2\alpha$$

$$T_{21} = \frac{1}{8\pi}\,(H_2H_1^*) = \frac{-E_0^2}{8\pi}\cos\alpha\sin\alpha$$

$$T_{31} = 0$$

Therefore,

$$\mathbf{p}_{\text{inc.}} = \frac{E_0^2}{8\pi}\,(\cos^2\alpha,\ \cos\alpha\sin\alpha,\ 0)$$

For the reflected wave, we have

$$T_{11} = \frac{1}{8\pi}\,(H_1H_1^* - E_0^2)$$

$$= \frac{1}{8\pi}\,(E_0^2 \sin^2\alpha - E_0^2)$$

$$= \frac{-E_0^2}{8\pi}\cos^2\alpha$$

$$T_{21} = \frac{1}{8\pi}\,(H_2H_1^*) = \frac{E_0^2}{8\pi}\,E_0^2\cos\alpha\sin\alpha$$

therefore,

$$\mathbf{p}_{\text{ref.}} = \frac{E_0^2}{8\pi}\,(\cos^2\alpha,\ -\cos\alpha\sin\alpha,\ 0)$$

Therefore,

$$\mathbf{p} = \mathbf{p}_{\text{inc.}} + \mathbf{p}_{\text{ref.}} = \frac{E_0^2}{4\pi}\cos^2\alpha\,(1,0,0)$$

$$= \frac{E_0^2}{4\pi} \cos^2 \alpha\, \hat{\mathbf{x}}$$

as obtained before. [*Note:* We could have added the fields, and evaluated the Maxwell stress tensor at the surface for the combined field $\mathbf{E}_T = \mathbf{E}_0 + \mathbf{E}_r$ and $\mathbf{H}_T = \mathbf{H}_0 + \mathbf{H}_r$. Since $e^{i\mathbf{k}\cdot\mathbf{r}} = e^{i\mathbf{k}'\cdot\mathbf{r}}$ on the surface, the result agrees with the above one. We did not use this method because the total fields are known to correspond to stationary waves, in which case concepts such as G and $T_{\alpha\beta}$ have an interpretation harder to grasp physically.]

(c) From the equation

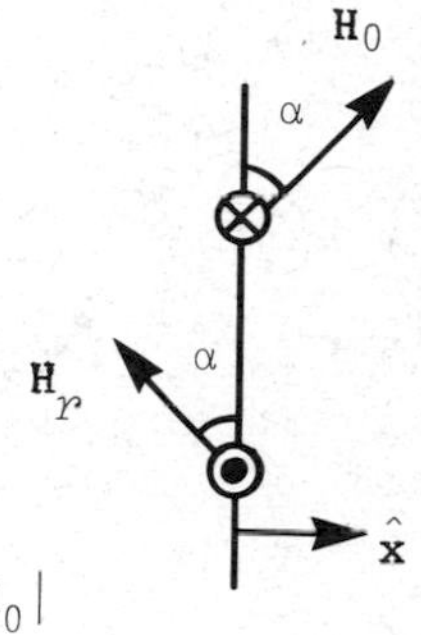

$$\nabla \times \mathbf{H} = \frac{4\pi}{c} \mathbf{j}$$

we obtain the boundary condition,

$$(-\hat{\mathbf{x}}) \times (\mathbf{H}_0 + \mathbf{H}_r) = \frac{4\pi}{c} \mathbf{j}_\ell$$

$$|\mathbf{H}_r| = |\mathbf{H}_0|$$

$j_\ell \to$ current/unit length. This equation is only true for total fields and currents. In terms of known quantities ($\mathbf{H}_0$ and $\mathbf{H}_r$), we must introduce a factor of 1/2 in the expression for the force per unit area:

$$\mathbf{p} = \tfrac{1}{2} \frac{\mathbf{j} \times (\mathbf{H}_0 + \mathbf{H}_r)}{c}$$

as for the electric case, when the "self-field" produced by the induced charges themselves are considered. Then (with $\mathbf{H}_{tot} = \mathbf{H}_0 + \mathbf{H}_r$),

$$\mathbf{p} = -\frac{1}{2c} \frac{c}{4\pi} [(\hat{\mathbf{x}} \times \mathbf{H}_t) \times \mathbf{H}_t]$$

$$= \frac{+1}{8\pi} \mathbf{H}_t \times (\hat{\mathbf{x}} \times \mathbf{H}_t)$$

$$= + \frac{1}{8\pi} [\hat{\mathbf{x}}(\mathbf{H}_t \times \mathbf{H}_t) - \mathbf{H}_t(\mathbf{H}_t \cdot \hat{\mathbf{n}})]$$

$$= + \frac{1}{8\pi} \mathbf{H}_t \cdot \mathbf{H}_t\, \hat{\mathbf{x}}$$

$$|\mathbf{p}| = + \frac{1}{8\pi} \frac{1}{2} (\mathbf{H}_t \cdot \mathbf{H}_t^*) = + \frac{1}{16\pi} (\mathbf{H}_0 + \mathbf{H}_r) \cdot (\mathbf{H}_0^* + \mathbf{H}_r^*)$$

$$= + \frac{1}{16\pi} \; (E_0^2 + 2E_0^2 \cos 2\alpha + E_0^2)$$

$$= + \frac{1}{16\pi} \cdot 2[E_0^2 (1 + \cos 2\alpha)]$$

$$= + \frac{E_0^2}{8\pi} \; 2 \cos^2 \alpha \doteq + \frac{E_0^2}{4\pi} \cos^2 \alpha$$

Therefore,

$$\mathrm{p} = \frac{E_0^2}{4\pi} \cos^2 \alpha \, \hat{\mathbf{x}} = 2\mu_{\mathrm{inc.}} \; \cos^2 \alpha \, \hat{\mathbf{x}}$$

in agreement with previous results.

Chapter 15

PROBLEM 15.1

(a) Write down the expressions for the retarded potentials making sure the variables on which the potentials and the charge and current densities depend are explicitly shown. By appropriate use of δ functions express these potentials as integrals over both space and time.
(b) Calculate the retarded potentials for the case of an arbitrarily moving point charge (Liénard-Wiechert potentials).
(c) Investigate the possibility that more than one point on the trajectory of the above point charge will contribute simultaneously to the potentials at a given point in space.

Solution:

(a) To satisfy the equations

$$\nabla^2\phi - \frac{1}{c^2}\frac{\partial^2\phi}{\partial t^2} = -\frac{1}{\varepsilon}\,\rho(\mathbf{r},t)$$

$$\nabla^2\mathbf{A} - \frac{1}{c^2}\frac{\partial^2\mathbf{A}}{\partial t^2} = -\mu\mathbf{j}(\mathbf{r},t)$$

we have the solutions,

$$\phi(\mathbf{r},t) = \frac{1}{4\pi\varepsilon_0}\int_{\substack{\text{all}\\\text{space}}} \frac{\rho(\mathbf{r}',t')dv'}{|\mathbf{r}-\mathbf{r}'|}$$

$$\mathbf{A}(\mathbf{r},t) = \frac{\mu_0}{4\pi}\int \frac{\mathbf{j}(\mathbf{r}',t')dv'}{|\mathbf{r}-\mathbf{r}'|}$$

where $t' = t - |\mathbf{r}-\mathbf{r}'|/c$. These are the retarded potentials. Now express these potentials as integrals over space and time. From the definition,

$$\int_a^b \delta(x-x')dx' = 1, \quad \text{if } a < x < b$$
$$= 0, \quad \text{otherwise}$$

We can write

$$\phi(\mathbf{r},t) = \frac{1}{4\pi\varepsilon_0}\int\int \frac{\rho(\mathbf{r}',t')}{|\mathbf{r}-\mathbf{r}'|}\,\delta\left(t' - t + \frac{1}{c}\,|\mathbf{r}-\mathbf{r}'|\right)dv'\,dt'$$

$$\mathbf{A}(\mathbf{r},t) = \frac{\mu_0}{4\pi}\int\int \frac{\mathbf{j}(\mathbf{r}',t')}{|\mathbf{r}-\mathbf{r}'|}\,\delta\left(t' - t + \frac{1}{c}\,|\mathbf{r}-\mathbf{r}'|\right)dv'\,dt'$$

(b) Take a moving point charge, say an electron,

$$\rho(\mathbf{r}',t') = e\delta[\mathbf{r}' - \mathbf{r}'(t')]$$

$$\mathbf{j}(\mathbf{r}',t') = e\mathbf{v}(t')\delta[\mathbf{r}' - \mathbf{r}'(t')]$$

$$\mathbf{A}(\mathbf{r},t) = \frac{\mu_0 e}{4\pi}\int dt' \int \frac{\delta[t' - t + (1/c)|\mathbf{r}-\mathbf{r}'|]}{|\mathbf{r}-\mathbf{r}'|}\,dv'\,\mathbf{v}(t')\delta[\mathbf{r}' - \mathbf{r}'(t')]$$

now,

$$\delta[\mathbf{r}' - \mathbf{r}'(t')]\,dv' = 1$$

so,

$$\mathbf{A} = \frac{\mu_0 e}{4\pi}\int dt'\,\frac{\mathbf{v}(t')}{|\mathbf{r}-\mathbf{r}'(t')|}\,\delta\left(t' - t + \frac{1}{c}\,|\mathbf{r}-\mathbf{r}'(t')|\right)$$

We can use the form

$$\int f(t')\delta[g(t')]\,dt' = \sum \frac{f(t')}{g'(t')}$$

Now let

$$\mathbf{R}(t') = \mathbf{r} - \mathbf{r}(t')$$

$$R(t') = |\mathbf{r} - \mathbf{r}(t')|$$

$$R(t') = |\mathbf{r} - \mathbf{r}'|$$

So let

$$\tau = t' + \frac{\mathrm{R}(t')}{c}$$

Then we have

$$\int \frac{\mathbf{v}(t')}{R(t')}\,\delta(\tau - t) = \sum \frac{f(t')}{g'(t')}$$

where

$$f(t') = \frac{\mathbf{v}(t')}{R(t')}$$

$$g(t') = \delta(\tau - t)$$

$$\frac{dg}{dt'} = \frac{d}{dt'}\ \tau = 1 + \frac{1}{c}\ \frac{dR(t')}{dt'}$$

and

$$R(t') = \sqrt{\mathbf{R}(t') \cdot \mathbf{R}(t')}$$

so

$$\frac{dR(t')}{dt'} = \tfrac{1}{2}\ \frac{1}{\sqrt{\mathbf{R}(t') \cdot \mathbf{R}(t')}} \left[\mathbf{R}(t') \cdot \frac{d\mathbf{R}(t')}{dt'} + \mathbf{R}(t') \cdot \frac{d\mathbf{R}(t')}{dt'}\right]$$

$$= \frac{\mathbf{R}(t')}{R(t')} \cdot \frac{d\mathbf{R}(t')}{dt'} = \frac{\mathbf{R}(t')}{R(t')} \cdot \frac{(-)d\mathbf{r}'}{dt'} = \frac{\mathbf{R}(t')}{R(t')} \cdot [-\mathbf{v}(t')]$$

where $\mathbf{R}(t') = \mathbf{r} - \mathbf{r}'$, so

$$\frac{dg}{dt'} = 1 + \frac{1}{c}\ \frac{\mathbf{R}(t') \cdot [-\mathbf{v}(t')]}{R(t')}$$

and

$$\int = \frac{\mathbf{v}(t')}{R(t')[1 - \mathbf{R}(t') \cdot \mathbf{v}(t')/cR(t')]}$$

so,

$$\mathbf{A}(\mathbf{r},t) = \frac{\mu_0 e}{4\pi} \left[\frac{\mathbf{v}(t')}{R(t')[1 - \mathbf{R}(t') \cdot \mathbf{v}(t')/cR(t')]}\right]\Bigg|_{t' = t_{ret}}$$

and ϕ is in exactly the same form except there is no $v(t')$ in the numerator and $\mu_0/4\pi \rightarrow 1/4\pi\varepsilon_0$ so,

$$\phi(\mathbf{r},t) = \frac{e}{4\pi\varepsilon_0} \left[\frac{1}{R(t')[1 - \mathbf{R}(t') \cdot \mathbf{v}(t')/cR(t')]}\right]\Bigg|_{t' = t_{ret}}$$

(c)

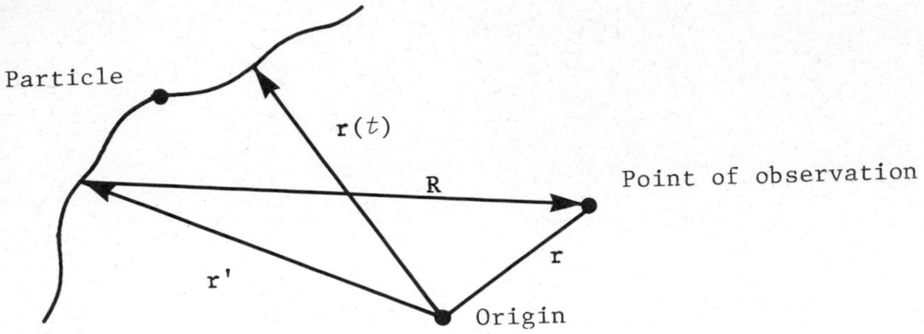

Is it possible to have two solutions of ϕ' to the equation

$$\mathbf{A} = \frac{\mu_0 e}{4\pi} \int dt' \frac{\mathbf{v}(t')}{|\mathbf{r} - \mathbf{r}'(t')|} \delta\left(t' - t + \frac{1}{c}[\mathbf{r} - \mathbf{r}'(t')]\right)$$

and its corresponding ϕ? The answer is no, because the path ACB is longer than the path AB, and if there were two solutions, then it would mean that they both took the same time to go different distances (since the particle is in the same media and travels at the same speed from $A \to B$ as from $C \to B$), then this would mean that the particle would have to travel at a speed $> c$ in the medium from $A \to C$ in order to take the same time to travel from $A \to C \to B$ as from $A \to B$. This can happen in some media (Cherenkov Radiation).

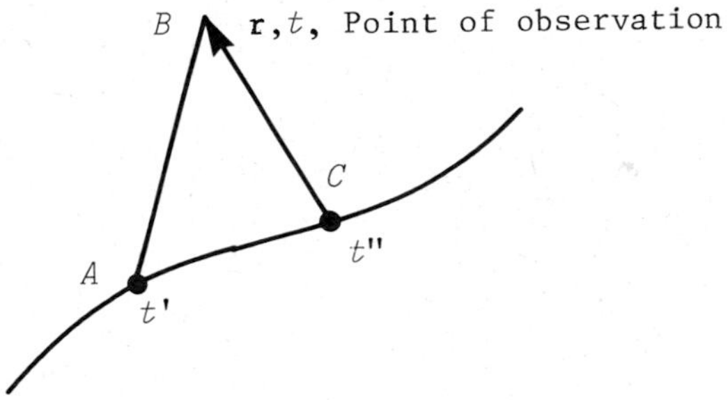

PROBLEM 15.2 Starting from the relativistic expression for the magnetic field by a uniformly moving point charge, develop an expression for the magnetic field of an infinitely long wire carrying a steady current I.

Solution:

$$\mathbf{B}_{\text{past}} = \frac{q}{4\pi\varepsilon_0 c^2} \frac{\mathbf{v} \times \mathbf{R}(1 - v^2/c^2)}{R^3[1 - (v^2/c^2)\sin^2\theta]^{3/2}}$$

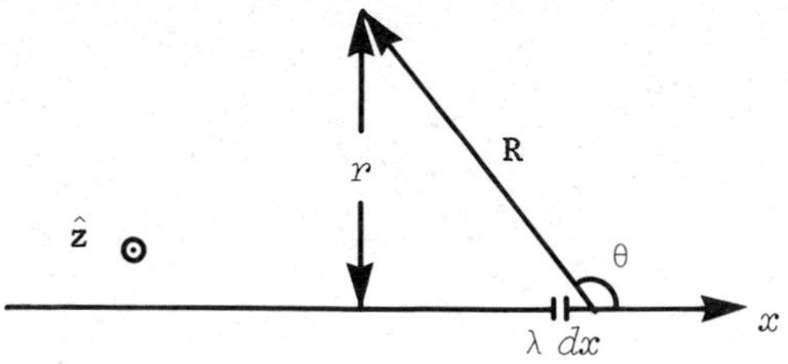

let

$$\lambda = \frac{\text{charge}}{\text{unit length}}$$

$q = \lambda\, dx$, then $I = \lambda v$ = current. B clearly points out of paper for all contributions.

$$\frac{r}{R} = \sin\theta, \quad R^2 = r^2 + x^2$$

$$d\mathbf{B}_{\text{wire}} = \frac{\hat{\mathbf{z}}\,\lambda\, dx\, v\, r(1 - v^2/c^2)}{4\pi\varepsilon_0 c^2[R^2 - (v^2/c^2)r^2]^{3/2}} = \frac{\hat{\mathbf{z}}\mu_0 I\, r(1 - v^2/c^2)}{4\pi} \frac{dx}{[(1 - v^2/c^2)r^2 + x^2]^{3/2}}$$

$$\mathbf{B}_{\text{wire}} = \hat{\mathbf{z}} \frac{\mu_0 I\, r(1 - v^2/c^2)}{4\pi} \left(\frac{1}{(1 - v^2/c^2)r^2} \frac{x}{[(1 - v^2/c^2)r^2 + x^2]^{\frac{1}{2}}} \right)_{-\infty}^{\infty} = \frac{\hat{\mathbf{z}}\mu_0 I}{2\pi r} \;!!$$

PROBLEM 15.3 As a rough measure of the relativistic "flattening" of the configuration of electric field lines from a moving charge, we might use the angle α between two conical surfaces which include between them half the total electric flux. That is, half the flux through a sphere shall be contained in the equatorial zone between $\theta' = \pi/2 + \alpha/2$ and $\theta' = \pi/2 - \alpha/2$. Consider only the extreme relativistic case, with $\gamma \gg 1$. Then only angles θ' such that $\theta' = \pi/2 - \varepsilon$, with $|\varepsilon| \ll 1$ need be considered. Show first that

$$E' = \frac{Q}{(r')^2} \frac{\gamma}{(1 + \gamma^2\varepsilon^2)^{3/2}}$$

Now let ε range from $-\alpha/2$ to $+\alpha/2$ and integrate to obtain the flux through the narrow equatorial belt.

Solution:

$$\sin^2\theta = \sin^2\left(\frac{\pi}{2} - \varepsilon\right) = \cos^2\varepsilon \simeq 1 - \varepsilon^2, \quad \varepsilon^2 \ll 1$$

$$\frac{1 - \beta^2}{[1 - \beta^2(1 - \varepsilon^2)]^{3/2}} = \frac{1 - \beta^2}{(1 - \beta^2 + \beta^2\varepsilon^2)^{3/2}} = \frac{1 - \beta^2}{(1 + \beta^2)[(1 + (\beta^2\varepsilon^2)/(1 - \beta^2)]^{3/2}}$$

$$\simeq \frac{\gamma}{(1+\gamma^2\varepsilon^2)^{3/2}} \quad \text{for } \beta^2 \simeq 1$$

$$\int \mathbf{E}\cdot d\mathbf{a} = \frac{Q\gamma}{r^2}\int_{\alpha/2}^{\alpha/2} \frac{2\pi r^2\, d\varepsilon}{(1+\gamma^2\varepsilon^2)^{3/2}} = \tfrac{1}{2}\,\pi Q$$

Let $\gamma\varepsilon = \tan\,\phi$, $\gamma\, d\varepsilon = \sec^2\,\phi\, d\phi$

$$\int \frac{\sec^2\,\phi\, d\phi}{(1+\tan^2\,\phi)^{3/2}} = \int \cos\,\phi\, d\phi = \sin\,\phi\Big|_{-\tan^{-1}\,\gamma\alpha/2}^{\tan^{-1}\,\gamma\alpha/2} = 1$$

$$2\sin\left(\tan^{-1}\frac{\gamma\alpha}{2}\right) = 1, \quad \tan^{-1}\frac{\gamma\alpha}{2} = \sin^{-1}(0.5) = 30°$$

$$\frac{\gamma\alpha}{2} = \tan\,30° = \frac{1}{\sqrt{3}}, \quad \alpha = \frac{2}{\gamma\sqrt{3}}$$

PROBLEM 15.4 A 30-BeV proton passes 10^{-7} cm away from a hydrogen atom. Estimate the peak magnitude of the electric field and the approximate duration of the electric field pulse to which the atom is subjected. Do the same for a 30-BeV electron passing at the same distance. Use the approximate equivalents for the rest masses, namely 1 BeV for the proton, and 0.5 MeV for the electron.

Solution:

$$E = T + Mc^2 = 30 \text{ BeV} + 1 \text{ BeV} = \frac{Mc^2}{\sqrt{1 - v^2/c^2}} = \gamma(1 \text{ BeV})$$

$$\gamma = 31$$

$$E'(\max) = \gamma\,\frac{Q}{r^2} = 31 \times \frac{4.8\times 10^{-10}}{(10^{-7})^2} = 1.5\times 10^6\ \frac{\text{statvolt}}{\text{cm}}$$

Since most of the flux is in a cone of angle $\alpha = 2/\sqrt{3}\,\gamma$, and $v \simeq c$,

$$\Delta t \simeq \frac{r\alpha}{v} \simeq \frac{r}{\gamma c} = \frac{10^{-7}}{31\times 3\times 10^{10}} \simeq 1\times 10^{-19} \text{ sec}$$

<u>Electron</u>

$$30.0005 \text{ BeV} = (0.0005 \text{ BeV})\gamma$$

$$\gamma \simeq 6\times 10^4 \simeq 2\times 10^3\ (\gamma \text{ proton})$$

$$E'(\max) \simeq 1.5\times 10^6 \times 2\times 10^3 = 3\times 10^9 \text{ statvolt/cm}$$

$$\Delta t \simeq 1\times 10^{-19}/2\times 10^3 \simeq 0.5\times 10^{-22} \text{ sec}$$

PROBLEM 15.5 Consider the trajectory of a charged particle which is moving with a speed 0.8 c in the x direction when it enters a large region in which there is a uniform electric field in the y direction. Show that the x velocity of the particle must actually decrease. What about the x component of momentum?

Solution:

$$\frac{dP_y}{dt} = F_y = qE_y, \quad \frac{dP_x}{dt} = F_x = 0$$

The x component of the momentum is conserved. However, since

$$P_x = \frac{m_0 v_x}{\sqrt{1 - v^2/c^2}}$$

and $v^2 = v_x^2 + v_y^2$ is increasing, or

$$E = \frac{m_0 c^2}{\sqrt{1 - v^2/c^2}}$$

is increasing, v_x must actually decrease.

PROBLEM 15.6 The deflection plates in a high-voltage cathode-ray oscilloscope are two rectangular plates, 4 cm long and 1.5 cm wide, spaced 0.8 cm apart. There is a difference of potential of 6000 V between the plates. An electron which has been accelerated through a potential difference of 250 kV enters this deflector from the left, moving parallel to the plates and halfway between them, initially. We want to find the position of the electron and its direction of motion when it leaves the deflecting field at the other end of the plates. We shall neglect the fringing field and assume the electric field between the plates is uniform right up to the end. The rest mass of the electron may be taken as 500 keV. First carry out the analysis in the "lab" frame by answering these questions: $\gamma = ?$; $\beta = ?$; p_x, in units of mc, = ?; time spent between the plates = ? (neglect the change in horizontal velocity); transverse momentum component acquired, in units of mc, = ?; transverse velocity at exit = ? direction of flight at exit? Now describe this whole process as it would appear in an inertial frame which moved with the electron at the moment it entered the deflecting region: What do the plates look like? What is the field between them? What happens to the electron in this coordinate system? Your main object in this exercise is to convince yourself that the two descriptions are completely consistent.

Solution:

$$E = kE + mc^2 = 250 + 500 = \gamma_0\, mc^2 = 500\, \gamma_0$$

$$\gamma_0 = 1.5, \quad \beta_0 = \sqrt{\frac{\gamma_0 - 1}{\gamma_0}} = 0.745$$

$$\frac{P_x}{mc} = \frac{\gamma m v_x}{mc} = \gamma\beta = 1.118$$

$$t = \frac{x}{v_x} = \frac{x}{\beta c} = \frac{4}{0.745 \times 3 \times 10^{10}} = 1.789 \times 10^{-10} \text{ s}$$

$$\frac{P_y}{mc} = \frac{F_y \Delta t}{mc} = \frac{qE_y \Delta t}{mc} = \frac{qE_y (x)}{\beta mc^2}$$

$$E_y = \frac{v}{y} = \frac{6000 \text{ V}}{0.8 \text{ cm}}, \quad qE_y = \frac{6000 \text{ eV}}{0.8 \text{ cm}}$$

$$\frac{P_y}{mc} = \frac{6000 \text{ eV} \times 4 \text{ cm}}{0.8 \text{ cm} \times 0.745 \times 500{,}000 \text{ eV}} = 8.05 \times 10^{-2} = \frac{\gamma m v_y}{mc}$$

Since $\beta^2 = \beta_x^2 + \beta_y^2$ and $\beta_x >> \beta_y$, $\gamma \cong \gamma_0$

$$v_y = \frac{8.05 \times 10^{-2} \times 3 \times 10^{10}}{1.5} = 1.611 \times 10^9 \text{ cm/sec}$$

$$y = \bar{v}_y \Delta t = \tfrac{1}{2} v_y \Delta t = \tfrac{1}{2} \times 1.611 \times 10^9 \times 1.789 \times 10^{-10} = 0.144 \text{ cm}$$

$$\theta = \tan^{-1} \frac{P_y}{P_x} = 4.12°$$

In the other frame, $E'_y = \gamma E_y$, $\Delta t' = \Delta t/\gamma$

$$P'_y = qE'_y \Delta t' = q(\gamma E_y)(\Delta t/\gamma) = qE_y \Delta t = P_y$$

$$y' = \tfrac{1}{2} v'_y \Delta t' = \tfrac{1}{2} \frac{P'_y}{m} \Delta t' = \tfrac{1}{2} \frac{P_y}{m} \frac{\Delta t}{\gamma} = \tfrac{1}{2} \frac{P_y}{mc} \frac{c\Delta t}{\gamma}$$

$$= \tfrac{1}{2} \times 8.05 \times 10^{-2} \times \frac{3 \times 10^{10} \times 1.789 \times 10^{-10}}{1.5} = 0.144 \text{ cm} = y$$

PROBLEM 15.7 Particle 1, with charge q_1, was at rest. Particle 2, with charge q_2, went past it, moving at speed v, the closest distance of approach being b. The first particle was so massive that the velocity it acquired from the Coulomb force caused only negligible displacement during the time of passage. Likewise, the second particle was so massive that any change in speed or deflection of its path from a straight line was negligible. How much transverse momentum was acquired by each particle, as a result of the near encounter? (Transverse here means perpendicular to the direction of motion of particle 2.) Answer this part first for particle 2, working in the frame where particle 1 is at rest. Gauss' law applied to a cylinder will make an integration unnecessary. Is your formula relativistically exact? Now consider the momentum acquired by particle 1. The field acting on it is that of a moving charge. By the appropriate integration the transverse momentum could be calculated. However, you could also justify the use of Gauss' law instead, or you could argue directly from your first result. Explain.

Solution: Assuming 1 remains at rest and 2 continues in a straight line during the interaction,

$$\Delta p_2 = \int \mathbf{F}\, dt = q_2 \int \mathbf{E}\, dt = \left(\frac{q_2}{v}\right) \int_{-\infty}^{\infty} \mathbf{E}\, dx = \frac{q_2}{v} \int_{-\infty}^{\infty} E_y\, dx$$

Consider E at the surface of a cylinder, radius b, infinitely long. Over this surface,

$$\int \mathbf{E} \cdot d\mathbf{r} = \int E_y\, (2\pi b\, dx) = 2\pi b \int_{-\infty}^{\infty} E_y\, dx$$

$$= 4\pi q_1$$

or

$$\int_{-\infty}^{\infty} E_y\, dx = \frac{2q_1}{b}$$

So,

$$\Delta p_2 = \frac{q_2}{v} \int_{-\infty}^{\infty} E_y\, dx = \frac{2q_1 q_2}{b\, v}$$

The same argument will work if we go to the frame in which q_2 is at rest or, more simply, since $\mathbf{p}_1 + \mathbf{p}_2 = \text{constant}$ (only internal forces),

$$\Delta p_1 = -\Delta P_2 = \frac{-2q_1 q_2}{vb}$$

PROBLEM 15.8

(a) For speeds small compared to c, the field transformations can be written in a very simple form,

$$\mathbf{E}' = \mathbf{E} + \frac{\mathbf{v}}{c} \times \mathbf{B}$$

$$\mathbf{B}' = \mathbf{B} - \frac{\mathbf{v}}{c} \times \mathbf{E}$$

where $\mathbf{v}$ is the velocity with which the "primed" coordinate system is moving as seen from the "unprimed" frame. Let $\mathbf{v} = \hat{\mathbf{x}}\beta c$ correspond to the particular situation where it holds that

$$E'_x = E_x, \quad E'_y = \gamma(E_y - \beta B_z), \quad E'_z = \gamma(E_z + \beta B_y)$$

$B'_x = B_x,\quad B'_y = \gamma(B_y + \beta E_z),\quad B'_z = \gamma(B_z - \beta E_y)$

[As usual $\beta \to v/c$ and $\gamma \to (1-\beta^2)^{-\frac{1}{2}}$.] All primed quantities are measured in the frame F', which is moving in the positive x direction with speed v as seen from F. Show that the above equations for **E'** and **B'** give results consistent with the above equations for the components of **E'** and **B'** in the approximation $(1-\beta^2)^{\frac{1}{2}} \simeq 1$. Let the primed frame be that of a jet aircraft flying in the direction of magnetic north in a region where the earth's magnetic field has a magnetude of 0.4 Gauss and makes an angle of 30° with the vertical, pointing downward as in the Northern Hemisphere.

(b) What is the direction of the extra component of the electric field, in aircraft coordinates, which arises from the motion through the magnetic field?

(c) What is its magnitude in statvolts/cm?

Solution:

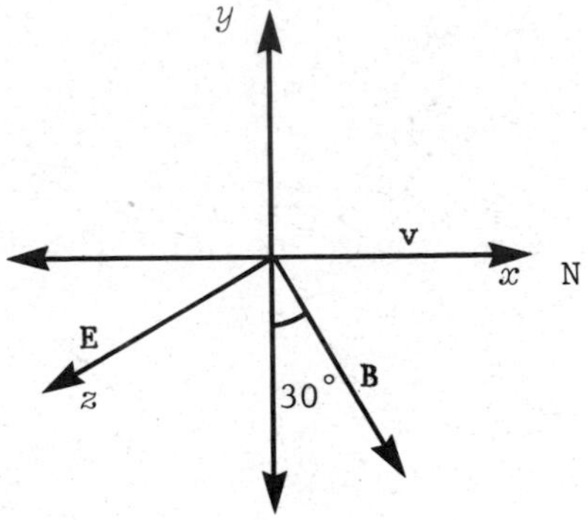

(a) $\mathbf{E} + \frac{\mathbf{v}}{c} \times \mathbf{B} = \mathbf{E} + \beta\hat{\mathbf{x}} \times \mathbf{B}$

$$= E_x\hat{\mathbf{x}} + (E_y - \beta B_z)\hat{\mathbf{y}} + (E_z + \beta B_y)\hat{\mathbf{z}}$$

$\mathbf{B} - \frac{\mathbf{v}}{c} \times \mathbf{E} = \mathbf{B} - \beta\hat{\mathbf{x}} \times \mathbf{E}$

$$= B_x\hat{\mathbf{x}} + (B_y + \beta E_z)\hat{\mathbf{y}} + (B_z - \beta E_y)\hat{\mathbf{z}}$$

These agree with the components of **E'** and **B'** in the above equations except for the factor $\gamma \simeq 1$.

$$\mathbf{E} = 0, \quad \mathbf{B} = (B \sin 30°)\hat{\mathbf{x}} - (B \cos 30°)\hat{\mathbf{y}}$$

$$\mathbf{E}' = \frac{\mathbf{v}}{c} \times \mathbf{B}' \simeq \frac{\mathbf{v}}{c} \times \mathbf{B}$$

$$E' = \beta\hat{x} \times B = -\beta B \cos 30°$$

So E' points west. Assume $v \simeq 500$ mph $\simeq 800$ km/hr $= 8 \times 10^7$ cm/3600 sec

$$\beta = \frac{v}{c} = 7.5 \times 10^{-7}$$

$$E'_z \simeq 7.5 \times 10^{-7} \times 0.4 \times 0.866 = 2.6 \times 10^{-7} \text{ statvolt/cm}$$

Suppose a wingspan of $\simeq 10$ m. Then the emf between wing tips is,

$$V = 2.6 \times 10^{-7} \frac{\text{statvolt}}{\text{cm}} \times 300 \frac{\text{V}}{\text{statvolt}} \times 10^3 \text{ cm} = 78 \text{ mV}$$

In the moving frame $F' = F'_\perp = e/r^2$, so

$$F_\perp = \frac{1}{\gamma} F'_\perp = \frac{e^2}{\gamma r^2}$$

Note that r does not change since it is perpendicular to the direction of relative motion.

Take the unprimed system to be that in which the electrons are at rest.

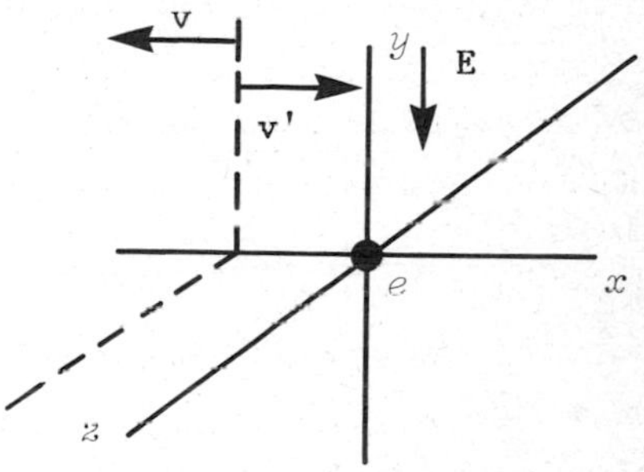

$$\mathbf{E} = -\frac{(e)}{r^2}\hat{\mathbf{y}}$$

$$\mathbf{B} = 0$$

$$\mathbf{E}' = \gamma\mathbf{E} = -\left(\frac{\gamma e}{r^2}\right)\hat{\mathbf{y}}$$

$$\mathbf{B}' = \frac{\mathbf{v}'}{c} \times \mathbf{E}' = \beta\hat{\mathbf{x}} \times \left(-\frac{\gamma e}{r^2}\right)\hat{\mathbf{y}} = -\left(\frac{\beta\gamma e}{r^2}\right)\hat{\mathbf{z}}$$

$$\mathbf{F}' = -e\left(\mathbf{E}' + \frac{\mathbf{v}'}{c}\ \mathbf{B}'\right)$$

$$= \left(\frac{\gamma e^2}{r^2}\right)\hat{\mathbf{y}} + \beta\mathbf{x} \times \left(\frac{\beta\gamma e^2}{r^2}\right)\hat{\mathbf{z}}$$

$$\mathbf{F}' = \frac{\gamma e^2}{r^2}\hat{\mathbf{y}}\ (1 - \beta^2) = \left(\frac{e^2}{\gamma r^2}\right)\hat{\mathbf{y}}$$

where F' is the force in the lab system.

$\lim_{v \to c} \frac{1}{\gamma} = 0$, so $\mathbf{F} \to 0$ as $v \to c$.

PROBLEM 15.9 In the neighborhood of the origin in the coordinate system x, y, z, there is an electric field **E** of magnitude 100 statvolt/cm pointing in a direction that makes angles of 30° with the x axis and 60° with the y axis. The frame F' has its axes parallel to those just described, but is moving, relative to the first frame, with a speed 0.6 c in the positive y direction. Find the direction and magnitude of the electric field which will be reported by an observer in the frame F'. What magnetic field does this observer report?

Solution:

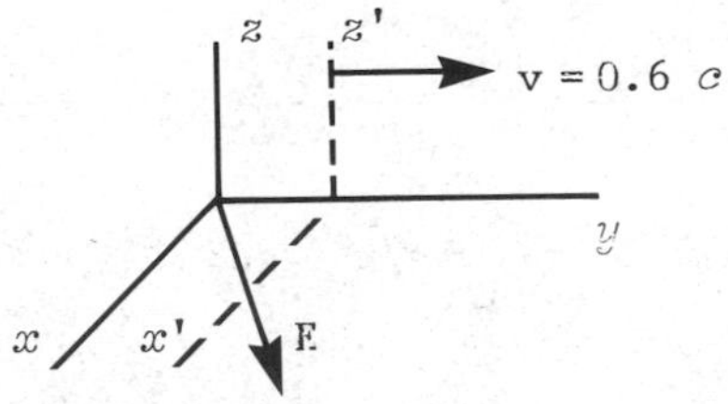

Since B = 0 in F, only a simple transformation is required.

$$E_{\parallel} = 100 \sin 30^\circ = 50 \text{ statvolt/cm}$$

$$E_{\perp} = 100 \cos 30^\circ = 86.6 \text{ statvolt/cm}$$

$$\gamma = \frac{1}{\sqrt{1-(0.06)^2}} = 1.25$$

$$E_{\parallel}' = E_{\parallel} = 50 \text{ statvolt/cm}$$

$$E_{\perp}' = \gamma E_{\perp} = 1.25 \times 86.6 = 108.3 \text{ statvolt/cm}$$

$$E' = \sqrt{(50)^2 + (108.3)^2} = 119.2 \text{ statvolt/cm}$$

$$\theta = \tan^{-1}(50/108.3) = 24.8^\circ$$

$$\mathbf{B}' = \frac{\mathbf{v}'}{c} \times \mathbf{E}' = -0.6\,\hat{\mathbf{y}} \times 108.3\,\hat{\mathbf{x}} = 65.0\,\hat{\mathbf{z}} \text{ G}$$

PROBLEM 15.10 A capacitor consists of two parallel rectangular plates with a vertical separation of 2 cm. The east-west dimension of the plates is 20 cm, the north-south dimension is 10 cm. The capacitor has been charged by connecting it temporarily to a battery of 300 V (1 statvolt). How many excess electrons are on the negative plate? What is the electric field strength between the plates? Now give the following quantities as they would be measured in a frame of reference which is moving eastward, relative to the laboratory in which the plates are at rest, with speed 0.6 c: the three dimensions of the capacitor; the number of excess electrons on the negative plate; the electric field strength between the plates. Answer the same questions for a frame of reference which is moving upward with speed 0.6 c.

Solution: In the usual parallel-plate approximations, $E = v/d = 1$ statvolt/2 cm = 0.5 statvolt/cm = $4\pi\sigma$ · number of electrons is equal to

$$\frac{E}{4\pi}\frac{\text{Area}}{e} = \frac{0.5}{4\pi} \times \frac{200 \text{ cm}^2}{4.8 \times 10^{-10} \text{ esu}} = 1.66 \times 10^{10}$$

In the system moving east: $\gamma = (1 - 0.6^2)^{-\frac{1}{2}} = 1.25$

Dimensions = $16 \times 10 \times 2$

Number of electrons = 1.66×10^{10}

$$E' = 4\pi\sigma' = \frac{4\pi Q}{\text{Area}'} = \frac{4\pi \times 1.66 \times 10^{10} \times 4.8 \times 10^{-10}}{160 \text{ cm}^2} = 0.625 \text{ statvolt/cm} = \gamma E$$

In the system moving up: $\gamma = 1.25$

Dimensions = $20 \times 10 \times 1.6$

Number of electrons = 1.66×10^{10}

$$E' = 4\pi Q(\text{Area}') = \frac{4\pi Q}{\text{Area}} = E = 0.5 \text{ statvolt/cm}$$

PROBLEM 15.11 A particle of charge e moves in a circular path of radius R in the $x-y$ plane with a constant angular velocity ω_0.

(a) Show that the exact expression for the angular distribution of power radiated into the mth multiple of ω_0 is

$$\frac{dP_m}{d\Omega} = \frac{e^2\omega_0^4R^2m^2}{2\pi c^3}\left[\left(\frac{dJ_m(m\beta \sin\theta)}{d(m\beta \sin\theta)}\right)^2 + \frac{\cos^2\theta}{\beta^2} J_m^2(m\beta \sin\theta)\right]$$

where $\beta = \omega_0 R/c$, and $J_m(x)$ is the Bessel function of order m.

(b) Assume nonrelativistic motion and obtain an approximate result for $dP_m/d\Omega$.

(c) Assume extreme relativistic motion and obtain the results for relativistic particle in instantaneously circular motion.

Solution:

(a) $$\left.\frac{d^2I}{d\omega d\Omega}\right|_{\omega = m\omega_0} = \frac{e^2\omega^2}{4\pi^2 c}\left|\int_{-\infty}^{\infty} \hat{\mathbf{n}} \times (\hat{\mathbf{n}} \times \beta) \exp[i\omega(t - (\hat{\mathbf{n}} \cdot \mathbf{r}(t))/c)]\, dt\right|^2_{\omega = m\omega_0}$$

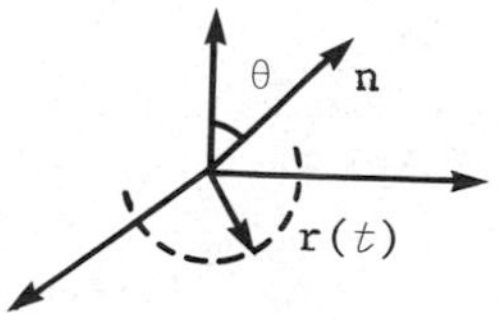

$$\hat{\mathbf{n}} \times (\hat{\mathbf{n}} \times \beta) = \beta(-\hat{\boldsymbol{\varepsilon}}_{\parallel} \sin\omega_0 t + \hat{\boldsymbol{\varepsilon}}_{\perp} \cos\omega_0 t \cos\theta)$$

$$\omega\left(t - \frac{\hat{\mathbf{n}}\cdot\mathbf{r}(t)}{c}\right) = m\omega_0\left(t - \frac{R}{c}\sin\omega_0 t \sin\theta\right)$$

$$\frac{d^2I}{d\omega d\Omega} = \frac{e^2\omega^2}{4\pi^2 c} \left| -\hat{\boldsymbol{\varepsilon}}_{\parallel} A_{\parallel}(\omega) + \hat{\boldsymbol{\varepsilon}}_{\perp} A_{\perp}(\omega)\right|^2$$

$$A_{\parallel}(\omega) = \beta \int_0^{2\pi} \sin\omega_0 t \exp^{[im\omega_0(t - (R/c)\sin\omega_0 t \sin\theta)]} dt$$

$$A_{\perp}(\omega) = \beta \int_0^{2\pi} \cos\omega_0 t \cos\theta \exp^{[im\omega_0(t - (R/c)\sin\omega_0 t \sin\theta)]} dt$$

$$\beta = \frac{\omega_0 R}{c}, \quad \phi = \omega_0 t$$

$$A_{\parallel}(\omega) = \frac{\beta}{\omega_0} \int_0^{2\pi} \sin\phi\, e^{[i(m\phi - m\beta\sin\phi\sin\theta)]} d\phi$$

$$= \frac{\beta i}{\omega_0} \frac{d}{d(m\beta\sin\theta)} \int_0^{2\pi} \left(e^{i(m\phi - m\beta\sin\phi\sin\theta)} d\phi\right)$$

$$= \frac{\beta i}{\omega_0} 2\pi \frac{d}{d(m\beta\sin\theta)} J_m(m\beta\sin\theta)$$

$$|A_{\parallel}(\omega)|^2 = \frac{\beta^2 4\pi^2}{\omega_0^2} \left(\frac{dJ_m(m\beta\sin\theta)}{d(m\beta\sin\theta)}\right)^2$$

$$A_{\perp}(\omega) = \frac{\beta}{\omega_0} \int_0^{2\pi} \cos\phi\cos\theta\, e^{i(m\phi - \beta m\sin\phi\sin\theta)} d\phi$$

integrating by parts, $u = e^{im\phi}$, $du = ime^{im\phi}\, d\phi$

$$dv = e^{-i\beta m\sin\phi\sin\theta} d(\sin\theta)$$

$$v = \frac{e^{-i\beta m\sin\phi\sin\theta}}{i\beta m\sin\phi}$$

$$A_\perp(\omega) = \frac{\beta}{\omega_0}\left(\cos\theta\, e^{im\phi} e^{-i\beta m \sin\phi \sin\theta}\Big|_0^{2\pi} - \frac{im\cos\theta}{i\beta m \sin\theta}\int_0^{2\pi} e^{i(m\phi - \sin\phi\sin\theta)}\, d\theta\right)$$

$$A_\perp(\omega) = \frac{-2\pi\cot\theta}{\omega_0} J_m(\beta m \sin\theta)$$

$$|A_\perp(\omega)|^2 = \frac{4\pi^2\cot^2\theta}{\omega_0^2} J_m^2(\beta m \sin\theta)$$

$$\frac{d^2I}{d\omega d\Omega} = \frac{e^2\omega_0^2}{4\pi^2 c}\frac{m^2\beta^2 4\pi^2}{\omega_0^2}\left[\left(\frac{dJ_m(\beta m\sin\theta)}{d(m\beta\sin\theta)}\right)^2 + \frac{\cot^2\theta}{\beta^2}J_m^2(m\beta\sin\theta)\right]$$

$$= \frac{e^2\omega_0^2 R^2 m^2}{c^3}\left[\left(\frac{dJ_m(m\beta\sin\theta)}{d(m\beta\sin\theta)}\right)^2 + \frac{\cot^2\theta}{\beta^2}J_m^2(m\beta\sin\theta)\right]$$

$$\frac{dP_m}{d\Omega} = \frac{1}{2\pi}\omega_0^2 \left.\frac{d^2I}{d\omega d\Omega}\right|_{\omega = m\omega_0}$$

$$\frac{dP_m}{d\Omega} = \frac{e^2\omega_0^4 R^2 m^2}{2\pi c^3}\left[\left(\frac{dJ_m(m\beta\sin\theta)}{d(m\beta\sin\theta)}\right)^2 + \frac{\cot^2\theta}{\beta^2}J_m^2(m\beta\sin\theta)\right]$$

$$\frac{dP_m}{d\Omega} = \frac{e^2\omega_0^2 m^2}{2\pi c}\left[\beta^2\left(\frac{dJ_m(m\beta\sin\theta)}{d(m\beta\sin\theta)}\right)^2 + \cot^2\theta\, J_m^2(m\beta\sin\theta)\right]$$

$$\frac{dJ_m(m\beta\sin\theta)}{d(m\beta\sin\theta)} = J_{m-1}(m\beta\sin\theta) - \frac{1}{\beta\sin\theta}J_m(m\beta\sin\theta)$$

$\beta \to 0$, so β, β^2 terms can be ignored.

$$\frac{dP_m}{d\Omega} = \frac{e^2\omega_0^2 m^2}{2\pi c}\left(\frac{J_m^2(m\beta\sin\theta)}{\sin^2\theta} + \cot^2\theta\, J_m^2(m\beta\sin\theta)\right)$$

$$x \ll 1,\quad J_m(x) \to \frac{1}{\Gamma(m+1)}\left(\frac{x}{2}\right)^m$$

Keep $m = 1$ terms,

$$\frac{dP_{m=1}}{d\Omega} = \frac{e^2\omega_0^2}{2\pi c}\left(\frac{1}{4} + \cos^2\theta\left(\frac{\sin^2\theta}{4}\right)\right)$$

$$\frac{dP}{d\Omega} \approx \frac{e^2\omega_0^2}{8\pi c}(1 + \cos^2\theta)$$

PROBLEM 15.12 Using the Liénard-Wiechert fields, discuss the time-average power radiated per unit solid angle in nonrelativistic motion of a particle with charge e, moving along the z axis with instantaneous position $z(t) = a\cos\omega_0 t$.

Solution: From the Liénard-Wiechert fields, we have seen that for nonrelativistic motion the instantaneous power radiated into a solid angle $d\Omega$ is:

$$\frac{dP}{d\Omega} = \frac{e^2}{4\pi c^3}|\mathbf{v}|^2 \sin^2\theta$$

where θ is the angle between $\mathbf{v}$ and direction to field point. Then, if the power has a periodic dependence T, time average power radiated into the solid angle is:

$$\left\langle\frac{dP}{d\Omega}\right\rangle = \frac{1}{T}\int_0^T \frac{dP}{d\Omega}\,dt$$

$$z(t) = a\cos\omega_0 t, \quad \therefore \quad |\dot{\mathbf{v}}|^2 = \omega_0^4 a^2 \cos^2\omega_0 t$$

$$\left\langle\frac{dP}{d\Omega}\right\rangle = \frac{1}{2\pi/\omega_0}\int_0^{2\pi/\omega_0} \frac{e^2}{4\pi c^3}\omega_0^4 a^2 \sin^2\theta\cos^2\omega_0 t\,dt$$

$$= \frac{\omega_0}{2\pi}\cdot\frac{e^2}{4\pi c^3}\omega_0^4 a^2 \sin^2\theta\int_0^{2\pi/\omega_0}\cos^2\omega_0 t\,dt$$

$$= \frac{1}{2\pi}\frac{e^3}{4\pi c^3}\omega_0^4 a^2 \sin\theta\int_0^{2\pi}\cos^2 x\,dx$$

$$\left\langle\frac{dP}{d\Omega}\right\rangle_t = \frac{e^2\omega_0^4 a^2}{8\pi c^3}\sin^2\theta$$

Angular distribution

with cylindrical symmetry about z

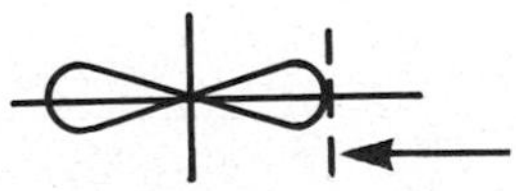

$$\frac{e^2 a^2 \omega_0^4}{8\pi d^3}$$

Total power radiated:

$$P = \int\left\langle\frac{dP}{d\Omega}\right\rangle_t d\Omega = \frac{e^2\omega_0^4 a^2}{8\pi c^3}\int_0^{2\pi}d\phi\int_0^{\pi}\sin^3\theta\,d\theta$$

$$P = \frac{e^2\omega_0^4 a^2}{3c^3}$$

PROBLEM 15.13 A nonrelativistic particle of charge ze, mass m, and kinetic energy E makes a head-on collision with a fixed central force field of finite range. The interaction is repulsive and described by a potential $V(r)$, which becomes greater than E at close distances.

(a) Show that the total energy radiated is given by

$$\Delta W = \frac{4}{3}\frac{z^2e^2}{m^2c^3}\sqrt{\frac{m}{2}}\int_{r_{min}}^{\infty}\left|\frac{dV}{dr}\right|^2 \frac{dr}{\sqrt{V(r_{min}) - V(r)}}$$

where r_{min} is the closest distance of approach in the collision.

(b) If the interaction is a Coulomb potential $V(r) = zZe^2/r$, show that the total energy radiated is

$$\Delta W = \frac{8}{45}\frac{zmv_0^5}{Zc^3}$$

where v_0 is the velocity of the charge at infinity.

Solution:

(a) The Larmor result for the total instantaneous power radiated is

$$P = \frac{2}{3}\frac{e^2}{c^3}|\dot{\mathbf{v}}|^2$$

$$|\dot{\mathbf{v}}|^2 = |\mathbf{a}|^2$$

which is the square of acceleration on the charge Ze, since the force field is described by a potential V.

$$\mathbf{F} = M\mathbf{a} = -\nabla V = -\hat{\mathbf{r}}\frac{\partial V(r)}{\partial r}$$

$$\mathbf{a} = -\frac{\hat{\mathbf{r}}}{M}\frac{dV(r)}{dr}$$

and

$$|\mathbf{a}|^2 = \frac{1}{M^2}\left|\frac{dV(r)}{dr}\right|^2$$

$$P = \frac{2}{3}\frac{z^2e^2}{M^2c^3}\left|\frac{dV(r)}{dr}\right|^2$$

Now

$$E_T = \frac{p^2}{2M} + V(r)$$

$$P = M\frac{dr}{dt} = \sqrt{2M[E_T - V(r)]}$$

$$\frac{dr}{dt} = \sqrt{\frac{2}{M}}\ \sqrt{E_T - V(r)}$$

$$dt = \sqrt{\frac{M}{2}}\ \frac{dr}{\sqrt{E_T - V(r)}}$$

The total energy radiated

$$\Delta W = 2\int_{t=\text{coll. time}}^{t=0} \frac{dE}{dt}\,dt = 2\int_{t_0}^{t_\infty} P(t)dt$$

Factor of 2 is for path in *and* path out. But we only have P as a function of r. [Also: $E_T = E(\text{Kinetic energy}) = V(r_{\min})$.]

$$\Delta W = 2\int_{t_0}^{t_\infty} \frac{2}{3}\frac{Z^2e^2}{M^2c^3}\left|\frac{dV}{dr}\right|^2 dt$$

$$\Delta W = \frac{4}{3}\frac{Z^2e^2}{M^2c^3}\sqrt{\frac{M}{2}}\int_{r_0}^{\infty}\left|\frac{dV}{dr}\right|^2 \frac{dr}{\sqrt{v(r_0) - v(r)}}$$

$$\Delta W = \frac{4}{3}\frac{Z^2e^2}{M^2c^3}\sqrt{\frac{M}{2}}\int_{r_0}^{\infty}\left|\frac{dV}{dr}\right|^2 \frac{dr}{\sqrt{v(r_0) - v(r)}}$$

r_0 = closest approach distance.

(b) Let $V(r) = \dfrac{zZe^2}{r}$

$$\frac{dV}{dr} = -\frac{zZe^2}{r^2}, \quad \left|\frac{dV}{dr}\right|^2 = \frac{z^2Z^2e^4}{r^4}$$

$$\Delta W = \frac{4}{3}\frac{z^2e^2}{M^2c^3}\sqrt{\frac{M}{2}}\ z^2Z^2e^4\int_{r_0}^{\infty}\frac{dr}{r^4}\ \frac{1}{\sqrt{v(r_0) - v(r)}}$$

$$\Delta W = \frac{4}{3}\frac{z^2e^2}{M^2c^3}\sqrt{\frac{M}{2}}\ \frac{z^2Z^2e^2}{\sqrt{zZe^2}}\sqrt{r_0}\int_{r_0}^{\infty}\frac{1}{r^2}\ \frac{dr/r^2}{\sqrt{1 - (r_0/r)}}$$

Now,

$$I = \int_{r_0}^{\infty} \frac{1}{r^2} \frac{dr/r^2}{\sqrt{1 - r_0(1/r)}} = -\int_{r_0}^{\infty} \left(\frac{1}{r}\right)^2 \frac{d(1/r)}{\sqrt{1 - (r_0)(1/r)}}$$

$$= \int_0^{1/r_0} \frac{u^2\,du}{\sqrt{1 - r_0 u}}$$

integrating by parts

$$= \frac{-2(8 + 4r_0 u + 3r_0^2 u^2)}{15(r_0)^3} \sqrt{1 - r_0 u} \Bigg|_0^{1/r_0}$$

$$I = \frac{-2}{15r_0^3}\{0 - 8\sqrt{1}\} = \frac{2.8}{15r_0^3}$$

$$\Delta W = \frac{8}{3} \frac{z^2e^2}{M^2c^3} \sqrt{\frac{M}{2}} \frac{z^2Z^2e^4}{\sqrt{zZe^2}} \frac{8}{15} \frac{\sqrt{r_0}}{r_0^3}$$

So

$$\tfrac{1}{2}Mv_0^2 = \frac{zZe^2}{r_0}$$

$$\sqrt{\frac{r_0}{zZe^2}} = \sqrt{\frac{2}{Mv_0^2}}$$

$$\frac{1}{r_0^3} = \left(\frac{mv_0^2}{2zZe^2}\right)^3$$

$$\Delta W = \frac{8}{3} \frac{Z^2e^2}{M^2c^3} \sqrt{\frac{M}{2}} \frac{1}{\sqrt{\tfrac{1}{2}Mv_0^2}} \frac{8}{15} z^2Ze^4 \left(\frac{Mv_0^2}{2zZe^2}\right)^3$$

$$\Delta W = \frac{8}{3} \frac{Z^2e^2}{M^2c^3} \frac{8}{15} M_0^3 v_0^5 \frac{1}{zZe^2} \frac{1}{8}$$

$$\Delta W = \frac{8}{45} \frac{zMv_0^5}{Zc^3}$$

PROBLEM 15.14 A charge e moves in simple harmonic motion along the z axis, $z(t') = a\cos(\omega_0 t')$.

(a) Show that the instantaneous power radiated per unit solid angle is

$$\frac{dP(t')}{d\Omega} = \frac{e^2 c\beta^4}{4\pi a^2} \frac{\sin^2\theta\cos^2(\omega_0 t')}{(1+\beta\cos\theta\sin\omega_0 t')^5}$$

where $\beta = a\omega_0/c$.

(b) By performing a time averaging, show that the average power per unit solid angle is

$$\frac{dP}{d\Omega} = \frac{e^2 c\beta^4}{32\pi a^2}\left(\frac{4+\beta^2\cos^2\theta}{(1-\beta^2\cos^2\theta)^{7/2}}\right)\sin^2\theta$$

(c) Make rough sketches of the angular distribution for nonrelativistic motion.

Solution:

(a) $Z(t') = a\cos\omega_0 t'$

Now we have seen that the angular distribution of the instantaneously radiated power is

$$\frac{dP(t)}{d\Omega} = \frac{e^2}{4\pi c}\frac{|\hat{n}\times\{(\hat{n}-\beta)\times\dot{\beta}\}|^2}{(1-\hat{n}\cdot\beta)^5}$$

$$\dot{Z}(t') = c\beta = -a\omega_0\sin\omega_0 t'$$

$$\ddot{Z}(t') = c\dot{\beta} = -a\omega_0^2\cos\omega_0 t'$$

$$\beta = \left(\frac{-a\omega_0^2}{c}\sin\omega_0 t'\right)\hat{k}, \quad \hat{n} = \hat{i}x + \hat{j}y + \hat{k}z$$

$$\hat{n} = \hat{i}\sin\theta\cos\theta + \hat{y}\sin\theta\sin\phi + \hat{k}\cos\theta$$

$$\dot{\beta} = \left(\frac{-a\omega_0^2}{c}\cos\omega_0 t'\right)\hat{k}, \quad \beta = \frac{+a\omega_0}{c}$$

$$\beta\cdot\dot{\beta} = 0$$

$$\hat{n}\times\dot{\beta} = -\left(\frac{-a\omega_0^2}{c}\cos\omega_0 t'\right)(\hat{j}\sin\theta\cos\theta - \hat{i}\sin\theta\sin\phi)$$

$$\hat{n}\times(\hat{n}\times\beta) = \left(\frac{a\omega_0^2}{c}\cos\omega_0 t'\right)(\hat{k}\sin^2\theta\cos^2\phi + \hat{k}\sin^2\theta\sin^2\phi$$

$$-\hat{i}\sin\theta\cos\theta\cos\phi - \hat{j}\sin\theta\cos\theta\sin\phi)$$

$$|\hat{n}\times(\hat{n}\times\beta)|^2 = \left(\frac{a\omega_0^2}{c}\cos\omega_0 t'\right)^2(\sin^4\theta + \sin^2\theta\cos^2\theta\cos^2\phi$$

$$+\sin^2\theta\cos^2\theta\sin^2\phi)$$

$$= \left(\frac{a\omega_0^2}{c}\cos\omega\; t'\right)^2 (\sin^2\theta + \cos^2\theta)\sin^2\theta$$

$$|\hat{\mathbf{n}}\times(\hat{\mathbf{n}}\times\dot{\beta})|^2 = \left(\frac{a\omega_0^2}{c}\cos\omega_0 t'\right)^2 \sin^2\theta = \frac{c^2}{a^2}\beta^4\sin^2\theta\cos^2\omega_0 t'$$

and

$$\hat{\mathbf{n}}\cdot\beta = \left(\frac{-a\omega_0}{c}\sin\omega_0 t'\right)\cos\theta$$

$$1-\hat{\mathbf{n}}\cdot\beta = 1+\frac{a\omega_0}{c}\sin\omega_0 t'\cos\theta = 1+\beta\cos\theta\sin\omega_0 t'$$

$$\frac{dP(t')}{d\Omega} = \frac{e^2c}{4\pi a^2}\,\frac{\beta^4\sin^2\theta\cos^2\omega_0 t'}{(1+\beta\cos\theta\sin\omega_0 t')^5}$$

(b) $$\left\langle\frac{dP(t')}{d\Omega}\right\rangle_{t'} = \frac{2\pi}{\omega_0}\int_0^{2\pi/\omega_0}\frac{dP(t')}{d\Omega}\,dt'$$

$$= \frac{2\pi}{\omega_0}\,\frac{e^2c\beta^4}{4\pi a^2}\sin^2\theta\int_0^{2\pi/\omega_0}\frac{\cos^2\omega_0 t'\,dt'}{[1+(\beta\cos\theta)\sin\omega_0 t']^5}$$

Let:

$$I = \int_0^{2\pi/\omega_0}\frac{\cos^2\omega_0 t'\,dt'}{[1+(\beta\cos\theta)\sin\omega_0 t']^5} = \frac{1}{\omega_0}\int_0^{2\pi}\frac{\cos^2 x\,dx}{[1+b\sin x]^5}$$

$x = \omega_0 t'$, $b = \beta\cos\theta$. Integrate by parts: $u = \cos x$, $du = -\sin x\,dx$

$$V = \int\frac{\cos x\,dx}{(1+b\sin x)^5}$$

$$V = -\int\frac{d(\sin x)}{(1+b\sin x)^5} = -\int\frac{d\omega}{(1+b\omega)^5} = \frac{1}{4b}\,\frac{1}{(1+b\omega)^4} = \frac{1}{4b}\,\frac{1}{(1+b\sin x)^4}$$

$$I = \frac{1}{\omega_0}\left(\left.\frac{\cos x}{4b(1+b\sin x)^4}\right|_0^{2\pi} + \frac{1}{4b}\int_0^{2\pi}\frac{\sin x\,dx}{(1+b\sin x)^4}\right)$$

$$= \frac{1}{\omega_0}\,\frac{1}{4b}\int_0^{2\pi}\frac{\sin x\,dx}{(1+b\sin x)^4}$$

Now

$$\int_0^{2\pi} \frac{dx}{(a+b\sin x)^3} = \frac{2\pi(a^2+b^2/2)}{(a^2-b^2)^{5/2}}$$

from

$$\int_0^{2\pi} \frac{dx}{(a+b\sin x)^2} = \frac{2\pi a}{(a^2-b^2)^{3/2}}$$

and

$$\frac{d}{db}\left(\frac{1}{(a+b\sin x)^3}\right) = -\frac{1}{(a+b\sin x)^6}\, 3(a+b\sin x)^2 \sin x = -3\frac{\sin x}{(a+b\sin x)^4}$$

$$\int_0^{2\pi} \frac{\sin x\, dx}{(a+b\sin x)^4} = -\frac{1}{3}\frac{d}{db}\int_0^{2\pi}\frac{dx}{(a+b\sin x)^3} = \frac{-1}{3}\frac{d}{db}\left(\frac{2\pi(a^2+b^2/2)}{(a^2-b^2)^{5/2}}\right)$$

$$\int_0^{2\pi} \frac{\sin x\, dx}{(a+b\sin x)^4} = -\frac{1}{3}\left(\frac{2\pi b}{(a^2-b^2)^{5/2}} - \frac{2\pi(a^2+b^2/2)}{(a^2-b^2)^5}\cdot\frac{5}{2}(a^2-b^2)^{3/2}(-2b)\right)$$

$$= -\frac{1}{3}\left(\frac{2\pi b(a^2-b^2)}{(a^2-b^2)^{7/2}} + \frac{2\pi b\cdot 5\cdot(a^2+b^2/2)}{(a^2-b^2)^{7/2}}\right)$$

$$= -\frac{2\pi}{3}\left(\frac{a^2b - b^3 + 5a^2b + 5b^3/2)}{(a^2-b^2)^{7/2}}\right)$$

$$\int^{2\pi} \frac{\sin x\, dx}{(1+b\sin x)^4} = -\frac{2\pi}{3}\left(\frac{6b+3b^3/2}{(1-b^2)^{7/2}}\right)$$

$$I = \frac{1}{\omega_0}\left(\frac{-\pi}{6b}\right)\left(\frac{6b+3b^3/2}{(1-b^2)^{7/2}}\right) = \frac{-\pi}{\omega_0}\left(\frac{1+\frac{1}{4}b^2}{(1-b^2)^{7/2}}\right)$$

$$\left\langle\frac{dP(t')}{d\Omega}\right\rangle_{t'} = \left(\frac{2\pi}{\omega_0}\right)^{-1}\frac{e^2c\beta^4}{4\pi a^2}\sin^2\theta\left(\frac{\pi}{4\omega_0}\right)\left(\frac{4+\beta^2\cos^2\theta}{(1-\beta^2\cos^2\theta)^{7/2}}\right)$$

$$\left\langle\frac{dP(t')}{d\Omega}\right\rangle_{t'} = \frac{e^2c\beta^4}{32\pi a^2}\left(\frac{4+\beta^2\cos^2\theta}{(1-\beta^2\cos^2\theta)^{7/2}}\right)\sin^2\theta$$

(c) For nonrelativistic motion, $\beta \ll 1$,

$$\frac{1}{(1-\beta^2\cos^2\theta)^{7/2}} \simeq 1 + \frac{7}{2}\beta^2\cos^2\theta$$

$$\frac{4+\beta^2\cos^2\theta}{(1-\beta^2\cos^2\theta)^{7/2}} \simeq 4 + 14\beta^2\cos^2\theta + \beta^2\cos^2\theta + O(\beta^4)$$

$$\simeq 4 + 15\beta^2\cos^2\theta$$

$$\left\langle\frac{dP}{d\Omega}\right\rangle \simeq \frac{e^2c\beta^4}{32\pi a^2}\,(4\sin^2\theta + 15\beta^2\cos^2\theta\sin^2\theta)$$

To lowest order in β

$$\left\langle\frac{dP}{d\Omega}\right\rangle \simeq \frac{e^2c\beta^4}{32\pi a^2}\,4\sin^2\theta = \frac{e^2c\beta^4}{8\pi a^2}\sin^2\theta$$

The angular distribution looks like

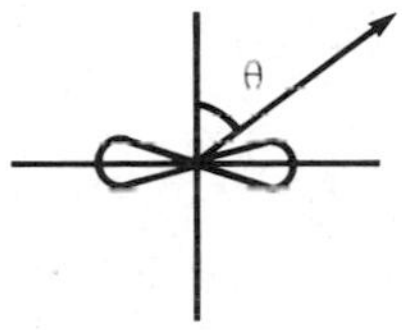

PROBLEM 15.15 Bohr's correspondence principle states that in the limit of large quantum numbers the classical power radiated in the fundamental is equal to the product of the quantum energy ($\hbar\omega_0$) and the reciprocal mean lifetime of the transition from pincipal quantum number n to $(n-1)$.

(a) Using nonrelativistic approximations, show that in a hydrogen-like atom the transition probability (reciprocal mean lifetime) for a transition from a circular orbit of principal quantum number n to $(n-1)$ is given classically by

$$\frac{1}{\tau} = \frac{2}{3}\frac{e^2}{\hbar c}\left(\frac{Ze^2}{\hbar c}\right)^4 \frac{mc^2}{\hbar}\frac{1}{n^5}$$

(b) For hydrogen compare the classical value from (a) with the correct quantum-mechanical results for the transitions $2p \rightarrow 1s$ (1.6×10^{-9} s), $4f \rightarrow 3d$ (7.3×10^{-8} s), $6h \rightarrow 5g$ (6.1×10^{-7} s).

Solution:

(a) $P_1 = \frac{k\omega_0}{\tau}$

$$\frac{dP_1}{d\Omega} = \frac{e^2\omega_0^4 R^2}{2\pi c^3}\left[\left(\frac{dJ_1(\beta \sin\theta)}{d(\beta\sin\theta)}\right)^2 + \frac{\cot^2\theta}{\beta^2} J_1^2(\beta\sin\theta)\right]$$

$$\beta = \frac{\omega_0 R}{c}, \quad \beta << 1$$

$$J_1(x) \simeq \frac{x}{2} - \frac{x^3}{8 \cdot 1! \cdot 2!}$$

$$\frac{dJ_1(x)}{dx} \simeq \tfrac{1}{2} - \frac{3x^2}{16}, \quad x = \beta\sin\theta$$

$$\frac{dP_1}{d\Omega} = \frac{e^2\omega_0^4 R^2}{2\pi c^3}\left[\tfrac{1}{4}\left(1 - \frac{3x^2}{8}\right)^2 + \frac{\cot^2\theta}{2\beta^2}\left(x - \frac{x^3}{8}\right)^2\right]$$

$$= \frac{e^2\omega_0^4 R^2}{2\pi c^3}\left[\tfrac{1}{4}\left(1 + \frac{8x^4}{16} - \frac{3x^2}{4}\right) + \tfrac{1}{2}\cot^2\theta\,\frac{1}{\beta^2}\left(x^2 - \frac{x^4}{4} + \frac{x^6}{64}\right)\right]$$

$$\simeq \frac{e^2\omega_0^4 R^2}{2\pi c^3}\left(\tfrac{1}{4} + \tfrac{1}{2}\cos^2\theta\sin^2\theta\right)$$

$$= \frac{e^2\omega_0^4 R^2}{8\pi c^3}\left(1 + 2\cos^2\theta\right)$$

$$P_1 = \int \frac{dP_1}{d\Omega}\,d\Omega = \frac{e^2\omega_0^4 R^2}{8\pi c^3}\left(4\pi + 4\pi\int_{-1}^{1}\cos^2\theta\, d(\cos\theta)\right)$$

$$= \frac{e^2\omega_0^4 R^2}{2c^3}(1 + 0) = \frac{e^2\omega_0^4 R^2}{2c^3}$$

$$\frac{1}{\tau} = \frac{1}{\hbar\omega_0}\,\frac{e^2\omega_0^4 R^2}{2c^3}$$

$$R = \frac{n^2}{z\alpha}\frac{\hbar}{mc}, \quad \alpha = \frac{e^2}{\hbar c}, \quad v = \omega_0 R = \frac{z\alpha c}{n}$$

$$\hbar\omega_0 = -\tfrac{1}{2}\, mc^2\alpha^2 \left(\frac{1}{n^2} - \frac{1}{(n-1)^2}\right)$$

$$= -\tfrac{1}{2}mc^2\alpha^2 \left(\frac{1 - n^2 - 2n - n^2}{[n(n-1)]^2}\right) = \tfrac{1}{2}\, mc^2\alpha^2 \left(\frac{2n-1}{[n(n-1)]^2}\right)$$

$$\frac{1}{\tau} = \frac{\omega_0}{\hbar}\, \frac{e^2}{2c^3} \left(\frac{z\alpha c}{n}\right)^2 = \frac{e^2}{2\hbar c^3} \left(\frac{z\alpha c}{n}\right)^3 \left(\frac{z\alpha mc}{n^2\hbar}\right)$$

$$= \frac{e^2}{2\hbar c^3} \left(\frac{ze^2}{\hbar c}\right)^4 \frac{mc^4}{\hbar n^5} = \frac{e^2}{2\hbar c} \left(\frac{ze^2}{\hbar c}\right)^4 \frac{mc^2}{n^5}$$

$$\frac{1}{\tau} = \frac{2}{3}\frac{e^2}{\hbar c} \left(\frac{ze^2}{\hbar c}\right)^4 \frac{mc^2}{\hbar}\, \frac{1}{n^5}$$

(b) $\alpha = \frac{e^2}{\hbar c} = \frac{1}{137}$, $z = 1$, $mc^2 = 0.51$ MeV, $\hbar = 6.58 \times 10^{-16}$ eV · sec

$$\frac{1}{\tau} = \frac{1.07 \times 10^{10}}{n^2}, \quad \tau = 9.35 \times 10^{-11}\, n^2$$

Classical	Quantum Mechancial
3.7×10^{-10} sec	1.6×10^{-9} sec
1.5×10^{-9} sec	7.3×10^{-8} sec
3.4×10^{-9} sec	6.1×10^{-7} sec

PROBLEM 15.16 A particle of mass m, charge q, moves in a plane perpendicular to a uniform, static, magnetic induction B.

(a) Calculate the total energy radiated per unit time, expressing it in terms of the constants already defined and the ratio γ of the particle's total energy to its rest energy.
(b) If at time $t = 0$ the particle has a total energy $E_0 = \gamma_0 mc^2$, show that it will have energy $E = \gamma mc^2 < E_0$ at a time t, where

$$t \approx \frac{3m^3c^5}{2q^4B^2}\left(\frac{1}{\gamma} - \frac{1}{\gamma_0}\right)$$

provided $\gamma \gg 1$.

(c) If the particle initially is nonrelativistic and has a kinetic energy ε_0 at $t = 0$, what is the kinetic energy at time t?

(d) If the particle is actually trapped in the magnetic dipole field of the earth and is spiraling back and forth along a line of force, does it radiate more energy near the equator, or while near its turning points? Why? Make quantitative statements if you can.

Solution:

(a) $$\left|\frac{e}{c}\,\mathbf{v}\times\mathbf{B}\right| = \frac{evB}{c} = \left|\frac{d\bar{p}}{dt}\right|$$

$$P(t') = \frac{2}{3}\,\frac{e^2}{m^2c^3}\,\gamma^2\left(\frac{d\bar{p}}{dt}\right)^2 = \frac{2}{3}\,\frac{e^2}{m^2c^3}\,\gamma^2\left(\frac{evB}{c}\right)^2$$

$$P(t') = \frac{2}{3}\,\frac{q^4v^2B^2\gamma^2}{m^2c^5} = \frac{2}{3}\,\frac{q^2\beta^2B^2\gamma^2}{m^2c^3}$$

$$\beta^2 = 1 - \frac{1}{\gamma^2}, \quad \beta \approx 1$$

$$P(t') \simeq \frac{2}{3}\,\frac{q^2B^2\gamma^2}{m_0^2c^3}$$

$$E = \gamma m_0c^2 = \gamma E_{\text{rest.}} = \gamma E_r, \quad E_{\text{rest.}} = m_0c^2$$

$$P(t') = \frac{2}{3}\,\frac{q^2B^2}{m_0^2c^3}\left(\frac{E}{E_r}\right)^2$$

(b) $$P = \frac{-dE}{dt}$$

$$\frac{-dE}{dt} = \frac{2}{3}\,\frac{q^2B^2E^2}{E_r^2m^2c^3} = \frac{2}{3}\,\frac{q^2B^2cE^2}{E_r^4}$$

$$\frac{-dE}{E^2} = \frac{2}{3}\,\frac{q^2cB^2}{E_r^4}\,dt$$

$$-\int_{E_0}^{E} \frac{dE}{E^2} = \int_0^t \frac{2}{3}\frac{q^2cB^2}{E_r^4}\,dt$$

$$\left.\frac{1}{E}\right|_{E_0}^{E} = \frac{1}{E} - \frac{1}{E_0} = \frac{2}{3}\frac{q^2cB^2t}{E_r^4}$$

$$t = \frac{3}{2}\frac{m_0^4c^8}{q^2B^2c}\left(\frac{1}{\gamma m_0c^2} - \frac{1}{\gamma_0 m_0c^2}\right) = \frac{3}{2}\frac{m^3c^5}{q^2B^2}\left(\frac{1}{\gamma} - \frac{1}{\gamma_0}\right)$$

$$t \simeq \frac{3m^3c^5}{2q^2B^2}\left(\frac{1}{\gamma} - \frac{1}{\gamma_0}\right), \quad \gamma \gg 1$$

(c) $$\frac{-dE}{dt} = \frac{2}{3}\frac{q^2B^2\beta^2\gamma^2}{m^2c^3}, \quad \gamma^2 = \frac{1}{1-\beta^2}$$

$\beta \ll 1, \quad \gamma^2 \simeq 1$

$$\frac{-dE}{dt} = \frac{2}{3}\frac{q^2v^2B^2}{m^2c^5} = \frac{2}{3}\frac{q^2B^2}{m^2c^5}\frac{2}{m}\left(\tfrac{1}{2}mv^2\right)$$

$$\frac{-dE}{dt} = \frac{4}{3}\frac{q^2B^2}{m^2c^5}\varepsilon(t)$$

$$\frac{dE}{\varepsilon} = -\frac{4}{3}\frac{q^2B^2}{m^3c^5}\,dt$$

$$\ln \varepsilon'\Big|_{\varepsilon_0}^{\varepsilon} = -\frac{4}{3}\frac{q^2B^2}{m^3c^5}\,t$$

$$\ln\left(\frac{\varepsilon}{\varepsilon_0}\right) = -\frac{4}{3}\frac{q^2B^2}{m^3c^5}\,t$$

$$\frac{\varepsilon}{\varepsilon_0} = \exp[-(4/3)(q^2B^2t/m^3c^5)]$$

$$\varepsilon(t) = \varepsilon(0)\exp\{-[(4/3)(q^2B^2/m^3c^5)t]\}$$

(d) Near the turning points the particle has the largest acceleration and $P(t') \propto (d\bar{p}/dt)^2$ so the particle radiates more energy at the endpoints.

PROBELM 15.17 In a certain reference frame a static, uniform, electric field E_0 is parallel to the x axis, and a static, uniform, magnetic induction $B_0 = 2E_0$ lies in the $x-y$ plane, making an angle θ with the x axis. Determine the relative velocity of a reference frame in which the electric and magnetic fields are parallel. What are the fields in that frame for $\theta << 1$ and $\theta \to (\pi/2)$?

Solution: Along z axis

$$E'_z = 0, \quad B_z = 0, \quad E'_1 = \gamma(E_1 - \beta B_2)$$

$$E'_2 = \gamma(E_2 + \beta B_1), \quad B'_1 = \gamma(B_1 + \beta E_2), \quad B'_2 = \gamma(B_2 - \beta E_1)$$

$$E'_1 = \gamma E_0(1 - 2\beta \sin\theta), \quad E'_2 = \gamma E_0(2\beta \cos\theta)$$

$$B'_1 = 2\gamma \cos\theta E_0, \quad B'_2 = \gamma(2 \sin\theta - \beta)E_0$$

$$\frac{E'_2}{E'_1} = \frac{B'_2}{B'_1} = \frac{\gamma E_0(2\beta \cos\theta)}{\gamma E_0(1 - 2\beta \sin\theta)} = \frac{\gamma E_0(\sin\theta - \beta)}{2\gamma E_0 \cos\theta}$$

So,

$$4\beta \cos^2\theta = (1 - 2\beta \sin\theta)(2 \sin\theta - \beta)$$

$$4\beta \cos^2\theta = 2 \sin\theta - \beta - 4\beta \sin^2\theta + 2\beta^2 \sin\theta$$

$$5\beta = 2\beta^2 \sin\theta + 2 \sin\theta$$

$$2\beta^2 \sin\theta - 5\beta + 2 \sin\theta = 0$$

$$B = \frac{5 \pm \sqrt{25 - 16 \sin^2\theta}}{4 \sin\theta}$$

$$= \frac{5 \pm 5\sqrt{1 - (4/5) \sin^2\theta}}{\sin\theta} = \frac{5}{\sin\theta}\left(1 \pm 1 \mp \frac{16}{50} \sin^2\theta\right)$$

for

$$\beta = \frac{5}{\sin\theta}\left(1 + 1 - \frac{16}{50} \sin^2\theta\right)$$

take limit $\theta \to 0$, $\to 0(\sin\theta \to 0)$ for

$$\beta = \frac{5}{\sin\theta}\left(1 - 1 + \frac{16}{50} \sin^2\theta\right), \quad \beta \to 0$$

so,

$$B' = 2E_0\hat{\mathbf{x}}$$

$$E' = E_0\hat{\mathbf{x}}$$

for

$$\theta \to \frac{\pi}{2}, \quad \sin\theta \to 1, \quad \beta \to \frac{5-3}{4} = \tfrac{1}{2}, \quad \gamma = 1.15$$

$$E'_1 = (1.15)E_0(1-1) \qquad B'_1 = 2(1.15)0$$

$$E'_2 = (1.15)E(0) \qquad B'_2 = 1.15(1+\tfrac{1}{2})E_0$$

$$\mathbf{B}' = 1.725\, E_0\hat{\mathbf{y}}$$

$$\mathbf{E}' = 0$$

PROBLEM 15.18 A particle of mass m and charge e moves in the laboratory in crossed, static, uniform, electric, and magnetic fields. **E** is parallel to the x axis; **B** is parallel to the y axis.

(a) For $|\mathbf{E}| < |\mathbf{B}|$ make the necessary Lorentz transformation to obtain explicitly parametric equations for the particle's trajectory.
(b) Repeat the calculation of (a) for $|\mathbf{E}| > |\mathbf{B}|$.

Solution:

(a) $\mathbf{E} = E_0\hat{\mathbf{x}}, \quad \mathbf{B} = B_0\hat{\mathbf{y}}$

Let velocity of moving frame be

$$u = \frac{ceB}{B^2}\,\hat{\mathbf{z}}$$

$$E'_1 = E'_2 = E'_3 = B'_2 = B'_3 = 0, \quad B'_1 = \frac{B_0^2 - E_0^2}{B_0} = B'_0$$

$$\frac{d\mathbf{v}'}{dt'} = \mathbf{v}\times\mathbf{u}', \quad \mathbf{u}' = \frac{c\mathbf{B}'}{\gamma mc} = \frac{e\gamma}{\gamma mc}\,\frac{B_0^2 - E_0^2}{B_0}$$

$$\frac{dz'}{dt'} = v'_x\, u', \quad \frac{dv'_x}{dt'} = -v'_z u', \quad \frac{dv'_y}{dt'} = 0$$

So

$$v'_x = -i(v'_{x0}\sin\omega' t' + v'_{z0}\cos\omega' t')$$

$$v'_y = v'_{y0}$$

$$v'_z = v'_{x0}\cos\omega' t' + v'_{z0}\sin\omega' t'$$

$$v'_z = \frac{v_z - u}{(1 - v_z u/c^2)}, \quad v'_x = v_x\sqrt{1-(u^2/c^2)}, \quad v'_y = v_y\sqrt{1-(u^2/c^2)},$$

$$t' = \left(t + \frac{uz}{c^2}\right)/(1 - u^2/c^2)^{\frac{1}{2}}$$

$$v_y = v_{y0}$$

$$\frac{v_z - u}{1 - (v_z u/c^2)} = v_{x0}\gamma^{-1} \cos \frac{e}{mc} \frac{B_0^2 - E_0^2}{B_0} \gamma \left(t + \frac{uz}{c^2}\right)$$

$$+ \frac{(v_{z0} - u)}{1 - (v_z u/c^2)} \sin \frac{e}{mc} \frac{B_0^2 - E_0^2}{B_0} \gamma \left(t + \frac{uz}{c^2}\right)$$

$$\gamma v_x = -i \left[\gamma v_{x0} \sin \frac{e}{mc} \frac{B_0^2 - E_0^2}{B_0} \gamma \left(t + \frac{uz}{c^2}\right) + \frac{v_{z0} - u}{1 - (v_z u/c^2)} \cos \frac{e}{mc} \frac{B_0^2 - E_0^2}{B_0} \gamma \left(t + \frac{uz}{c^2}\right)\right]$$

$$\gamma^3 \left(\frac{cv - v_0^2 v/c}{(v^2 - v_0^2)^{1/2}(c^2 - v_0^2)^{1/2}}\right) \frac{dv}{dt} = \frac{eE_0}{m_0}$$

$$\left(1 - \frac{v^2}{c^2}\right)^{-3} \frac{v(c - v_0^2/c)}{(v^2 - v_0^2)^{1/2}(c^2 - v_0^2)^{1/2}} \frac{dv}{dt} = \frac{eE_0}{m_0}$$

$$\left(1 - \frac{v^2}{c^2}\right)^{-3/2} \frac{(v/c)(c^2 - v_0^2)^{1/2}}{(v^2 - v_0^2)^{1/2}} \frac{dv}{dt} = \frac{eE_0}{m_0}$$

$$c^2(c^2 - v_0^2)^{-3/2} \frac{v(c^2 - v_0^2)^{1/2}}{(v^2 - v_0^2)^{1/2}} \frac{dv}{dt} = \frac{eE_0}{m_0}$$

$$\int_{v_0}^{v} \frac{v\,dv}{(c^2 - v^2)^{3/2}(v^2 - v_0^2)^{1/2}} = \frac{E_0 e}{m_0 c^2} \int_0^t \frac{dt}{(c^2 - v_0^2)^{1/2}}, \quad u^2 = v^2 - v_0^2$$

$$\int_0^{(u^2 - v_0^2)^{1/2}} \frac{du}{(c^2 - v_0^2 - u^2)^{3/2}} = \left. \frac{u}{(c^2 - v_0^2)(c^2 - v_0^2 - u^2)^{1/2}} \right|_0^{(u^2 - v_0^2)^{1/2}}$$

$$\frac{(u^2 - v_0^2)^{1/2}}{(c^2 - u^2)^{1/2}} = \frac{eE_0(c^2 - v_0^2)^{1/2}}{m_0 c^2} t = at$$

$$\frac{v^2 - v_0^2}{c^2 - v^2} = a^2 t^2, \quad v^2 + v_0^2 + a^2 t^2 c^2 - \alpha^2 t^2 v^2, \quad v = \left(\frac{v_0^2 + a^2 c^2 t^2}{1 + \alpha^2 t^2}\right)^{1/2}$$

$$v_x = \left(\frac{c^2 - ((v_0^2 + a^2t^2c^2)/(c^2 - v_0^2))}{c^2 - v_0^2}\right)^{\frac{1}{2}} v_0 = \left(\frac{(c^2 - v_0^2)/(1 + a^2t^2)}{c^2 - v_0^2}\right)^{\frac{1}{2}} = \frac{v_0}{(1 + a^2t^2)^{\frac{1}{2}}}$$

$$v^2 = v_x^2 + v_z^2, \quad v_z = \frac{v_0^2 + a^2c^2t}{1 + a^2t^2} - \frac{v_0^2}{1 + a^2t^2} = \frac{a^2tc^2}{1 + a^2t^2}$$

$$x = v_0 \int_0^t \frac{dt}{(1 + a^2t^2)^{\frac{1}{2}}} = \frac{v_0}{a} \int_0^t \frac{dt}{(a^2 + t^2)^{\frac{1}{2}}} = \frac{v_0}{a} \sinh^{-1} at$$

$$x = c(a^2 + t^2)^{\frac{1}{2}} \Big|_0^t = \frac{c}{a} [(1 + a^2t^2)^{\frac{1}{2}} - 1]$$

PROBLEM 15.19 A charged particle oscillates harmonically with frequency ω according to the equation $x = a \cos \omega t$. If $\omega a/c = 1/3$, then $v \simeq c$ at the center of oscillation. Calculate the average rate of radiation in all directions using the exact formula and compare with the result in the dipole approximation.

Solution: Our equation is

$$\frac{dW}{dt} = \frac{e^2}{4\pi c^3} \int d\Omega \left(\frac{(\bar{a})^2}{(1 - \beta \cdot \vec{n})^3} + \frac{2(\vec{n} \cdot \vec{a})(\vec{a} \cdot \beta^2)}{(1 - \beta \cdot \vec{n})^4} - \frac{(1 - \beta^2)(\vec{n} \cdot \vec{a})^2}{(1 - \beta \cdot \vec{n})^5}\right)$$

with

$$\mathbf{r} = A \cos \omega t \ \hat{\mathbf{z}}, \quad \beta = \frac{-A\omega}{c} \sin \omega t \ \hat{\mathbf{z}} = \beta \hat{\mathbf{z}},$$

$$\vec{a} = -A\omega^2 \cos \omega t \ \hat{\mathbf{z}} = a\hat{\mathbf{z}}$$

Then, in spherical coordinates,

$$\frac{dW}{dt} = \frac{e^2a^2}{4\pi c^3} \left(\int \frac{d\Omega}{(1 - \beta \cos \theta)^5} + 2\beta \int \frac{\cos \theta \, d\Omega}{(1 - \beta \cos \theta)^4} - (1 - \beta^2) \int \frac{\cos^2 \theta \, d\Omega}{(1 - \beta \cos \theta)^5}\right)$$

$$= \frac{2\pi e^2a^2}{4\pi c^3} \left(\int_0^\pi \frac{\sin \theta \, d\theta}{(1 - \beta \cos \theta)^5} + 2\beta \int_0^\pi \frac{\cos \theta \sin \theta \, d\theta}{(1 - \beta \cos \theta)^4} - (1 - \beta^2) \int_0^\pi \frac{\cos^2 \theta \sin \theta \, d\theta}{(1 - \beta \cos \theta)^5}\right)$$

$$= \frac{e^2a^2}{2c^3} \left(\frac{1}{\beta} \int_{-\beta}^{+\beta} \frac{dz}{(1 - z)^5} + \frac{2\beta}{\beta^2} \int_{-\beta}^{+\beta} \frac{z}{(1 - z)^4} dz - \frac{(1 - \beta^2)}{\beta^3} \int_{-\beta}^{+\beta} \frac{z^2 \, dz}{(1 - z)^5}\right)$$

$$= \frac{e^2 a^2}{2c^3} \left\{ \frac{1}{\beta} I_1 + \frac{2}{\beta} I_2 - \frac{1 - \beta^2}{\beta^3} I_3 \right\}$$

where

$$I_1 = \int_{-\beta}^{+\beta} \frac{dz}{(1 - z)^5}; \quad I_2 = \int_{-\beta}^{+\beta} \frac{z\,dz}{(1 - z)^4}; \quad I_3 = \int_{-\beta}^{+\beta} \frac{z^2\,dz}{(1 - z)^5}$$

$$I_1 = \left. \frac{1}{2(1 - z)^2} \right|_{-\beta}^{+\beta} = \frac{2\beta}{(1 - \beta^2)^2}$$

$$I_2 = - \left. \frac{1 - 3z}{6(1 - z)^3} \right|_{-\beta}^{+\beta} = \frac{8\beta^3}{3(1 - \beta^2)^3}$$

$$I_3 = \left. \frac{1 - 4z + 6z^2}{12(1 - z)^4} \right|_{-\beta}^{+\beta} = \frac{2\beta^3(1 + 5\beta^2)}{3(1 - \beta^2)^4}$$

Using these results,

$$\frac{1}{\beta} I_1 + \frac{2}{\beta} I_2 - \frac{1 - \beta^2}{\beta^3} I_3 = \frac{1}{\beta} (I_1 + 2I_2) - \frac{1 - \beta^2}{\beta^3} I_3$$

$$= \frac{2}{(1 - \beta^2)^2} + \frac{18\beta^2}{3(1 - \beta^2)^3} - \frac{2(1 + 5\beta^2)}{3(1 - \beta^2)^3}$$

$$= \frac{2}{3(1 - \beta^2)^3} [3(1 - \beta^2) + 8\beta^2 - 1 - 5\beta^2]$$

$$= \frac{2}{3(1 - \beta^2)^3} (3 - 3\beta^2 + 8\beta^2 - 1 - 5\beta^2)$$

$$= \frac{4}{3(1 - \beta^2)^3}$$

Therefore,

$$\frac{dW}{dt} = \frac{e^2 a^2}{2c^3} \frac{4}{3(1 - \beta^2)^3} = \frac{2e^2 a^2}{3c^3} \frac{1}{(1 - \beta^2)^3} \qquad (1)$$

where

$$a = -A\omega^2 \cos \omega t, \qquad \beta = \frac{-A\omega}{c} \sin \omega t$$

so,

$$\left\langle\frac{dW}{dt}\right\rangle = \frac{2e^2}{3c^3} \left\langle\frac{a^2}{(1-\beta^2)^3}\right\rangle = \frac{2e^2A^4\omega^4}{3c^3} \left\langle\frac{\cos^2 \omega t}{[1-(A\omega/c)^2 \sin^2 \omega t]^3}\right\rangle$$

Now if $D = A\omega/c \leq 1$,

$$\left\langle\frac{\cos^2 \omega t}{[1-(A\omega/c)^2 \sin^2 \omega t]^3}\right\rangle = \left\langle\frac{1/(1+\tan^2 x)}{[1-(D^2 \tan^2 x/1+\tan^2 x)]^3}\right\rangle$$

$$= \frac{1}{\pi}\int_{-\pi/2}^{\pi/2} \frac{(1+\tan^2 x)^2\,dx}{(1+E^2 \tan^2 x)^3}, \quad \text{where } E^2 = 1 - D^2$$

$$= \frac{1}{\pi}\int_{-\infty}^{+\infty} \frac{1+t^2\,dt}{(1+E^2t^2)^3} = \frac{2}{\pi}\int_0^{\infty} \frac{1+t^2\,dt}{(1+E^2t^2)^3}$$

Now, to obtain the integral, we proceed as follows,

$$\int_0^\infty \frac{1+t^2\,dt}{(1+E^2t^2)^3} = \int_0^\infty \frac{1\,dt}{(1+E^2t^2)^3} + \int_0^\infty \frac{t^2}{(1+E^2t^2)^3}$$

but,

$$\int_0^\infty \frac{t^2\,dt}{(1+E^2t^2)^3} = \int_0^\infty \frac{t \cdot t\,dt}{(1+E^2t^2)^3} = -\frac{1}{2(2E^2)} \left.\frac{t}{(1+E^2t^2)^2}\right|_0^\infty \int_0^\infty \frac{1}{4E^2}\frac{1\,dt}{(1+E^2t^2)^2}$$

$$= \frac{1}{4E^2}\int_0^\infty \frac{1\,dt}{(1+E^2t^2)^2}$$

Therefore,

$$\int_0^\infty \frac{1+t^2\,dt}{(1+E^2t^2)^3} = \int_0^\infty \frac{1\,dt}{(1+E^2t^2)^3} \quad \frac{1}{4E^2}\int_0^\infty \frac{1\,dt}{(1+E^2t^2)^3}$$

Using

$$\frac{1}{(1+a^2x^2)^{m+1}} = \frac{-a^2x^2}{(1+a^2x^2)^{m+1}} + \frac{1}{(1+a^2x^2)^m}$$

we obtain,

$$\int_0^\infty \frac{1\,dt}{(1+a^2t^2)^{m+1}} = +\int_0^\infty \frac{dt(-a^2t)}{(1+a^2t^2)^{m+1}} + \int_0^\infty \frac{dt\;1}{(1+a^2t^2)^m}$$

$$= \frac{1}{2m}\,\frac{t}{(1+a^2t^2)^m}\bigg|_0^\infty - \int_0^\infty \frac{1}{2m}\,\frac{1\,dt}{(1+a^2t^2)^m} + \int_0^\infty \frac{1\,dt}{(1+a^2t^2)^m}$$

$$= \left(1 - \frac{1}{2m}\right)\int_0^\infty \frac{1\,dt}{(1+a^2t^2)^m}$$

$$= \frac{2m-1}{2m}\int_0^\infty \frac{1\,dt}{(1+a^2t^2)^m}, \qquad m \geq 1$$

that is,

$$\int_0^\infty \frac{1}{(1+E^2t^2)^{m+1}} = \frac{2m-1}{2m}\left(\frac{2(m-1)-1}{2(m-1)}\right)\cdots\left(\frac{1}{2}\right)\int_0^\infty \frac{1\,dt}{1+E^2t^2}$$

$$= \left(\frac{2m-1}{2m}\right)\frac{2(m-1)-1}{2(m-1)}\cdots\frac{1}{2E}\,\arctan(Et)\bigg|_0^\infty$$

$$= \frac{2m-1}{2m}\,\frac{2m-3}{2m-2}\cdots\tfrac{1}{2}\,\frac{\pi}{2E}$$

so

$$\left\langle\frac{\cos^2\omega t}{[1-(A\omega/c)^2\sin^2\omega t]^3}\right\rangle = \frac{2}{\pi}\int_0^\infty \frac{1+t^2\,dt}{(1+E^2t^2)^3}$$

$$= \frac{2}{\pi}\left(\int_0^\infty \frac{1\,dt}{(1+E^2t^2)^3}\quad \frac{1}{4E^2}\int_0^\infty \frac{1\,dt}{(1+E^2t^2)^2}\right)$$

$$= \frac{2}{\pi}\left(\frac{3}{4}\cdot\frac{1}{2}\cdot\frac{1}{E}\cdot\frac{\pi}{2}\right) + \frac{2}{\pi}\cdot\frac{1}{4E^2}\left(\frac{1}{2}\cdot\frac{1}{E}\cdot\frac{\pi}{2}\right)$$

$$= \frac{1}{2E}\left(\frac{3}{4}+\frac{1}{4E^2}\right) = \frac{1}{2}\,\frac{1}{E^3}\left(\frac{3E^2+1}{4}\right)$$

but $E^2 = 1 - D^2 = 1 - (A\omega/c)^2$, therefore,

$$\left\langle \frac{\cos^2 \omega t}{[1 - (A\omega/c)^2 \sin^2 \omega t]^3} \right\rangle = \frac{1}{2} \frac{3[1 - (A\omega/c)^2 + 1]}{4E^3}$$

$$= \frac{1}{2E^3} \left(\frac{3 - 3(A\omega/c)^2 + 1}{4} \right)$$

$$= \frac{1}{2} \left(1 - \frac{3}{4} (A\omega/c)^2 \right) \frac{1}{E^2}$$

that is,

$$\left\langle \frac{\cos^2 \omega t}{[1 - (A\omega/c)^2 \sin^2 t]^3} \right\rangle = \frac{1}{2} \frac{1 - (3/4)(A\omega/c)^2}{[\sqrt{1 - (A\omega/c)^2}]^3}$$

So that, finally

$$\frac{dW}{dt} = \frac{e^2 A^4 \omega^4}{3c^3} \frac{1 - (3/4)(A\omega/c)^2}{[\sqrt{1 - (A\omega/c)^2}]^3} \tag{2}$$

Ususally it is not worthwhile working integrals out explicitly, unless you suspect you may be unable to do them, or that they will yield some interesting or useful physical result. We note that

$$p_{\text{dip.}} = \frac{e^2 A^2 \omega^4}{3c^3}$$

is the average rate of radiation in all directions for the (nonrelativistic) dipole approximation. So, we can write our result in the form

$$p = p_{\text{dip.}} \frac{1 - (3/4)(A\omega/c)^2}{[1 - (A\omega/c)^2]^{3/2}} \tag{3}$$

So that if $\omega A/c \ll 1$,

$$p \simeq p_{\text{dip.}} \left[1 - \frac{3}{4} \left(\frac{A\omega}{c} \right)^2 \right] \left[1 + \frac{3}{2} \left(\frac{A\omega}{c} \right)^2 - \dots \right]$$

$$\simeq p_{\text{dip.}} \left[1 + \frac{3}{4} \left(\frac{A\omega}{c} \right)^2 + \dots \right]$$

$$\therefore \quad p \simeq p_{\text{dip.}} \left[1 + \frac{3}{4} \left(\frac{A\omega}{c} \right)^2 + \dots \right] \tag{4}$$

We see that in the nonrelativistic limit, $A\omega/c \to 0$, the equation reduces to the expected form. In our particular case, with $\omega A/c = 1/3$, we obtain [from Eq. (1)]

$$p = 1.094\ p_{dip.}$$

which is our final result.

PROBLEM 15.20 Show that the rate of radiation from a charge e, moving in a circle of radius a with constant angular velocity ω, is given by

$$\frac{dW}{dt} = \frac{2e^2}{3c^3} \frac{a^2\omega^4}{[1 - (a\omega/c)^2]^2}$$

Solution: Our exact expression is

$$\frac{dW}{dt} = \frac{e^2}{4\pi c^3} \int \frac{d\Omega}{(1 - \beta \cdot n)^5} [a^2(1 - \beta \cdot n)^2 + 2(n \cdot a)(1 - \beta \cdot n)(a \cdot \beta) - (n \cdot a)^2(1 - \beta^2)]$$

where a = the acceleration at time t, and $\beta = v/c$. For uniform circular motion, $a \perp v$, so that our expression reduces to

$$\frac{dW}{dt} = \frac{e^2}{4\pi c^3} \int \frac{d\Omega}{(1 - \beta \cdot n)^5} [a^2(1 - \beta \cdot \vec{n})^2 - (\vec{n} \cdot \vec{a})\ (1 - \beta^2)]$$

$$= \frac{e^2}{4\pi c^3} \left[a^2\left(\int \frac{d\Omega}{(1 - \beta \cdot n)^3}\right) - (1 - \beta^2)\left(\int \frac{d\Omega(\vec{n} \cdot \vec{a})^2}{(1 - \beta \cdot n)^5}\right)\right]$$

Since $a \perp v$, that is, $a \perp \beta$, at each time t, set up a system of spherical coordinates with the z axis along the v direction and the x axis along the direction of a. Then,

$$\vec{n} \cdot \vec{a} = \vec{a} \sin\theta \cos\phi$$

$$\vec{n} \cdot \beta = \beta \cos\theta$$

Therefore,

$$\frac{dW}{dt} = \frac{e^2a^2}{4\pi c^3} \left(\int \frac{d\Omega}{(1 - \beta\cos\theta)^3} - (1 - \beta^2) \int \frac{d\Omega \sin^2\theta \cos^2\phi}{(1 - \beta\cos\theta)^5}\right)$$

$$= \frac{e^2a^2}{4\pi c^3} [I_1 - (1 - \beta^2)I_2]$$

$$I_1 = \int \frac{d\Omega}{(1-\beta\cos\theta)^3} = 2\pi \int_0^{\pi} \frac{\sin\theta\, d\theta}{(1-\beta\cos\theta)^3} = \frac{2\pi}{\beta} \int_{-\beta}^{+\beta} \frac{dz}{(1-z)^3}$$

$$= \frac{2\pi}{\beta} \frac{2\beta}{(1-\beta^2)^2} = \frac{4\pi}{(1-\beta^2)^2}$$

$$I_2 = \pi \int_0^{\pi} \frac{\sin\theta\,(1-\cos^2\theta)d\theta}{(1-\beta\cos\theta)^5} = \pi\left(\int_0^{\pi} \frac{\sin\theta\, d\theta}{(1-\beta\cos\theta)^5} - \int_0^{\pi} \frac{\cos^2\theta\sin\theta\, d\theta}{(1-\beta\cos\theta)^5}\right)$$

$$= \pi\left(\frac{1}{\beta}\int_{-\beta}^{+\beta} \frac{dz}{(1-z)^5} - \frac{1}{\beta^3}\int_{-\beta}^{+\beta} \frac{z^2}{(1-z)^5}\right)$$

We now use,

$$\int_{-\beta}^{\beta} \frac{dz}{(1-z)^5} = \frac{1}{4} \frac{1}{(1-z)^4}\Bigg|_{-\beta}^{+\beta} = \frac{2\beta(1+\beta^2)}{(1-\beta^2)^4}$$

and

$$\int_{-\beta}^{\beta} \frac{z^2}{(1-z)^5}\, dz = \frac{1-4z+6z^2}{12(1-z)^4}\Bigg|_{-\beta}^{+\beta} = \frac{2\beta^3}{3(1-\beta^2)^4}\,(1+5\beta^2)$$

Therefore,

$$I_2 = \frac{2\pi}{(1-\beta^2)^4}\left((1+\beta^2) - \frac{1}{3} - \frac{5}{3}\beta^2\right) = \frac{2\pi}{(1-\beta^2)^4}\left[-\beta^2\left(\frac{2}{3}\right) + \frac{2}{3}\right]$$

$$= \frac{4\pi}{3} \frac{1-\beta^2}{(1-\beta^2)^4} = \frac{4\pi}{3} \frac{1}{(1-\beta^2)^3}$$

So,

$$\frac{dW}{dt} = \frac{e^2a^2}{4\pi c^3}\left(\frac{4\pi}{(1-\beta^2)^2} - (1-\beta^2)\,\frac{4\pi}{3}\cdot\frac{1}{(1-\beta^2)^3}\right)$$

$$= \frac{2e^2a^2}{3c^3} \frac{1}{(1-\beta^2)^2}$$

Using $a^2 = R^2\omega^4$, $v = R\omega$, with R radius of circle, this becomes

$$\frac{dW}{dt} = \frac{2e^2}{3c^3} \frac{R^2\omega^4}{[1 - (R\omega/c)^2]^2}$$

PROBLEM 15.21 Show that the formula $\mathbf{H} = \mathbf{n}' \times \mathbf{E}$, which is generally valid for a point charge in arbitrary motion, reduces to $\mathbf{H} = \boldsymbol{\beta} \times \mathbf{E}$ for a point charge in uniform motion.

Solution: Consider the diagram

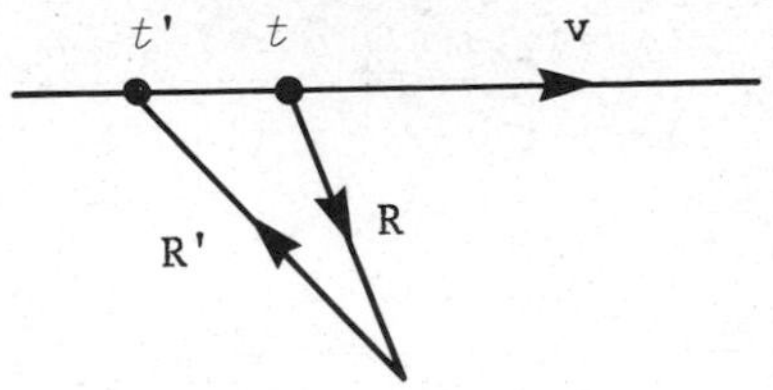

We have $\mathbf{R}' = \mathbf{v}(t - t') + \mathbf{R} = \mathbf{v}\frac{R'}{c} + \mathbf{R} = \vec{\beta}R' + \mathbf{R}$ where $\boldsymbol{\beta} = \mathbf{v}/c$. Therefore,

$$\mathbf{R} = \mathbf{R}' - \boldsymbol{\beta}R'$$

or

$$\mathbf{n}R = \mathbf{n}'R' - \boldsymbol{\beta}R' \tag{1}$$

Using Eq. (1),

$$0 = \mathbf{n} \times \mathbf{n}R = \mathbf{n} \times \mathbf{n}'R' - \mathbf{n} \times \boldsymbol{\beta}R', \quad \mathbf{n} \times \mathbf{n}' = \mathbf{n} \times \boldsymbol{\beta} \tag{2}$$

For a charge with uniform motion,

$$\mathbf{E} = \frac{\ell}{S^3}(1 - \beta^2)\mathbf{R}$$

therefore, using Eq. (2),

$$\mathbf{H} = \mathbf{n}' \times \mathbf{E} = \frac{\ell}{S^3}(1 - \beta^2)(\mathbf{n}' \times \mathbf{n}R) = \frac{\ell R}{S^3}(1 - \beta^2)(\mathbf{n}' \times \mathbf{n}) = \frac{\ell R}{S^3}(1 - \beta^2)(\boldsymbol{\beta} \times \mathbf{n})$$

$$= \boldsymbol{\beta} \times \frac{\ell R\mathbf{n}}{S^3}(1 - \beta^2) - \boldsymbol{\beta} \times \mathbf{E}$$

that is,

$$\mathbf{H} = \boldsymbol{\beta} \times \mathbf{E}$$